Introduction to the Standard Model and Beyond

The Standard Model of particle physics is an amazingly successful theory describing the fundamental particles and forces of nature. This text, written for a two-semester graduate course on the Standard Model, develops a practical understanding of the theoretical concepts it's built upon, to prepare students to enter research. The author takes a historical approach to demonstrate to students the process of discovery which is often overlooked in other textbooks, presenting quantum field theory and symmetries as the necessary tools for describing and understanding the Standard Model. He develops these tools using a basic understanding of quantum mechanics and classical field theory, such as Maxwell's electrodynamics, before discussing the important role that Noether's theorem and conserved charges play in the theory. Worked examples feature throughout the text, while homework exercises are included for the first five parts, with solutions available online for instructors. Inspired by the author's own teaching experience, suggestions for independent research topics have been provided for the second half of the course, which students can then present to the rest of the class.

Stuart Raby is a professor of physics at The Ohio State University. He is among the original proponents of the supersymmetric extension of the Standard Model and a pioneer of supersymmetric grand unified theories. His work focuses on the experimental consequences of physics beyond the Standard Model, ranging from collider experiments to proton decay searches, dark matter candidates and questions in cosmology. His first book, *Supersymmetric Grand Unified Theories*, was published by Springer in 2017.

Introduction to the Standard Model and Beyond

Quantum Field Theory, Symmetries and Phenomenology

STUART RABY

The Ohio State University

CAMBRIDGE
UNIVERSITY PRESS

University Printing House, Cambridge CB2 8BS, United Kingdom

One Liberty Plaza, 20th Floor, New York, NY 10006, USA

477 Williamstown Road, Port Melbourne, VIC 3207, Australia

314–321, 3rd Floor, Plot 3, Splendor Forum, Jasola District Centre, New Delhi – 110025, India

103 Penang Road, #05–06/07, Visioncrest Commercial, Singapore 238467

Cambridge University Press is part of the University of Cambridge.

It furthers the University's mission by disseminating knowledge in the pursuit of
education, learning and research at the highest international levels of excellence.

www.cambridge.org
Information on this title: www.cambridge.org/9781108494199
DOI: 10.1017/9781108644129

First published 2021

Printed in the United Kingdom by TJ Books Limited, Padstow Cornwall

A catalogue record for this publication is available from the British Library.

ISBN 978-1-108-49419-9 Hardback

Additional resources for this publication at www.cambridge.org/raby

Contents

Part VII Road to the Standard Model: Quantum Chromodynamics

Part VIII Road to the Standard Model: Electroweak Theory

Color plates section can be found between pages 320 and 321

Preface

The goal of the book is to develop a practical understanding of the theoretical concepts inherent in the Standard Model and to enable students to enter a research program in this exciting field. Note that the topics discussed in this book are also useful for students with interests in nuclear and astro/particle physics.

This book bears a strong resemblance to similar texts. However, it provides an introduction to field theory and the Standard Model which I believe may be more accessible to both experimentalists and theorists. Also, by following the development of the theory historically, it may allow students to understand the process of discovery, which seems to be missing in more recent texts. The development of the Standard Model relied heavily on symmetries, which can be both continuous and discrete, global and local, softly broken and exact. Therefore the book focuses on an understanding of symmetries in particle physics; their consequences and tests thereof.

This book is designed as a text for a one-year course on the Standard Model. The prerequisites include advanced courses on classical mechanics, electrodynamics and quantum mechanics. However, it does not require quantum field theory as a prerequisite. Homework problems and solutions are included in the book and have been used in the first half of the course. However, in the latter half of the course we have found that, firstly, there are too many directions in particle physics for which there is insufficient time to discuss in class, and, secondly, students have relished the independent research they do in discovering the details of some of these directions on their own. Thus, I provide a list of possible topics for independent research at the end of the book. Students typically present two research papers and lectures on these topics in class. In developing the material in the book we acknowledge some texts which have had a significant influence on our presentation. These include *Elementary Particle Physics* (Wiley, 1966) by S. Gasiorowicz, *Particle Physics and Introduction to Field Theory* (CRC Press, 1981) by T. D. Lee and *Lie Algebras in Particle Physics*, 2nd edition (CRC Press, 2018) by H. Georgi.

Note, the material in the book is presented step by step: introducing subjects as foundations which are then built upon and used over and over. For students who have already had a field theory course, the material in Part II may be encouraged as a refresher. Chapters 8, 9 and 10, in particular, on the symmetries of complex scalars and fermions, should be emphasized. Then instruction might cover the material in Parts III–X.

What is Particle Physics?

The 20th century witnessed the most amazing discovery of what we now call the Standard Model of Particle Physics. The Standard Model describes the results of hundreds of thousands of experiments, with of order 28 arbitrary parameters that are fit by experiment. Particle physics can be defined as humankind's search for an understanding of the basic building blocks of matter and their interactions. As such it began with the earliest musing about space, time and atoms by Aristotle and Democritus, with the study of gravitational phenomena by Galileo and Newton and the study of electric and magnetic phenomena by the likes of Coulomb, Ampere, Faraday and Maxwell. By the end of the 19th century, physicists were wondering whether all of nature was understood and, according to Maxwell in his Introductory Lecture on Experimental Physics, held at Cambridge in October 1871,

> ...the opinion seems to have got abroad, that in a few years all the great physical constants will have been approximately estimated, and that the only occupation which will then be left to men of science will be to carry on these measurements to another place of decimals. ...But we have no right to think thus of the unsearchable riches of creation, or of the untried fertility of those fresh minds into which these riches will continue to be poured. ...But the history of science shews that even during the phase of her progress in which she devotes herself to improving the accuracy of the numerical measurement of quantities with which she has long been familiar, she is preparing the materials for the subjugation of the new regions, which would have remained unknown if she had been contented with the rough methods of her early pioneers. I might bring forward instances gathered from every branch of science, shewing how the labour of careful measurement has been rewarded by the discovery of new fields of research, and by the development of new scientific ideas.

Soon after Maxwell made these comments, a period of highly significant scientific breakthroughs began with the discovery of radio waves by Hertz (1886–1889), followed quickly by those X-rays by Roentgen (1895) and nuclear radiation by Becquerel (1896). It was these discoveries which put into motion the dramatic events of the 20th century.

Beginning with the discovery of the electron by Thomson in 1897 and the seminal work of Rutherford, Geiger and Marsden in 1911, physicists have investigated the atom using particle beams (electrons, protons and alpha particles) as probes. They developed new detection methods: the geiger counter, scintillators, cloud and then bubble chambers. This new paradigm for probing matter and new detectors led to many discoveries.

The mystery of the atom was unraveled. With Bohr's quantum theory and Pauli's exclusion principle, a complete theory of the periodic table of the elements was developed, leading to the fields of physical chemistry and, eventually, microbiology.

The nucleus was shattered and the strong interactions, holding protons and neutrons inside the core, were investigated. Hence the beginning of nuclear physics.

These new detectors were then pointed to the heavens, unveiling a shower of energetic charged particles emanating from distant galaxies. New particles, such as the muon, the pion and the positron (also known as an anti-electron) were discovered in the cosmic ray shower. You know, it would take a lead umbrella to stop this eternal shower.

By the late 1950s, a major step forward came with the advent of particle accelerators at Brookhaven in New York and CERN in Geneva. With controlled particle beams (made of strongly interacting particles, such as pions, kaons and protons), experiments revealed a virtual particle zoo, with many more species of strongly interacting particles – all called hadrons. It became evident that the hadronic zoo was immense and it was unlikely that hadrons were elementary or fundamental particles at all.

To make a long story short, by 1974, the chaos of discovery lead to the Standard Model describing all observed particle phenomena in terms of three fundamental forces (four, including gravity) and the fundamental building blocks of matter, i.e. quarks and leptons.

Only now, after the dust of this chaotic discovery settles, are we able with hindsight to recognize the underlying principles which define the theory we call the Standard Model. It is these principles, and their logical extension, which I will attempt to describe in this course. Particle physics has relied on amazing new methods of experiment to explore nature. Moreover, it has relied on *symmetries* and their consequences to describe nature. It is this hand-in-hand method which has propelled us into the present state of our understanding. Once again, we are at a moment in history where the role of scientists appears to be just to measure the fundamental constants to the next decimal point. But like Maxwell, more than 100 years ago, we can expect that new experimental techniques and measurements will lead to the next epoch of discovery.

Acknowledgements

This book is based on a Standard Model course that I have given at Ohio State University over the course of some 30 years. The material presented here has greatly benefited from the many graduate students who have taken the course and asked many tough questions. I would not have been able to write this book without their help. I also want to thank Professor Leonard Susskind for his strong support which kept me in the field. Finally, I want to thank my wife, Michele, whose guidance over these 50 years has always kept me grounded.

PART I

GETTING STARTED

1   Notation

Before we begin discussing the theory we must agree on some notation. The space–time coordinates are represented by the Lorentz four-vector $x^\mu = (x^0, x^i) \equiv (ct, \vec{x})$ $i = 1, 2, 3$. The index μ takes on values 0, 1, 2, 3. The index μ can be raised or lowered using the space–time metric $g_{\mu\nu}$ with $x_\mu \equiv g_{\mu\nu} x^\nu$ and

$$g_{\mu\nu} \equiv \begin{pmatrix} 1 & & & \\ & -1 & & \\ & & -1 & \\ & & & -1 \end{pmatrix} \text{ in Cartesian coordinates.} \tag{1.1}$$

We define the Lorentz-invariant four-vector length by

$$x^2 \equiv x^\mu x^\nu g_{\mu\nu}. \tag{1.2}$$

The energy–momentum four-vector is given by $P^\mu = (P^0, \vec{p}c) \equiv (E, \vec{p}c)$ and the particle mass is given via the Lorentz scalar

$$P^2 \equiv P^\mu P^\nu g_{\mu\nu} \equiv E^2 - \vec{p}^2 c^2 = m^2 c^4. \tag{1.3}$$

In order to discuss dimensions of various quantities, it is convenient to consider the Dirac equation, which describes the relativistic quantum mechanics of spin $1/2$ particles. We first introduce the Dirac gamma matrices, satisfying $\{\gamma^\mu, \gamma^\nu\} = 2g^{\mu\nu}$. Then Dirac defined the Hamiltonian given by the 4×4 matrix

$$H = \vec{\alpha} \cdot \vec{p}c + \beta m c^2 + V(\hat{x}), \tag{1.4}$$

where $\gamma^0 \equiv \beta$ $\gamma^i = \beta \alpha^i$.

The energy eigenvalue is given by

$$E^2 = H_0^2 \equiv \vec{p}^2 c^2 + m^2 c^4, \tag{1.5}$$

where H_0 is the free particle Hamiltonian. In the Schrödinger equation, the energy-momenta are replaced by operators in Hilbert space. We have

$$P^\mu = (P^0, \vec{p}c) \equiv i\hbar c \, \partial^\mu = i\hbar c \, \frac{\partial}{\partial x_\mu}$$

$$\equiv i\hbar c \left(\frac{\partial}{c \partial t}, -\frac{\partial}{\partial x^i} \right),$$

(with $\nabla^i \equiv \frac{\partial}{\partial x^i}$). The potential for an electron in the field of Z protons is given by

$$V = e\phi = -\frac{Ze^2}{4\pi\epsilon_0 r} \text{ central force.} \tag{1.6}$$

Finally, we obtain the Schrödinger equation

$$H\psi = i\hbar \frac{\partial}{\partial t}\psi, \qquad (1.7)$$

$$\hbar c \left[\vec{\alpha} \cdot (-i\vec{\nabla}) + \beta \frac{mc}{\hbar} - \frac{Z\alpha}{r} \right] \psi = i\hbar \frac{\partial \psi}{\partial t}, \qquad (1.8)$$

where we define the dimensionless fine structure constant $\alpha \equiv \frac{e^2}{4\pi\epsilon_0 \hbar c}$. [1]

We can now consider the dimensions of various quantities. Each term in the above equation necessarily has the same dimension. We have, where the bracket $[x]$ represents the dimension of the quantity x and ℓ stands for length,

$$[\vec{\nabla}] = \ell^{-1} \qquad \left[\frac{\hbar}{mc} \right] = \ell \quad \text{Compton wavelength}$$

$$[\vec{x}] = \ell \qquad [\alpha] = \ell^0$$

$$[ct] = \ell$$

$$[E] \sim \hbar\, c\, \ell^{-1} \qquad [\vec{p}] = \hbar\, \ell^{-1}.$$

$$[m] \sim \frac{\hbar}{c}\, \ell^{-1}$$

We will define the unit of energy commonly used in atomic, nuclear, astro and particle physics. We have the electron volt given by $1 \text{ eV} = 1.6 \times 10^{-19}$ joules, where 1 joule = 1 coulomb volt. Since the charge on an electron, $e = 1.6 \times 10^{-19}$ coulombs, 1 eV is, by definition, the energy an electron has after traversing a voltage difference of 1 volt. Then $1 \text{ keV} = 10^3 \text{ eV}; 1 \text{ MeV} = 10^6 \text{ eV}; 1 \text{ GeV} = 10^9 \text{ eV}$: and $1 \text{ TeV} = 10^3 \text{ GeV}$.

Note, also in the **mks** system we have

$$1 \text{ kg } \frac{\text{m}^2}{\text{s}^2} \equiv 1 \text{ joule}$$

$$\Rightarrow 1 \text{ kg} = (3 \times 10^8)^2 \frac{\text{joule}}{c^2}$$

$$= \frac{9 \times 10^{16}}{1.6} \frac{10^{19} \text{ eV}}{c^2} = 5.625 \times 10^{29} \frac{\text{MeV}}{c^2}.$$

In these units the proton and electron masses are given by

$$m_p \simeq 1.67 \times 10^{-24} \text{ gm} \simeq 1.67 \times 10^{-27} \text{ kg}$$

$$\simeq 938 \; MeV/c^2$$

$$m_e \simeq 9 \times 10^{-31} \text{ kg} \simeq 50 \times 10^4 \text{ eV}/c^2$$

$$\simeq \frac{1}{2} \text{ MeV}/c^2.$$

---

[1] Note, given $e = 1.6 \times 10^{-19}$ C and $\frac{1}{4\pi\epsilon_0} = 9 \times 10^9$ N m^2/C^2, we find an approximate value for the fine structure constant, $\alpha \approx \frac{1}{137.5}$. Given fixed values of e, $\hbar$ and c, better experimental measurements of α can determine ϵ_0 to higher accuracy.

The values of $\hbar$ and c will be important to remember. We have $\hbar = [E\ t] = 6.6 \times 10^{-22}$ MeV $\cdot$ s and $c = 3 \times 10^{8}$m/s. Thus we have $\hbar\, c \approx 200 \times 10^{-13}$ MeV cm. Define the unit of length 1fermi $\equiv 1$ fm $= 10^{-13}$ cm and we have $\hbar\, c \approx 200$ MeV fm.

Cross-sections, σ, are in units of a barn $\equiv 1$ b $= 10^{-24}$ cm^2. Strong-interaction cross-sections $\sigma_{st} \sim 30$ mb $= 3 \times 10^{-26}$ cm^2 and weak cross-sections are of order $\sigma_{weak} \sim 1$ picobarn $\equiv 1$ pb $= 10^{-36}$ cm^2. Finally a femtobarn $\equiv 1$ fb $= 10^{-39}$ cm^2.

Note, following the standard conventions, in this book we will redefine the dimensions of space and time such that $\hbar = c = \epsilon_0 = 1$. We can always revert to the **mks** units by rescaling the end result by the appropriate powers of $\hbar$ and c. Therefore $(200$ MeV$)^{-1} \doteq 1$ fm $\doteq 3 \times 10^{-24}$ s, where the notation $\doteq$ means conversions using $\hbar$ and c. Finally, a cross-section of $(200$ MeV$)^{-2} \doteq 10$ mb and 1 TeV$^{-2} \simeq 400$ pb.

PART II

SYMMETRIES AND QUANTUM FIELD THEORY

2 Poincaré Invariance

In particle physics it has always been true that symmetries are very important. One might go so far as to state that the Standard Model could not have been constructed without understanding and making use of the many, both exact and broken, symmetries of nature. The symmetries of nature define the observable properties of elementary particles. In this chapter, we begin our discussion of the symmetry which gives us the basic labels we use to define all particle states, i.e. momentum, energy, mass and spin. They are given in terms of a complete set of commuting operators of the Poincaré group,

$$\{\hat{P}_\mu,\ \hat{M}^2,\ \hat{W}^2,\ \hat{W}_3\}. \tag{2.1}$$

And the single-particle states are given by their eigenvalues,

$$|p, m, s, \sigma\rangle. \tag{2.2}$$

However, before we get started on studying the Poincaré group, let's take a moment to discuss the concept of a group.

2.1 What Is a Group?

ASIDE: The group G is defined by a set of elements, $g_i \in G$, and a multiplication rule, $g_1 \circ g_2$. In addition, in order to be a group one needs to satisfy the following group properties:

1. closure: if $g_1 \in G$, $g_2 \in G$, then $g_1 \circ g_2 \in G$,
2. associativity: if $g_{1,2,3} \in G$, then $g_1 \circ (g_2 \circ g_3) = (g_1 \circ g_2) \circ g_3$,
3. identity: $I \in G$, such that $I \circ g_i = g_i \circ I = g_i$,
4. inverse: if $g_1 \in G$, then $g_1^{-1} \in G$, such that $g_1^{-1} \circ g_1 = g_1 \circ g_1^{-1} = I$.

Some simple examples of groups are:

(1) $\{\Re\}$, i.e. the set of real numbers with zero removed, with the multiplication rule, $\circ$, given simply by multiplication. If $x, y \in \Re$, then $x \cdot y \in \Re$. Think of the identity element and inverse.

(2) $\{\mathbb{Z}\}$, the set of integers, with the multiplication rule, $\circ$, given simply by addition, $+$. If $m \in \mathbb{Z}$, then $n \circ m \equiv n + m \in \mathbb{Z}$. Think of the identity element and inverse.

(3) $\mathbb{Z}_2 = \{1, -1\}$, with the multiplication rule, $\circ$, given simply by multiplication.

Now let's consider a group that all of you are familiar with, i.e. the group $SU(2)$. The group $SU(2)$ is given by the following set, i.e.

$$SU(2) \equiv \{u \mid \text{complex } 2 \times 2 \text{ matrices, satisfying } u^\dagger u = \mathbb{I}_{2\times2}, \det(u) = 1\}, \quad (2.3)$$

with the group product–matrix multiplication, satisfying the following group properties. Given u_1, $u_2 \in SU(2)$,

1. closure: $u_1 \cdot u_2 = u_3 \in SU(2)$,

2. unique identity: $\mathbb{I}_{2\times2} \in SU(2)$,

3. unique inverse: if $u \in SU(2)$, then $u^{-1} \equiv u^\dagger \in SU(2)$.

This defines the group $SU(2)$.

The general form of an element of $SU(2)$ is given by the 2×2 matrix

$$u = \begin{pmatrix} \alpha & \beta \\ -\beta^* & \alpha^* \end{pmatrix}, \quad (2.4)$$

with $|\alpha|^2 + |\beta|^2 = 1$. The parameter space of the group $SU(2)$ is thus the three-dimensional surface of a four-dimensional sphere. Also,

$$u^\dagger = \begin{pmatrix} \alpha^* & -\beta \\ \beta^* & \alpha \end{pmatrix} \quad (2.5)$$

and

$$u^\dagger u = \begin{pmatrix} |\alpha|^2 + |\beta|^2 & 0 \\ 0 & |\alpha|^2 + |\beta|^2 \end{pmatrix} \equiv \mathbb{I}_{2\times2}. \quad (2.6)$$

Now consider infinitesimal transformations

$$\alpha = 1 + \frac{i}{2}\theta_3, \quad \beta = \frac{1}{2}(\theta_2 + i\theta_1). \quad (2.7)$$

Then we have

$$u = \mathbb{I}_{2\times2} + i\vec{\theta} \cdot \vec{J}, \quad (2.8)$$

where

$$J_i = \sigma_i/2 \quad (2.9)$$

and σ_i, $i = 1, 2, 3$ are the Pauli matrices

$$\sigma_1 = \begin{pmatrix} 0 & 1 \\ 1 & 0 \end{pmatrix}, \tag{2.10}$$

$$\sigma_2 = \begin{pmatrix} 0 & -i \\ i & 0 \end{pmatrix},$$

$$\sigma_3 = \begin{pmatrix} 1 & 0 \\ 0 & -1 \end{pmatrix}.$$

We can then exponentiate this infinitesimal transformation and obtain a finite rotation given by

$$u = e^{i\vec{\theta} \cdot \vec{J}}. \tag{2.11}$$

θ_i are the three rotation angles and J_i are the three generators (or elements of the Lie algebra) of $SU(2)$. We then obtain the commutation relations

$$[J_i, \ J_j] = i\epsilon_{ijk} \ J_k, \tag{2.12}$$

which we recognize as the Lie algebra of angular momentum.

We can now represent the action of angular momentum on single-particle states in terms of the unitary operator

$$\mathcal{U}(\vec{\theta}) \equiv \mathcal{U}(u(\vec{\theta})) = e^{i\vec{\theta} \cdot \hat{\vec{J}}}. \tag{2.13}$$

Since $u(\vec{\theta})$ satisfies the group property

$$u^{-1}(\vec{\theta}) \ u(\vec{\theta'}) \ u(\vec{\theta}) = u(\vec{\theta''}) \tag{2.14}$$

we have also that

$$\mathcal{U}(u^{-1}(\vec{\theta}) \ u(\vec{\theta'}) \ u(\vec{\theta})) = \mathcal{U}(u(\vec{\theta''})). \tag{2.15}$$

Now consider infinitesimal transformations, which will allow us to show that the angular momentum operators on single-particle states satisfy the same commutation relations as angular momentum operators on spinors. We have

$$\mathcal{U}(\vec{\theta}) \approx 1 + i\vec{\theta} \cdot \hat{\vec{J}} - \frac{1}{2}(\vec{\theta} \cdot \hat{\vec{J}})^2 + \ldots \tag{2.16}$$

and from the defining representation we have

$$u^{-1}(\vec{\theta}) \ u(\vec{\theta'}) \ u(\vec{\theta}) = u(\vec{\theta''}), \tag{2.17}$$

$$(1 - i\vec{\theta} \cdot \vec{J} + \ldots) \ (1 + i\vec{\theta'} \cdot \vec{J} + \ldots) \ (1 + i\vec{\theta} \cdot \vec{J} + \ldots) = (1 + i\vec{\theta''} \cdot \vec{J} + \ldots),$$

$$(1 + i\vec{\theta'} \cdot \vec{J} + \theta_i \ \theta'_j[J_i, \ J_j] - \frac{1}{2}(\vec{\theta'} \cdot \vec{J})^2 + \ldots) = (1 + i\vec{\theta''} \cdot \vec{J} + \ldots).$$

Using $[J_i, \ J_j] = i\epsilon_{ijk} \ J_k$ we obtain

$$\theta''_k = \theta'_k + \epsilon_{kij} \ \theta_i \ \theta'_j. \tag{2.18}$$

Similarly we have

$$\mathcal{U}(u^{-1}(\vec{\theta})\ u(\vec{\theta'})\ u(\vec{\theta})) = \mathcal{U}(u(\vec{\theta''})), \tag{2.19}$$

$$\mathcal{U}(u^{-1}(\vec{\theta}))\ \mathcal{U}(u(\vec{\theta'}))\ \mathcal{U}(u(\vec{\theta})) = \mathcal{U}(u(\vec{\theta''})),$$

$$(1 - i\vec{\theta}\cdot\hat{\vec{J}} + \dots)\ (1 + i\vec{\theta'}\cdot\hat{\vec{J}} + \dots)\ (1 + i\vec{\theta}\cdot\hat{\vec{J}} + \dots) = (1 + i\vec{\theta''}\cdot\hat{\vec{J}} + \dots),$$

and we obtain

$$[\hat{J}_i,\ \hat{J}_j] = i\epsilon_{ijk}\ \hat{J}_k. \tag{2.20}$$

Note, the second line in Eqn. 2.19 is the property that a representation of the Lie algebra must satisfy the homomorphism that

$$\mathcal{U}(u(\vec{\theta_1}))\ \mathcal{U}(u(\vec{\theta_2})) \equiv \mathcal{U}(u(\vec{\theta_1})u(\vec{\theta_2})). \tag{2.21}$$

This guarantees that every representation satisfies the group multiplication properties, and, as a consequence, the generators in every representation of the Lie algebra satisfy the same commutation relations. All the properties we associate with particles are defined as consequences of symmetries. We will talk about energy, momentum and mass; about spin, electric charge, etc. All of these quantities are eigenvalues of conserved charges, which themselves are the generators of symmetries of nature.

2.2 The Lie Algebra of the Poincaré Group

So now let's discuss the Poincaré group $\equiv \mathcal{P}$, i.e. the group of inhomogeneous Lorentz transformations. The generators are

$$P_\mu,\ M_{\mu\nu} = -M_{\nu\mu}. \tag{2.22}$$

We begin the discussion with the defining representation, i.e. the position four-vector, where $x^\mu \to x'^\mu$, with $x'^\mu \equiv \Lambda^\mu{}_\nu x^\nu + a^\mu$. $\Lambda^\mu{}_\nu$ is a Lorentz transformation and a^μ corresponds to a translation. Lorentz transformations satisfy

$$g_{\mu\nu}\Lambda^\mu{}_\alpha\Lambda^\nu{}_\beta \equiv g_{\alpha\beta}, \tag{2.23}$$

such that for $a^\mu = 0$

$$x'^\mu x'^\nu g_{\mu\nu} = \Lambda^\mu{}_\alpha\Lambda^\nu{}_\beta g_{\mu\nu} x^\alpha x^\beta = g_{\alpha\beta} x^\alpha x^\beta. \tag{2.24}$$

We use the following convention for the metric,

$$g_{\mu\nu} = \begin{pmatrix} 1 & & & \\ & -1 & & \\ & & -1 & \\ & & & -1 \end{pmatrix} \quad \text{and } g^\mu{}_\nu \equiv \delta^\mu{}_\nu, \tag{2.25}$$

where $\delta^\mu{}_\nu$ is the Kronecker delta and we use the metric to raise and lower Lorentz indices, i.e. $x_\mu \equiv g_{\mu\nu} x^\nu$.

We then find

$$\Lambda_\nu{}^\alpha \Lambda^\nu{}_\beta \equiv \delta^\alpha{}_\beta, \tag{2.26}$$

and thus the identities

$$\Lambda_\nu{}^\alpha \equiv (\Lambda^T)^\alpha{}_\nu \equiv (\Lambda^{-1})^\alpha{}_\nu, \tag{2.27}$$

where the middle term is the transpose matrix and the last term is the inverse matrix. Thus in 4×4 matrix notation we have $\Lambda^T \Lambda = \mathbb{I}_{4\times 4}$, which then implies $\det \Lambda = \pm 1$.

The identity transformation of the Poincaré group is given by $\Lambda = \mathbb{I}_{4\times 4}, \; a^\mu = 0$. Now let's define the group product. Acting twice with the Poincaré transformation we have

$$
\begin{aligned}
x^\mu \to x''^\mu &= \Lambda'^\mu{}_\nu x'^\nu + a'^\mu \\
&= \Lambda'^\mu{}_\nu (\Lambda^\nu{}_\alpha x^\alpha + a^\nu) + a'^\mu \\
&\equiv \Lambda''^\mu{}_\alpha x^\alpha + a''^\mu,
\end{aligned}
$$

where

$$\Lambda'' = \Lambda' \Lambda \quad a'' = \Lambda' a + a'. \tag{2.28}$$

This is, by definition, a semi-direct product group.

Let's now summarize some properties of the group matrices. We have

- $\det \Lambda = \pm 1$.
- Also, $1 \equiv g_{00} = (\Lambda^0{}_0)^2 - (\Lambda^i{}_0)^2$, where the index i is summed from 1 to 3. Thus $(\Lambda^0{}_0)^2 = 1 + (\Lambda^i{}_0)^2 \geq 1$, which implies $\Lambda^0{}_0 \geq 1$ or $\Lambda^0{}_0 \leq -1$.

Now consider the set of transformations continuously connected to the identity (called the Proper Orthochronos transformations, satisfying

$$
\begin{aligned}
P.O. = \{\Lambda, a | \Lambda (\text{real } 4 \times 4 \text{ matrices}), a(4 \text{ vector}), g_{\mu\nu} \Lambda^\mu{}_\alpha \Lambda^\nu{}_\beta \equiv g_{\alpha\beta}, \\
\det \Lambda = 1, \; \Lambda^0{}_0 \geq 1\}.
\end{aligned} \tag{2.29}
$$

These form a subgroup of $\mathcal{P}$.

We can then obtain all transformations satisfying $\Lambda^0{}_0 \leq -1$ by the discrete transformation, $\{T\Lambda\}$, with $\Lambda \in P.O.$ where

$$
T = \begin{pmatrix} -1 & & & \\ & 1 & & \\ & & 1 & \\ & & & 1 \end{pmatrix}. \tag{2.30}
$$

We can also obtain all Lorentz matrices satisfying $\det \Lambda = -1$ by the discrete transformations given by

$$\{P\Lambda, \; T\Lambda\}, \tag{2.31}$$

with

$$P = \begin{pmatrix} 1 & & & \\ & -1 & & \\ & & -1 & \\ & & & -1 \end{pmatrix}. \tag{2.32}$$

The transformation matrices P, T define parity and time-reversal transformations, respectively.

Single-particle states form irreducible unitary representations of the Poincaré group. We label the quantum numbers of these states by the eigenvalues of a complete set of commuting *conserved* operators. The action of the Poincaré group on the vector space of single-particle states can be represented by unitary operators $U(a, \Lambda)$ which satisfy the group multiplication law (**A**):

$$U(a', \Lambda')U(a, \Lambda) = U(a' + \Lambda'a, \Lambda'\Lambda), \tag{2.33}$$

where (**A**) implies (**B**):

$$U(a, \Lambda) \equiv U(a, 1)U(0, \Lambda). \tag{2.34}$$

We have $U(a, 1) \equiv U(a)$ defines pure translations and $U(0, \Lambda) \equiv U(\Lambda)$ defines pure Lorentz transformations.

By studying the infinitesimal transformations and making use of the group multiplication law we can obtain the commutation relations of the Poincaré generators. Consider the infinitesimal transformations for a translation given in terms of a_μ and for a Lorentz transformation given by $\Lambda^\mu{}_\nu \simeq \delta^\mu{}_\nu + \alpha^\mu{}_\nu$.

$$\Lambda^T \Lambda = \mathbb{I}_{4 \times 4} \Rightarrow \Lambda_\nu{}^\alpha \Lambda^\nu{}_\beta = \delta^\alpha{}_\beta$$
$$\equiv (\delta_\nu{}^\alpha + \alpha_\nu{}^\alpha)(\delta^\nu{}_\beta + \alpha^\nu{}_\beta)$$
$$= \delta^\alpha{}_\beta + \alpha_\beta{}^\alpha + \alpha^\alpha{}_\beta = \delta^\alpha{}_\beta + 0(\alpha^2)$$
$$\Rightarrow \quad \alpha_\beta{}^\alpha + \alpha^\alpha{}_\beta = 0;$$
$$\text{multiplying by } g_{\gamma\alpha}: \quad \alpha_{\beta\gamma} + \alpha_{\gamma\beta} = 0$$
$$\Rightarrow \alpha_{\beta\gamma} = -\alpha_{\gamma\beta}.$$

We've thus identified the group parameters, a_μ, $\alpha_{\mu\nu} = -\alpha_{\nu\mu}$. The unitary operators for a finite Poincaré transformations are then given by

$$U(a) \equiv e^{i\hat{P}^\mu a_\mu},$$
$$U(\Lambda) \equiv e^{\frac{i}{2}\alpha_{\mu\nu}\hat{M}^{\mu\nu}},$$

where $\hat{P}^\mu$, $\hat{M}^{\mu\nu}$ are Hermitian operators acting on the single-particle vector space.

ASIDE:

Note, Poincaré transformations (in fact Lorentz boosts) in the defining representation are NOT unitary.

We have, for example, a finite boost in the z direction given by

$$\Lambda \Rightarrow \Lambda^{\mu}{}_{\nu} = \begin{pmatrix} \cosh\beta & 0 & 0 & \sinh\beta \\ 0 & 1 & 0 & 0 \\ 0 & 0 & 1 & 0 \\ \sinh\beta & 0 & 0 & \cosh\beta \end{pmatrix},$$

$$\Lambda^{T} \Rightarrow (\Lambda^{T})^{\mu}{}_{\nu} = \Lambda_{\nu}{}^{\mu} = g_{\nu\alpha}\Lambda^{\alpha}{}_{\beta}g^{\beta\mu} = \begin{pmatrix} \cosh\beta & 0 & 0 & -\sinh\beta \\ 0 & 1 & 0 & 0 \\ 0 & 0 & 1 & 0 \\ -\sinh\beta & 0 & 0 & \cosh\beta \end{pmatrix}.$$

An infinitesimal boost in the z direction is thus of the form

$$\Lambda^{\mu}{}_{\nu}(z\ boost) \approx \begin{pmatrix} 1 & 0 & 0 & 0 \\ 0 & 1 & 0 & 0 \\ 0 & 0 & 1 & 0 \\ 0 & 0 & 0 & 1 \end{pmatrix} + \beta \begin{pmatrix} 0 & 0 & 0 & 1 \\ 0 & 0 & 0 & 0 \\ 0 & 0 & 0 & 0 \\ 1 & 0 & 0 & 0 \end{pmatrix}. \tag{2.35}$$

We will now prove that a finite boost in the z direction is given by $\Lambda \equiv e^{i\alpha_{03}M^{03}}$ where $\alpha_{03} \equiv \beta$, and

$$iM^{03} \equiv \begin{pmatrix} 0 & 0 & 0 & 1 \\ 0 & 0 & 0 & 0 \\ 0 & 0 & 0 & 0 \\ 1 & 0 & 0 & 0 \end{pmatrix}. \tag{2.36}$$

Note: In general a Lorentz transformation is given by

$$\Lambda^{\mu}{}_{\nu} = (e^{\frac{i}{2}\alpha_{\lambda\rho}M^{\lambda\rho}})^{\mu}{}_{\nu} \approx \delta^{\mu}{}_{\nu} + i\alpha_{\lambda\rho}(M^{\lambda\rho})^{\mu}{}_{\nu}\ \ (\text{for}\ \ \lambda < \rho) + \cdots. \tag{2.37}$$

Then a boost in the z direction is given approximately by $\Lambda^{\mu}{}_{\nu}(z\ boost) \approx \delta^{\mu}{}_{\nu} + i\alpha_{03}(M^{03})^{\mu}{}_{\nu}$, where iM^{03} is given in Eqn. 2.36.

Note,

$$(iM^{03})^2 \equiv \begin{pmatrix} 1 & 0 & 0 & 0 \\ 0 & 0 & 0 & 0 \\ 0 & 0 & 0 & 0 \\ 0 & 0 & 0 & 1 \end{pmatrix}.$$

Then we have

$$e^{i\alpha_{03}M^{03}} = 1 + \sum_{n=1}^{\infty} \frac{(\alpha_{03})^n}{n!}(iM^{03})^n$$

$$= \left(\sum_{m=0}^{\infty} \frac{(\alpha_{03})^{2m}}{(2m)!}\right) \begin{pmatrix} 1 & 0 & 0 & 0 \\ 0 & 0 & 0 & 0 \\ 0 & 0 & 0 & 0 \\ 0 & 0 & 0 & 1 \end{pmatrix} + \left(\sum_{m=0}^{\infty} \frac{(\alpha_{03})^{2m+1}}{(2m+1)!}\right) iM^{03}$$

$$+ \begin{pmatrix} 0 & 0 & 0 & 0 \\ 0 & 1 & 0 & 0 \\ 0 & 0 & 1 & 0 \\ 0 & 0 & 0 & 0 \end{pmatrix}$$

$$\Rightarrow e^{i\alpha_{03}M^{03}} = \begin{pmatrix} 0 & 0 & 0 & 0 \\ 0 & 1 & 0 & 0 \\ 0 & 0 & 1 & 0 \\ 0 & 0 & 0 & 0 \end{pmatrix} + \cosh(\alpha_{03}) \begin{pmatrix} 1 & 0 & 0 & 0 \\ 0 & 0 & 0 & 0 \\ 0 & 0 & 0 & 0 \\ 0 & 0 & 0 & 1 \end{pmatrix}$$

$$+ \sinh(\alpha_{03}) \begin{pmatrix} 0 & 0 & 0 & 1 \\ 0 & 0 & 0 & 0 \\ 0 & 0 & 0 & 0 \\ 1 & 0 & 0 & 0 \end{pmatrix} \equiv \Lambda,$$

with $\beta \equiv \alpha_{03}$.
Q.E.D.
However, $\Lambda = e^{i\beta M^{03}}$ is NOT unitary, since $(M^{03})^{\mu}{}_{\nu}$ is NOT Hermitian. Note, boosts in the four-dimensional vector representation are not unitary, whereas rotations are unitary.

Now, return to unitary operators, which provide the action of the Poincaré group on single-particle states. For translations we have $U(a) = e^{i\hat{P}^{\mu} a_{\mu}}$. Given the group multiplication law $U(a)U(a') \equiv U(a + a')$ we consider infinitesimal transformations and obtain

$$(1 + i\hat{P} \cdot a - \frac{1}{2}(\hat{P} \cdot a)^2 + \dots)(1 + i\hat{P} \cdot a' - \frac{1}{2}(\hat{P} \cdot a')^2 + \dots) = 1 + i\hat{P} \cdot (a + a')$$

$$- \frac{1}{2}(\hat{P} \cdot a + \hat{P} \cdot a')^2 + \dots. \tag{2.38}$$

To order a_{μ}^0 and a_{μ}^1 both sides agree, so consider order a_{μ}^2. We have

$$-\frac{1}{2}(\hat{P} \cdot a)^2 - \frac{1}{2}(\hat{P} \cdot a')^2 - \hat{P} \cdot a \hat{P} \cdot a' = -\frac{1}{2}(\hat{P} \cdot a)^2 - \frac{1}{2}(\hat{P} \cdot a')^2 - \frac{1}{2}(\hat{P} \cdot a' \hat{P} \cdot a + \hat{P} \cdot a \hat{P} \cdot a') \tag{2.39}$$

or

$$0 = \frac{1}{2}[\hat{P} \cdot a, \hat{P} \cdot a'] = \frac{1}{2} a_{\mu} a'_{\nu} [\hat{P}^{\mu}, \hat{P}^{\nu}]. \tag{2.40}$$

This has to be valid for all values of a_μ, a'_ν. Thus we have

$$[\hat{P}^\mu, \hat{P}^\nu] = 0. \tag{2.41}$$

This is the first step in obtaining the Lie algebra of the Poincaré group. What about the commutators, $[\hat{P}^\mu, \hat{M}^{\nu\gamma}] = ?$ or $[\hat{M}^{\mu\nu}, \hat{M}^{\gamma\sigma}] = ?$ In order to find these commutation relations, we once more use the group multiplication law given by

$$U(a', \Lambda')U(a, \Lambda) = U(a' + \Lambda'a, \Lambda). \tag{2.42}$$

Using Eqns. 2.33 and 2.34 we have

$$U(a')U(\Lambda')U(a)U(\Lambda) = \underbrace{U(a' + \Lambda'a)}_{U(a')U(\Lambda'a)}\ \underbrace{U(\Lambda'\Lambda)}_{U(\Lambda')U(\Lambda)} \tag{2.43}$$

or

$$U(\Lambda')U(a)U^{-1}(\Lambda') = U(\Lambda'a). \tag{2.44}$$

Thus,

$$U(\Lambda')e^{i\hat{P}^\mu a_\mu}U^{-1}(\Lambda') = e^{i\hat{P}^\nu \Lambda'^{\ \mu}_\nu a_\mu}. \tag{2.45}$$

Now we drop the prime and evaluate the expression to linear order in a_μ. We have

$$U(\Lambda)\hat{P}^\mu U^{-1}(\Lambda) = \hat{P}^\nu \Lambda_\nu^{\ \mu} \equiv (\Lambda^{-1})^\mu_{\ \nu}\hat{P}^\nu. \tag{2.46}$$

Thus we find that $\hat{P}^\mu$ transforms as a four-vector.

Now use

$$U(\Lambda) = e^{\frac{i}{2}\alpha_{\mu\nu}\hat{M}^{\mu\nu}}$$

$$\simeq 1 + \frac{i}{2}\alpha_{\mu\nu}\hat{M}^{\mu\nu} + 0(\alpha^2)$$

and

$$\Lambda^\mu_{\ \nu} \simeq \delta^\mu_{\ \nu} + \alpha^\mu_{\ \nu}$$

to obtain

$$(1 + \frac{i}{2}\alpha_{\mu\nu}\hat{M}^{\mu\nu})\hat{P}^\sigma(1 - \frac{i}{2}\alpha_{\mu\nu}\hat{M}^{\mu\nu}) = \hat{P}^\alpha(\delta_\alpha^{\ \sigma} + \alpha_\alpha^{\ \sigma}), \tag{2.47}$$

which implies

$$\frac{i}{2}\alpha_{\mu\nu}[\hat{M}^{\mu\nu}, \hat{P}^\sigma] = \hat{P}^\mu \alpha_\mu^{\ \sigma}$$

$$= \hat{P}^\mu g^{\nu\sigma}\alpha_{\mu\nu}$$

$$= \frac{1}{2}(\hat{P}^\mu g^{\nu\sigma} - \hat{P}^\nu g^{\mu\sigma})\alpha_{\mu\nu}$$

or

$$[\hat{M}^{\mu\nu}, \hat{P}^\sigma] = -i(\hat{P}^\mu g^{\nu\sigma} - \hat{P}^\nu g^{\mu\sigma})$$

Finally, for homework you will show that, using the following relation,

$$U(\Lambda)U(\Lambda')U^{-1}(\Lambda) = U(\Lambda\Lambda'\Lambda^{-1}), \tag{2.48}$$

we obtain the commutation relation

$$[\hat{M}^{\mu\nu}, \hat{M}^{\rho\sigma}] = i(\hat{M}^{\mu\rho}g^{\nu\sigma} + \hat{M}^{\nu\sigma}g^{\mu\rho} - \hat{M}^{\nu\rho}g^{\mu\sigma} - \hat{M}^{\mu\sigma}g^{\nu\rho}). \qquad (2.49)$$

We have used the fact that $U^{-1}(\Lambda) \equiv U(\Lambda^{-1})$, since $U(\Lambda^{-1})U(\Lambda) = U(\Lambda^{-1}\Lambda) \equiv \mathbb{I}$.

$U(a)$ and $U(\Lambda)$ act on the single-particle vector space, defined by the maximal set of commuting operators in the Lie algebra of the Poincaré group, i.e. $\{\hat{P}^\mu, \hat{M}^{\mu\nu}\} \subset \mathcal{L}_{\mathcal{P}}$. These are given by the set

$$\{\hat{P}^\mu, \ \hat{M}^2 \equiv \hat{\mathcal{P}}_\mu\hat{P}^\mu, \ \text{ and a subset of } \hat{M}^{\mu\nu}\}, \qquad (2.50)$$

where $\left[\hat{M}^{\mu\nu}, \hat{M}^2\right] = \left[\hat{P}^\mu, \hat{M}^2\right] = 0$ and $\left[\hat{P}^\mu, \hat{P}^\nu\right] = 0$.

For now just consider (we shall define spin later) the states $|P_\mu, m\rangle$ where

$$\hat{P}^\mu|P_\mu, m\rangle = P^\mu|P_\mu, m\rangle,$$
$$\hat{M}^2|P_\mu, m\rangle = m^2|P_\mu, m\rangle$$
$$\text{and } U(a)|P_\mu, m\rangle \equiv e^{iP^\mu a_\mu}|P_\mu, m\rangle.$$

What about $U(\Lambda)|P_\mu, m\rangle$? Consider $\hat{P}^\mu U(\Lambda)|P_\mu, m\rangle$ and use the relation

$$U^{-1}(\Lambda)\hat{P}^\mu U(\Lambda) = (\Lambda\hat{P})^\mu, \qquad (2.51)$$

which gives

$$\hat{P}^\mu U(\Lambda) = U(\Lambda)(\Lambda\hat{P})^\mu. \qquad (2.52)$$

$$\hat{P}^\mu U(\Lambda)|P_\mu, m\rangle = U(\Lambda)(\Lambda\hat{P})^\mu|P_\mu, m\rangle = (\Lambda P)^\mu U(\Lambda)|P_\mu, m\rangle, \qquad (2.53)$$

which implies that

$$U(\Lambda)|P_\mu, m\rangle \equiv |(\Lambda P)_\mu, m\rangle. \qquad (2.54)$$

2.3 Exercises

2.1 A particle with mass m decays into two particles with masses, m_1, m_2 and four momenta, p_1, p_2, respectively. Find the energy and momenta of particles 1 and 2 in the rest frame of the decaying particle.

2.2 (a) What is the minimum energy per proton for two beams of protons with equal and opposite momenta to produce a pair of top quarks ($m_t \sim 175$ GeV)? (b) What is the minimum energy required in the same problem if a beam of protons hits a fixed hydrogen target?

2.3 Using the group property $U(\Lambda)U(\Lambda')U^{-1}(\Lambda) = U(\Lambda\Lambda'\Lambda^{-1})$ where $U(\Lambda) = e^{\frac{i}{2}\alpha_{\mu\nu}\hat{M}^{\mu\nu}}$ and $\Lambda_{\mu\nu} \simeq g_{\mu\nu} + \alpha_{\mu\nu} + O(\alpha^2)$, show that

$$\left[\hat{M}^{\mu\nu}, \ \hat{M}^{\rho\sigma}\right] = i\left(\hat{M}^{\mu\rho}g^{\nu\sigma} + \hat{M}^{\nu\sigma}g^{\mu\rho} - \hat{M}^{\nu\rho}g^{\mu\sigma} - \hat{M}^{\mu\sigma}g^{\nu\rho}\right). \qquad (2.55)$$

2.4 B and $\bar{B}$ = anti-B mesons are produced in the process $e^+e^- \to B\bar{B}$ at a total center of momentum energy $\sqrt{s} = 90$ GeV. Given $m_B = m_{\bar{B}} = 5277.6$ MeV and $\tau_B = \tau_{\bar{B}} = 11.8 \times 10^{-13}$ s, ($m_B = B$ mass, $\tau_B = B$ lifetime), calculate the distance a B travels in its lifetime.

2.5 The Thomson cross-section for light scattering on electrons is given by $\sigma_T = \frac{8\pi\alpha^2}{3m_e^2}$ where $\alpha \simeq 1/137$ and $m_e \simeq \frac{1}{2}$ MeV is the electron mass. Calculate σ_T and give your answer in cm^2.

3 Spin

What about spin? Energy, momentum, mass and the spin of a particle all come as representations of the Poincaré group, i.e. the symmetry of space–time. In this chapter we see how the quantum number of spin enters into our ensemble of particle properties.

3.1 Massive Particles

We will now follow Wigner's theory of induced representations of the Poincaré group. Define $|\tilde{P}\rangle$ as the state of a massive particle in its rest frame, where the four-vector of a massive particle at rest is given by

$$\tilde{P} \equiv (m, \vec{0}). \tag{3.1}$$

Then, using a pure boost, we can take the particle at rest to a new Lorentz frame where it is in motion, i.e.

$$|P\rangle \equiv U\left(L(P)\right)|\tilde{P}\rangle \equiv |L(P)\tilde{P}\rangle, \tag{3.2}$$

where $L(P)$ is a pure Lorentz boost.

We define Wigner's "little group" (LG) as the *subset* of Lorentz transformations which leave $\tilde{P}^{\mu}$ invariant, i.e. $\Lambda^{0\mu}{}_{\nu}$ is an element of the LG if

$$\Lambda^{0\mu}{}_{\nu}\tilde{P}^{\nu} \equiv \tilde{P}^{\mu}. \tag{3.3}$$

We use the LG to define the spin of a particle. Recalling the infinitesimal transformation $\Lambda^{0\mu}{}_{\nu} = \delta^{\mu}{}_{\nu} + \alpha^{0\mu}{}_{\nu}$, we have

$$\alpha^{0}{}_{\mu\nu}\tilde{P}^{\nu} = 0, \tag{3.4}$$

which implies that

$$\alpha^{0}{}_{i0} = \alpha^{0}{}_{0i} \equiv 0. \tag{3.5}$$

However,

$$\alpha^{0}{}_{ij} = -\alpha^{0}{}_{ji} \tag{3.6}$$

are arbitrary and define a set of three independent parameters. We can then define the following convenient form for $\alpha^0{}_{\mu\nu}$. We have

$$\alpha^0{}_{\mu\nu} = \epsilon_{\mu\nu\rho\sigma}\tilde{P}^\rho n^\sigma = -\alpha^0{}_{\nu\mu}$$
$$= \epsilon_{\mu\nu0\sigma}n^\sigma m,$$

where n^i are three independent parameters. Note, $\epsilon_{\mu\nu\rho\sigma}$ is the four-index tensor, anti-symmetric under interchange of any two adjacent indices, with $\epsilon_{0123} \equiv 1$ and n^σ is an arbitrary Lorentz four-vector.[1]

The LG operator acting on single-particle states is now given by

$$\mathcal{U}(\Lambda^0) \simeq 1 + \frac{i}{2}\alpha^0{}_{\mu\nu}\hat{M}^{\mu\nu}$$
$$= 1 + \frac{i}{2}\epsilon_{\mu\nu\rho\sigma}\tilde{P}^\rho n^\sigma \hat{M}^{\mu\nu},$$
$$\mathcal{U}(\Lambda^0)|\tilde{P}, m\rangle = (1 + \frac{i}{2}\epsilon_{\mu\nu\rho\sigma}n^\sigma \hat{M}^{\mu\nu}\hat{P}^\rho)|\tilde{P}, m\rangle.$$

We now define the Pauli–Lubanski spin operator,

$$\hat{W}_\sigma \equiv -\frac{1}{2}\epsilon_{\mu\nu\rho\sigma}\hat{M}^{\mu\nu}\hat{P}^\rho. \tag{3.7}$$

Then

$$\mathcal{U}(\Lambda^0)|\tilde{P}, m\rangle \equiv (1 - in^\sigma \hat{W}_\sigma)|\tilde{P}, m\rangle,$$
$$|\Lambda^0\tilde{P}, m\rangle = |\tilde{P}, m\rangle'.$$

What are some of the properties of $\hat{W}$? We have

$$\hat{W}_\sigma \hat{P}^\sigma \equiv 0, \tag{3.8}$$

which implies

$$\hat{W}_0|\tilde{P}, m\rangle \equiv 0, \tag{3.9}$$

and

$$[\hat{W}_\sigma, \hat{P}^\mu] = 0. \tag{3.10}$$

In addition, it can be shown that

$$[\hat{M}_{\mu\nu}, \hat{W}_\sigma] = -i(\hat{W}_\mu g_{\nu\sigma} - \hat{W}_\nu g_{\mu\sigma}), \tag{3.11}$$

which implies that $\hat{W}_\sigma$ transforms as a Lorentz four-vector. We also have

$$\left[\hat{W}_\lambda, \hat{W}_\sigma\right] = i\epsilon_{\lambda\sigma\alpha\beta}\hat{W}^\alpha \hat{P}^\beta. \tag{3.12}$$

Thus for massive particles at rest we find [the notation $\doteq$ signifies the equality is valid when acting on rest frame states]

$$\hat{W}_\sigma|\tilde{P}, m\rangle = -\frac{1}{2}\epsilon_{\mu\nu0\sigma}\hat{M}^{\mu\nu}m|\tilde{P}, m,\rangle, \tag{3.13}$$

---

[1] Note, $\epsilon_{0ijk} \equiv \epsilon_{ijk}$.

which we write as, acting on rest frame states,

$$\hat{W}_\sigma \doteq -\frac{1}{2}\epsilon_{\mu\nu0\sigma}\hat{M}^{\mu\nu}m \tag{3.14}$$

or

$$\frac{\hat{W}_i}{m} \doteq \hat{J}_i \equiv -\frac{1}{2}\epsilon_{imn}\hat{M}^{mn}, \tag{3.15}$$

$$\hat{W}_0 \doteq 0. \tag{3.16}$$

Note:

$$[\hat{J}_i,\ \hat{J}_j] = i\epsilon_{ijk}\ \hat{J}_k, \tag{3.17}$$

i.e. the Lie algebra of the groups $SO(3)$ or $SU(2)$.

Thus $\hat{W}_\mu$ is a Lorentz covariant generalization of the rotation operators and $\hat{W}_\mu \hat{W}^\mu$ is a Lorentz scalar. Moreover, in the rest frame

$$\hat{W}_\mu \hat{W}^\mu |\tilde{P}, m\rangle \doteq -m^2 \hat{\vec{J}}^2 |\tilde{P}, m\rangle$$
$$= -m^2 s(s+1)|\tilde{P}, m\rangle,$$

where $s = 0, \frac{1}{2}, 1, \frac{3}{2}, \ldots$ is the spin of the state.

Summary for Massive Single-Particle States

Thus finally for massive particles we can define the complete set of commuting operators

$$\{\hat{P}^\mu, \hat{M}^2 \equiv \hat{P}^2, \hat{W}^2/m^2, \hat{W}_3/m\} \tag{3.18}$$

or, equivalently,

$$\hat{P}^\mu, \hat{M}^2, \hat{\vec{J}}^2, \hat{J}_3, \tag{3.19}$$

with eigenstates

$$|P^\mu, m, s, \sigma\rangle. \tag{3.20}$$

$$\hat{J}_3|P^\mu, m, s, \sigma\rangle|_{\vec{p}=0} = \sigma|P^\mu, m, s, \sigma\rangle|_{\vec{p}=0}, \tag{3.21}$$

and

$$\hat{\vec{J}}^2|P^\mu, m, s, \sigma\rangle|_{\vec{p}=0} = s(s+1)|P^\mu, m, s, \sigma\rangle|_{\vec{p}=0}. \tag{3.22}$$

$$(\hat{J}_1 \pm i\hat{J}_2)|P^\mu, m, s, \sigma\rangle|_{\vec{p}=0} = \sqrt{(s \mp \sigma)(s \pm \sigma + 1)}\ |P^\mu, m, s, \sigma \pm 1\rangle|_{\vec{p}=0}. \tag{3.23}$$

Now we want to know how Lorentz transformations act on states in motion. We use a Lorentz boost given by $L(p)$ which takes $\tilde{P} = (m, \vec{0})$ into

$$P^\mu = (E, \vec{p}) = L^\mu{}_\nu(P)\tilde{P}^\nu. \tag{3.24}$$

Then define

$$|P^\mu, m, s, \sigma\rangle \equiv U\left(L(P)\right)|\tilde{P}^\mu, m, s, \sigma\rangle. \qquad (3.25)$$

How does $|P^\mu, m, s, \sigma\rangle$ transform under Poincaré transformations? We have

$$U(a)|P^\mu, m, s, \sigma\rangle = e^{iP^\mu a_\mu}|P^\mu, m, s, \sigma\rangle, \qquad (3.26)$$

with

$$E \equiv P^0 = (\vec{p}^{\,2} + m^2)^{1/2}. \qquad (3.27)$$

We then have

$$U(\Lambda)|P^\mu, m, s, \sigma\rangle = \underbrace{U(\Lambda)U\left(L(P)\right)}_{U\left(\Lambda L(P)\right)}|\tilde{P}, m, s, \sigma\rangle$$

$$= U\left(L(\Lambda P)\right)\underbrace{U\left(L^{-1}(\Lambda P)\right)U\left(\Lambda L(P)\right)}_{U\left(L^{-1}(\Lambda P)\Lambda L(P)\right)}|\tilde{P}, m, s, \sigma\rangle.$$

Consider the equation

$$L^{-1}(\Lambda P)\Lambda L(P)\tilde{P} \equiv L^{-1}(\Lambda P)(\Lambda P)$$
$$\equiv \tilde{P}.$$

Hence, $L^{-1}(\Lambda P)\Lambda L(P)$ is an element of the LG and

$$R(\Lambda P, P) \equiv L^{-1}(\Lambda P)\Lambda L(P) \qquad (3.28)$$

is called a Wigner rotation. Figure 3.1 shows a simple example of a Wigner rotation using Mathematica. Now we have

$$U\left(R(\Lambda P, P)\right)|\tilde{P}^\mu, m, s, \sigma\rangle \equiv \sum_{\sigma'}|\tilde{P}^\mu, m, s, \sigma'\rangle\langle\tilde{P}^\mu, m, s, \sigma'|U(R)|\tilde{P}^\mu, m, s, \sigma\rangle$$

$$\equiv \sum_{\sigma'} D^{(s)}_{\sigma'\sigma}\left(R(\Lambda P, P)\right)|\tilde{P}^\mu, m, s, \sigma'\rangle, \qquad (3.29)$$

where

$$D^{(s)}_{\sigma'\sigma} \equiv (e^{-i\theta\hat{n}\cdot\vec{J}^{(s)}})_{\sigma'\sigma} \qquad (3.30)$$

are the Wigner rotation matrices in the spin s representation.

Finally, we have

$$U(\Lambda)|P^\mu, m, s, \sigma\rangle \equiv \sum_{\sigma'} D^{(s)}_{\sigma'\sigma}\left(R(\Lambda P, P)\right)|(\Lambda P)^\mu, m, s, \sigma'\rangle. \qquad (3.31)$$

We see that, in general, Lorentz transformations rotate the components of spin.

```
sx := Sinh[x]
cx := Sqrt[1 + sx^2]

Lx := {{cx, sx, 0, 0}, {sx, cx, 0, 0}, {0, 0, 1, 0}, {0, 0, 0, 1}}
Lix := {{cx, -sx, 0, 0}, {-sx, cx, 0, 0}, {0, 0, 1, 0}, {0, 0, 0, 1}}

Simplify[Lix.Lx]
```

$$\{\{1, 0, 0, 0\}, \{0, 1, 0, 0\}, \{0, 0, 1, 0\}, \{0, 0, 0, 1\}\}$$

```
sz := Sinh[z]
cz := Sqrt[1 + sz^2]

Lz := {{cz, 0, 0, sz}, {0, 1, 0, 0}, {0, 0, 1, 0}, {sz, 0, 0, cz}}
Liz := {{cz, 0, 0, -sz}, {0, 1, 0, 0}, {0, 0, 1, 0}, {-sz, 0, 0, cz}}

Simplify[Liz.Lz]
```

$$\{\{1, 0, 0, 0\}, \{0, 1, 0, 0\}, \{0, 0, 1, 0\}, \{0, 0, 0, 1\}\}$$

```
Lambdaxz := Lx.Lz
Lambdaxz
```

$$\left\{\left\{\sqrt{1 + \text{Sinh}[x]^2}\ \sqrt{1 + \text{Sinh}[z]^2}\ ,\ \text{Sinh}[x],\ 0,\ \sqrt{1 + \text{Sinh}[x]^2}\ \text{Sinh}[z]\right\},\right.$$
$$\left\{\text{Sinh}[x]\ \sqrt{1 + \text{Sinh}[z]^2}\ ,\ \sqrt{1 + \text{Sinh}[x]^2}\ ,\ 0,\ \text{Sinh}[x]\ \text{Sinh}[z]\right\},$$
$$\left.\{0, 0, 1, 0\},\ \left\{\text{Sinh}[z],\ 0,\ 0,\ \sqrt{1 + \text{Sinh}[z]^2}\ \right\}\right\}$$

```
tx := sx / cx
tz := sz / cz
gam := 1 / (tx^2 + tz^2 / cx^2)
beta := cz * cx - 1
Lixz := {{cz * cx, -cz * sx, 0, -sz},
    {-cz * sx, 1 + gam * beta * tx^2, 0, gam * beta * tx * tz / cx},
    {0, 0, 1, 0},
    {-sz, gam * beta * tx * tz / cx, 0, 1 + gam * beta * tz^2 / cx^2}}
Lxz := {{cz * cx, cz * sx, 0, sz},
    {cz * sx, 1 + gam * beta * tx^2, 0, gam * beta * tx * tz / cx},
    {0, 0, 1, 0},
    {sz, gam * beta * tx * tz / cx, 0, 1 + gam * beta * tz^2 / cx^2}}

Simplify[Lixz.Lxz]
```

$$\{\{1, 0, 0, 0\}, \{0, 1, 0, 0\}, \{0, 0, 1, 0\}, \{0, 0, 0, 1\}\}$$

```
R := Lixz.Lambdaxz
FunctionExpand[Factor[FullSimplify[R]]]
```

$$\left\{\{1, 0, 0, 0\},\ \left\{0,\ \frac{\sqrt{\text{Cosh}[z]^2}\ \text{Sinh}[x]^2 + \sqrt{\text{Cosh}[x]^2}\ \text{Sinh}[z]^2}{\text{Cosh}[z]^2\ \text{Sinh}[x]^2 + \text{Sinh}[z]^2},\right.\right.$$
$$\left.0,\ \frac{\left(-1 + \sqrt{\text{Cosh}[x]^2}\ \sqrt{\text{Cosh}[z]^2}\right)\text{Sinh}[x]\ \text{Sinh}[z]}{\text{Cosh}[z]^2\ \text{Sinh}[x]^2 + \text{Sinh}[z]^2}\right\},\ \{0, 0, 1, 0\},$$
$$\left\{0,\ \frac{\text{Sinh}[x]\ \text{Sinh}[z]\left(-1 + \sqrt{\text{Cosh}[x]^2}\ \sqrt{\text{Cosh}[z]^2} - \text{Cosh}[z]^2\ \text{Sinh}[x]^2 - \text{Sinh}[z]^2\right)}{\sqrt{\text{Cosh}[x]^2}\ \sqrt{\text{Cosh}[z]^2}\ \left(\text{Cosh}[z]^2\ \text{Sinh}[x]^2 + \text{Sinh}[z]^2\right)},\right.$$
$$\left.\left.0,\ \frac{\sqrt{\text{Cosh}[z]^2}\ \text{Sinh}[x]^2 + \sqrt{\text{Cosh}[x]^2}\ \text{Sinh}[z]^2}{\text{Cosh}[z]^2\ \text{Sinh}[x]^2 + \text{Sinh}[z]^2}\right\}\right\}$$

```
sx := 100
sz := 200

N[Simplify[R]]
```

```
{{1., 0., 0., 0.}, {0., 0.0149987, 0., 0.999888},
 {0., 0., 1., 0.}, {0., -0.999888, 0., 0.0149987}}
```

Fig. 3.1   A Mathematica notebook example of a Wigner rotation.

3.2 Massless Particles

For massless particles, there is no rest frame. Nevertheless we still have

$$\hat{W}_\sigma \equiv -\frac{1}{2}\epsilon_{\mu\nu\rho\sigma}\hat{M}^{\mu\nu}\hat{P}^\rho \tag{3.32}$$

(see Eqn. 3.4). Moreover, $\hat{W}_\sigma$ is still a LG generator, which leaves invariant the massless particle state

$$\tilde{P} = (p, 0, 0, p) \tag{3.33}$$

in the so-called, standard form.

What is the algebra of the LG? We derive the algebra as follows:

$$\hat{W}_\sigma \hat{P}^\sigma |\tilde{P}\rangle \equiv 0, \tag{3.34}$$

which implies that

$$(\hat{W}_0 + \hat{W}_3)|\tilde{P}\rangle = 0. \tag{3.35}$$

In addition, we have

$$\hat{P}_\sigma \hat{P}^\sigma |\tilde{P}\rangle = 0. \tag{3.36}$$

Finally, we have

$$\hat{W}_\sigma \hat{W}^\sigma |\tilde{P}\rangle = C|\tilde{P}\rangle \tag{3.37}$$

where the constant C is a Lorentz scalar with dimensions of mass.

We now want to prove that $C = 0$.

Consider the Lorentz boost in the z direction

$$\Lambda = \begin{pmatrix} \cosh\beta & 0 & 0 & \sinh\beta \\ 0 & 1 & 0 & 0 \\ 0 & 0 & 1 & 0 \\ \sinh\beta & 0 & 0 & \cosh\beta \end{pmatrix}. \tag{3.38}$$

Then

$$\lim_{\beta\to-\infty} \Lambda\tilde{P} = \lim_{\beta\to-\infty} e^\beta p \to 0. \tag{3.39}$$

But

$$\hat{W}_\sigma \hat{W}^\sigma |\lim_{\beta\to-\infty} \Lambda\tilde{P}\rangle = C|\lim_{\beta\to-\infty} \Lambda\tilde{P}\rangle \to 0 \tag{3.40}$$

and therefore $C = 0$ and

$$\hat{W}_\sigma \hat{W}^\sigma |\tilde{P}\rangle = 0. \tag{3.41}$$

(See Weinberg, 1996, ch. 2, for more details.)

Now let us use the Pauli–Lubanski commutation relations (Eqn. 3.12). We have

$$[\hat{W}_1, \hat{W}_2]|\tilde{P}\rangle = (i\epsilon_{1203}\hat{W}^0 p + i\epsilon_{1230}\hat{W}^3 p)|\tilde{P}\rangle$$
$$= i\epsilon_{1203}\, p(\hat{W}^0 - \hat{W}^3)|\tilde{P}\rangle$$
$$= i\epsilon_{1203}\, p(\hat{W}_0 + \hat{W}_3)|\tilde{P}\rangle \equiv 0.$$

In addition, we have

$$[\hat{W}_3, \hat{W}_1]|\tilde{P}\rangle = ip\hat{W}_2|\tilde{P}\rangle,$$
$$[\hat{W}_3, \hat{W}_2]|\tilde{P}\rangle = -ip\hat{W}_1|\tilde{P}\rangle,$$

where the right-hand side of the first equation follows from

$$i\epsilon_{31\mu\nu}\hat{W}^\mu \tilde{P}^\nu \tag{3.42}$$

and, for example, $\epsilon_{3120} = -1$.

The commutation relation $[\hat{W}_1, \hat{W}_2] = 0$ implies that these operators generate translations in the 1–2 plane, while $\hat{W}_3/p$ generates rotations about the z axis. The eigenvalues of translations, in general, are continuous, and, in particular, $\hat{W}_1^2 + \hat{W}_2^2$ takes on continuous values. But $\hat{W}^\mu \hat{W}_\mu |\tilde{P}\rangle = 0$. Therefore, we have

$$[(\hat{W}_0)^2 - (\hat{W}_3)^2) - ((\hat{W}_1)^2 + (\hat{W}_2)^2)]|\tilde{P}\rangle = 0. \tag{3.43}$$

But

$$((\hat{W}_0)^2 - (\hat{W}_3)^2)|\tilde{P}\rangle \equiv (\hat{W}_0 + \hat{W}_3)(\hat{W}_0 - \hat{W}_3)|\tilde{P}\rangle = 0. \tag{3.44}$$

Hence

$$[(\hat{W}_1)^2 + (\hat{W}_2)^2]|\tilde{P}\rangle = 0. \tag{3.45}$$

The only conclusion is thus

$$\hat{W}_1|\tilde{P}\rangle = \hat{W}_2|\tilde{P}\rangle = 0. \tag{3.46}$$

Combining the results of Eqns. 3.41 and 3.46 we find

$$\hat{W}_\mu|\tilde{P}\rangle \propto \hat{P}_\mu|\tilde{P}\rangle. \tag{3.47}$$

Since $\hat{W}_\mu$ and $\hat{P}_\mu$ are both Lorentz four-vectors, this implies that, in general (for all massless states),

$$\hat{W}_\mu = -\lambda\hat{P}_\mu, \tag{3.48}$$

where λ is a Lorentz scalar. We then define

$$\lambda \equiv \frac{-\hat{W}_\mu n^\mu}{\hat{P}_\mu n^\mu} = \frac{-\frac{1}{2}\epsilon_{\mu\nu\rho\sigma}\hat{M}^{\nu\rho}\hat{P}^\sigma n^\mu}{\hat{P}_\mu n^\mu}, \tag{3.49}$$

where n^μ is an arbitrary four-vector. Now let $n^\mu \equiv (1, \vec{0})$, then

$$\lambda = \frac{\overbrace{-\dfrac{1}{2}\epsilon_{ijk}\hat{M}^{ij}\hat{P}^k}^{\hat{J}_k}}{\hat{P}^0},$$

$$\lambda \equiv \frac{\hat{\vec{J}} \cdot \hat{\vec{P}}}{|\hat{\vec{P}}|}. \tag{3.50}$$

The parameter λ is called helicity, i.e. the component of angular momentum in the direction of motion of the particle.

A zero mass state has only one spin component. However, if we require **parity** invariance, then $\vec{p} \to -\vec{p}$, $\vec{J} \to \vec{J}$ and $\lambda \to -\lambda$, i.e. two spin states form a parity doublet. As an example, the two helicity states of a photon, $\lambda = \pm 1$, are related by parity.

Note: for massive states, under a Lorentz boost,

$$\xrightarrow{\Rightarrow \; spin}_{\vec{p}} \quad \text{can go to a new Lorentz frame such that} \quad \xleftarrow{\Rightarrow \; spin}_{\vec{p}},$$

$$\lambda > 0 \qquad\qquad\qquad\qquad\qquad\qquad\qquad\qquad \lambda < 0$$

which implies that both spins are necessary. Note, also: λ, as defined with the particular unit vector, n^μ, is not Lorentz invariant. However, λ is invariant under pure rotations.

We have shown that massless states only have helicity quantum numbers, while massive states form complete representations of the rotation group. Nevertheless, even for massive states, it is still sometimes useful to represent them in a helicity eigenstate basis. Let us now define the helicity basis for the spin of massive particles. Choose $|\tilde{P}^\mu, m, s, \lambda\rangle$ with λ defined as the z component of spin. Then boost the state in the z direction, with

$$|P'^\mu, m, s, \lambda\rangle \equiv U(L(P'))|\tilde{P}^\mu, m, s, \lambda\rangle, \tag{3.51}$$

where

$$P'^\mu \equiv (E, 0, 0, p)$$
$$= L(P')^\mu{}_\nu \, \tilde{P}^\nu.$$

Now, rotate the three-momentum into an arbitrary $\vec{p}$ direction. Since helicity is a scalar under rotations, λ doesn't change.

In general, we have

$$|P^\mu, m, s, \lambda\rangle \equiv U(R_{PP'})|P'^\mu, m, s, \lambda\rangle, \tag{3.52}$$

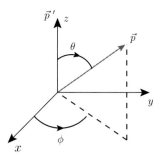

Fig. 3.2 This figure defines the momentum vector in spherical coordinates.

with

$$R_{PP'} \equiv R(\phi, \theta, -\phi)$$
$$= e^{-i\phi \hat{J}_3} e^{-i\theta \hat{J}_2} e^{i\phi \hat{J}_3},$$

where $\vec{p}$ points in the (θ, ϕ) direction with respect to the z axis (see Fig. 3.2).

ASIDE: More about the Lorentz group.
Define

$$\hat{J}_i \equiv -\frac{1}{2}\epsilon_{imn}\hat{M}^{mn} \quad \text{rotations,}$$
$$\hat{K}_i \equiv \hat{M}_{i0} \quad \text{boosts.}$$

These have the following commutation relations

$$\left[\hat{J}_i, \hat{P}_k\right] = i\epsilon_{ikl}\,\hat{P}_l,$$
$$\left[\hat{J}_i, \hat{P}_0\right] = 0,$$
$$\left[\hat{K}_i, \hat{P}_k\right] = i\hat{P}_0\,g_{ik},$$
$$\left[\hat{K}_i, \hat{P}_0\right] = -i\hat{P}_i$$

and

$$\left[\hat{J}_m, \hat{J}_n\right] = i\epsilon_{mnk}\,\hat{J}_k,$$
$$\left[\hat{J}_m, \hat{K}_n\right] = i\epsilon_{mnk}\,\hat{K}_k,$$
$$\left[\hat{K}_m, \hat{K}_n\right] = -i\epsilon_{mnk}\,\hat{J}_k.$$

This is the Lie algebra of the group $SO(3,1)$, while the first line ($[\hat{J}_m, \hat{J}_n]$) is the Lie algebra of the group $SO(3) \approx SU(2)$.

We can now define the complexified Lie algebra

$$\hat{j}_m^{\pm} \equiv \frac{1}{2}(\hat{J}_m \pm i\hat{K}_m), \tag{3.53}$$

which satisfy

$$\left[\hat{j}_m^\pm, \hat{j}_n^\pm\right] = i\epsilon_{mnk}\,\hat{j}_k^\pm,$$

$$\left[\hat{j}_m^\pm, \hat{j}_n^\mp\right] = 0.$$

This is the Lie algebra of the group $SU(2) \times SU(2)$. Using this notation, a general Lorentz transformation can be written in the form

$$U(\Lambda) \equiv e^{-ia\cdot\hat{J}+ib\cdot\hat{K}}$$
$$= e^{\hat{j}^+\cdot(b-ia)}e^{-\hat{j}^-\cdot(b+ia)}.$$

In the notation of $SU(2) \times SU(2)$ with generators $\{\hat{j}_m^+,\ \hat{j}_m^-\}$ we have the fundamental spinor representations given by $(\frac{1}{2},0)$ and $(0,\frac{1}{2})$, while the four-vector representation is given by $(\frac{1}{2},\frac{1}{2})$. Note, $(\frac{1}{2},0)$ is an eigenvector of $\hat{j}_m^-$ with eigenvalue 0. Thus, on this state we have $\hat{J}_m = i\hat{K}_m$. Hence, in this representation we have

$$J_m = \frac{1}{2}\sigma_m, \quad j_m^+ = J_m = \frac{1}{2}\sigma_m, \tag{3.54}$$

where σ_m, $m = 1,2,3$ are the Pauli matrices. For the $(0,\frac{1}{2})$ state we have $\hat{J}_m = -i\hat{K}_m$ and

$$J_m = \frac{1}{2}\sigma_m, \quad j_m^- = J_m = \frac{1}{2}\sigma_m. \tag{3.55}$$

4 Completeness and Normalization

In this chapter we discuss the completeness and normalization relation for single-particle states. These are the states which would be in the beam of particles in a scattering experiment, and the particles observed asymptotically in a detector. We then define the abstract notion of cross-sections and decay rates.

4.1 Single-Particle States

Single-particle states form a complete set of states in the subspace of a single particle. Therefore, given the simplified notation

$$|\vec{p}, \lambda\rangle \equiv |P^\mu, m, s, \lambda\rangle, \tag{4.1}$$

with m = mass, s = spin and λ = helicity, we have the completeness relation

$$\sum_\lambda \int dLips(p)|\vec{p}, \lambda\rangle\langle\vec{p}, \lambda| \equiv \mathbb{I}, \tag{4.2}$$

where $dLips(p)$ is called Lorentz invariant phase space and is defined below,

$$dLips(p) \equiv \frac{d^4p}{(2\pi)^3}\delta(P^2 - m^2)\theta(P^0)$$

$$= \frac{d^4p}{(2\pi)^3 2E_p}\left(\delta(P^0 - E_p) + \delta(P^0 + E_p)\right)\theta(P^0) \tag{4.3}$$

$$= \frac{d^3\vec{p}}{(2\pi)^3 2E_p}, \tag{4.4}$$

with $E_p \equiv (\vec{p}^{\,2} + m^2)^{1/2}$ (for more information on $Lips(p)$ see (pilkuhn, 1967)

Now let's confirm that indeed $Lips(p)$ is Lorentz invariant. We will look at each term separately. The first term d^4p is invariant, since under a Lorentz transformation we have $P' = \Lambda P$ and the Jacobian of the transformation is given by $\det \Lambda \equiv 1$. Hence, d^4p is Lorentz invariant. The term $\theta(P^0)$ is also invariant since if $P^0 > 0$, then also $P^{0\prime} > 0$. Finally, $\delta(P^2 - m^2)$ is Lorentz invariant and in the second step we used

$$\delta(f(x)) \equiv \sum_i \frac{\delta(x - x_i)}{\left|\frac{\partial f}{\partial x}\right|_{x=x_i}}, \tag{4.5}$$

where $f(x_i) \equiv 0$.

We now derive the normalization of states, which is already determined by the completeness relation. Let us define the identity operator, $\mathbb{I}$, satisfying

$$\mathbb{I}\,|\vec{p}\,',\lambda'\rangle = \left(\sum_\lambda \int dLips(p)|\vec{p},\lambda\rangle\langle\vec{p},\lambda|\right)|\vec{p}\,',\lambda'\rangle \equiv |\vec{p}\,',\lambda'\rangle, \qquad (4.6)$$

with the solution given by the Lorentz invariant normalization of single-particle states,

$$\langle\vec{p},\lambda|\vec{p}\,',\lambda'\rangle = (2\pi)^3 2E_p \delta^3(\vec{p}-\vec{p}')\delta_{\lambda\lambda'}. \qquad (4.7)$$

Note, we can also identify the operator $\mathbb{I} \equiv \hat{N}$ as the number operator such that

$$\hat{N}|\vec{p},\lambda\rangle \equiv |\vec{p},\lambda\rangle \qquad (4.8)$$

implies that the single-particle state has one particle in all space. Using this we can then find the number of particles per unit volume. We have

$$\lim_{\vec{p}\,'\to\vec{p}}\langle\vec{p}\,',\lambda|\hat{N}|\vec{p},\lambda\rangle \equiv (2\pi)^3 2E_p \delta^3(\vec{0}). \qquad (4.9)$$

The volume of space, V, is given by

$$V \equiv \int_{\vec{p}\,'\to\vec{p}} d^3\vec{x}\ e^{i(\vec{p}-\vec{p}\,')\cdot\vec{x}} \equiv (2\pi)^3 \delta^3(\vec{0}), \qquad (4.10)$$

which implies the result

$$\langle\vec{p},\lambda|\hat{N}|\vec{p},\lambda\rangle \equiv 2E_p V, \qquad (4.11)$$

and thus the number of particles per unit volume in the single-particle states (normalized in this way) is given by

$$\langle\vec{p},\lambda|\frac{\hat{N}}{V}|\vec{p},\lambda\rangle \equiv 2E_p. \qquad (4.12)$$

Note, one would expect the number of particles per unit volume to have dimensions of ℓ^{-3}. However our normalization of states is not consistent with this. Nevertheless, don't worry, since the dimensions of observable quantities will all work out in the end.

4.2 Definition of Cross-Section and Lifetime

What do we observe in a scattering experiment? We observe the quantity $\frac{N_m}{T}$, defined as the number of counts per unit time in a detector of a particle [of state m] with four-momentum $[\vec{p}, \vec{p} + \vec{\Delta p}]$ and mass m. Consider the laboratory frame, (Fig. 4.1) we have

$$\frac{dN_m}{T} \propto N_2 \times F_1, \qquad (4.13)$$

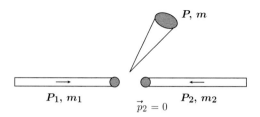

Fig. 4.1   The laboratory frame for scattering.

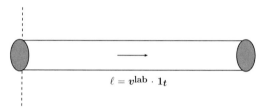

Fig. 4.2   The beam of incident particles moving through an imaginary line.

where N_2 is the number of particles in the target, F_1 is the incident flux and $\frac{dN_m}{T}$ is the number of particles scattered into a solid angle per second. We then define the differential cross-section, $d\sigma_{12\to m}$, as the proportionality constant, by the relation

$$\frac{dN_m}{T} \equiv N_2 F_1 d\sigma_{12\to m}. \tag{4.14}$$

The differential cross-section or transition rate is equivalent to the probability of detecting an event in the detector with some solid angle per unit time per target particle per incident flux.

The definition of the cross-section in simple scattering events involves some *idealizations* which are not totally *realistic*. Let us consider the idealizations vs. reality below:

- The idealization – we consider a plane wave beam, i.e. a momentum eigenstate, incident on an infinite-plane target, perpendicular to the beam.
- The reality –
  (1) the target is within the beam;
  (2) it is a "thin" target, so there is no multiple scattering or absorption of the emitted particles;
  (3) the plane wave beam is in reality a localized wave packet in both the transverse and longitudinal directions.

Consider the beam of particles (see Fig. 4.2) with speed v^{lab} and moving a distance ℓ in one unit of time, 1_t. The density of particles in the beam is $\rho_1^{\text{lab}} = N_1/V$ and the incident flux is given by

$$F_1 = \frac{\rho_1 \ell}{1_t} = \frac{\text{the number of particles per unit area}}{\text{unit time}}$$
$$= \rho_1^{\text{lab}} v^{\text{lab}}. \tag{4.15}$$

Then we have

$$\frac{dN_m}{T} \equiv \rho_2(\text{rest})\rho_1^{\text{lab}} v^{\text{lab}} d\sigma_{12\to\text{m}} V, \tag{4.16}$$

where $N_2 \equiv \rho_2(\text{rest}) V$.

Therefore, we find the differential cross-section given, in general, by

$$d\sigma_{12\to m} \equiv \frac{(dN_m/TV)}{\rho_1^{\text{lab}}\rho_2(\text{rest})v^{\text{lab}}}, \tag{4.17}$$

and the integrated cross-section, integrated over all of phase space, is given by

$$\sigma_{12\to m} \equiv \frac{(N_m/TV)}{\rho_1^{\text{lab}}\rho_2(\text{rest})v^{\text{lab}}}. \tag{4.18}$$

We now show that the term

$$\rho_1^{\text{lab}}\rho_2(\text{rest})v^{\text{lab}} \tag{4.19}$$

is a Lorentz invariant quantity.

Proof

$$\rho_1^{\text{lab}} \equiv \rho_1(\text{rest}) \frac{E_1^{\text{lab}}}{m_1} \tag{4.20}$$

due to a Lorentz contraction. Then we have

$$\rho_1^{\text{lab}}\rho_2(\text{rest})v^{\text{lab}} = \rho_1(\text{rest}) \frac{E_1^{\text{lab}}}{m_1} \cdot \rho_2(\text{rest}) \frac{|\vec{p_1}|^{\text{lab}}}{E_1^{\text{lab}}} \tag{4.21}$$

$$\equiv \rho_1(\text{rest})\rho_2(\text{rest}) \frac{|\vec{p_1}|^{\text{lab}}}{m_1}.$$

Digression on kinematics:

$$P^\mu = (E, \vec{p}) = (P^0, p^i),$$
$$x^\mu = (t, \vec{x}) = (x^0, x^i),$$
$$P \cdot x = P^\mu x^\nu g_{\mu\nu} = Et - \vec{p} \cdot \vec{x}.$$

In the laboratory frame we have

$$P_1 = (E_1^{\text{lab}}, \vec{p}_1^{\ \text{lab}}), \quad P_2 = (m_2, \vec{0}). \tag{4.22}$$

Define

$$s \equiv (P_1 + P_2)^2 = m_1^2 + m_2^2 + 2m_2 E_1^{\text{lab}}, \tag{4.23}$$

with

$$E_1^{\text{lab}} \equiv \sqrt{m_1^2 + (\vec{p}_1^{\ \text{lab}})^2}. \tag{4.24}$$

Using Eqn. 4.23 we have

$$(s - m_1^2 - m_2^2)^2 = 4m_2^2 \left((\vec{p}_1^{\text{ lab}})^2 + m_1^2 \right) \tag{4.25}$$

$$m_2 |\vec{p}_1^{\text{ lab}}| = \frac{1}{2} \sqrt{(s - m_1^2 - m_2^2)^2 - 4m_1^2 m_2^2}$$

$$\equiv \frac{1}{2} \sqrt{\lambda(s_1 m_1^2, m_2^2)}.$$

which defines the function, $\lambda(s, m_1^2, m_2^2)$.

In the center of momentum system (CMS) we have

$$\vec{p}_1 + \vec{p}_2 \equiv 0. \tag{4.26}$$

Thus,

$$\vec{p}_1 \equiv \vec{p} = -\vec{p}_2, \quad p \equiv |\vec{p}| \tag{4.27}$$

and

$$s = (E_1 + E_2)^2, \tag{4.28}$$

where the total CMS energy is

$$E_{CM} \equiv E_1 + E_2 = \sqrt{s}. \tag{4.29}$$

We also have

$$s \equiv (P_1 + P_2)^2 = (m_1^2 + m_2^2 + 2P_1 \cdot P_2), \tag{4.30}$$

with

$$P_1 \cdot P_2 = E_1 E_2 - \vec{p}_1 \cdot \vec{p}_2 = E_1 E_2 + (\vec{p})^2. \tag{4.31}$$

Thus,

$$(s - m_1^2 - m_2^2) = 2(E_1 E_2 + p^2) \tag{4.32}$$

and using $E_1 = \sqrt{p^2 + m_1^2}$, $E_2 = \sqrt{p^2 + m_2^2}$ and $p \equiv |\vec{p}|$ we obtain the identity

$$p s^{1/2} \equiv \frac{1}{2} \sqrt{\lambda}. \tag{4.33}$$

Let's consider some of the properties of λ. We have

$$\lambda(s, m_1^2, m_2^2) \equiv s^2 + m_1^4 + m_2^4 - 2s m_1^2 - 2s m_2^2 - 2m_1^2 m_2^2, \tag{4.34}$$

which is totally symmetric under interchange of s, m_1^2, m_2^2. Also,

$$\lambda \equiv [s - (m_1 + m_2)^2][s - (m_1 - m_2)^2]. \tag{4.35}$$

Returning to (Eqn. 4.21) and using the result of Eqn. 4.25, we see that

$$\rho_1^{(\text{lab})} \rho_2(\text{rest}) v^{\text{lab}} \equiv \rho_1(\text{rest}) \rho_2(\text{rest}) \frac{\sqrt{\lambda}}{2m_1 m_2} \tag{4.36}$$

is Lorentz invariant. □

We can now rewrite Eqn. 4.17 as

$$d\sigma_{12 \to m} \equiv \frac{(dN_m/TV)}{\rho_1(\text{rest})\rho_2(\text{rest})} \frac{2m_1 m_2}{\sqrt{\lambda}}, \tag{4.37}$$

which is Lorentz covariant.

The four-volume element, TV, is Lorentz invariant, but dN_m depends on the direction of momentum in some frame. The total number of events integrated over all momenta,

$$N_m \equiv \int dN_m, \tag{4.38}$$

is however Lorentz invariant. Thus, the total cross-section,

$$\sigma_{12 \to m}, \tag{4.39}$$

is also Lorentz invariant.

Finally, with our Lorentz covariant normalization of states (for two-particle scattering), we obtain $\rho_1 \equiv 2E_1$ and $\rho_1(rest) \equiv 2m_1$, which gives

$$d\sigma_{12 \to m} \equiv \frac{dN_m/TV}{2\sqrt{\lambda}}. \tag{4.40}$$

The dimensions of a cross-section is $[\sigma] = \ell^2, \frac{1}{m^2}$.[1]

It is useful now to define the quantity known as *luminosity*. It depends on the incident flux in the scattering experiment. We have

$$\mathcal{L}_{12}(t) \equiv N_2 F_1. \tag{4.41}$$

Then the number of events in state m per unit time in some solid angle is given by

$$\frac{dN_m}{T} \equiv \mathcal{L}_{12}(t) d\sigma_{12 \to m} \tag{4.42}$$

and the integrated number of events (integrated over all solid angles) is given by

$$\frac{N_m}{T} \equiv \mathcal{L}_{12}(t) \sigma_{12 \to m}. \tag{4.43}$$

Luminosity has dimensions $\frac{1}{\ell^2 s}$. We also define the *integrated luminosity* (integrated over time) such that

$$N_m \equiv \int_0^T dt \frac{N_m}{T} \equiv \underbrace{[\int_0^T dt \mathcal{L}_{12}(t)]}_{\mathcal{L}_T} \sigma_{12 \to m}. \tag{4.44}$$

---

[1] This may appear a bit strange, since it means that the quantity N_m/TV is dimensionless. Moreover, TV is the volume of space–time, which is infinite. But recall, we are using plane wave states, which exist over all of space, and in our idealized scattering experiment the process occurs over an infinite amount of time. Nevertheless, when we calculate these quantities using the quantum mechanics of plane wave states, all the dimensions will come out as stated. Just wait!

The integrated luminosity has dimensions,

$$[\mathcal{L}_T] \equiv \frac{1}{\sigma}. \tag{4.45}$$

Thus, given $\mathcal{L}_T$ and $\sigma_{12 \to m}$, we can obtain the total number of events observed in the experiment.

4.3 Decay Rates

Define

$$\frac{dN_m}{T} \equiv N_a \; d\Gamma_{a \to m}, \tag{4.46}$$

where $\frac{dN_m}{T}$ is equivalent to the number of hits of state m in a cone of phase space per unit of time $[t, t + \Delta t]$. N_a is the number of a particles at time t and $d\Gamma_{a \to m}$ is the partial differential decay rate. Integrating over phase space we obtain

$$\frac{N_m}{T} \equiv N_a \Gamma_{a \to m}, \tag{4.47}$$

where $\Gamma_{a \to m}$, the probability per unit time for a to decay into some state m, is the partial decay rate or *partial width*. Finally, summing over all final states we obtain the total decay rate,

$$\Gamma_a \equiv \sum_m \Gamma_{a \to m}. \tag{4.48}$$

Note: $\frac{\Delta N_a}{\Delta t}$ is the change in the number of a particles per unit of time $[t, t + \Delta t]$. Thus, if when a decays it goes into some state m, we have the relation

$$\frac{\Delta N_a}{\Delta t} + \left(\sum_m \frac{N_m}{T} \right) \equiv 0. \tag{4.49}$$

As a consequence, we have

$$\frac{\Delta N_a}{\Delta t} = -N_a \; \Gamma_a,$$
$$\frac{dN_a(t)}{dt} = -N_a(t)\Gamma_a,$$
$$N_a(t) = N_a(0)e^{-t/\tau_a},$$

where

$$\tau_a \equiv \frac{\hbar}{\Gamma_a} \tag{4.50}$$

is called the *lifetime of a*.

Note: T is *not* Lorentz invariant. Hence, Γ_a is also *not* Lorentz invariant. We have

$$\Gamma_a = \left(\frac{\sum_m N_m}{T}\right) / N_a = \frac{\left(\sum_m N_m\right)}{\rho_a} / TV \equiv \frac{(N/TV)}{\rho_a}. \tag{4.51}$$

But $\rho_a(\vec{p}_a) \equiv \rho_a(\vec{0})\frac{E_a}{m_a}$. Thus,

$$\Gamma_a(\vec{p}_a) \equiv \Gamma_a(\vec{0})\,\frac{m_a}{E_a}, \tag{4.52}$$

where

$$\Gamma_a(\vec{0}) \equiv \Gamma_a(rest). \tag{4.53}$$

Usually the lifetime or width, as given in tables, is defined in the rest frame. We have

$$d\Gamma_{a \to m}\,[\text{rest frame}] \equiv \frac{(dN_m/TV)}{2m_a}. \tag{4.54}$$

A decay rate has dimensions $[\Gamma_a] = \ell^{-1}, m$.

In this chapter we have obtained some abstract notions of cross-sections and decay rates. However, in order to calculate these quantities we will require quantum mechanics and the concept of a scattering matrix, S matrix, which we discuss in the next chapter. The S matrix gives the quantum mechanical probability for a scattering or decay process to occur.

5 Quantum Mechanics

In order to actually calculate the value for either a cross-section or a decay rate one requires a theory and formalism. That formalism is given by relativistic quantum field theory. But before we embark on field theory let's see how far we can go with just quantum mechanics. In this chapter we define the scattering (or S) matrix and discuss some of its properties. We use it to derive the general formula for cross-sections and decay rates. These formulae will be used in field theory calculations.

5.1 Scattering Probability

The number of particles/time entering a detector is determined by the probability for a transition from an initial state i to a final state f given by P_{fi}. P_{fi} is given in terms of the S matrix S_{fi} by

$$P_{fi} \equiv |S_{fi}|^2. \tag{5.1}$$

In general, in quantum mechanics, we have

$$i\frac{d}{dt}\psi(t) = H\psi(t) \text{ [Schrödinger equation]},$$

$$\psi(t) \equiv \sum_i a_i(t)|i\rangle \quad \text{where } \{|i\rangle\} \text{ form a complete set of states},$$

$$\langle i|j\rangle \equiv \delta_{ij},$$

$$\sum_i |i\rangle\langle i| \equiv 1.$$

Then

$$i\frac{d}{dt}a_i(t) = \sum_j H_{ij}a_j(t), \tag{5.2}$$

$$H_{ij} \equiv \langle i|H|j\rangle.$$

A general solution to the Schrödinger equation is given in terms of the evolution operator, $\mathcal{U}(t, t')$, satisfying

$$i\frac{d}{dt}\mathcal{U}(t, t_i) = H\mathcal{U}(t, t_i) \tag{5.3}$$

and

$$\mathcal{U}(t_2, t_1) = \mathcal{U}(t_2, t)\mathcal{U}(t, t_1),$$
$$\mathcal{U}(t_1, t_1) = 1. \tag{5.4}$$

We then have

$$\psi(t) = \mathcal{U}(t, t_1)\psi(t_1) \tag{5.5}$$

or, equivalently,

$$a_j(t) = \sum_i \mathcal{U}_{ji}(t, t_1)a_i(t_1). \tag{5.6}$$

5.2 *S* Matrix

From the Schrödinger equation, we have

$$\frac{d}{dt}|\psi(t)|^2 \equiv 0, \tag{5.7}$$

where $|\psi(t)|^2 \equiv \langle\psi(t)|\psi(t)\rangle$. Therefore

$$|\psi(t)|^2 \quad = \quad |\psi(t_1)|^2 \tag{5.8}$$

implies that

$$\sum_i |a_i(t)|^2 \quad = \quad \sum_i |a_i(t_1)|^2. \tag{5.9}$$

Then, using Eqn. 5.6, we have

$$\sum_i |a_i(t)|^2 = \left(\sum_{jk} \left(\sum_i \mathcal{U}_{ij}(t, t_1)\mathcal{U}_{ik}^*(t, t_1) \right) a_j(t_1)a_k^*(t_1) \right)$$
$$= \sum_i |a_i(t_1)|^2,$$

which implies that

$$\sum_i \mathcal{U}_{ij}(t, t_1)\mathcal{U}_{ik}^*(t, t_1) \equiv \delta_{jk} \tag{5.10}$$

or

$$\mathcal{U}\mathcal{U}^\dagger = \mathcal{U}^\dagger\mathcal{U} \equiv \mathbf{1}, \tag{5.11}$$

i.e. $\mathcal{U}$ is a unitary evolution operator.

For a free particle with energy E_a, we have

$$\mathcal{U}(t_2, t_1) \equiv \exp[-iE_a(t_2 - t_1)]. \tag{5.12}$$

We can now define the *S* matrix. The Hamiltonian for an interacting system can be written in terms of the free particle Hamiltonian and the interaction term, with

$$H = H_0 + H_{int}. \tag{5.13}$$

Forces are typically short range, except for electromagnetic forces. However, even for electromagnetic interactions, screening effects tend to also make these forces short range. Therefore, let's assume that the time during which particles interact is in the range

$$-\frac{T}{2} < t < \frac{T}{2}. \tag{5.14}$$

Outside of this time range, we assume that the particles travel freely. Thus, for the incoming and outgoing particles we have

$$\mathcal{U}_i\left(0, -\frac{T}{2}\right) = \exp[-i\frac{T}{2}\sum_{i=1}^{n_i} E_i],$$

$$\mathcal{U}_f\left(\frac{T}{2}, 0\right) = \exp[-i\frac{T}{2}\sum_{j=1}^{n_f} E_j].$$

The S matrix is now given by

$$S_{fi} = \mathcal{U}_f^{-1}\left(\frac{1}{2}T, 0\right) \mathcal{U}_{fi}\left(\frac{1}{2}T, -\frac{1}{2}T\right) \mathcal{U}_i^{-1}\left(0, -\frac{1}{2}T\right), \tag{5.15}$$

where the initial and final factors cancel the free particle time dependence.[1] The S matrix satisfies the following important properties

(1) $S_{fi} \equiv S$ matrix is time independent;
(2) S is unitary,

$$S_{fi}S_{if'}^{\dagger} = S_{if}^{\dagger}S_{fi'} \equiv \mathbf{1}. \tag{5.16}$$

It is worth commenting on some additional properties of the S matrix.

- The term "free particles" includes all bound states. (e.g. $pp \to d$ [deuteron] $+ \pi^+$ [positively charged pion]), where the deuteron and pion are treated as free particles.

  The S matrix does not distinguish elementary and composite particles since eigenstates of $H_{\text{free}} \equiv H_0$ must form a complete set of states. Since we cannot expand bound states in terms of unbound states they must be included.
- Stable particles include protons, electrons, γs (photons); all others decay. When considering unstable particles, i.e. resonances, T has a natural limit. T is less than the lifetime of most short-lived particles. For example, in the process $pp \to pp$, $pp\pi^0$ followed by $\pi^0 \to \gamma\gamma$, the lifetime of the π^0 is $\tau_{\pi^0} \sim 10^{-16}$ s or ($\Delta E = 1/\tau_{\pi^0} \sim 6.6 \times 10^{-6}$ MeV. These two final states pp, $pp\pi^0$ are connected via unitarity on timescales $< 10^{-16}$ s only, unless the 2γs are explicitly incorporated into the S matrix.

Typical strong-interaction timescales (or equivalently, energy scales) for low energy pp collisions are of order $10-100$ MeV, i.e. corresponding to much shorter

---

[1] The cancelation of the free particle time dependence, as explicitly done here, will also be clear when we re-derive the S matrix in field theory.

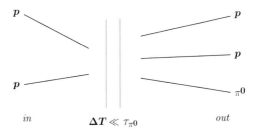

Fig. 5.1   Scattering process $pp \to pp\pi^0$.

collision timescales than the π^0 lifetime. Hence, we can treat the π^0 as a stable particle when discussing strong-interaction processes (see Fig. 5.1).

The S_{fi} matrix element is written in terms of a transition amplitude T_{fi} and an identity element corresponding to no scattering. We have

$$S_{fi} \equiv \langle f|i\rangle + i(2\pi)^4 \delta^4(P_i - P_f)T_{fi}. \tag{5.17}$$

The four-momentum delta function is always present due to the conservation of energy and momentum, i.e. $S_{fi} \equiv 0$ unless $E, \vec{p}$ are conserved. We shall see, when we discuss the field theoretic calculation of S, that the above form of the S matrix is output (not input) from the calculation.

Let's now evaluate the transition probability $P_{fi} = |S_{fi}|^2$ for $i \neq f$. We have

$$P_{fi} \equiv |S_{fi}|^2 = VT(2\pi)^4 \delta^4(P_i - P_f)|T_{fi}|^2,$$
$$f \neq i$$

where

$$(2\pi)^4 \delta^4(P_i - P_f) \equiv \int e^{ix\cdot(P_i - P_f)} d^4x \tag{5.18}$$

and we used

$$[(2\pi)^4 \delta^4(P_i - P_f)]^2 = VT[(2\pi)^4 \delta^4(P_i - P_f)]. \tag{5.19}$$

We are now in a position to express the scattering cross-sections and decay rates in terms of transition amplitudes. In particular, for two-particle scattering $i = ab, f \equiv \{n\}$, we have (Eqn. 4.40)

$$\sigma_{ab\to n} = \frac{N_n/TV}{2\sqrt{\lambda(s, m_a^2, m_b^2)}}, \tag{5.20}$$

where

$$N_n \text{ (number of events in detector)} = \int |S_{(ab\to n)}|^2 dLips(P_1, \ldots, P_n) \tag{5.21}$$

and

$$dLips(P_1, \ldots P_n) \equiv \frac{1}{(2\pi)^{3n}} \prod_{i=1}^{n} \frac{d^3\vec{p}_i}{2E_i}. \tag{5.22}$$

Hence, we have

$$\sigma_{ab \to n} = \frac{1}{2[\lambda(s, m_a^2, m_b^2)]^{1/2}} \int dLips(s; P_1, \ldots, P_n)|T_{fi}|^2, \tag{5.23}$$

with $s \equiv (P_a + P_b)^2$. Note: T_{fi} is a Lorentz scalar. The quantity $dLips(s; P_1, \ldots, P_n)$ is defined as follows,

$$dLips(s; P_1, \ldots, P_n) \equiv (2\pi)^4 \delta^4(P_i - P_f) dLips(P_1, \ldots, P_n), \tag{5.24}$$

with $P_f \equiv \sum_{j=1}^n P_j$ and $P_i = P_a + P_b$.

The differential cross-section is given by

$$d\sigma_{ab \to n} = \frac{|T_{fi}|^2}{2[\lambda]^{1/2}} dLips(s; P_1, \ldots, P_n). \tag{5.25}$$

For homework, you will show that the two-to-two scattering differential cross-section is given by

$$d\sigma_{ab \to cd} = \frac{|T_{fi}|^2}{2[\lambda]^{1/2}} \frac{p' d\Omega}{16\pi^2 s^{1/2}}$$

$$= \frac{|T_{fi}|^2}{64\pi^2 s} \frac{p'}{p} d\Omega,$$

using $ps^{1/2} = \frac{1}{2}\sqrt{\lambda}$ or

$$\frac{d\sigma}{d\Omega}(ab \to cd) = \left(\frac{|T_{fi}|}{8\pi}\right)^2 \frac{p'}{sp}. \tag{5.26}$$

Note, up until now, we have ignored spin in scattering processes. Now if a, b, c, d have spin s_a, s_b, s_c, s_d and helicity $\lambda_a, \lambda_b, \lambda_c, \lambda_d$ then we have

$$T_{fi} \equiv T(s, \Omega[\propto t]; \lambda_a, \lambda_b, \lambda_c, \lambda_d). \tag{5.27}$$

Let's assume that the initial states have their spin polarized, and we measure the spins of the final states. Given the initial states, a, b, with spin, J_a, J_b, we use the density matrices for the initial states, $P(M_a), P(M_b)$ (for a further discussion of spin density matrices, see Chapter 6, Section 6.2). Then we have the differential cross-section given by

$$\frac{d\sigma}{d\Omega}(a, b \to c, M_c; d, M_d) = \frac{1}{4s} \frac{p'}{p} \sum_{M_a, M_b} P(M_a) P(M_b) \left| \frac{T(s, \Omega, M_a, \ldots, M_d)}{4\pi} \right|^2. \tag{5.28}$$

However, if we don't measure the spin of the final states, then we sum over M_c, M_d. We have

$$\frac{d\sigma}{d\Omega}(ab, \to cd) = \frac{1}{4s} \frac{p'}{p} \sum_{M_a, \ldots, M_d} P(M_a) P(M_b) \left| \frac{T(s, \Omega, M_a, \ldots, M_d)}{4\pi} \right|^2. \tag{5.29}$$

For decays, $a \to 1, \ldots, n$, we have at rest

$$\Gamma_{a \to n} = \frac{N/TV}{2m_a} \tag{5.30}$$

or

$$\Gamma_{a \to n} = \frac{1}{2m_a} \int dLips(m_a^2; P_1, \ldots, P_n)(T_{fa})^2. \tag{5.31}$$

5.3 Treatment of Unstable Particles

In this section, we would like to discuss the treatment of unstable particles. Consider the scattering process $\overset{i}{\overbrace{pp}} \to \overset{c}{\overbrace{pp}} \overset{d}{\overbrace{\pi^0}}$ where the overscript identifies the incoming and outgoing states. The reaction time associated with a strong interaction, $\overset{i}{\overbrace{pp}}$, is $\tau_R \sim 10^{-23}$ s and the timescale for the decay of the pion $\overset{d}{\overbrace{\pi^0}}$ to $\overset{1\ 2}{\overbrace{2\gamma}}$ is given by $\tau_{\pi^0} \sim 10^{-16}$ s. The S matrix element for the intermediate process $i \to cd$ is given by

$$S(i \to cd) = \mathcal{U}_d^{\prime\,-1}(t,0)\,\mathcal{U}_c^{-1}(t,0)\,\mathcal{U}_{cd,i}(t,-\tfrac{1}{2}T)\,\mathcal{U}_i^{-1}(0,-\tfrac{1}{2}T), \tag{5.32}$$

which, by construction, is time independent. We have

$$P_d = P_1 + P_2, \qquad E_d' = m \equiv m_{\pi^0}, \tag{5.33}$$

$\mathcal{U}_d'(t,0) \equiv \exp(-imt - \tfrac{1}{2}\Gamma t)$ and $1 \geq P(i,cd) = |\mathcal{U}_{i \to cd}(t,-\tfrac{1}{2}T)|^2 \equiv |\mathcal{U}_{cd,i}|^2$. Note, the relevant timescales are given by

$$\tau_{\pi^0} > t > T \gg \tau_R. \tag{5.34}$$

$P(i,cd)$ is the probability that d has been produced at time t, but *not* decayed. $P(i \to cd)$ goes to zero at $t \to \infty$ since d eventually decays.

For the complete process $i \to c12$ we have

$$\mathcal{U}_{c12,i} = \int_{-\frac{1}{2}T}^{T'} dt\,\mathcal{U}_{12,d}(T',t)\,\mathcal{U}_c(T',t)\,\mathcal{U}_{cd,i}(t,-\tfrac{1}{2}T). \tag{5.35}$$

In a perturbative treatment of the decay process, we treat d as stable under H_0. We have

$$S(d \to 12) = \mathcal{U}_{12}^{-1}(T',0)\,\mathcal{U}_{12,d}(T',t)\,\mathcal{U}_d^{-1}(0,t), \tag{5.36}$$

where $\mathcal{U}_d^{-1}(0,t) = e^{-iE_d t}$ and the decay is proportional to the interaction Hamiltonian, H_{int}. $S(d \to 12)$ is perturbatively time independent. Note, $T' \gg \tau_{\pi^0} > t$.

Now we can make some educated rearrangements in the formula for the S matrix element. We have (using the results of Eqns. 5.32 and 5.35)

$$S(i \to c12) = \mathcal{U}_{12}^{-1}(T',0)\, \mathcal{U}_c^{-1}(T',0)\, \mathcal{U}_{c12,i}(T', -\tfrac{1}{2}T)\, \mathcal{U}_i^{-1}(0, -\tfrac{1}{2}T)$$

$$\equiv \int_{-\frac{1}{2}T}^{T'} dt \underbrace{\left[\mathcal{U}_{12}^{-1}(T',0)\, \mathcal{U}_{12,d}(T',t)\, \mathcal{U}_d^{-1}(0,t)\right]}_{[S(d\to12)]}$$

$$\times\, \mathcal{U}_d(0,t)\, \mathcal{U}_d'(t,0)$$

$$\times \underbrace{\left[\mathcal{U}_d'^{\,-1}(t,0)\, \mathcal{U}_c^{-1}(T',0)\, \mathcal{U}_c(T',t)\, \mathcal{U}_{cd,i}(t, -\tfrac{1}{2}T)\, \mathcal{U}_i^{-1}(0, -\tfrac{1}{2}T)\right]}_{[S(i\to cd)]}.$$

Hence, using the fact that the S matrix elements are time independent, we find

$$S(i \to c12) \equiv S(d \to 12) \times \underbrace{\int_{-\frac{1}{2}T}^{T'} dt\, \mathcal{U}_d(0,t)\, \mathcal{U}_d'(t,0)}_{-i\Phi(E_d)}\ S(i \to cd). \qquad (5.37)$$

The function $\Phi(E)$ describes the propagation of the unstable state. We have

$$\Phi(E) = i \int_{-\frac{1}{2}T}^{T'} dt\, e^{iEt} e^{(-imt - \frac{1}{2}\Gamma t)}$$

$$\simeq i \int_0^\infty dt\, e^{it(E - m + \frac{i}{2}\Gamma)}$$

$$= \frac{1}{m - E - i\frac{\Gamma}{2}}$$

and in terms of the transition amplitude we have (see Fig. 5.2)

$$S(i \to c12) = S(d \to 12)(-i\Phi(E_d))S(i \to cd)$$
$$\equiv i(2\pi)^4 \delta^4(P_i - P_c - P_1 - P_2)T_{c12,i}.$$

The present analysis is in the framework of non-relativistic quantum mechanics. For relativistic particles (which allows for particle propagation forward and backward in time) we have

$$\Phi(s) = \frac{i}{m^2 - s - im\Gamma}$$

$$\simeq \frac{i}{2m}\left[\Phi(E) + \Phi(-E)\right] \qquad \{\, E = \text{particle},\ -E = \text{antiparticle}\}$$

$$\equiv \frac{\frac{i}{2m}}{m - E - i\frac{\Gamma}{2}} + \frac{\frac{i}{2m}}{m + E - i\frac{\Gamma}{2}}$$

$$= \frac{i + \frac{\Gamma}{2m}}{(m - i\frac{\Gamma}{2})^2 - E^2},$$

with $s = E^2$ in the center of momentum system (CMS) of 12 and Γ is the *total width of the particle*. Note: this will be re-derived using relativistic quantum field theory. Thus for a relativistic particle, we have

$$T_{c12,i} \equiv T_{12,d}\Phi(s_d)T_{cd,i}. \qquad (5.38)$$

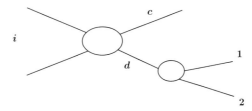

Fig. 5.2 Diagrammatic representation of an unstable particle decay.

This is the Breit–Wigner amplitude for resonance production and decay.

Now let us evaluate the cross-section for this process (Fig. 5.2 with $i = \{a,\ b\}$) We have

$$\sigma(ab \to c12) = \frac{1}{4qs^{1/2}} \int |T(ab \to cd)|^2 |T(d \to 12)|^2 \times \frac{dLips(s; P_c, P_1, P_2)}{(m^2 - s_d)^2 + m^2\Gamma^2}, \quad (5.39)$$

where q is the momentum of the incoming particles in the CMS. The cross-section for this process takes a particularly simple and intuitive form in the so-called narrow resonance approximation,

$$\sigma(ab \to c12) \overset{\underset{lim}{m\Gamma \to 0}}{\to} \sigma(ab \to cd)BR(d \to 12), \quad (5.40)$$

where $BR(d \to 12) \equiv \frac{\Gamma(d \to 12)}{\Gamma}$ is the probability for the state d to decay to the particular final state 12 (also known as the branching fraction). Hence, the cross-section for $ab \to c12$ equals the cross-section for $ab \to cd$ times the probability for $d \to 12$.

The proof of this statement follows. Note: the following analysis demonstrates the manipulations of phase space formulae, which can also be useful in other contexts. We begin with the formulae

$$dLips(P) \equiv \frac{1}{(2\pi)^3} \frac{d^3\vec{p}}{2E} = \frac{1}{(2\pi)^3} d^4 P \delta(P^2 - m^2)\theta(P^0) \quad (5.41)$$

and

$$dLips(s; P_c, P_1, P_2) \equiv (2\pi)^4 \delta^4(P - P_c - P_1 - P_2)dLips(P_c, P_1, P_2), \quad (5.42)$$

with $s = P^2$. Define $P_d \equiv P_1 + P_2$ and $P_d^2 \equiv S_d$.

Now multiply Eqn. 5.42 by $1 \equiv d^4 P_d\ \delta^4(P_d - P_1 - P_2)$. We obtain

$$dLips(s; P_c, P_1, P_2) \equiv \delta^4(P - P_c - P_d)dLips(P_c)d^4 P_d\ dLips(s_d; P_1, P_2). \quad (5.43)$$

But

$$d^4 P_d = d^4 P_d\ \delta(P_d^2 - s_d)\ ds_d = (2\pi)^3\ dLips(P_d)\ ds_d, \quad (5.44)$$

since $P_d^0 = P_1^0 + P_2^0 > 0$. This implies that

$$dLips(s; P_c, P_1, P_2) = \frac{1}{2\pi} dLips(s; P_c, P_d)\ dLips(s_d; P_1, P_2)\ ds_d. \quad (5.45)$$

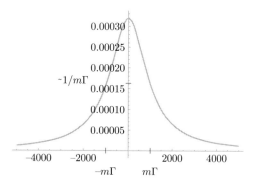

Fig. 5.3 Breit–Wigner amplitude (using Mathematica).

After these manipulations, we can now write (Eqn. 5.39)

$$\sigma(ab \to c12) = \frac{1}{4qs^{1/2}} \int |T(ab \to cd)|^2 \frac{ds_d}{2\pi} \, dLips(s; P_c, P_d) \qquad (5.46)$$

$$\times \frac{|T(d \to 12)|^2 \, dLips(s_d; P_1, P_2)}{(m^2 - s_d)^2 + m^2 \Gamma^2}$$

$$= \int \frac{ds_d}{\pi} \sigma(ab \to cd)|_{s_d} \frac{m\Gamma_{s_d}(d \to 12)}{(m^2 - s_d)^2 + m^2 \Gamma^2},$$

where we used Eqn. 5.25 and

$$\Gamma_{s_d}(d \to 12) \equiv \frac{1}{2m} \int dLips(s_d; P_1, P_2)|T(d \to 12)|^2$$

$$[\text{valid for narrow resonance}] \quad \frac{\Gamma}{m} \to 0$$

$$\simeq \Gamma_{m^2}(d \to 12)$$

$$\sigma(ab \to cd)|_{s_d} \simeq \sigma_{m_d^2}(ab \to cd).$$

Note: $\Gamma = \sum_f \Gamma(d \to f)$ is the total width, while $\Gamma(d \to 12)$ is the partial width.

Now let us specialize to the narrow resonance approximation $\frac{\Gamma}{m} \to 0$. Consider the integrand in Eqn. 5.46 (see Fig. 5.3). In the limit that the decay rate is much smaller than the mass, the Breit-Wigner amplitude squared becomes very peaked with a narrow width. The integral can be approximated by the following.

$$\lim_{m\Gamma \to 0} \left\{ \int_{s_1}^{s_2} \frac{m\Gamma \frac{ds}{\pi}}{(m^2 - s)^2 + m^2 \Gamma^2} = \lim_{m\Gamma \to 0} \left\{ -\frac{1}{\pi} \tan^{-1} \left(\frac{m^2 - s}{m\Gamma} \right) \right\} \right\}\Big|_{s_1}^{s_2}$$

$$= \begin{cases} 1 & \text{for } s_1 < m^2 < s_2 \\ 0 & \text{otherwise} \end{cases}.$$

Given $x = \frac{m^2 - s}{m\Gamma}$, we define $tan(y) = x$ or $y = tan^{-1}(x)$ (see Fig. 5.4). Hence, the integrand is a test function, which in the limit $m\Gamma \to 0$ becomes a delta function. We have

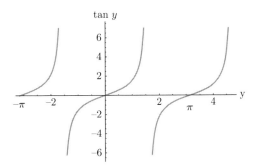

The parameter $\tan(y)$ (using Mathematica).

$$\lim_{m\Gamma \to 0}\left(\frac{m\Gamma/\pi}{(m^2-s)^2+m^2\Gamma^2}\right)=\delta(m^2-s). \tag{5.47}$$

Finally, we see the result in Eqn. 5.40.

$$\lim_{m\Gamma \to 0}\sigma(ab \to c12) \to \sigma(ab \to cd)BR(d \to 12). \tag{5.48}$$

5.4 Exercises

5.1 For two-particle phase space:

(1) Given $p=p_1'+p_2'$ with $p^2=s$, $p=(\sqrt{s},\;\vec{o})$, $p_1'=(E',\;\vec{p}\,')$ and defining $p'=|\vec{p}\,'|$, show that $dLips(s;p_1',\;p_2')=\frac{p'd\Omega}{16\pi^2 s^{1/2}}$, where $d\Omega=d\cos\theta d\phi$ and $p'=\sqrt{\frac{\lambda(s,\,m_1^2,\,m_2^2)}{4s}}$. The three-vector $\vec{p}\,'$ is defined in Fig. 5.5.

(2) Consider two body scattering $1+2 \to 3+4$ with $p_1+p_2=p_3+p_4$. In the CMS, let $p_1=(e,\;\vec{p})$, $p=|\vec{p}|$ and $p_3=(E',\;\vec{p}\,')$, $p'=|\vec{p}\,'|$ with $\vec{p}\cdot\vec{p}\,'=pp'\cos\theta$. Define the Lorentz invariants, $s=(p_1+p_2)^2$, $t=(p_3-p_1)^2$, $u=(p_1-p_4)^2$, with $s+t+u \equiv \sum_{i=1}^{4} m_i^2$. Show that the differential cross-section is given by $d\sigma=\frac{1}{4s}\frac{p'}{p}\left|\frac{T(s,\Omega)}{4\pi}\right|^2 d\Omega=\frac{\pi}{4p^2 s}\left|\frac{T(s,\;t)}{4\pi}\right|^2 dt$, where $\Omega=\{\theta,\phi\}$ and the second equality assumes T is independent of ϕ.

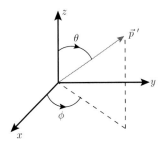

Definition of the polar angles, θ, ϕ.

5.2 For three-body phase space, show that in the CMS system, where $\vec{p}_1 + \vec{p}_2 + \vec{p}_3 = 0$, $dLips(s; p_1, \ p_2, \ p_3) = \frac{1}{8}(2\pi)^{-5}d\Omega' dE_1 dE_2$, where Ω' is all the angles in Ω_1, Ω_2, except for θ_{12} defined by $\vec{p}_1 \cdot \vec{p}_2 = p_1 p_2 \cos \theta_{12}$. Thus, the three-body phase space is constant on the energy surface, $E_1 - E_2$.

5.3 Now calculate two-body phase space in the laboratory frame described by the incoming four-momenta $- p = (m, 0), k = (E, \vec{k})$. Calculate $dLips(s, k', p')$, with the outgoing four-momenta given by $p' = (E'_p, \vec{p}\,')$ and $k' = (E', \vec{k}\,')$ and, in the laboratory frame, the scattering angle between $\vec{k}$ and $\vec{k}\,'$ is given by θ such that $\vec{k} \cdot \vec{k}\,' = kk' \cos(\theta)$.

6 Unitarity and Partial Waves

In this chapter we derive the unitarity relations for the transition amplitude. These have important consequences. In particular, for the partial wave decomposition of two-body scattering cross-sections we define the partial wave cross-sections and derive the unitarity bounds. These played an important role in deriving the Standard Model as we shall see later.

6.1 Unitary S Matrix

Unitarity of the S matrix has explicit consequences for transition amplitudes that we would like to explore. Recall

$$S_{fi} \equiv \underbrace{\langle f|i\rangle}_{\delta_{fi}} + i(2\pi)^4\delta^4(P_i - P_f)T_{fi}. \tag{6.1}$$

Given this expression we then find

$$\sum_f S_{if}^\dagger S_{fi'} = \delta_{ii'} = \sum_f S_{fi}^* S_{fi'}$$

$$= \sum_f \left(\delta_{fi} - i(2\pi)^4\delta^4(P_i - P_f)T_{fi}^*\right) \times \left(\delta_{fi'} + i(2\pi)^4\delta^4(P_{i'} - P_f)T_{fi'}\right)$$

$$= \delta_{ii'} + i(2\pi)^4\delta^4(P_{i'} - P_i)T_{ii'} - i(2\pi)^4\delta^4(P_i - P_{i'})T_{i'i}^*$$

$$+ (2\pi)^4\delta^4(P_i - P_{i'})\sum_f \int (2\pi)^4\delta^4(P_i - P_f)dLips(P_f)T_{fi}^*T_{fi'}.$$

We therefore obtain the unitarity relation for transition amplitudes,

$$-i(T_{ii'} - T_{i'i}^*) = \sum_f \int dLips(s; P_f)T_{fi}^*T_{fi'}. \tag{6.2}$$

We will now discuss several different examples of unitarity.

1. Consider first the simplest case, $i = i'$, i.e. a single-particle state. We have (using Eqn. 5.31)

$$2\mathrm{Im}T_{ii} = \sum_f \int dLips(s; P_f)|T_{fi}|^2 \equiv 2m_i\Gamma_i$$

or

$$\mathrm{Im}T_{ii} \equiv m_i\Gamma_i, \tag{6.3}$$

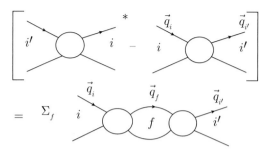

Fig. 6.1 Diagrammatic representation of the partial wave expansion, assuming two-body intermediate states.

where Γ_i is the total decay rate for the state, i.[1]

2. Now consider elastic scattering at zero angle. We have $i = i'$ where i is a two-particle state. Then we find (using Eqns. 4.33 and 5.23 with q the momentum in the center of momentum system [CMS])

$$2\mathrm{Im}T_{ii} = \sum_f \int dLips(s; P_f)|T_{fi}|^2,$$

$$2\mathrm{Im}T_{ii} \equiv 4qs^{1/2}\sigma_i(\text{total}),$$

where $\sigma_i(\text{total}) \equiv \sum_f \sigma(i \to f)$ is the total cross-section including both elastic and inelastic channels. This is known as the *optical theorem*, which states that the imaginary part of the forward elastic scattering amplitude is proportional to the total cross-section.

3. Now consider two-body scattering in the CMS. We will focus on a partial wave expansion, assuming that three or more particle final states can be neglected in the intermediate sum on f. For simplicity we shall consider spin-zero particles. We have (see Fig. 6.1)

$$i\left(T_{ii'}^*(\theta) - T_{i'i}(\theta)\right) \equiv \sum_f \frac{q_f}{16\pi^2} s^{1/2} \int d\Omega'\, T_{fi}(\theta')\, T_{fi'}^*(\theta''). \tag{6.4}$$

We have assumed that $T(\theta)$ is independent of the angle ϕ. This is valid for spin-zero particles. The other angles relevant for the analysis are defined in Fig. 6.2. We have Ω' which represents the angle between $\vec{q}_f$ with respect to $\vec{q}_i$ and Ω'' is the angle between $\vec{q}_f$ and $\vec{q}_{i'}$. Given the spherical coordinates defined in Fig. 6.2 we find

---

[1] For a single particle in field theory, the quantity T_{ii} represents the inverse propagator. For an unstable scalar particle with four-momentum, p, we have $T_{ii} = p^2 - m_i^2 + im_i\Gamma_i$. This is then the Breit–Wigner form of the unstable particle propagator.

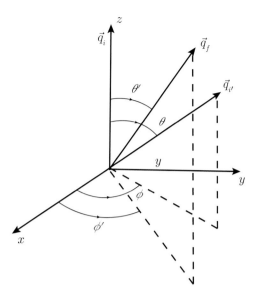

Fig. 6.2 Diagrammatic representation of the angles used in the expansion.

$$q_{i'} \equiv |\vec{q}_{i'}|,$$
$$\vec{q}_{i'} \equiv (\cos\phi\sin\theta, \sin\phi\sin\theta, \cos\theta)q_{i'},$$
$$\vec{q}_f \equiv (\cos\phi'\sin\theta', \sin\phi'\sin\theta', \cos\theta')q_f,$$
$$\vec{q}_f \cdot \vec{q}_{i'} \equiv \cos\theta'' q_f q_{i'}.$$

We also have

$$\cos\theta'' = \cos\phi\cos\phi'\sin\theta\sin\theta' + \sin\phi\sin\phi'\sin\theta\sin\theta' + \cos\theta\cos\theta',$$
$$\cos\theta'' \equiv \cos(\phi - \phi')\sin\theta\sin\theta' + \cos\theta\cos\theta'.$$

We now expand the scattering amplitude in terms of partial waves (a generalization of relations valid for non-relativistic quantum mechanics). We have

$$\frac{1}{4\pi}T_{fi}(s,\theta) \equiv 2s^{1/2}\sum_{L=0}^{\infty}(2L+1)P_L(x)T_{fi}^L(s), \tag{6.5}$$

where $x \equiv \cos\theta$ and $T_{fi}^L(s)$ is the partial wave amplitude. $P_L(x)$ are Legendre polynomials which form a complete set of functions of x. Using the following identities for Legendre polynomials (Eqn. 6.6),

$$\int_{-1}^{1} dx(L+\frac{1}{2})P_L(x)P_{L'}(x) = \delta_{L'}, \tag{6.6}$$

$$\int d\phi' P_L(x'') \equiv 2\pi P_L(x)P_L(x'),$$

we can invert the relation (Eqn. 6.5) and find

$$T_{fi}^L(s) = \frac{1}{16\pi s^{1/2}}\int_{-1}^{1} T_{fi}(s,\theta)\, P_L(x)dx. \tag{6.7}$$

Let's now substitute the partial wave expansion into the unitarity relation, Eqn. 6.4. We find

$$i(T^{L^*}_{ii'} - T^{L}_{i'i}) = 2 \sum_f q_f T^{L}_{fi} T^{L^*}_{fi'},$$ (6.8)

which is valid for all s (implicit).

Consider elastic scattering with $i = i'$ (*not* forward scattering), for example, $\pi\pi \to \pi\pi, \pi\pi\pi\pi, K\bar{K}$. At low energies, $s < 16m_\pi^2$, $f = i$ is the only open channel. We then have

$$\mathrm{Im} T^{L}_{ii} = \sum_f q_f T^{L}_{fi} T^{L^*}_{fi}.$$ (6.9)

Now define

$$T^{L}(s) \equiv T^{L}_{ii}(s),$$
$$T^{L}_{f}(s) \equiv T^{L}_{fi}(s),$$
$$q \equiv q_i.$$

Then

$$\mathrm{Im} T^{L}(s) = \sum_f q_f |T^{L}_{f}(s)|^2$$ (6.10)

$$= q|T^{L}(s)|^2$$

for $s < s_{\mathrm{threshold}}$. In this case we have

$$\mathrm{Im}(q T^{L}(s)) = |q T^{L}(s)|^2.$$ (6.11)

Now let

$$q T^{L}(s) \equiv r_L e^{i\delta_L}.$$ (6.12)

Plugging this into Eqn. 6.11 we find

$$r_L = \sin \delta_L$$ (6.13)

and

$$T^{L}(s) = \frac{1}{q} e^{i\delta_L(s)} \sin \delta_L(s),$$ (6.14)

where $e^{i\delta_L(s)}$ is the phase shift.

Now we define the partial wave S matrix element by

$$S^{L}_{fi} \equiv \delta_{fi} + 2i T^{L}_{fi} \sqrt{q_f q_i},$$ (6.15)

which reproduces the partial wave unitarity constraint (Eqn. 6.8).

If, for example, only the elastic channel is open we have

$$S^{L}(s) = 1 + 2i e^{i\delta_L(s)} \times \underbrace{\sin \delta_L(s)}_{\dfrac{e^{i\delta_L(s)} - e^{-i\delta_L(s)}}{2i}}$$ (6.16)

and thus

$$S^{L}(s) \equiv e^{2i\delta_L(s)}.$$ (6.17)

In general, using Eqns. 5.26, 6.5 and 6.14, we have

$$\frac{d\sigma(i \to f)}{d\Omega} = \frac{q'}{q} \left| \sum_{L=0}^{\infty} (2L+1) P_L(\cos\theta) T_{fi}^L(s) \right|^2$$

or $\qquad \sigma(s) \equiv \sum_{L=0}^{\infty} \sigma_L(s) = 4\pi \frac{q'}{q} \sum_{L=0}^{\infty} (2L+1) \left| T_{fi}^L(s) \right|^2.$

In the elastic region ($s < s_{\text{threshold}}$) we have

$$\sigma_L(s) = \frac{4\pi}{q^2} (2L+1) \sin^2 \delta_L(s) \tag{6.18}$$

and thus we obtain the so-called unitarity bound given by

$$\sigma_L^{\max}(s) = \frac{4\pi}{q^2} (2L+1). \tag{6.19}$$

Note: q is the CMS momentum of the incoming particle. So $\sqrt{s} = E_1 + E_2 = \sqrt{m_1^2 + q^2} + \sqrt{m_2^2 + q^2}$, and for $q \gg m_1, m_2$ we have

$$s \sim 4q^2. \tag{6.20}$$

Hence,

$$\sigma_L^{\max}(s) \sim \frac{16\pi}{s} (2L+1), \tag{6.21}$$

i.e. cross-sections decrease as $1/s$.

Let us now consider the effect of inelastic channels. We have

$$\text{Im} T^L(s) = q|T^L(s)|^2 + \sum_r q_r |T_r^L|^2, \tag{6.22}$$

where the last term includes all other two-body inelastic channels $f \neq i = i'$. We now have the elastic part of the S matrix, S^L (Eqn. 6.15), given by

$$S^L \equiv 1 + 2i T^L q,$$

and plugging this into Eqn. 6.22 we find

$$\text{Re}(1 - S^L) \equiv 2q \text{Im} T^L$$
$$= \frac{1}{2} |S^L - 1|^2 + \frac{1}{2} \sum_r |S_r^L|^2,$$

where $S_r^L \equiv 2i T_r^L \sqrt{q_r q}$.

Let $S^L \equiv \eta_L e^{2i\delta_L}$, where η_L is called the inelasticity parameter. We then have

$$1 - \eta_L \cos 2\delta_L = \frac{1}{2}(1 + \eta_L^2 - 2\eta_L \cos 2\delta_L) + \frac{1}{2} \sum_r |S_r^L|^2,$$

$$\eta_L^2 = 1 - \sum_r |S_r^L|^2.$$

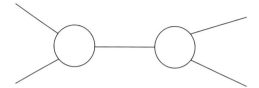

Fig. 6.3   Diagrammatic representation of a resonance in the s channel.

But $0 \le \sum_r |S_r^L|^2 \le 1$ by unitarity. Hence,

$$0 \le \eta_L \le 1. \tag{6.23}$$

Finally,

$$\begin{aligned} \sigma_L(\text{elastic}) &= 4\pi(2L+1)|T^L|^2 \\ &= \frac{\pi}{q^2}(2L+1)|1 - \eta_L e^{2i\delta_L}|^2, \end{aligned}$$

$$\begin{aligned} \sigma_L(\text{inelastic}) &= \frac{4\pi}{q}(2L+1)\sum_r q_r|T_r^L|^2 \\ &= \frac{\pi}{q^2}(2L+1)(1 - \eta_L^2), \end{aligned}$$

$$\begin{aligned} \sigma_L(\text{total}) &= \sigma_L(\text{elastic}) + \sigma_L(\text{inelastic}) \\ &= \frac{2\pi}{q^2}(2L+1)(1 - \eta_L \cos 2\delta_L). \end{aligned}$$

Experimentally, one measures $\sigma_L(\text{inelastic})$ and obtains η_L. One then measures $\sigma_L(\text{elastic})$ and determines δ_L. The phase shift δ_L (in the elastic region) gives us information about resonances (see Fig. 6.3).

$$\begin{aligned} T^L(s) &= \frac{1}{q} \sin \delta_L(s) e^{i\delta_l(s)} \\[1mm] &= \frac{1}{q} \frac{\sin \delta_L(s)}{e^{-i\delta_L(s)}} \\[1mm] &= \frac{1}{q} \frac{\sin \delta_L(s)}{\cos \delta_L(s) - i \sin \delta_L(s)}. \end{aligned}$$

$$\begin{aligned} T^L(s) &= \frac{1}{q} \frac{1}{\cot \delta_L(s) - i} \\[1mm] &= \frac{1}{q} \frac{m_r \Gamma_r}{m_r^2 - s - im_r \Gamma_r} \qquad \text{Breit–Wigner} \end{aligned}$$

We see that

$$\cot(\delta_L(s)) = \frac{m_r^2 - s}{m_r \Gamma_r}, \tag{6.24}$$

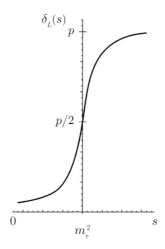

Fig. 6.4   Phase shift as a function of s (using Mathematica).

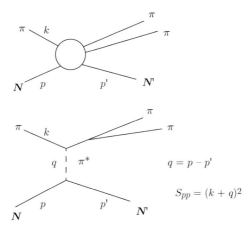

Fig. 6.5   A possible set up for measuring the $\pi\pi$ scattering amplitude.

such that on resonance, $s = m_r^2$, we have $\delta_L(s) = \frac{\pi}{2}$. Hence the phase shift goes through $\frac{\pi}{2}$ at resonance (see Fig. 6.4). We also see that

$$\cot \delta_L(s) \to \infty \Leftarrow s \to -\infty$$
$$\cot \delta_L(s) \to -\infty \Leftarrow s \to \infty$$

The phase shift analysis of scattering amplitudes has been used to identify new particle resonances. For example, in the scattering process $\pi\pi \longrightarrow \rho \longrightarrow \pi\pi$ where $\pi\pi \to \pi\pi\pi\pi$ is negligible, experimenters have identified the ρ meson as a $\pi\pi$ resonance, with $M_\rho = 770$ MeV and $m_\pi \simeq 140$ MeV. Many more resonances have been found in this way, leading to an explosion in the number of, apparently, new fundamental strongly interacting particles. See Fig. 6.5 for an example of how $\pi\pi$ scattering amplitudes can be measured experimentally.

6.2 Spin $ab \to cd_M$ with $d_M \to 1, \ldots, n$

The transition amplitude for this process, where M is the spin in a certain direction and J_d is total spin of state d, is given by

$$T = \sum_{M=-J_d}^{J_d} T(d_M \to 1, \ldots, n) \; \Phi(s_d) \; T(ab \to cd_M). \tag{6.25}$$

We then obtain

$$|T|^2 \equiv \frac{1}{(m^2 - s_d)^2 + m^2 \Gamma^2} \sum_{M,M'} X_{MM'}, \tag{6.26}$$

with

$$X_{MM'} = T(d_M \to 1, \ldots, n)T^*(d_{M'} \to 1, \ldots, n) \times T(ab \to cd_M)T^*(ab \to cd_{M'}). \tag{6.27}$$

We define the quantities

$$R(\text{prod})_{MM'} \equiv T(ab \to cd_M)T^*(ab \to cd_{M'})$$
$$R(\text{decay})_{MM'} \equiv T(d_M \to 1, \ldots, n)T^*(d_{M'} \to 1, \ldots, n).$$

Then we have

$$|T|^2 = \frac{1}{(m^2 - s_d)^2 + m^2 \Gamma^2} \underbrace{\sum_{MM'} R(\text{decay})_{MM'} R(\text{prod})_{MM'}}_{\text{Trace}(R(\text{decay})R^T(\text{prod}))}. \tag{6.28}$$

Define the production spin density matrix, ρ, by

$$\rho(s,t)_{MM'} \equiv \left[\sum_M |T(s,t,M)|^2 \right]^{-1} R(\text{prod})_{MM'},$$

$$\text{where} \quad T(s,t,M) \equiv T(ab \to cd_M),$$

$$\rho = \rho^\dagger \quad \text{and} \quad Tr(\rho) \equiv 1.$$

Now define $P(M) \equiv \rho_{MM}$ in a basis where ρ is diagonal. $P(M)$ is the probability that the state d is produced with spin M. $P(M)$ is real and positive. For unpolarized states $P(M) \equiv \frac{1}{2J+1}$. such that $Tr(\rho) \equiv \sum_{M=-J}^{J} P(M) = 1$.

Now consider the generalization of the scattering amplitude for the process $ab \to cd$ for the case that the initial states a, b with spin J_a, J_b are polarized. For the initial states we can use the density matrices $P(M_a)$, $P(M_b)$ and we assume that for the final states we measure their spin. In this case the differential cross-section is given by

$$\frac{d\sigma}{d\Omega}(a, b \to c, M_c; d, M_d) = \frac{1}{4s} \frac{p'}{p} \sum_{M_a, M_b} P(M_a) P(M_b) \left| \frac{T(s, \Omega, M_a, \ldots, M_d)}{4\pi} \right|^2. \tag{6.29}$$

However, if we don't measure the spin of the final states, then we sum over M_c, M_d. We have

$$\frac{d\sigma}{d\Omega}(ab, \to cd) = \frac{1}{4s}\frac{p'}{p} \sum_{M_a,\ldots,M_d} P(M_a)P(M_b) \left| \frac{T(s, \Omega, M_a, \ldots, M_d)}{4\pi} \right|^2. \tag{6.30}$$

6.3 Application of the Formalism on Lorentz Transformations

When reading this section, it will be useful to refer to Lee and Yang (1958) and Gasiorowicz (1966), ch. 14.

How do we know spin of the Λ^0 baryon? We are sort of jumping ahead here, since we haven't discussed N (nucleon) and π (pion) states yet, but the determination uses the formalism we've discussed. Consider the decay of the $\Lambda^0 \to N\pi$ (Fig. 6.6). In the Λ^0 rest frame we define the direct product state for the $N\pi$ system. We have

$$|\vec{p}_N, \lambda_N; \vec{p}_\pi, \lambda_\pi\rangle|_{(\vec{p}_\pi = -\vec{p}_N = -\vec{p})} \equiv |W, \vec{p}; \lambda, 0\rangle, \tag{6.31}$$

where the nucleon and pion helicities at this point historically were known to have values, $\lambda_N = \lambda = \pm\frac{1}{2}; \quad \lambda_\pi = 0$, with the energy of the state given by W. For the Λ^0 decaying at rest, we have

$$W \equiv E_N + E_\pi \equiv m_{\Lambda^0}. \tag{6.32}$$

Now the state with definite values of $\vec{J}^2, J_3$, is given by[2]

$$|W; J, M; \lambda_1, \lambda_2\rangle = N_J \int_0^{2\pi} d\phi \int_0^\pi d\theta \sin\theta D_{M,\lambda_1-\lambda_2}^{(J)*} (R(\phi, \theta, 0)) |W, \hat{p}; \lambda_1, \lambda_2\rangle, \tag{6.35}$$

Fig. 6.6 Diagrammatic representation of the decay of a Λ^0 baryon into $N\pi$. The possible values of $J_{\Lambda^0} = 1/2, 3/2, 5/2, \ldots$.

[2] This is the rotation which takes the state back to the z direction. Start by defining the state $|W, \vec{p^0}; \lambda_1\lambda_2\rangle$, with $\vec{p^0} \equiv (0, 0, |\vec{p}|)$, which satisfies

$$J_3|W; \vec{p^0}; \lambda_1\lambda_2\rangle = (\lambda_1 - \lambda_2)|W; \vec{p^0}; \lambda_1\lambda_2\rangle. \tag{6.33}$$

since, in the CMS, $\vec{p}_1 = -\vec{p}_2 = \vec{p^0}$, and the helicities are, by definition, the angular momenta in the direction of motion. Then

$$|W, \vec{p}; \lambda_1\lambda_2\rangle = U(R(\phi, \theta, 0)) |W, \vec{p^0}; \lambda_1\lambda_2\rangle. \tag{6.34}$$

with $N_J \equiv \left(\frac{2J+1}{4\pi}\right)^{1/2}$, where

$$|W, \hat{p}; \lambda_1 \lambda_2\rangle \equiv \frac{1}{(2\pi)^3} \left(\frac{p}{4W}\right)^{1/2} |W, \vec{p}; \lambda_1 \lambda_2\rangle, \qquad (6.36)$$

$\hat{p} \equiv \vec{p}/|\vec{p}|$ and

$$\langle W, \hat{p}\,'; \lambda_1' \lambda_2' | W, \hat{p}; \lambda_1 \lambda_2\rangle \equiv \delta^2(\hat{p}\,' - \hat{p}) \delta_{\lambda_1' \lambda_1} \delta_{\lambda_2' \lambda_2}, \qquad (6.37)$$

if $|W, \vec{p}, \lambda_1 \lambda_2\rangle$ is normalized as the product of single-particle states.

The normalization

$$\langle W; J', M'; \lambda_1' \lambda_2' | W; J, M; \lambda_1 \lambda_2\rangle \equiv \delta_{JJ'} \delta_{MM'} \delta_{\lambda_1 \lambda_1'} \delta_{\lambda_2 \lambda_2'} \qquad (6.38)$$

follows from the identity

$$\int_0^{2\pi} d\alpha \int_{-1}^{1} d\cos\beta \; D_{M\lambda}^{(J)*}(R(\alpha, \beta, 0)) D_{M'\lambda}^{(J')}(R(\alpha, \beta, 0)) = \frac{4\pi}{2J+1} \delta_{JJ'} \delta_{MM'}. \qquad (6.39)$$

These states are angular momentum eigenstates satisfying

$$J_3 |W; J, M; \lambda_1 \lambda_2\rangle = M |W; J, M; \lambda_1 \lambda_2\rangle$$
$$\vec{J}^{\,2} |W; J, M; \lambda_1 \lambda_2\rangle = J(J+1) |W; J, M; \lambda_1 \lambda_2\rangle.$$

We also have the inverse relation

$$|W, \hat{p}; \lambda_1 \lambda_2\rangle = \sum_{J,M} N_J D_{M, \lambda_1 - \lambda_2}^{(J)}(R(\phi, \theta, o)) |W; J, M; \lambda_1 \lambda_2\rangle. \qquad (6.40)$$

Under rotations, we have

$$U(R(\alpha, \beta, \gamma)) |W; J, M; \lambda_1 \lambda_2\rangle \equiv \sum_{M'} |W; J, M'; \lambda_1 \lambda_2\rangle D_{M'M}^{(J)}(R(\alpha, \beta, \gamma)), \quad (6.41)$$

where

$$D_{M'M}^{(J)}(R(\alpha, \beta, \gamma)) \equiv \left(e^{-i\alpha J_3^{(J)}} e^{-i\beta J_2^{(J)}} e^{i\gamma J_3^{(J)}}\right)_{M'M} \qquad (6.42)$$

and α, β, γ are Euler angles.

The rotation matrices satisfy the relations

1. $D_{M'M}^{(J)}(R(\alpha, \beta, \gamma)) = e^{-i\alpha M' - i\gamma M} d_{M'M}^{(J)}(\beta),$

$$d_{M'M}^{(J)}(\beta) \equiv \left(e^{-i\beta J_2^{(J)}}\right)_{M'M}.$$

They satisfy:[3]

(1) $d_{M',M}^{(J)}(\beta)$ are real,

(2) $d_{M',M}^{(J)}(-\beta) = d_{M,M'}^{(J)}(\beta),$

(3) $d_{M',M}^{(J)}(\beta) = d_{-M,-M'}^{(J)}(\beta),$

(4) $d_{M',M}^{(J)}(\beta) = (-1)^{M-M'} d_{M,M'}^{(J)}(\beta).$

---

[3] See Haber (n.d.).

2. $\displaystyle\sum_M D^{(J)}_{M'M}(R)D^{(J)}_{MM''}(R') \equiv D^{(J)}_{M'M''}(RR').$

3. $D^{*(J)}_{MM'}(R) = D^{(J)}_{M'M}(R^{-1}).$

The Λ^0 is in a state $\Psi_\alpha \equiv \sum_M C^\alpha_M |\Psi_{J,M}\rangle$ and $J \equiv J_{\Lambda^0}$. Thus, the decay amplitude is

$$
\begin{aligned}
\langle \pi N|S|\Psi_\alpha\rangle &\equiv \sum_M C^\alpha_M \langle \pi N|S|\Psi_{J,M}\rangle \\
&= \sum_M C^\alpha_M N_J D^{(J)^*}_{M\lambda}(R_{\vec{p}})\langle W; J, M; \lambda, 0|S|\Psi_{J,M}\rangle,
\end{aligned}
$$

where we have used the inverse relation, Eqn. 6.40. To simplify notation, we define $S_\lambda \equiv \langle W; J, M; \lambda, 0|S|\Psi_{J,M}\rangle$.

Angular momentum conservation guarantees that

$$
J_{\pi N} \equiv J_{\Lambda^0} \equiv J, \quad M_{\pi N} = M_{\Lambda^0} \equiv M. \tag{6.43}
$$

The decay amplitude squared is then given by

$$
|\langle \pi N|S|\Lambda\rangle|^2 = \frac{2J+1}{4\pi} \sum_{MM'} C^\alpha_M C^{\alpha}_{M'}{}^* |S_\lambda|^2 D^{(J)^*}_{M\lambda}(R_{\vec{p}})D^{(J)}_{M'\lambda}(R_{\vec{p}}). \tag{6.44}
$$

Using $D^{(J)}_{M\lambda}(R_{\vec{p}}) \equiv e^{-iM\phi}d^J_{M\lambda}(\theta)$, the angular distribution[4] is given by

$$
\begin{aligned}
\frac{1}{2J+1}|\langle \pi N|S|\Lambda\rangle|^2 &\equiv W_\lambda(\theta,\phi) \\
&= \frac{1}{4\pi} \sum_{MM'} \rho_{MM'} |S_\lambda|^2 \; e^{-i(M'-M)\phi} d^J_{M\lambda}(\theta)d^J_{M'\lambda}(\theta),
\end{aligned}
$$

where $\rho_{MM'} \equiv \sum_\alpha C^\alpha_M C^{*\alpha}_{M'}$ is the density matrix for the Λ^0 beam, satisfying $Tr\rho = 1$, and $\rho_{MM} \equiv P(M)$ is the probability of measuring Λ^0 in the spin state M.

Now define

$$
W(\theta) \equiv \frac{1}{2\pi}\int_0^{2\pi} d\phi \sum_\lambda W_\lambda(\theta;\phi) = \frac{1}{4\pi} \sum_\lambda \sum_M \rho_{MM} |S_\lambda|^2 \left(d^J_{M\lambda}(\theta)\right)^2. \tag{6.45}
$$

$W(\theta)$ gives the number of events at angle θ with respect to the z direction. Then we can obtain the averages over the decay angle θ given by

$$
\langle\cos\theta\rangle \equiv \frac{\int_{-1}^{1} d\cos\theta \cos\theta W(\theta)}{\int_{-1}^{1} d\cos\theta W(\theta)}, \tag{6.46}
$$

which is the average value of $\cos\theta$ and

$$
\langle P_L(\cos\theta)\rangle \equiv \frac{\int_{-1}^{1} d\cos\theta P_L(\cos\theta) W(\theta)}{\int_{-1}^{1} d\cos\theta W(\theta)}. \tag{6.47}
$$

---

[4] $\frac{1}{2J+1}$ = average over Λ^0 spin.

Given the Clebsch–Gordan coefficients,

$$C(J, J', \ell; M, M', M + M') \equiv C(J, M; J', M' | \ell, M + M'), \qquad (6.48)$$

etc., we have the identities

$$\left(d^J_{M,\lambda}(\theta) d^J_{-M,-\lambda}(\theta)\right) \equiv \sum_{\ell=0}^{2J} (-1)^{M-\lambda} C(J, J, \ell; M, -M, 0) \, C(J, J, \ell; \lambda, -\lambda, 0) P_\ell(\cos\theta)$$

$$(6.49)$$

and

$$C(J, J, 0; M, -M, 0) = (-1)^{M-J} \frac{1}{\sqrt{2J+1}}$$

$$C(J, J, 1; M, -M, 0) = (-1)^{J-M} \sqrt{\frac{3}{2J+1}} \frac{M}{\sqrt{J(J+1)}}.$$

Using the following results,

$$\int_{-1}^{1} d\cos\theta \, W(\theta) = \frac{1}{4\pi} \sum_\lambda \sum_M P(M) |S_\lambda|^2 \int_{-1}^{1} d\cos\theta$$

$$\times \sum_\ell (-1)^{M-\lambda} C(J, J, \ell; M, -M, 0) C(J, J, \ell; \lambda, -\lambda, 0) P_\ell(\cos\theta)$$

$$= \frac{2}{4\pi(2J+1)} \sum_M P(M) \sum_\lambda |S_\lambda|^2, \qquad (6.50)$$

where $\sum_M P(M) \equiv 1$ and

$$\int_{-1}^{1} d\cos\theta \, P_\ell(\cos\theta) = 2\delta_{\ell 0},$$

$$\int_{-1}^{1} d\cos\theta \, P_\ell(\cos\theta) P_L(\cos\theta) = \frac{2\delta_{\ell L}}{2\ell + 1}.$$

Thus, we obtain

$$\langle P_L(\cos\theta) \rangle = \frac{\frac{2J+1}{2L+1} \sum_M \sum_\lambda P(M) |S_\lambda|^2 (-1)^{M-\lambda} C(J, J, L; M, -M, 0) C(J, J, L; \lambda, -\lambda, 0)}{\left(\sum_\lambda |S_\lambda|^2\right)}.$$

$$(6.51)$$

Now let's define a normalized S matrix element, $\bar{S}_\lambda \equiv \dfrac{S_\lambda}{\left(\sum_\lambda |S_\lambda|^2\right)^{1/2}}$ such that

$$\sum_\lambda |S_\lambda|^2 \equiv 1. \text{ We also define the parameter}$$

$$\alpha \equiv |\bar{S}_{1/2}|^2 - |\bar{S}_{-1/2}|^2 \qquad (6.52)$$

with $|\alpha| \le 1$.

Given these preliminaries we can now obtain the interesting, and extremely clever, part. Let's define and calculate two test functions, $T_{JM}^{\pm}$, given as follows:

$$T_{JM}^{+} \equiv \frac{2}{2J+1} \sum_{L \ even} (-1)^{M-1/2}(2L+1)\frac{C(J,J,L;M,-M,0)}{C(J,J,L;1/2,-1/2,0)} \langle P_L(\cos\theta)\rangle$$

$$= 2 \sum_{L \ even} \sum_{M',\lambda} (-1)^{M-1/2}(-1)^{M'-\lambda}P(M')|\bar{S}_\lambda|^2 \qquad (6.53)$$

$$\times \ C(J,J,L;M',-M',0)C(J,J,L;M,-M,0) \times \frac{C(J,J,L;\lambda,-\lambda,0)}{C(J,J,L;1/2,-1/2,0)}.$$

Now use

$$C(J,J,L;-M,M,0) \equiv (-1)^{2J-L} \times C(J,J,L;M,-M,0) \qquad (6.54)$$

to do the λ sum. We have

$$\sum_{\lambda=\pm 1/2} (-1)^{-1/2-\lambda}|\bar{S}_\lambda|^2 \frac{C(J,J,L;\lambda,-\lambda,0)}{C(J,J,L;1/2,-1/2,0)}$$

$$= -|\bar{S}_{1/2}|^2 + |\bar{S}_{-1/2}|^2(-1)^{2J-L}$$

$$\equiv -1 \quad \text{for } L \text{ even and } J = 1/2, 3/2, 5/2, \ldots.$$

Hence,

$$T_{JM}^{+} = 2\sum_{M'}\sum_{L \ even}(-1)^{M+M'+1}P(M')C(J,J,L;M',-M',0)C(J,J,L;M,-M,0)$$

$$= \sum_{M'}\sum_{L}(1+(-1)^L)(-1)^{M+M'+1}P(M')C(J,J,L;M',-M',0)C(J,J,L;M,-M,0)$$

$$= \sum_{M',L} P(M')(-1)^{M+M'+1}$$

$$\times \{C(J,J,L;M',-M',0)C(J,J,L;M,-M,0)$$

$$+ (-1)^{2J}C(J,J,L;-M',M',0)C(J,J,L;M,-M,0)\},$$

where in the last line we used the result of Eqn. 6.54. But, using

$$\sum_{L} C(J,J,L;M,-M,0)C(J,J,L;\mu,-\mu,0) \equiv \delta_{\mu M} \qquad (6.55)$$

we obtain

$$T_{JM}^{+} \equiv -(P(M)-P(-M)). \qquad (6.56)$$

Also define

$$T_{JM}^{-} \equiv \frac{2}{2J+1} \sum_{L \ odd}(-1)^{M+1/2}(2L+1) \times \frac{C(J,J,L;M,-M,0)}{C(J,J,L;1/2,-1/2,0)}\langle P_L(\cos\theta)\rangle,$$

$$\qquad (6.57)$$

and, via similar analysis, we obtain

$$T_{JM}^{-} = -\alpha\left(P(M)-P(-M)\right). \qquad (6.58)$$

But since $|\alpha| \leq 1$ and $0 \leq P(M) \leq 1$, we have

$$
\begin{aligned}
|T_{JM}^-| &= |\alpha| \, |P(M) - P(-M)| \\
&\leq |P(M) - P(-M)| \\
&= |T_{JM}^+| \leq 1.
\end{aligned}
$$

Experimentally, we have

$$W(\theta) \propto 1 + A \cos\theta, \tag{6.59}$$

which implies that

$$\langle P_L(\cos\theta)\rangle = 0 \quad \text{for } L \geq 2 \tag{6.60}$$

and $\langle \cos\theta \rangle \simeq 0.19$ (Crawford et al., 1959). Now using the definitions, Eqns. 6.53 and 6.57, we have

$$T_{JJ}^+ \equiv \frac{2}{2J+1}(-1)^{J-1/2}\frac{C(J,J,0;J,-J,0)}{C(J,J,0;1/2,-1/2,0)}\underbrace{\langle P_0(\cos\theta)\rangle}_{\langle 1\rangle \equiv 1},$$

$$T_{JJ}^- \equiv \frac{2}{2J+1}(-1)^{J+1/2}\frac{3C(J,J,1;J,-J,0)}{C(J,J,1;1/2,-1/2,0)}\underbrace{\langle P_1(\cos\theta)\rangle}_{\langle\cos\theta\rangle}.$$

But recall

$$|T_{JJ}^-| \leq |T_{JJ}^+|, \tag{6.61}$$

which gives

$$|\langle\cos\theta\rangle| \leq \frac{1}{3}\left(\frac{C(J,J,0;J,-J,0)}{C(J,J,0;1/2,-1/2,0)}\right)\left(\frac{C(J,J,1;1/2,-1/2,0)}{C(J,J,1;J,-J,0)}\right)$$

$$|\langle\cos\theta\rangle| \leq \frac{1}{3}\left(\frac{C(J,J,1;1/2,-1/2,0)}{C(J,J,1;J,-J,0)}\right) \equiv \frac{1}{6J}$$

or

$$0.19 \leq \frac{1}{6J} \tag{6.62}$$

and

$$J \leq \frac{1}{1.14}. \tag{6.63}$$

Thus, $0 < J < 1$, which gives the solution that

$$J \equiv J_{\Lambda^0} = \frac{1}{2}. \tag{6.64}$$

ASIDE: Proof of Eqn. 6.37. The normalization of the two-particle state is given by

$$
\begin{aligned}
\langle W, \vec{p}'; \lambda_1'\lambda_2'|W, \vec{p}; \lambda_1\lambda_2\rangle &= (2\pi)^3 2E_N \delta^3(\vec{p}_N' - \vec{p}_N)(2\pi)^3 2E_\pi \delta^3(\vec{p}_\pi' - \vec{p}_\pi)\delta_{\lambda_1'\lambda_1}\delta_{\lambda_2'\lambda_2} \\
&= (2\pi)^6 4E_N E_\pi \delta^3(\vec{p}_N' - \vec{p}_N)\,\delta^3(\vec{p}_\pi' - \vec{p}_\pi)\delta_{\lambda_1'\lambda_1}\delta_{\lambda_2'\lambda_2}.
\end{aligned}
$$

Now introduce new momenta, $P = P_N + P_\pi$, $K = \frac{1}{2}(P_N - P_\pi)$ and $k = |\vec{k}|$. Then

$$\langle W, \vec{p}'; \lambda_1' \lambda_2' | W, \vec{p}; \lambda_1 \lambda_2 \rangle = (2\pi)^6 4 E_N E_\pi \delta^3(\vec{p}' - \vec{p})\, \delta^3(\vec{k}' - \vec{k}) \delta_{\lambda_1' \lambda_1} \delta_{\lambda_2' \lambda_2}$$

$$= (2\pi)^6 \frac{4 E_N E_\pi}{k^2} \delta^3(\vec{p}' - \vec{p})\, \delta(k' - k) \delta^2(\hat{k}' - \hat{k}) \delta_{\lambda_1' \lambda_1} \delta_{\lambda_2' \lambda_2}$$

$$= (2\pi)^6 \frac{4 E_N E_\pi}{k^2} \frac{dP^0}{dk} \delta^4(P' - P)\, \delta^2(\hat{k}' - \hat{k}) \delta_{\lambda_1' \lambda_1} \delta_{\lambda_2' \lambda_2}.$$

In the last line we used $\delta(k' - k) = \delta(P^{0\prime} - P^0) \frac{dP^0}{dk}$.

In the CMS we have $P^0 = E_N + E_\pi \equiv W$, $\frac{dP^0}{d|\vec{p}_N|} = \frac{dE_N}{dp} + \frac{dE_\pi}{dp} = \frac{p}{E_N} + \frac{p}{E_\pi} \equiv \frac{pW}{E_N E_\pi}$ and $\vec{k} = \vec{p}$, $k = p$. Thus $\frac{dP^0}{dk} \equiv \frac{kW}{E_N E_\pi}$. Finally, we obtain

$$\langle W, \vec{p}'; \lambda_1' \lambda_2' | W, \vec{p}; \lambda_1 \lambda_2 \rangle = (2\pi)^6 \frac{4W}{p} \delta^2(\hat{p}' - \hat{p}) \delta_{\lambda_1' \lambda_1} \delta_{\lambda_2' \lambda_2} \delta^4(P - P'). \quad (6.65)$$

Hence the state $|W, \hat{p}; \lambda_1 \lambda_2 \rangle$ is normalized as given in Eqn. 6.37 where the factor $\delta^4(P - P')$ is implicit.

7 Introduction to Field Theory

In order to calculate the S matrix we need the formalism of relativistic quantum field theory. In the next few chapters we shall derive the basic constituents of field theory by defining multi-particle states for spin 0, $1/2$ and 1 particles. These multi-particle states will be constructed in terms of creation and annihilation operators acting on a Fock vacuum for free particles. The creation and annihilation operators will satisfy commutation (anti-commutation) relations for integral (half-integral) spin states. Given the multi-particle plane-wave states we then define wave functions and operators creating these non-trivial particle states from the vacuum. This leads us directly to quantum fields satisfying commutation or anti-commutation relations, and to Hamiltonians and then Lagrangians describing the systems. So this is our bottom-up approach constructing field theory from the free-particle eigenstates of the Poincaré group.

7.1 Multi-Particle States: Fock Space

Consider single-particle states labeled by the eigenvalues of the operators $\hat{P}^\mu$, $\hat{P}^2$, $\hat{W}^2$, ($\hat{W}_3$ or helicity, λ). We have the states

$$|\vec{p}, \lambda\rangle \equiv |P^\mu, m, s, \lambda\rangle. \tag{7.1}$$

We shall be interested in scalar (boson), fermion and vector (boson) fields or $s = 0, 1/2, 1$ states, either massive or massless.

Let's begin by discussing $s = 0$ states. We define operators, $\{a^\dagger(\vec{p}), a(\vec{p})\}$, for different momenta $\vec{p}$, which create or annihilate single-particle states from a vacuum. These are called Fock states and the Fock vacuum is defined such that $a(\vec{p})|0\rangle \equiv 0$ for all momenta $\vec{p}$ and with normalization $\langle 0|0\rangle \equiv 1$. We then have a single-particle state

$$a^\dagger(\vec{p})|0\rangle \equiv |\vec{p}\rangle. \tag{7.2}$$

These states are normalized such that

$$E_p \equiv \sqrt{\vec{p}^2 + m^2},$$
$$\langle \vec{p}\,'|\vec{p}\rangle \equiv (2\pi)^3 2E_p \delta^3(\vec{p} - \vec{p}\,')$$
$$\equiv \langle 0|a(\vec{p}\,')a^\dagger(\vec{p})|0\rangle$$
$$\equiv \langle 0|[a(\vec{p}\,'), a^\dagger(\vec{p})]|0\rangle,$$

which implies that[1]

$$[a(\vec{p}\,'), a^\dagger(\vec{p})] \equiv (2\pi)^3 \, 2E_p \, \delta^3(\vec{p} - \vec{p}\,'). \tag{7.4}$$

This is our first quantum commutation relation. We also define

$$[a(\vec{p}\,'), a(\vec{p})] \equiv [a^\dagger(\vec{p}\,'), a^\dagger(\vec{p})] = 0. \tag{7.5}$$

With these definitions, we can now define a number operator $\hat{N}$ which counts the number of particles in the system. We have

$$\hat{N} \equiv \int \frac{d^3\vec{p}}{(2\pi)^3 2E_p} \, a^\dagger(\vec{p})a(\vec{p}),$$

$$\hat{N}|0\rangle \equiv 0,$$

$$\hat{N}|\vec{p}\rangle \equiv \int \frac{d^3\vec{p}'}{(2\pi)^3 \, 2E_{p'}} \, a^\dagger(\vec{p}\,') \quad \times \quad \underbrace{a(\vec{p}\,')a^\dagger(\vec{p})|0\rangle}_{(2\pi)^3 \, 2E_{p'}\delta^3(\vec{p} - \vec{p}\,')|0\rangle}$$

implying that

$$\hat{N}|\vec{p}\rangle \equiv |\vec{p}\rangle. \tag{7.6}$$

We now consider a multi-particle state, which can be written in the form

$$a^\dagger(\vec{p}_1)a^\dagger(\vec{p}_2)\ldots a^\dagger(\vec{p}_n)|0\rangle \equiv |\vec{p}_1, \ldots, \vec{p}_n\rangle, \tag{7.7}$$

with

$$\vec{p}_1 \neq \vec{p}_2 \neq \cdots \neq \vec{p}_n \tag{7.8}$$

and normalization $\mathbf{1}^n \equiv \langle \vec{p}_1, \ldots, \vec{p}_n | \vec{p}_1, \ldots, \vec{p}_n \rangle$ where $\mathbf{1}$ is defined as follows

$$\mathbf{1} \equiv (2\pi)^3 \, 2E_p \, \delta^3(\vec{0}). \tag{7.9}$$

Then we have

$$\hat{N}|\vec{p}_1, \ldots, \vec{p}_n\rangle = n|\vec{p}_1, \ldots, \vec{p}_n\rangle. \tag{7.10}$$

You might ask, what happens if $\vec{p}_i = \vec{p}_j, \quad i \neq j$? In general, we have

$$a^\dagger(\vec{p}_i)|n(\vec{p}_1), \ldots, n(\vec{p}_i), \ldots\rangle = N_+ \left(n(\vec{p}_i)\right)|n(\vec{p}_1), \ldots, n(\vec{p}_i) + 1, \ldots\rangle, \tag{7.11}$$

where N_+ is the normalization factor and

$$n(\vec{p}_i\rangle \equiv \text{number of particles in state } \vec{p}_i. \tag{7.12}$$

Now let's evaluate the normalization factor, N_+, in a simple case. We have

$$a^\dagger(\vec{p})|n(\vec{p})\rangle = N_+ \left(n(\vec{p})\right)|n(\vec{p}) + 1\rangle. \tag{7.13}$$

---

[1] Note: Peskin and Schroeder (1995) choose to define

$$\sqrt{2E_p} \, a^\dagger(\vec{p})|0\rangle \equiv |\vec{p}\rangle. \tag{7.3}$$

Hence,

$$\langle n(\vec{p}) + 1 | n(\vec{p}) + 1 \rangle |N_+|^2 = \langle n(\vec{p}) | \underbrace{a(\vec{p}) a^\dagger(\vec{p})}_{a^\dagger(\vec{p}) a(\vec{p}) + \mathbf{1}} | n(\vec{p}) \rangle \tag{7.14}$$

and thus

$$\underbrace{\langle n(\vec{p}) + 1 | n(\vec{p}) + 1 \rangle}_{\mathbf{1}^{(n(\vec{p})+1)}} |N_+|^2 = (n(\vec{p}) + 1) \, \mathbf{1} \underbrace{\langle n(\vec{p}) | n(\vec{p}) \rangle}_{\mathbf{1}^{n(\vec{p})}} \tag{7.15}$$

and

$$|N_+ \left(n(\vec{p}) \right)|^2 \equiv n(\vec{p}) + 1. \tag{7.16}$$

Hence (up to a phase) we have

$$N_+ \left(n(\vec{p}) \right) \equiv \sqrt{n(\vec{p}) + 1}. \tag{7.17}$$

Carrying this process in reverse, we find the result which should be familiar from the quantum mechanics of an harmonic oscillator, indicative of Bose–Einstein statistics. We have

$$|n(\vec{p})\rangle \equiv \frac{1}{\sqrt{n(\vec{p})!}} \left(a^\dagger(\vec{p}) \right)^{n(\vec{p})} |0\rangle. \tag{7.18}$$

As a consequence of the commutation relations, we also have

$$a^\dagger(\vec{p}_1) a^\dagger(\vec{p}_2) |0\rangle \equiv a^\dagger(\vec{p}_2) a^\dagger(\vec{p}_1) |0\rangle, \tag{7.19}$$

i.e. the wave functions are symmetric under interchange of the particles

$$|\vec{p}_1, \vec{p}_2\rangle = |\vec{p}_2, \vec{p}_1\rangle. \tag{7.20}$$

The energy and momentum operators for this system are easily obtained:

$$\hat{H} \equiv \int \frac{d^3\vec{p}}{(2\pi)^3 2E_p} \, E_p \, a^\dagger(\vec{p}) a(\vec{p}), \tag{7.21}$$

$$\hat{\vec{P}} \equiv \int \frac{d^3\vec{p}}{(2\pi)^3 2E_p} \, \vec{p} \, a^\dagger(\vec{p}) a(\vec{p}), \tag{7.22}$$

such that

$$[\hat{H}, a^\dagger(\vec{p}, t)] = E_p a^\dagger(\vec{p}, t)$$

$$\equiv -i \frac{\partial}{\partial t} a^\dagger(\vec{p}, t) \qquad \text{(Heisenberg equations of motion)}.$$

In addition, we have

1. $\hat{H}|0\rangle \equiv 0$;
2. $\hat{H}|\vec{p}\rangle = E_p|\vec{p}\rangle$;
3. $a^\dagger(\vec{p}, t)$ (Heisenberg-picture operators) $= e^{iE_p t} a^\dagger(\vec{p})$ (Schrödinger-picture operators).

Also

$$[\hat{\vec{P}}, a^\dagger(\vec{p})] \equiv \vec{p}\, a^\dagger(\vec{p}) \tag{7.23}$$

implies

1. $\hat{\vec{P}}\,|0\rangle \equiv 0;$
2. $\hat{\vec{P}}\,|\vec{p}\rangle \equiv \vec{p}\,|\vec{p}\rangle.$

We can also define a wave packet.

$$\begin{aligned}|f(\vec{x})\rangle &\equiv \int \frac{d^3\vec{p}}{(2\pi)^3 2E_p}\; f_{\vec{p}}(\vec{x})|\vec{p}\rangle \\ &\equiv \phi_{f+}(\vec{x})|0\rangle,\end{aligned}$$

where

$$\phi_{f+}(\vec{x}) \equiv \int \frac{d^3\vec{p}}{(2\pi)^3 2E_p}\; f_{\vec{p}}(\vec{x})a^\dagger(\vec{p}) \tag{7.24}$$

is a field operator, and

$$f_{\vec{p}}(\vec{x}) \equiv \langle \vec{p}\,|\phi_{f+}(\vec{x})\,|0\rangle \tag{7.25}$$

is the single-particle wave function. For a plane wave we have

$$f_{\vec{p}}(\vec{x}) \equiv e^{-i\vec{p}\cdot\vec{x}} \tag{7.26}$$

and

$$\phi_+(\vec{x}) \equiv \int \frac{d^3\vec{p}}{(2\pi)^3 2E_p}\; e^{-i\vec{p}\cdot\vec{x}}\, a^\dagger(\vec{p}). \tag{7.27}$$

We could also use the following field operator to create the plane wave state,

$$\phi(\vec{x}) \equiv \phi_+(\vec{x}) + \phi_+^\dagger(\vec{x})$$
$$\phi(\vec{x}) \equiv \int \frac{d^3\vec{p}}{(2\pi)^3 2E_p}\; \left(e^{-i\vec{p}\cdot\vec{x}}a^\dagger(\vec{p}) + e^{i\vec{p}\cdot\vec{x}}a(\vec{p})\right),$$

which satisfies

$$\phi^\dagger(\vec{x}) = \phi(\vec{x}) \quad \text{i.e. Hermitian.} \tag{7.28}$$

Let's evaluate the commutator,

$$\left[\hat{\vec{P}}, \phi(\vec{x})\right] = \int \frac{d^3\vec{p}\,'}{(2\pi)^3 2E_{p'}}\; \vec{p}\,' \int \frac{d^3\vec{p}}{(2\pi)^3 2E_p}$$

$$\times \left(e^{-i\vec{p}\cdot\vec{x}}\;\left[a^\dagger(\vec{p}\,')a(\vec{p}\,'), a^\dagger(\vec{p})\right] + e^{i\vec{p}\cdot\vec{x}}\left[a^\dagger(\vec{p}\,')a(\vec{p}\,'), a(\vec{p})\right]\right).$$

But $([AB, C] \equiv A[B, C] + [A, C]B)$ and

$$\left[a^\dagger(\vec{p}\,')a(\vec{p}\,'), a^\dagger(\vec{p})\right] = (2\pi)^3 2E_p\delta^3(\vec{p} - \vec{p}\,')a^\dagger(\vec{p}),$$
$$\left[a^\dagger(\vec{p}\,')a(\vec{p}\,'), a(\vec{p})\right] = -(2\pi)^3 2E_p\delta^3(\vec{p} - \vec{p}\,')a(\vec{p}).$$

Therefore, we have

$$\left[\hat{\vec{P}}, \phi(\vec{x})\right] = \int \frac{d^3\vec{p}}{(2\pi)^3 2E_p} \; \vec{p}\left(e^{-i\vec{p}\cdot\vec{x}} a^\dagger(\vec{p}) - e^{i\vec{p}\cdot\vec{x}} \; a(\vec{p})\right)$$

$$\equiv i\frac{\partial}{\partial\vec{x}}\phi(\vec{x})$$

or

$$\left[\hat{P}^i, \phi(\vec{x})\right] = -i\frac{\partial}{\partial x_i}\phi(\vec{x}) \qquad (x_i \equiv -x^i). \tag{7.29}$$

Thus, $\hat{\vec{P}}$ is the generator of translations. Note, we have

$$e^{-i\hat{\vec{P}}\cdot\vec{a}} \; \phi(\vec{x})e^{i\hat{\vec{P}}\cdot\vec{a}} \equiv \phi(\vec{x}+\vec{a})$$

$$\text{(order by order)} \qquad \equiv e^{\vec{a}\cdot\frac{\partial}{\partial\vec{x}}}\phi(\vec{x})$$

and the unitary translation operator,

$$\mathcal{U}(\vec{a}) \equiv e^{-i\hat{\vec{P}}\cdot\vec{a}} \equiv e^{i\hat{P}^i\cdot a_i}. \tag{7.30}$$

We can also express the transformation of the operator $\phi(\vec{x})$ under translations, in the standard form, by

$$\phi'(\vec{x}) \equiv \mathcal{U}(\vec{a}) \; \phi(\vec{x}) \; \mathcal{U}^{-1}(\vec{a}) \equiv \phi(\vec{x}\,'), \tag{7.31}$$

where $\vec{x}\,' = \vec{x} + \vec{a}$.

It is not surprising that $\hat{H}$ is a generator of time translations. Consider

$$\phi(x) \equiv \phi(\vec{x}, t) \qquad \text{Heisenberg operator}$$

$$\equiv \int \frac{d^3\vec{p}}{(2\pi)^3 2E_p} \left(e^{-i\vec{p}\cdot\vec{x}} \; a^\dagger(\vec{p}, t) + e^{i\vec{p}\cdot\vec{x}} \; a(\vec{p}, t)\right)$$

and for a free particle we have

$$\phi(x) \equiv \int \frac{d^3\vec{p}}{(2\pi)^3 2E_p} \left(e^{ip\cdot x} \; a^\dagger(\vec{p}) + e^{-ip\cdot x} \; a(\vec{p})\right), \tag{7.32}$$

where $a^\dagger(\vec{p}, t) \equiv e^{iE_p t} \; a^\dagger(\vec{p})$ and $p \cdot x \equiv E_p t - \vec{p} \cdot \vec{x}$. Then

$$\left[\hat{H}, \phi(x)\right] \equiv -i\frac{\partial}{\partial t}\phi(x) \tag{7.33}$$

or

$$e^{i\hat{H}a_0}\phi(x)e^{-i\hat{H}a_0} \equiv e^{a_0\frac{\partial}{\partial t}} \phi(x)$$

$$\equiv \phi(x^0 + a^0, \vec{x}).$$

In general, we have $\hat{P}^0 \equiv \hat{H}$ and thus

$$\left[\hat{P}^\mu, \phi(x)\right] = -i\partial^\mu\phi(x), \tag{7.34}$$

where the notation $\partial^\mu \equiv \frac{\partial}{\partial x_\mu}$. The unitary operator for space–time translations, as discussed in the first chapter, is now given by

$$\mathcal{U}(a) \equiv e^{i\hat{P}^\mu a_\mu}, \tag{7.35}$$

such that

$$\phi'(x) \equiv \mathcal{U}(a)\phi(x)U^{-1}(a) \equiv \phi(x'),$$
$$x' = x + a.$$

7.2 What about Lorentz Transformations?

We want

$$\mathcal{U}(\Lambda) \, |\vec{p}\rangle \equiv |\vec{\Lambda p}\rangle \tag{7.36}$$

and in analogy with translations,

$$\phi'(x) \equiv \mathcal{U}(\Lambda)\phi(x)\mathcal{U}^{-1}(\Lambda) \equiv \phi(x'), \tag{7.37}$$

where

$$x' \equiv \Lambda x$$
$$(x'^{\mu} \equiv \Lambda^{\mu}{}_{\nu}x^{\nu}),$$
$$\mathcal{U}(\Lambda) \equiv e^{\frac{i}{2}\alpha_{\mu\nu}\hat{M}^{\mu\nu}}.$$

For small α we have $\Lambda^{\mu}{}_{\nu} = g^{\mu}{}_{\nu} + \alpha^{\mu}{}_{\nu}$ or $\mathcal{U}(g + \alpha) \simeq 1 + \frac{i}{2}\alpha_{\mu\nu}\hat{M}^{\mu\nu} + 0(\alpha^2)$. Thus, Eqn. 7.37 gives

$$\phi'(x) \simeq \phi(x) + \frac{i}{2}\alpha_{\mu\nu}\left[\hat{M}^{\mu\nu}, \phi(x)\right] + \dots$$
$$= \phi(x^{\mu} + \alpha^{\mu}{}_{\nu}x^{\nu} + \dots)$$
$$= \phi(x) + \alpha^{\mu}{}_{\nu}x^{\nu}\partial_{\mu}\phi(x) + \dots$$

or

$$\left[\hat{M}^{\mu\nu}, \phi(x)\right] = -i(x^{\nu}\partial^{\mu} - x^{\mu}\partial^{\nu})\phi(x). \tag{7.38}$$

Note: In quantum mechanics, observables are given in terms of the matrix elements of Hermitian operators. Moreover, causality, in a Lorentz invariant theory, requires that the commutator, $[\phi(x), \phi(y)]$, vanish for space-like separations, $(x - y)^2 < 0$, i.e. two measurements made at space-like separations should be physically independent. Hence, we shall now show that Lorentz covariance and causality requires that we use the field operator

$$\phi(x) \equiv \phi_{+}(x) + \phi_{+}^{\dagger}(x) \tag{7.39}$$

to describe physical observables. The proof of the relation

$$[\phi(x), \phi(y)] = 0 \text{ for } (x - y)^2 < 0 \tag{7.40}$$

follows.

Proof   Define the operator (which in fact we shall show that, for free fields, it is a c-number),

$$C(x, y) \equiv [\phi(x), \phi(y)].$$

Then under translations we have

$$\mathcal{U}(-y)[\phi(x), \phi(y)]\mathcal{U}^{-1}(-y) = [\phi(x - y), \phi(0)].$$

Thus, $C(x, y) = C(x - y)$ is only a function of $x - y$. Under Lorentz transformations, we have

$$\mathcal{U}(\Lambda)[\phi(z), \phi(0)]\mathcal{U}^{-1}(\Lambda) = [\phi(z'), \phi(0)] \quad \text{for } z' \equiv \Lambda z. \tag{7.41}$$

Hence, $C(z) = C(z')$ is only a function of z^2.

Calculate

$$\begin{aligned}
[\phi(x), \phi(y))] &\equiv \int \frac{d^3\vec{p}}{(2\pi)^3 2E_p} \int \frac{d^3\vec{p}\,'}{(2\pi)^3 2E_{p'}} \\
&\quad \times \left[\left(e^{ip\cdot x} a^\dagger(\vec{p}) + e^{-ip\cdot x} a(\vec{p}) \right), \left(e^{ip'\cdot y} a^\dagger(\vec{p}\,') + e^{-ip'\cdot y} a(\vec{p}\,') \right) \right] \\
&= \int \frac{d^3\vec{p}}{(2\pi)^3 2E_p} \frac{d^3\vec{p}\,'}{(2\pi)^3 2E_{p'}} \left(e^{-ip\cdot x + ip'\cdot y} \left[a(\vec{p}), a^\dagger(\vec{p}\,') \right] \right. \\
&\quad \left. - e^{ip\cdot x - ip'\cdot y} \left[a(\vec{p}\,'), a^\dagger(\vec{p}) \right] \right) \\
&= \int \frac{d^3\vec{p}}{(2\pi)^3 2E_p} \left(e^{-ip\cdot(x-y)} - e^{ip\cdot(x-y)} \right) \\
&= \int \frac{d^4 p}{(2\pi)^3} \theta(p^0) \delta(p^2 - m^2) \left(e^{-ip\cdot(x-y)} - e^{ip\cdot(x-y)} \right) \\
&\equiv D(x - y) - D(y - x) \equiv C(x, y).
\end{aligned}$$

The function $D(z)$ is explicitly a Lorentz invariant function of z satisfying $D(z') = D(z)$ for $z' = \Lambda z$. Now let $z = x - y$ with $z^2 < 0$. Since the points z and $-z$ in Fig. 7.1 are space-like, one can be obtained from the other by a Lorentz transformation. Hence

$$D(z) = D(-z) \tag{7.42}$$

and

$$C(x, y) = [\phi(x), \phi(y)] = 0, \quad \text{for } (x - y)^2 < 0. \tag{7.43}$$

$\square$

Note, however that

$$C(x, y) = [\phi(x), \phi(y)] \neq 0 \quad \text{for } (x - y)^2 > 0, \tag{7.44}$$

since for $z^2 > 0$ the points z and $-z$ are not related by a Lorentz transformation.

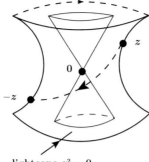

lightcone $z^2 = 0$

Fig. 7.1 Space–time coordinates with space-like and time-like regions specified. We can move from the point z to the point at $-z$ by a Lorentz transformation.

7.3 Let's Complete the Analogy of a Scalar Field with the Quantum-Harmonic Oscillator

The Hamiltonian for the quantum oscillator is given by

$$H_{osc} \equiv \frac{1}{2}p^2 + \frac{w^2}{2}q^2, \tag{7.45}$$

with the position, q, and momentum, p, satisfying the commutation relations,

$$[q, p] = i \quad [q, q] = 0 = [p, p]. \tag{7.46}$$

The parameters q, p are Hermitian operators acting on a Hilbert space. Define creation and annihilation operators in Hilbert space by

$$p = -i \left(\frac{w}{2}\right)^{1/2} (a - a^\dagger), \tag{7.47}$$

$$q = \frac{1}{(2w)^{1/2}} (a + a^\dagger),$$

or

$$a = \frac{i}{(2w)^{1/2}} (p - iwq),$$

$$a^\dagger = \frac{-i}{(2w)^{1/2}} (p + iwq),$$

satisfying

$$[a, a^\dagger] = \frac{1}{2w} \underbrace{[p - iwq, p + iwq]}_{2w} \equiv 1 \tag{7.48}$$

and

$$[a, a] = [a^\dagger, a^\dagger] = 0. \tag{7.49}$$

Following Eqns. 7.45 and 7.47 we have

$$
\begin{aligned}
H_{osc} &\equiv -\frac{1}{2}\frac{w}{2}\left(a - a^\dagger\right)^2 + \frac{1}{2}\frac{w^2}{2w}(a + a^\dagger)^2 \\
&= \frac{w}{2}\left(aa^\dagger + a^\dagger a\right) \equiv wa^\dagger a + \frac{w}{2}.
\end{aligned}
$$

The ground state of the oscillator $|0\rangle$ satisfies $a|0\rangle \equiv 0$. In the coordinate representation

$$
p \equiv -i\frac{\partial}{\partial q} \tag{7.50}
$$

implies that

$$
\langle q|p|q'\rangle \equiv -i\frac{\partial}{\partial q}\delta(q - q'). \tag{7.51}
$$

Then

$$
\langle q|a|0\rangle = \frac{1}{(2w)^{1/2}}\left(\frac{\partial}{\partial q} + wq\right)\langle q|0\rangle \equiv 0, \tag{7.52}
$$

or the ground state wave function is given by

$$
\langle q|0\rangle \equiv e^{-\frac{w}{2}q^2}. \tag{7.53}
$$

Excited states are given by

$$
|n\rangle = \frac{1}{\sqrt{n!}}(a^\dagger)^n|0\rangle. \tag{7.54}
$$

We then have

$$
H\,|n\rangle = \left(n + \frac{1}{2}\right)w\,|n\rangle, \tag{7.55}
$$

with

$$
H|0\rangle = \frac{w}{2}|0\rangle, \tag{7.56}
$$

i.e. the ground-state energy is non-zero due to zero-point quantum fluctuations.

Now consider the scalar field theory. We have

$$
\hat{H} = \int \frac{d^3\vec{p}}{(2\pi)^3 2E_p}\left(E_p a^\dagger(\vec{p})a(\vec{p}) + \mathbf{1}\frac{E_p}{2}\right), \tag{7.57}
$$

where recall,

$$
\mathbf{1} \equiv (2\pi)^3 2E_p\delta^3(\vec{0}), \tag{7.58}
$$

and we have added the last term which corresponds to the vacuum energy due to zero-point fluctuations. The Hamiltonian is given by a sum over an infinite set of harmonic oscillators in momentum space. The zero-point energy for a field theory is divergent. Including a momentum cut-off, Λ, the energy per unit volume diverges as Λ^4. This is not a problem, unless one includes gravity. Without gravity, only energy differences are relevant. However, once one includes gravity then the vacuum energy curves space–time. This vacuum energy is directly related to the cosmological constant by $G_N\Lambda^4$. With Λ of order the Planck scale, the cosmological constant is about a factor of 10^{120} larger than observation.

Continuing with the analogy with the non-relativistic harmonic oscillator we define the operators

$$P(\vec{p}) \equiv -i\left(\frac{E_p}{2}\right)^{1/2}\left(a(\vec{p}) - a^\dagger(-\vec{p})\right),$$

$$Q(\vec{p}) \equiv \frac{1}{(2E_p)^{1/2}}\left(a(\vec{p}) + a^\dagger(-\vec{p})\right).$$

Note: unlike the non-relativistic case, Q is not Hermitian. We have $Q^\dagger(\vec{p}) \equiv Q(-\vec{p})$. We now have

$$a(\vec{p}) \equiv \frac{i}{(2E_p)^{1/2}}\left(P(\vec{p}) - iE_pQ(\vec{p})\right). \tag{7.59}$$

Then we find

$$\hat{H} \equiv \int \frac{d^3\vec{p}}{(2\pi)^3 2E_p}\left(\frac{1}{2}|P(\vec{p})|^2 + \frac{E_p^2}{2}|Q(\vec{p})|^2\right) \tag{7.60}$$

is given by the sum over an infinite set of oscillators. The Fock ground state satisfies

$$a(\vec{p})|0\rangle \equiv 0. \tag{7.61}$$

Q, P satisfy the commutation relations,

$$[Q(\vec{p}), P(\vec{p}\,')] = -\frac{i}{2}\left[a(\vec{p}) + a^\dagger(-\vec{p}), a(\vec{p}\,') - a^\dagger(-\vec{p})\right]$$

$$= i(2\pi)^3 2E_p\delta^3(\vec{p} + \vec{p}\,').$$

Therefore, in the Schrödinger picture, we have

$$P(\vec{p}) \equiv -i\frac{\delta}{\delta Q(-\vec{p})}. \tag{7.62}$$

It is a functional derivative satisfying

$$\frac{\delta Q(\vec{p})}{\delta Q(\vec{p}\,')} \equiv (2\pi)^3 2E_p\delta^3(\vec{p} - \vec{p}\,'). \tag{7.63}$$

We can now evaluate the Fock ground state in the Schrödinger picture. Starting with Eqn. 7.61 we have

$$\langle[Q(\vec{p})]|a(\vec{p})|0\rangle = \frac{1}{(2E_p)^{1/2}}\left(\frac{\delta}{\delta Q(-\vec{p})} + E_pQ(\vec{p})\right)\langle[Q(\vec{p})]|0\rangle \equiv 0, \tag{7.64}$$

which gives

$$\langle[Q(\vec{p})]|0\rangle \equiv \exp\left(-\int \frac{d^3\vec{p}}{(2\pi)^3 2E_p} E_p|Q(\vec{p})|^2\right). \tag{7.65}$$

This is the vacuum wave functional in the Schrödinger picture.

Now consider the Lagrangian for the harmonic oscillator. We have

$$L = \frac{1}{2}\dot{q}^2 - \frac{w^2}{2}q^2, \tag{7.66}$$

with canonical momentum,

$$p \equiv \frac{\partial L}{\partial \dot{q}} \equiv \dot{q}. \tag{7.67}$$

Lagrange's equations of motion are

$$\partial_t \frac{\partial L}{\partial \dot{q}} - \frac{\partial L}{\partial q} = 0. \tag{7.68}$$

Then the Hamiltonian is obtained from the Lagrangian via a Legendre transformation,

$$H_{osc} \equiv \dot{q}p - L \equiv H(p,q),$$
$$= p^2 - \left(\frac{p^2}{2} - \frac{w^2 q^2}{2} \right),$$
$$= \frac{1}{2}p^2 + \frac{w^2}{2}q^2.$$

Similarly, for field theory, we have a Lagrangian,

$$L \equiv \int d^3\vec{x} \underbrace{\left[\frac{1}{2} \left(\frac{\partial \phi}{\partial t} \right)^2 - \frac{1}{2} \left(\frac{\partial \phi}{\partial \vec{x}} \right)^2 - \frac{m^2}{2}\phi^2 \right]}_{} \equiv \int d^3\vec{x}\mathcal{L}(\phi,\dot{\phi},\vec{\nabla}\phi),$$

$$\mathcal{L}(\phi,\dot{\phi},\vec{\nabla}\phi) = \frac{1}{2}\left(\partial_\mu \phi\right)^2 - \frac{m^2}{2}\phi^2 \equiv \frac{1}{2}g^{\mu\nu}\partial_\mu\phi\partial_\nu\phi - \frac{m^2}{2}\phi^2.$$

We can also describe this as an infinite number of coupled harmonic oscillators, each located within some small volume element τ_i with $\vec{x} \subseteq \tau_i$ and

$$\phi(\vec{x},t) \to \phi_i(t), \tag{7.69}$$

i.e. the field averaged over the volume element τ_i. The gradient term then becomes a finite difference term coupling the different oscillators and we sum over all i. Thus, each ϕ_i is a quantum oscillator and part of an infinite system of coupled oscillators.

Lagrange's equations of motion are given by

$$\partial_\mu \frac{\partial \mathcal{L}}{\partial \partial_\mu \phi} - \frac{\partial \mathcal{L}}{\partial \phi} = 0 \tag{7.70}$$

$$(\Box + m^2)\phi = 0 \tag{7.71}$$

where $\Box$ is the D'Alembertian given by

$$\Box \equiv g^{\mu\nu}\partial_\mu\partial_\nu. \tag{7.72}$$

The canonical coordinates for the real scalar field are given by

$$\phi(x) \equiv \int \frac{d^3\vec{p}}{(2\pi)^3 2E_p} \left[e^{ip\cdot x}a^\dagger(\vec{p}) + e^{-ip\cdot x}a(\vec{p}) \right]. \tag{7.73}$$

We have

$$\frac{\partial \phi}{\partial \vec{x}} = -i \int \frac{d^3\vec{p}}{(2\pi)^3 2E_p} \; \vec{p}\left[e^{ip\cdot x}a^\dagger(\vec{p}) - e^{-ip\cdot x}a(\vec{p}) \right]. \tag{7.74}$$

At time $t = 0$, we have

$$\phi(\vec{x}) \equiv \int \frac{d^3\vec{p}}{(2\pi)^3 \sqrt{2E_p}} \, e^{i\vec{p}\cdot\vec{x}} Q(\vec{p}). \tag{7.75}$$

Similarly, at $t = 0$, the canonical momentum is given by

$$\frac{\partial \mathcal{L}}{\partial \dot{\phi}(\vec{x})} \equiv \Pi_\phi(\vec{x}) = \dot{\phi}(\vec{x})$$

$$\equiv \int \frac{d^3\vec{p}}{(2\pi)^3 \sqrt{2E_p}} \, e^{i\vec{p}\cdot\vec{x}} P(\vec{p})$$

and

$$\frac{1}{2} \int d^3\vec{x} (\dot{\phi}(\vec{x}))^2 \equiv \int \frac{d^3\vec{p}}{(2\pi)^3 2E_p} \, \frac{1}{2}|P(\vec{p})|^2. \tag{7.76}$$

The Hamiltonian density is then given by

$$\mathcal{H} = \Pi_\phi \, \dot{\phi} - \mathcal{L} = \frac{1}{2} \left(\dot{\phi}^2 + (\vec{\nabla}\phi)^2 + m^2\phi^2 \right), \tag{7.77}$$

with the Hamiltonian,

$$H \equiv \int d^3\vec{x} \, \mathcal{H}$$

$$= \int d^3\vec{x} \left(\frac{1}{2}\Pi_\phi^2(\vec{x}) + \frac{1}{2} \left((\vec{\nabla}\phi)^2 + m^2\phi^2 \right) \right).$$

We then have

$$\frac{1}{2} \int d^3\vec{x} \left[(\vec{\nabla}\phi)^2 + m^2\phi^2 \right] |_{t=0} \equiv \int \frac{d^3\vec{p}}{(2\pi)^3 2E_p} \, \frac{E_p^2}{2} \, |Q(\vec{p})|^2, \tag{7.78}$$

which implies

$$H \equiv \int \frac{d^3\vec{p}}{(2\pi)^3 2E_p} \, \left(\frac{1}{2}|P(\vec{p})|^2 + \frac{E_p^2}{2} \, |Q(\vec{p})|^2 \right). \tag{7.79}$$

This returns us to the analogy between the non-relativistic quantum oscillator and the field-theory description in momentum space of an infinite set of coupled quantum oscillators.

The scalar field satisfies canonical, equal-time, commutation relations,

$$[\phi(\vec{x}, t), \Pi_\phi(\vec{y}, t)] \equiv i\delta^3(\vec{x} - \vec{y}). \tag{7.80}$$

7.4 Some Useful Identities

Now we want to derive two relations which become useful once we discuss perturbation theory for an interacting scalar field and the Feynman rules in Chapter 12. The first one is as follows, we show that

$$a^\dagger(\vec{p}) \equiv \int d^3\vec{x} \, \phi(\vec{x}, t) \, i \overleftrightarrow{\partial_t} \, e^{-ip\cdot x}, \tag{7.81}$$

Fig. 7.2 The Feynman diagram associated with a scalar propagator from space–time point y to x.

where $\overleftrightarrow{\partial_t}$ means differentiate to the right with a plus sign and then to the left with a minus sign. The proof is quite simple. We have

$$a^\dagger(\vec{p}) \equiv \int d^3\vec{x}\; \phi(\vec{x},t)\; i\, \overleftrightarrow{\partial_t}\; e^{-ip\cdot x}$$

$$= \int \frac{d^3\vec{p}\,'}{(2\pi)^3 2E_{p'}} \int d^3\vec{x} \quad \left\{ \left[e^{ip'\cdot x} a^\dagger(\vec{p}\,') + e^{-ip'\cdot x} a(\vec{p}\,') \right] i\, \overleftrightarrow{\partial_t}\; e^{-ip\cdot x} \right\}$$

$$\left\{ (E_p + E_{p'})\, e^{i(p'-p)\cdot x}\, a^\dagger(\vec{p}\,') + (E_p - E_{p'}) e^{-i(p'+p)\cdot x}\, a(\vec{p}\,') \right\}$$

Then using $\int d^3\vec{x}\; e^{i(p'-p)\cdot x} = (2\pi)^3 \delta^3(\vec{p} - \vec{p}\,')$, we obtain the result.

For the second important relation we shall evaluate the matrix element in the Fock vacuum of the time-ordered product of two scalar fields, also known as the Feynman propagator, see Fig. 7.2. We have

$$i\Delta_F(x-y) \equiv \langle 0|T\left(\phi(x)\phi(y)\right)|0\rangle$$

$$\equiv \langle 0|\left\{ \theta(x^0 - y^0)\phi(x)\phi(y) + \theta(y^0 - x^0)\phi(y)\phi(x) \right\}|0\rangle$$

$$= \int \frac{d^4p}{(2\pi)^4} \frac{i}{p^2 - m^2 + i\epsilon}\, e^{ip\cdot(x-y)}.$$

The proof of the last equality is:

$$i\Delta_F(x-y) = \theta(x^0 - y^0) \int \frac{d^3\vec{p}\,'}{(2\pi)^3 2E_{p'}} \frac{d^3\vec{p}}{(2\pi)^3 2E_p}\, e^{-ip'\cdot x} e^{ip\cdot y}$$

$$\times \underbrace{\langle 0|a(\vec{p}\,')a^\dagger(\vec{p})|0\rangle}_{(2\pi)^3 2E_p \delta^3(\vec{p} - \vec{p}\,')}$$

$$+\; \theta(y^0 - x^0) \int \frac{d^3\vec{p}\,'}{(2\pi)^3 2E_{p'}} \frac{d^3\vec{p}}{(2\pi)^3 2E_p}\, e^{-ip\cdot y} e^{ip'\cdot x}$$

$$\times \underbrace{\langle 0|a(\vec{p})a^\dagger(\vec{p}\,')|0\rangle}_{(2\pi)^3 2E_p \delta^3(\vec{p} - \vec{p}\,')}$$

$$= \int \frac{d^3\vec{p}}{(2\pi)^3 2E_p} \left\{ e^{-ip\cdot(x-y)}\, \theta(x^0 - y^0) + e^{ip\cdot(x-y)}\, \theta(y^0 - x^0) \right\}.$$

Now we define an integral realization of the Heaviside step function, $\theta(x^0)$. We have (see Fig. 7.3)

$$\theta(x^0) \equiv \frac{i}{2\pi} \int_{-\infty}^{+\infty} \frac{dE\; e^{-iEx^0}}{E + i\epsilon}. \tag{7.82}$$

Integral over the real axis with a pole located at $-i\epsilon$ in the complex E plane.

For $x^0 < 0$ we can extend the integral into the complex E plane by adding the semicircle at $\mathrm{Im}E \to +\infty$. In this case, by Cauchy's theorem we have

$$\theta(x^0 < 0) \equiv 0. \tag{7.83}$$

For $x^0 > 0$, on the other hand, we can extend the integral into the complex E plane by adding the semicircle at $\mathrm{Im}E \to -\infty$. We then have

$$\theta(x^0 < 0) \equiv 1. \tag{7.84}$$

Hence, $\theta(x^0)$ is the step function.

We now plug this representation of the step function into the integral where we replace the integral parameter E by p^0. We have

$$i\Delta_F(x-y)$$
$$= i \int \frac{d^3\vec{p}}{(2\pi)^3 2E_p} \frac{dp^0}{(2\pi)} \left\{ \frac{e^{-i(p^0+E_p)(x^0-y^0)+i\vec{p}\cdot(\vec{x}-\vec{y})}}{p^0 + i\epsilon} + \frac{e^{i(p^0+E_p)(x^0-y^0)-i\vec{p}\cdot(\vec{x}-\vec{y})}}{p^0 + i\epsilon} \right\}$$

We also define $d^4p \equiv d^3\vec{p}\,dp^0$ and then obtain

$$= i \int \frac{d^4p}{(2\pi)^4} \frac{1}{2E_p} \left\{ \frac{e^{-ip\cdot(x-y)}}{p^0 - E_p + i\epsilon} + \frac{e^{ip\cdot(x-y)}}{p^0 - E_p + i\epsilon} \right\}$$

$$= i \int \frac{d^4p}{(2\pi)^4} \frac{e^{ip\cdot(x-y)}}{2E_p} \underbrace{\left\{ \frac{-1}{p^0 + E_p - i\epsilon} + \frac{1}{p^0 - E_p + i\epsilon} \right\}}_{\dfrac{2(E_p - i\epsilon)}{(p^0)^2 - (E_p - i\epsilon)^2}}$$

$$= i \int \frac{d^4p}{(2\pi)^4} \frac{e^{ip\cdot(x-y)}}{p^2 - m^2 + i\epsilon},$$

where $E_p^2 \equiv \vec{p}^2 + m^2$ and $p^2 \equiv (p^0)^2 - \vec{p}^2$ is not necessarily equal to m^2. We have also used $i\epsilon E_p \sim i\epsilon$ in the limit $\epsilon \to 0^+$.

To complete this chapter let's discuss why we took the plane wave for the scalar field to be a Lorentz scalar function. The plane wave is defined by the expression

$$\langle 0|\phi(x)|\vec{p}\rangle \equiv e^{-ip\cdot x}. \tag{7.85}$$

But we can see how it transforms under Lorentz transformations. We have (using 7.37 is correct)

$$\langle 0|\phi(x)|\vec{p}\rangle \equiv e^{-ip\cdot x}$$
$$= \langle 0|U^{-1}(\Lambda)\mathcal{U}(\Lambda)\phi(x)U^{-1}(\Lambda)U(\Lambda)|\vec{p}\rangle$$
$$= \langle 0|\phi(x')|\vec{p}\,'\rangle \equiv e^{-ip'\cdot x'},$$

where $x' = \Lambda x$, $p' = \Lambda p$, and thus

$$p' \cdot x' = p \cdot x \tag{7.86}$$

is necessarily a Lorentz scalar.

7.5 Exercises

7.1 Redefine the Hamiltonian,

$$\hat{H} = \int \frac{d^3\vec{p}}{(2\pi)^3 2E_p} \left(E_p a^\dagger(\vec{p})a(\vec{p}) + \frac{E_p}{2}\mathbb{I} \right), \tag{7.87}$$

where we've added an infinite c-number constant $\int \frac{d^3\vec{p}}{(2\pi)^3 2E_p} \left(\frac{E_p}{2} \right) \mathbb{I}$, which we shall see is interpreted as the zero-point vacuum energy, and

$$\mathbb{I} \equiv (2\pi)^3 2E_p \delta^3(\vec{0}). \tag{7.88}$$

Define

$$P(\vec{p}) \equiv -i \left(\frac{E_p}{2} \right)^{1/2} \left(a(\vec{p}) - a^\dagger(-\vec{p}) \right), \tag{7.89}$$

$$Q(\vec{p}) \equiv \left(\frac{1}{2E_p} \right)^{1/2} \left(a(\vec{p}) + a^\dagger(-\vec{p}) \right). \tag{7.90}$$

Note $Q^\dagger(\vec{p}) = Q(-\vec{p})$.

Show that

$$\hat{H} = \int \frac{d^3\vec{p}}{(2\pi)^3 2E_p} \left(\frac{1}{2}|P(\vec{p})|^2 + \frac{E_p^2}{2}|Q(\vec{p})|^2 \right). \tag{7.91}$$

Thus, $\hat{H}$ is a "sum" over an infinite set of harmonic oscillators where $P(\vec{p})$ is the canonical momentum and $Q(\vec{p})$ is the coordinate.

7.2 Given

$$\phi(x) \equiv \int \frac{d^3\vec{p}}{(2\pi)^3 2E_p} \left[e^{ip\cdot x} a^\dagger(\vec{p}) + e^{-ip\cdot x} a(\vec{p}) \right], \tag{7.92}$$

(a) show that

$$\frac{1}{2} \int d^3\vec{x} \; \dot{\phi}(x)^2|_{t=0} = \int \frac{d^3\vec{p}}{(2\pi)^3 2E_p} \frac{1}{2}|P(\vec{p})|^2, \tag{7.93}$$

where $\dot{\phi}(x) \equiv \frac{\partial \phi}{\partial t}$ and $x^0 = t$. (b) show that

$$\frac{1}{2} \int d^3\vec{x} \left[\left(\vec{\nabla}\phi \right)^2 + m^2\phi^2 \right] |_{t=0} = \int \frac{d^3\vec{p}}{(2\pi)^3 2E_p} \frac{E_p^2}{2} |Q(\vec{p})|^2 . \tag{7.94}$$

(c) Thus, show that

$$\hat{H} = \int d^3\vec{x} \left(\frac{1}{2}\dot{\phi}(x)^2 + \frac{1}{2} \left[\left(\vec{\nabla}\phi \right)^2 + m^2\phi^2 \right] \right) . \tag{7.95}$$

7.3 Define

$$\Pi_\phi(\vec{x}, t) \equiv \dot{\phi}(\vec{x}, t) \equiv \dot{\phi}(x). \tag{7.96}$$

Show that (using the commutation relations of the creation and annihilation operators)

$$[\phi(\vec{x}, t), \ \Pi_\phi(\vec{y}, t)] = i\delta^3(\vec{x} - \vec{y}). \tag{7.97}$$

8 Complex Scalar Field

In this chapter we introduce the complex scalar field. With it comes a new continuous global symmetry and a new conserved charge, which we alternately call particle number, or, just charge. We discuss Noether's theorem in more detail; showing that for every continuous global symmetry of the Lagrangian there is a conserved current. Moreover, the conserved charge is the generator of the symmetry. We then define the discrete symmetries of charge conjugation, parity and time reversal.

8.1 Complex Scalar

We define the complex scalar field, ϕ, by (with $\phi_{1,2}$ two independent real scalar fields, BUT both having the same mass, m)

$$\phi(x) = \frac{1}{\sqrt{2}} \left(\phi_1(x) + i\phi_2(x) \right) \tag{8.C}$$

$$\equiv \int \frac{d^3\vec{p}}{(2\pi)^3 2E_p} \left[e^{ip\cdot x} b^\dagger(\vec{p}) + e^{-ip\cdot x} a(\vec{p}) \right],$$

where the commutation relations can be derived using the commutation relations for that of a real scalar field, Eqns. 7.4 and 7.5. We obtain

$$\left[a(\vec{p}\,'), a^\dagger(\vec{p}) \right] = (2\pi)^3 \, 2E_p \, \delta^3(\vec{p} - \vec{p}\,'), \tag{8.1}$$
$$\left[b(\vec{p}\,'), b^\dagger(\vec{p}) \right] = (2\pi)^3 \, 2E_p \, \delta^3(\vec{p} - \vec{p}\,')$$

and

$$a(\vec{p}) \, |0\rangle = b(\vec{p}) \, |0\rangle \equiv 0. \tag{8.2}$$

What is the significance of the two different operators a, b? We will now show that

$$|1, \vec{p}\rangle \equiv a^\dagger(\vec{p})|0\rangle,$$
$$|-1, \vec{p}\rangle \equiv b^+(\vec{p})|0\rangle,$$

where ± 1 are the quantum numbers of the charge $\hat{Q}$, i.e.

$$\hat{Q}|1, \vec{p}\rangle = |1, \vec{p}\rangle,$$
$$\hat{Q}|-1, \vec{p}\rangle = -|-1, \vec{p}\rangle.$$

If we define $|1, \vec{p}\rangle$ as the particle, then $|-1, \vec{p}\rangle$ is the anti-particle.

Let's now define the charge operator, $\hat{Q}$. We have the Lagrangian, given by

$$\mathcal{L} = \frac{1}{2}(\partial_\mu \phi_1)^2 + \frac{1}{2}(\partial_\mu \phi_2)^2 - \frac{1}{2}m^2(\phi_1^2 + \phi_2^2) \tag{8.3}$$
$$= |\partial_\mu \phi|^2 - m^2|\phi|^2.$$

Given the Lagrangian we obtain the equations of motion,

$$\partial_\mu \frac{\partial \mathcal{L}}{\partial(\partial_\mu \phi)} - \frac{\partial \mathcal{L}}{\partial \phi} = 0$$

or

$$(\Box + m^2)\phi = (\Box + m^2)\phi^\dagger = 0.$$

Using the equations of motion it is easy to see that there exists a conserved current,

$$j^\mu \equiv i\phi^\dagger \overleftrightarrow{\partial^\mu} \phi. \tag{8.4}$$

Proof

$$\partial_\mu j^\mu = i\partial_\mu \left(\phi^\dagger \partial^\mu \phi - (\partial^\mu \phi^\dagger)\phi\right)$$
$$= i\left(\phi^\dagger \Box \phi - (\Box \phi^\dagger)\phi\right) \equiv 0$$

$\Box$

Also, for current conservation, $j^\mu \equiv (j^0, \vec{j})$, means that it satisfies the continuity equation, $\partial_\mu j^\mu \equiv \partial_t j^0 + \vec{\nabla} \cdot \vec{j} = 0$. Finally, the charge operator, $\hat{Q}$, is given by

$$\hat{Q} = \int d^3\vec{x} j^0 \tag{8.5}$$
$$= \int \frac{d^3\vec{p}}{(2\pi)^3 2E_p} \left(a^\dagger(\vec{p})a(\vec{p}) - b^\dagger(\vec{p})b(\vec{p})\right) \quad + \quad \text{infinite constant.}$$

It satisfies the conservation law,

$$\frac{d\hat{Q}}{dt} \equiv \int d^3\vec{x} \partial_t j^0 = -\int_V d^3\vec{x} \, \vec{\nabla} \cdot \vec{j}$$
$$= -\int_\Sigma \vec{j} \cdot d\vec{s} = 0,$$

where we take V to be the volume of space and $\sum$ is the surface at infinity.[1]

After subtracting the infinite constant, we find

$$\hat{Q}|1, \vec{p}\rangle = |1, \vec{p}\rangle,$$
$$\hat{Q}|-1, \vec{p}\rangle = -|-1, \vec{p}\rangle.$$

The existence of a conserved current is no accident. It is the result of the symmetry:

$$\phi' = e^{-i\alpha}\phi, \tag{8.6}$$

[1] We assume that at any finite time all fields vanish at spatial infinity.

where α is a constant parameter. Note: $[\hat{Q}, \phi] = -\phi$. This is also easy to verify, since

$$\hat{Q} \equiv \int d^3x \; i\phi^\dagger \overleftrightarrow{\partial^0} \phi = \int d^3\vec{x} \, i \left(\phi^\dagger \, \Pi_{\phi^\dagger} - \Pi_\phi \phi \right). \tag{8.7}$$

As a consequence we have

$$\phi' = e^{i\alpha\hat{Q}} \, \phi e^{-i\alpha\hat{Q}} \equiv e^{-i\alpha}\phi, \tag{8.8}$$

i.e. $\hat{Q}$ is the generator of the symmetry.

ASIDE: In general, Noether's theorem states that to every continuous symmetry of the Lagrangian (and more generally of the action), there is a corresponding conserved current and a conserved charge. Moreover, the conserved charge is the generator of the symmetry.

Consider the Lagrangian density

$$\mathcal{L} = \mathcal{L}(\phi, \partial_\mu \phi). \tag{8.9}$$

Proof

$$0 = \delta\mathcal{L} \text{ (small variation)} \; = \frac{\partial \mathcal{L}}{\partial(\partial_\mu\phi)} \, \delta(\partial_\mu\phi) + \frac{\partial \mathcal{L}}{\partial\phi} \, \delta\phi + (\phi \to \phi^\dagger). \tag{8.10}$$

But $\delta(\partial_\mu\phi) \equiv \partial_\mu(\delta\phi)$. Therefore, using the equations of motion, we have

$$0 = \partial_\mu \left(\frac{\partial \mathcal{L}}{\partial(\partial_\mu\phi)} \delta\phi \right) - \left[\partial_\mu \frac{\partial \mathcal{L}}{\partial(\partial_\mu\phi)} - \frac{\partial \mathcal{L}}{\partial\phi} \right] \delta\phi + (\phi \to \phi^\dagger) \tag{8.11}$$

or

$$\alpha j^\mu = \frac{\partial \mathcal{L}}{\partial(\partial_\mu\phi)} \, \delta\phi + (\phi \to \phi^\dagger) \tag{8.12}$$

and

$$\alpha \partial_\mu j^\mu \equiv 0. \tag{8.13}$$

$\square$

For example, in the case $\phi' = e^{-i\alpha}\phi$, the infinitesimal transformation is given by $\delta\phi = \phi' - \phi = -i\alpha\phi$. Then, we have

$$\alpha j^\mu = \frac{\partial \mathcal{L}}{\partial(\partial_\mu\phi)} \, \delta\phi + \frac{\partial \mathcal{L}}{\partial(\partial_\mu\phi^\dagger)} \, \delta\phi^\dagger$$

$$= \alpha \left(-i(\partial_\mu\phi^\dagger)\phi + i(\partial^\mu\phi)\phi^\dagger \right)$$

or

$$j^\mu \equiv i\phi^\dagger \overleftrightarrow{\partial^\mu} \phi. \tag{8.14}$$

$$x \qquad\qquad y$$

Fig. 8.1 Feynman graph representation of a charged scalar propagator. The arrow represents the direction of motion of the particle.

Note, we can also derive the relations, which will become useful for the Lehmann–Symanzik–Zimmerman (LSZ) formalism in Chapter 12,

$$b^\dagger(\vec{p}) = \int d^3\vec{x} \; \phi(x) i \overleftrightarrow{\partial_t} e^{-ip\cdot x} \tag{8.15}$$

$$a^\dagger(\vec{p}) = \int d^3\vec{x} \; \phi^\dagger(x) i \overleftrightarrow{\partial_t} e^{-ip\cdot x},$$

and the Feynman propagator, given by

$$i\Delta_F(x-y) \equiv \langle 0|T\left(\phi(x)\phi^\dagger(y)\right)|0\rangle$$

$$\equiv \langle 0|\phi(x)\phi^\dagger(y)|0\rangle\theta(x^0-y^0) + \langle 0|\phi^\dagger(y)\phi(x)|0\rangle\theta(y^0-x^0)$$

$$= \int \frac{d^3\vec{p}\,'}{(2\pi)^3 2E_{p'}} \int \frac{d^3\vec{p}}{(2\pi)^3 2E_p}$$

$$\times \left\{ e^{-ip'\cdot x}e^{ip\cdot y} \; \overbrace{\langle 0|a(\vec{p}\,')a^\dagger(\vec{p})|0\rangle}^{(2\pi)^3\,2E_p\,\delta^3(\vec{p}-\vec{p}\,')} \; \theta(x^0-y^0) \right.$$

$$\left. + e^{-ip\cdot y}e^{ip'\cdot x} \; \overbrace{\langle 0|b(\vec{p})b^\dagger(\vec{p}\,')|0\rangle}^{(2\pi)^3\,2E_p\,\delta^3(\vec{p}-\vec{p}\,')} \; \theta(y^0-x^0) \right\}$$

$$= \int \frac{d^3\vec{p}}{(2\pi)^3 2E_p} \left\{ e^{-ip\cdot(x-y)}\theta(x^0-y^0) + e^{ip\cdot(x-y)}\theta(y^0-x^0) \right\}$$

$$\equiv \int \frac{d^4p}{(2\pi)^4} \frac{i}{p^2-m^2+i\epsilon} e^{ip\cdot(x-y)}.$$

For a complex scalar field, we represent the propagator by the Feynman graph, Fig. 8.1.

Note the arrow. When time flows in the direction the arrow is pointing ($x^0 > y^0$), then a particle is created at y^0 and annihilated at x^0; but when time flows in the direction opposite to which the arrow is pointing ($y^0 > x^0$), then the anti-particle is created at x^0 and annihilated at y^0.

8.2 Discrete Symmetries of the Charged Scalar Field

In this section we shall define the discrete operations of charge conjugation, $\mathbf{C}$, parity, $\mathbf{P}$, and time reversal, $\mathcal{T}$. These symmetries are defined in an intuitively obvious way. At the end we will show that the free-field theory is invariant under these symmetries. Note, however, that the weak interactions are not invariant under either parity, $\mathbf{P}$, or the product, $\mathbf{CP}$. Nevertheless, we will show that $\mathbf{CP}\mathcal{T}$ is always a symmetry of any local Lorentz-invariant field theory.

Let's first consider charge conjugation. By definition, we assume that there exists an operator with the following properties:

- $\mathbf{C}|0\rangle = |0\rangle$;
- $\mathbf{C}a(\vec{p})\mathbf{C}^{-1} = b(\vec{p})$;
- $\mathbf{C}b(\vec{p})\mathbf{C}^{-1} = a(\vec{p})$;
- $\mathbf{C}\mathbf{C}^\dagger = \mathbf{C}^\dagger\mathbf{C} = 1$, which implies that $\mathbf{C}^{-1} = \mathbf{C}^\dagger$.

Then it can be shown that

$$\mathbf{C}\phi(x)\mathbf{C}^{-1} = \phi^\dagger(x),$$
$$\mathbf{C}\hat{Q}\mathbf{C}^{-1} = -\hat{Q}.$$

As a consequence we have

$$\mathbf{C}a^\dagger(\vec{p})|0\rangle = \mathbf{C}a^\dagger(\vec{p})\mathbf{C}^{-1}\mathbf{C}\,|0\rangle$$
$$= b^\dagger(\vec{p})|0\rangle$$

or

$$\mathbf{C}|1,\vec{p}\rangle = |-1,\vec{p}\rangle. \tag{8.16}$$

Thus, $\mathbf{C}$ is the charge conjugation operator. In free-field theory, we can find the operator $\mathbf{C}$. We have

$$\mathbf{C} = \exp\left[i\pi \int \frac{d^3\vec{p}}{(2\pi)^3 2E_p}\ \left(b^\dagger(\vec{p}) - a^\dagger(\vec{p})\right)\left(b(\vec{p}) - a(\vec{p})\right)\right]. \tag{8.17}$$

Now let's consider parity. If ψ is a physical state of the system, then we denote the space-inverted state by ψ'. If the theory is invariant under space inversions, then ψ', ψ are related by unitary transformations, i.e.

$$\psi' = \mathbf{P}\psi, \tag{8.18}$$

and the expectation value of observables in ψ are equal to the expectation value of the space-inverted observables in ψ'. Thus, we have

$$\langle\psi|\vec{\hat{P}}|\psi\rangle = -\langle\psi'|\vec{\hat{P}}|\psi'\rangle$$
$$= -\langle\psi|\mathbf{P}^{-1}\vec{\hat{P}}\mathbf{P}|\psi\rangle,$$

$$\langle \psi | \vec{\hat{J}} | \psi \rangle = \langle \psi' | \vec{\hat{J}} | \psi' \rangle$$
$$= \langle \psi | \mathbf{P}^{-1} \vec{\hat{J}} \mathbf{P} | \psi \rangle,$$
$$\langle \psi | \hat{Q} | \psi \rangle = \langle \phi' | \hat{Q} | \psi' \rangle$$
$$= \langle \psi | \mathbf{P}^{-1} \hat{Q} \mathbf{P} | \psi \rangle.$$

Hence we obtain the transformation laws,

$$\mathbf{P}^{-1} \vec{\hat{P}} \mathbf{P} = -\vec{\hat{P}}, \tag{8.19}$$

$$\mathbf{P}^{-1} \vec{\hat{J}} \mathbf{P} = \vec{\hat{J}},$$

$$\mathbf{P}^{-1} \hat{Q} \mathbf{P} = \hat{Q},$$

since the state $|\psi\rangle$ is arbitrary.

Given the momentum operator,

$$\vec{\hat{P}} \equiv \int \frac{d^3 \vec{p}}{(2\pi)^3 2E_p} \vec{p} \left(a^\dagger(\vec{p}) a(\vec{p}) + b^\dagger(\vec{p}) b(\vec{p}) \right), \tag{8.20}$$

we find

$$\mathbf{P}^{-1} a(\vec{p}) \, \mathbf{P} = \eta a(-\vec{p}),$$

$$\mathbf{P}^{-1} b(\vec{p}) \, \mathbf{P} = \eta^* b(-\vec{p}),$$

with $|\eta| = 1$ defined so that

$$\phi_\mathbf{P}(\vec{x}, t) \equiv \mathbf{P}^{-1} \phi(\vec{x}, t) \mathbf{P} = \eta \phi(-\vec{x}, t). \tag{8.21}$$

Note: $\mathbf{P}^{-1} H \mathbf{P} = H$ only if parity is conserved.

If $\mathbf{P}|0\rangle = |0\rangle$, then it is possible to choose $\eta = \pm 1$ (Feinberg and Weinberg, 1959) As a result we have

$$\mathbf{P}|\vec{p}\rangle = \eta \, |-\vec{p}\rangle,$$

$$\mathbf{P}^2 |\vec{p}\rangle = \eta^2 \, |\vec{p}\rangle \equiv |\vec{p}\rangle.$$

By definition, a scalar (or bosonic) state with phase η given by

$$\eta = +1 \qquad \text{is a scalar,}$$

$$\eta = -1 \qquad \text{is a pseudo-scalar.}$$

Now let's define the time-reversal operator with the following properties:

$$\mathcal{T}^+ \mathcal{T} = \mathcal{T} \mathcal{T}^+ = 1 \qquad \text{which implies that} \quad \mathcal{T}^{-1} \equiv \mathcal{T}^\dagger,$$

$$\mathcal{T} \phi(\vec{x}, t) \mathcal{T}^{-1} = \eta \, \phi^\dagger(\vec{x}, -t) \qquad |\eta| = 1 \tag{8.22}$$

$$\equiv \eta \, \phi^\dagger(\vec{x}, t').$$

This is then equivalent to

$$\mathcal{T} a(\vec{p}) \mathcal{T}^{-1} = \eta a^\dagger(-\vec{p}), \tag{8.23}$$

$$\mathcal{T} b(\vec{p}) \mathcal{T}^{-1} = \eta^* b^\dagger(-\vec{p}).$$

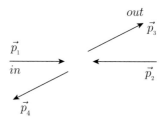

Fig. 8.2 A scattering process going forward in time.

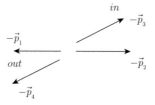

Fig. 8.3 The time-reversed process.

Time reversal when acting on states takes incoming states into outgoing states and vice versa. It is given by

$$\mathcal{T}|0\rangle = |0\rangle^* \equiv \langle 0| \tag{8.24}$$

and

$$\begin{aligned}
\mathcal{T}|\vec{p}\rangle &= \mathcal{T}a^\dagger(\vec{p})\mathcal{T}^{-1}\mathcal{T}|0\rangle \\
&= \eta^* a(-\vec{p})|0\rangle^* \\
&\equiv \eta^* \langle 0|a(-\vec{p}) \\
&= \eta^* \langle -\vec{p}|,
\end{aligned} \tag{8.25}$$

which graphically looks like the following:

$$\overset{\vec{p}}{\underset{\text{incoming}}{\longrightarrow}} \quad \mathcal{T} \quad \overset{-\vec{p}}{\underset{\text{outgoing}}{\longleftarrow}} . \tag{8.26}$$

For scattering, the action of time reversal takes a movie of the scattering going forward in time to the same movie played in reverse. Consider scattering with two incoming states and two outgoing states, called two-to-two scattering, shown in Fig. 8.2. In the time-reversed scattering process, we have Fig. 8.3.

Now, with respect to time reversal, we want to make sure that the Heisenberg equations of motion for the field moving forward in time is the same for the time-reversed field moving backward in time. This guarantees that, assuming the Hamiltonian is invariant under time reversal, there is no way to distinguish between going forward or backward in time. The time-reversed process should be just like watching the movie in reverse. Thus, we want to check that, if we assume

$$\mathcal{T}H_0\mathcal{T}^{-1} \equiv H_0, \tag{8.27}$$

the equations of motion are form invariant under time reversal. The Heisenberg equations are given by

$$[H, \phi(\vec{x}, t)] = -i\frac{\partial}{\partial t}\phi(\vec{x}, t). \tag{8.28}$$

The $\mathcal{T}$-transformed field is given by Eqn. 8.22,

$$\mathcal{T}\phi(\vec{x}, t)\mathcal{T}^{-1} = \eta\phi^\dagger(\vec{x}, -t). \tag{8.29}$$

Therefore, we have

$$\mathcal{T}[H, \phi(\vec{x}, t)]\mathcal{T}^{-1} = -i\frac{\partial}{\partial t} \mathcal{T}\phi(\vec{x}, t)\mathcal{T}^{-1}$$

$$= -i\frac{\partial}{\partial t}\eta\phi^\dagger(\vec{x}, -t)$$

$$\mathcal{T}H\phi(\vec{x}, t)\mathcal{T}^{-1} - \mathcal{T}\phi(\vec{x}, t)H\mathcal{T}^{-1} = i\eta \frac{\partial}{\partial t'} \phi^\dagger(\vec{x}, t'), \tag{8.30}$$

where $t' \equiv -t$. In order to retain form invariance, we need

$$[H, \phi^\dagger(\vec{x}, t')] = -i \frac{\partial}{\partial t'} \phi^\dagger(\vec{x}, t') \tag{8.31}$$

or the left-hand side needs to be $-[H, \phi^\dagger(\vec{x}, t')]\eta$.

The solution to this problem requires a very special property for the operation $\mathcal{T}$. We define that when $\mathcal{T}$ acts on a product of fields we have the following relation.

$$\mathcal{T}AB\mathcal{T}^{-1} \equiv \mathcal{T}B\mathcal{T}^{-1}\mathcal{T}A\mathcal{T}^{-1}, \tag{8.32}$$

such that

$$\mathcal{T}H\phi(\vec{x}, t)\mathcal{T}^{-1} \equiv \mathcal{T}\phi(\vec{x}, t)\mathcal{T}^{-1}\mathcal{T}H\mathcal{T}^{-1}$$

$$= \eta\phi^\dagger(\vec{x}, t')H,$$

where we have assumed that $\mathcal{T}H\mathcal{T}^{-1} = H$, i.e. $\mathcal{T}$ is a symmetry operation. Then Eqn. 8.30 gives

$$[H, \phi^\dagger(\vec{x}, t')] = -i\frac{\partial}{\partial t'}\phi^\dagger(\vec{x}, t'), \tag{8.33}$$

i.e., the time-reversed field satisfies the same equation of motion as ϕ.

Now we want to check to see if the discrete operations, **C**, **P**, $\mathcal{T}$, are symmetries of the free-field theory. As in Eqn. 8.2, the Lagrangian is

$$\mathcal{L} = |\partial_\mu\phi|^2 - m^2|\phi|^2 \tag{8.34}$$

and the Hamiltonian is given by

$$H = \int d^3\vec{x}(|\partial_t\phi(\vec{x}, t)|^2 + |\vec{\nabla}\phi(\vec{x}, t)|^2 + m^2|\phi(\vec{x}, t)|^2). \tag{8.35}$$

Consider first **C**. Under **C**, $\phi^C(x) \equiv \mathbf{C}\phi(x)\mathbf{C}^{-1} = \phi^\dagger(x)$. Clearly, we have

$$\mathcal{L}(\phi^C) = \mathcal{L}(\phi) \tag{8.36}$$

and, similarly, $\mathbf{C}H\mathbf{C}^{-1} = H$, hence **C** is a symmetry operation.

Now consider **P**. We have (Eqn. 8.21)

$$\mathbf{P}^{-1}H\mathbf{P} \equiv \int d^3\vec{x}(|\partial_t\phi(-\vec{x},t)|^2 + |\vec{\nabla}\phi(-\vec{x},t)|^2 + m^2|\phi(-\vec{x},t)|^2) \tag{8.37}$$

$$= \int d^3\vec{x}\,'(|\partial_t\phi(\vec{x}\,',t)|^2 + |\vec{\nabla}\phi(\vec{x}\,',t)|^2 + m^2|\phi(\vec{x}\,',t)|^2) \tag{8.38}$$

$$\equiv H \tag{8.39}$$

where $\vec{x}\,' = -\vec{x}$. Hence $\mathbf{P}$ is a symmetry operation.

Now let's explicitly show that the free Hamiltonian is invariant under time reversal. We have

$$\mathcal{T}H(t)\mathcal{T}^{-1} = \int d^3\vec{x}\;\mathcal{T}\left[(\partial_t\phi^\dagger\,\partial_t\phi) + (\vec{\nabla}\phi^\dagger \cdot \vec{\nabla}\phi) + m^2(\phi^\dagger\phi)\right]\mathcal{T}^{-1}$$

$$= \int d^3\vec{x}\left[\partial_t\phi^\dagger(\vec{x},-t)\partial_t\phi(\vec{x},-t)\right.$$

$$\left. + \vec{\nabla}\phi^\dagger(\vec{x},-t) \cdot \vec{\nabla}\phi(\vec{x},-t) + m^2\phi^\dagger(\vec{x},-t)\phi(\vec{x},-t)\right]$$

$$\equiv H(t') \equiv H,$$

since H is time independent.

8.3 Exercises

8.1 Given

$$\phi(x) = \int \frac{d^3\vec{p}}{(2\pi)^3 2E_p}\left[e^{ip\cdot x}b^\dagger(\vec{p}) + e^{-ip\cdot x}a(\vec{p})\right., \tag{8.40}$$

Show (using the commutation relations in Eqn. 8.1, with $[a,a] = [a^\dagger,a^\dagger] = 0$ $[b,b] = [b^\dagger,b^\dagger] = 0$) that

$$[\phi(x),\,\Pi_\phi(y)]\,|_{x^0=y^0} = i\delta^3(\vec{x}-\vec{y}), \tag{8.41}$$

where

$$\Pi_\phi(x) \equiv \frac{\delta\mathcal{L}}{\delta\dot{\phi}(x)} = \dot{\phi}^*(x) \equiv \frac{\partial\phi^*}{\partial t}. \tag{8.42}$$

8.2 Given

$$\hat{Q} = \int d^3\vec{x}j^0, \tag{8.43}$$

where

$$j^0 \equiv i\phi^*\overset{\leftrightarrow}{\partial}_t\phi, \tag{8.44}$$

show that

$$\left[\hat{Q},\,\phi\right] = -\phi. \tag{8.45}$$

8.3 Prove

$$\phi' \equiv e^{i\alpha\hat{Q}}\,\phi e^{-i\alpha\hat{Q}} = e^{-i\alpha}\phi. \tag{8.46}$$

8.4 Show that Eqn. 8.22 is a consequence of Eqn. 8.23.

9 Spin-1/2 Particles

In this chapter we introduce the Dirac equation for free fermions and derive the fermion field, i.e. the solution to the Dirac equation. We then consider the symmetries of the Dirac equation. In this chapter we focus on Lorentz invariance, while in the next chapter we will consider the phase symmetries of the Dirac Lagrangian.

9.1 Dirac Equation

All the matter particles which make up atoms, i.e. electrons, protons and neutrons, have spin $1/2$ and are called fermions. Thus, we will spend some extra time concentrating on the physics of spin-1/2 particles. Let's first agree on a representation of the Pauli matrices. We have

$$\sigma^1 = \begin{pmatrix} 0 & 1 \\ 1 & 0 \end{pmatrix} \quad \sigma^2 = \begin{pmatrix} 0 & -i \\ i & 0 \end{pmatrix} \quad \sigma^3 = \begin{pmatrix} 1 & 0 \\ 0 & -1 \end{pmatrix}. \tag{9.1}$$

In addition, the relativistic treatment of fermions requires the Dirac gamma matrices. Here we present them in, what we'll call, the standard representation. We have these 4×4 matrices given by

$$\gamma^0 = \begin{pmatrix} 1 & 0 \\ 0 & -1 \end{pmatrix} \quad \gamma^i = \begin{pmatrix} 0 & \sigma^i \\ -\sigma^i & 0 \end{pmatrix} \quad \gamma^5 \equiv i\gamma^0 \gamma^1 \gamma^2 \gamma^3 = \begin{pmatrix} 0 & 1 \\ 1 & 0 \end{pmatrix}, \tag{9.2}$$

which satisfy the anti-commutation relations,

$$\{\gamma^\mu, \gamma^\nu\}_{\alpha\beta} = 2g^{\mu\nu}\delta_{\alpha\beta}; \quad \{\gamma^\mu, \gamma^5\} = 0, \tag{9.3}$$

with $\gamma^\mu = (\gamma^0, \gamma^i)$, $\alpha^i \equiv \gamma^0\gamma^i = \begin{pmatrix} 0 & \sigma^i \\ \sigma^i & 0 \end{pmatrix}$ and $\beta \equiv \gamma^0$. Here are some useful properties of these gamma matrices. We will use the notation, $*$ is complex conjugation, $\dagger$ is Hermitian conjugation and T is the transpose. The Pauli matrices are Hermitian. $\gamma^{0*} = \gamma^{0\dagger} = \gamma^0$, $\gamma^{i*} = \begin{pmatrix} 0 & \sigma^{i*} \\ -\sigma^{i*} & 0 \end{pmatrix}$, $\gamma^{iT} = \begin{pmatrix} 0 & -\sigma^{iT} \\ \sigma^{iT} & 0 \end{pmatrix}$ and $\alpha^{i\dagger} = \alpha^i$.

Consider now the Dirac equation for a free fermion. We have the Lagrangian

$$\mathcal{L} = \psi^\dagger \gamma^0 (i\gamma^\mu \partial_\mu - m)\psi. \tag{9.4}$$

In standard notation, we define $\psi^\dagger\gamma^0 \equiv \overline{\psi}$, $\not{\partial} \equiv \gamma^\mu\partial_\mu$ and the kinetic term $\gamma^\mu\partial_\mu$ is sometimes replaced with $\frac{1}{2}\overset{\leftrightarrow}{\not{\partial}}$, which makes the Lagrangian Hermitian, and ψ and $\overline{\psi}$ are treated on an equal footing.

Lagrange equations of motion are as follows.

$$\frac{\partial\mathcal{L}}{\partial\overline{\psi}} = 0 = (i\not{\partial} - m)\psi, \tag{9.5}$$

$$\partial_\mu\frac{\partial\mathcal{L}}{\partial(\partial_\mu\psi)} - \frac{\partial\mathcal{L}}{\partial\psi} = 0 = \overline{\psi}(i\overset{\leftarrow}{\partial}_\mu\gamma^\mu + m). \tag{9.6}$$

This then gives us $i(\gamma^0\partial_t\psi + \gamma^i\partial_i\psi) - m\psi = 0$ or in compact notation

$$i\dot{\psi} = (-i\alpha^i\nabla^i + \beta m)\psi. \tag{9.7}$$

The canonical momentum is given by[1]

$$\Pi_{\psi_\alpha} = \frac{\partial\mathcal{L}}{\partial\dot{\psi}_\alpha} = i\psi_\alpha^\dagger, \qquad \Pi_{\overline{\psi}_\alpha} = \frac{\partial\mathcal{L}}{\partial\dot{\overline{\psi}}_\alpha} = 0. \tag{9.8}$$

We then obtain the Hamiltonian density given by

$$\mathcal{H} = \Pi_{\psi_\alpha}\dot{\psi}_\alpha - \mathcal{L} = \psi^\dagger(-i\alpha^i\nabla^i + \beta m)\psi \tag{9.9}$$

and the equal-time anti-commutation relations

$$\{\psi_\alpha(\vec{x}, t),\ \psi_\beta^\dagger(\vec{y}, t)\} = \delta^3(\vec{x} - \vec{y})\delta_{\alpha\beta}, \tag{9.10}$$

$$\{\psi_\alpha(\vec{x}, t),\ \psi_\beta(\vec{y}, t)\} = 0.$$

As a check that our anti-commutation relations are correct, one just needs to check that Lagrange's equations of motion are reproduced by the Heisenberg equations,

$$[H,\ \psi_\alpha] = -i\dot{\psi}_\alpha, \tag{9.11}$$

where $H \equiv \int d^3\vec{x}\mathcal{H}$.

We now want to identify the general solution to the wave equation, Eqn. 9.7. We look for a solution of the form[2]

$$\psi_\alpha(\vec{x}, t) = \int \frac{d^3\vec{p}}{(2\pi)^3 2E_p} S_\alpha(\vec{p}, t)e^{i\vec{p}\cdot\vec{x}}. \tag{9.12}$$

$S_\alpha(\vec{p}, t)$ is the wave function in momentum space with the spin index, $\alpha = 1, \ldots, 4$. We then have

$$(-i\vec{\alpha}\cdot\vec{\nabla} + \beta\, m)\psi(\vec{x}, t) = \int \frac{d^3\vec{p}}{(2\pi)^3 2E_p}(\vec{\alpha}\cdot\vec{p} + \beta\, m)S(\vec{p}, t)e^{i\vec{p}\cdot\vec{x}}$$

$$\equiv i\dot{\psi}. \tag{9.13}$$

This then gives the equation for $S(\vec{p}, t)$,

$$(\vec{\alpha}\cdot\vec{p} + \beta m)S(\vec{p}, t) = i\dot{S}(\vec{p}, t). \tag{9.14}$$

---

[1] This notation is a bit sloppy, since $\Pi_{\overline{\psi}_\alpha} = 0$ is a constraint equation. Derivation of the anti-commutation relations is discussed by Dirac in his Belfer Lectures.

[2] We follow the very pedagogical analysis of Lee (1981).

The general solution for $S(\vec{p}, t)$ can be expanded in terms of a complete set of spinor wave functions,

$$\{u(\vec{p}, s),\ v(-\vec{p}, s)\}, \tag{9.15}$$

with $s = \pm 1/2$, which are solutions to the equations

$$(\vec{\alpha} \cdot \vec{p} + \beta m) \left\{ \begin{array}{c} u(\vec{p}, s) \\ v(-\vec{p}, s) \end{array} \right\} = E_p \left\{ \begin{array}{c} u(\vec{p}, s) \\ -v(-\vec{p}, s) \end{array} \right\}. \tag{9.16}$$

Note: $(\vec{\alpha} \cdot \vec{p} + \beta m)$ is a traceless, Hermitian matrix, with real eigenvalues. Also, Eqns. 9.16 are equivalent to the equations

$$(\not{p} - m)u(\vec{p}, s) = 0, \quad (\not{p} + m)v(\vec{p}, s) = 0. \tag{9.17}$$

We also define the matrices,

$$\Sigma^{\mu\nu} \equiv \frac{i}{4}[\gamma^\mu,\ \gamma^\nu]. \tag{9.18}$$

It is easy to verify that

$$\frac{1}{2}\Sigma_i \equiv \frac{1}{2}\epsilon_{ijk}\Sigma^{jk} = \left(\begin{array}{cc} \frac{\sigma^i}{2} & 0 \\ 0 & \frac{\sigma^i}{2} \end{array} \right). \tag{9.19}$$

In a later section, we shall identify $\Sigma^{\mu\nu}$ with the generator of Lorentz transformations on spinors and Σ_i as the generator of rotations. For now, we obtain an independent set of equations for the spinor wave functions. We have

$$\frac{\vec{\Sigma} \cdot \hat{p}}{2} \left\{ \begin{array}{c} u(\vec{p}, s) \\ v(-\vec{p}, s) \end{array} \right\} = s \left\{ \begin{array}{c} u(\vec{p}, s) \\ v(-\vec{p}, s) \end{array} \right\}, \tag{9.20}$$

where $\hat{p} \equiv \vec{p}/|\vec{p}|$ and $s = \pm 1/2$ will be identified with helicity. Note: $\vec{\Sigma} \cdot \vec{p}$ is also a traceless Hermitian matrix and

$$\left[(\vec{\alpha} \cdot \vec{p} + \beta m),\ \frac{\vec{\Sigma} \cdot \hat{p}}{2} \right] \equiv 0. \tag{9.21}$$

Thus these two operators (at a given $\vec{p}$) can be simultaneously diagonalized with real eigenvalues. The four eigenstates $\{u(\vec{p}, s),\ v(-\vec{p}, s)\}$ with eigenvalues, $\pm E_p$, $s = \pm 1/2$ are a complete set spanning the four-dimensional vector space of states with momentum $\vec{p}$. They are orthogonal and they can be normalized such that

$$u^\dagger(\vec{p}, s)u(\vec{p}, s') = v^\dagger(-\vec{p}, s)v(-\vec{p}, s') \equiv 2E_p\delta_{ss'}, \tag{9.22}$$
$$u^\dagger(\vec{p}, s)v(-\vec{p}, s') = v^\dagger(-\vec{p}, s)u(\vec{p}, s') \equiv 0.$$

We expand $S(\vec{p}, t)$ in this basis. We have

$$S(\vec{p}, t) = \sum_{s=\pm 1/2} (b(\vec{p}, s, t)u(\vec{p}, s) + d^\dagger(-\vec{p}, s, t)v(-\vec{p}, s)), \tag{9.23}$$

where b, d are operators on Fock space. Plugging this result into Eqns. 9.12 and 9.23 we then have the quantized spinor fields given by

$$\psi_\alpha(\vec{x},t) = \int \frac{d^3\vec{p}}{(2\pi)^3 2E_p} \sum_s (b(\vec{p},s,t)u(\vec{p},s)e^{i\vec{p}\cdot\vec{x}} + d^\dagger(\vec{p},s,t)v(\vec{p},s)e^{-i\vec{p}\cdot\vec{x}}), \quad (9.24)$$

$$\psi_\alpha^\dagger(\vec{x},t) = \int \frac{d^3\vec{p}}{(2\pi)^3 2E_p} \sum_s (b^\dagger(\vec{p},s,t)u^\dagger(\vec{p},s)e^{-i\vec{p}\cdot\vec{x}} + d(\vec{p},s,t)v^\dagger(\vec{p},s)e^{i\vec{p}\cdot\vec{x}}).$$

If we now define the equal-time anti-commutation relations for the Dirac creation and annihilation operators by

$$\{b(\vec{p},s,t), b^\dagger(\vec{p}\,',s',t)\} = (2\pi)^3 2E_p \delta^3(\vec{p}-\vec{p}\,')\delta_{ss'}, \quad (9.25)$$

$$\{d(\vec{p},s,t), d^\dagger(\vec{p}\,',s',t)\} = (2\pi)^3 2E_p \delta^3(\vec{p}-\vec{p}\,')\delta_{ss'}, \quad (9.26)$$

we reproduce the canonical anti-commutation relations of the fields. Moreover, our single-particle states will be normalized in the standard way.

Note: the above expansions are valid for either free or interacting fields. The only difference will be in the time dependence of the fields. For free fields, the Hamiltonian is

$$H_0 = \int d^3\vec{x}\, \psi^\dagger(-i\vec{\alpha}\cdot\vec{\nabla} + \beta m)\psi, \quad (9.27)$$

and plugging in the fields given in Eqn. 9.24 we obtain

$$H_0 = \int \frac{d^3\vec{p}}{(2\pi)^3 2E_p}\, E_p \sum_s (b^\dagger(\vec{p},s)b(\vec{p},s) - d(\vec{p},s)d^\dagger(\vec{p},s)) \quad (9.28)$$

$$= \int \frac{d^3\vec{p}}{(2\pi)^3 2E_p}\, E_p \sum_s (b^\dagger(\vec{p},s)b(\vec{p},s) + d^\dagger(\vec{p},s)d(\vec{p},s) - (2\pi)^3 2E_p \delta^3(\vec{0})).$$

The last term represents the negative energy in the Dirac sea. Given the free Hamiltonian, we obtain Heisenberg equations of motion for the creation and annihilation operators. We have[3]

$$\underbrace{[H_0,\ b(\vec{p},s,t)]}_{\equiv -E_p b(\vec{p},s,t)} = -i\dot{b}(\vec{p},s,t). \quad (9.29)$$

Similarly for $d(\vec{p},s,t)$. Hence, we obtain the solution to the free-field time dependence given by

$$b(\vec{p},s,t) = e^{-iE_p t}b(\vec{p},s), \quad (9.30)$$

$$d(\vec{p},s,t) = e^{-iE_p t}d(\vec{p},s).$$

Plugging this into the expression for the spinor fields, Eqn 9.24, we obtain the free fermion field

$$\psi_\alpha(\vec{x},t) = \int \frac{d^3\vec{p}}{(2\pi)^3 2E_p} \sum_s (b(\vec{p},s)u(\vec{p},s)e^{-ip\cdot x} + d^\dagger(\vec{p},s)v(\vec{p},s)e^{ip\cdot x}). \quad (9.31)$$

---

[3] Using the identity $[AB,\ C] \equiv A\{B,\ C\} - \{A,\ C\}B$.

For interacting fields, the time dependence is much more complicated.

ASIDE: Define

$$\bar{u}(\vec{p}, s) \equiv u^\dagger(\vec{p}, s)\gamma^0 \tag{9.32}$$

and similarly for $v(\vec{p}, s)$. We then have

$$\sum_s u(\vec{p}, s)\bar{u}(\vec{p}, s) = \not{p} + m, \tag{9.33}$$

$$\sum_s v(\vec{p}, s)\bar{v}(\vec{p}, s) = \not{p} - m. \tag{9.34}$$

The projection operator onto states with helicity s is given by

$$\frac{1}{2}\left[\mathbb{I} + 2s\vec{\Sigma} \cdot \hat{\vec{p}}\right]. \tag{9.35}$$

We then have

$$u(\vec{p}, s)\bar{u}(\vec{p}, s) = (\not{p} + m)\,\frac{1}{2}\left[\mathbb{I} + 2s\vec{\Sigma} \cdot \hat{\vec{p}}\right] \tag{9.36}$$

$$v(\vec{p}, s)\bar{v}(\vec{p}, s) = (\not{p} - m)\,\frac{1}{2}\left[\mathbb{I} - 2s\vec{\Sigma} \cdot \hat{\vec{p}}\right].$$

For massless fermions, the spin projection operator becomes an helicity projection operator given by

$$\left[\mathbb{I} + \gamma_5(2s)\right]. \tag{9.37}$$

We can also derive some useful identities for traces of gamma matrices. We have

$$\mathrm{Tr}(\gamma_\mu\gamma_\nu) = 4g_{\mu\nu}, \tag{9.38}$$

$$\mathrm{Tr}(\gamma_\mu\gamma_\alpha\gamma_\nu\gamma_\beta) = 4(g_{\mu\alpha}g_{\nu\beta} + g_{\mu\beta}g_{\nu\alpha} - g_{\mu\nu}g_{\alpha\beta}),$$

$$\mathrm{Tr}(\gamma_\mu\gamma_\alpha\gamma_\nu\gamma_\beta\gamma_5) = 4i\epsilon_{\mu\alpha\nu\beta}.$$

Now let's define our states in Fock space. We have

$$b|0\rangle = d|0\rangle \equiv 0, \tag{9.39}$$

$$b^\dagger(\vec{p}, s)|0\rangle \equiv |e^-(\vec{p}, s)\rangle, \tag{9.40}$$

$$d^\dagger(\vec{p}, s)|0\rangle \equiv |e^+(\vec{p}, s)\rangle. \tag{9.41}$$

Note, as in the case of a complex scalar field, we will be able to define a conserved charge operator, $\hat{Q}$, with eigenvalues, ± 1. A two-particle state of fermions satisfies

$$b^\dagger(\vec{p}, s)b^\dagger(\vec{p}\,', s')|0\rangle = -b^\dagger(\vec{p}\,', s')b^\dagger(\vec{p}, s)|0\rangle, \tag{9.42}$$

which implies that these states satisfy Fermi–Dirac statistics.

9.2 Causality for Fermions

Fermionic operators satisfy anti-commutation relations, instead of commutation relations as for scalars. For free fields we have for $(x - y)^2 < 0$,

$$\{\psi_\alpha(x), \ \psi_\beta(y)\} = 0, \tag{9.43}$$
$$\{\psi_\alpha(x), \ \overline{\psi}_\beta(y)\} = 0.$$

Hermitian observables, for fermions, are bilinear in the fields. We define an arbitrary Hermitian observable,

$$O_i(x) \equiv \overline{\psi}(x)\Gamma_i\psi(x), \tag{9.44}$$

where Γ_i are a complete set of 4×4 gamma matrices.

By causality, we require[4] that

$$[O_i(x), \ O_j(y)] = 0 \qquad \text{for} \quad (x - y)^2 < 0. \tag{9.46}$$

We can then check that causality is satisfied using the identity

$$[AB, \ CD] \equiv A\{B, \ C\}D - AC\{B, \ D\} - C\{A, \ D\}B + \{A, \ C\}DB. \tag{9.47}$$

9.3 Phase Factor Conventions for Spinors

The homogeneous equations for the spinors, $\{u, \ v\}$, allow for them to be determined up to arbitrary phases. We choose the phases such that

(1) $u(\vec{p}, s) = \gamma_2 v^*(\vec{p}, s),$
    $v(\vec{p}, s) = \gamma_2 u^*(\vec{p}, s);$

(2) $\gamma^0 u(\vec{p}, s) = u(-\vec{p}, -s),$
    $\gamma^0 v(\vec{p}, s) = -v(-\vec{p}, -s);$

(3) $\Sigma_2 u^*(\vec{p}, s) = e^{i\theta(\vec{p}, s)} u(-\vec{p}, s);$
    $\Sigma_2 v^*(\vec{p}, s) = -e^{-i\theta(-\vec{p}, s)} v(-\vec{p}, s),$
    where $e^{i\theta(\vec{p}, s)} = -e^{i\theta(-\vec{p}, s)}.$

---

[4] A similar identity for commutation relations is given by

$$[AB, \ CD] \equiv A[B, \ C]D + AC[B, \ D] - C[D, \ A]B - [C, \ A]BD. \tag{9.45}$$

ASIDE: The spinors are defined in terms of the homogeneous equations, Eqns. 9.16, 9.20, and the normalization conditions, Eqn. 9.22. In this aside, we demonstrate the proof of the phase conventions.

Proof (1) We use the identities $\gamma_2 \Sigma_i^* \gamma_2 = \Sigma_i$, $(\gamma_2)^2 = -1$, $\gamma_2 \alpha_i^* \gamma_2 = -\alpha_i$, $\gamma_2 \beta^* \gamma_2 = \beta$.
Start with Eqn. 9.16:

$$(\vec{\alpha} \cdot \vec{p} + \beta m) \begin{Bmatrix} u(\vec{p}, s) \\ v(-\vec{p}, s) \end{Bmatrix} = E_p \begin{Bmatrix} u(\vec{p}, s) \\ -v(-\vec{p}, s) \end{Bmatrix}. \tag{9.48}$$

Take the complex conjugate

$$(\vec{\alpha}^* \cdot \vec{p} + \beta^* m) \begin{Bmatrix} u^*(\vec{p}, s) \\ v^*(-\vec{p}, s) \end{Bmatrix} = E_p \begin{Bmatrix} u^*(\vec{p}, s) \\ -v^*(-\vec{p}, s) \end{Bmatrix}. \tag{9.49}$$

Then multiply by $-\gamma_2$

$$(-\vec{\alpha} \cdot \vec{p} + \beta m) \begin{Bmatrix} \gamma_2 u^*(\vec{p}, s) \\ \gamma_2 v^*(-\vec{p}, s) \end{Bmatrix} = -E_p \begin{Bmatrix} \gamma_2 u^*(\vec{p}, s) \\ -\gamma_2 v^*(-\vec{p}, s) \end{Bmatrix} \tag{9.50}$$

and change $\vec{p} \to -\vec{p}$, we have

$$(\vec{\alpha} \cdot \vec{p} + \beta m) \begin{Bmatrix} \gamma_2 u^*(-\vec{p}, s) \\ \gamma_2 v^*(\vec{p}, s) \end{Bmatrix} = E_p \begin{Bmatrix} -\gamma_2 u^*(-\vec{p}, s) \\ \gamma_2 v^*(\vec{p}, s) \end{Bmatrix}. \tag{9.51}$$

Now check Eqn. 9.20,

$$\frac{\vec{\Sigma} \cdot \hat{p}}{2} \begin{Bmatrix} u(\vec{p}, s) \\ v(-\vec{p}, s) \end{Bmatrix} = s \begin{Bmatrix} u(\vec{p}, s) \\ v(-\vec{p}, s) \end{Bmatrix}. \tag{9.52}$$

Follow the same steps and we obtain, first,

$$\frac{\vec{\Sigma}^* \cdot \hat{p}}{2} \begin{Bmatrix} u^*(\vec{p}, s) \\ v^*(-\vec{p}, s) \end{Bmatrix} = s \begin{Bmatrix} u^*(\vec{p}, s) \\ v^*(-\vec{p}, s) \end{Bmatrix}, \tag{9.53}$$

and then

$$\frac{-\vec{\Sigma} \cdot \hat{p}}{2} \begin{Bmatrix} \gamma_2 u^*(\vec{p}, s) \\ \gamma_2 v^*(-\vec{p}, s) \end{Bmatrix} = s \begin{Bmatrix} \gamma_2 u^*(\vec{p}, s) \\ \gamma_2 v^*(-\vec{p}, s) \end{Bmatrix}, \tag{9.54}$$

and finally

$$\frac{\vec{\Sigma} \cdot \hat{p}}{2} \begin{Bmatrix} \gamma_2 u^*(-\vec{p}, s) \\ \gamma_2 v^*(\vec{p}, s) \end{Bmatrix} = s \begin{Bmatrix} \gamma_2 u^*(-\vec{p}, s) \\ \gamma_2 v^*(\vec{p}, s) \end{Bmatrix}. \tag{9.55}$$

Clearly, up to normalization, we have

$$\begin{Bmatrix} u(\vec{p}, s) \\ v(\vec{p}, s) \end{Bmatrix} = \begin{Bmatrix} N_u \gamma_2 v^*(\vec{p}, s) \\ N_v \gamma_2 u^*(\vec{p}, s) \end{Bmatrix}. \tag{9.56}$$

Consider just the equation, $v(\vec{p}, s) = N_v \gamma_2 u^*(\vec{p}, s)$. From this we obtain $v^\dagger(\vec{p}, s) = -u^T(\vec{p}, s) \gamma_2 N_v^*$. Thus, $2E_p \equiv v^\dagger(\vec{p}, s) v(\vec{p}, s) = u^T(\vec{p}, s) u^*(\vec{p}, s) |N_v|^2 \equiv 2E_p |N_v|^2$. Thus, $|N_v| = 1$ and we choose $N_v = 1$. Similarly, we can choose $N_u = 1$. $\square$

Proof   (2) We use the identities $\gamma^0\alpha^i\gamma^0 = -\alpha^i$, $\gamma^0\Sigma_i\gamma^0 = \Sigma_i$, $(\gamma^0)^2 = 1$. Take the equations

$$(\vec{\alpha}\cdot\vec{p} + \beta m)u(\vec{p},s) = E_p u(\vec{p},s) \tag{9.57}$$

and

$$\frac{\vec{\Sigma}\cdot\hat{p}}{2}\, u(\vec{p},s) = s u(\vec{p},s). \tag{9.58}$$

Multiplying by γ^0, we obtain

$$(-\vec{\alpha}\cdot\vec{p} + \beta m)\gamma^0 u(\vec{p},s) = E_p\gamma^0 u(\vec{p},s) \tag{9.59}$$

and

$$\frac{-\vec{\Sigma}\cdot\hat{p}}{2}\,\gamma^0 u(\vec{p},s) = -s\gamma^0 u(\vec{p},s). \tag{9.60}$$

This implies that

$$u(-\vec{p},-s) = N\gamma^0 u(\vec{p},s), \tag{9.61}$$

where $|N| = 1$. We choose $N = 1$. So we have

$$u(-\vec{p},-s) = \gamma^0 u(\vec{p},s). \tag{9.62}$$

Now for $v(\vec{p},s)$ we use the previous results, i.e.

$$v(\vec{p},s) = \gamma_2 u^*(\vec{p},s). \tag{9.63}$$

We have

$$u^*(-\vec{p},-s) = \gamma^0 u^*(\vec{p},s) \tag{9.64}$$

and multiplying by γ_2 we find

$$v(-\vec{p},-s) \equiv \gamma_2 u^*(-\vec{p},-s) = \gamma_2\gamma^0 u^*(\vec{p},s) = -\gamma^0 v(\vec{p},s). \tag{9.65}$$

$\square$

Proof   (3) - We use the identities $\Sigma_2\Sigma_i^*\Sigma_2 = -\Sigma_i$, $\Sigma_2\alpha^{i*}\Sigma_2 = -\alpha^i$, where $\Sigma_2 \equiv \begin{pmatrix} \sigma_2 & 0 \\ 0 & \sigma_2 \end{pmatrix}$. Start with the equations

$$(\vec{\alpha}\cdot\vec{p} + \beta m)u(\vec{p},s) = E_p u(\vec{p},s) \tag{9.66}$$

and

$$\frac{\vec{\Sigma}\cdot\hat{p}}{2}\, u(\vec{p},s) = s u(\vec{p},s). \tag{9.67}$$

Take the complex conjugate and then multiply by Σ_2. We have

$$(-\vec{\alpha}\cdot\vec{p} + \beta m)\Sigma_2 u^*(\vec{p},s) = E_p\Sigma_2 u^*(\vec{p},s) \tag{9.68}$$

and

$$-\frac{\vec{\Sigma}\cdot\hat{p}}{2}\,\Sigma_2 u^*(\vec{p},s) = s\Sigma_2 u^*(\vec{p},s). \tag{9.69}$$

This implies that

$$\Sigma_2 u^*(\vec{p}, s) \propto u(-\vec{p}, s),$$

with the proportionality given by a phase. We have

$$\Sigma_2 u^*(\vec{p}, s) = e^{i\theta(\vec{p}, s)} u(-\vec{p}, s). \tag{9.70}$$

Now let's derive the properties of this phase factor. Take the complex conjugate of Eqn. 9.70. We have

$$- \Sigma_2 u(\vec{p}, s) = e^{-i\theta(\vec{p}, s)} u^*(-\vec{p}, s). \tag{9.71}$$

Now take Eqn. 9.70 again and take $\vec{p} \to -\vec{p}$. We get

$$\Sigma_2 u^*(-\vec{p}, s) = e^{i\theta(-\vec{p}, s)} u(\vec{p}, s). \tag{9.72}$$

This gives

$$u(\vec{p}, s) = e^{-i\theta(-\vec{p}, s)} \Sigma_2 u^*(-\vec{p}, s),$$

$$u^*(\vec{p}, s) = -e^{i\theta(-\vec{p}, s)} \Sigma_2 u(-\vec{p}, s)$$

and finally

$$\Sigma_2 u^*(\vec{p}, s) = -e^{i\theta(-\vec{p}, s)} u(-\vec{p}, s). \tag{9.73}$$

Combining this last equation with Eqn. 9.70 we have

$$e^{i\theta(\vec{p}, s)} = -e^{i\theta(-\vec{p}, s)}. \tag{9.74}$$

Similarly for $v(\vec{p}, s)$ we take the following steps, starting with phase relation (1). We have $\Sigma_2 v^*(\vec{p}, s) = \Sigma_2 (\gamma_2 u^*(\vec{p}, s))^* = -\Sigma_2 \gamma_2 u(\vec{p}, s) = -\gamma_2 (\Sigma_2 u(\vec{p}, s)) = \gamma_2 u^*(-\vec{p}, s) e^{-i\theta(\vec{p}, s)} = e^{-i\theta(\vec{p}, s)} v(-\vec{p}, s) = -e^{-i\theta(-\vec{p}, s)} v(-\vec{p}, s)$ or

$$\Sigma_2 v^*(\vec{p}, s) = -e^{-i\theta(-\vec{p}, s)} v(-\vec{p}, s). \tag{9.75}$$

$\square$

Given the choice of phase conventions for the Dirac spinors, $\{u, v\}$, we can now find an explicit expression for the spinors, consistent with the phase conventions, and recall $s = $ helicity. We have

$$u(\vec{p}, s) \equiv \sqrt{E + m} \begin{pmatrix} \chi_{\hat{p}}(s) \\ \frac{\vec{\sigma} \cdot \vec{p}}{E + m} \chi_{\hat{p}}(s) \end{pmatrix} \tag{9.76}$$

and

$$v(\vec{p}, s) \equiv \gamma_2 u^*(\vec{p}, s) = \sqrt{E + m} \begin{pmatrix} \frac{\vec{\sigma} \cdot \vec{p}}{E + m} \sigma_2 \chi_{\hat{p}}^*(s) \\ \sigma_2 \chi_{\hat{p}}^*(s) \end{pmatrix}, \tag{9.77}$$

with the two-component spinor, $\chi_{\hat{p}}(s)$, satisfying

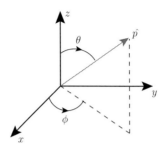

Fig. 9.1 Definition of the polar angles, θ, ϕ.

(1) $\vec{\sigma} \cdot \vec{p}\, \chi_{\hat{p}}(s) = 2s\, |\vec{p}| \chi_{\hat{p}}(s)$,

(2) $\chi_{\hat{p}}(-s) = \chi_{-\hat{p}}(s)$,

(3) $\sigma_2 \chi_{\hat{p}}^*(s) = N(\hat{p}, s) \chi_{\hat{p}}(-s)$,

with $N(\hat{p}, s) = -N(\hat{p}, -s) = N(-\hat{p}, -s)$.

Finally, the two-component spinors are given by

$$\chi_{\hat{p}}(s = +1/2) \equiv i \begin{pmatrix} \cos\frac{\theta}{2}\, e^{-i\frac{\phi}{2}} \\ \sin\frac{\theta}{2}\, e^{i\frac{\phi}{2}} \end{pmatrix}, \tag{9.78}$$

$$\chi_{\hat{p}}(s = -1/2) \equiv \begin{pmatrix} -\sin\frac{\theta}{2}\, e^{-i\frac{\phi}{2}} \\ \cos\frac{\theta}{2}\, e^{i\frac{\phi}{2}} \end{pmatrix},$$

where the polar angles, θ, ϕ, are defined in Fig. 9.1.

9.4 Lorentz Transformations of Fermions

Recall, for scalar fields we have

$$\phi'(x) = e^{i\frac{\alpha_{\mu\nu}}{2}\hat{M}^{\mu\nu}}\, \phi(x) e^{-i\frac{\alpha_{\mu\nu}}{2}\hat{M}^{\mu\nu}} = \phi(x') \qquad x' = \Lambda x. \tag{9.79}$$

From this we obtain the commutation relations of the scalar field with $\hat{M}^{\mu\nu}$, i.e.

$$[\hat{M}^{\mu\nu},\ \phi(x)] = i(x^\mu \partial^\nu - x^\nu \partial^\mu)\phi(x). \tag{9.80}$$

Now consider the Dirac Lagrangian,

$$\mathcal{L} = \overline{\psi}(i\slashed{\partial} - m)\psi. \tag{9.81}$$

We look for a symmetry such that

$$\psi'(x) = S\psi(x') \qquad x'^\mu = \Lambda^\mu{}_\nu x^\nu \tag{9.82}$$

and

$$\overline{\psi}'(x) \equiv \psi^\dagger(x') S^\dagger \gamma^0, \tag{9.83}$$

where S is a 4×4 matrix which is determined as a solution of the equation

$$\mathcal{L}'(x) \equiv \mathcal{L}(x'). \tag{9.84}$$

Note, in this case the action, defined by $S = \int d^4 x \mathcal{L}(x)$, is invariant.

We want

$$\begin{aligned}
\mathcal{L}'(x) &= \overline{\psi}'(x)(i\slashed{\partial} - m)\psi'(x) \\
&= \psi^\dagger(x') S^\dagger \gamma^0 (i\gamma^\mu \partial_\mu - m) S \psi(x') \\
&= \overline{\psi}(x')(i\slashed{\partial}' - m)\psi(x') \equiv \mathcal{L}(x').
\end{aligned} \tag{9.85}$$

In order to obtain the last line in the equation, we use the relation,

$$\partial'_\mu = (\Lambda^{-1})^\nu{}_\mu \partial_\nu.$$

We then obtain the relations that the matrix S must satisfy in order for this transformation to be a symmetry of the action. We have

$$\begin{aligned}
S^\dagger \gamma^0 S &= \gamma^0, \\
S^\dagger \gamma^0 \gamma^\mu S &= \gamma^0 \gamma^\nu (\Lambda^{-1})^\mu{}_\nu
\end{aligned} \tag{9.86}$$

such that

$$S^\dagger \gamma^0 \gamma^\mu \partial_\mu S = \gamma^0 \gamma^\nu (\Lambda^{-1})^\mu{}_\nu \partial_\mu = \gamma^0 \gamma^\nu \partial'_\nu \equiv \gamma^0 \slashed{\partial}'. \tag{9.87}$$

It can be shown that the solution to Eqns. 9.86 is given by

$$S = e^{i \frac{\alpha_{\mu\nu}}{2} \Sigma^{\mu\nu}}, \tag{9.88}$$

where $\Sigma^{\mu\nu}$ was defined in Eqn. 9.18.

We now have the full Lorentz transformation of fermion fields given by

$$\psi'_\alpha(x) \equiv e^{i \frac{\alpha_{\mu\nu}}{2} \hat{M}^{\mu\nu}} \psi_\alpha(x) e^{-i \frac{\alpha_{\mu\nu}}{2} \hat{M}^{\mu\nu}} = [S(\alpha_{\mu\nu})\psi(x')]_\alpha. \tag{9.89}$$

From this we find the action of Lorentz transformations on fermions given by

$$[\hat{M}^{\mu\nu}, \ \psi_\alpha(x)] = i(x^\mu \partial^\nu - x^\nu \partial^\mu)\psi_\alpha(x) + (\Sigma^{\mu\nu}\psi(x))_\alpha. \tag{9.90}$$

Rotations and boosts are given by the operators

$$\hat{J}_i \equiv -\frac{1}{2}\epsilon_{imn}\hat{M}^{mn} \quad \text{rotations}, \tag{9.91}$$

$$\hat{K}_i \equiv \hat{M}_{i0} \quad \text{boosts}.$$

So acting on fermion fields we have

$$[\hat{J}_i(t), \ \psi_\alpha(\vec{x}, t)] = -\left[\left(\ell_i + \frac{\Sigma_i}{2}\right)\psi(\vec{x}, t)\right]_\alpha, \tag{9.92}$$

where $\ell_i = -i\epsilon_{ijk}x^j \frac{\partial}{\partial x^k} \equiv (-i\vec{x} \times \vec{\nabla})_i$ and $\Sigma_i = \epsilon_{ijk}\Sigma^{jk}$.[5]

---

[5] It is easy to show that $[\ell_i, \ \ell_j]f = i\epsilon_{ijk}\ell_k f$ where f is an arbitrary test function of $\vec{x}$ and also $[\frac{\Sigma_i}{2}, \frac{\Sigma_j}{2}] = i\epsilon_{ijk}\frac{\Sigma_k}{2}$, i.e. they satisfy the same Lie algebra.

The momentum and angular momentum operators on Fock space are given by

$$\hat{P}^i \equiv i \int d^3\vec{x}\, \psi^\dagger(\vec{x}, t)\partial^i \psi(\vec{x}, t), \tag{9.93}$$

$$\hat{J}_i \equiv \int d^3\vec{x}\, \psi^\dagger(\vec{x}, t)\left[\ell_i + \frac{\Sigma_i}{2}\right]\psi(\vec{x}, t). \tag{9.94}$$

In terms of creation and annihilation operators, we also have

$$\hat{\vec{P}} = \int \frac{d^3\vec{p}}{(2\pi)^3 2E_p}\, \vec{p}\sum_s (b^\dagger(\vec{p}, s)b(\vec{p}, s) + d^\dagger(\vec{p}, s)d(\vec{p}, s)). \tag{9.95}$$

We can check that

$$[\hat{P}^i,\ \psi_\alpha(\vec{x}, t)] = -i\partial^i\psi_\alpha(\vec{x}, t) \tag{9.96}$$

and with $\hat{P}^0 = H_0$ we have

$$[\hat{P}^\mu,\ \psi_\alpha(\vec{x}, t)] = -i\partial^\mu\psi_\alpha(\vec{x}, t). \tag{9.97}$$

We also have

$$[H_0,\ \hat{\vec{P}}] = [H_0, \hat{\vec{J}}] = 0. \tag{9.98}$$

Thus, momentum and angular momentum are conserved.

Given this extensive introduction, we now want to show that the states $|e^\pm(\vec{p}, s)\rangle$ are eigenstates of $\hat{\vec{J}} \cdot \hat{\vec{P}}$ with eigenvalue $s|\vec{p}|$.[6] Hence, s is the helicity of the state. We have

$$\hat{\vec{P}}\,|0\rangle = \hat{\vec{J}}\,|0\rangle = 0. \tag{9.99}$$

In addition,

$$\hat{\vec{P}}\,|e^\pm(\vec{p}, s)\rangle = \vec{p}\,|e^\pm(\vec{p}, s)\rangle. \tag{9.100}$$

Now we will show that

$$\hat{\vec{J}} \cdot \hat{\vec{P}}\,|e^\pm(\vec{p}, s)\rangle = |\vec{p}|s|e^\pm(\vec{p}, s)\rangle. \tag{9.101}$$

Proof

$$\hat{P}^k\psi^\dagger(\vec{x}, t)|0\rangle = i\nabla_k\psi^\dagger(\vec{x}, t)|0\rangle \tag{9.102}$$

and

$$\hat{J}_i\,\psi^\dagger(\vec{x}, t)\,|0\rangle = \left(-i(\vec{x} \times \vec{\nabla})_i\psi^\dagger(\vec{x}, t) + \psi^\dagger(\vec{x}, t)\frac{\Sigma_i}{2}\right)|0\rangle. \tag{9.103}$$

In both cases, we used the commutation relations to obtain the results. Therefore,

$$\hat{\vec{J}} \cdot \hat{\vec{P}}\,\psi^\dagger(\vec{x}, t)|0\rangle = \frac{i}{2}\nabla_k\psi^\dagger(\vec{x}, t)\Sigma_k|0\rangle. \tag{9.104}$$

Now let's derive a useful identity. We have

$$\psi^\dagger(\vec{x}, t) = \int \frac{d^3\vec{p}}{(2\pi)^3 2E_p}\sum_s (b^\dagger(\vec{p}, s, t)u^\dagger(\vec{p}, s)e^{-i\vec{p}\cdot\vec{x}} + d(\vec{p}, s, t)v^\dagger(\vec{p}, s)e^{i\vec{p}\cdot\vec{x}}), \tag{9.105}$$

---

[6] See Peskin and Schroeder (1995).

and integrating over $\vec{x}$ we obtain

$$\int d^3\vec{x}\, e^{i\vec{p}\cdot\vec{x}}\psi_\alpha^\dagger(\vec{x},t) = \int \frac{d^3\vec{p}\,'}{(2\pi)^3 2E_{p'}} \sum_s (b^\dagger(\vec{p}\,',s,t)u^\dagger(\vec{p}\,',s)(2\pi)^3\delta^3(\vec{p}-\vec{p}\,')$$
$$+ \, d(\vec{p}\,',s,t)v^\dagger(\vec{p}\,',s)(2\pi)^3\delta^3(\vec{p}+\vec{p}\,'))$$
$$= \frac{1}{2E_p}\sum_s(b^\dagger(\vec{p},s,t)u^\dagger(\vec{p},s) + d(-\vec{p},s,t)v^\dagger(-\vec{p},s)).$$

We then have the following identities, which we shall refer to again when we consider the Lehmann–Symanzik–Zimmerman (LSZ) formalism in Chapter 12.

$$\int d^3\vec{x}e^{i\vec{p}\cdot\vec{x}}\psi^\dagger(\vec{x},t)u(\vec{p},s) = b^\dagger(\vec{p},s,t), \tag{9.106}$$

$$\int d^3\vec{x}e^{-i\vec{p}\cdot\vec{x}}\psi^\dagger(\vec{x},t)v(\vec{p},s) = d(\vec{p},s,t),$$

$$\int d^3\vec{x}e^{i\vec{p}\cdot\vec{x}}v^\dagger(\vec{p},s)\psi(\vec{x},t) = d^\dagger(\vec{p},s,t).$$

Using this result, we find that

$$\hat{\vec{J}}\cdot\hat{\vec{P}}\,|e^-(\vec{p}s)\rangle \equiv \hat{\vec{J}}\cdot\hat{\vec{P}}\,b^\dagger(\vec{p},s,t)|0\rangle$$
$$= \frac{i}{2}\int d^3\vec{x}\,e^{i\vec{p}\cdot\vec{x}}\nabla_k\psi^\dagger(\vec{x},t)\Sigma_k u(\vec{p},s)|0\rangle$$
$$= \int d^3\vec{x}e^{i\vec{p}\cdot\vec{x}}\psi^\dagger(\vec{x},t)\underbrace{\left(\frac{\vec{\Sigma}\cdot\vec{p}}{2}\right)u(\vec{p},s)}_{|\vec{p}|su(\vec{p},s)}|0\rangle,$$

where the last equality was obtained by an integration by parts. Therefore, we have

$$\hat{\vec{J}}\cdot\hat{\vec{P}}\,|e^-(\vec{p},s)\rangle = |\vec{p}|sb^\dagger(\vec{p},s,t)|0\rangle \equiv |\vec{p}|s|e^-(\vec{p},s)\rangle. \tag{9.107}$$

$\square$

Thus, the operators, $\frac{\hat{\vec{J}}\cdot\hat{\vec{P}}}{|\hat{\vec{P}}|}$ and $\hat{\vec{P}}$ are simultaneously diagonalized with eigenvalues helicity s and momentum $\vec{p}$. A similar analysis can show that

$$\hat{\vec{J}}\cdot\hat{\vec{P}}\,|e^+(\vec{p},s)\rangle \equiv \hat{\vec{J}}\cdot\hat{\vec{P}}\,d^\dagger(\vec{p},s,t)|0\rangle = |\vec{p}|sd^\dagger(\vec{p},s,t)|0\rangle \equiv |\vec{p}|s|e^+(\vec{p},s)\rangle. \tag{9.108}$$

9.5 Exercises

9.1 Given the anti-commutation relations (Eqn. 9.10), define

$$Q_\Gamma \equiv \int d^3\vec{x}\psi^\dagger(\vec{x},t)\Gamma\psi(\vec{x},t), \tag{9.109}$$

where $\psi^\dagger \Gamma \psi \equiv \psi_\alpha^\dagger \Gamma_{\alpha\beta} \psi_\beta$ and $\Gamma_{\alpha\beta}$ may contain a differential operator acting on $\vec{x}$. Show that

$$[Q_\Gamma,\ Q_{\Gamma'}] = Q_{[\Gamma,\ \Gamma']}. \tag{9.110}$$

9.2 Given

$$H_0 \equiv \int d^3\vec{x}\ \psi^\dagger \Gamma_H \psi, \tag{9.111}$$

with

$$\Gamma_H \equiv \left(-i\vec{\alpha} \cdot \vec{\nabla} + \beta m\right), \tag{9.112}$$

show that

$$[H_0,\ \psi_\alpha(\vec{x}, t)] = -i\dot{\psi}_\alpha(\vec{x}, t) \tag{9.113}$$

reproduces the Dirac equation.

9.3 (a) Given the Dirac Lagrangian with external sources coupled to fermions

$$\mathcal{L} = \bar{\psi}(i\slashed{\partial} - m)\psi + \bar{\eta}\psi + \bar{\psi}\eta, \tag{9.114}$$

where η, $\bar{\eta}$ are fermionic classical sources, we obtain the equations of motion

$$(i\slashed{\partial}_x - m)\psi(x) = -\eta(x). \tag{9.115}$$

Find the Green's function which solves this equation,

$$\psi(x) = -\int d^4y\, S_F(x - y)\eta(y). \tag{9.116}$$

(b) Given the free Dirac field,

$$\psi_\alpha(\vec{x}, t) = \int \frac{d^3\vec{p}}{(2\pi)^3 2E_p} \sum_s (b(\vec{p}, s, t)u(\vec{p}, s)e^{-ip\cdot x} + d^\dagger(\vec{p}, s, t)v(\vec{p}, s)e^{ip\cdot x}), \tag{9.117}$$

show that

$$i[S_F(x - y)]_{\alpha\beta} \equiv \langle 0|T\left(\psi_\alpha(x)\bar{\psi}_\beta(y)\right)|0\rangle \tag{9.118}$$

$$= i\int \frac{d^4p}{(2\pi)^4}e^{-ip\cdot(x-y)}\left(\frac{\slashed{p} + m}{p^2 - m^2 + i\epsilon}\right)_{\alpha\beta},$$

where

$$T\left(\psi_\alpha(x)\bar{\psi}_\beta(y)\right) \equiv \theta(x^0 - y^0)\psi_\alpha(x)\bar{\psi}_\beta(y) - \theta(y^0 - x^0)\bar{\psi}_\beta(y)\psi_\alpha(x). \tag{9.119}$$

Note, the minus sign is for fermions.

Weyl Spinors

Phase symmetries of the Dirac Lagrangian lead to two new quantum numbers, charge (or particle number) and chirality. These two symmetries play an important role in the Standard Model. The $U(1)$ phase symmetry associated with particle number can, in some cases, be a local symmetry. In this case the conserved charge couples to a $U(1)$ gauge boson, such as the photon. Chirality is important in many ways. Chiral symmetry is only exact when fermions are massless. Nevertheless, we shall see that it is an approximate symmetry at energies above the mass of the fermion; spontaneous breaking of chiral symmetry is responsible for the nucleon mass and for light pions and kaons; and, finally, the Standard Model electroweak interactions are chiral. For all of these reasons it is extremely important to explore and elaborate on the $U(1)$ phase symmetry associated with chirality.

10.1 Phase Symmetries of the Dirac Lagrangian

In the previous chapter we discussed the Lorentz invariance of the Dirac Lagrangian. Now we want to discuss the phase symmetries of the Dirac theory. There are two such symmetries, commonly referred to as charge or particle number and chiral symmetry. Consider first particle number. We have

$$\psi' = e^{i\alpha} \, \psi \text{ and infinitesimally } \delta\psi = i\alpha\psi, \tag{10.1}$$
$$\psi'^{\dagger} = e^{-i\alpha} \, \psi^{\dagger} \text{ and infinitesimally } \delta\psi^{\dagger} = -i\alpha\psi^{\dagger}$$

where α is a space–time constant phase. The Lagrangian is invariant under this transformation and we have

$$0 = \delta\mathcal{L} \equiv \alpha \, \partial_{\mu} \, j^{\mu}, \tag{10.2}$$

where $\jmath^{\mu}$ is the conserved current. We find

$$-\alpha j^{\mu} = \frac{\partial\mathcal{L}}{\partial(\partial_{\mu}\psi)}\delta\psi + \frac{\partial\mathcal{L}}{\partial(\partial_{\mu}\psi^{\dagger})}\delta\psi^{\dagger} = i\bar{\psi}\gamma^{\mu}(i\alpha\psi)$$

or

$$j^{\mu} = \bar{\psi}\gamma^{\mu}\psi. \tag{10.3}$$

The conserved charge is then given by

$$\hat{N} \equiv -\hat{Q} = \int d^3\vec{x}\, j^0 = \int d^3\vec{x}\, \psi^\dagger \psi \tag{10.4}$$

$$= \int \frac{d^3\vec{p}}{(2\pi)^3 2E_p} \sum_s (b^\dagger(\vec{p},s)b(\vec{p},s) - d^\dagger(\vec{p},s)d(\vec{p},s)),$$

where the expression is, by definition, normal ordered, i.e. annihilation operators are always on the right and any infinite constant has been subtracted. With this definition of the charge operator, we have

$$\hat{Q}\, \underbrace{b^\dagger(\vec{p},s)|0\rangle}_{|e^-_{\vec{p},s}\rangle} = -|e^-_{\vec{p},s}\rangle, \tag{10.5}$$

$$\hat{Q}\, \underbrace{d^\dagger(\vec{p},s)|0\rangle}_{|e^+_{\vec{p},s}\rangle} = |e^+_{\vec{p},s}\rangle$$

The operator $\hat{N}$ counts the number of particles $-$ the number of anti-particles, whereas $\hat{Q}$ gives the electric charge of an electron in units of $e = 1.6 \times 10^{-19}$ C.

Now consider chiral symmetry. We have

$$\psi' = e^{i\alpha\gamma_5}\psi \text{ and infinitesimally } \delta\psi = i\alpha\gamma_5\psi, \tag{10.6}$$

$$\psi^{\dagger'} = \psi^\dagger e^{-i\alpha\gamma_5} \text{ and infinitesimally } \delta\psi^\dagger = -i\alpha\psi^\dagger\gamma_5,$$

where α is a space–time constant phase and, recall, $\gamma_5 = \begin{pmatrix} 0 & 1 \\ 1 & 0 \end{pmatrix}$ with $\{\gamma_5,\ \gamma_\mu\} = 0$. In the case of a chiral rotation, the kinetic term is invariant. We have

$$\bar{\psi}'\slashed{\partial}\psi' = \psi'^\dagger \gamma_0 \gamma_\mu \partial^\mu \psi' \tag{10.7}$$

$$= \psi^\dagger e^{-i\alpha\gamma_5} \gamma_0 \gamma_\mu \partial^\mu e^{i\alpha\gamma_5}\psi$$

$$= \bar{\psi}\gamma_\mu \partial^\mu \psi = \bar{\psi}\slashed{\partial}\psi.$$

But the mass term is NOT invariant, since

$$\bar{\psi}'\psi' = \psi'^\dagger \gamma_0 \psi' \tag{10.8}$$

$$= \psi^\dagger e^{-i\alpha\gamma_5} \gamma_0 e^{i\alpha\gamma_5}\psi$$

$$= \bar{\psi}e^{2i\alpha\gamma_5}\psi \equiv \cos 2\alpha\, \bar{\psi}\psi + i\sin 2\alpha\, \bar{\psi}\gamma_5\psi.$$

The chiral current is given by the expression

$$\alpha j^{\mu 5} = \frac{\partial \mathcal{L}}{\partial(\partial_\mu \psi)}\delta\psi + \frac{\partial \mathcal{L}}{\partial(\partial_\mu \psi^\dagger)}\delta\psi^\dagger = i\gamma^\mu(i\alpha\gamma_5\psi)$$

or

$$j^{\mu 5} = -\bar{\psi}\gamma^\mu\gamma_5\psi. \tag{10.9}$$

It is not conserved, since

$$\alpha\partial_\mu j^{\mu 5} \equiv \delta\mathcal{L} = -m\delta(\bar{\psi}\psi) = -2i\alpha m\bar{\psi}\gamma_5\psi$$

or

$$\partial_\mu j^{\mu 5} = -2im\bar{\psi}\gamma_5\psi. \tag{10.10}$$

However, in the massless limit $(m \to 0)$ we have the conserved chiral charge

$$\hat{Q}_5 \equiv \int d^3\vec{x} j_5^0. \tag{10.11}$$

10.2 The Massless Limit of the Dirac Theory, $m = 0$

The Dirac equation becomes

$$\gamma^\mu \partial_\mu \psi = 0 \tag{10.12}$$

or equivalently

$$\gamma^\mu \partial_\mu \gamma_5 \psi = 0. \tag{10.13}$$

We can therefore define the fields, $\psi_{L,R}$, given by

$$\psi_L \equiv \frac{1}{2}(1 - \gamma_5)\psi \equiv P_L\psi, \tag{10.14}$$

$$\psi_R \equiv \frac{1}{2}(1 + \gamma_5)\psi \equiv P_R\psi, \tag{10.15}$$

where $P_{L,R}$ are projection operators satisfying $P_L^2 = P_L$, $P_R^2 = P_R$, $P_L P_R = 0$. The fields, $\psi_{L,R}$, are eigenvalues of γ_5 (and thus chirality or handedness) with

$$\gamma_5 \psi_L = -\psi_L \qquad \text{negative chirality,} \tag{10.16}$$
$$\gamma_5 \psi_R = \psi_R \qquad \text{positive chirality.}$$

In general, we also have

$$\psi \equiv \psi_L + \psi_R \tag{10.17}$$

with

$$\gamma^\mu \partial_\mu \psi_L = 0 = \gamma^\mu \partial_\mu \psi_R. \tag{10.18}$$

It will be useful to study the solutions of the Dirac equation in momentum space. We have the equations for the Dirac spinors, Eqns. 9.16, 9.20, which now have the form

$$\vec{\alpha} \cdot \vec{p} \left\{ \begin{array}{c} u(\vec{p}, s) \\ v(-\vec{p}, s) \end{array} \right\} = |\vec{p}| \left\{ \begin{array}{c} u(\vec{p}, s) \\ -v(-\vec{p}, s) \end{array} \right\}, \tag{10.19}$$

$$\frac{\vec{\Sigma} \cdot \hat{p}}{2} \left\{ \begin{array}{c} u(\vec{p}, s) \\ v(-\vec{p}, s) \end{array} \right\} = s \left\{ \begin{array}{c} u(\vec{p}, s) \\ v(-\vec{p}, s) \end{array} \right\}. \tag{10.20}$$

Note, the matrices α^i and Σ_i satisfy the relation

$$\gamma_5 \Sigma_i \equiv \alpha^i. \tag{10.21}$$

Hence, we have

$$\gamma_5 \vec{\Sigma} \cdot \hat{p} \left\{ \begin{array}{c} u(\vec{p}, s) \\ v(-\vec{p}, s) \end{array} \right\} = \left\{ \begin{array}{c} u(\vec{p}, s) \\ -v(-\vec{p}, s) \end{array} \right\}, \tag{10.22}$$

$$\vec{\Sigma} \cdot \hat{p} \left\{ \begin{array}{c} u(\vec{p}, s) \\ v(-\vec{p}, s) \end{array} \right\} = 2s \left\{ \begin{array}{c} u(\vec{p}, s) \\ v(-\vec{p}, s) \end{array} \right\}. \tag{10.23}$$

Plugging the second equation into the first and using $(2s)^2 = 1$, we have

$$\gamma_5 \left\{ \begin{array}{c} u(\vec{p}, s) \\ v(-\vec{p}, s) \end{array} \right\} = 2s \left\{ \begin{array}{c} u(\vec{p}, s) \\ -v(-\vec{p}, s) \end{array} \right\}. \tag{10.24}$$

We can then conclude that the chiral fields, $\psi_{L,R}$, contain the following spinors:

$$\psi_L \supset \{u(\vec{p}, s = -1/2), \ v(-\vec{p}, s = 1/2)\}, \tag{10.25}$$
$$\psi_R \supset \{u(\vec{p}, s = 1/2), \ v(-\vec{p}, s = -1/2)\}.$$

Therefore we have

$$\tag{10.26}$$

$$\psi_L(\vec{x}, t) = \int \frac{d^3\vec{p}}{(2\pi)^3 2E_p} \left(b(\vec{p}, -1/2)u(\vec{p}, -1/2)e^{-ip\cdot x} + d^\dagger(\vec{p}, 1/2)v(\vec{p}, 1/2)e^{ip\cdot x} \right),$$

$$\psi_R(\vec{x}, t) = \int \frac{d^3\vec{p}}{(2\pi)^3 2E_p} \left(b(\vec{p}, 1/2)u(\vec{p}, 1/2)e^{-ip\cdot x} + d^\dagger(\vec{p}, -1/2)v(\vec{p}, -1/2)e^{ip\cdot x} \right).$$

The interpretation of this result is that the field

$\psi_L^\dagger$	creates a left-handed particle/negative helicity
	annihilates a right-handed anti-particle/positive helicity,
$\psi_R^\dagger$	creates a right-handed particle/positive helicity
	annihilates a left-handed anti-particle/negative helicity. (10.27)

The field ψ_L could, for example, be used to represent a neutrino, since, as far as we know, a neutrino has only these quantum numbers.

Now let's rewrite the Lagrangian in terms of these chiral fields. We have

$$\mathcal{L} = \bar{\psi} i \partial\!\!\!/ \psi - m \bar{\psi} \psi. \tag{10.28}$$

Using $\psi = \psi_L + \psi_R$ and $\bar{\psi}_L = \psi_L^\dagger \gamma_0 \equiv \psi_L^\dagger \gamma_0 P_R$, we have

$$\mathcal{L} = \bar{\psi}_L i \partial\!\!\!/ \psi_L + \bar{\psi}_R i \partial\!\!\!/ \psi_R - m(\bar{\psi}_L \psi_R + \bar{\psi}_R \psi_L). \tag{10.29}$$

Note that, in the limit $m \to 0$, ψ_L and ψ_R decouple.

Let's now define a new basis of gamma matrices such that γ_5 is diagonal. This is called the chiral representation of the gamma matrices. We define the similarity transformation by

$$\gamma'_\mu = S^{-1}\gamma_\mu S, \quad \gamma'_5 = S^{-1}\gamma_5 S, \tag{10.30}$$

with

$$S = \frac{1}{\sqrt{2}} \begin{pmatrix} 1 & 1 \\ -1 & 1 \end{pmatrix} \tag{10.31}$$

$$S^{-1} = \frac{1}{\sqrt{2}} \begin{pmatrix} 1 & -1 \\ 1 & 1 \end{pmatrix},$$

where $S^{-1}S = 1$.

Since it is a similarity transformation we have $\{\gamma'_\mu, \gamma'_\nu\} = 2g_{\mu\nu}\mathbb{I}$ and $\{\gamma'_\mu, \gamma'_5\} = 0$. The gamma matrices in the chiral representation are now

$$\gamma'_0 = \begin{pmatrix} 0 & 1 \\ 1 & 0 \end{pmatrix}, \quad \gamma'^i = \gamma^i = \begin{pmatrix} 0 & \sigma^i \\ -\sigma^i & 0 \end{pmatrix}, \quad \gamma'_5 = \begin{pmatrix} -1 & 0 \\ 0 & 1 \end{pmatrix}. \tag{10.32}$$

Thus,

$$P'_L = \begin{pmatrix} 1 & 0 \\ 0 & 0 \end{pmatrix}, \quad P'_R = \begin{pmatrix} 0 & 0 \\ 0 & 1 \end{pmatrix}, \quad \alpha'^i = S^{-1}\alpha^i S = \begin{pmatrix} -\sigma^i & 0 \\ 0 & \sigma^i \end{pmatrix}. \tag{10.33}$$

In the chiral basis of gamma matrices, the fermion fields are now given by

$$\psi' = S^{-1}\psi \equiv \psi'_L + \psi'_R, \tag{10.34}$$

where

$$\psi'_L = P'_L \psi' \equiv \begin{pmatrix} \phi_L \\ 0 \end{pmatrix}, \tag{10.35}$$

$$\psi'_R = P'_R \psi' \equiv \begin{pmatrix} 0 \\ \phi_R \end{pmatrix}.$$

Thus,

$$\psi' \equiv \begin{pmatrix} \phi_L \\ \phi_R \end{pmatrix}, \tag{10.36}$$

where $\phi_{L,R}$ are two-component spinors. We can now write the Dirac Lagrangian in terms of these two-component spinors. It takes several steps, starting from

$$\mathcal{L} = \bar{\psi}(i\slashed{\partial} - m)\psi = \bar{\psi}S(S^{-1}(i\slashed{\partial} - m)S)S^{-1}\psi = \psi'^\dagger(i\partial_t + i\vec{\alpha}' \cdot \vec{\nabla} - \gamma'_0 m)\psi' \tag{10.37}$$

and ending with

$$\mathcal{L} = \phi_L^\dagger(i\partial_t - i\vec{\sigma} \cdot \vec{\nabla})\phi_L + \phi_R^\dagger(i\partial_t + i\vec{\sigma} \cdot \vec{\nabla})\phi_R - m(\phi_L^\dagger\phi_R + \phi_R^\dagger\phi_L). \tag{10.38}$$

It is convenient to define the new Lorentz covariant generalization of the Pauli matrices given by

$$\sigma^\mu = (1, \sigma^i), \quad \bar{\sigma}^\mu = (1, -\sigma^i) \tag{10.39}$$

with

$$\gamma'^\mu = \begin{pmatrix} 0 & \sigma^\mu \\ \bar{\sigma}^\mu & 0 \end{pmatrix}. \tag{10.40}$$

Then the Dirac Lagrangian takes the simple form

$$\mathcal{L} = \phi_L^\dagger(i\bar{\sigma}^\mu\partial_\mu)\phi_L + \phi_R^\dagger(i\sigma^\mu\partial_\mu)\phi_R - m(\phi_L^\dagger\phi_R + \phi_R^\dagger\phi_L). \tag{10.41}$$

It is clear from the structure of the Lagrangian that the three terms are separately Lorentz invariant.

The phase symmetries of the Dirac Lagrangian now take a very simple form as well. We have

$$\psi \to e^{i\alpha}\psi \tag{10.42}$$

which now becomes

$$\phi_{L,R} \to e^{i\alpha}\phi_{L,R}. \tag{10.43}$$

On the other hand, the chiral symmetry

$$\psi \to e^{i\beta\gamma_5}\psi \tag{10.44}$$

now becomes

$$\phi_L \to e^{-i\beta}\phi_L, \qquad \phi_R \to e^{i\beta}\phi_R. \tag{10.45}$$

The chiral symmetry is broken by the mass term. In the case of, say, a free massless neutrino, a consistent Dirac Lagrangian can take the simple form

$$\mathcal{L} = \phi_L^\dagger(i\bar{\sigma}^\mu\partial_\mu)\phi_L. \tag{10.46}$$

10.3 Lorentz Transformations and Weyl Spinors

The Lorentz generators acting on spinor indices are given by

$$\Sigma'^{\mu\nu} = \frac{i}{4}[\gamma'^\mu, \gamma'^\nu] = \frac{i}{4}\begin{pmatrix} (\sigma^\mu\bar{\sigma}^\nu - \sigma^\nu\bar{\sigma}^\mu) & 0 \\ 0 & (\bar{\sigma}^\mu\sigma^\nu - \bar{\sigma}^\nu\sigma^\mu) \end{pmatrix}. \tag{10.47}$$

A finite Lorentz transformation on spinors is then given by

$$S = e^{i\frac{\alpha_{\mu\nu}}{2}\Sigma'^{\mu\nu}} = \begin{pmatrix} e^{-\frac{1}{8}\alpha_{\mu\nu}(\sigma^\mu\bar{\sigma}^\nu - \sigma^\nu\bar{\sigma}^\mu)} & 0 \\ 0 & e^{-\frac{1}{8}\alpha_{\mu\nu}(\bar{\sigma}^\mu\sigma^\nu - \bar{\sigma}^\nu\sigma^\mu)} \end{pmatrix}. \tag{10.48}$$

Thus Lorentz transformations are reducible in the chiral basis. The irreducible spinor representations are the two-component chiral spinors, also known as Weyl spinors.

It turns out that we can simplify the Dirac Lagrangian one more time by proving that, instead of using both right and left chiral spinors, we can work solely in terms of left-handed spinors. Consider the equation of motion for a massless right-handed chiral spinor. We have

$$i\sigma^\mu\partial_\mu\phi_R = 0, \tag{10.49}$$

whereas the left-handed chiral spinor satisfies the equation of motion,

$$i\bar{\sigma}^{\mu}\partial_{\mu}\phi_L = 0. \tag{10.50}$$

Now we want to show that the field $-i\sigma_2\phi_R^*$ is a left-handed chiral spinor, i.e. it satisfies the same equations of motion as in Eqn. 10.50. Start with Eqn. 10.49 and take the complex conjugate. We have

$$i\sigma^{\mu*}\partial_{\mu}\phi_R^* = 0. \tag{10.51}$$

Then, using the identity,

$$\sigma_2\,\sigma^{\mu*}\sigma_2 \equiv \bar{\sigma}^{\mu}, \tag{10.52}$$

we have

$$i\bar{\sigma}^{\mu}\partial_{\mu}(-i\sigma_2\phi_R^*) = 0. \tag{10.53}$$

Thus, if ϕ_R is a right-handed spinor, then

$$\chi_2 \equiv -i\sigma_2\phi_R^* \tag{10.54}$$

is a left-handed spinor.[1] Note: we also have the inverse relation,

$$\phi_R = i\sigma_2\chi_2^*. \tag{10.55}$$

Now define the left-handed spinor as

$$\chi_1 \equiv \phi_L \tag{10.56}$$

and the Dirac four-component spinor can be written solely in terms of two independent left-handed chiral fields, $\chi_{1,2}$,

$$\psi_D = \begin{pmatrix} \chi_1 \\ i\sigma_2\chi_2^* \end{pmatrix}. \tag{10.57}$$

Then the Dirac Lagrangian for a massive Dirac fermion is given by

$$\mathcal{L} = \chi_1^{\dagger}(i\bar{\sigma}^{\mu}\partial_{\mu})\chi_1 + \chi_2^{\dagger}(i\bar{\sigma}^{\mu}\partial_{\mu})\chi_2 + m(\chi_{2\alpha}\epsilon^{\alpha\beta}\chi_{1\beta} + h.c.), \tag{10.58}$$

where $h.c.$ stands for Hermitian conjugate and we have used the identities,[2]

$$\phi_R^{\dagger}(i\sigma^{\mu}\partial_{\mu})\phi_R \equiv \partial_{\mu}\chi_2^{\dagger}(-i\bar{\sigma}^{\mu})\chi_2 = \chi_2^{\dagger}(i\bar{\sigma}^{\mu}\partial_{\mu})\chi_2, \tag{10.59}$$

$$\phi_L^{\dagger}\phi_R + \phi_R^{\dagger}\phi_L \equiv -\chi_2^T(i\sigma_2)\chi_1 + h.c. \tag{10.60}$$

and

$$\epsilon^{\alpha\beta} \equiv (i\sigma_2)^{\alpha\beta} \equiv \begin{pmatrix} 0 & 1 \\ -1 & 0 \end{pmatrix}. \tag{10.61}$$

Note, a Lorentz scalar is obtained with a bilinear product of two left Weyl spinors and anti-symmetrizing the spin indices, α, β.

[1] This was a shortcut to obtain the result. We could also show that the Lorentz generators in the ϕ_R representation go over to the generators in the ϕ_L representation upon taking the same steps.

[2] An integration by parts is implied.

To be absolutely clear, let's consider a concrete example of a free electron and Majorana neutrino. For the fields we have

$$\psi_e = \begin{pmatrix} \chi_e \\ i\sigma_2\chi_{\bar{e}}^* \end{pmatrix}, \qquad \psi_\nu = \begin{pmatrix} \chi_\nu \\ 0 \end{pmatrix}. \tag{10.62}$$

The Lagrangian is given by

$$\mathcal{L} = \chi_e^\dagger(i\bar{\sigma}^\mu\partial_\mu)\chi_e + \chi_{\bar{e}}^\dagger(i\bar{\sigma}^\mu\partial_\mu)\chi_{\bar{e}} + m_e(\chi_{\bar{e}_\alpha}\epsilon^{\alpha\beta}\chi_{e_\beta} + h.c.) \tag{10.63}$$
$$+\chi_\nu^\dagger(i\bar{\sigma}^\mu\partial_\mu)\chi_\nu + m_\nu(\chi_{\nu_\alpha}\epsilon^{\alpha\beta}\chi_{\nu_\beta} + h.c.).$$

The field, χ_e, annihilates a left-handed electron and creates a right-handed anti-electron, while the field, $\chi_{\bar{e}}$, annihilates a left-handed anti-electron and creates a right-handed electron.

For the electron, we have electron number conservation due to the phase symmetry

$$\chi_e \to e^{i\alpha}\chi_e, \quad \chi_{\bar{e}} \to e^{-i\alpha}\chi_{\bar{e}} \tag{10.64}$$

and the chiral symmetry,

$$\chi_e \to e^{i\beta}\chi_e, \quad \chi_{\bar{e}} \to e^{i\beta}\chi_{\bar{e}}, \tag{10.65}$$

is broken by the mass term. For the neutrino, we could define neutrino number in terms of the symmetry,

$$\chi_\nu \to e^{i\alpha}\chi_\nu. \tag{10.66}$$

However, this symmetry is explicitly broken by the mass term. This is what defines a Majorana mass term, i.e. it violates the fermion number by two units.

Finally, note that for a massless Dirac fermion, both chirality and helicity are conserved, and moreover the eigenvalues, handedness and helicity, are identical. However for a massive Dirac fermion, chirality is broken, thus handedness is not conserved. Nevertheless, angular momentum, and consequentially helicity, is always conserved.

11 Spin-1 Particles

Spin-1 particles are used to represent massive vector mesons, such as the ρ meson in strong interactions. They are also the particles associated with the strong and electroweak interactions of the Standard Model. In this chapter we present the fields for both massive and massless spin-1 particles. The important new ingredient in the discussion of massless spin-1 particles is the local symmetry, known as gauge invariance, and the additional step we must take in order to quantize the theory, known as choosing a gauge. This is the last step before we derive the Feynman rules for calculating S matrix elements in the next chapter.

11.1 Massive Spin-1 Particles

Let's first consider the case, $m \neq 0$. The Lagrangian for a real vector field is given by

$$\mathcal{L} = -\frac{1}{4}F_{\mu\nu}^2 + \frac{1}{2}m^2 A_{\mu}^2, \tag{11.1}$$

where $A_{\mu} \equiv A_{\mu}^{\dagger}$, $A_{\mu}^2 \equiv g_{\mu\nu}A^{\mu}A^{\nu}$, $F_{\mu\nu} \equiv \partial_{\mu}A_{\nu} - \partial_{\nu}A_{\mu}$ and $F_{\mu\nu}^2 \equiv g_{\mu\alpha}g_{\nu\beta}F^{\mu\nu}F^{\alpha\beta}$. Lagrange's equations of motion are thus

$$\partial_{\mu}\frac{\partial\mathcal{L}}{\partial(\partial_{\mu}A_{\nu})} - \frac{\partial\mathcal{L}}{\partial A_{\nu}} = 0 \tag{11.2}$$

or

$$\partial_{\mu}F^{\mu\nu} + m^2 A^{\nu} = 0. \tag{11.3}$$

This implies the two equations

$$\partial_{\mu}A^{\mu} = 0, \qquad (\Box + m^2)A^{\mu} = 0. \tag{11.4}$$

Under a Lorentz transformation the vector field transforms as

$$A'_{\mu}(x) \equiv \mathcal{U}(\Lambda)A_{\mu}(x)\mathcal{U}^{-1}(\Lambda) \equiv A_{\nu}(x')\Lambda^{\nu}{}_{\mu}, \tag{11.5}$$

where

$$x' \equiv \Lambda x,$$
$$(x'^{\mu} \equiv \Lambda^{\mu}{}_{\nu}x^{\nu}),$$
$$\mathcal{U}(\Lambda) \equiv e^{\frac{i}{2}\alpha_{\mu\nu}\hat{M}^{\mu\nu}}.$$

Then we have

$$F'_{\mu\nu}(x) = \Lambda^{\alpha}{}_{\mu}\Lambda^{\beta}{}_{\nu}F_{\alpha\beta}(x'), \tag{11.6}$$

and the action is invariant.

The canonical momenta are given by

$$\Pi^{\mu} \equiv \frac{\partial\mathcal{L}}{\partial\dot{A}_{\mu}} = F^{\mu 0} = \partial^{\mu}A^0 - \partial^0 A^{\mu}. \tag{11.7}$$

Hence

$$\Pi^0 = 0 \tag{11.8}$$

$$\Pi^i = -(\nabla^i A^0 + \dot{A}^i). \tag{11.9}$$

Since the field, $\dot{A}^0$, does not appear in $\mathcal{L}$, we can solve the equations of motion of A^0 in terms of the fields, A^i and Π^i. Thus, there are only three physical degrees of freedom for the massive vector field. The Lagrangian density and Hamiltonian are then given by

$$\mathcal{L} = \frac{1}{2}(\vec{\Pi})^2 - \frac{1}{4}F_{ij}F^{ij} - \frac{1}{2}m^2(\vec{A})^2 + \frac{1}{2}m^2(A^0)^2 \tag{11.10}$$

and

$$H = \int d^3\vec{x}(\Pi^i\dot{A}_i - \mathcal{L}) \tag{11.11}$$

$$= \int d^3\vec{x}[\Pi^i(\Pi^i + \nabla^i A^0) - \frac{1}{2}(\Pi^i)^2 + \frac{1}{4}F_{ij}F^{ij} + \frac{1}{2}m^2(A_i)^2 - \frac{1}{2}m^2(A^0)^2].$$

Then, using the Lagrange's equations of motion, Eqn. 11.3,

$$\partial_i F^{i0} = -m^2 A^0 = \vec{\nabla}\cdot\vec{\Pi}, \tag{11.12}$$

we find

$$H \equiv \int d^3\vec{x}[\frac{1}{2}(\vec{\Pi})^2 + \frac{(\vec{\nabla}\cdot\vec{\Pi})^2}{2m^2} + \frac{1}{4}F_{ij}F^{ij} + \frac{1}{2}m^2(\vec{A})^2]. \tag{11.13}$$

The canonical equal time commutation relations are given by

$$[A^i(\vec{x},t),\ \Pi_j(\vec{y},t)] = i\delta^i{}_j\delta^3(\vec{x}-\vec{y}). \tag{11.14}$$

One can then check that the Heisenberg equations,

$$[H,\ A_i] = -i\dot{A}_i, \tag{11.15}$$

reproduce Lagrange's equations of motion.

The Fock-space representation of the free vector field is given by

$$A_{\mu}(x) = \int \frac{d^3\vec{p}}{(2\pi)^3 2E_p}\sum_{\lambda=1}^{3}[\epsilon_{\mu}(\vec{p},\lambda)a(\vec{p},\lambda)e^{-ip\cdot x} + \epsilon^*_{\mu}(\vec{p},\lambda)a^{\dagger}(\vec{p},\lambda)e^{ip\cdot x}], \tag{11.16}$$

where the annihilation, $a(\vec{p},\lambda)$, and creation, $a^{\dagger}(\vec{p},\lambda)$, operators satisfy the commutation relations,

$$[a(\vec{p},\lambda),\ a^{\dagger}(\vec{p}',\lambda')] = (2\pi)^3 2E_p\delta^3(\vec{p}-\vec{p}')\delta_{\lambda\lambda'}. \tag{11.17}$$

The polarization vectors, $\epsilon_\mu(\vec{p}, \lambda)$ with $\lambda = 0, \pm 1$, satisfy the relation

$$p^\mu \epsilon_\mu(\vec{p}, \lambda) = 0. \tag{11.18}$$

This implies that for a particle at rest we have $\epsilon_0(\vec{0}, \lambda) = 0$.

The polarization vectors, $\epsilon_i(\vec{0}, \lambda)$, are eigenstates of the angular momentum operators in the spin-1 representation. In this case the generators are given by

$$(S_i)_{jk} \equiv -i\epsilon_{ijk}, \quad \{i, j, k\} = 1\text{--}3. \tag{11.19}$$

They satisfy the Lie algebra,

$$[S_i, \; S_j] = i\epsilon_{ijk} S_k. \tag{11.20}$$

If we define λ as the eigenvalues of

$$(S_3)_{jk} = \begin{pmatrix} 0 & -i & 0 \\ i & 0 & 0 \\ 0 & 0 & 0 \end{pmatrix}, \tag{11.21}$$

we find the three eigenfunctions, given by

$$\epsilon_i(\vec{0}, 0) = \begin{pmatrix} 0 \\ 0 \\ 1 \end{pmatrix} \equiv \delta_{i3}, \tag{11.22}$$

$$\epsilon_i(\vec{0}, 1) = \frac{1}{\sqrt{2}} \begin{pmatrix} 1 \\ i \\ 0 \end{pmatrix} \equiv \frac{1}{\sqrt{2}} (\delta_{i1} + i\delta_{i2}),$$

$$\epsilon_i(\vec{0}, -1) = \frac{1}{\sqrt{2}} \begin{pmatrix} 1 \\ -i \\ 0 \end{pmatrix} \equiv \frac{1}{\sqrt{2}} (\delta_{i1} - i\delta_{i2}).$$

Given the states with spin in the z direction, we can then define helicity eigenstates with eigenvalues, $\lambda = 0, \pm 1$. The polarization vectors are normalized such that

$$\epsilon_i(\vec{0}, \lambda)\epsilon_i^*(\vec{0}, \lambda') = \delta_{\lambda\lambda'} \tag{11.23}$$

or, in general,

$$\epsilon^\mu(\vec{p}, \lambda)\epsilon_\mu^*(\vec{p}, \lambda') = -\delta_{\lambda\lambda'}. \tag{11.24}$$

We shall need to know the value of the following two-index Lorentz tensor,

$$P_{\mu\nu}(\vec{p}) \equiv \sum_{\lambda=1}^{3} \epsilon_\mu(\vec{p}, \lambda)\epsilon_\nu^*(\vec{p}, \lambda), \tag{11.25}$$

for arbitrary values of the momentum, $\vec{p}$. We know that

$$P_{00}(\vec{0}) = P_{0i}(\vec{0}) = 0. \tag{11.26}$$

Moreover, using the results of Eqn. 11.22, we can prove that

$$P_{ij}(\vec{0}) = [\delta_{i3}\delta_{j3} + \delta_{i1}\delta_{j1} + \delta_{i2}\delta_{j2}] \equiv -g_{ij}. \tag{11.27}$$

In general, $P_{\mu\nu}(\vec{p})$ transforms as a Lorentz tensor, satisfies

$$p^{\mu} P_{\mu\nu} = 0 \tag{11.28}$$

and is only a function of the Lorentz four-vector, p^{μ}. Hence, in general, we can write

$$P_{\mu\nu}(\vec{p}) = A g_{\mu\nu} + B p_{\mu} p_{\nu} + C \epsilon_{\mu\nu\lambda\rho} p^{\lambda} p^{\rho}, \tag{11.29}$$

where A, B, C are constants and the last term vanishes by symmetry. Then

$$p^{\mu} P_{\mu\nu}(\vec{p}) = A p_{\nu} + B p^2 p_{\nu} = 0$$

implies that

$$B = -\frac{A}{m^2} \text{ and } p^2 = m^2.$$

Now

$$P_{\mu\nu}(\vec{p}) = A \left(g_{\mu\nu} - \frac{p_{\mu} p_{\nu}}{m^2} \right)$$

and using $P_{ij}(\vec{0}) = -g_{ij}$, we have $A = -1$. Hence, finally, we have

$$P_{\mu\nu}(\vec{p}) = \left(-g_{\mu\nu} + \frac{p_{\mu} p_{\nu}}{m^2} \right). \tag{11.30}$$

We now want to evaluate the propagator for massive vector bosons. Recall, the Lagrangian is given by

$$\mathcal{L} = -\frac{1}{4} F_{\mu\nu}^2 + \frac{1}{2} m^2 A_{\mu}^2 - j_{\mu} A^{\mu}, \tag{11.31}$$

where we have added a classical source, j_{μ}, coupled to the vector field, A^{μ}, satisfying $\partial^{\mu} j_{\mu} = 0$. Lagrange's equations of motion are then given by

$$\partial_{\mu} F^{\mu\nu} + m^2 A^{\nu} = j^{\nu} \tag{11.32}$$

or

$$K^{\mu\nu}(x) A_{\nu}(x) = j^{\mu}(x). \tag{11.33}$$

The kernel, $K^{\mu\nu}(x)$, is a differential operator given by

$$K^{\mu\nu}(x) \equiv [g^{\mu\nu}(\Box + m^2) - \partial^{\mu} \partial^{\nu}]. \tag{11.34}$$

This differential equation can be solved using the Green's function, which is the inverse operator for the kernel. This is best found in momentum space. We define the Fourier transforms,

$$A_{\nu}(x) \equiv \int \frac{d^4 p}{(2\pi)^4} \, e^{ip \cdot x} \tilde{A}_{\nu}(p), \tag{11.35}$$

$$j^{\mu}(x) \equiv \int \frac{d^4 p}{(2\pi)^4} \, e^{ip \cdot x} \tilde{j}^{\mu}(p).$$

Then in momentum space we find

$$\tilde{K}^{\mu\nu}(p) \tilde{A}_{\nu}(p) = [-g^{\mu\nu}(p^2 - m^2) + p^{\mu} p^{\nu}] \tilde{A}_{\nu}(p) = \tilde{j}^{\mu}(p). \tag{11.36}$$

Hence,

$$\tilde{A}_\nu(p) = \tilde{G}_{\nu\lambda}(p)\tilde{j}^\lambda(p), \tag{11.37}$$

where

$$\tilde{K}^{\mu\nu}\tilde{G}_{\nu\lambda} = \delta^\mu{}_\lambda. \tag{11.38}$$

We want to show that

$$\tilde{G}_{\nu\lambda}(p) = \frac{P_{\nu\lambda}(p)}{p^2 - m^2}, \tag{11.39}$$

where $P_{\mu\nu}(p)$ is the two-index tensor we calculated earlier. The proof is straightforward, we have

$$\tilde{K}^{\mu\nu}(p)P_{\nu\lambda}(p) \equiv [-g^{\mu\nu}(p^2 - m^2) + p^\mu p^\nu]\left(-g_{\nu\lambda} + \frac{p_\nu p_\lambda}{m^2}\right) \equiv \delta^\mu{}_\lambda(p^2 - m^2). \tag{11.40}$$

Thus dividing by $(p^2 - m^2)$ we obtain the result in Eqn. 11.39.

The vector field is then given by

$$A_\nu(x) = \int \frac{d^4p}{(2\pi)^4}\, e^{ip\cdot x}\frac{P_{\nu\lambda}(p)}{p^2 - m^2}\, \tilde{j}^\lambda(p) \equiv \int d^4y\, G_{\nu\lambda}(x - y)j^\lambda(y). \tag{11.41}$$

The Green's function is given by

$$G_{\nu\lambda}(x - y) \equiv \int \frac{d^4p}{(2\pi)^4}\, e^{ip\cdot(x-y)}\frac{\left(-g_{\nu\lambda} + \frac{p_\nu p_\lambda}{m^2}\right)}{p^2 - m^2 + i\epsilon}. \tag{11.42}$$

The Feynman propagator is, in fact, given by the Green's function with the appropriate factor of $i\epsilon$. We claim, without proof, that

$$\langle 0|T(A_\nu(x)A_\lambda(y))|0\rangle \equiv iG_{\nu\lambda}(x - y). \tag{11.43}$$

Finally, we have a useful identity given by

$$a^\dagger(\vec{p}, \lambda) = -\int d^3\vec{x}A_\mu(\vec{x}, t)i\overleftrightarrow{\partial_t}e^{-ip\cdot x}\epsilon^\mu(\vec{p}, \lambda). \tag{11.44}$$

11.2 Massless Spin-1 Particles

The massless limit of the massive theory is not well defined. The problem is that the straightforward derivation of the propagator fails. And the failure is due to a symmetry, called gauge invariance. The Lagrangian for the theory is

$$\mathcal{L} = -\frac{1}{4}F_{\mu\nu}^2 - j^\mu A_\mu \tag{11.45}$$

with the equations of motion given by

$$\partial_\mu F^{\mu\nu} = j^\nu. \tag{11.46}$$

This implies that the gauge field A_μ can only couple to a conserved current satisfying

$$\partial_\mu j^\mu = 0. \tag{11.47}$$

This is all well and good. But let's now calculate the propagator as we did in the massive case. The kernel is given by

$$K^{\mu\nu}(x-y) = \delta^4(x-y)(g^{\mu\nu}\Box_y - \partial_y^\mu \partial_y^\nu). \tag{11.48}$$

We want the inverse, but it doesn't exist, since

$$\int d^4 y K^{\mu\nu}(x-y)\partial_\nu^y \Lambda(y) \equiv 0 \tag{11.49}$$

for an arbitrary function, $\Lambda(y)$. This is a consequence of gauge invariance, i.e. the transformation

$$A_\mu(x) \rightarrow A_\mu(x) + \partial_\mu \Lambda(x) \tag{11.50}$$

leaves the Lagrangian invariant. In order to calculate a Green's function, we must first choose a gauge. For example, let's choose the Coulomb gauge given by

$$\vec{\nabla} \cdot \vec{A}(x) = 0. \tag{11.51}$$

In the Coulomb gauge, the equations of motion become

$$\partial_\mu F^{\mu 0} = \vec{\nabla} \cdot \vec{E} = j^0, \tag{11.52}$$

where

$$E^i = F^{i0} = \partial^i A^0 - \partial^0 A^i. \tag{11.53}$$

We thus have

$$\vec{\nabla} \cdot \vec{E} = -\vec{\nabla}^2 A^0 = j^0. \tag{11.54}$$

Hence, A^0 is a dependent field. The other field equations are given by

$$\partial_\mu F^{\mu j} = \partial_0 F^{0j} + \partial_i F^{ij} = j^j. \tag{11.55}$$

After one step this gives

$$-\Box A_j + \nabla^j \dot{A}^0 = j^j. \tag{11.56}$$

Then, using Eqn. 11.54, we have

$$\dot{A}^0 = -(\vec{\nabla}^2)^{-1}\dot{j}^0 \equiv (\vec{\nabla}^2)^{-1}\nabla^i j^i \tag{11.57}$$

we find

$$\Box A_j = -\left(\delta^{ji} - \frac{\nabla^j \nabla^i}{\vec{\nabla}^2}\right) j^i. \tag{11.58}$$

Thus, A_j only couples to the transverse current.

Now consider canonical quantization of the massless spin-1 particle. We have

$$\mathcal{L} = \frac{1}{2}(\vec{E}^2 - \vec{B}^2) - j^\mu A_\mu. \tag{11.59}$$

The canonical momentum is given by

$$\Pi^\mu \equiv \frac{\partial \mathcal{L}}{\partial \dot{A}_\mu} = F^{\mu 0}. \tag{11.60}$$

Hence,

$$\Pi^0 \equiv 0 \tag{11.61}$$

is a constraint equation.

$$\Pi^i = E^i = \partial^i A^0 - \partial^0 A^i \tag{11.62}$$

implies that

$$\dot{A}_i = E^i - \partial^i A^0. \tag{11.63}$$

The Hamiltonian density is then given by

$$\mathcal{H} \equiv \Pi^i \dot{A}_i - \mathcal{L} = \frac{1}{2}\vec{\Pi}^2 + \Pi^i \nabla^i A^0 + \frac{1}{2}\vec{B}^2 + j^\mu A_\mu \tag{11.64}$$

and the Hamiltonian is

$$H = \int d^3\vec{x}(\mathcal{H} + j^\mu A_\mu) \tag{11.65}$$

$$= \int d^3\vec{x} \left(\frac{1}{2}\vec{\Pi}^2 + \frac{1}{2}\vec{B}^2 - A_0(\vec{\nabla} \cdot \vec{\Pi} - j^0) + \vec{j} \cdot \vec{A} \right).$$

Using the equation of motion $\vec{\nabla} \cdot \vec{\Pi} - j^0 = 0$, the field A^0 does not appear in the Hamiltonian. In addition, Heisenberg's equations of motion must reproduce Lagrange's equations, Eqn. 11.58. Given the Heisenberg equations,

$$[H, \ A_i(x)] = -i\dot{A}_i(x) \tag{11.66}$$

and

$$[H, \ \Pi^i(x)] = -i\dot{\Pi}^i(x) \tag{11.67}$$

and the equal-time commutation relations given by

$$[A_i(\vec{x}, t), \ \Pi^j(\vec{y}, t)] = i \left(\delta_i{}^j - \frac{\partial_i \partial^j}{\vec{\nabla}^2} \right) \delta^3(\vec{x} - \vec{y}), \tag{11.68}$$

we indeed reproduce Lagrange's equations.

The gauge field is given by

$$A_i(x) = \int \frac{d^3\vec{p}}{(2\pi)^3 2E_p} \sum_{\lambda = \pm 1} [a(\vec{p}, \lambda)\epsilon_i(\vec{p}, \lambda)e^{-ip\cdot x} + a^\dagger(\vec{p}, \lambda)\epsilon_i^*(\vec{p}, \lambda)e^{ip\cdot x}]. \tag{11.69}$$

The polarization vectors satisfy,

$$\vec{p} \cdot \vec{\epsilon}(\vec{p}, \lambda) = 0, \tag{11.70}$$

which implies that there are only two helicity states, and

$$E_p^2 = \vec{p}^2. \tag{11.71}$$

For a free field, we have

$$\Pi^i(x) = \dot{A}_i(x), \quad A^0(x) = 0, \tag{11.72}$$

and the canonical commutation relations for the annihilation and creation operators, $a(\vec{p}, \lambda)$, $a^\dagger(\vec{p}, \lambda)$, is given by

$$[a(\vec{p}, \lambda),\ a^\dagger(\vec{p}\,', \lambda')] = (2\pi)^3 2E_p \delta^3(\vec{p} - \vec{p}\,')\delta_{\lambda\lambda'}. \tag{11.73}$$

The polarization vectors satisfy

$$\hat{p} \cdot \vec{S}\ \epsilon(\vec{p}, +1) = \epsilon(\vec{p}, +1), \tag{11.74}$$
$$\hat{p} \cdot \vec{S}\ \epsilon(\vec{p}, -1) = -\epsilon(\vec{p}, -1),$$

where $\vec{S}$ is the spin operator.

The massless spin-1 propagator, or, for example, the photon propagator, depends on the gauge choice. In the Coulomb gauge we have

$$\langle 0|T(A_i(x)A_j(y))|0\rangle \equiv iG_{ij}(x - y), \tag{11.75}$$

with

$$\nabla^i G_{ij} = 0. \tag{11.76}$$

We have

$$G_{ij}(x - y) = \int \frac{d^4p}{(2\pi)^4} e^{ip\cdot(x-y)} \frac{(\delta_{ij} - \frac{p_i p_j}{\vec{p}^2})}{p^2 + i\epsilon}. \tag{11.77}$$

Note, as you might expect, it is not manifestly Lorentz covariant. It is however Lorentz invariant if we apply a Lorentz transformation followed by a gauge transformation to return to the Coulomb gauge.

There are other gauge choices which preserve Lorentz covariance. For example, consider adding a term to the Lagrangian which violates gauge invariance. We have

$$\mathcal{L} = -\frac{1}{4}F_{\mu\nu}^2 - \frac{1}{2\xi}(\partial_\mu A^\mu)^2 - j_\mu A^\mu. \tag{11.78}$$

When $(\partial_\mu A^\mu) = 0$ we have added zero to $\mathcal{L}$. The equations of motion are now

$$\partial_\mu F^{\mu\nu} + \frac{1}{\xi}\partial^\nu(\partial_\alpha A^\alpha) = j^\nu. \tag{11.79}$$

For a conserved current, j_μ, the term $(\partial_\alpha A^\alpha)$ is a free field satisfying the equation $\Box(\partial_\alpha A^\alpha) = 0$.

The calculation of the Green's function now proceeds without a problem. The equations of motion can be written in the form

$$\int d^4y K^{\nu\alpha}(x - y)A_\alpha(y) = j^\nu(y), \tag{11.80}$$

with

$$K^{\nu\alpha}(x - y) = \delta^4(x - y)(\Box g^{\nu\alpha} - \partial^\nu \partial^\alpha(1 - 1/\xi)). \tag{11.81}$$

The Green's function is the solution to the equation

$$\int d^4y K^{\nu\lambda}(x - y)G_{\lambda\mu}(y - z) \equiv \delta^4(x - y)\delta^\nu{}_\mu. \tag{11.82}$$

Finally, we have

$$\langle 0|T(A_\lambda(x)A_\mu(y))|0\rangle \equiv iG_{\lambda\mu}(x-y) = i \int \frac{d^4p}{(2\pi)^4} e^{ip\cdot(x-y)} \frac{(-g_{\lambda\mu} + (1-\xi)\frac{p_\lambda p_\mu}{p^2})}{p^2 + i\epsilon}.$$
(11.83)

Note, ξ is a free parameter which labels the gauge choice. Moreover, gauge invariance requires that no physical quantity (meaning no gauge-invariant quantity) depends on ξ. This is true, as long as A_μ couples to a conserved current whose vertices satisfy $p_\mu j^\mu = 0$.

12 The S Matrix in Field Theory

The S matrix is defined as the overlap from an in-coming state, α, at minus infinity in time to an outgoing state, β, at plus infinity in time.[1] We have

$$S_{\beta\alpha} \equiv (\psi_\beta(t \to +\infty)|\psi_\alpha(t \to -\infty)), \tag{12.1}$$

where

$$\psi_\beta(t \to +\infty) \ \{\text{set of outgoing states}\},$$
$$\psi_\alpha(t \to -\infty) \ \{\text{set of incoming states}\}.$$

The "in" states include

$$\psi_\alpha \supset \{\psi_0, \ a^\dagger(p)_{in}\psi_0, \ a^\dagger(p_1)_{in}a^\dagger(p_2)_{in}\psi_0, \ \dots\}, \tag{12.2}$$

where ψ_0 is the exact Heisenberg vacuum state and the creation operators, $a^\dagger(p)_{in}$, create single-particle states. These are a complete set of states obtained using free-particle creation operators, i.e. asymptotic states. Similarly the "out" states include

$$\psi_\beta \supset \{\psi_0, \ a^\dagger(p)_{out}\psi_0, \ a^\dagger(p_1)_{out}a^\dagger(p_2)_{out}\psi_0, \ \dots\}, \tag{12.3}$$

which are also a complete set of states obtained using free particle creation operators. Since both the "in" and "out" states form a complete set, they must be related by a unitary transformation. We have

$$\psi_\alpha^{in} = \sum_\beta S_{\beta\alpha}\psi_\beta^{out} \tag{12.4}$$

or

$$\left(\psi_\beta^{out}|\ \psi_\alpha^{in}\right) \equiv S_{\beta\alpha} \qquad S \text{ matrix.} \tag{12.5}$$

Our fields for both bosons and fermions were expanded, at any given time, in terms of a complete set of states with given momentum and spin, i.e. a general Fourier decomposition which is valid for both free or interacting fields. Using these fully interacting, Heisenberg picture, fields we can now define the "in" and "out" creation and annihilation operators in terms of the asymptotic limit as time, $t \to \pm\infty$. The summary of these results follows.

---

[1] We follow the analysis of Gasiorowicz (1966, chs. 6 and 7) and Itzykson and Zuber (1980, ch. 5).

SUMMARY:

- For a *real scalar* we have the fully interacting field given by (Eqn. 7.81)

$$\phi(\vec{x},t) = \int \frac{d^3\vec{p}}{(2\pi)^3 2E_p} [\overbrace{e^{-i\vec{p}\cdot\vec{x}}a^\dagger(\vec{p},t)}^{e^{ip\cdot x}a^\dagger(\vec{p},t)e^{-iE_p t}} + e^{i\vec{p}\cdot\vec{x}}a(\vec{p},t)]. \qquad (12.6)$$

Define

$$\tilde{a}^\dagger(\vec{p},t) \equiv \int d^3\vec{x}\phi(\vec{x},t)\, i\overleftrightarrow{\partial}_t\, e^{-ip\cdot x} \Rightarrow a^\dagger(\vec{p},t)e^{-iE_p t} \qquad (12.7)$$

such that at $t \to \pm\infty$ we have $\tilde{a}^\dagger(\vec{p},t) \to a^\dagger(\vec{p})$ for either "in" or "out" states. Note, we have explicitly canceled the free-particle time dependence for the asymptotic state.

In a more precise derivation of the "in" or "out" fields, we would write, for example,

$$\phi(x) = \phi_{in}(x) + \int d^4x' \Delta_{ret}(x - x', m)J(x'), \qquad (12.8)$$

where the interacting field equation is given by $(\Box + m^2)\phi(x) = J(x)$, $\mathcal{L}_I = \phi(x)J(x)$ is the interaction term, $\phi_{in}(x)$ satisfies free-field equations and the retarded Green's function, $\Delta_{ret}(x)$, vanishes in the limit $x \to -\infty$.

- For a *complex scalar field* we have similarly (Eqns. 8.C and 8.15)

$$\phi(\vec{x},t) = \int \frac{d^3\vec{p}}{(2\pi)^3 2E_p} [e^{-i\vec{p}\cdot\vec{x}}b^\dagger(\vec{p},t) + e^{i\vec{p}\cdot\vec{x}}a(\vec{p},t)]. \qquad (12.9)$$

Define

$$\tilde{b}^\dagger(\vec{p},t) \equiv \int d^3\vec{x}\, \phi(\vec{x},t)\, i\overleftrightarrow{\partial}_t e^{-ip\cdot x}, \qquad (12.10)$$

$$\tilde{a}^\dagger(\vec{p},t) \equiv \int d^3\vec{x}\, \phi^\dagger(\vec{x},t)\, i\overleftrightarrow{\partial}_t e^{-ip\cdot x}. \qquad (12.11)$$

- For a *fermion field* we have (Eqn. 9.106)

$$\psi_\alpha(\vec{x},t) = \int \frac{d^3\vec{p}}{(2\pi)^3 2E_p} \sum_s [b(\vec{p},s,t)u(\vec{p},s)e^{i\vec{p}\cdot\vec{x}} + d^\dagger(\vec{p},s,t)v(\vec{p},s)e^{-i\vec{p}\cdot\vec{x}}].$$
$$\qquad (12.12)$$

Define

$$\tilde{b}^\dagger(\vec{p},s,t) \equiv \int d^3\vec{x}\psi^\dagger(\vec{x},t)u(\vec{p},s)e^{-ip\cdot x} \qquad (12.13)$$
$$= b^\dagger(\vec{p},s,t)e^{-iE_p t},$$

$$\tilde{d}^\dagger(\vec{p},s,t) \equiv \int d^3\vec{x}\, v^\dagger(\vec{p},s)\psi(\vec{x},t)e^{-ip\cdot x} \qquad (12.14)$$
$$= d^\dagger(\vec{p},s,t)e^{-iE_p t} .$$

- For a *vector field* we have (Eqn. 11.44)

$$A_\mu(x) = \int \frac{d^3\vec{p}}{(2\pi)^3 2E_p} \sum_{\lambda=1}^{3} [\epsilon_\mu(\vec{p},\lambda)a(\vec{p},\lambda,t)e^{i\vec{p}\cdot\vec{x}} + \epsilon_\mu^*(\vec{p},\lambda)a^\dagger(\vec{p},\lambda,t)e^{-i\vec{p}\cdot\vec{x}}].$$

$$\text{(12.15)}$$

Define

$$\tilde{a}^\dagger(\vec{p},\lambda,t) = -\int d^3\vec{x} A_\mu(\vec{x},t) i \overleftrightarrow{\partial_t} e^{-ip\cdot x} \epsilon^\mu(\vec{p},\lambda).$$

$$\text{(12.16)}$$

Note, by definition, the asymptotic limit of all these creation and, similarly, annihilation operators are time independent, i.e. the free-field time dependence of the operators has been explicitly removed.

Now that we have the creation, and annihilation, operators for "in" and "out" states, we can evaluate the S matrix. We shall use the formalism known as the LSZ (Lehmann–Symanzik–Zimmerman) reduction formalism (Lehmann et al., 1957). In the end we shall have the S matrix element given in terms of the matrix elements of the time-ordered product of interacting fields in the exact interacting Heisenberg-picture vacuum. To get started, let's consider a simple S matrix element for two "in" states scattering to an arbitrary state, α. We have

$$S_{\alpha;pq} = \left(\psi_\alpha^{out}|\psi_{pq}^{in}\right) \tag{12.17}$$

$$= \left(\psi_\alpha^{out}|a_{in}^\dagger(\vec{p})\psi_q^{in}\right)$$

$$= \lim_{t\to-\infty} \left(\psi_\alpha^{out}|\tilde{a}^\dagger(\vec{p},t)\psi_q^{in}\right)$$

$$= \lim_{t\to-\infty} \left(\psi_\alpha^{out}| \int_{x^0=t} d^3\vec{x}\phi^\dagger(x) i\overleftrightarrow{\partial_0} e^{-ip\cdot x}\psi_q^{in}\right).$$

Now we use the identity

$$\int_{x^0\to-\infty} d^3\vec{x}[...] \equiv \int_{x^0\to+\infty} d^3\vec{x}[...] - \int d^4x \frac{\partial}{\partial x^0}[...], \tag{12.18}$$

where [...] refers to some arbitrary integrand. The first term gives

$$\tag{12.19}$$

$$\lim_{t\to+\infty} \left(\psi_\alpha^{out}| \int_{x^0=t} d^3\vec{x}\ \phi^\dagger(x) i\overleftrightarrow{\partial_0} e^{-ip\cdot x}\psi_q^{in}\right) = \lim_{t\to+\infty} \left(\psi_\alpha^{out}|\tilde{a}^\dagger(\vec{p},t)\psi_q^{in}\right)$$

$$= \left(\psi_\alpha^{out}|a_{out}^\dagger(\vec{p})\psi_q^{in}\right)$$

$$= \left(a_{out}(\vec{p})\psi_\alpha^{out}|\psi_q^{in}\right)$$

$$\equiv \mathbf{1}_{\vec{p}',\vec{p}} \left(\psi_{\alpha'}^{out}|\psi_q^{in}\right),$$

where $\mathbf{1}_{\vec{p}',\vec{p}} = \langle\vec{p}\,'|\vec{p}\rangle$. Note, the state $\psi_{\alpha'}^{out}$ is the state ψ_α^{out} with the state $|\vec{p}\rangle$ removed. We have

$$S_{\alpha;pq} = \mathbf{1}_{\vec{p}\,',\vec{p}} S_{\alpha';q} + \bar{S}_{\alpha;qp}, \tag{12.20}$$

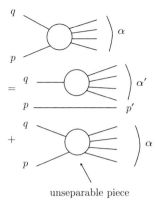

First two terms in the S matrix, i.e. the non-interacting piece and the rest.

where the first term describes no interaction, i.e. the initial state just passes through to the outgoing state without any interaction, while the second term describes the unseparable piece, i.e. the interacting piece, see Fig. 12.1.

Now for a stable single-particle state, we have $\psi_q^{in} \equiv \psi_q^{out}$. We thus have

$$\left(a_{out}(\vec{p})\psi_\alpha^{out}|\psi_q^{in}\right) \equiv \left(a_{out}(\vec{p})\psi_\alpha^{out}|\psi_q^{out}\right) = \left(\psi_\alpha^{out}|\psi_{pq}^{out}\right) \equiv \mathbf{1}_{\alpha;pq}. \tag{12.21}$$

Hence, we now have

$$(S-\mathbf{1})_{\alpha;pq} \equiv \quad \text{second term}$$

$$(S-\mathbf{1})_{\alpha;pq} \equiv -i\int d^4x \frac{\partial}{\partial x^0}\left[(\psi_\alpha^{out}|\phi^\dagger(x)\psi_q)\overleftrightarrow{\partial_0}\,e^{-ip\cdot x}\right]$$

$$= -i\int d^4x\left[(\psi_\alpha^{out}|\phi^\dagger(x)\psi_q)\,\frac{\partial^2}{\partial x^{0^2}}e^{-ip\cdot x} - (\psi_\alpha^{out}|\,\frac{\partial^2}{\partial x^{0^2}}\,\phi^\dagger(x)\psi_q)e^{-ip\cdot x}\right].$$

Now, using the identity

$$\frac{\partial^2}{\partial x^{0^2}}e^{-ip\cdot x} \equiv -E_p^2 e^{-ip\cdot x}$$

$$= -(\vec{p}^{\,2}+m^2)e^{-ip\cdot x}$$

$$= (\vec{\nabla}^2-m^2)e^{-ip\cdot x},$$

we have

$$(S-\mathbf{1})_{\alpha;\mathbf{pq}} \equiv -i\int d^4x\left[(\psi_\alpha^{(out)}|\phi^\dagger(x)\psi_q)(\vec{\nabla}^2-m^2)e^{-ip\cdot x}\right.$$

$$\left. -(\psi_\alpha^{(out)}|\partial_0^2\phi^\dagger(x)\psi_q)e^{-ip\cdot x}\right] \tag{12.22}$$

$$= i\int d^4x(\psi_\alpha^{(out)}|\phi^\dagger(x)\psi_q)\,(\overleftarrow{\Box}_x+m^2)e^{-ip\cdot x},$$

where to obtain the last term we integrated by parts and we define the differential operator acting to the left, $\overleftarrow{K}_x$, by

$$\overleftarrow{K}_x \equiv (\overleftarrow{\Box}_x+m^2). \tag{12.23}$$

Note, we can also use the equivalent expression

$$(S-1)_{\alpha;pq} \equiv \int d^4x \; (\psi_\alpha^{out}|\phi^\dagger(x)\psi_q) \, \frac{(p^2-m^2)}{i} \, e^{-ip\cdot x}. \qquad (12.24)$$

For fermions, we have

$$(\psi_\alpha^{out}|\psi_{pq}^{in}) \equiv (\psi_\alpha^{out}|b^{\dagger\,in}(\vec{p},s)\psi_q) \qquad (12.25)$$

and the corresponding S matrix element given by

$$(S-1)_{\alpha;pq} \equiv i \int d^4x \; (\psi_\alpha^{out}|\bar{\psi}(x)\psi_q)(i\overleftarrow{\partial}_x + m)u(\vec{p},s)e^{-ip\cdot x}. \qquad (12.26)$$

Hopefully by now the reader gets the general idea. So now let's consider the analysis to its conclusion for a particular two-body elastic scattering process, depicted in Fig. 12.2. The scattering amplitude for this process is given by

$$S \equiv \mathbf{1} + i(2\pi)^4\delta^4(p+q-p'-q')T. \qquad (12.27)$$

After the first step described previously we have[2]

$$(S-1)_{p'q';pq} \equiv i \int d^4x (\psi_{p'q'}^{out}|\phi^\dagger(x)\psi_p) \overleftarrow{K}_x \, e^{-iq\cdot x} \qquad (12.28)$$

where $K_x = \Box_x + m_s^2$ or we could have started with the fermion and obtained

$$(S-1)_{p'q';pq} \equiv i \int d^4x (\psi_{p'q'}^{out}|\bar{\psi}(x)\psi_q) \overleftarrow{D}_x \, u(\vec{p},s)e^{-ip\cdot x}, \qquad (12.29)$$

where

$$\overleftarrow{D}_x = i \overleftarrow{\partial}_x + m_f. \qquad (12.30)$$

We can start with either expression and then continue with the LSZ formalism. Let's start with the first expression and then continue. We have, at the next step,

$$(S-1)_{p'q';pq} \equiv i \int d^4x (\psi_{p'} a^{out}(q')|\phi^\dagger(x) \; \psi_p)\overleftarrow{K}_x \, e^{-iq\cdot x}. \qquad (12.31)$$

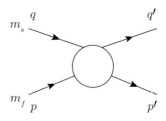

Fig. 12.2 Two-body elastic scattering process with one fermion and one scalar.

[2] Note, we are being a bit sloppy here, since the factor of $(-\mathbf{1})_{p'q';pq}$ in the following expression for the S matrix only appears at the final step in the LSZ reduction formalism. But for simplicity, since we know it is coming, we place it here.

Now we use the identities

$$(\psi_{p'} a^{out}(q')|\phi^\dagger(x)\psi_p) = \lim_{t\to+\infty} (\psi_{p'} \tilde{a}(q',t)|\phi^\dagger(x)\psi_p)$$

$$= \lim_{y^0=t\to+\infty} i \int d^3\vec{y}\, e^{iq'\cdot y}\, \overset{\leftrightarrow}{\partial}_{y^0}\, (\psi_{p'}\phi(y)\phi^\dagger(x)\psi_p)$$

and

$$\lim_{y^0\to+\infty} \phi(y)\,\phi^\dagger(x) \equiv \lim_{y^0\to+\infty} (\theta(y^0-x^0)\phi(y)\phi^\dagger(x) + \theta(x^0-y^0)\phi^\dagger(x)\phi(y))$$

$$\equiv \lim_{y^0\to+\infty} T(\phi(y)\phi^\dagger(x)). \tag{12.32}$$

We then have

$$(\psi_{p'} a^{out}(q')|\phi^\dagger(x)\psi_p) = \lim_{y^0=t\to-\infty} i \int d^3\vec{y} e^{iq'\cdot y}\overset{\leftrightarrow}{\partial}_{y^0} (\psi_{p'} T(\phi(y)\phi^\dagger(x))\psi_p)$$

$$+i \int d^4y \partial_{y^0} [e^{iq'\cdot y}\, \overset{\leftrightarrow}{\partial}_{y^0} (\psi_{p'} T(\phi(y)\phi^\dagger(x))\psi_p)].$$

The first term gives

$$(\psi_{p'}\phi^\dagger(x)a^{in}(q')\psi_p) \equiv 0, \tag{12.33}$$

since $a^{in}(q')|\psi(p)) \equiv 0$. Thus only the second term contributes and we have

$$(\psi_{p'} a^{out}(q')|\phi^\dagger(x)\psi_p) = i \int d^4y \partial_{y^0} [e^{iq'\cdot y}\overset{\leftrightarrow}{\partial}_{y^0} (\psi_{p'} T(\phi(y)\phi^\dagger(x))\psi_p)]$$

$$= i \int d^4y \{e^{iq'\cdot y}\partial^2_{y^0}(\psi_{p'} T(\dots)\psi_p) - (\partial^2_{y^0}e^{iq'\cdot y})(\psi_{p'} T(\dots)\psi_p)\}$$

$$= i \int d^4y\, e^{iq'\cdot y}\vec{K}_y(\psi_{p'} T(\phi(y)\phi^\dagger(x))\psi_p). \tag{12.34}$$

Putting it all together, we finally arrive, after step 2, with the expression

$$(S-\mathbf{1})_{p'q';pq} \equiv i^2 \int d^4x \int d^4y e^{iq'\cdot y}\overrightarrow{K_y}(\psi_{p'} T(\phi(y)\phi^\dagger(x))\psi_p)\overleftarrow{K_x}\, e^{-iq\cdot x}. \tag{12.35}$$

This procedure is continued until all external states are accounted for. In the end we obtain the expression

$$(\mathbf{S}-\mathbf{1})_{p'q';pq} \equiv i^4 \int d^4x_1\dots d^4x_4 e^{iq'\cdot x_1}e^{ip'\cdot x_2}\bar{u}(\overrightarrow{p'},s')\, \overrightarrow{K_{x_1}}\overrightarrow{D_{x_2}} \tag{12.36}$$

$$\times (\psi_0 T(\phi(x_1)\psi(x_2)\bar{\psi}(x_3)\phi^\dagger(x_4))\psi_0)\, \overleftarrow{K_{x_4}}\overleftarrow{D_{x_3}}\, u(\overrightarrow{p},s)e^{-iq\cdot x_4}e^{-ip\cdot x_3}$$

where

$$\overrightarrow{D_x} \equiv -i\,\overrightarrow{\not{\partial}}_x + m_f \tag{12.37}$$

$$\overleftarrow{D_x} \equiv i\,\overleftarrow{\not{\partial}}_x + m_f. \tag{12.38}$$

Let's now summarize the results of the LSZ reduction formalism. For two-to-two scattering, the transition amplitude is described by the vacuum-to-vacuum matrix element, where ψ_0 is the exact Heisenberg-picture vacuum, of the time-ordered product of interacting Heisenberg-picture fields; one for each incoming or outgoing particle. In addition, there is the appropriate kinetic operator, one for each field, acting on the four-point function. Finally, there is the appropriate wave function for each incoming and outgoing particle, an integration over the position of each field in the four-point function and a factor of i^4. There is now a simple generalization of this result for all possible spin 0, 1/2 or 1 fields. In general, the LSZ rules are as follows: First, considering incoming states, we have[3]

- scalars $\phi^\dagger(x)\overleftarrow{K}_x\, e^{-iq\cdot x}$,
- anti-scalars $e^{-iq\cdot x}\overrightarrow{K}_x\, \phi(x)$,
- fermions $\bar{\psi}(x)\overleftarrow{D}_x\, u(\vec{q},s)e^{-iq\cdot x}$,
- anti-fermions $e^{-iq\cdot x}\bar{v}(\vec{q},s)\overrightarrow{D}_x\, \psi(x)$,
- vectors $e^{-iq\cdot x}(-\epsilon^\mu(\vec{q},\lambda))(\Box_x + m_V^2)A_\mu(x)$.

For outgoing states, we have

- scalars $e^{iq\cdot x}\,\overrightarrow{K}_x\, \phi(x)$,
- anti-scalars $\phi^\dagger(x)\overleftarrow{K}_x\, e^{iq\cdot x}$,
- fermions $e^{iq\cdot x}\,\bar{u}(\vec{q},s)\overrightarrow{D}_x\, \psi(x)$,
- anti-fermions $\bar{\psi}(x)\overleftarrow{D}_x\, v(\vec{q},s)e^{iq\cdot x}$,
- vectors $e^{iq\cdot x}(-\epsilon^{\mu*}(\vec{q},\lambda))(\Box_x + m_V^2)A_\mu(x)$.

Note, the time-ordered product of the fields, at different space–time points, are all to be taken inside the matrix element in the exact vacuum. Then, the kinetic operators act on the n-point function. The wave functions are on the outside (with the appropriate momentum and helicity for each state), and, finally, there is an integral over each space–time point and a factor of i^n. For photons, $m_V = 0$ and the polarization vectors require a choice of gauge. In the Feynman gauge we have

$$\sum_{\lambda=\pm 1} \epsilon_\mu(\vec{q},\lambda)\epsilon_\nu^*(\vec{q},\lambda) = -g_{\mu\nu}, \qquad (12.39)$$

using the normalization given by

$$\epsilon^{\mu*}(\vec{q},\lambda)\epsilon_\mu(\vec{q},\lambda') \equiv -\delta_{\lambda\lambda'}. \qquad (12.40)$$

Given the LSZ result, we see that the S matrix element is the residue of the poles in the external lines of the n-point Green's function:

$$G_n(x_1,\dots,x_n) \equiv (\psi_0 T(\phi(x_1)\cdots\phi(x_n))\psi_0), \qquad (12.41)$$

where $\phi(x_i)$ represents a generic field. Now, we must evaluate G_n. We do this using perturbation theory. In Appendix A we show, using the interaction picture, that

$$G_n(x_1,\dots,x_n) = \frac{\langle 0|T\left(e^{-i\int dt' H_I(t')}\phi_I(x_1)\cdots\phi_I(x_n)\right)|0\rangle}{\langle 0|T\left(e^{-i\int dt' H_I(t')}\right)|0\rangle}, \qquad (12.42)$$

---

[3] Note, we implicitly sum over all spinor indices in the following equations.

where H_I is the interaction Hamiltonian and the interaction-picture fields satisfy free-field equations. This is the Gell-Mann–Low theorem.

Finally, using Wick's theorem (for a derivation, see Appendix B), you will be able to evaluate the last expression, since

$$T(\phi_I(x_1)\cdots\phi_I(x_n)) = N(\phi_I(x_1)\cdots\phi_I(x_n))$$
$$+ \{N\left(\underbracket{\phi_I(x_1)\phi_I}(x_2)\cdots\phi_I(x_n)\right)$$

$+$ sum over all possible permutations of a single contraction

$+$ sum over all double contractions

$+ \ldots$, until every possible field is contracted$\}$.

Note, contractions are by definition only for fields in pairs. We have for real scalars

$$\underbracket{\phi_I(x_1)\phi_I}(x_2) \equiv i\Delta_F(x_1 - x_2), \tag{12.43}$$

where Δ_F is the Feynman propagator. Note, this is a pure C number. For fermions, we have

$$\underbracket{\psi_I(x_1)\bar{\psi}_I}(x_2) \equiv iS_F(x_1 - x_2), \tag{12.44}$$

$$\underbracket{\psi_I(x_1)\psi_I}(x_2) \equiv 0. \tag{12.45}$$

In general, non-trivial contractions give the Feynman Green's functions. In addition, Wick's theorem requires the fields to be moved to the appropriate place so that a contracted pair is side by side. In addition, in order to accomplish this, one must commute scalars with scalars or fermions, and one must anti-commute fermions with fermions. This process is accomplished as if scalars are classical numbers and fermions are classical Grassmann variables. Given this result we see that the only non-vanishing term is given by the last term in Wick's theorem, i.e. the term with all fields contracted in pairs in all possible ways.

$$\langle 0|T(\phi_I(x_1)\cdots\phi_I(x_n))|0\rangle = N(\text{fully contracted}). \tag{12.46}$$

Note, if there are an odd number of fields, then the result is identically zero.

PART III

QUANTUM ELECTRODYNAMICS

Quantum Electrodynamics

In this chapter, we apply our field-theory knowledge to a specific example, quantum electrodynamics (QED), which is the first well-understood relativistic quantum field theory of the Standard Model. In fact, it has been tested to very high order in perturbation theory. Moreover, for almost 20 years it was the only successful and established part of the Standard Model. The weak interactions were described by an effective field theory, the four fermion interactions proposed by Fermi in 1932. And the strong nuclear forces were believed to be intractable using field theory. Therefore, the only tool one had for understanding the strong interactions was to abstract some of the known properties of field theory and propose them as axioms for the S matrix, i.e. Poincaré invariance, unitarity, crossing symmetry and analyticity. We will use our discussion of QED to study the electromagnetic properties of electrons and muons. We will then use QED as a probe to study the properties of strongly interacting particles. Here, we describe the $U(1)$ local symmetry of QED and elaborate on the Feynman rules for the theory. This hopefully makes it clear how to use the formalism for the S matrix discussed in the previous chapter. We then discuss the property known as crossing symmetry.

13.1 Introduction to QED

The Lagrangian of QED is given by

$$\mathcal{L} = -\frac{1}{4}F_{\mu\nu}^2 + \bar{\psi}(i\slashed{D} - m)\psi, \tag{13.1}$$

where the covariant derivative, D_μ, is given by

$$iD_\mu \equiv i\partial_\mu - eA_\mu\, Q \tag{13.2}$$

and Q is the charge of the field in units of $e = 1.6 \times 10^{-19}$ C. For the possible fields, $f = \{e, \nu, p, n, u, d\}$, i.e. electron, neutrino, proton, neutron, up or down quark, we have $Q_e = -1$, $Q_\nu = 0$, $Q_p = +1$, $Q_n = 0$, $Q_u = 2/3$, $Q_d = -1/3$.

The Lagrangian is gauge invariant. It is invariant under a continuous, local symmetry, where $\alpha(x)$ is space–time dependent. The fields transform as follows:

$$\psi'(x) = e^{i\alpha(x)Q}\,\psi(x), \tag{13.3}$$

$$A'_\mu(x) = A_\mu(x) - \frac{1}{e}\partial_\mu\alpha(x). \tag{13.4}$$

With these transformations it is easy to see that

$$iD'_\mu\psi'(x) = e^{i\alpha(x)Q}\big(i\partial_\mu - eA'_\mu(x)Q - (\partial_\mu\alpha(x)Q)\big)\psi(x) \tag{13.5}$$

$$= e^{i\alpha(x)Q}iD_\mu\psi(x). \tag{13.6}$$

Hence, the fermionic term in $\mathcal{L}$ is invariant, and we have

$$\bar{\psi}'(x)(i\slashed{D}' - m)\psi'(x) = \bar{\psi}(x)(i\slashed{D} - m)\psi(x). \tag{13.7}$$

We also know that the gauge term is invariant, since

$$F'_{\mu\nu} = F_{\mu\nu}. \tag{13.8}$$

Now we can divide the Lagrangian into a free and interacting part, $\mathcal{L} = \mathcal{L}_0 + \mathcal{L}_I$, with

$$\mathcal{L}_0 = -\frac{1}{4}F_{\mu\nu}^2 + \bar{\psi}(i\slashed{\partial} - m)\psi \tag{13.9}$$

and

$$\mathcal{L}_I = -eA_\mu\bar{\psi}\gamma^\mu Q\psi. \tag{13.10}$$

13.2 Feynman Rules in Coordinate Space

For homework, you will evaluate the Bhabha scattering amplitude, discussed in Section 13.4, starting from the results of the Lehmann–Symanzik–Zimmerman (LSZ) formalism and obtaining the scattering amplitude in position space. Then performing the space–time integrals and obtaining the scattering amplitude in momentum space. This process should make it clear how the Feynman rules in momentum space are obtained. For now, we just give the general Feynman rules first in coordinate space and then in momentum space.

1. For every photon line we have (in the $\xi = 1$ gauge)

$$\langle 0|T(A_\mu(x)A_\nu(0))|0\rangle = \underbrace{A_\mu(x)A_\nu(0)} \tag{13.11}$$

$$= -ig_{\mu\nu}\Delta_F(x), \tag{13.12}$$

with

$$\Delta_F(x) = \int \frac{d^4p}{(2\pi)^4}\frac{e^{ip\cdot x}}{p^2 + i\epsilon}. \tag{13.13}$$

This is represented by the Feynman diagram, Fig. 13.1.

Fig. 13.1   Photon propagator.

Fig. 13.2   Fermion propagator.

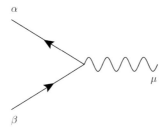

Fig. 13.3   Photon–fermion vertex.

2. For every fermion line we have

$$\langle 0|T(\psi_\alpha(x)\bar{\psi}_\beta(0))|0\rangle = \underbrace{\psi_\alpha(x)\bar{\psi}_\beta(0)} \tag{13.14}$$
$$= i(S_F(x))_{\alpha\beta},$$

with

$$S_F(x) = \int \frac{d^4p}{(2\pi)^4}\, e^{-ip\cdot(x-y)}\left(\frac{\not{p}+m}{p^2-m^2+i\epsilon}\right)_{\alpha\beta} \tag{13.15}$$

and the Feynman graph given in Fig. 13.2.

3. At every vertex given in Fig. 13.3 we have the factor

$$-ie(\gamma^\mu)_{\alpha\beta}Q. \tag{13.16}$$

4. (a) For every external photon we have

$$\begin{aligned}
\text{outgoing} &\quad -\epsilon_\mu^*(\vec{q},\lambda)e^{iq\cdot x}, \\
\text{incoming} &\quad -\epsilon_\mu(\vec{q},\lambda)e^{-iq\cdot x},
\end{aligned} \tag{13.17}$$

with

$$\sum_{\lambda=\pm 1}\epsilon_\mu(\vec{q},\lambda)\epsilon_\nu^*(\vec{q},\lambda) = -g_{\mu\nu}. \tag{13.18}$$

(b) For every external fermion we have

$$\text{incoming fermion} \quad u(\vec{q}, s)e^{-iq \cdot x}, \tag{13.19}$$

$$\text{outgoing fermion} \quad \bar{u}(\vec{q}, s)e^{iq \cdot x},$$

$$\text{incoming anti-fermion} \quad \bar{v}(\vec{q}, s)e^{-iq \cdot x},$$

$$\text{outgoing anti-fermion} \quad v(\vec{q}, s)e^{iq \cdot x}.$$

In addition, there is a kinetic operator, appropriate to each field, acting on the coordinates of each external line.

5. We have a factor of

$$i^n \int \prod_{i=1}^{n} d^4 x_i, \tag{13.20}$$

with n the number of external lines.

6. Sum over all possible ways of connecting lines to form *connected* Feynman diagrams, i.e. those diagrams that cannot be divided into two or more parts by drawing a line through the diagram that does not intersect any propagator line.

7. A relative sign for each graph is determined by Wick contractions.

8. We have a factor of $\frac{1}{n!}$, where, in this case, n is the order of the perturbation series.

13.3 Feynman Rules in Momentum Space

1. For every internal photon line we have the photon propagator,

$$\frac{-ig_{\mu\nu}}{p^2 + i\epsilon}, \tag{13.21}$$

with the appropriate Feynman diagram, Fig. 13.1.

2. For every internal fermion line we have

$$i \left(\frac{\not{p} + m}{p^2 - m^2 + i\epsilon} \right)_{\alpha\beta}, \tag{13.22}$$

with the Feynman propagator, Fig. 13.2.

3. At every vertex we have the factor

$$- ie(\gamma^\mu)_{\alpha\beta} Q \tag{13.23}$$

with the diagram in Fig. 13.3.

4. There is an overall factor in front,

$$(2\pi)^4 \delta^4 \left(\sum_{i_{out}} p_i^{out} - \sum_{i_{in}} p_i^{in} \right), \tag{13.24}$$

which enforces energy and momentum conservation. This is a direct consequence of Poincaré invariance.

5. Draw all possible connected diagrams, i.e. those diagrams that cannot be divided into two or more parts by drawing a line through the diagram that does not intersect any propagator line. Assign momentum to each internal line, such that four-momentum is conserved at every vertex.

6. Integrate over the momentum for every closed loop with the integration measure,

$$\int \frac{d^4 p}{(2\pi)^4}, \tag{13.25}$$

i.e. these momenta are not constrained by energy–momentum conservation.

7. Insert a factor of

$$(-1)^L, \tag{13.26}$$

where L is the number of closed fermion loops.

8. (a) Insert a factor of

$$\frac{1}{n!}, \tag{13.27}$$

where n is the order of the perturbation.

(b) Insert a combinatoric factor that counts the number of graphs with the same topology, i.e. this counts the number of independent Wick contractions giving the same Feynman graph.

(c) Insert a factor of ± 1 for the relative sign of the graph resulting from the Wick contraction of fermion fields.

9. Insert the appropriate wave functions for the incoming and outgoing lines.

(a) For every external photon we have

$$\text{outgoing} \quad \epsilon_\mu^*(\vec{q}, \lambda), \tag{13.28}$$

$$\text{incoming} \quad \epsilon_\mu(\vec{q}, \lambda).$$

(b) For every external fermion we have

$$\begin{array}{ll} \text{incoming fermion} & u(\vec{q}, s), \\ \text{outgoing fermion} & \bar{u}(\vec{q}, s), \\ \text{incoming anti-fermion} & \bar{v}(\vec{q}, s), \\ \text{outgoing anti-fermion} & v(\vec{q}, s). \end{array} \tag{13.29}$$

Note: there are no longer any external lines associated with the incoming or out-going particles. These have been eliminated by the action of the kinetic operators.[1] In addition, there are no disconnected Feynman diagrams, since these are also elimi-nated, either by the kinetic operators acting on both ends of a disconnected external line or because vacuum-to-vacuum diagrams are canceled, order by order in per-turbation theory, by terms coming from the denominator of the Gell-Mann–Low theorem.

13.3.1 Bubbles and Disconnected Diagrams at Second Order in Perturbation Theory

The perturbation expansion will produce many diagrams to second order in perturbation theory. These include two vacuum-to-vacuum bubble diagrams (**a** and **b**) and two disconnected diagrams (**c** and **d**) in Fig. 13.4, as well as the two connected diagrams, Figs. 13.5 and 13.6. To second order, the two bubbles in **a** and **b** cancel and we are just left with the zeroth-order discon-nected diagram. Diagram **c** corresponds to a wave-function correction, while diagram **d** actually vanishes due to a symmetric integral. Finally, the two remaining disconnected diagrams are eliminated in this order in perturbation the-ory when the kinetic operators are corrected by both wave function and mass renormalization.

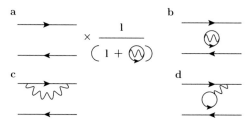

Fig. 13.4 Feynman diagrams contributing to Bhabha scattering at second order in perturbation theory. **a** and **b** include vacuum-to-vacuum bubble diagrams, **c** and **d** are propagator corrections.

---

[1] This statement is only correct when one takes into account that the kinetic operators for the "in" and "out" states receive corrections order by order in perturbation theory. In addition to the bare mass receiving corrections in order to define the on-shell mass of the physical states, the terms proportional to $\Box$ (for scalars) or $\not{\partial}$ (for fermions) also receive corrections. These corrections redefine the propagator for the states, as well as the inverse propagators, i.e. the kinetic operators. These complications are carefully covered in any field-theory text. In the end, all propagators for both "in" and "out" states are, using the standard terminology, "amputated."

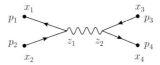

Fig. 13.5 An s channel diagram, where $s = (p_1 + p_2)^2$.

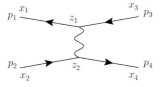

Fig. 13.6 A t channel diagram, where $t = (p_3 - p_1)^2$.

13.4 Bhabha Scattering

As an example of the Feynman rules in momentum space, consider Bhabha scattering, i.e.

$$e^+ e^- \longrightarrow e^+ e^- \tag{13.30}$$

$$p_1 p_2 \qquad p_3 p_4$$

$$s_1 s_2 \qquad s_3 s_4.$$

There are two Feynman diagrams in position space, given in Figs. 13.5 and 13.6.

For each graph there is a combinatoric factor of 2 (due to interchanging $z_1 \leftrightarrow z_2$) and a relative sign difference. These graphs are derived from the four-point Green's function

$$(\psi_0 T(\bar{\psi}(x_2)\psi(x_1)\psi(x_4)\bar{\psi}(x_3))\psi_0), \tag{13.31}$$

which at second order in perturbation theory is given by

$$\frac{1}{2!} \langle 0|T(\bar{\psi}(x_2)\psi(x_1)\psi(x_4)\bar{\psi}(x_3)\bar{\psi}(z_1)\gamma^\mu \psi(z_1)\bar{\psi}(z_2)\gamma^\nu \psi(z_2)A_\mu(z_1)A_\nu(z_2))|0\rangle. \tag{13.32}$$

Note: the fields in the last expression are implicitly interaction-picture operators. The two inequivalent contractions responsible for graphs, Figs. 13.5 and 13.6, are given in Eqn. 13.33.

$$A_\mu(z_1)A_\nu(z_2) \times \tag{13.33}$$

$$\left\{ \bar{\psi}(x_2)\psi(x_1)\bar{\psi}(x_4)\bar{\psi}(x_3)\bar{\psi}(z_1)\gamma^\mu\psi(z_1)\bar{\psi}(z_2)\gamma^\nu\psi(z_2) \quad (1) \right.$$

$$+$$

$$\left. \bar{\psi}(x_2)\psi(x_1)\psi(x_4)\bar{\psi}(x_3)\bar{\psi}(z_1)\gamma^\mu\psi(z_1)\bar{\psi}(z_2)\gamma^\nu\psi(z_2) \quad (2) \right\}.$$

Also, the denominator in the Gell-Mann–Low theorem would contribute a two-loop vacuum-to-vacuum amplitude, which is infinite. This, however, does not contribute to this order in perturbation theory, since the resulting amplitude is disconnected and is killed by the kinetic operators.[2] In addition, there is also a diagram contributing to a one-loop correction of each Fermion propagator, but this is also killed by the kinetic operators, since it is a disconnected diagram.

In momentum space we find

$$(S-1) = (2\pi)^4\delta^4(p_1 + p_2 - p_3 - p_4) \times T \tag{13.34}$$

(see Eqn. 5.17), where the transition amplitude is given by

$$T = [-\bar{v}(\vec{p}_1, s_1)\gamma^\mu u(\vec{p}_2, s_2)\bar{u}(\vec{p}_4, s_4)\gamma^\nu v(\vec{p}_3, s_3)] \tag{13.35}$$

$$\times [-ig_{\mu\nu} \frac{1}{(p_1 + p_2)^2 + i\epsilon} (ie)^2]\} \quad \{\text{contraction } (1) - s \text{ channel}\}$$

$$+\{[+\bar{u}(\vec{p}_4, s_4)\gamma^\nu u(\vec{p}_2, s_2)\ \bar{v}(\vec{p}_1, s_1)\gamma^\mu v(\vec{p}_3, s_3)] \tag{13.36}$$

$$\times [-ig_{\mu\nu} \frac{1}{(p_1 - p_3)^2 + i\epsilon} (ie)^2] \quad \{\text{contraction } (2) - t \text{ channel}\}$$

and

$$T \equiv T(-, p_4, s_4; +, p_3, s_3; -, p_2, s_2; +, p_1, s_1). \tag{13.37}$$

13.5 Crossing Symmetry

Before we continue to discuss the phenomenology of QED, there is one more property of relativistic quantum field theory that we wish to discuss here. This is the property known as crossing symmetry, and it is a property of the perturbation expansion of any relativistic quantum field theory. Here, however, we will use the example of Bhabha scattering to demonstrate the symmetry.

[2] It also cancels an identical disconnected diagram coming from the numerator.

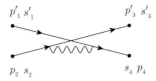

A u channel diagram, where $u = (p'_1 - p_4)^2$.

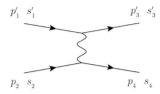

A t channel diagram, where $t = (p'_1 - p'_3)^2$.

In Figs. 13.5 and 13.6 we have the two Feynman diagrams for Bhabha scattering, evaluated to second order in perturbation theory, Eqns. 13.35 and 13.36. A crossed process is given by Eqn. 13.38

$$e^- e^- \qquad \longrightarrow \qquad e^- e^- \qquad\qquad (13.38)$$
$$p'_1 p_2 \qquad\qquad p'_3 p_4$$
$$s'_1 s_2 \qquad\qquad s'_3 s_4$$

with the Feynman diagrams of Figs. 13.7 and 13.8. It is obtained diagrammatically by reversing the incoming and outgoing anti-particle lines. It can be shown, order by order in perturbation theory, that the mathematical expression for these crossed diagrams, Figs. 13.7 and 13.8, can be obtained from the original diagrams, Figs. 13.5 and 13.6, by continuing the momenta as follows,

$$p_1 \to -p'_3 \quad \text{and} \quad p_3 \to -p'_1. \qquad\qquad (13.39)$$

We now want to make this property of relativistic quantum field theory plausible. We will show that, to this order, the amplitudes for

$$T(-, p_4, s_4; +, p_3, s_3; -, p_2, s_2; +, p_1, s_1) \qquad\qquad (13.40)$$

become, upon crossing,

$$T(-, p_4, s_4; +, -p'_1, s_3; -, p_2, s_2; +, -p'_3, s_1)$$
$$\equiv T(-, p_4, s_4; -, p'_3, s'_3; -, p_2, s_2; -, p'_1, s'_1), \qquad\qquad (13.41)$$

where

$$s'_3 \equiv -s_1, \quad s'_1 \equiv -s_3. \qquad\qquad (13.42)$$

The final amplitude is given, up to a phase, by the Feynman rules for the crossed process, Eqn. 13.41.

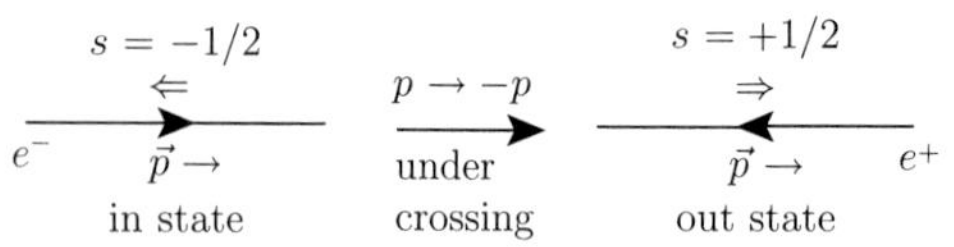

$$s = -1/2 \qquad\qquad\qquad s = +1/2$$
$$\Leftarrow \qquad\qquad p \to -p \qquad\qquad \Rightarrow$$

$$e^- \quad \vec{p} \to \qquad\qquad \text{under} \qquad\qquad \vec{p} \to \quad e^+$$

$$\text{in state} \qquad\qquad \text{crossing} \qquad\qquad \text{out state}$$

Fig. 13.9 A graphical representation of crossing on spinors where s is helicity.

Proof First consider the photon propagators in Figs. 13.5 and 13.6. They are given by $\frac{1}{s+i\epsilon}$ and $\frac{1}{t+i\epsilon}$, respectively. Upon crossing, we have

$$s = (p_1 + p_2)^2 \to (p_2 - p_3')^2 \equiv (p_1' - p_4)^2 \equiv u, \qquad (13.43)$$

i.e. $s \to u$, and

$$t = (p_1 - p_3)^2 \to (p_1' - p_3')^2 \equiv t, \qquad (13.44)$$

i.e. $t \to t$. Thus, the propagators for Bhabha scattering go into the propagators for elastic $e^- e^-$ scattering.

Now consider the spinors. We will show that, upon crossing,

$$u(E_1, \vec{p}_1, s_1) \to u(-E_3', -\vec{p}_3', s_1) \equiv \left(\frac{1}{2 s_1 i N(\hat{p}_3, s_1)} \right) v(E_3', \vec{p}_3', s_1), \qquad (13.45)$$

where $\left(\frac{1}{2 s_1 i N(\hat{p}_3, s_1)} \right)$ is a phase factor. Hence, from the Feynman rules, the wave function for an incoming particle with momentum, p_1, and helicity, s_1, becomes, upon crossing, the wave function for an outgoing anti-particle with momentum, p_3', and helicity, $-s_1$ (see Fig. 13.9).

Now let's derive this result. The spinors satisfy the equation

$$(\not{p} - m)u(\vec{p}, s) = 0, \qquad (\not{p} + m)v(\vec{p}, s) = 0. \qquad (13.46)$$

Thus, if we take $p_\mu \to -p_\mu$, we find

$$u(E, \vec{p}, s) \to u(-E, -\vec{p}, s) \propto v(E, \vec{p}, s), \qquad (13.47)$$

i.e. they satisfy the same equations.

They also satisfy the equation

$$\frac{\vec{\Sigma} \cdot \hat{p}}{2} \left\{ \begin{array}{c} u(\vec{p}, s) \\ v(-\vec{p}, s) \end{array} \right\} = s \left\{ \begin{array}{c} u(\vec{p}, s) \\ v(-\vec{p}, s) \end{array} \right\}. \qquad (13.48)$$

Therefore, upon crossing, we have[3]

$$-\frac{\vec{\Sigma} \cdot \hat{p}}{2} \left\{ \begin{array}{c} u(-E, -\vec{p}, s) \\ v(E, \vec{p}, s) \end{array} \right\} = s \left\{ \begin{array}{c} u(-E, -\vec{p}, s) \\ v(E, \vec{p}, s) \end{array} \right\}. \qquad (13.49)$$

Thus, $v(\vec{p}, s)$ is an eigenstate of helicity with eigenvalue $-s$ and

$$u(-E, -\vec{p}, s) = N v(E, \vec{p}, s), \qquad (13.50)$$

where N is a phase factor.

---

[3] We only use crossing on the u spinor in the following equation. For the v spinor we just take $\vec{p} \to -\vec{p}$.

The normalization factor can be explicitly derived given our phase conventions for the spinors. We have, Eqn. 9.76,

$$u(\vec{p}, s) \equiv \sqrt{E + m} \begin{pmatrix} \chi_{\hat{p}}(s) \\ \frac{\vec{\sigma} \cdot \vec{p}}{E + m} \chi_{\hat{p}}(s) \end{pmatrix} \tag{13.51}$$

and, Eqn. 9.77,

$$v(\vec{p}, s) \equiv \gamma_2 u^*(\vec{p}, s) = \sqrt{E + m} \begin{pmatrix} \frac{\vec{\sigma} \cdot \vec{p}}{E + m} \sigma_2 \chi_{\hat{p}}^*(s) \\ \sigma_2 \chi_{\hat{p}}^*(s) \end{pmatrix}. \tag{13.52}$$

Then,

$$u(-E, -\vec{p}, s) \equiv \sqrt{m - E} \begin{pmatrix} \chi_{-\hat{p}}(s) \\ \frac{\vec{\sigma} \cdot \vec{p}}{E - m} \chi_{-\hat{p}}(s) \end{pmatrix}. \tag{13.53}$$

But $\chi_{-\hat{p}}(s) = \chi_{\hat{p}}(-s)$. Therefore, we have

$$u(-E, -\vec{p}, s) \equiv \frac{\sqrt{m^2 - E^2}}{\sqrt{E + m}} \begin{pmatrix} \chi_{\hat{p}}(-s) \\ \frac{\vec{\sigma} \cdot \vec{p}}{E^2 - m^2} \chi_{\hat{p}}(-s)(E + m) \end{pmatrix}. \tag{13.54}$$

Then, using $\sqrt{m^2 - E^2} = i|\vec{p}|$ and $\vec{\sigma} \cdot \vec{p}\chi_{\hat{p}}(s) = 2s|\vec{p}|\chi_{\hat{p}}(s)$, we find

$$u(-E, -\vec{p}, s) \equiv \frac{1}{2si} \sqrt{E + m} \begin{pmatrix} \frac{\vec{\sigma} \cdot \vec{p}}{E + m} \chi_{\hat{p}}(-s) \\ \chi_{\hat{p}}(-s) \end{pmatrix}. \tag{13.55}$$

Finally, we use the identity, $N(\hat{p}, s)\chi_{\hat{p}}(-s) \equiv \sigma_2\, \chi_{\hat{p}}^*(s)$ and we have

$$u(-E, -\vec{p}, s) = N\, v(E, \vec{p}, s), \tag{13.56}$$

with

$$N = \left(\frac{1}{2siN(\hat{p}, s)} \right). \tag{13.57}$$

Similarly,

$$v(-E, -\vec{p}, s) = N'u(E, \vec{p}, s). \tag{13.58}$$

□

Thus in momentum space, we find the amplitude for the crossed diagram,

$$T = (S - 1) = (2\pi)^4 \delta^4(p_1' + p_2 - p_3' - p_4) \times T$$
$$\{[-\bar{u}(\vec{p}_3\,', s_3')\gamma^\mu u(\vec{p}_2, s_2)\bar{u}(\vec{p}_4, s_4)\gamma^\nu u(\vec{p}_1\,', s_1')]\} \tag{13.59}$$
$$\times [-ig_{\mu\nu} \frac{1}{(p_1' - p_4)^2 + i\epsilon} (ie)^2]\} \quad \{\text{contraction } (1) - u \text{ channel}\}$$
$$+\{[+\bar{u}(\vec{p}_4, s_4)\gamma^\nu u(\vec{p}_2, s_2)\, \bar{u}(\vec{p}_3\,', s_3')\gamma^\mu u(\vec{p}_1\,', s_1')]\} \tag{13.60}$$
$$\times [-ig_{\mu\nu} \frac{1}{(p_1' - p_3')^2 + i\epsilon} (ie)^2]\} \quad \{ \text{ contraction } (2) - t \text{ channel}\}$$

and

$$T \equiv T(-, p_4, s_4; +, -p_1', s_3; -, p_2, s_2; +, -p_3', s_1) \tag{13.61}$$
$$= T(-, p_4, s_4; -, p_3', s_3'; -, p_2, s_2; -, p_1', s_1').$$

13.6 Exercises

13.1 Calculate the scattering amplitude for the Bhabha scattering process in Eqn. 13.30 starting from the LSZ formalism (as in the example in Eqn. 12.36), then using the Gell-Mann–Low formula and Wick's theorem to first derive the amplitude in position space. Then do all the space–time integrals to obtain the amplitude in momentum space.

13.2 Calculate the differential cross-section for Bhabha scattering (in the limit $E_i \gg m_e$; i.e. neglect the electron mass) for unpolarized initial states and assuming that only the momentum of the final states are measured.

13.3 Calculate the differential cross-section for an incoming electron with negative helicity and positron with positive helicity to produce a μ^- with negative helicity and a μ^+ with positive helicity, i.e. $(e_{-1/2}^- e_{+1/2}^+ \rightarrow \mu_{-1/2}^- \mu_{+1/2}^+)$. Compare the result for a similar scattering process with $(e_{-1/2}^- e_{+1/2}^+ \rightarrow \mu_{-1/2}^- \mu_{-1/2}^+)$. Do not neglect the muon mass! Explain the results in terms of angular momentum and/or chirality conservation.

14 Magnetic Moments in QED

In this chapter we consider some of the successes of quantum electrodynamics (QED). In particular, we focus on the theoretical calculations of the anomalous magnetic moments of the electron and muon and compare with the present experimental data. We also briefly discuss the Lamb shift. Although there are apparently some discrepancies between theory and experiment of the anomalous magnetic moment, these results nevertheless set the bar for theoretical calculations in the Standard Model. They also allow us to see how both strong and weak interaction physics enters into the electromagnetic phenomenon of electrons and muons at higher orders of perturbation theory.

14.1 The Dirac Magnetic Moment

Consider the interaction Hamiltonian, $\mathcal{H}_I$, responsible for the process in Fig. 14.1. We have

$$\mathcal{H}_I = e A^\mu \bar{\psi} \gamma_\mu Q \psi. \tag{14.1}$$

Now consider the matrix element of $\mathcal{H}_I$ taken between single particle states, i.e.

$$\langle p', s' | \mathcal{H}_I | q, \lambda; p, s \rangle = e \epsilon^\mu(\vec{q}, \lambda) \bar{u}(\vec{p}\,', s') \gamma_\mu Q u(\vec{p}, s). \tag{14.2}$$

Before we continue, let's derive a useful identity (Gordon identity) valid on-shell. We have (where $\doteq$ signifies the equality is valid in matrix elements between spinors)

$$
\begin{aligned}
2m\gamma_\mu &\doteq \slashed{p}' \gamma_\mu + \gamma_\mu \slashed{p} \\
&= \frac{1}{2}(\{\gamma_\mu, \slashed{p}\} + [\gamma_\mu, \slashed{p}]) \\
&\quad + \frac{1}{2}(\{\gamma_\mu, \slashed{p}'\} - [\gamma_\mu, \slashed{p}']) \\
&= (p' + p)_\mu - \frac{1}{2}[\gamma_\mu, \gamma_\nu]q^\nu
\end{aligned}
$$

or using $\Sigma_{\mu\nu} \equiv \frac{i}{4}[\gamma_\mu, \gamma_\nu]$ we have

$$\gamma_\mu \doteq \frac{(p' + p)_\mu}{2m} + \frac{i}{m}\Sigma_{\mu\nu}q^\nu. \tag{14.3}$$

Plugging Eqn 14.3 into Eqn 14.2 we have two terms.

In the non-relativistic limit with $p'_\mu \approx p_\mu \approx m u_\mu$, where u_μ is the four-vector velocity, the first term just represents the charge current. We are interested in

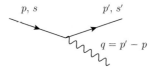

Fig. 14.1 Soft photon interacts with an electron.

the second term, which will give the magnetic-moment interaction. For electrons, $Q_e = -1$, and we have, for the second term,

$$\langle p', s'|\mathcal{H}_I|q, \lambda; p, s\rangle_{2nd} = i\frac{eQ_e}{m} \epsilon^\mu(\vec{q}, \lambda)q^\nu \bar{u}(\vec{p}\,', s')\Sigma_{\mu\nu}u(\vec{p}, s). \qquad (14.4)$$

Now, using[1] $-\partial^\nu A^\mu(x) \approx e^{-iq\cdot x}i\epsilon^\mu(\vec{q}, \lambda)q^\nu$, we have

$$\langle p', s'|\mathcal{H}_I|q, \lambda; p, s\rangle_{2nd} = -\frac{eQ_e}{m}\langle p', s'|\partial^\nu A^\mu \bar{\psi}\Sigma_{\mu\nu}\psi|q, \lambda; p, s\rangle$$

$$= \frac{eQ_e}{2m}\langle p', s'|F^{\mu\nu}\bar{\psi}\Sigma_{\mu\nu}\psi|q, \lambda; p, s\rangle.$$

Now, using[2] $F^{ij} \equiv -\epsilon_{ijk}\,B_k, \quad \Sigma_{ij} \equiv \frac{1}{2}\epsilon_{ijk}\Sigma_k$, we have

$$F_{ij}\Sigma^{ij} \equiv -\vec{B}\cdot\vec{\Sigma}. \qquad (14.5)$$

Hence,[3]

$$\langle p', s'|\mathcal{H}_I|q, \lambda; p, s\rangle_{2nd} = -\langle p', s'|\frac{eQ_e}{2m}\vec{B}\cdot\bar{\psi}\vec{\Sigma}\psi|q, \lambda; p, s\rangle. \qquad (14.6)$$

Recall, the spin operator on fermions is given by

$$\vec{S} = \frac{\vec{\Sigma}}{2} \qquad (14.7)$$

and we finally obtain

$$\langle p', s'|\mathcal{H}_I|q, \lambda; p, s\rangle_{2nd} = -\langle p', s'|\frac{g\,eQ_e}{2m}\vec{B}\cdot\bar{\psi}\,\vec{S}\psi|q, \lambda; p, s\rangle$$

$$= -\langle p', s'|\bar{\psi}\vec{\mu}_e\psi\cdot\vec{B}|q, \lambda; p, s\rangle, \qquad (14.8)$$

where the gyromagnetic ratio, g, is 2. This gives the Dirac magnetic moment, $\vec{\mu}$, of a point-like particle, such as the electron, muon or tau lepton, with

$$\vec{\mu}_e = \frac{geQ_e\hbar}{2m_e}\vec{S}. \qquad (14.9)$$

The quantity

$$\mu_B = \frac{e\hbar}{2m_e} \approx 5.8 \times 10^{-11}\text{MeV/T} \qquad (14.10)$$

---

[1] For a more transparent derivation, see Section 15.2.

[2] The magnetic field is given in terms of the vector potential by the expression, $\vec{B} = \vec{\nabla} \times \vec{A}$, where the four-vector potential is by definition given by the expression, $A^\mu \equiv (A^0,\, A^i)$.

[3] Note, in the non-relativistic limit, $q_0 \approx 0$, thus the term proportional to Σ_{0i} is suppressed.

is the Bohr magneton, while

$$\mu_N = \frac{e\hbar}{2m_p} \approx 3.2 \times 10^{-14} \text{MeV/T} \tag{14.11}$$

is the nuclear magneton.

14.2 Anomalous Magnetic Moments

The Dirac magnetic moment, discussed in the previous section, is just the first order in a perturbation series. The one-loop correction, given by the diagram in Fig. 14.2, is of the form

$$\langle p', s' | \mathcal{H}_I | q, \lambda; p, s \rangle_{2nd} = a^{(2)} \frac{e}{2m} \langle p', s' | F_{\mu\nu} \bar{\psi} \Sigma^{\mu\nu} Q\psi | q, \lambda; p, s \rangle. \tag{14.12}$$

Thus, to order e^2, the gyromagnetic ratio is given by $g = 2 + 2a^{(2)}$ or

$$\frac{g-2}{2} = a^{(2)}, \tag{14.13}$$

where $a^{(2)}$ is known as the anomalous magnetic moment. The one-loop calculation was originally done by Julian Schwinger (Schwinger, 1948) and he obtained the result

$$a^{(2)} = \frac{\alpha}{2\pi}. \tag{14.14}$$

This one-loop calculation can be found in Appendix C.

QED is one of the most successful parts of the Standard Model and early on it was used to learn a great deal about the so-called elementary particles, such as the proton and neutron. Let's review some of its successes. Consider the latest experimental measurements and theoretical calculations of the anomalous magnetic moments of the electron and the muon.

Experimentally (Hanneke et al., 2011), we have

$$a_e^{expt.} \equiv \frac{g_e - 2}{2} = 1159652180.73(28) \times 10^{-12}, \tag{14.15}$$

where the parenthesis corresponds to the uncertainty in the last two digits. Theoretically, the calculation of a_e (for a summary of results, see Aoyama et al., 2018),

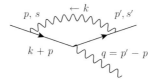

Fig. 14.2   One-loop quantum correction to soft-photon interaction with an electron where k is the loop momentum.

is given by

$$a_e^{theory} = 0.5 \left(\frac{\alpha}{\pi}\right) - (0.328478965579193\ldots)\left(\frac{\alpha}{\pi}\right)^2 + (1.181241456587\ldots)\left(\frac{\alpha}{\pi}\right)^3$$
$$- (1.912245764\ldots)\left(\frac{\alpha}{\pi}\right)^4 + (6.675(192))\left(\frac{\alpha}{\pi}\right)^5$$
$$+ (0.03053(23) + 1.6927(120)) \times 10^{-12}, \tag{14.16}$$

which is a tabulation of thousands of Feynman diagrams, and the last term includes the weak and hadronic loops, respectively. Of course, an accurate prediction for a_e depends on the value of the electromagnetic fine-structure constant, α. At the time of the Aoyama et al. paper, given the best experimentally determined value for α, the largest uncertainty in a_e^{theory} came from the uncertainty in α. Therefore, Aoyama et al. turned the analysis around and used the theoretical value of a_e, Eqn. 14.16, along with the experimental measurement, Eqn. 14.15, to obtain, at the time, the most accurate determination of α given by

$$\alpha^{-1} = 137.0359991491(15)(14)(330), \tag{14.17}$$

where the uncertainties come from the tenth-order QED, the hadronic correction and the experiment. Since the analysis of Aoyama et al. in 2018, a new, more accurate, measurement of the fine-structure constant (Parker et al., 2018) now leads to a *discrepancy with the standard model* (Davoudiasl and Marciano, 2018) (see Fig. 14.3). The discrepancy is

$$\Delta a_e = a_e^{exp} - a_e^{SM} = [-87\pm28(\text{exp})\pm23(\alpha)\pm2(\text{theory})]\times 10^{-14} = (-87\pm36)\times 10^{-14} \tag{14.18}$$

where in the last term the uncertainties are added in quadrature. In a more recent measurement of the fine structure constant (Morel et al., 2020) the discrepancy with the Standard Model has become smaller with the opposite sign.

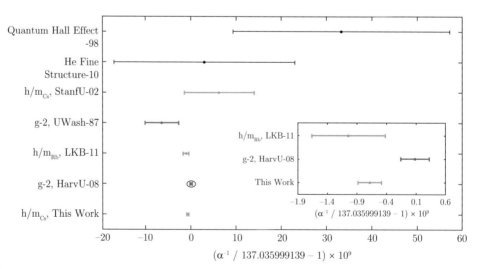

Fig. 14.3 Recent measurements of the fine-structure constant. The figure and the last point come from Parker et al. (2018). The circled point comes from combining the theoretical expression for a_e and the experimental measurement.

14.3   Measurement of a_e

The most precise measurement of a_e comes from an experiment using a Penning trap (Hanneke et al., 2011) (for a summary, see Sturm et al., 2013). This is so ingenious, it is worth our while to describe the theory behind the experiment. A Penning trap, simply stated, uses a magnetic field to confine a single electron. So let's consider the case of an electron in an external magnetic field,

$$\vec{B} = B_z \hat{z}, \tag{14.19}$$

and recall,

$$\vec{\mu}_e = -\frac{g_e e}{2m_e} \vec{S}. \tag{14.20}$$

The dynamics can be described by the Schrödinger equation with the Hamiltonian,

$$H = \frac{1}{2m_e}(\vec{p} + e\vec{A})^2 - \vec{\mu}_e \cdot \vec{B}. \tag{14.21}$$

It is convenient to choose a gauge with

$$A^x = -B_z y, \quad A^y = A^z = 0. \tag{14.22}$$

We now use the Schrödinger equation, given by

$$H\psi = E\psi, \tag{14.23}$$

$$\left\{ \frac{1}{2m_e}[(-i\partial_x - eB_z y)^2 - \partial_y^2 - \partial_z^2] + \frac{g_e e}{2m_e} B_z \frac{\sigma_3}{2} \right\} \psi = E\psi,$$

with

$$\psi_\pm = e^{i(k_x x + k_z z)} \chi(y) |\pm\rangle. \tag{14.24}$$

We then have

$$\left\{ \frac{1}{2m_e}[(k_x - eB_z y)^2 - \partial_y^2 + k_z^2] + \frac{g_e e}{2m_e} B_z S_\pm \right\} \psi_\pm = E\psi_\pm, \tag{14.25}$$

$$\left\{ \frac{-\partial_y^2}{2m_e} + \frac{\omega_c^2 m_e}{2}(y_0 - y)^2 + \frac{k_z^2}{2m_e} + \frac{g_e}{2}\omega_c S_\pm \right\} \chi(y)|\pm\rangle = E\chi(y)|\pm\rangle,$$

where

$$\omega_c \equiv \frac{eB_z}{m_e} \tag{14.26}$$

is the cyclotron frequency and

$$k_x \equiv eB_z y_0, \quad S_\pm = \pm\frac{1}{2}. \tag{14.27}$$

This is the Hamiltonian for an harmonic oscillator in the y direction with energy levels given by

$$E_{n\pm} = \left(n + \frac{1}{2} + \frac{g_e}{2}S_\pm\right)\omega_c + \frac{k_z^2}{2m_e} \tag{14.28}$$

Energy levels of a single electron in a Penning trap.

with $n = 0, 1, 2, 3, \ldots$. In a more suggestive form we have

$$\frac{E_{n_\pm}}{\omega_c} = \left(n + \frac{1}{2} \left(1 \pm \frac{g_e}{2} \right) \right) + \frac{k_z^2}{2 m_e \omega_c}$$

and using $\frac{g_e}{2} = 1 + a_e$ we have

$$\frac{E_{n_\pm}}{\omega_c} = \left(n + \frac{1}{2} \left(\begin{array}{c} 2 + a_e \\ -a_e \end{array} \right) \right) + \frac{k_z^2}{2 m_e \omega_c}. \tag{14.29}$$

Hence, experimentally, a_e is given by

$$a_e \equiv \frac{\Delta E}{\omega_c} = \frac{E_{0_+} - E_{1_-}}{\omega_c} \tag{14.30}$$

(see Fig. 14.4).

14.4 Anomalous Magnetic Moment of the Muon

The experimental value of the muon's anomalous magnetic moment is given by
Bennett et al. (2006) and Mohr et al. (2016),

$$a_\mu^{expt.} = 116592089(63) \times 10^{-11}, \tag{14.31}$$

which is the final result of the E821 experiment at Brookhaven National Laboratory
(BNL). The theoretical calculation of a_μ has several contributions,

$$a_\mu^{theory} = a_\mu(QED) + a_\mu(EW) + a_\mu(hadron). \tag{14.32}$$

The latest result for $a_\mu(QED)$, up to $(\frac{\alpha}{\pi})^5$ with the value for α obtained from a_e
(as summarized in Aoyama et al., 2018) is given by

$$a_\mu(QED) = 1165847188.41(7)(17)(6)(28) \times 10^{-12}, \tag{14.33}$$

where the uncertainties are due to the lepton–mass ratios, the numerical evaluation
of the eighth-order QED term, the numerical evaluation of the tenth-order QED
term and the fine-structure constant from left to right.[4] The electroweak term

---

[4] If instead, the latest value of α (Parker et al., 2018) is used, one finds the correction to $a_\mu(QED)$
is of order 0.04×10^{-11}, which is a small contribution Δa_μ.

Fig. 14.5 Some one-loop contributions to the electroweak term.

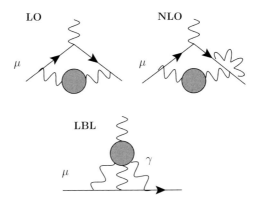

Fig. 14.6 Some HVP leading-order (LO), next-to-leading order (NLO) and LBL diagrams.

includes up to two loop contributions of $W^\pm, Z^0$ and Higgs boson, Fig. 14.5. The result is

$$a_\mu(EW) = 15.36(10) \times 10^{-10}. \tag{14.34}$$

Finally, the hadronic contribution includes several terms corresponding to hadronic vacuum polarization (HVPLO, HVPNLO, HVPNNLO), and light-by-light scattering (LBL), Fig. 14.6. (For further discussion, see Davier et al., 2017; Jegerlehner, 2017; Teubner, 2017.) Note, the leading order contribution to the hadronic vacuum polarization can be obtained directly from experiment, using the measurement of $\sigma(e^+e^- \to hadrons)$, analyticity of the scattering amplitude and the optical theorem, which directly relates the cross-section to the full scattering amplitude. We have $a_\mu^{had.LO} = \frac{m_\mu^2}{12\pi^3} \int_{s_{th}}^\infty ds \frac{1}{s} \hat{K}(s) \sigma_{had.}(s)$ (Fig. 14.7). The weight function $\frac{\hat{K}(s)}{s} = \frac{O(1)}{s}$, so that lower energies dominate in the integral. Similarly, one also uses the process $\sigma(\tau \to \nu_\tau + hadrons)$ with consistent results.

On the other hand, the LBL contribution has the largest uncertainty due to the fact that it can only be obtained in a model-dependent way, with pions and other hadrons in the loop, and not directly from quantum chromodynamics (QCD). Although recent lattice QCD results are roughly consistent with the models.

Combining all the theoretical results gives the values for

$$a_\mu^{theory} = \begin{cases} 116591783(51) \times 10^{-11} & \text{(Jegerlehner, 2017)} \\ 116591812(47) \times 10^{-11} & \text{(Teubner, 2017)} \\ 116591820(52) \times 10^{-11} & \text{(Davier et al., 2017)} \end{cases} \tag{14.35}$$

$$2 \, \text{Im} \, \text{WWW} \underset{had.}{\bigodot} \text{WWW} = \Sigma_{\boldsymbol{had.}} \int \mathbf{d\Phi} \, \big| \text{WWW} \big|^2$$

Fig. 14.7 HVPLO can be obtained directly from experiment using analyticity and the optical theorem. We have $a_\mu^{had.LO} = \frac{m_\mu^2}{12\pi^3} \int_{s_{th}}^{\infty} ds \frac{1}{s} \hat{K}(s) \sigma_{had.}(s)$. The weight function $\frac{\hat{K}(s)}{s} = \frac{O(1)}{s}$, so that lower energies dominate in the integral.

The difference between measurement and theory ranges from 269×10^{-11} to 306×10^{-11}, corresponding to a discrepancy of 3.3σ to 3.8σ discrepancy.

14.5 How Is a_μ Measured?

The anomalous magnetic moment of the muon was measured at Brookhaven in the BNL821 experiment (Bennett et al., 2006). They used a polarized beam of muons in a storage ring located in a plane perpendicular to a magnetic field $\vec{B}$ (see Fig. 14.8). The momentum of the muons, $\vec{p}$, changes with the cyclotron frequency, ω_c. The spin of the muons, $\vec{s}_\mu$, changes with the frequency, $\omega_s = \omega_c(1+a_\mu)$, where a_μ is the anomalous magnetic moment of the muon.[5] Thus, $\vec{s}_\mu$ precesses around the $\vec{p}$ direction with frequency,

$$\omega_a \equiv \omega_s - \omega_c = a_\mu \omega_c. \tag{14.36}$$

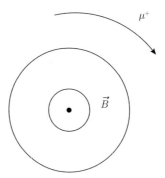

Fig. 14.8 A cartoon of the BNL821 muon storage ring with the magnetic field, $\vec{B}$, coming out of the paper.

---

[5] The spin of the muon in an external magnetic field satisfies the equation of motion $\frac{d\vec{S}}{dt} = \vec{\mu} \times \vec{B}$.

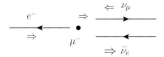

Fig. 14.9   Electrons with maximum energy in muon decay at rest.

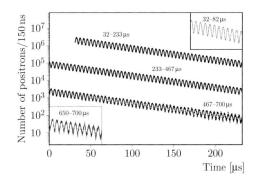

Fig. 14.10 $N_e(t)$, i.e. the number of positrons with energy $E_e > 2$ GeV as a function of time. Reprinted with permission from Brown et al. (2001). Copyright (2001) by the American Physical Society.

Now, there is an interesting factoid about muon decays. Consider a μ^- in its rest frame decaying into an electron, a muon neutrino and an electron anti-neutrino. The electrons can be emitted with varying values of their energy. However, angular momentum conservation requires that electrons with maximum energy (i.e. when $\vec{p}_{\nu_\mu} \simeq \vec{p}_{\nu_e}$) move along the muon spin direction (see Fig. 14.9). Then, when you boost the muon to the laboratory frame, the electrons with maximum energy will oscillate in time with frequency, ω_a. This is measured in Fig. 14.10 as $N_e(t)$, i.e. the number of electrons with energy $E_e > 2$ GeV as a function of time. The frequency of oscillations is just ω_a, which gives a_μ using Eqn. 14.36.

A new muon g-2 E-989 experiment at Fermilab recently began taking data in 2017 (Gohn, 2019). The experiment uses the muon storage ring from BNL821 and expects to have 21 times the statistics of the previous experiment and a factor of 4 improvement in the uncertainty. Results are eagerly awaited.

14.6 Strong-Interaction Contribution to Electromagnetic Processes

In the calculation of a_μ, the strong-interaction contribution was represented by the cross-section $\sigma(e^+e^- \to hadrons)$. In general, we have processes such as in Fig. 14.11. The imaginary part of the photon polarization tensor (i.e. the one-particle irreducible photon two-point function) is related to the total cross-section

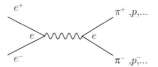

Fig. 14.11 The production of hadrons in e^+e^- scattering.

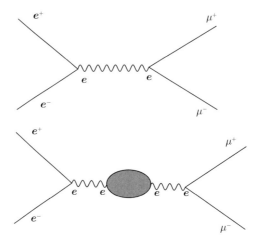

Fig. 14.12 Intermediate strong-interaction contribution $e^+e^- \to \mu^+\mu^-$ scattering.

for $e^+e^- \to hadrons$. The real part of the two-point function can be written in terms of the imaginary part using dispersion relations (like the Kramers–Kronig relation in electrodynamics). We will say more about dispersion relations later in the book.

For now, we argue that the hadronic contribution to the photon propagator can be measured in $e^+e^- \to \mu^+\mu^-$ scattering. Consider the Feynman diagrams in Fig. 14.12. The second term is of order e^4; however, it includes strong interactions in the loop. Note, the cross-section for $e^+e^- \to hadrons$ is given by

$$\sigma_h(s) \equiv \sigma_{e^+e^- \to hadrons} \simeq \sum_{hadrons} \int |T_{e^+e^- \to hadrons}|^2 dLips. \qquad (14.37)$$

We can then define an effective photon propagator

$$\frac{1}{q^2} \to \frac{1}{q^2} + \frac{1}{4\pi^2\alpha} \int_{4m_\pi^2}^\infty \frac{\sigma_h(s)ds}{q^2 - s + i\epsilon}. \qquad (14.38)$$

Experimentally, the following interference effect is measured, Fig. 14.13. The cross-over occurs near the mass of the ϕ meson, which has spin, parity and charge-conjugation properties, $J^{PC} = 1^{--}$. The ϕ meson contribution to the cross-section is given by (see Fig. 14.14)

$$\sigma_h(s) \propto \frac{g_{\phi\gamma}^2 \Gamma_{\phi \to \pi^+\pi^-\pi^0}}{(s - m_\phi^2)^2 + m_\phi^2 \Gamma_\phi^2}, \qquad (14.39)$$

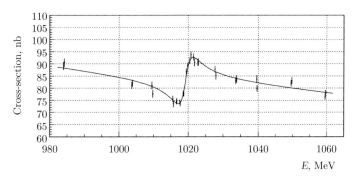

Fig. 14.13 $\sigma_{e^+e^-\to\mu^+\mu^-}$ as a function of $\sqrt{s}$ in MeV. Reprinted with permission from Achasov et al. (2001). Copyright (2001) by the American Physical Society.

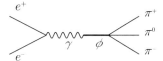

Fig. 14.14 $\sigma_h(s) \simeq \sigma_{e^+e^-\to\pi^+\pi^-\pi^0}$ as a function of $\sqrt{s}$ in MeV.

which is enhanced at $\sqrt{s} = m_\phi \simeq 1019$ MeV.

After including all known corrections to the photon propagator, one can parameterize the effects of new physics (beyond the Standard Model) by the substitution

$$\frac{e^2}{q^2} \to \left(\frac{e^2}{q^2} \pm \frac{2\pi}{\Lambda_\pm^2}\right), \tag{14.40}$$

where the second term corresponds to a four-fermion contact interaction. Experimentally (Tanabashi et al., 2018), looking at processes such as $e^+e^- \to e^+e^-, \mu^+\mu^-$, one finds, roughly, $\Lambda_+ \geq 8.3$ TeV, $\Lambda_- \geq 9.5$ TeV.

14.7 Lamb Shift

QED is extremely well tested. We've considered the anomalous magnetic moment of the electron. Now let's briefly consider radiative corrections to the spectrum of hydrogen. One of the most significant results regards the so-called Lamb shift. The energy levels of a bound electron are given in terms of the relativistic Dirac equation with

$$E\psi_e = (\vec{\alpha} \cdot \vec{p} + \beta m + A_0(r))\psi_e. \tag{14.41}$$

The energy levels are labeled using the notation

$$nL_J, \tag{14.42}$$

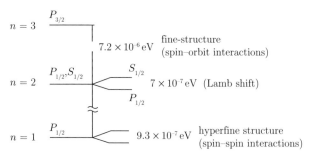

Fig. 14.15 Energy levels of the relativistic description of the hydrogen atom.

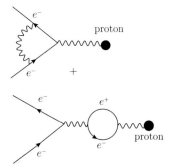

Fig. 14.16 The one-loop radiative corrections to the Lamb shift.

where n is the principal quantum number of the state, L is the orbital angular momentum and J is the total angular momentum. The $2\,P_{1/2}$ and $2\,S_{1/2}$ levels are degenerate solutions to the Dirac equation, see Fig. 14.15. However, due to radiative corrections, which uses the Bethe–Salpeter equations for bound electrons, (Bethe and Salpeter, 1957), the two energy levels are split (see Fig. 14.16). The splitting was measured in 1947 by Lamb and Retherford (1947, 1950) and the result is in good agreement with the theoretical calculation.

In summary, we have seen that QED has been tested quite successfully to many orders of perturbation theory. Nevertheless there are now some discrepancies between theory and experiment. There are two possibilities. These discrepancies may actually go away with further experimental tests of the theory. On the other hand, they may provide clues to new physics beyond the Standard Model. In either case, it is quite clear that the electron and muon are extremely well described as fundamental (elementary), point-like particles.

The Size of the Proton

In the elastic scattering of electrons on protons and deuterium it was found that the proton and neutron were not point-like particles. In fact, they had a size of the order of a fermi. Thus, unlike the electron or muon, they were not elementary particles. In this chapter we focus on the structure of the nucleon, as measured in elastic e–p and e–D scattering experiments.

15.1 Elastic Electron–Proton Scattering

Consider elastic electron–proton scattering as given in Fig. 15.1. The scattering amplitude for a point-like proton is given by the expression

$$T = -e^2 \bar{u}(k')\gamma^\mu u(k)\frac{g_{\mu\nu}}{q^2}\bar{u}(p')\gamma^\nu u(p). \tag{15.1}$$

This process was measured at Stanford Linear Accelerator Center (SLAC) in 1956 by R. Hofstadter and collaborators (Chambers and Hofstadter, 1956). It was found that the proton is not point-like at all. In fact, the proton and neutron have a finite size.

Let's first describe the relevant kinematics for elastic electron–proton scattering in the laboratory frame with $p = (m_p, \vec{0})$. The relevant three momenta are defined in Fig. 15.2.

We also define the Lorentz invariant momentum transfer

$$
\begin{aligned}
q^2 = (k - k')^2 &= 2m_e^2 - 2k \cdot k' \tag{15.2}\\
&= 2m_e^2 - 2EE' + 2|\vec{k}||\vec{k}'|\cos(\theta)\\
&\simeq -4EE'\left(\frac{1 - \cos\theta}{2}\right) \quad \text{for } E \gg m_e\\
&= -4EE'\sin^2\frac{\theta}{2}.
\end{aligned}
$$

The transition amplitude is now expressed by

$$T \equiv \frac{e^2}{q^2}[j^\mu(k', k)]_{electron}[J_\mu(p, p')]_{proton}, \tag{15.3}$$

where

$$j^\mu(k', k) \equiv -\bar{u}(\vec{k}', s')\gamma^\mu u(\vec{k}, s) \equiv \langle \vec{k}', s'|j^\mu_{EM}(0)|\vec{k}, s\rangle, \tag{15.4}$$

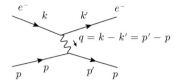

Fig. 15.1 Elastic electron–proton scattering.

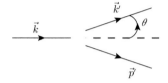

Fig. 15.2 Three momenta for elastic electron–proton scattering.

$$J^\mu(p',p) \equiv \langle \vec{p}\,', s_p' | j_{EM}^\mu(0) | \vec{p}, s_p \rangle, \tag{15.5}$$

and

$$j_{EM}^\mu(0) \equiv \bar{\psi} \gamma^\mu Q \psi \tag{15.6}$$

is the electromagnetic current.

   Note:

$$\partial_\mu j_{EM}^\mu \equiv 0 \tag{15.7}$$

implies that

$$q_\mu j^\mu(k',k) = q_\mu J^\mu(p',p) \equiv 0. \tag{15.8}$$

Proof

$$
\begin{aligned}
0 &\equiv \langle \vec{p}\,', s_p' | \partial_\mu j_{EM}^\mu(0) | \vec{p}, s_p \rangle \\
&\equiv \langle \vec{p}\,', s_p' | i[\hat{P}_\mu, j_{EM}^\mu(0)] | \vec{p}, s_p \rangle \\
&\equiv i(p'-p)_\mu \langle \vec{p}\,', s_p' | j_{EM}^\mu(0) | \vec{p}, s_p \rangle \\
&\equiv i q_\mu J^\mu(p',p).
\end{aligned}
$$

$\square$

   The differential cross-section one obtains for a point-like spin-1/2 particle is given by

$$\left(\frac{d\sigma}{d\Omega} \right)_{\text{point-like spin } (1/2,\,1/2)} = \left(\frac{d\sigma}{d\Omega} \right)_{(1/2,0)} \left(1 + 2\tau \tan^2 \frac{\theta}{2} \right), \tag{15.9}$$

where $\tau = |q^2|/4m_p^2$,

$$\left(\frac{d\sigma}{d\Omega} \right)_{(1/2,0)} \equiv \frac{\alpha^2 \cos^2 \frac{\theta}{2}}{4E^2 \sin^4 \frac{\theta}{2}[1 + \frac{2E}{m_p} \sin^2 \frac{\theta}{2}]}, \tag{15.10}$$

and the term with τ is the magnetic-moment contribution.

These results are for a point-like proton, but a real proton has structure. We parameterize that structure by what is called the proton form factor, i.e.

$$J_\mu(p',p) \equiv \bar{u}(\vec{p}\,', s_p')\Gamma_\mu(p',p)u(\vec{p}, s_p), \tag{15.11}$$

where $\Gamma_\mu(p',p)$ is constructed using four-momenta and a complete set of γ-type matrices, i.e.

- $\mathbb{I}$,   scalar,
- γ_μ,   vector,
- $\Sigma_{\mu,\nu} = \frac{i}{4}[\gamma_\mu, \gamma_\nu]$,   tensor,   also $\Sigma_{\mu\nu}\gamma^5$,
- $\gamma_\mu\gamma^5$,   axial vector,
- γ^5,   pseudo-scalar.

For a point particle, we have

$$J_\mu(p',p) = \bar{u}(\vec{p}\,', s')\gamma_\mu u(\vec{p}, s). \tag{15.12}$$

In a later chapter we shall discuss the charge conjugation and parity transformations of fermions and show that quantum electrodynamics (QED) is invariant under parity. Parity invariance puts constraints on the possible matrices, Γ_μ. Now, we'll briefly discuss parity transformations on the current and gauge field. Under a parity transformation, we have

$$\vec{p} \to -\vec{p} \equiv \vec{p}_P,$$
$$s \to -s \equiv s_P.$$

The four-momentum then transforms as

$$p^\mu \to (\mathbf{P}p)^\mu \tag{15.13}$$

where $\mathbf{P} = \text{diag}(1, -1, -1, -1)$. Then, the parity-transformed vertex function, $J_\mu(p',p)$, is given by

$$\begin{aligned} J_\mu^P(p',p) &\equiv \bar{u}(\vec{p}_P', s_p')\gamma_\mu u(\vec{p}_P, s_P) \\ &= \bar{u}(\vec{p}\,', s')\gamma_0\gamma_\mu\gamma_0 u(\vec{p}, s) \\ &\equiv [\mathbf{P}J(p',p)]_\mu, \end{aligned} \tag{15.14}$$

where the second equality follows from phase convention number 2, i.e.

$$u(\vec{p}_P, s_P) \equiv u(-\vec{p}, -s) \equiv \gamma_0 u(\vec{p}, s). \tag{15.15}$$

We thus see that J_μ transforms as a four-vector under parity.

Consider the vertex function,

$$J_{\mu 5}(p',p) \equiv \bar{u}(\vec{p}\,', s')\gamma_\mu\gamma^5 u(\vec{p}, s). \tag{15.16}$$

It transforms under parity as

$$J_{\mu 5}^P(p',p) \equiv \bar{u}(\vec{p}\,', s')\gamma_0\gamma_\mu\gamma^5\gamma_0 u(\vec{p}, s) \equiv -[\mathbf{P}J_5(p',p)]_\mu, \tag{15.17}$$

i.e. $J_{\mu 5}$ transforms as an axial (or pseudo) vector.

Now we give a simplified description of the parity invariance of QED. The interaction Hamiltonian density is given by

$$\mathcal{H}_I = eA^\mu \bar{\psi}\gamma_\mu Q\psi = eA^\mu j_\mu. \qquad (15.18)$$

Under parity $j_\mu \to [\mathbf{P}j]_\mu$. Therefore, parity invariance of QED requires

$$A_P^\mu = [\mathbf{P}A]^\mu, \qquad (15.19)$$

i.e. the photon is odd under parity. Since $\mathbf{P}^2 = 1$, $\mathcal{H}_I$ is invariant under parity. Hence, in general, $J_\mu(p', p)$ must transform as a vector, since it turns out that parity is conserved by both electromagnetic and strong interactions.

The most general vertex function, transforming as a Lorentz four-vector under parity is given by

$$J_\mu(p', p) = \bar{u}(\vec{p}\,', s')[\gamma_\mu k_1(q^2) + 2i\Sigma_{\mu\nu}q^\nu k_2(q^2) + 2i\Sigma_{\mu\nu}(p' + p)^\nu k_3(q^2)$$
$$+ q_\mu \, k_4(q^2) + (p' + p)_\mu k_5(q^2)]u(\vec{p}, s).$$

The k_i are, a priori, arbitrary functions only of q^2, since $(p + p')^2 = 2m^2 + 2p \cdot p'$ and $q^2 = (p' - p)^2 = 2m^2 - 2p \cdot p'$; thus $(p + p')^2 = 4m^2 - q^2$. We now show that several terms in the expression for J_μ are redundant. Recall:

$$\bar{u}(\vec{p}\,', s')(p' + p)_\mu u(\vec{p}, s) = \bar{u}(\vec{p}\,', s')[2m\gamma_\mu - 2i\Sigma_{\mu\nu}q^\nu]u(\vec{p}, s)$$

and

$$\bar{u}(\vec{p}\,', s')2i\Sigma_{\mu\nu}(p' + p)^\nu u(\vec{p}, s) = \bar{u}(\vec{p}\,', s')[-q_\mu]u(\vec{p}, s).$$

Thus, terms with $(p' + p)$ can be re-expressed in terms of the $k_{1,2,4}$ terms. Hence, without loss of generality, we have

$$J_\mu(p', p) = \bar{u}(\vec{p}\,', s')[\gamma_\mu F_1(q^2) + i\frac{\kappa}{m}F_2(q^2)\Sigma_{\mu\nu}q^\nu + q_\mu F_3(q^2)]u(\vec{p}, s). \qquad (15.20)$$

Now consider the constraint of current conservation, i.e.

$$\partial_\mu j^\mu = 0 \implies q^\mu J_\mu(p', p) = 0. \qquad (15.21)$$

Hence,

$$0 = \bar{u}(\vec{p}\,', s')[\slashed{q}F_1(q^2) + q^2 F_3(q^2)]u(\vec{p}, s). \qquad (15.22)$$

But we have

$$\bar{u}(\vec{p}\,', s')\slashed{q}\, u(\vec{p}, s) \equiv \bar{u}(\vec{p}\,', s')(\slashed{p}' - \slashed{p})u(\vec{p}, s)$$
$$= u(p'', s')(m - m)u(\vec{p}, s) \equiv 0.$$

Thus, $\partial_\mu j^\mu = 0$ implies that $F_3(q^2) = 0$.

F_1 and F_2 are the electromagnetic form factors and we now have the final expression

$$J_\mu(p', p) = \bar{u}(\vec{p}\,', s')[\gamma_\mu F_1(q^2) + i\frac{\kappa}{m}F_2(q^2)\Sigma_{\mu\nu}\, q^\nu]u(\vec{p}, s), \qquad (15.23)$$

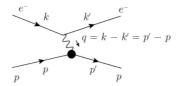

Elastic electron–proton scattering with the blob on the proton vertex, representing the form factor.

where, we shall show later,

$$F_1(0) \equiv Q, \qquad \text{charge of the particle} \qquad (15.24)$$

$$\kappa F_2(0) \equiv \frac{g-2}{2}, \qquad \text{anomalous magnetic moment.}$$

The amplitude for electron–proton elastic scattering is now given by

$$T_{ep \to ep} = -\frac{e^2}{q^2} \bar{u}(\vec{k}\,', s_e') \gamma^\mu u(\vec{k}, s_e) J_\mu(p', p). \qquad (15.25)$$

(see Fig. 15.3, where the blob represents the nucleon form factor).

To calculate the differential cross-section for unpolarized beams summed over the final spins we must first evaluate the expression

$$\sum_{spins} |T_{ep \to ep}|^2$$

$$= \frac{e^4}{(q^2)^2} \sum_{s_e, s_e'} (\bar{u}(\vec{k}\,', s_e') \gamma^\mu u(\vec{k}, s_e) \bar{u}(\vec{k}, s_e) \gamma^\nu u(\vec{k}\,', s_e')) \sum_{s_p, s_p'} (J_\mu(p', p) J_\nu^\dagger(p', p)).$$

$$(15.26)$$

We then obtain the differential cross-section, i.e. the Rosenbluth formula,

$$\frac{d\sigma}{d\Omega} = \left(\frac{d\sigma}{d\Omega}\right)_{1/2,0} \left\{ [F_1^2 + \tau \kappa^2 F_2^2] + \left[2\tau (F_1 + \kappa F_2)^2 \tan^2 \frac{\theta}{2}\right] \right\}. \qquad (15.27)$$

Note, $F_1(0) + \kappa F_2(0) = \frac{g}{2} \equiv \mu$, the total magnetic moment of the nucleon in units of $\frac{e}{m}\vec{S}$.

The observed values of the proton and neutron form factors at zero momentum transfer are given in Table 15.1.

Table 15.1 Proton and neutron form factors at zero momentum.			
Proton	$F_1^p(0) = 1$	$F_2^p(0) \equiv 1$	$\kappa_p = +1.79$
Neutron	$F_1^n(0) = 0$	$F_2^n(0) \equiv 1$	$\kappa_n = -1.91$

15.2 Physical Interpretation of the Form Factor

Now let's derive the properties of the form factor that we stated without proof in the last section, Eqn. 15.24. The interaction Hamiltonian is given by

$$H_I = e \int d^3\vec{x} A^\mu(x) \times \underbrace{j_\mu^{EM}(x)}_{e^{i\hat{P}\cdot x}\, j_\mu^{EM}(0)\, e^{-i\hat{P}\cdot x}} . \qquad (15.28)$$

In an external electromagnetic field, the nucleon interaction is given by

$$\langle \vec{p}\,', s' | H_I | \vec{p}, s \rangle = e \int d^3\vec{x} A^\mu_{ext.}(\vec{x}) J_\mu(p', p) \underbrace{e^{i(p'-p)\cdot x}}_{e^{iq\cdot x}}, \qquad (15.29)$$

where $J_\mu(p', p)$ is given in Eqn. 15.23 and $\gamma_\mu \doteq \frac{(p'+p)_\mu}{2m} + \frac{i}{m}\Sigma_{\mu\nu}q^\nu$. Hence, we now have

$$J_\mu(p', p) = \bar{u}(\vec{p}\,', s')\, [\frac{(p'+p)_\mu}{2m} F_1(q^2) + \frac{i}{m}\Sigma_{\mu\nu}q^\nu(F_1(q^2) + \kappa F_2(q^2))]u(\vec{p}, s). \qquad (15.30)$$

In the non-relativistic limit, with $E' \simeq E$ or $q^0 \simeq 0$, $\frac{\vec{p}\,'}{m} \simeq \frac{\vec{p}}{m} = \vec{v}$ and the four-vector velocity, $u^\mu \equiv (1, \vec{v})$, we have

$$J_\mu(p', p) \simeq 2m[\delta_{ss'}F_1(q^2)u_\mu + \frac{i}{m}\left(\frac{\bar{u}(\vec{p}\,', s')\,\Sigma_{\mu i}\,u(\vec{p}, s)}{2m}\right)q^i\,(F_1(q^2) + \kappa F_2(q^2))]. \qquad (15.31)$$

Consider the term containing

$$\left(\frac{\bar{u}(\vec{p}\,', s')\Sigma_{0i}u(\vec{p}, s)}{2m}\right)q^i$$

$$= \frac{1}{2m}\sqrt{E'+m}\left(\chi_{\hat{p}}^\dagger{}'(s') \quad -\chi_{\hat{p}}^\dagger{}'(s')\frac{\vec{\sigma}\cdot\vec{p}\,'}{E'+m}\right)\frac{i}{2}\begin{pmatrix} 0 & \vec{q}\cdot\vec{\sigma} \\ \vec{q}\cdot\vec{\sigma} & 0 \end{pmatrix}$$

$$\begin{pmatrix} \chi_{\hat{p}}(s) \\ \frac{\vec{\sigma}\cdot\vec{p}}{E+m}\chi_{\hat{p}}(s) \end{pmatrix}\sqrt{E+m}$$

$$\sim \frac{i}{2}\chi_{\hat{p}}^\dagger{}'(s')\left(\vec{q}\cdot\vec{\sigma}\frac{\vec{\sigma}\cdot\vec{p}}{2m} - \frac{\vec{\sigma}\cdot\vec{p}\,'}{2m}\vec{q}\cdot\vec{\sigma}\right)\chi_{\hat{p}}(s)$$

$$\sim -\frac{i}{m}(s-s')^2|\vec{p}|^2\,\underbrace{\chi_{\hat{p}}^\dagger{}'(s')\,\chi_{\hat{p}}(s)}_{\delta_{ss'}} \equiv 0.$$

Thus, this term only contributes to order $(\frac{|\vec{p}|}{m})^3$ and can be neglected. Note also that Σ_{0i} is not the spin operator.

Thus, we have

$$\langle \vec{p}\,', s' | H_I | \vec{p}, s \rangle = 2m \{ e \left(\int d^3 \vec{x} A_0^{ext.}(x) e^{-i\vec{q}\cdot\vec{x}} \right) \delta_{ss'} F_1(q^2) \qquad (15.32)$$

$$+ e \left(\int d^3 \vec{x} \, A_i^{ext.}(x) e^{-i\vec{q}\cdot\vec{x}} \right) \delta_{ss'} F_1(q^2) v^i$$

$$- \frac{e}{m} \left(\int d^3 \vec{x} \, B_i^{ext.}(x) e^{-i\vec{q}\cdot\vec{x}} \right)$$

$$\times (F_1(q^2) + \kappa F_2(q^2)) \left(\frac{\bar{u}(\vec{p}\,', s') \frac{\Sigma_i}{2} u(\vec{p}, s)}{2m} \right) \},$$

where $\vec{B} = \vec{\nabla} \times \vec{A}$ and $q^2 \approx -\vec{q}\,^2$.

Note:

$$\int \frac{d^3 \vec{q}}{(2\pi)^3} \int d^3 \vec{x} A_0^{ext.}(\vec{x}) e^{-i\vec{q}\cdot\vec{x}} \int d^3 \vec{x}\,' \rho(\vec{x}\,') e^{i\vec{q}\cdot\vec{x}\,'} \equiv \int d^3 \vec{x} A_0^{ext.}(\vec{x}) \rho(\vec{x}),$$

where $\rho(\vec{x})$ is the charge density. Now define

$$F_1(-\vec{q}\,^2) \equiv \int d^3 \vec{x}\,' \rho(\vec{x}\,') e^{i\vec{q}\cdot\vec{x}\,'}.$$

Hence,

$$\rho(\vec{x}) \equiv \int \frac{d^3 \vec{q}}{(2\pi)^3} e^{-i\vec{q}\cdot\vec{x}} F_1(-\vec{q}\,^2). \qquad (15.33)$$

Finally, we see that the total charge, Q, is given by

$$Q \equiv F_1(0) = \int d^3 \vec{x} \rho(\vec{x}). \qquad (15.34)$$

Similarly, we can show that the magnetic-moment distribution is given by

$$\mu(\vec{x}) \equiv \int \frac{d^3 \vec{q}}{(2\pi)^3} e^{-i\vec{q}\cdot\vec{x}} (F_1(q^2) + \kappa F_2(q^2)) \qquad (15.35)$$

and the total magnetic moment of the nucleon is given by $\vec{\mu}_N = \frac{g}{2} \frac{e\hbar}{m_N} \vec{S}$ with

$$\frac{g}{2} \equiv F_1(0) + \kappa F_2(0) \equiv \int d^3 \vec{x} \mu(\vec{x}). \qquad (15.36)$$

Now, consider the meaning of the q^2 dependence of the form factors. Given a Taylor series in q^i, we have

$$F_1(-\vec{q}\,^2) = F_1(0) + \frac{q^i q^j}{2} \frac{\partial^2 F_1}{\partial q^i \partial q^j} |_{\vec{q}=0} + \dots, \qquad (15.37)$$

where the first-order derivative term vanishes, since

$$\frac{\partial F_1}{\partial q^i} |_{\vec{q}=0} = -i \int d^3 \vec{x} \rho(\vec{x}) x^i \equiv 0, \qquad (15.38)$$

by spherical symmetry. The second-derivative term gives

$$\frac{\partial^2 F_1}{\partial q^i \partial q^j}|_{\vec{q}=0} = -\int d^3\vec{x}\, x^i x^j \rho(\vec{x}) \tag{15.39}$$

$$= -\frac{1}{3}\int d^3\vec{x}(\vec{x}^{\,2})\rho(\vec{x})\delta^{ij}$$

$$\equiv -\frac{1}{3}\langle \vec{x}^{\,2}\rangle\delta^{ij},$$

where $\langle \vec{x}^{\,2}\rangle$ is the root-mean-square charge radius of the nucleon. Thus,

$$F_1^p(-\vec{q}^{\,2}) = 1 - \frac{1}{6}\vec{q}^{\,2}\langle \vec{x}^{\,2}\rangle_p + \dots, \tag{15.40}$$

$$F_1^n(-\vec{q}^{\,2}) = -\frac{1}{6}\vec{q}^{\,2}\langle \vec{x}^{\,2}\rangle_n + \dots.$$

More accurately, to order $\frac{\vec{q}^{\,2}}{m^2}$ in the non-relativistic limit (see Sachs, 1962), we can define the electric and magnetic form factors, G_E, G_M, given by

$$G_E(q^2) = F_1(q^2) - \tau\kappa F_2(q^2) \tag{15.41}$$

$$G_M(q^2) = F_1(q^2) + \kappa F_2(q^2).$$

The differential cross-section is then given by

$$\frac{d\sigma}{d\Omega} = \left(\frac{d\sigma}{d\Omega}\right)_{1/2,0}\left(\frac{G_E^2 + \tau G_M^2}{1 + \tau} + 2\tau G_M^2\tan^2\frac{\theta}{2}\right). \tag{15.42}$$

The form factors for both the proton and neutron were obtained in experiments at SLAC (see Bumiller et al., 1961) with elastic electron–proton and electron–deuteron scattering. The neutron form factors were obtained from the second process by subtracting the proton contribution. The results imply that the nucleons are NOT point-like particles. In fact, the nucleons have a charge radius of order 1 fermi, i.e. we have

$$\langle \vec{x}^{\,2}\rangle_p|_{electric} \equiv -6\frac{dG_E^p}{dq^2}|_{q^2=0} = [(0.80\pm0.01)\times10^{-13}\text{ cm}]^2, \tag{15.43}$$

$$\langle \vec{x}^{\,2}\rangle_p|_{magnetic} \equiv -6\frac{dG_M^p}{dq^2}|_{q^2=0} = [(0.80\pm0.02)\times10^{-13}\text{ cm}]^2.$$

In addition, the form factors have a so-called dipole dependence on q^2 with

$$G_E^p(q^2) = \frac{G_M^p(q^2)}{\mu_p} = \frac{G_M^n(q^2)}{\mu_n} \sim \left(\frac{1}{1-\frac{q^2}{0.71}}\right)^2, \tag{15.44}$$

$$G_E^n(q^2) = 0,$$

with q^2 measured in GeV^2. Finally

$$\mu_p \equiv G_M^p(0) = \frac{g_p}{2} = 2.79, \tag{15.45}$$

$$\mu_n \equiv G_M^n(0) = \frac{g_n}{2} = -1.91.$$

Clearly, both the proton and neutron are not point-like particles. Moreover, they have a size of order 1 fermi and they have non-trivial charge and magnetic-moment distributions. It is difficult to consider them as elementary particles.

15.3 Form Factors in the Crossed Channel

The momentum transfer in electron–nucleon scattering, Fig. 15.3, satisfies $q^2 = t \leq 0$. Now consider the crossed process, $e^+e^- \to N\bar{N}$, Fig. 15.4, which is obtained by crossing symmetry, taking $p \to -p$, $k' \to -k'$ and $q = p' - p \to q = (p' + p)$. Now, $q^2 = s > 0$. Thus $e^+e^- \to N\bar{N}$ measures the nucleon form factors in the $q^2 > 0$ region. Moreover, *it must be the same function*. Vector mesons were discovered and they were expected in order to explain the simple behavior of the q^2 dependence of the form factors.

For simplicity, consider first the process $e^+e^- \to \pi^+\pi^-$, i.e. no spinors (Fig. 15.5). The differential cross-section is given by

$$\frac{d\sigma}{d\Omega}(e^+e^- \to \pi^+\pi^-) = \frac{\alpha^2}{s^{5/2}}|\vec{p}_\pi|^3|F_\pi(s)|^2 \sin^2\theta^*, \tag{15.46}$$

with $|\vec{p}_\pi| = \frac{1}{2}(s - 4m_\pi^2)^{1/2}$ and F_π is the pion form factor. The form factor represents the fact that pions are not point particles. They have structure and, phenomenologically, that structure is given in terms of pion resonances which can appear in this two-body channel, as in Fig. 15.6. Each vector meson is strongly interacting and decays, in this case, into two pions. Note the spin, parity and charge-conjugation properties of the vector mesons are the same as that of the photon, i.e. $J^{PC} = 1^{--}$. In Table 15.2 we list the vector mesons which couple to the photon and their mass, widths, isospin and dominant decay modes. We will discuss these states in more detail as we go on.

A phenomenological description of the pion form factor is given in terms of a sum over resonances, as follows (Fig. 15.7). We have

$$F_\pi(q^2) = \sum_V \frac{g_{V\pi\pi}}{2\gamma_V}\left[\frac{m_V^2}{m_V^2 - q^2 - im_V\Gamma_V}\right], \tag{15.47}$$

with

$$\Gamma_V \equiv \frac{2}{3}\frac{g_{V\pi\pi}^2}{4\pi}\frac{|\vec{p}_\pi|^3}{m_V^2}. \tag{15.48}$$

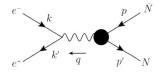

Fig. 15.4   Elastic electron–proton scattering in the crossed channel.

Table 15.2 Strong-interaction vector mesons which couple directly to photons

	Mass (MeV)	Γ (MeV)	Isospin	Decay branching ratios
ρ	770	150	1	$\pi\pi \sim 100\%$
ω	783	10	0	$\pi^+\pi^-\pi^0$ 90%, $\pi^+\pi^-$ 1%, $\pi^0\gamma$ 9%
ϕ	1020	4	0	K^+K^- 47%, $k_L k_S$ 35%, $\pi^+\pi^-\pi^0$ 16%
ρ'	~ 1450	~ 230	1	$4\,\pi, \pi\pi$
J/ψ	3095	0.069	0	Multi-hadronic
ψ'	3684	0.225	0	Multi-hadronic
Υ	9460	0.052	0	Multi-hadronic

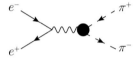

Fig. 15.5 $e^+e^- \to \pi^+\pi^-$ scattering.

Fig. 15.6 $e^+e^- \to \pi^+\pi^-$ scattering pictured in terms of a sum over π–π resonances.

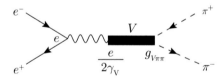

Fig. 15.7 Phenomenological description of $e^+e^- \to \pi^+\pi^-$.

$\Gamma_{\rho\pi\pi} = (150 \pm 23)$ MeV and $m_\rho = (775 \pm 7)$ MeV are measured, for example, in experiments at Orsay (Benaksas et al., 1972), in the process $e^+e^- \to \pi^+\pi^-$ at $q^2 \simeq m_\rho^2$. The phenomenological coupling constant measured is $\frac{g_{\rho\pi\pi}^2}{4\pi} = 2.8 \pm 0.5$. This is clear evidence of strong-interaction physics, i.e. this cannot be used in a perturbation expansion.

In addition, by measuring the differential cross-section, $\frac{d\sigma}{d\Omega}(e^+e^- \to \pi^+\pi^-)$ at $s \simeq m_\rho^2$, they find $\frac{\gamma_\rho^2}{4\pi} = 0.64 \pm 0.1$. This leads to the empirical result

$$2\gamma_\rho \simeq g_{\rho\pi\pi}. \tag{15.49}$$

This is no accident since, if the ρ dominates the form factor, the condition $F_\pi(0) = 1$ (i.e. charge = 1) implies $2\gamma_\rho \simeq g_{\rho\pi\pi}$.

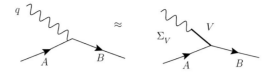

Fig. 15.8   Vector Meson Dominance relation.

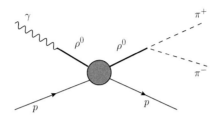

Fig. 15.9 ρ-Meson photo-production.

For the process $e^+e^- \to N\bar{N}$ we can use a similar phenomenological analysis of the form factors. We have

$$F_N(q^2) = \sum_V \frac{g_{VN\bar{N}}}{2\gamma_V} \left[\frac{m_V^2}{m_V^2 - q^2 - im_V\Gamma_V} \right] \simeq \left(\frac{1}{1 - q^2/m_0^2} \right)^2, \qquad (15.50)$$

which is an approximate dipole form.

This brings us to the phenomenological description of photons interacting with strongly interacting particles, known as the Vector Meson Dominance (VMD) model. According to this model we obtain the formula

$$|T(\gamma A \to B)|^2 = |\sum_{V=\rho,\omega,\phi} \frac{e}{2\gamma_V} T(V_{tr} A \to B)|^2, \qquad (15.51)$$

where V_{tr} is the transverse part of the massive vector field and the states A, B are strong-interaction states, i.e. baryons or mesons. Graphically, it is represented by Fig. 15.8. For example, applying VMD, we can obtain the relation

$$\frac{d\sigma}{dt}(\gamma A \to \rho A) = \frac{e^2}{4\gamma_\rho^2} \frac{d\sigma}{dt}(\rho A \to \rho A), \qquad (15.52)$$

which describes the photo-production of vector mesons. An example would be the photo-production of ρ mesons seen in the process $\gamma p \to \rho p \to \pi^+\pi^- p$, Fig. 15.9. One can also measure $\frac{d\sigma}{dt}(\rho p \to \rho p)$ in the process of Fig. 15.10. Photon–vector–meson couplings can be measured. As an example, see Table 15.3.

Table 15.3 $\gamma - V$ couplings measured either in photo-production experiments or $e^+e^- \to \pi^+\pi^-$ (Leith, 1978).

Experiment	$\frac{\gamma_\rho^2}{4\pi}$	$\frac{\gamma_\omega^2}{4\pi}$	$\frac{\gamma_\phi^2}{4\pi}$
Storage rings ($e^+e^- \to \pi^+\pi^-$) at $q^2 = m_V^2$	0.64 ± 0.1	4.6 ± 0.5	2.8 ± 0.2
Photo-production ($\gamma A \to V A$) at $q^2 = 0$	0.61 ± 0.03	9.6 ± 2.1	5.9 ± 2.4

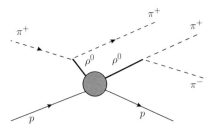

Fig. 15.10 ρ-Meson–proton scattering.

Finally, consider photon–nucleus interactions, $\sigma_T(\gamma A)$, where A is the atomic mass of the nucleus and σ_T is the total cross-section. According to VMD, we have $\sigma_T(\gamma p) = \sigma_T(\gamma n)$, since the photon couples to the nucleon through the vector mesons, and thus we might expect that $\sigma_T(\gamma A) = A\,\sigma_T(\gamma p)$. However, vector mesons have a short mean free path inside the nucleus, since they interact strongly. Thus, a photon will only see a fraction of the nucleus (near the surface): the so-called shadowing effect. Experiments find at 5.5 GeV, $\sigma_T(\gamma A) = A^{0.95\pm0.02}\sigma_T(\gamma p)$, which implies that some shadowing is observed. To summarize, VMD provides rough agreement with the data (for a review, see Schildknecht, 2006).

15.4 Exercises

15.1 Given the amplitude for elastic electron–proton scattering given in Eqn. 15.25, calculate the Rosenbluth differential cross-section in Eqn. 15.27.

15.2 Given the Rosenbluth differential cross-section for elastic electron–proton scattering, find the cross-section for the process $e^+e^- \to p\bar{p}$ using crossing symmetry.

PART IV

DISCRETE SYMMETRIES AND THEIR CONSEQUENCES

16 Charge Conjugation and Parity

Symmetries have played a major role in the development of the Standard Model. So in the next couple of chapters we will focus on both discrete and continuous symmetries. We have already seen that continuous global symmetries provide conserved currents and charges, while continuous local symmetries, as in quantum electrodynamics (QED), give us new forces. Symmetries can either be space–time symmetries or internal symmetries. Poincaré invariance is an example of the former and QED is an example of the latter. Note, Einstein's theory of general relativity can be derived as a consequence of local Poincaré invariance. In the case of QED, the symmetry is unbroken and, as a consequence, the phase of an electron is a non-observable. To summarize, an exact (unbroken) symmetry, whether global or local, implies a non-observable quantity, a symmetry transformation and a conservation law. As another example, absolute spatial direction may be the non-observable, which implies a rotation symmetry and conservation of angular momentum.

On the other hand, a broken symmetry implies that the non-observable becomes observable. Examples of broken symmetries are absolute right-handedness, i.e. $\vec{r} \rightarrow -\vec{r}$, or parity, and absolute sign of the charge, i.e. $e \rightarrow -e$, or charge conjugation of electric charge. In these cases we know that **P**, **C**, **CP** are broken by the weak interactions. For example, absolute charge can be fixed using the following experiment:

$$\frac{\Gamma(k_L \rightarrow e^+ \pi^- \nu)}{\Gamma(k_L \rightarrow e^- \pi^+ \bar{\nu})} = 1.00648 \pm 0.00035, \tag{16.1}$$

and therefore by rate counting one can distinguish between e^+, e^- since **CP** is violated in k_L decays. This might be very useful in any encounter with aliens from another galaxy. They may be made of the same particles which make up us or they may be made of anti-particles. It could get very risky to decide whether or not to shake "hands" when first meeting. If they are made of anti-matter, then at first touch you both would annihilate, which wouldn't be pretty. So before shaking hands, you ask them to do the experiment. If what they call electrons appears with a higher rate in k_L decays, then in fact you can conclude that what they call electrons, you would call positrons, and shaking hands is prohibited.

In this chapter, we focus on defining **C** and **P** symmetry for fermions, and as an aside we define lepton number. QED is defined in terms of a continuous local and internal, $U(1)$, phase symmetry. The Lagrangian density is

$$\mathcal{L} = -\frac{1}{4}F_{\mu\nu}^2 + \bar{\psi}(i\slashed{\partial} - e\slashed{A}Q - m)\psi \tag{16.2}$$

and the action is $S = \int d^4x \mathcal{L}$. As we have seen previously, the global phase symmetry $\psi \to e^{i\alpha Q}\psi$, $A_\mu \to A_\mu$ leads via Noether's theorem to a conserved current $j^\mu = \bar{\psi}\gamma^\mu Q\psi$ and a conserved charge, $L \equiv -\int d^3\vec{x}\, j^0$, which we identify as lepton number for $\psi = \{\psi_e,\ \psi_\mu,\ \psi_\tau\}$. The Lagrangian is invariant under the local symmetry $\psi \to e^{i\alpha(x)Q}\psi$, $A_\mu \to A_\mu - \frac{1}{e}\partial_\mu\alpha$.

16.1 Charge Conjugation

Let's now define charge conjugation. We define an operator $\mathbf{C}$ such that $\mathbf{C}\mathbf{C}^\dagger = \mathbf{C}^\dagger\mathbf{C} = 1$ and the transformation on fermions is given by

$$\mathbf{C}\psi\mathbf{C}^\dagger = \eta_C \psi^C, \tag{16.3}$$

with $|\eta_C| = 1$. We want this transformation to satisfy the condition

$$\mathbf{C}j^\mu\mathbf{C}^\dagger = -j^\mu, \tag{16.4}$$

such that it changes the sign of the charge. Then invariance of the QED interaction term, $ej^\mu A_\mu$, requires that

$$\mathbf{C}A_\mu\mathbf{C}^\dagger = -A_\mu. \tag{16.5}$$

So what is $\mathbf{C}$? It satisfies

$$\mathbf{C}\bar{\psi}\gamma^\mu\psi\mathbf{C}^\dagger = (\mathbf{C}\bar{\psi}\mathbf{C}^\dagger)\gamma^\mu(\mathbf{C}\psi\mathbf{C}^\dagger) = -\bar{\psi}\gamma^\mu\psi. \tag{16.6}$$

Define $\mathbf{C}$ such that

$$\mathbf{C}\psi\mathbf{C}^\dagger = \eta_C\gamma_2\psi^{\dagger T}, \quad \mathbf{C}\psi^\dagger\mathbf{C}^\dagger = -\eta_C^*\psi^T\gamma_2. \tag{16.7}$$

Then it can be seen that

$$\begin{aligned}
\mathbf{C}j^\mu\mathbf{C}^\dagger &= (-\eta_C^*\psi^T\gamma_2\gamma_0)\gamma^\mu(\eta_C\gamma_2\psi^{\dagger T}) \\
&= \psi_\beta(\gamma_2\gamma^{\mu\dagger}\gamma_2\gamma_0)_{\beta\alpha}\psi_\alpha^\dagger \\
&= -\psi_\alpha^\dagger(\gamma_2\gamma^{\mu\dagger}\gamma_2\gamma_0)_{\beta\alpha}\psi_\beta + const. \\
&= -\psi^\dagger(\gamma_2\gamma^{\mu\dagger}\gamma_2\gamma_0)^T\psi + const. \\
&= -\bar{\psi}\gamma^\mu\psi + const.,
\end{aligned}$$

where we've used the identities $(\gamma_2\gamma^{\mu\dagger}\gamma_2)^T = \gamma^\mu$ and $\gamma_0\gamma^\mu\gamma_0 = \gamma^{\mu\dagger}$. Note, $j^\mu \equiv \frac{1}{2}[\bar{\psi}, \gamma^\mu\psi]$ satisfies $\mathbf{C}j^\mu\mathbf{C}^\dagger = -j^\mu$ without having to subtract an infinite constant. In addition, we have $\int d^3\vec{x}\, j^0(x) = \int \frac{d^3\vec{p}}{(2\pi)^3 2E_p}(b^\dagger b - d^\dagger d) + const.$. We can then define the normal ordered current $j^\mu \equiv\, :\bar{\psi}\gamma^\mu\psi:$ where we explicitly subtract the infinite constant. The normal ordered current also satisfies $\mathbf{C}j^\mu\mathbf{C}^\dagger = -j^\mu$ without any additional subtraction.

What does the operator $\mathbf{C}$ do on states? We have

$$\mathbf{C}\psi(x)\mathbf{C}^\dagger = \eta_c \psi^C(x), \tag{16.8}$$

$$\text{LHS} = \int \frac{d^3\vec{p}}{(2\pi)^3 2E_p} \sum_s [\mathbf{C}b(\vec{p},s)\mathbf{C}^\dagger u(\vec{p},s)e^{-ip\cdot x} + \mathbf{C}d^\dagger(\vec{p},s)\mathbf{C}^\dagger v(\vec{p},s)e^{ip\cdot x}],$$

$$\text{RHS} = \eta_c \int \frac{d^3\vec{p}}{(2\pi)^3 2E_p} \sum_s [b^\dagger(\vec{p},s)\gamma_2 u^*(\vec{p},s)e^{ip\cdot x} + d(\vec{p},s)\gamma_2 v^*(\vec{p},s)e^{-ip\cdot x}].$$

Recall, $\gamma_2 u^*(\vec{p},s) = v(\vec{p},s)$ and $\gamma_2 v^*(\vec{p},s) = u(\vec{p},s)$. Plugging this into Eqn. 16.8 we find

$$\mathbf{C}d^\dagger(\vec{p},s)\mathbf{C}^\dagger = \eta_C b^\dagger(\vec{p},s), \tag{16.9}$$

$$\mathbf{C}b^\dagger(\vec{p},s)\mathbf{C}^\dagger = \eta_C^* d^\dagger(\vec{p},s).$$

Therefore, we find that, as expected, particles go into anti-particles, and vice versa, under charge conjugation.

Now that we have defined the charge conjugation transformation, such that the interaction term is invariant, we need to check that the free Lagrangian is also invariant. Let's first check the kinetic term for fermions. We have

$$\begin{aligned}
\mathbf{C}\bar{\psi}i\slashed{\partial}\psi\mathbf{C}^\dagger &= \mathbf{C}\psi^\dagger\mathbf{C}^\dagger\gamma_0 i\slashed{\partial}\mathbf{C}\psi\mathbf{C}^\dagger \\
&= -\psi^T\gamma_2\gamma_0 i\slashed{\partial}\gamma_2\psi^{\dagger T} \\
&= -\psi_\alpha^\dagger(\gamma_2 i\overleftarrow{\slashed{\partial}}^\dagger\gamma_2\gamma_0)_{\beta\alpha}\psi_\beta \\
&= -\psi^\dagger(\gamma_2 i\overleftarrow{\slashed{\partial}}^\dagger\gamma_2\gamma_0)^T\psi \\
&= -\bar{\psi}i\overleftarrow{\slashed{\partial}}\psi.
\end{aligned}$$

But the action

$$S = \int d^4x\,\bar{\psi}i\slashed{\partial}\psi \rightarrow -\int d^4x\,\bar{\psi}i\overleftarrow{\slashed{\partial}}\psi = \int d^4x\,\bar{\psi}i\slashed{\partial}\psi + (\text{surface terms} = 0).$$

The mass term is also invariant. Therefore, the action is invariant. Note the Lagrangian density $\mathcal{L} \supset \bar{\psi}(\frac{1}{2}i\overleftrightarrow{\slashed{\partial}} - m)\psi$ is invariant under charge conjugation without having to consider surface terms. It is also Hermitian. The gauge kinetic term is also invariant under charge conjugation. Finally, the Lagrangian density $\mathcal{L} = \bar{\psi}(\frac{1}{2}i\overleftrightarrow{\slashed{\partial}} - m)\psi - \frac{e}{2}A_\mu[\bar{\psi},\gamma^\mu\psi] - \frac{1}{4}F_{\mu\nu}^2$ also satisfies $\mathbf{C}\mathcal{L}\mathbf{C}^\dagger \equiv \mathcal{L}$.

16.2 Parity

We define the parity transformation on fermion fields by

$$\mathbf{P}\psi(\vec{x},t)\mathbf{P}^\dagger = \eta_P\gamma_0\psi(\vec{x}\,',t'), \tag{16.10}$$

where $\vec{x}\,' = -\vec{x}$, $t' = t$ and $|\eta_P| = 1$.

$$\text{LHS} = \int \frac{d^3\vec{p}}{(2\pi)^3 2E_p} \sum_s [\mathbf{P}b(\vec{p}, s)\mathbf{P}^\dagger u(\vec{p}, s)e^{-ip\cdot x} + \mathbf{P}d^\dagger(\vec{p}, s)\mathbf{P}^\dagger v(\vec{p}, s)e^{ip\cdot x}],$$

$$\text{RHS} = \eta_P \int \frac{d^3\vec{p}}{(2\pi)^3 2E_p} \sum_s [b(\vec{p}, s)\gamma_0 u(\vec{p}, s)e^{-ip\cdot x'} + d^\dagger(\vec{p}, s)\gamma_0 v(\vec{p}, s)e^{ip\cdot x'}]$$

$$= \eta_P \int \frac{d^3\vec{p}}{(2\pi)^3 2E_p} \sum_s [b(-\vec{p}, -s)\gamma_0 u(-\vec{p}, -s)e^{-ip\cdot x} + d^\dagger(-\vec{p}, -s)$$

$$\gamma_0 v(-\vec{p}, -s)e^{ip\cdot x}]$$

$$= \eta_P \int \frac{d^3\vec{p}}{(2\pi)^3 2E_p} \sum_s [b(-\vec{p}, -s)u(\vec{p}, s)e^{-ip\cdot x} - d^\dagger(-\vec{p}, -s)v(\vec{p}, s)e^{ip\cdot x}]$$

where in the next-to-last equation we changed variables $\vec{p} \to -\vec{p}$ and in the last equation we use the relation

$$\gamma_0 u(\vec{p}, s) = u(-\vec{p}, -s),$$
$$\gamma_0 v(\vec{p}, s) = -v(-\vec{p}, -s).$$

Hence, we find

$$\mathbf{P}b^\dagger(\vec{p}, s)\mathbf{P}^\dagger = \eta_P^* b^\dagger(-\vec{p}, -s) \tag{16.11}$$
$$\mathbf{P}d^\dagger(\vec{p}, s)\mathbf{P}^\dagger = -\eta_P d^\dagger(-\vec{p}, -s).$$

Except for the minus sign in the second line, this is what we expect for a parity operation.

Let's now check the invariance of QED under parity. We have

$$\mathbf{P}\bar{\psi}(i\slashed{\partial} - e\slashed{A} - m)\psi\mathbf{P}^\dagger = \mathbf{P}\bar{\psi}\mathbf{P}^\dagger(i\slashed{\partial} - e\mathbf{P}\slashed{A}\mathbf{P}^\dagger - m)\mathbf{P}\psi\mathbf{P}^\dagger$$

$$= \psi^\dagger(x')(i\slashed{\partial} - e\mathbf{P}\slashed{A}\mathbf{P}^\dagger - m)\gamma_0\psi(x')$$

$$= \bar{\psi}(x')(i\gamma^{\mu\dagger}\partial_\mu - e\gamma^{\mu\dagger}\mathbf{P}A_\mu(x)\mathbf{P}^\dagger - m)\psi(x').$$

But $\gamma^{0\dagger} = \gamma^0$ and $\gamma^{i\dagger} = -\gamma^i$, hence $\gamma^{\mu\dagger}\partial_\mu \equiv \gamma^\mu\partial'_\mu$. Let

$$\mathbf{P}A_\mu(x)\mathbf{P}^\dagger \equiv (\mathbf{P}A(x'))_\mu, \tag{16.12}$$

where $(\mathbf{P}x)_\mu \equiv x'_\mu$. Then we have $\gamma^{\mu\dagger}\mathbf{P}A_\mu(x)\mathbf{P}^\dagger = \gamma^{\mu\dagger}(\mathbf{P}A(x'))_\mu \equiv \gamma^\mu A_\mu(x')$. Finally, we have

$$\mathbf{P}\bar{\psi}(x)(i\slashed{\partial} - e\slashed{A}(x) - m)\psi(x)\mathbf{P}^\dagger = \bar{\psi}(x')(i\slashed{\partial}' - e\slashed{A}(x') - m)\psi(x'), \tag{16.13}$$

and the action is invariant, since $x \to x'$ is a change of variables.

16.3 Transformation of the Photon in the Coulomb Gauge

The gauge field transforms as follows:

$$\mathbf{P}A_\mu(x)\mathbf{P}^\dagger = (\mathbf{P}A(x'))_\mu \tag{16.14}$$
$$\mathbf{C}A_\mu(x)\mathbf{C}^\dagger = -A_\mu(x). \tag{16.15}$$

The gauge-field Lagrangian density is invariant under charge conjugation and satisfies $\mathcal{L}(x) = \mathcal{L}(x')$ under parity. Thus, the gauge action is invariant.

In the Coulomb gauge we have

$$A_i(x) = \int \frac{d^3\vec{p}}{(2\pi)^3 2E_p} \sum_{\lambda=\pm 1} [a(\vec{p},\lambda)\epsilon_i(\vec{p},\lambda)e^{-ip\cdot x} + h.c.]$$

with $\vec{p} \cdot \vec{\epsilon}(\vec{p},\lambda) \equiv 0$ and we can choose a phase convention, such that $\epsilon_i(-\vec{p},-\lambda) = \epsilon_i(\vec{p},\lambda)$. Then we can see that the creation operator transforms as follows:

$$\mathbf{P}a^\dagger(\vec{p},\lambda)\mathbf{P}^\dagger = -a^\dagger(-\vec{p},-\lambda) \tag{16.16}$$
$$\mathbf{C}a^\dagger(\vec{p},\lambda)\mathbf{C}^\dagger = -a^\dagger(\vec{p},\lambda).$$

16.4 Applications of **C** and **P** Invariance

Discrete symmetries lead to selection rules. In this section we discuss several of these.

1. Furry's theorem: An even number of photons cannot scatter into an odd number of photons, or vice versa.

Proof   Consider a state of n photons, i.e. $|n\gamma\rangle = \Pi_{i=1}^{n}a^\dagger(\vec{p}_i,\lambda_i)|0\rangle$. Then $\mathbf{C}|n\gamma\rangle = (-1)^n|n\gamma\rangle$. Since the action of QED is invariant, the Hamiltonian, the unitary evolution operator and consequently the S matrix are also invariant, i.e. $\mathbf{C}S\mathbf{C}^\dagger = S$. Thus,

$$\langle n'\gamma|S|n\gamma\rangle = \langle n'\gamma|\mathbf{C}S\mathbf{C}^\dagger|n\gamma\rangle$$
$$= (-1)^{n+n'}\langle n'\gamma|S|n\gamma\rangle$$

or

$$\langle n'\gamma|S|n\gamma\rangle = 0 \ \ \text{if} \ \ n + n' \ \ \text{is odd.} \tag{16.17}$$

This selection rule is valid to all orders in the electromagnetic interactions.   □

Note, if the strong interactions are also invariant under **C**, i.e. $\mathbf{C}H_{st}\mathbf{C}^\dagger = H_{strong}$, then the selection rule is valid to all orders in H_{strong} as well.

2. The π^0 is an eigenstate of charge conjugation with $\eta_C = +1$.

Proof   Consider the decay $\pi^0 \to 2\gamma$. The π^0 is a strong-interaction eigenstate and the 2 γ decay is its dominant decay mode. This implies that

$$\mathbf{C}|\pi^0\rangle = |\pi^0\rangle. \tag{16.18}$$

If the π^0 decay is solely due to electromagnetic processes and assuming the strong interactions are invariant under charge conjugation, then the decay $\pi^0 \to$ an odd number of photons should be forbidden. Experimentally, the branching ratio $B(\pi^0 \rightarrow 3\gamma) < 5 \times 10^{-6}$. This is consistent with the invariance under **C** of both the strong and electromagnetic interactions. Note, we shall show much later that in quantum chromodynamics (QCD), **C** invariance can be proven just as in QED. $\qquad\qquad\qquad\qquad\qquad\qquad\qquad\qquad\qquad\qquad\qquad\qquad\quad$ □

3. Positronium states (e^+e^- bound states) can be labeled by $^{2S+1}L_J$ where L is the orbital angular momentum, J is the total angular momentum and $S = 0, 1$ is the spin of the state. We want to show that their charge-conjugation (C) and parity (P) quantum numbers depend solely on L and S, i.e. $P = -(-1)^L$ and $C = (-1)^{L+S}$.

Proof QED is C and P invariant, thus positronium states have well-defined C, P quantum numbers. For small $\alpha = \frac{e^2}{4\pi} \sim \frac{1}{137}$, a non-relativistic description is valid since excitation energies, $\Delta E \sim m_e v^2$, satisfy $\frac{\Delta E}{m_e} \sim v^2 \sim \alpha^2 \ll 1$. In this limit the effects of spin–orbital couplings and the presence of virtual photons are negligible. The positronium-state vector is then given by

$$|^{2S+1}L_J\rangle = \sum_{m,\sigma_z,\sigma_z'} \int d^3\vec{p}\, Y_{LM}(\hat{p})\chi_m^S(\sigma_z,\sigma_z')C_{Mm}(p)b^\dagger(\vec{p},\sigma_z)d^\dagger(-\vec{p},\sigma_z')|0\rangle,$$

$$(16.19)$$

where

(1) $\sigma_z, \sigma_z' = \pm\frac{1}{2}$ are spin components in the z direction;

(2) $\chi_m^S(\sigma_z,\sigma_z') = (-1)^{S+1}\chi_m^S(\sigma_z',\sigma_z)$ is the spin wave function. For $S = 1$, we have the triplet state given by

$$\chi_1^1 = (\uparrow_-\uparrow_+) \quad m = 1,$$
$$\chi_0^1 = (\uparrow_-\downarrow_+ + \downarrow_-\uparrow_+)/\sqrt{2} \quad m = 0,$$
$$\chi_{-1}^1 = (\downarrow_-\downarrow_+) \quad m = -1,$$

and for $S = 0$ we have

$$\chi_0^0 = (\uparrow_-\downarrow_+ - \downarrow_-\uparrow_+)/\sqrt{2} \quad m = 0;$$

(3) $Y_{LM}(-\hat{p}) = (-1)^L Y_{Lm}(\hat{p})$ are spherical harmonics for the spatial wave functions;

(4) $J_z = M + m$;

(5) $C_{Mm}(p)$ is the appropriate Clebsch–Gordan coefficient $\begin{pmatrix} J & L & S \\ J_z & M & m \end{pmatrix}$, multiplied by the radial wave function, which is a function of $p = |\vec{p}|$.

Now evaluate the action of **P** and **C** on these states. First, we have

$$\mathbf{P}|^{2S+1}L_J\rangle = \ldots\ldots \mathbf{P}b^\dagger(\vec{p},\sigma_z)d^\dagger(-\vec{p},\sigma_z')|0\rangle$$

$$= \ldots\ldots -b^\dagger(-\vec{p},\sigma_z)d^\dagger(\vec{p},\sigma_z')|0\rangle$$

$$= -\sum_{m,\sigma_z,\sigma_z'}\int d^3\vec{p}\, Y_{LM}(\hat{p})\chi_m^S(\sigma_z,\sigma_z')C_{Mm}(p)b^\dagger(-\vec{p},\sigma_z)d^\dagger(\vec{p},\sigma_z')|0\rangle$$

$$= -\sum_{m,\sigma_z,\sigma_z'}\int d^3\vec{p}\, Y_{LM}(-\hat{p})\chi_m^S(\sigma_z,\sigma_z')C_{Mm}(p)b^\dagger(\vec{p},\sigma_z)d^\dagger(-\vec{p},\sigma_z')|0\rangle$$

$$\equiv -(-1)^L|^{2S+1}L_J\rangle.$$

To summarize the parity eigenvalue, factor $(-1)^L$ comes from $\hat{p} \to -\hat{p}$ and -1 comes from the relative parity of e^+, e^-.

Now evaluate the action of **C** on these states. We have

$$\mathbf{C}|^{2S+1}L_J\rangle = \ldots\ldots \mathbf{C}b^\dagger(\vec{p},\sigma_z)d^\dagger(-\vec{p},\sigma_z')|0\rangle$$

$$= \ldots\ldots d^\dagger(\vec{p},\sigma_z)b^\dagger(-\vec{p},\sigma_z')|0\rangle$$

$$= \ldots\ldots -b^\dagger(-\vec{p},\sigma_z')d^\dagger(\vec{p},\sigma_z)|0\rangle$$

$$= -\sum_{m,\sigma_z,\sigma_z'}\int d^3\vec{p}\, Y_{LM}(\hat{p})\chi_m^S(\sigma_z,\sigma_z')C_{Mm}(p)b^\dagger(-\vec{p},\sigma_z')d^\dagger(\vec{p},\sigma_z)|0\rangle$$

$$= -\sum_{m,\sigma_z,\sigma_z'}\int d^3\vec{p}\, Y_{LM}(-\hat{p})\chi_m^S(\sigma_z',\sigma_z)C_{Mm}(p)b^\dagger(\vec{p},\sigma_z)d^\dagger(-\vec{p},\sigma_z')|0\rangle$$

$$\equiv (-1)^{L+S}|^{2S+1}L_J\rangle.$$

To summarize the charge-conjugation eigenvalue, under the transformation $e^+ \leftrightarrow e^-$, the factor $(-1)^L$ comes from $\hat{p} \to -\hat{p}$ and $(-1)^{S+1}$ comes from $\sigma_z \leftrightarrow \sigma_z'$, and (-1) comes from fermi statistics.   □

These results lead to selection rules. In particular, **C** invariance requires that

- if $L+S$ even, then the state cannot decay into an odd number of photons
- if $L+S$ odd, then the state cannot decay into an even number of photons.

These decays are forbidden to leading order in α which is consistent with the data.

(4) Consider the decay of a spin 0 particle to two photons. Since parity is conserved in QED, we show that for a scalar particle, ($P = +1$), the polarization directions of the 2γs are parallel; and for a pseudo-scalar particle, ($P = -1$), the polarization directions of the 2γs are perpendicular.[1]

Proof A two-photon state, in the center of momentum system (CMS), can be written (in the Coulomb gauge) as

$$|2\gamma\rangle = \int d^3\vec{p}\, \chi_{ij}(\vec{p})a_i^\dagger(\vec{p})\, a_j^\dagger(-\vec{p})|0\rangle,$$

where $a_i^\dagger(\vec{p}) \equiv \sum_{\lambda=\pm 1}a^\dagger(\vec{p},\lambda)\epsilon_i(\vec{p},\lambda)$ and $\vec{p}\cdot\vec{a}^\dagger(\vec{p}) \equiv 0$.

---

[1] This result was originally found by Yang (1950).

To conserve angular momentum in the decay, the 2γ state has $J = 0$. Thus, $\chi_{ij}(\vec{p})$ must transform as a second-rank tensor under rotations. In general, we have

$$\chi_{ij}(\vec{p}) = A\delta_{ij} + B\epsilon_{ijk}p_k + Cp_ip_j, \tag{16.20}$$

where A, B, C are functions of $p = |\vec{p}|$. Note, C does not contribute to the state $|2\gamma\rangle$, since the creation operators are annihilated by $\vec{p}$. Thus,

$$\chi_{ij}(\vec{p}) = A\delta_{ij} + B\epsilon_{ijk}p_k. \tag{16.21}$$

Under parity, we have $\mathbf{P}a^\dagger(\vec{p}, \lambda)\mathbf{P}^\dagger = -a^\dagger(-\vec{p}, -\lambda)$ and $\epsilon_i(\vec{p}, \lambda) = \epsilon_i(-\vec{p}, -\lambda)$. Hence, $\mathbf{P}a_i^\dagger(\vec{p})\mathbf{P}^\dagger = -a_i^\dagger(-\vec{p})$. Therefore,

$$\begin{aligned}
\mathbf{P}|2\gamma\rangle &= \int d^3\vec{p}\,\chi_{ij}(\vec{p})\mathbf{P}a_i^\dagger(\vec{p})a_j^\dagger(-\vec{p})|0\rangle \\
&= \int d^3\vec{p}\,\chi_{ij}(\vec{p})a_i^\dagger(-\vec{p})a_j^\dagger(\vec{p})|0\rangle \\
&= \int d^3\vec{p}\,\chi_{ij}(-\vec{p})a_i^\dagger(\vec{p})a_j^\dagger(-\vec{p})|0\rangle \\
&\equiv \pm|2\gamma\rangle
\end{aligned}$$

for $\chi_{ij}(-\vec{p}) = \pm\chi_{ij}(\vec{p})$.

Hence, we have

$$\chi_{ij}(\vec{p}) = \begin{cases} A\,\delta_{ij} & P = +1, \\ B\,\epsilon_{ijk}p_k & P = -1. \end{cases} \tag{16.22}$$

Therefore, for $P = +1$ the polarization vectors of the 2γs are parallel, and for $P = -1$, they are perpendicular. $\qquad\square$

As an example, consider the decay, $\pi^0 \to 2\gamma$. Assume, for now, that $J_{\pi^0} = 0$ (which we will determine later), then use the polarizations of the 2γs to determine the parity of the π^0. Actually, we use the decay $\pi^0 \to e^+e^-e^+e^-$ as in Fig. 16.1. Note, the plane of the e^+e^- pairs contains the polarization vector of the off-shell photon. This is due to the coupling $a_i^\dagger j_i^{EM}$ and chirality ($\approx$ helicity) conservation (see Fig. 16.2).

The theoretical calculation of the orientation of the two planes, $(e^+e^-)_a$ and $(e^+e^-)_b$ (Kroll and Wada, 1955), is given by the distribution

$$W(\phi) = 1 \pm 0.18\cos(2\phi), \tag{16.23}$$

where ϕ is the angle between the two planes and the $(+, -)$ signs correspond to (scalar, pseudo-scalar), respectively.

The process was measured experimentally in 1962 by Samios et al. (1962). They parameterized the result by

Fig. 16.1 π^0 decay to two e^+e^- pairs.

Fig. 16.2 Virtual γ decay to e^+e^- pair. The double arrow for the virtual photon is the polarization direction (or the electric-field direction of the photon) while the double arrows for the electron and positron are their helicity (or chirality) with the momentum in the direction of the current.

$$W(\phi) = 1 + \alpha\cos(2\phi) \tag{16.24}$$

and found $\alpha = -0.12 \pm 0.15$, which is consistent with π^0 being a pseudo-scalar and inconsistent with π^0 being a scalar by 2σ. With a slightly different evaluation of the data the authors found a 3.6σ evidence for π^0 being pseudo-scalar. There are several other experimental measurements using different processes which confirm that the parity of a π^0 is $P_{\pi^0} = -1$.

It is now instructive to consider a different theoretical approach for studying the parity of a π^0. It is an effective Lagrangian approach which is especially useful when studying processes containing strongly interacting particles. The coupling of a π^0 to two photons can be described by one of two different effective Lagrangians, depending on whether $P_{\pi^0} = \pm 1$. For $P_{\pi^0} = +1$ we can use the effective Lagrangian

$$\mathcal{L}_{eff} = \frac{g}{2}\phi F_{\mu\nu}F^{\mu\nu}, \tag{16.25}$$

where g is a phenomenological coupling constant and $\phi = \phi^\dagger$ is the π^0 field. For $P_{\pi^0} = -1$ we can use

$$\mathcal{L}_{eff} = \frac{g}{2}\phi F_{\mu\nu}F_{\rho\sigma}\epsilon^{\mu\nu\rho\sigma}. \tag{16.26}$$

Under parity, it was shown for homework that $\mathbf{P}F_{\mu\nu}(x)^2\mathbf{P}^\dagger = F_{\mu\nu}(x')^2$ and $\mathbf{P}F_{\mu\nu}(x)^*F^{\mu\nu}(x)\mathbf{P}^\dagger = -F_{\mu\nu}(x')^*F^{\mu\nu}(x')$ where $^*F^{\mu\nu} \equiv \frac{1}{2}\epsilon^{\mu\nu\rho\sigma}F_{\rho\sigma}$. Therefore, in order for either Lagrangian to be invariant under parity, it must be so that $\mathbf{P}\phi(x)\mathbf{P}^\dagger = \eta_{\pi^0}\phi(x')$ with $\eta_{\pi^0} \equiv P_{\pi^0} = \pm 1$. Then, since $E^i = \partial^i A^0 - \dot{A}^i$ and $B^i = \epsilon_{ijk}\partial_j A^k$, we see that the photon polarization vector is parallel to $\vec{E}$ and perpendicular to $\vec{B}$. Finally, $F_{\mu\nu}^2 = 2(\vec{E}^2 - \vec{B}^2)$ and $F_{\mu\nu}{}^*F^{\mu\nu} = 2\vec{E}\cdot\vec{B}$. Thus, for $P_{\pi^0} = +1$ the polarization vectors of the two photons are parallel and for $P_{\pi^0} = -1$, they are perpendicular. This agrees with our previous result.

Now recall that $C_{\pi^0} = +1$, since π^0 decays to two photons. Again, we have assumed that the π^0 has spin, $J = 0$. We will show how this is obtained later. Thus, in summary, we have for a π^0: $J^{PC} = 0^{-+}$.

5. Show that a spin 1 particle cannot decay into two photons, i.e. Landau–Yang theorem (Landau, 1948; Yang, 1950).

Proof We define the two-photon state in the CMS. We have

$$|2\gamma\rangle_i = \int d^3\vec{p}\,\chi_{ijk}(\vec{p})a_j^\dagger(\vec{p})a_k^\dagger(-\vec{p})|0\rangle, \tag{16.27}$$

where $\chi_{ijk}(\vec{p})$ is a rank three tensor under rotations. In general, we have

$$\chi_{ijk}(\vec{p}) = A\epsilon_{ijk} + Bp_i\delta_{jk} + Cp_j\delta_{ik} + Dp_k\delta_{ij} + B'p_i\epsilon_{jkl}p_l$$
$$+ C'p_j\epsilon_{ikl}p_l + D'p_k\epsilon_{ijl}p_l + Ep_ip_jp_k.$$

The quantities $A, B, \ldots, D', E$ are functions of $p = |\vec{p}|$. Since $\vec{p} \cdot \vec{a}^\dagger(\vec{p}) = 0$, the terms with C, D, C', D', E do not contribute to the state vector. Thus, we have

$$\chi_{ijk}(\vec{p}) = A\epsilon_{ijk} + Bp_i\delta_{jk} + B'p_i\epsilon_{jkl}p_l. \tag{16.28}$$

Note, the wave function satisfies the relation

$$\chi_{ijk}(\vec{p}) \equiv -\chi_{ikj}(-\vec{p}). \tag{16.29}$$

Given this property and Bose statistics we have

$$|2\gamma\rangle_i = \int d^3\vec{p}\,\chi_{ijk}(\vec{p})a_j^\dagger(\vec{p})a_k^\dagger(-\vec{p})|0\rangle$$
$$= \int d^3\vec{p}\,\chi_{ijk}(\vec{p})a_k^\dagger(-\vec{p})a_j^\dagger(\vec{p})|0\rangle$$
$$= \int d^3\vec{p}\,\chi_{ikj}(-\vec{p})a_j^\dagger(\vec{p})a_k^\dagger(-\vec{p})|0\rangle,$$

where the last equality follows from the change of variables, $\vec{p} \to -\vec{p}$ and $j \leftrightarrow k$.

Averaging the results of the first and last step, we find

$$|2\gamma\rangle_i \equiv \frac{1}{2}\int d^3\vec{p}\,[\chi_{ijk}(\vec{p}) + \chi_{ikj}(-\vec{p})]a_j^\dagger(\vec{p})a_k^\dagger(-\vec{p})|0\rangle \equiv 0. \tag{16.30}$$

Thus 2γs cannot be in a state with total angular momentum $J = 1$. $\qquad\square$

Consider the decay $\pi^0 \to 2\gamma$, we conclude that $J_{\pi^0} \neq 1$. Later we will show that $J_{\pi^0} \neq 2$, hence $J_{\pi^0} = 0$.

Let's now use these results in order to learn some properties of positronium. The positronium state ${}^{2S+1}L_{J=1}$ cannot decay to two photons but it can decay into three photons, where $\mathbf{C}_{3\gamma} = -1$. Now, the positronium ground state decays into two photons, hence $\mathbf{C} = +1 = (-1)^{L+S}$. A ground-state wave function is typically symmetric in space; hence, $L = 0$ and thus $S = 0$. So the ground state is ${}^1S_0 \to 2\gamma$. On the other hand, the spin-flipped excitation decays via ${}^3S_1 \to 3\gamma$. As a result it turns out that the decay rates satisfy $\Gamma_{{}^1S_0} \gg \Gamma_{{}^3S_1}$, since the latter rate is suppressed by phase space as well as an extra factor of the fine-structure constant, α.

16.5 Exercises

16.1 Calculate the transformation of the following operators under $\mathbf{C}$ and $\mathbf{P}$ using

$$\mathbf{C}A_\mu(x)\mathbf{C}^\dagger = -A_\mu(x), \qquad \mathbf{P}A_\mu(x)\mathbf{P}^\dagger = PA_\mu(x'), \tag{16.31}$$
$$\mathbf{C}\psi(x)\mathbf{C}^\dagger = \eta_c\gamma_2\psi^\dagger(x), \qquad \mathbf{P}\psi(x)\mathbf{P}^\dagger = \eta_p\gamma^0\psi(x'),$$

where P is an operation on four vectors defined by $x' = (-\vec{x}, t) \equiv \mathrm{P}x$.

$$\bar{\psi}\Gamma\psi, \tag{16.32}$$

with

$$\Gamma = \{\mathbb{I},\ \gamma_5,\ \gamma_\mu\gamma_5,\ \gamma_\mu,\ \Sigma_{\mu\nu},\ \gamma_5\Sigma_{\mu\nu}\} \tag{16.33}$$

and

$$F_{\mu\nu}^2 = F_{\mu\nu}F_{\alpha\beta}g^{\mu\alpha}g^{\nu\beta},\ F^*F = \frac{1}{2}F_{\mu\nu}F_{\alpha\beta}\epsilon^{\mu\nu\alpha\beta}. \tag{16.34}$$

16.2 The magnetic-dipole interaction is given by

$$F_{\mu\nu}\bar{\psi}\Sigma^{\mu\nu}\psi. \tag{16.35}$$

Show that this interaction term is invariant under **C** and **P**. Determine whether the electric-dipole interaction given by

$$F_{\mu\nu}\bar{\psi}\gamma_5\Sigma^{\mu\nu}\psi \tag{16.36}$$

is invariant under **C**, **P** or **CP**.

16.3 Given

$$\mathbf{C}b^\dagger(\vec{p}, s)\mathbf{C}^\dagger = \eta_c^* d^\dagger(\vec{p}, s) \tag{16.37}$$
$$\mathbf{C}d^\dagger(\vec{p}, s)\mathbf{C}^\dagger = \eta_c b^\dagger(\vec{p}, s),$$

and letting $\eta_c = 1$, show that

$$\mathbf{C} \equiv \exp\left[i\pi \sum_s b^\dagger(\vec{p}, s)b(\vec{p}, s)\right] \times \exp\left[\frac{\pi}{2} \sum_s \left(d^\dagger(\vec{p}, s)b(\vec{p}, s) - b^\dagger(\vec{p}, s)d(\vec{p}, s)\right)\right]. \tag{16.38}$$

Hint: define the transformation

$$M_\theta \equiv \exp\left[\theta \sum_s \left(d^\dagger(\vec{p}, s)b(\vec{p}, s) - b^\dagger(\vec{p}, s)d(\vec{p}, s)\right)\right], \tag{16.39}$$

where θ is a continuous real parameter. Show via differentiation with respect to θ that

$$M_\theta b^\dagger(\vec{p}, s)M_\theta^\dagger = \cos\theta b^\dagger(\vec{p}, s) + \sin\theta d^\dagger(\vec{p}, s), \tag{16.40}$$
$$M_\theta d^\dagger(\vec{p}, s)M_\theta^\dagger = -\sin\theta b^\dagger(\vec{p}, s) + \cos\theta d^\dagger(\vec{p}, s).$$

Note also that in the expressions above the integral over momentum is implicit, i.e.

$$\sum_s \equiv \sum_s \int \frac{d^3\vec{p}}{(2\pi)^3 2E_p}. \tag{16.41}$$

Time-Reversal Invariance

In this chapter we discuss time-reversal symmetry and some of its experimental consequences. In particular, we derive the reciprocity relation (or detailed balance) and discuss how it was used to obtain the spin of the pion. Since the $\pi^0 \to 2\gamma$, we know from before that $J_{\pi^0} \neq 1$. Hence, $J_{\pi^0} = 0$ or 2.

17.1 Time-Reversal Invariance of QED

First, recall the main results from Chapter 8, Section 8.2 in Chapter 8, Eqns. 8.22, 8.23, 8.25, 8.32. The bottom line is that time reversal takes incoming states to outgoing states and vice versa. In addition, we have

$$\langle f | AB | i \rangle \equiv \langle f | T^\dagger T A B T^\dagger T | i \rangle \tag{17.1}$$
$$\equiv \langle i_T | T B T^\dagger T A T^\dagger | f_T \rangle.$$

Time-reversal invariance implies that $T H T^\dagger = H$. We also saw that $T^2 \phi T^{\dagger 2} = \phi$.

Now consider quantum electrodynamics (QED). The Lagrangian density is given by

$$\mathcal{L} = -\frac{1}{4} F_{\mu\nu}^2 + \bar{\psi}(i\slashed{\partial} - m)\psi - e A_\mu \bar{\psi}\gamma^\mu Q \psi.$$

We need to define a transformation that leaves the kinetic term invariant and takes $t \to -t$. Let

$$T \psi(\vec{x}, t) T^\dagger = A \psi^*(\vec{x}, -t) \equiv \psi^\dagger(\vec{x}, -t) A^T \tag{17.2}$$

and choose the matrix A such that the kinetic term is invariant. Then we have

$$T \psi^\dagger(\vec{x}, t) T^\dagger = A^* \psi(\vec{x}, -t)$$

and

$$T \bar{\psi}(\vec{x}, t) T^\dagger = \gamma_0 A^* \psi(\vec{x}, -t), \tag{17.3}$$

where we used $\gamma_0^T = \gamma_0$.

Now calculate the transformation of the kinetic term. We have

$$T \bar{\psi}(\vec{x}, t)(i\slashed{\partial} - m)\psi(\vec{x}, t) T^\dagger \equiv T \psi(\vec{x}, t) T^\dagger (i \overleftarrow{\slashed{\partial}} - m)^T T \bar{\psi}(\vec{x}, t) T^\dagger$$
$$= \bar{\psi}(\vec{x}, -t)\gamma_0 A^T (-i \overrightarrow{\slashed{\partial}} - m)^T \gamma_0 A^* \psi(\vec{x}, -t)$$
$$= \bar{\psi}(\vec{x}, t')(i\slashed{\partial}' - m)\psi(\vec{x}, t'),$$

where $t' = -t$ and in the next-to-last line we performed an integration by parts which is valid in the action. The last equation provides the conditions that the matrix A must satisfy.

We have $\delta S = 0$ if and only if

$$\gamma_0 A^T \gamma_0 A^* = 1 \quad \text{or} \quad A^\dagger \gamma_0 A = \gamma_0,$$
$$\gamma_0 A^T \gamma_0^2 A^* = \gamma_0 \quad \text{or} \quad A^\dagger A = 1,$$
$$-\gamma_0 A^T \gamma_i^T \gamma_0 A^* = \gamma_i \quad \text{or} \quad A^\dagger \alpha_i A = -\alpha_i^T.$$

Recall, $\alpha_i = \gamma_0 \gamma_i$, $\alpha_i^\dagger = \alpha_i$. The solution to these equations for the matrix A is given by

$$A = \Sigma_2 \equiv \begin{pmatrix} \sigma_2 & 0 \\ 0 & \sigma_2 \end{pmatrix}. \tag{17.4}$$

Now consider the action of $\mathcal{T}$ on states. We have

$$\psi_{free}(\vec{x},t) = \int \frac{d^3\vec{p}}{(2\pi)^3 2E_p} \sum_s (b(\vec{p},s)u(\vec{p},s)e^{-ip\cdot x} + d^\dagger(\vec{p},s)v(\vec{p},s)e^{ip\cdot x}).$$

Then, under time reversal, we have

$$\text{LHS} \quad \mathcal{T}\psi(\vec{x},t)\mathcal{T}^\dagger = \ldots\cdots\sum_s (\mathcal{T}b(\vec{p},s)\mathcal{T}^\dagger\ldots\cdots + \mathcal{T}d^\dagger(\vec{p},s)\mathcal{T}^\dagger\ldots\ldots),$$

$$\text{RHS} \quad \Sigma_2\,\psi^*(\vec{x},-t) = \ldots\cdots\sum_s (b^\dagger(\vec{p},s)\Sigma_2 u^*(\vec{p},s)e^{ip\cdot x'} + d(\vec{p},s)\Sigma_2 v^*(\vec{p},s)e^{-ip\cdot x'})$$

$$= \int \frac{d^3\vec{p}}{(2\pi)^3 2E_p} \sum_s (b^\dagger(-\vec{p},s)\Sigma_2 u^*(-\vec{p},s)e^{-ip\cdot x}$$
$$+ d(-\vec{p},s)\Sigma_2 v^*(-\vec{p},s)e^{ip\cdot x}),$$

where $x' = (\vec{x},-t)$ and ψ^* means complex conjugation on numbers, but Hermitian conjugation on operators. In the next-to-last line we changed variables $\vec{p} \to -\vec{p}$. Using one of the phase conventions derived earlier, Chapter 9, Section 9.3. We find

$$\mathcal{T}b(\vec{p},s)\,\mathcal{T}^\dagger = e^{i\theta(-\vec{p},s)}b^\dagger(-\vec{p},s), \tag{17.5}$$
$$\mathcal{T}d^\dagger(\vec{p},s)\,\mathcal{T}^\dagger = e^{-i\theta(-\vec{p},s)}d(-\vec{p},s).$$

Note, helicity is unchanged under time reversal, but, since $\vec{p} \to -\vec{p}$, spin changes sign under $\mathcal{T}$. Finally, we have

$$\mathcal{T}^2\,\psi\,\mathcal{T}^{\dagger 2} = \Sigma_2 \mathcal{T}\psi^*\mathcal{T}^\dagger = \Sigma_2\Sigma_2^T\psi = -\psi. \tag{17.6}$$

Let's now evaluate the transformation of the interaction term. We have

$$\mathcal{T}j_{EM}^\mu\mathcal{T}^\dagger = \mathcal{T}\bar{\psi}\gamma^\mu Q\psi\mathcal{T}^\dagger \tag{17.7}$$
$$= \mathcal{T}\,\psi\mathcal{T}^\dagger(\gamma^\mu Q)^T\mathcal{T}\bar{\psi}\mathcal{T}^\dagger$$
$$= \psi^\dagger(\vec{x},-t)A^T(\gamma^\mu Q)^T\gamma_0 A^*\psi(\vec{x},-t)$$
$$= \bar{\psi}(\vec{x},-t)(A^\dagger\gamma_0\gamma^\mu A\gamma_0)^T Q\psi(\vec{x},-t)$$
$$\equiv j_{\mu\,EM}(\vec{x},-t).$$

Note, in the last line we used the identity, $(A^\dagger \gamma_0 \gamma^\mu A \gamma_0)^T \equiv \gamma_\mu$.

Thus, for invariance of the interaction term $\propto A_\mu j^\mu_{EM}$ we require

$$\mathcal{T} A_\mu(\vec{x}, t) \mathcal{T}^\dagger = A^\mu(\vec{x}, -t). \tag{17.8}$$

This defines the transformation of photons under time reversal. We have

$$\mathcal{T} A_i(\vec{x}, t) \mathcal{T}^\dagger = -A_i(\vec{x}, -t) \tag{17.9}$$

and[1]

$$\mathcal{T} a(\vec{p}, \lambda) \mathcal{T}^\dagger = -a^\dagger(-\vec{p}, \lambda). \tag{17.10}$$

This then defines the action of time reversal on photon states. Also

$$\mathcal{T}^2 A_\mu(\vec{x}, t) \mathcal{T}^{\dagger^2} = A_\mu(\vec{x}, t). \tag{17.11}$$

Finally, let's check that the gauge kinetic term is invariant. We have

$$\mathcal{T} F_{\mu\nu} F^{\mu\nu} \mathcal{T}^\dagger \equiv \mathcal{T}(\partial_\mu A_\nu - \partial_\nu A_\mu)(\partial^\mu A^\nu - \partial^\nu A^\mu)\mathcal{T}^\dagger$$
$$= (\partial^\mu A_\nu(x') - \partial^\nu A_\mu(x'))(\partial_\mu A^\nu(x') - \partial_\nu A^\mu(x')),$$

where $x' = (\vec{x}, -t)$, $\partial^\mu = (\partial^0, \partial^i)$, $\partial_\mu = (\partial^0, -\partial^i)$ and $\partial'_\mu = (-\partial^0, -\partial^i) \equiv -\partial^\mu$. Hence, we have

$$\mathcal{T} F_{\mu\nu} F^{\mu\nu} \mathcal{T}^\dagger = (\partial'_\mu A_\nu(x') - \partial'_\nu A_\mu(x'))(\partial'^\mu A^\nu(x') - \partial'^\nu A^\mu(x')) \tag{17.12}$$
$$\equiv F_{\mu\nu}(x') F^{\mu\nu}(x').$$

Thus, the action of QED is invariant under time reversal.

Theorem: We want to prove that the S matrix satisfies,

$$\mathcal{T} S \mathcal{T}^\dagger = S, \tag{17.13}$$

assuming that the theory is time-reversal invariant.

Proof Let $H = H_0 + H_I$ such that $\mathcal{T} H_0 \mathcal{T}^\dagger = H_0$ and $\mathcal{T} H_I \mathcal{T}^\dagger = H_I$. The S matrix is defined using the interaction picture via the Schrödinger equation for the evolution operator, $U(t, t_0)$,

$$H_I(t) U(t, t_0) \equiv i\frac{\partial}{\partial t} U(t, t_0), \tag{17.14}$$

with $U(t_0, t_0) = 1$ and $H_I(t) \equiv e^{iH_0 t} H_I e^{-iH_0 t}$. Then, we have

$$\mathcal{T} H_I(t) \mathcal{T}^\dagger = \mathcal{T} e^{iH_0 t} H_I e^{-iH_0 t} \mathcal{T}^\dagger$$
$$= \mathcal{T} e^{-iH_0 t} \mathcal{T}^\dagger \mathcal{T} H_I \mathcal{T}^\dagger \mathcal{T} e^{iH_0 t} \mathcal{T}^\dagger$$
$$= e^{-iH_0 t} H_I e^{iH_0 t} \equiv H_I(-t).$$

---

[1] We use the phase convention for the polarization vector, such that $\epsilon^*(\vec{p}, \lambda) = \epsilon(\vec{p}, -\lambda)$.

Now consider the action of time reversal on the evolution operator. We have

$$\mathcal{T}H_I(t)U(t,t_0)\mathcal{T}^\dagger \equiv i\frac{\partial}{\partial t}\mathcal{T}U(t,t_0)\mathcal{T}^\dagger,$$

$$\mathcal{T}U(t,t_0)\mathcal{T}^\dagger \ H_I(-t) = i\frac{\partial}{\partial t}\mathcal{T}U(t,t_0)\mathcal{T}^\dagger,$$

which implies that

$$H_I(-t)\mathcal{T}U^\dagger(t,t_0)\mathcal{T}^\dagger = i\frac{\partial}{\partial(-t)}\ \mathcal{T}U^\dagger(t,t_0)\mathcal{T}^\dagger. \tag{17.15}$$

Hence, we have

$$\mathcal{T}U^\dagger(t,t_0)\mathcal{T}^\dagger = U(-t,-t_0)$$

or

$$\mathcal{T}U(t,t_0)\mathcal{T}^\dagger = U^\dagger(-t,-t_0). \tag{17.16}$$

Now $S \equiv U(\infty,-\infty)$ and $\mathcal{T}U(\infty,-\infty)\mathcal{T}^\dagger = U^\dagger(-\infty,\infty) \equiv U(\infty,-\infty)$, since $U^\dagger(t,t_0)U(t,t_0) = 1$ and $U(t_0,t)U(t,t_0) = 1$. Thus, we have $\mathcal{T}S\mathcal{T}^\dagger = S$. □

17.2 Phenomenological Consequences of Time-Reversal Invariance

(1) Reciprocity relation (or detailed balance)

If we have a reaction $a+b+\cdots \rightleftarrows a'+b'+\ldots$, with the following set of quantum numbers for the initial and final states, respectively, given by $\{\vec{p}_i, s_i\}$, $\{\vec{p}_j\,', s_j'\}$. Using the short-hand notation for the initial and final states, $|\vec{p}_i, s_i\rangle$, $|\vec{p}_j\,', s_j'\rangle$, where s_i, s_j' are helicities. We want to show that

$$|\langle \vec{p}_j\,', s_j'|S|\vec{p}_i, s_i\rangle| = |\langle -\vec{p}_i, s_i|S| - \vec{p}_j\,', s_j'\rangle|. \tag{17.17}$$

For two-to-two scattering the process and the time-reversed process appears in Fig. 17.1.

Proof   Under time reversal, the states transform as

$$\mathcal{T}|\vec{p}_i, s_i\rangle = e^{i\theta}\langle -\vec{p}_i, s_i|, \tag{17.18}$$

$$\mathcal{T}|\vec{p}_j\,', s_j'\rangle = e^{i\theta'}\langle -\vec{p}_j\,', s_j'|.$$

Hence we have

$$\langle \vec{p}_j\,', s_j'|S|\vec{p}_i, s_i\rangle = \langle \vec{p}_j\,', s_j'|\mathcal{T}^\dagger\mathcal{T}S\mathcal{T}^\dagger\mathcal{T}|\vec{p}_i, s_i\rangle \tag{17.19}$$

$$= e^{i(\theta-\theta')}\ \langle -\vec{p}_i, s_i|S| - \vec{p}_j\,', s_j'\rangle.$$

□

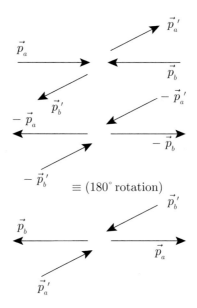

Fig. 17.1   Detailed balance for two-to-two scattering.

(2) Consider two body reactions $a + b \to a' + b'$ in the center of momentum system (CMS). We have $\sqrt{s} = E_a + E_b = E_{a'} + E_{b'}$ and $\vec{p}_a = -\vec{p}_b \equiv \vec{p}, \vec{p}_{a'} = -\vec{p}_{b'} \equiv \vec{p}\,'$. Now define the momenta in polar coordinates by $\vec{p} = (p, \theta, \phi)$ and $\vec{p}\,' = (p', \theta', \phi')$. Then the unpolarized differential scattering cross-section (where the helicities of the final states are not measured) satisfies

$$(2s_a + 1)(2s_b + 1)d\sigma(ab \to a'b') = \frac{1}{4s}\frac{p'}{p} \sum_{spin} |\frac{T_{ij}}{4\pi}|^2 d\Omega', \qquad (17.20)$$

where

$$T_{ij} = \langle \vec{p}_j\,', s_j'|S|\vec{p}_i, s_i \rangle \equiv \langle \vec{p}\,', s_a', s_b'|S|\vec{p}, s_a, s_b \rangle. \qquad (17.21)$$

For the time-reversed scattering process we have

$$(2s_{a'} + 1)(2s_{b'} + 1)d\sigma(a'b' \to ab) = \frac{1}{4s}\frac{p}{p'} \sum_{spin} |\frac{T_{ji}}{4\pi}|^2 d\Omega, \qquad (17.22)$$

where[2]

$$T_{ji} \equiv \langle \vec{p}, s_a, s_b|S|\vec{p}\,', s_{a'}, s_{b'} \rangle. \qquad (17.23)$$

But by the reciprocity relation applied to two-to-two scattering, we have

$$|T_{ij}| = |T_{ji}|. \qquad (17.24)$$

---

[2] Note, we've made use of the 180° rotation, as depicted in Fig. 17.1.

Hence, we find

$$\frac{d\sigma(ab \to a'b')}{d\sigma(a'b' \to ab)} = \left(\frac{p'}{p}\right)^2 \frac{(2s_{a'} + 1)(2s_{b'} + 1)}{(2s_a + 1)(2s_b + 1)} \frac{d\Omega'}{d\Omega}, \tag{17.25}$$

which then gives the final relation

$$\frac{\sigma(ab \to a'b')}{\sigma(a'b' \to ab)} = \left(\frac{p'}{p}\right)^2 \frac{(2s_{a'} + 1)(2s_{b'} + 1)}{(2s_a + 1)(2s_b + 1)} \times \left\{ \begin{array}{cc} 1/2 & a \neq b,\ a' = b' \\ 2 & a = b,\ a' \neq b' \\ 1 & \text{otherwise} \end{array} \right\}. \tag{17.26}$$

Note, the final factors in this expression are symmetry factors for identical particles (which have been neglected previously). They account for double counting in the phase-space integrals.

(3) Finally, let's discuss one method for determining the pion spin. Consider the scattering process

$$\pi^+ D \to pp \tag{17.27}$$

where D is deuterium with spin 1, and $s_a = s_\pi$, $s_b = s_D = 1$, $s_{a'} = s_{b'} = 1/2$. We also consider the time-reversed process,

$$pp \to \pi^+ D. \tag{17.28}$$

From the reciprocity relation, we have

$$\frac{\sigma(\pi^+ D \to pp)}{\sigma(pp \to \pi^+ D)} = \frac{1}{2} \left(\frac{p_{pp}}{p_{\pi^+ D}}\right)^2 \frac{(2)(2)}{3(2s_\pi + 1)}. \tag{17.29}$$

These processes were measured experimentally (Durbin et al., 1951), and they found

$$s_\pi = 0. \tag{17.30}$$

(4) Some additional tests of the reciprocity relation have been made. For example, the processes

$$Mg^{24} + D \leftrightarrows Mg^{25} + p$$
$$Mg^{24} + \alpha \leftrightarrows Al^{27} + p$$

have been used to show that the reciprocity relation is valid to within $\sim 1/2\%$. These processes test time-reversal invariance of the strong interactions.

18 CPT Theorem

Although the symmetries of **C**, **P** and $\mathcal{T}$ are separately broken, we show that the combined symmetry **CP$\mathcal{T}$** is preserved in any local Lorentz-invariant field theory. We then discusses the consequences of **CP$\mathcal{T}$**.

For fermions we have found

$$\mathcal{T}\psi(\vec{x}, t)\mathcal{T}^\dagger = \eta_t \Sigma_2 \psi^*(\vec{x}, -t)$$

$$\mathbf{P}\mathcal{T}\psi(\vec{x}, t)\mathcal{T}^\dagger \mathbf{P}^\dagger = \eta_t \sigma_2 \underbrace{\mathbf{P}\psi^*(\vec{x}, -t)\mathbf{P}^\dagger}_{\eta_p^* \gamma_0 \psi^*(-\vec{x}, -t)}$$

$$\Theta\psi(\vec{x}, t)\Theta^\dagger = \eta_t \eta_p^* \Sigma_2 \gamma_0 \underbrace{\mathbf{C}\psi^*(-\vec{x}, -t)\,\mathbf{C}^\dagger}_{\eta_c^* \gamma_2^* \psi(-\vec{x}, -t)}$$

$$= -\eta_t \eta_p^* \eta_c^* \,\Sigma_2 \gamma_0 \gamma_2 \,\psi(-\vec{x}, -t),$$

where we defined $\Theta \equiv \mathbf{CP}\mathcal{T}$. Now, defining $\eta \equiv \eta_t \eta_p^* \eta_c^*$ and using the identity $-\Sigma_2 \gamma_0 \gamma_2 \equiv \gamma_5$, we obtain

$$\Theta\psi(x)\,\Theta^\dagger \equiv \eta\gamma_5\psi(-x). \tag{18.1}$$

18.1 CPT Theorem

Theorem 18.1   *CPT theorem: Any Lorentz-invariant Lagrangian density, $\mathcal{L}(x)$, made of products of local operators and their derivatives, satisfies*

$$\Theta\mathcal{L}(x)\Theta^\dagger = \mathcal{L}(-x), \tag{18.2}$$

provided we choose the transformations (which follow from previous definitions up to a phase)

$$\Theta\psi(x)\Theta^\dagger = \eta\gamma_5\psi(-x)$$

$$\Theta\phi(x)\Theta^\dagger = \phi(-x)$$

$$\Theta A_\mu(x)\Theta^\dagger = -A_\mu(-x).$$

Proof   Consider the bilinear in fermions, $O(x) \equiv g : \psi_b^\dagger(x)\,\Gamma\,\psi_{b'}(x) :$, which is normal ordered and Γ is one element of a complete set of 4×4 matrices,

$\Gamma = \{\gamma_0,\ \gamma_0\gamma_5,\ \gamma_0\gamma_\mu,\ \gamma_0\gamma_\mu\gamma_5,\ \gamma_0\Sigma_{\mu\nu}\}$ which define scalar (S), pseudo-scalar (P), vector (V), axial-vector (A) and tensor (T) matrices, respectively. Note, they satisfy

$$\Gamma\gamma_5 = -\gamma_5\Gamma \quad \text{for S, P, T,}$$
$$\Gamma\gamma_5 = \gamma_5\Gamma \quad \text{for V, A.}$$

Thus, we have

$$\Theta O(x)\Theta^\dagger = g : \eta\psi_{b'}(-x)\gamma_5\Gamma^T\gamma_5\psi_b^*(-x)\eta^* : \qquad (18.3)$$
$$= -g : \psi_b^\dagger(-x)\gamma_5\Gamma\gamma_5\ \psi_{b'}(-x) :$$
$$\equiv \left\{ \begin{array}{cc} O(-x) & \text{S, P, T} \\ -O(-x) & \text{V, A} \end{array} \right\}.$$

Note, in addition we have $\frac{\partial}{\partial x_\mu} \equiv -\frac{\partial}{\partial(-x)_\mu}$. We then define the quantity

$$F_{\mu_1\dots\mu_N}(x) \quad \text{local product of terms} \quad (O(x),\ \partial_\mu), \qquad (18.4)$$

which satisfies

$$\Theta F_{\mu_1\dots\mu_N}(x)\ \Theta^\dagger = (-1)^N F_{\mu_1\dots\mu_N}(-x). \qquad (18.5)$$

Consider also that products of bosonic fields, given by

$$T_{\mu_1\dots\mu_N}(x) = (A_\mu(x),\ \partial_\mu,\ \phi(x)), \qquad (18.6)$$

satisfy

$$\Theta T_{\mu_1\dots\mu_N}(x)\Theta^\dagger = (-1)^N T_{\mu_1\dots\mu_N}(-x). \qquad (18.7)$$

Therefore, letting

$$\mathcal{L}(x) = \sum_{\mu_i} T_{\mu_1\dots\mu_N}(x)F^{\mu_1\dots\mu_N}(x), \qquad (18.8)$$

then $\mathcal{L}(x)$ satisfies

$$\Theta\mathcal{L}(x)\Theta^\dagger = \mathcal{L}(-x). \qquad (18.9)$$

$\square$

18.2 Applications of the CPT Theorem

(A) Consider a particle at rest, e.g. the proton state given by $|p, m\rangle$, where p stands for proton and m is the spin in the z direction. We then have the transformations

$$\mathbf{C}|p, m\rangle = |\bar{p}, m\rangle\eta_c,$$
$$\mathbf{P}|p, m\rangle = |p, m\rangle\eta_p,$$
$$\mathcal{T}|p, m\rangle = \langle p, -m|\eta_t^*,$$

and thus

$$\Theta|p,m\rangle = \mathbf{CPT}|p,m\rangle$$
$$= \mathbf{CPT}b^\dagger(\vec{0},m)\mathbf{T}^\dagger\mathbf{T}|0\rangle$$
$$= \langle 0|\mathbf{CP}\eta_t^* b(\vec{0},-m)\mathbf{P}^\dagger\mathbf{C}^\dagger$$
$$= \langle 0|\eta_p\eta_t^*\mathbf{C}\ b(\vec{0},-m)\mathbf{C}^\dagger$$
$$= \eta_c\eta_p\eta_t^*\langle 0|\ d(\vec{0},-m) \equiv \langle \bar{p},-m|\eta^*.$$

(B) In addition, we have $\Theta H\Theta^\dagger = H$ and $\Theta S\Theta^\dagger = S$.

The following three results are a consequence of $\mathbf{CPT}$ invariance.

(1) The mass of a particle and its anti-particle are equal.

$$m_a \equiv \langle a,m|H|a,m\rangle$$
$$= \langle a,m|\Theta^\dagger\Theta H\Theta^\dagger\Theta|a,m\rangle$$
$$= \langle \bar{a},-m|H|\bar{a},-m\rangle \equiv m_{\bar{a}}.$$

(2) The charge of a particle and its anti-particle are equal and opposite. Recall, the charge operator is given by $\hat{Q} \equiv \int d^3\vec{x}\ j^0(x)$ and under Θ we have, Eqn. 18.3, $\Theta j^0(x)\Theta^\dagger = -j^0(-x)$. If charge is conserved, then $\Theta\hat{Q}\Theta^\dagger = -\hat{Q}$.

Therefore, we have

$$Q_a \equiv \langle a,m|\hat{Q}|a,m\rangle$$
$$= \langle a,m|\Theta^\dagger\Theta\ \hat{Q}\ \Theta^\dagger\Theta|a,m\rangle$$
$$= -\langle \bar{a},-m|\hat{Q}|\bar{a},-m\rangle \equiv -Q_{\bar{a}}.$$

(3) The lifetime of a particle and its anti-particle are equal.

Proof   Consider that a state a decays into a set of final states, $\{b\}$, and its anti-particle, $\bar{a}$, decays into the $\mathbf{CPT}$ conjugate states, $\{\bar{b}\}$. Consider, also, that a, b are eigenstates of H_{strong} and, for example, $H_I = H_{weak}$. Then the total decay rate (or inverse lifetime) is given by the expression

$$\tau_a^{-1} \equiv \frac{1}{2m_a}\sum_b(2\pi)^4\delta^4(p_b - p_a)|\langle b_{free}|U(\infty,0)H_I|a,m\rangle|^2, \qquad (18.10)$$

where $\sum_b \equiv \sum_b \int dLips(\vec{p}_b)$. We also have the lifetime for the anti-particle given by

$$\tau_{\bar{a}}^{-1} \equiv \frac{1}{2m_a}\sum_{\bar{b}}(2\pi)^4\delta^4(p_{\bar{b}} - p_{\bar{a}})|\langle \bar{b}_{free}|U(\infty,0)H_I|\bar{a},m\rangle|^2. \qquad (18.11)$$

$U(\infty,0)$ includes H_{strong} interactions only and allows the decay products of a at $t = 0$ to evolve to $t \to \infty$. τ_a^{-1} is independent of m and is calculated to first order in H_I with $\vec{p}_a \equiv 0$.

Under the action of Θ we have

$$\tau_a^{-1} \equiv \frac{1}{2m_a} \sum_b (2\pi)^4 \delta^4(p_b - p_a) |\langle b_{free} | \Theta^\dagger \Theta \, U(\infty, 0) H_I \Theta^\dagger \Theta | a, m \rangle|^2$$

$$= \frac{1}{2m_a} \sum_b (2\pi)^4 \delta^4(p_b - p_a) |\langle \bar{a}, -m | H_I U^\dagger(-\infty, 0) | \bar{b}_{free} \rangle|^2$$

$$= \frac{1}{2m_a} \sum_b (2\pi)^4 \delta^4(p_b - p_a) |\langle \bar{b}_{free} | U(-\infty, 0) H_I | \bar{a}, -m \rangle|^2.$$

Note, $U^\dagger(0, \infty) \equiv U(\infty, 0)$, and the S matrix satisfies, $S \equiv U(\infty, 0) U(0, -\infty)$ and $S^\dagger = U(-\infty, 0) U(0, \infty)$. In addition, since the mass of particles and anti-particles are identical, we have $p_a = p_{\bar{a}}$, $p_b = p_{\bar{b}}$ and for every state b, there exists a **CPT** conjugate state $\bar{b}$. We thus find

$$\tau_a^{-1} \equiv \frac{1}{2m_a} \sum_{\bar{b}} (2\pi)^4 \delta^4(p_{\bar{b}} - p_{\bar{a}}) |\langle \bar{b}_{free} | S^\dagger U(\infty, 0) H_I | \bar{a}, -m \rangle|^2$$

$$= \dots \dots \, |\sum_{\bar{b}'} \langle \bar{b}_{free} | S^\dagger | \bar{b}'_{free} \rangle \langle \bar{b}'_{free} | U(\infty, 0) H_I | \bar{a}, -m \rangle|^2$$

$$= \dots \sum_{\bar{b}', \bar{b}''} (\langle \bar{b}_{free} | S^\dagger | \bar{b}'_{free} \rangle \langle \bar{b}_{free} | S^\dagger | \bar{b}''_{free} \rangle^*)$$

$$\times \langle \bar{b}'_{free} | U(\infty, 0) H_I | \bar{a}, -m \rangle \langle \bar{b}''_{free} | U(\infty, 0) H_I | \bar{a}, -m \rangle^*,$$

where we've used the completeness relation $\sum_{\bar{b}'} |\bar{b}'_{free}\rangle \langle \bar{b}'_{free}| \equiv 1$.

But by unitarity of the S matrix, we have

$$\sum_{\bar{b}} (2\pi)^4 \delta^4(p_{\bar{b}} - p_{\bar{a}}) \langle \bar{b}''_{free} | S | \bar{b}_{free} \rangle \langle \bar{b}_{free} | S^\dagger | \bar{b}'_{free} \rangle \equiv \delta_{\bar{b}', \bar{b}''} (2\pi)^4 \delta^4(p_{\bar{b}'} - p_{\bar{a}}). \quad (18.12)$$

Hence,

$$\tau_a^{-1} = \frac{1}{2m_a} \sum_{\bar{b}'} (2\pi)^4 \delta^4(p_{\bar{b}'} - p_{\bar{a}}) |\langle \bar{b}'_{free} | U(\infty, 0) H_I | \bar{a}, -m \rangle|^2 \equiv \tau_{\bar{a}}^{-1}, \quad (18.13)$$

or equivalently, $\Gamma_a \equiv \Gamma_{\bar{a}}$. $\square$

Note, since the sum over all intermediate states was required in the proof, i.e. $\Gamma_a = \sum_b \Gamma_{a \to b}$ and $\Gamma_{\bar{a}} = \sum_{\bar{b}} \Gamma_{\bar{a} \to \bar{b}}$, and we found that $\Gamma_a = \Gamma_{\bar{a}}$, this does not mean that $\Gamma_{a \to b} = \Gamma_{\bar{a} \to \bar{b}}$. In general, these can be different, i.e. $\Gamma_{a \to b} \neq \Gamma_{\bar{a} \to \bar{b}}$ if **C** or **CP** are violated.

PART V

FLAVOR SYMMETRIES

Global Symmetries

In this chapter we begin our discussion of continuous global symmetries of nature. These are necessarily approximate symmetries of nature. Nevertheless, the elaboration of these symmetries and their consequences were crucial for developing the theory we call the Standard Model.

19.1 Baryon and Lepton Numbers

By the late 1950s, the particle zoo had increased dramatically with the advent of government labs at Brookhaven National Laboratory in New York and CERN in Geneva. Some of the states which were now included were states with

- $B = 1$ $p, n, \Lambda, \Sigma, \ldots$
- $B = -1$ $\bar{p}, \bar{n}, \bar{\Lambda}, \bar{\Sigma}, \ldots$
- $B = 0$ $e^{\pm}, \mu^{\pm}, \nu, \bar{\nu}, \gamma, \pi^{\pm}, \pi^0, K^{\pm}$

The baryon number, B, is conserved. It is defined as a phase symmetry of baryonic states, for example, for protons we have

$$|p\rangle \to e^{iB\theta}|p\rangle = e^{i\theta}|p\rangle. \tag{19.1}$$

The symmetry leaves the action unchanged or, equivalently,

$$H_\theta = e^{iB\theta} H e^{-iB\theta} \equiv H. \tag{19.2}$$

Note, if B is conserved, then since the proton is the lowest mass state with $B = 1$, it cannot decay. However, energy, momentum and charge conservation would allow the possible decay processes such as $p \to e^+ + \gamma$, or $e^+ + \pi^0$, etc. Experimentally, $\tau_p \times BR^{-1}(p \to e^+ \pi^0) > 10^{34}$ years. Processes like $\pi^- p \to \pi^0 n$, $\pi^0 n p \bar{p}$ are seen and conserve the baryon number. What about the possible oscillation process, $n \to \bar{n}$. Does this conserve or violate the baryon number?

The lepton number, L_e, L_μ, L_τ (prior to the discovery of neutrino oscillations), appeared to be conserved for each type of lepton, separately. For example,

- $L_e = 1$ e^-, ν_e
- $L_e = -1$ $e^+, \bar{\nu}_e$
- $L_e = 0$ $\mu^{\pm}, \nu_\mu, \bar{\nu}_\mu, \gamma, p, n, \bar{p}, \bar{n}, \pi^{\pm}, \pi^0, K^{\pm}, \ldots$

Now, as far as we know, only the total lepton number, $L = L_e + L_\mu + L_\tau$, may be conserved.

We have the observed decay processes, $\pi^\pm$ decay,

$$\pi^+ \rightarrow \begin{cases} \mu^+ \ \nu_\mu \\ e^+ \ \nu_e \end{cases}, \tag{19.3}$$

$$\pi^- \rightarrow \begin{cases} \mu^- \ \bar{\nu}_\mu \\ e^- \ \bar{\nu}_e \end{cases}, \tag{19.4}$$

nuclear beta decay,

$$n \rightarrow p + e^- + \bar{\nu}_e,$$

and the crossed process

$$\bar{\nu}_e + p \rightarrow e^+ + n.$$

Note, it wasn't originally known that there was more than one kind of neutrino. In fact, an experiment in 1962 by Lederman and co-workers (Danby et al., 1962) proved that, at the time, there were two kinds of neutrinos. The experiment used the decay $\pi^+ \rightarrow \mu^+ + \nu$ with the subsequent flux of νs to try to observe the reaction $\nu + n \rightarrow p + e^-$. This process was not observed, but $\nu + n \rightarrow p + \mu^-$ was observed. They concluded that there were two neutrino types, ν_μ and ν_e, and only ν_μ was evident in the decay, $\pi^+ \rightarrow \mu^+ + \nu_\mu$. In addition, apparently, L_μ and L_e were separately conserved. Consistent with this we find in muon decay

$$\mu^+ \rightarrow e^+ \nu_e \bar{\nu}_\mu,$$

$$\mu^- \rightarrow e^- \bar{\nu}_e \nu_\mu.$$

L_μ, L_e conservation forbids the processes $\mu^- \rightarrow e^- + \gamma$ (experimentally $BR(\mu^- \rightarrow e^- \gamma) \leq 4.2 \times 10^{-13}$)(Mori, 2017) and $\nu_\mu + n \rightarrow p + e^-$. These are also forbidden by $L_\mu - L_e$.

In nuclear decays, the process $(A, Z) \rightarrow (A, Z + 2) + e^- + e^-$ is forbidden by L_e alone. It has never been observed. Finally, the process $(A, Z) \rightarrow (A - 2, Z) + m\gamma$ has never been observed. This corresponds to two neutrons annihilating, which violates the baryon number.[1]

19.2 Isotopic Spin

Isotopic spin was introduced in the 1930s by Heisenberg to describe the approximate charge independence of the strong interactions. We can take the proton and neutron Dirac spinors and define a new nucleon spinor by

$$\psi_N(x) = \begin{pmatrix} \psi_p(x) \\ \psi_n(x) \end{pmatrix}. \tag{19.5}$$

---

[1] Later in the book we shall show that both baryon and lepton numbers are violated in the Standard Model by gauge anomalies.

This is used to describe a phenomenological description of strong interactions valid for length scales ≥ 1 fm. Now instead of just defining a discrete symmetry which interchanges protons and neutrons, we define a continuous, global $U(2)$ symmetry acting on ψ_N. By definition, the group $U(2)$ is given by the set

$$U(2) = \{u | 2 \times 2 \text{ complex matrices, satisfying } u^\dagger u = \mathbb{I}_{2\times2}\}, \qquad (19.6)$$

with the group product given by matrix multiplication. Under this transformation we can consider $\psi \equiv \psi_N = \begin{pmatrix} \psi_1 (\equiv \psi_p) \\ \psi_2 (\equiv \psi_n) \end{pmatrix}$ and $\psi_i \rightarrow \psi_i' = u_{ij}\psi_j$. ψ_i, $i = 1, 2$ satisfy the anti-commutation relations

$$\{\psi_i(\vec{x}, t),\ \psi_j^\dagger(\vec{y}, t)\} = \delta_{ij}\delta^3(\vec{x} - \vec{y}). \qquad (19.7)$$

In order for the transformation to preserve these commutation relations we have

$$\{\psi_i'(\vec{x}, t),\ \psi_j^{\dagger'}(\vec{y}, t)\} \equiv \delta_{ij}\delta^3(\vec{x} - \vec{y})$$
$$= u_{ik}u_{lj}^\dagger \underbrace{\{\psi_k(\vec{x}, t),\ \psi_l^\dagger(\vec{y}, t)\}}_{\delta_{lk}\delta^3(\vec{x}-\vec{y})}.$$

This requires the condition $(uu^\dagger)_{ij} = \delta_{ij}$ or $uu^\dagger = \mathbb{I}$. Hence, u are unitary 2×2 matrices.

Let's now define the transformation in terms of operators on Fock space. There exists a unitary operator, $\mathcal{U}$, such that[2]

$$\mathcal{U}\psi(x)\mathcal{U}^\dagger \equiv u\psi(x). \qquad (19.8)$$

Given

$$\psi_p(x) = \int \frac{d^3\vec{k}}{(2\pi)^3 2E_k} \sum_s (b_p(\vec{k}, s)u(\vec{k}, s)e^{-ik\cdot x} + d_p^\dagger(\vec{k}, s)v(\vec{k}, s)e^{ik\cdot x},$$

$$\psi_n(x) = \int \frac{d^3\vec{k}}{(2\pi)^3 2E_k} \sum_s (b_n(\vec{k}, s)u(\vec{k}, s)e^{-ik\cdot x} + d_n^\dagger(\vec{k}, s)v(\vec{k}, s)e^{ik\cdot x},$$

we have

$$\mathcal{U} \begin{pmatrix} b_p(\vec{k}, s) \\ b_n(\vec{k}, s) \end{pmatrix} \mathcal{U}^\dagger = u \begin{pmatrix} b_p(\vec{k}, s) \\ b_n(\vec{k}, s) \end{pmatrix}$$

$$\mathcal{U} \begin{pmatrix} d_p^\dagger(\vec{k}, s) \\ d_n^\dagger(\vec{k}, s) \end{pmatrix} \mathcal{U}^\dagger = u \begin{pmatrix} d_p^\dagger(\vec{k}, s) \\ d_n^\dagger(\vec{k}, s) \end{pmatrix}.$$

Note, $u \in U(2)$ has four arbitrary continuous parameters, since u is a complex 2×2 matrix, which implies eight real parameters. However, $uu^\dagger = \mathbb{I}$ gives four constraints. Therefore u is defined in terms of just four real parameters. Now define $\det u \equiv e^{-2i\theta}$, then $u \equiv e^{-i\theta}\ u_I$ such that $\det u_I = 1$. u_I is an element of the group $SU(2)$, where

$$SU(2) \equiv \{u_I | 2 \times 2 \text{ unitary matrices; } u_I^\dagger u_I = \mathbb{I}, \det u_I = 1\}. \qquad (19.9)$$

---

[2] We shall explicitly construct it later.

The general form for u_I was given in Eqns. 2.4, 2.11. We have $u_I \equiv e^{i\frac{\vec{\tau}}{2}\cdot\vec{\theta}}$ where τ_i are Pauli matrices, $\theta_i, i = 1, 2, 3$ are three real parameters.

We can now see that $U(2) \equiv SU(2) \otimes U(1)$, i.e. a direct product of two independent groups. The group $U(1)$ is given by $\vec{\theta} = 0$, such that $u = e^{-i\theta}\mathbb{I}$. The transformation of the nucleon field under $U(1)$ is given by

$$\mathcal{U}(\theta, \vec{\theta} = 0) \begin{pmatrix} d_p^\dagger(\vec{k}, s) \\ d_n^\dagger(\vec{k}, s) \end{pmatrix} \mathcal{U}^\dagger(\theta, \vec{\theta} = 0) = e^{-i\theta} \begin{pmatrix} d_p^\dagger(\vec{k}, s) \\ d_n^\dagger(\vec{k}, s) \end{pmatrix}.$$

Then define $\mathcal{U}(\theta) \equiv e^{i\hat{B}\theta}$ with $\hat{B} = \hat{B}^\dagger$, an operator on Fock space. Let's calculate

$$-i\frac{d}{d\theta}[\mathcal{U}(\theta, \vec{\theta} = 0) \begin{pmatrix} d_p^\dagger(\vec{k}, s) \\ d_n^\dagger(\vec{k}, s) \end{pmatrix} \mathcal{U}^\dagger(\theta, \vec{\theta} = 0)]_{\theta=0} \equiv -\begin{pmatrix} d_p^\dagger(\vec{k}, s) \\ d_n^\dagger(\vec{k}, s) \end{pmatrix},$$

which gives the relation

$$\left[\hat{B}, \begin{pmatrix} d_p^\dagger(\vec{k}, s) \\ d_n^\dagger(\vec{k}, s) \end{pmatrix}\right] = -\begin{pmatrix} d_p^\dagger(\vec{k}, s) \\ d_n^\dagger(\vec{k}, s) \end{pmatrix}. \tag{19.10}$$

Similarly,

$$\left[\hat{B}, \begin{pmatrix} b_p(\vec{k}, s) \\ b_n(\vec{k}, s) \end{pmatrix}\right] = -\begin{pmatrix} b_p(\vec{k}, s) \\ b_n(\vec{k}, s) \end{pmatrix} \tag{19.11}$$

or

$$\left[\hat{B}, \begin{pmatrix} b_p^\dagger(\vec{k}, s) \\ b_n^\dagger(\vec{k}, s) \end{pmatrix}\right] = \begin{pmatrix} b_p^\dagger(\vec{k}, s) \\ b_n^\dagger(\vec{k}, s) \end{pmatrix}. \tag{19.12}$$

We now know how $\hat{B}$ acts on proton and neutron states, since $|p\rangle = b_p^\dagger|0\rangle$ and $|n\rangle = b_n^\dagger|0\rangle$. Thus, assuming $\hat{B}|0\rangle = 0$, we find $\hat{B}|p\rangle = |p\rangle$ and $\hat{B}|n\rangle = |n\rangle$, while $|\bar{p}\rangle = d_p^\dagger|0\rangle$ and $|\bar{n}\rangle = d_n^\dagger|0\rangle$. Thus, $\hat{B}|\bar{p}\rangle = -|\bar{p}\rangle$ and $\hat{B}|\bar{n}\rangle = -|\bar{n}\rangle$. Hence, the operator $\hat{B}$ is the generator of the baryon number.

Now consider the $SU(2)$ transformations. We have

$$\mathcal{U}_I(\vec{\theta})\psi(x)\mathcal{U}_I^\dagger(\vec{\theta}) \equiv u_I\psi(x). \tag{19.13}$$

This transformation takes protons into neutrons (this is *isospin*). Heisenberg assumed that the strong interactions (i.e. nuclear forces) were isospin invariant, i.e. $\mathcal{U}_I H_{strong}\mathcal{U}_I^\dagger = H_{strong}$ or $[\mathcal{U}_I, H_{strong}] = 0$.

ASIDE: Note, we have discussed the action of $\mathcal{U}$ on free (or interaction-picture) states. However, if we write $H_{strong} = H_0 + H_I$ and $\mathcal{U}_I H_0 \mathcal{U}_I^\dagger = H_0$, $\mathcal{U}_I H_I \mathcal{U}_I^\dagger = H_I$, then it is easy to see that the Heisenberg states transform in the same way as the interaction-picture states. If, for example, the interaction-picture state is given by $|p\rangle_0 \equiv b_p^\dagger|0\rangle$, then the fully interacting state is given by

$|p\rangle = \lim_{t\to-\infty} e^{iH_{strong}t} e^{-iH_0 t}|p\rangle_0$. Then we have

$$\mathcal{U}_I |p\rangle = \lim_{t\to-\infty} \mathcal{U}_I e^{iH_{strong}t} e^{-iH_0 t}|p\rangle_0$$

$$\equiv \lim_{t\to-\infty} e^{iH_{strong}t} e^{-iH_0 t} \mathcal{U}_I |p\rangle_0$$

and we evaluated $\mathcal{U}_I |p\rangle_0$. We also have $[\mathcal{U}_I, S] = 0$.

For homework, you found that

$$\mathcal{U}_I(\vec{\theta}) \equiv e^{-i\vec{I}\cdot\vec{\theta}}, \tag{19.14}$$

where

$$[I_i, \psi_N(x)] = -\frac{\tau_i}{2}\psi_N(x). \tag{19.15}$$

This is the solution to the equation

$$\mathcal{U}_I(\vec{\theta})\psi_N(x)\mathcal{U}_I^\dagger(\vec{\theta}) \equiv u_I \psi_N(x), \tag{19.16}$$

with $u_I \equiv e^{i\frac{\vec{\tau}}{2}\cdot\vec{\theta}}$. I_i are the generators of isospin rotations, satisfying the Lie algebra

$$[I_i,\ I_j] = i\epsilon_{ijk}I_k. \tag{19.17}$$

We also have

$$I_i \equiv \int d^3\vec{x}\,\psi_N^\dagger(x)\frac{\tau_i}{2}\psi_N(x) \tag{19.18}$$

and

$$j_i^\mu(x) \equiv \bar{\psi}_N(x)\gamma^\mu\frac{\tau_i}{2}\psi_N(x) \tag{19.19}$$

satisfy $\partial_\mu j_i^\mu(x) = 0$, if isospin is conserved.

In terms of creation and annihilation operators we have

$$I_i \equiv \int \frac{d^3\vec{k}}{(2\pi)^3 2E_k} \sum_s (b_l^\dagger(\vec{k},s)\frac{(\tau_i)_{lm}}{2}b_m(\vec{k},s) - d_m^\dagger(\vec{k},s)\underbrace{\frac{(\tau_i)_{lm}}{2}}_{\frac{(\tau_i^T)_{ml}}{2}}d_l(\vec{k},s)). \tag{19.20}$$

On nucleon states we have

$$\mathcal{U}_I(\vec{\theta})\begin{pmatrix}|\bar{p}\rangle\\|\bar{n}\rangle\end{pmatrix} = u_I(\vec{\theta})\begin{pmatrix}|\bar{p}\rangle\\|\bar{n}\rangle\end{pmatrix},$$

$$\mathcal{U}_I(\vec{\theta})\begin{pmatrix}|p\rangle\\|n\rangle\end{pmatrix} = \begin{pmatrix}|p\rangle\\|n\rangle\end{pmatrix} u_I^\dagger(\vec{\theta}).$$

Finally, $[\mathcal{U}_I,\ H_{strong}] = 0$ for arbitrary $\vec{\theta}$ implies that $[I_i,\ H_{strong}] = 0$.

19.3 Tests of Approximate Isospin Symmetry

In the previous section, we discussed the generators of isospin symmetry and their commutation relations. Now we want to connect this to phenomenology. If isospin is a symmetry of the theory, then particle states should come in representations of isospin, with eigenvalues given in terms of a complete set of commuting operators. For isospin, which is apparently a symmetry of the strong interactions, we have the complete set of commuting operators given by

$$\{H_{strong}, \vec{I}^2, I_3\}, \tag{19.21}$$

with eigenstates,

$$H_{strong} |I, m, E_k, \vec{k}\rangle = E_k |I, m, E_k, \vec{k}\rangle,$$
$$I_3 |I, m, E_k, \vec{k}\rangle = m |I, m, E_k, \vec{k}\rangle,$$
$$\vec{I}^2 |I, m, E_k, \vec{k}\rangle = I(I+1) |I, m, E_k, \vec{k}\rangle.$$

Now let's show that

$$I_i \left(\begin{array}{c} |p\rangle \\ |n\rangle \end{array} \right) = \left(\begin{array}{c} |p\rangle \\ |n\rangle \end{array} \right) \left(\frac{\tau_i}{2}\right). \tag{19.22}$$

Proof:

$$\left(\begin{array}{c} |p\rangle \\ |n\rangle \end{array} \right) = \left(\begin{array}{c} b_1^\dagger |0\rangle \\ b_2^\dagger |0\rangle \end{array} \right).$$

Also, using Eqn. 19.20, we have

$$I_i b_l^\dagger |0\rangle = \int \frac{d^3\vec{k}}{(2\pi)^3 2E_k} \sum_s (b_j^\dagger(\vec{k}, s) \frac{(\tau_i)_{jm}}{2} b_m(\vec{k}, s) - \ldots) b_l^\dagger |0\rangle = b_j^\dagger |0\rangle \left(\frac{\tau_i}{2}\right)_{jl}, \tag{19.23}$$

where we used the anti-commutation relations

$$\{b_m(\vec{k}, s), b_l^\dagger(\vec{p}, s')\} = \delta_{ml}(2\pi)^3 2E_k \delta^3(\vec{k} - \vec{p})\delta_{ss'}.$$

Note, we use the standard representation of the Pauli matrices,

$$\tau_3 = \left(\begin{array}{cc} 1 & 0 \\ 0 & -1 \end{array} \right), \quad \tau_1 = \left(\begin{array}{cc} 0 & 1 \\ 1 & 0 \end{array} \right), \quad \tau_2 = \left(\begin{array}{cc} 0 & -i \\ i & 0 \end{array} \right).$$

Hence, we have

$$I_3|p\rangle = \frac{1}{2}|p\rangle, \quad I_3|n\rangle = -\frac{1}{2}|n\rangle.$$

We also define raising and lowering operators,

$$I_\pm \equiv I_1 \pm iI_2, \tag{19.24}$$

such that $I_1 = \frac{I_+ + I_-}{2}$, $I_2 = \frac{I_+ - I_-}{2i}$ and $\tau_\pm \equiv \frac{\tau_1 \pm i\tau_2}{2}$ with $\tau_+ = \begin{pmatrix} 0 & 1 \\ 0 & 0 \end{pmatrix}$ and

$\tau_- = \begin{pmatrix} 0 & 0 \\ 1 & 0 \end{pmatrix}$.

We then find that

$$I_+ |l\rangle = |j\rangle (\tau_+)_{jl}. \tag{19.25}$$

Using $(\tau_+)_{jl} = \delta_{j1}\delta_{l2}$, $(\tau_-)_{jl} = \delta_{j2}\delta_{l1}$, we have

$$I_+|1\rangle = 0, \quad I_+|2\rangle = |1\rangle, \tag{19.26}$$

with $|p\rangle \equiv |1\rangle$, $|n\rangle \equiv |2\rangle$. Similarly, we have

$$I_- |l\rangle = |j\rangle (\tau_-)_{jl}. \tag{19.27}$$

Therefore we have

$$I_-|1\rangle = |2\rangle, \quad I_-|2\rangle = 0. \tag{19.28}$$

Finally, the commutator

$$[I_+,\ I_-] = 2I_3 \tag{19.29}$$

can be used to find

$$\begin{aligned}
\vec{I}^{\,2}|p\rangle &\equiv (I_1^2 + I_2^2 + I_3^2)|p\rangle \\
&= \left[\frac{1}{2}(I_+ I_- + I_- I_+) + I_3^2 \right]|p\rangle \\
&= \frac{3}{4}|p\rangle.
\end{aligned}$$

Hence, the proton has isospin $I_p = 1/2$. Let's now prove that $I_n = I_p = 1/2$. We have

$$I_p(I_p + 1) = \langle p|\vec{I}^{\,2}|p\rangle = \langle p|\vec{I}^{\,2}I_+|n\rangle.$$

But, $\left[\vec{I}^{\,2}, I_i\right] = 0$. Hence

$$I_p(I_p + 1) = \langle p|\vec{I}^{\,2}I_+|n\rangle = \langle p|I_+\vec{I}^{\,2}|n\rangle = \langle I_-\ p|\vec{I}^{\,2}|n\rangle = \langle n|\vec{I}^{\,2}|n\rangle = I_n(I_n + 1) \tag{19.30}$$

or the nucleon multiplet has $I_N = I_p = I_n = 1/2$.

We can use the same analysis to prove that the mass of the proton and neutron must be equal, assuming isospin is a symmetry of the strong interactions, i.e. $[H_{strong}, I_i] = 0$. We have

$$m_p = \langle p|H_{strong}|p\rangle = \langle p|H_{strong}\ I_+|n\rangle = \langle n|H_{strong}|n\rangle = m_n. \tag{19.31}$$

Clearly, isospin is broken by electromagnetic interactions, since protons and neutrons do not have the same charge. We might, however, expect that protons would be heavier than neutrons, due to electromagnetic mass corrections. But exactly the opposite is true. We will return to this issue later.

19.4 Isospin of Pions

According to Yukawa (1935), pions mediate the strong nuclear force between protons and neutrons. There are three types of pions, $\pi^\pm$, π^0, which, to a good approximation, are degenerate. They interact with nucleons as in Fig. 19.1. Since there are three types of pions which are almost degenerate, we assume they form an isospin triplet. We have

$$I_i|\pi_j\rangle = i\epsilon_{ijk}|\pi_k\rangle. \tag{19.32}$$

π^i transforms as a vector as in Fig. 19.2.

The fields, $\pi_i(x)$, are real pseudo-scalar fields. We can then define the combinations which are eigenstates of I_3. We have

$$|\pi^0\rangle \equiv |\pi_3\rangle, \quad |\pi^\pm\rangle \equiv \frac{1}{\sqrt{2}}|\pi_1 \pm i\pi_2\rangle. \tag{19.33}$$

Then it is easy to see that

$$I_3|\pi^0\rangle = 0, \quad I_3|\pi^\pm\rangle = \pm|\pi^\pm\rangle$$

and

$$I_\pm|\pi^0\rangle = \mp\sqrt{2}|\pi^\pm\rangle,$$

$$I_+|\pi^\pm\rangle = (I_1 + iI_2)\frac{1}{\sqrt{2}}|\pi_1 \pm i\pi_2\rangle$$

$$= \pm\frac{i^2}{\sqrt{2}}|\pi_3\rangle - \frac{i^2}{\sqrt{2}}|\pi_3\rangle = \left\{ \begin{array}{c} 0 \\ \sqrt{2}\,|\pi_3\rangle \end{array} \right. .$$

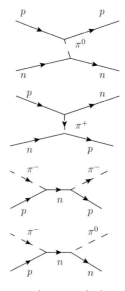

Fig. 19.1   Some strong interactions between nucleons and pions.

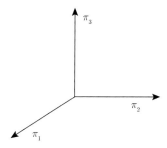

Fig. 19.2 Pions transform as an ordinary vector under isospin.

Thus, we have

$$I_+|\pi^+\rangle = 0, \quad I_+|\pi^-\rangle = \sqrt{2}\,|\pi^0\rangle,$$
$$I_-|\pi^-\rangle = 0, \quad I_-|\pi^+\rangle = -\sqrt{2}\,|\pi^0\rangle.$$

Assuming that pions are strong-interaction eigenstates and that the strong interactions commute with isospin, we have $m_{\pi^0} = m_{\pi^\pm}$. The states π^+ and π^- are **C** or **CP** conjugates, thus **CP$\mathcal{T}$** invariance requires $m_{\pi^+} = m_{\pi^-}$. Again, since electromagnetic interactions break isospin, all pions are not degenerate in mass. Nevertheless, we have $m_{\pi^\pm} > m_{\pi^0}$. Finally, we have $I_\pi = 1$.

19.5 Exercises

19.1 Calculate the differential cross-sections for the processes

(a) $e^+e^- \to \mu^+\mu^-$ and
(b) $e^+e^- \to \pi^+\pi^-$.

Assume unpolarized initial states and sum over the final spins. Explain the difference in the angular dependence θ (measured between the incoming electron and outgoing positively charged particle) for the two processes and the difference in the threshold behavior, i.e. the momentum dependence of the differential cross-sections near threshold. (Hint: both are consequences of angular momentum conservation.)

20 Testing Isospin and G Parity

20.1 Testing Isospin with Scattering Cross-Sections

The cross-section for πN scattering has a resonance at $\sqrt{s} \simeq 1232$ MeV. The resonance appears in the processes $\pi^+ p \to \pi^+ p$, $\pi^- p \to \pi^0 n$, $\pi^- p \to \pi^- p$, $\pi^+ n \to \pi^0 p$, and $\pi^- n \to \pi^- n$ at approximately the same value of $\sqrt{s}$. Thus, it is interpreted as a strong-interaction resonance, i.e. an eigenstate of H_{strong} with mass, $m = m_\Delta$, where the delta resonance comes in four charge types, $\Delta^{++}, \Delta^+, \Delta^0, \Delta^-$. Since there are four states, approximately degenerate in mass, we can assume that they are an isospin multiplet with $I_\Delta = 3/2$ and baryon number $B_\Delta = 1$. For homework, using the fact that the pions are an iso-triplet and the nucleon is an iso-doublet, you used isospin invariance and the Wigner–Eckart theorem to evaluate the ratio of πN scattering cross-sections, on resonance at $\sqrt{s} = 1232$ MeV. You found that the cross-sections were in the ratio,

$$\sigma(\pi^+ p \to \pi^+ p) : \sigma(\pi^- p \to \pi^0 n) : \sigma(\pi^- p \to \pi^- p) = 9 : 2 : 1.$$

Note, we can define the Wigner–Eckart reduced matrix element

$$S_{3/2} \equiv \langle \pi^+ p | S | \pi^+ p \rangle, \tag{20.1}$$

since, in terms of isospin, the state

$$|\pi^+ p\rangle \equiv |1, 1; \tfrac{1}{2}, \tfrac{1}{2}\rangle \equiv |\tfrac{3}{2}, \tfrac{3}{2}\rangle$$

is unique. On the other hand, the state

$$|\pi^- p\rangle \equiv |1, -1; \tfrac{1}{2}, \tfrac{1}{2}\rangle = \left(\tfrac{3}{2}, -\tfrac{1}{2}|1, -1; \tfrac{1}{2}, \tfrac{1}{2}\right)|\tfrac{3}{2}, -\tfrac{1}{2}\rangle + \left(\tfrac{1}{2}, -\tfrac{1}{2}|1, -1; \tfrac{1}{2}, \tfrac{1}{2}\right)|\tfrac{1}{2}, -\tfrac{1}{2}\rangle.$$

Also, isospin invariance of the S matrix implies that

$$S_{3/2} \equiv \langle \tfrac{3}{2}, \pm\tfrac{3}{2}|S|\tfrac{3}{2}, \pm\tfrac{3}{2}\rangle = \langle \tfrac{3}{2}, \pm\tfrac{1}{2}|S|\tfrac{3}{2}, \pm\tfrac{1}{2}\rangle$$

is independent of I_3. Finally, we have

$$\frac{\sigma(\pi^- p \to \pi^- p)}{\sigma(\pi^+ p \to \pi^+ p)}\Big|_{\sqrt{s}=m_\Delta} = |\left(\tfrac{3}{2}, -\tfrac{1}{2}|1, -1; \tfrac{1}{2}, \tfrac{1}{2}\right)|^4 = \frac{1}{9}.$$

Experimentally, we can find graphs of these cross-sections in the Particle Data Book (Tanabashi et al., 2018). Using an eyeball determination we find

$\sigma(\pi^+ p \to \pi^+ p)_{\sqrt{s}=m_\Delta} \sim 200$ mb and $\sigma(\pi^- p \to \pi^- p)_{\sqrt{s}=m_\Delta} \sim 28$ mb, which is consistent with the theory to within 20%.

Another example is given by the processes, $p + D \to \pi^0 + {}^3He$, $\pi^+ + {}^3H$. Deuterium $D \equiv {}^2H$ is a proton–neutron bound state with $I_D = 0$, while the states ${}^3H = (pnn)$ and ${}^3He = (ppn)$ seem to form an iso-doublet with

$$\begin{pmatrix} {}^3He \\ {}^3H \end{pmatrix} = |\tfrac{1}{2}, \pm\tfrac{1}{2}\rangle. \tag{20.2}$$

Note, the deuterium ground state has $J = 1$ with a symmetric spatial wave function ($L = 0$), a symmetric spin wave function ($S = 1$) and an anti-symmetric isospin wave function ($I = 0$). Thus the ground-state wave function is totally anti-symmetric as required by the Pauli principle for an NN bound state.

Let's now test our hypothesis of isospin conservation by calculating the ratio

$$R = \frac{\sigma(p\,D \to \pi^+\,{}^3H)}{\sigma(p\,D \to \pi^0\,{}^3He)}. \tag{20.3}$$

We can identify the isospin quantum number of the initial and final states. We have

$$|p\,D\rangle = |\tfrac{1}{2}, \tfrac{1}{2}\rangle,$$

$$|\pi^0\,{}^3He\rangle = |1,0; \tfrac{1}{2}, \tfrac{1}{2}\rangle = |\tfrac{1}{2}, \tfrac{1}{2}\rangle \left(\tfrac{1}{2}, \tfrac{1}{2}|1,0; \tfrac{1}{2}, \tfrac{1}{2}\right) + \dots (I = 3/2),$$

$$|\pi^+\,{}^3H\rangle = |1,1; \tfrac{1}{2}, -\tfrac{1}{2}\rangle = |\tfrac{1}{2}, \tfrac{1}{2}\rangle \left(\tfrac{1}{2}, \tfrac{1}{2}|1,1; \tfrac{1}{2}, -\tfrac{1}{2}\right) + \dots (I = 3/2).$$

Hence,

$$R = \frac{\sigma(p\,D \to \pi^+\,{}^3H)}{\sigma(p\,D \to \pi^0\,{}^3He)} = \frac{|(\tfrac{1}{2}, \tfrac{1}{2}|1,1; \tfrac{1}{2}, -\tfrac{1}{2})|^2}{|(\tfrac{1}{2}, \tfrac{1}{2}|1,0; \tfrac{1}{2}, \tfrac{1}{2})|^2} \equiv \frac{2/3}{1/3} = 2. \tag{20.4}$$

Experimentally, we find

$$R - \begin{cases} 1.91 \pm 0.25 \\ 2.26 \pm 0.11 \end{cases},$$

which agrees with the prediction of isospin conservation.

To summarize, isospin conservation (even if it is only approximate) defines multiplets that are

- degenerate in mass
- cross-section relations which can be tested experimentally.

There are many more relations, even among atomic nuclei, providing convincing evidence that isospin is an approximate symmetry of the strong, nuclear forces.

20.2 G-Parity Invariance of the Strong Interactions

In this section, we want to define the transformation called G parity (Lee and Yang, 1956a) and then discuss the consequences of G-parity invariance of the strong interactions. The first step in this discussion is to show that a doublet and anti-doublet under isospin are the same, up to a change of basis. Consider the transformations of the anti-doublet $\{\bar{p}, \bar{n}\}$ and doublet $\{p, n\}$. We have

$$\mathcal{U}_I(\vec{\theta}) \begin{pmatrix} |\bar{p}\rangle \\ |\bar{n}\rangle \end{pmatrix} = u_I(\vec{\theta}) \begin{pmatrix} |\bar{p}\rangle \\ |\bar{n}\rangle \end{pmatrix},$$

$$\mathcal{U}_I(\vec{\theta}) \begin{pmatrix} |p\rangle \\ |n\rangle \end{pmatrix} = \begin{pmatrix} |p\rangle \\ |n\rangle \end{pmatrix} u_I^\dagger(\vec{\theta}),$$

where $u_I(\vec{\theta}) \overset{|\vec{\theta}|\ll 1}{\sim} 1 + i\frac{\vec{\tau}}{2} \cdot \vec{\theta}$ and $u_I^*(\vec{\theta}) \overset{|\vec{\theta}|\ll 1}{\sim} 1 + i\left(-\frac{\vec{\tau}^T}{2} \cdot \vec{\theta}\right)$. We now want to show that

$$\mathcal{U}_I(\vec{\theta}) \begin{pmatrix} |\bar{n}\rangle \\ -|\bar{p}\rangle \end{pmatrix} = \begin{pmatrix} |\bar{n}\rangle \\ -|\bar{p}\rangle \end{pmatrix} u_I^\dagger(\vec{\theta}), \tag{20.5}$$

i.e. if $\begin{pmatrix} |p\rangle \\ |n\rangle \end{pmatrix}$ transforms as an $SU(2)$ doublet, (**2**), and $\begin{pmatrix} |\bar{p}\rangle \\ |\bar{n}\rangle \end{pmatrix}$ transforms as a

(**2̄**), then $\begin{pmatrix} |\bar{n}\rangle \\ -|\bar{p}\rangle \end{pmatrix}$ also transforms as a (**2**).

Proof　We have

$$u_I(\vec{\theta}) \equiv e^{i\frac{\vec{\theta}}{2}\cdot\vec{\tau}} \equiv e^{i\frac{\tau_0}{2}\alpha}, \tag{20.6}$$

where $\tau_0 \equiv \hat{\theta}\cdot\vec{\tau}$, $\vec{\theta} \equiv \alpha\hat{\theta}$ and $\alpha \equiv |\vec{\theta}|$. We then have $\tau_0^2 = \tau_i\tau_j\hat{\theta}_i\hat{\theta}_j = \frac{1}{2}\{\tau_i, \tau_j\}\hat{\theta}_i\hat{\theta}_j = 1$. Given this, we obtain

$$u_I(\vec{\theta}) \equiv \cos\frac{\alpha}{2} + i\vec{\tau}\cdot\hat{\theta}\sin\frac{\alpha}{2}. \tag{20.7}$$

Note the identity, $\tau_2\vec{\tau} = -\vec{\tau}^*\tau_2 \equiv -\vec{\tau}^T\tau_2$, which then gives

$$\tau_2\left(\frac{\vec{\tau}}{2}\right)\tau_2 = \left(-\frac{\vec{\tau}^T}{2}\right) \tag{20.8}$$

and

$$\tau_2 u_I(\vec{\theta}) = u_I^*(\vec{\theta})\tau_2. \tag{20.9}$$

Thus,

$$\tau_2\mathcal{U}_I(\vec{\theta}) \begin{pmatrix} |\bar{p}\rangle \\ |\bar{n}\rangle \end{pmatrix} = \tau_2 u_I(\vec{\theta}) \begin{pmatrix} |\bar{p}\rangle \\ |\bar{n}\rangle \end{pmatrix},$$

which gives

$$\mathcal{U}_I(\vec{\theta})\tau_2 \begin{pmatrix} |\bar{p}\rangle \\ |\bar{n}\rangle \end{pmatrix} = u_I^*(\vec{\theta})\tau_2 \begin{pmatrix} |\bar{p}\rangle \\ |\bar{n}\rangle \end{pmatrix} \equiv \tau_2 \begin{pmatrix} |\bar{p}\rangle \\ |\bar{n}\rangle \end{pmatrix} u_I^\dagger(\vec{\theta}). \tag{20.10}$$

Hence, $i\tau_2\begin{pmatrix} |\bar{p}\rangle \\ |\bar{n}\rangle \end{pmatrix}$ transforms as a $(\mathbf{2})$ with $i\tau_2\begin{pmatrix} |\bar{p}\rangle \\ |\bar{n}\rangle \end{pmatrix} \equiv \begin{pmatrix} |\bar{n}\rangle \\ -|\bar{p}\rangle \end{pmatrix}$. $\square$

Similarly, for the nucleon field, $\psi_N(x)$, we have

$$\mathcal{U}_I(\vec{\theta})\psi_N\mathcal{U}_I^\dagger(\vec{\theta}) = u_I\psi_N$$

and thus

$$\mathcal{U}_I(\vec{\theta})\psi_N^\dagger\mathcal{U}_I^\dagger(\vec{\theta}) = \psi_N^\dagger u_I^\dagger = u_I^*\psi_N^\dagger.$$

Recall, the charge-conjugate field is given by $\psi_N^C \equiv \gamma_2\psi_N^\dagger$. Therefore, we obtain

$$\mathcal{U}_I(\vec{\theta})(i\tau_2\psi_N^C)\mathcal{U}_I^\dagger(\vec{\theta}) = u_I(i\tau_2\psi_N^C). \tag{20.11}$$

Hence, both fields $\psi_N = \begin{pmatrix} \psi_p \\ \psi_n \end{pmatrix}$ and $i\tau_2\psi_N^C = \begin{pmatrix} \psi_n^C \\ -\psi_p^C \end{pmatrix}$ transform as iso-doublets.

Since both $\begin{pmatrix} p \\ n \end{pmatrix}$ and $\begin{pmatrix} \bar{n} \\ -\bar{p} \end{pmatrix}$ are iso-doublets, perhaps the discrete transformation which takes

$$\begin{pmatrix} p \\ n \end{pmatrix} \to \begin{pmatrix} \bar{n} \\ -\bar{p} \end{pmatrix} \tag{20.12}$$

commutes with $SU(2)$ and H_{strong} and is also a symmetry operation. This is the symmetry called G parity. First, define the transformation

$$\mathbf{C}\begin{pmatrix} p \\ n \end{pmatrix} = \begin{pmatrix} \bar{p} \\ \bar{n} \end{pmatrix} \tag{20.13}$$

and then look for the transformation $\begin{pmatrix} \bar{p} \\ \bar{n} \end{pmatrix} \to \begin{pmatrix} \bar{n} \\ -\bar{p} \end{pmatrix}$. Consider an isospin rotation about the 2 axis by an angle $\theta_2 = \pi$ or $\alpha = \pi$ and $\hat{\theta}_i = \delta_{i2}$. Then we have

$$e^{-i\pi I_2}\begin{pmatrix} \bar{p} \\ \bar{n} \end{pmatrix} \equiv i\tau_2\begin{pmatrix} \bar{p} \\ \bar{n} \end{pmatrix} = \begin{pmatrix} \bar{n} \\ -\bar{p} \end{pmatrix}. \tag{20.14}$$

We now define the operator which generates G parity given by

$$\mathbf{G} \equiv e^{-i\pi I_2}\mathbf{C} \neq \mathbf{C}e^{-i\pi I_2}. \tag{20.15}$$

We finally have

$$\mathbf{G}\begin{pmatrix} p \\ n \end{pmatrix} = \begin{pmatrix} \bar{n} \\ -\bar{p} \end{pmatrix}. \tag{20.16}$$

Now we also find

$$\mathbf{G}\begin{pmatrix} \bar{n} \\ -\bar{p} \end{pmatrix} = e^{-i\pi I_2}\begin{pmatrix} n \\ -p \end{pmatrix} = i\tau_2 e^{-i\pi I_2}\begin{pmatrix} p \\ n \end{pmatrix} \tag{20.17}$$

$$= (i\tau_2)(i\tau_2)^*\begin{pmatrix} p \\ n \end{pmatrix} = -\begin{pmatrix} p \\ n \end{pmatrix}.$$

Thus, we have

$$\mathbf{G}^2 \begin{pmatrix} p \\ n \end{pmatrix} = - \begin{pmatrix} p \\ n \end{pmatrix}, \tag{20.18}$$

$$\mathbf{G}^2 \begin{pmatrix} \bar{n} \\ -\bar{p} \end{pmatrix} = - \begin{pmatrix} \bar{n} \\ -\bar{p} \end{pmatrix}.$$

We now want to prove the following theorem,

$$[\mathbf{G}, I_i] \equiv 0. \tag{20.19}$$

If this theorem is true, then we can simultaneously diagonalize the operators

$$\{ \mathbf{G}, \ I_3, \ \vec{I}^{\,2}, \ H_{strong} \}.$$

Proof   We have

$$\mathcal{U}_I \mathbf{G} \begin{pmatrix} p \\ n \end{pmatrix} = \mathcal{U}_I \begin{pmatrix} \bar{n} \\ -\bar{p} \end{pmatrix} = u_I^* \begin{pmatrix} \bar{n} \\ -\bar{p} \end{pmatrix},$$

$$\mathbf{G}\mathcal{U}_I \begin{pmatrix} p \\ n \end{pmatrix} = u_I^* \mathbf{G} \begin{pmatrix} p \\ n \end{pmatrix} = u_I^* \begin{pmatrix} \bar{n} \\ -\bar{p} \end{pmatrix}.$$

This gives

$$[\mathcal{U}_I, \mathbf{G}] \begin{pmatrix} p \\ n \end{pmatrix} = 0. \tag{20.20}$$

Similarly, we have

$$[\mathcal{U}_I, \mathbf{G}] \begin{pmatrix} \bar{n} \\ -\bar{p} \end{pmatrix} = 0. \tag{20.21}$$

Hence, on any state $|A\rangle$ constructed out of the vacuum with many $ps, ns, \bar{p}s, \bar{n}s$ we have $[\mathcal{U}_I, \mathbf{G}]|A\rangle = 0$. Therefore, in words, we have $[\mathcal{U}_I, \mathbf{G}] = 0$ when acting on any state made of any number of nucleons and anti-nucleons.

What about a general hadronic state? Any hadronic state, including states with $\rho s, \pi s, K$ can be reached by starting with just Ns and $\bar{N}$s by strong interactions, such as, for example, $N + N \rightarrow N + N + \pi$. But we assumed that $[I_i, H_{strong}] = 0$ and $[\mathbf{C}, H_{strong}] = 0$, therefore we have $[\mathbf{G}, H_{strong}] = 0$. In addition, we now have

$$[[\mathbf{G}, \mathcal{U}_I], H_{strong}] = 0. \tag{20.22}$$

Thus, if $[\mathbf{G}, \mathcal{U}_I] = 0$ on the initial state in a scattering process, it must still be zero on the final state. Hence, we conclude that

$$[\mathbf{G}, \mathcal{U}_I] = 0. \tag{20.23}$$

Finally, since $\left[\mathbf{G}, \mathcal{U}_I(\vec{\theta})\right] = 0$ for any value of $\vec{\theta}$ and $\mathcal{U}_I(\vec{\theta}) \equiv e^{-i\vec{\theta} \cdot \vec{I}}$, we find

$$[I_i, \mathbf{G}] = 0. \tag{20.24}$$

$\square$

Now we want to prove that on any state with total isospin, I, we have $\mathbf{G}$ on that state having an eigenvalue $\mathbf{G}^2 \equiv (-1)^{2I}$. The proof is not difficult. We have already seen that (Eqn. 20.18) $\mathbf{G}^2|N\rangle = -|N\rangle$ and $\mathbf{G}^2 i\tau_2|\bar{N}\rangle = -i\tau_2|\bar{N}\rangle$. Thus, on any iso-doublet, we have $\mathbf{G}^2|\frac{1}{2}\rangle = -|\frac{1}{2}\rangle$. And on a general multi-nucleon state $|A\rangle$, we have

$$\mathbf{G}^2|A\rangle = \left\{ \begin{array}{ll} -|A\rangle & I_A = \frac{1}{2}, \frac{3}{2}, \ldots \quad \text{odd number of } Ns \\ |A\rangle & I_A = 0, 1, 2, \ldots \quad \text{even number of } Ns \end{array} \right. \tag{20.25}$$

Hence,

$$G^2 \equiv (-1)^{2I}. \tag{20.26}$$

20.3   Some Applications of Isospin and G-Parity Conservation

(1) What is the G parity of a pion?

$$\mathbf{G}|\pi^0\rangle \equiv e^{-i\pi I_2}\mathbf{C}|\pi^0\rangle \tag{20.27}$$
$$\equiv e^{-i\pi I_2}|\pi^0\rangle = -|\pi^0\rangle,$$

where the last equality follows from $180°$ rotation about the 2 isospin axis. Since $[I_i, \mathbf{G}] = 0$, the G parity of the entire pion iso-triplet multiplet is identical. Therefore, we have

$$\mathbf{G}|\pi\rangle = -|\pi\rangle \tag{20.28}$$

for $\pi^\pm$, π^0. Using G parity we obtain selection rules. For example, no process can take an even number of pions into an odd number of pions, or vice versa. On the other hand, since

$$\mathbf{G}|\pi N\rangle = -|\pi\bar{N}\rangle, \tag{20.29}$$

we can have processes, such as $\pi + N \to n\pi + N$, with n arbitrary, without violating G parity.

(2) Vector mesons are observed as strong-interaction resonances in e^+e^- scattering. For example, we have $e^+e^- \to$ virtual $\gamma \to \{\rho^0, \omega^0, \phi, J/\psi, \psi', \Upsilon, \ldots\}$ (see Table 15.2). Since $\mathbf{C}A_\mu\mathbf{C}^\dagger = -A_\mu$ and both quantum electrodynamics and the strong interactions preserve charge conjugation, this implies that all vector mesons satisfy $\mathbf{C}|V\rangle = -|V\rangle$. What can we say about their spin and parity? On-shell photons (at $q^2 = 0$) are pure vector ($J = 1$), with parity $P = -1$ for A_i. However, off-shell photons with $q^2 > 0$ have scalar components ($J = 0$), for A_0 and $P = +1$. Consider the process in Fig. 20.1. Assume V is a scalar and prove this false. The amplitude is proportional to

$$\langle 0|j_{EM}^\mu(0)|V(q)\rangle \equiv cq^\mu, \tag{20.30}$$

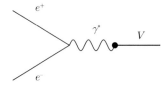

Fig. 20.1 $e^+e^- \to V$ where V is a vector meson.

where q^μ is the only available vector if V is a scalar field. But recall, gauge invariance requires $\partial_\mu j^\mu_{EM} = 0$. This then implies that

$$q_\mu \langle 0|j^\mu_{EM}(0)|V(q)\rangle \equiv cq^2 = 0 \qquad (20.31)$$

for any $q^2 \neq 0$. Thus, the only solution is that the constant $c = 0$. Therefore, we conclude that V cannot be a scalar, i.e. $J^{PC}_V = 1^{--}$.

(a) V cannot decay into any number of π^0s, since $\mathbf{C}|n\pi^0\rangle = |n\pi^0\rangle$ and $\mathbf{C}|V\rangle = -|V\rangle$.

(b) Experimentally, we find the dominant decay mode of the ρ meson is $\rho \to \pi^+\pi^-$. Using G parity we have $\mathbf{G}|\pi\pi\rangle = |\pi\pi\rangle$, which implies that $\mathbf{G}|\rho\rangle = |\rho\rangle$ by G-parity conservation. This implies the selection rule that a ρ cannot decay into an odd number of pions.

(c) What is the isospin of ρ? We have $\mathbf{G}^2|\rho\rangle = |\rho\rangle$. This implies that $I_\rho = 0, 1, 2, \ldots$. It is easy to see that $I_\rho \neq 0$. Since $\mathbf{G} = e^{-i\pi I_2}\mathbf{C}$ and if $I_\rho = 0$, we find $\mathbf{G}|\rho_0\rangle = -|\rho_0\rangle$. But this contradicts (b). So $I_\rho \neq 0$. Also $I_\rho \geq 2$ is not allowed, since there exist only three states, $\rho^\pm$, ρ^0, which are approximately degenerate. Therefore, we conclude that $I_\rho = 1$. Let's now check that this is consistent. We have $\mathbf{G}|\rho^0\rangle = e^{-i\pi I_2}\mathbf{C}|\rho^0\rangle = -e^{-i\pi I_2}|\rho^0\rangle = |\rho^0\rangle$, i.e. this is self-consistent.

(d) Note, the states ω, ϕ, J/ψ, etc. have no other degenerate states nearby in mass. Therefore, we have $I_{\omega, \phi, J/\psi,...} = 0$ and $\mathbf{G}_{\omega, \phi, J/\psi,...} = -1$. This gives the selection rule that the states ω, ϕ, J/ψ cannot decay into an even number of pions.

20.4 Exercises

20.1 Define the generator of isospin,

$$I_i = \frac{1}{2}\int d^3\vec{x}\,\psi^\dagger_N(x)\tau_i\psi_N(x), \qquad (20.32)$$

where $\psi_N(x)$, defined in Eqn. 19.5, is an iso-doublet field containing two four-component fields, ψ_p, ψ_n, representing the proton and neutron, respectively; τ_i, $i = 1, 2, 3$ are Pauli matrices acting in isospin space.

(a) Show that

$$[I_i, \ I_j] = i\epsilon_{ijk}I_k \qquad (20.33)$$

and
  (b)

$$[I_i, \; \psi_N(x)] = -\frac{\tau_i}{2}\psi_N(x). \tag{20.34}$$

(c) Define rotation operators

$$U_I(\vec{\theta}) \equiv \exp(-i\vec{I} \cdot \vec{\theta}) \tag{20.35}$$

where θ_i are three real parameters. Show that

$$U_I(\vec{\theta})\psi_N(x)U_I^\dagger(\vec{\theta}) = u_I\psi_N(x), \tag{20.36}$$

where

$$u_I(\vec{\theta}) = \exp(i\frac{\vec{\tau}}{2} \cdot \vec{\theta}) \tag{20.37}$$

is the representation of the isospin rotation group in the two-dimensional space. (Hint: let $\vec{\theta} = \alpha\hat{\theta}$, where $\alpha = |\vec{\theta}|$ and $\hat{\theta} = \vec{\theta}/\alpha$ and differentiate with respect to α.)
  (d) Assume

$$[I_i, \; H_{strong}] = 0. \tag{20.38}$$

Show that this implies that the proton and neutron, which are eigenstates of H_{strong}, have equal mass.

20.2 The pions, $\pi^\pm$, π^0, transform as a vector under isospin, where

$$I_3|\pi^\pm\rangle = \pm|\pi^\pm\rangle, \quad I_3|\pi^0\rangle = 0 \tag{20.39}$$
$$I^\pm|\pi^0\rangle = \mp\sqrt{2}|\pi^\pm\rangle, \quad I^+|\pi^+\rangle = 0, \quad I^-|\pi^+\rangle = -\sqrt{2}|\pi^0\rangle$$

and

$$\pi^\pm \equiv \frac{1}{\sqrt{2}}(\pi_1 \pm i\pi_2), \quad \pi^0 \equiv \pi_3 \tag{20.40}$$

and

$$I^\pm \equiv (I_1 \pm iI_2). \tag{20.41}$$

π_i are real scalar fields and $I^\pm$ are raising and lowering operators.
  The pion–nucleon scattering cross-section, at a laboratory pion energy $\sim$ 300 MeV, is dominated by the resonance, $\Delta(1232)$ with isospin 3/2, i.e. $\pi + N \to \Delta(1232) \to \pi + N$. Use conservation of isospin, $[I_i, \; S] = 0$ (where S is the scattering operator), and the Wigner–Eckart theorem to show that the ratio of cross-sections,

$$\sigma(\pi^+p \to \pi^+p) : \sigma(\pi^-p \to \pi^0n) : \sigma(\pi^-p \to \pi^-p) = 9 : 2 : 1. \tag{20.42}$$

21 Evidence for New Particles, Quantum Numbers and Interactions

In $\pi^- p$ scattering experiments using bubble-chamber detectors, the process $\pi^- + p \to \pi^- + p$ scattering occurs frequently with a cross-section of order 20 mb, characteristic of strong interactions. But new final states were also produced at the same rate. For example, it was found that $\pi^- + p \to \Lambda^0 + K^0$ (lambda baryons and kaons), where the Λ^0 has the dominant decay mode $\Lambda^0 \to \pi + N$ and the K^0 decays, for example, by $K^0 \to \pi^+ + \pi^-$. The ρ meson is also produced via strong interactions, but its lifetime is so short (of order 10^{-23} s) that it is only observed as a resonance peak in a scattering cross-section. The new states Λ^0, K^0, on the other hand, travel a visible distance in the bubble chamber before decaying. They were observed as Vs in a bubble-chamber photo. Their longer lifetime corresponds to a weaker decay interaction than the strong interaction responsible for ρ-meson decay. Weak interactions were already known to be responsible for beta decay processes, such as $n \to p + e^- + \bar{\nu}$ or $\pi^+ \to \mu^+ + \nu_\mu$. Perhaps this was another example of weak interactions. But what would explain the fact that these new states were produced via strong interactions, and then decay only via weak interactions?

The decay mode of the Λ^0 is used to argue that the baryon number of the Λ^0 is $B_{\Lambda^0} = +1$ and thus $B_{K^0} = 0$. In addition, there is a state $\bar{\Lambda}^0$, which is degenerate with the Λ^0, and decays via $\bar{\Lambda}^0 \to \pi + \bar{N}$. Thus, it has the baryon number $B_{\bar{\Lambda}^0} = -1$. It is easy to see that the states Λ^0, $\bar{\Lambda}^0$ are NOT an iso-doublet.

Proof If $\begin{pmatrix} \Lambda^0 \\ \bar{\Lambda}^0 \end{pmatrix}$ were a doublet, this implies that $I_+ \bar{\Lambda}^0 = \Lambda^0$ and $I_- \Lambda^0 = \bar{\Lambda}^0$. Now we use the fact that isospin commutes with the baryon number. We then have

$$1 = \langle \Lambda^0 | B | \Lambda^0 \rangle = \langle \Lambda^0 | B I_+ | \bar{\Lambda}^0 \rangle$$
$$= \langle \Lambda^0 | I_+ B | \bar{\Lambda}^0 \rangle = \langle I_- \Lambda^0 | B | \bar{\Lambda}^0 \rangle$$
$$= \langle \bar{\Lambda}^0 | B | \bar{\Lambda}^0 \rangle = -1,$$

which is a contradiction. Therefore, we conclude $I_{\Lambda^0} = I_{\bar{\Lambda}^0} = 0$. $\square$

What about K^0? It turns out that there are four approximately degenerate states $\{K^+, K^-, K^0, \bar{K}^0\}$. Since K^0 is produced in the process $\pi^- + p \to \Lambda^0 + K^0$, and $I_\pi = 1$, $I_p = \frac{1}{2}$, $I_{\Lambda^0} = 0$, we can conclude that $I_{K^0} = \frac{1}{2}$ or $\frac{3}{2}$. So which is it?

In order to bring some understanding to these new particles, their production cross-sections and their weak decay rates, Abraham Pais noted that these new particles were always produced in association, i.e. the processes $\pi^- + p \to$

K^0+n or $\pi^0+\Lambda^0$ were never observed. Gell-Mann then ascribed this phenomenon to the existence of a new quantum number, he called "strangeness" (S). The new particles carried this new quantum number, with $S_{K^0} = +1$, $S_{\Lambda^0} = -1$ and $S_\pi = S_p = 0$. Assuming that S is conserved by the strong interactions, these new particles were necessarily produced in pairs. Their decays, on the other hand, went via weak interactions, which violated S conservation. Then, under charge conjugation, we can propose the transformation

$$\mathbf{C}|K^0\rangle = \eta_{K^0}|\bar{K}^0\rangle, \qquad (21.1)$$

with $S_{\bar{K}^0} = -1$. Assuming that isospin and strangeness commute, i.e. $[I_i, S] = 0$, then all states in an isospin multiplet necessarily have the same value of strangeness. Consistent with data, we have $S = +1$ for $\{K^0, K^+\}$ and $S = -1$ for $\{\bar{K}^0, K^-\}$. As far as isospin is concerned, we have two iso-doublets given by $\begin{pmatrix} K^+ \\ K^0 \end{pmatrix}$ and $\begin{pmatrix} \bar{K}^0 \\ -K^- \end{pmatrix}$.

These were not the only new particles produced associatively by the strong interactions. We also have the Σ baryons, produced in processes such as $\pi^- + p \to \Sigma^- + K^+$ or $\Sigma^0 + K^0$, with the dominant decay modes of the Σs given by $\Sigma^- \to \pi^- + n$ and $\Sigma^0 \to \Lambda^0 + \gamma$. There are in fact an iso-triplet of Σs, i.e. $\{\Sigma^+, \Sigma^0, \Sigma^-\}$, with $I_\Sigma = 1, S_\Sigma = -1$ and $B_\Sigma = +1$. Note, processes like $\pi^- + p \to K^0 + n$ and $\pi^- + p \to \Sigma^+ + K^-$ are forbidden by strangeness conservation.

Once kaons were produced via πN interactions, then beams of charged kaons were made such that KN interactions could be observed in bubble-chamber experiments. For example, the process $K^+ + p \to K^0 + \pi^+ + p$ was observed. The Δ^{++} state was seen as a resonance in the $\pi^+ p$ channel and a new resonance, K^{+*} was observed in the $K^0\pi^+$ channel. Clearly, there was evidence for an immense particle zoo, requiring some principle to bring order to this chaos.

21.1 Bringing Order to Chaos

This order was provided by a combination of detailed experimental measurements and the inspiration of Gell-Mann. The spin, parity assignments for the new particles were obtained by experiment. All the new particles were listed in tables with their spin, parity, isospin and strangeness quantum numbers. In 1956, Gell-Mann then organized all the new particles graphically in a two-dimensional diagram with a horizontal axis given by the third component of isospin, I_3, and the vertical axis given by a new quantum number, called hypercharge, Y, with $Y \equiv S + B$. In Figs. 21.1–21.4, we present the new organization for the spin 1/2, spin 3/2, spin 0 and spin 1 particles, respectively.[1] Note, the electric charge of the particle states is then

---

[1] For a reference to the early understanding of the particle zoo, see (Gell-Mann, 1956; Gell-Mann and Rosenbaum, 1957).

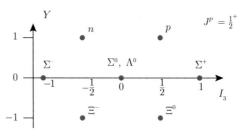

Fig. 21.1 Baryon octet with spin, parity, $J^P = \frac{1}{2}^+$.

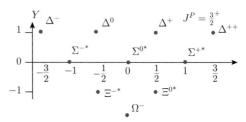

Fig. 21.2 Baryon decuplet with spin, parity, $J^P = \frac{3}{2}^+$.

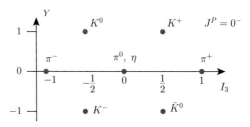

Fig. 21.3 $J^P = 0^-$ meson octet.

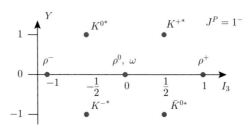

Fig. 21.4 $J^P = 1^-$ vector-meson octet.

given in terms of the Gell-Mann–Nishijima formula,

$$Q = I_3 + \frac{Y}{2}. \tag{21.2}$$

To summarize, global isospin and strangeness symmetries are only approximate. They are broken by electrodynamics and weak interactions. Nevertheless, they lead to useful relations.

Table 21.1 Changes in isospin in weak interactions. Note, in both cases the change $|\Delta I_3| = \frac{1}{2}$ and $|\Delta I| = \frac{3}{2}$ or $\frac{1}{2}$.

	$\Lambda^0 \to p + \pi^-$	$K^\pm \to \pi^\pm + \pi^0$
I	$0 \to \frac{1}{2}\ 1$	$\frac{1}{2} \to 1\ 1$
I_3	$0 \to \frac{1}{2} + (-1)$	$\pm\frac{1}{2} \to \pm 1 + 0$

Table 21.2 $|\Delta I|$ and $|\Delta I_3|$ in weak interactions. Note, only hadrons carry isospin.

$n \to p + e^- + \bar{\nu}_e$	$K^\pm \to \pi^0 + e^\pm + \left(\begin{matrix} \nu_e \\ \bar{\nu}_e \end{matrix} \right)$
$|\Delta I| = 0$	$|\Delta I| = \frac{1}{2}$
$|\Delta I_3| = 1$	$|\Delta I_3| = \frac{1}{2}$

- Isospin symmetry leads to an approximate degeneracy of states in an isospin multiplet.
- We obtain relations between scattering and decay amplitudes for states in the same isospin multiplet.
- Discrete G parity implies selection rules.
- Broken symmetries provide a framework for understanding symmetry breaking (see Tables 21.1 and 21.2).

We close this chapter with two examples of the last statement.

(1) Consider the electromagnetic interaction. It is easy to check that all the above states have electric charge given by the charge operator,

$$Q = I_3 + \frac{Y}{2} \equiv \int d^3 \vec{x} j_{EM}^0(x). \tag{21.3}$$

This implies that $j_{EM}^\mu \equiv j_3^\mu + \frac{1}{2} j_Y^\mu$. Note, by definition, we have $I_3 |\gamma\rangle = Y |\gamma\rangle \equiv 0$. Now since $H_{EM} \equiv e j_{EM}^\mu A_\mu$, we have that

$$[I_3,\ H_{EM}] = [Y,\ H_{EM}] = 0.$$

On the other hand, $[I_\pm,\ H_{EM}] \neq 0$. We conclude that the electromagnetic symmetry breaking of isospin nevertheless preserves I_3 and Y. Note, I_3 conservation of electromagnetic interactions allows for the processes $p \leftrightarrow p + \gamma$ and $n \leftrightarrow n + \gamma$. But the reaction $\Lambda \leftrightarrow n + \gamma$ violates both I_3 and Y. It is, however, allowed by the weak interactions which violate both I_3 and Y. Experimentally, the branching ratio $BR(\Lambda \to n + \gamma) = (1.02 \pm 0.33) \times 10^{-3}$.

(2) Consider the weak interaction processes in Table 21.1.

Note that in both cases the change $|\Delta I_3| = \frac{1}{2}$ and $|\Delta I| = \frac{3}{2}$ or $\frac{1}{2}$. Similarly, consider the processes in Table 21.2.

Let's postulate that the weak interactions satisfy $|\Delta I| \leq 1$. Assume, therefore that the weak interaction Hamiltonian, H_{weak}, is a tensor operator under isospin with $I_{H_{weak}} \leq 1$. Use this assumption to evaluate the ratio

$$R \equiv \frac{\Gamma(\Lambda^0 \to \pi^- \ p)}{\Gamma(\Lambda^0 \to \pi^0 \ n)}. \tag{21.4}$$

For the top process, we have

$$|\pi^- p\rangle = |\frac{3}{2}, -\frac{1}{2}\rangle \left(\frac{3}{2}, -\frac{1}{2}|1, -1; \frac{1}{2}, \frac{1}{2}\right) + |\frac{1}{2}, -\frac{1}{2}\rangle \left(\frac{1}{2}, -\frac{1}{2}|1, -1; \frac{1}{2}, \frac{1}{2}\right)$$

and

$$|\Lambda^0\rangle = |0, 0\rangle.$$

To zeroth order in the weak interactions, i.e. $(H_{weak})^0 : \langle \pi^- p|S_0|\Lambda^0\rangle = 0$. To first order in the weak interactions, $S_1 \sim H_{weak}$, we have

$$\langle \pi^- p|S_1|\Lambda^0\rangle \stackrel{|\Delta I| \leq 1}{=} \langle \frac{1}{2}, -\frac{1}{2}|S_1|0, 0\rangle \left(\frac{1}{2}, -\frac{1}{2}|1, -1; \frac{1}{2}, \frac{1}{2}\right)^* = \langle \frac{1}{2}, -\frac{1}{2}|S_1|0, 0\rangle(\sqrt{2/3}). \tag{21.5}$$

Similarly, for the bottom process, we have

$$|\pi^0 n\rangle = |\frac{3}{2}, -\frac{1}{2}\rangle \left(\frac{3}{2}, -\frac{1}{2}|1, 0; \frac{1}{2}, -\frac{1}{2}\right) + |\frac{1}{2}, -\frac{1}{2}\rangle \left(\frac{1}{2}, -\frac{1}{2}|1, 0; \frac{1}{2}, -\frac{1}{2}\right).$$

Hence,

$$\langle \pi^0 n|S_1|\Lambda^0\rangle \stackrel{|\Delta I| \leq 1}{=} \langle \frac{1}{2}, -\frac{1}{2}|S_1|0, 0\rangle \left(\frac{1}{2}, -\frac{1}{2}|1, 0; \frac{1}{2}, -\frac{1}{2}\right)^* = \langle \frac{1}{2}, -\frac{1}{2}|S_1|0, 0\rangle(-\sqrt{1/3}). \tag{21.6}$$

We thus find

$$R =\equiv \frac{\Gamma(\Lambda^0 \to \pi^- p)}{\Gamma(\Lambda^0 \to \pi^0 n)} = 2. \tag{21.7}$$

Experimentally, we have $BR(\Lambda^0 \to \pi^- p) = (64.1 \pm 0.5)\%$ and $BR(\Lambda^0 \to \pi^0 n) = (35.7 \pm 0.5)\%$ or $R = \frac{64.1}{35.7} \sim 2$, which is good to 20%. Note, this is not an exact rule. In cases where both $|\Delta I| = \frac{1}{2}$ and $\frac{3}{2}$ contribute, it is found experimentally that $|\Delta I| = \frac{1}{2}$ dominates. This is the so-called $|\Delta I| = \frac{1}{2}$ rule. In these cases, $|\Delta I| = \frac{3}{2}$ is allowed, but suppressed in amplitude by a factor of order $1/20$. This fact needs to be understood at the quark level.

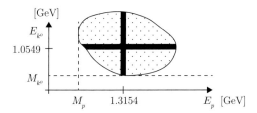

Fig. 21.5 This is a Dalitz plot of the differential cross-section for the process $K^+p \to K^0\pi^+p$.

21.2 Exercises

21.1 Consider the decay $K \to 2\pi$.

(a) What is the orbital angular momentum of the pions?

(b) Use Bose statistics to identify the allowed isospin states of the two pions, i.e. $I = 0, 1, 2$?

(c) If in weak decays, only $|\Delta I| \leq 1$ is allowed, is the process $K^+ \to \pi^+\pi^0$ permitted?

(d) Given

$$\tau_{K^+} = 1.24 \times 10^{-8} \text{ s}, \quad BR(K^+ \to \pi^+\pi^0) = (21.17 \pm 0.16)\% \quad (21.8)$$
$$\tau_{K^0} = 0.892 \times 10^{-10} \text{ s}, \quad BR(K^0 \to \pi^+\pi^-) = (68.61 \pm 0.28)\%,$$

calculate the ratio of the $|\Delta I| = 3/2$ to $\Delta I = 1/2$ amplitudes for these processes.

21.2 The scattering process $K^+p \to K^0\pi^+p$ is measured in the laboratory with a K^+ beam with energy $E_{K^+}^{lab} = 3$ GeV incident on a hydrogen bubble chamber. The K^0 decays into $\pi^+\pi^-$. The energy and momenta of the final states are measured and a cartoon of the observed differential cross-section (in the center of momentum system {CMS})

$$\frac{d^2\sigma}{dE_p dE_{K^0}} \quad (21.9)$$

is plotted in Fig. 21.5. In actuality, this experiment was performed at an energy of 3.5 GeV (de Baere et al., 1967). This is known as a Dalitz plot, where each point represents an event and the dark bands represent clusters of points. From this plot we learn that there are two resonances produced, $K^+p \to K^{*+}p$, followed by the decay $K^{*+} \to K^0\pi^+$, and $K^+p \to K^0\Delta^{++}$ followed by the decay $\Delta^{++} \to \pi^+p$.

Using the location of the bands given in the figure, find the mass of the Δ^{++} and K^{*+}. Note, $m_p = 0.938$ GeV, $m_{K^+} = 0.494$ GeV, $m_{K^0} = 0.498$ GeV and $m_{\pi^+} = 0.140$ GeV.

22 Representation Theory for $SU(2)$

In the previous chapter, we discussed the isospin symmetry of the strong interactions and identified the quantum numbers of many strong-interaction states. This chapter focusses on the representation theory for the group $SU(2)$. This is a precursor for our discussion of the flavor group, $SU(3)$, and the non-relativistic quark model of baryons and mesons.

The group $SU(2)$ is defined by the set of elements and a multiplication law. We have

$$SU(2) = \{u \mid u = 2 \times 2 \text{ complex matrices}; u^\dagger u = 1, \det u = 1\}. \tag{22.1}$$

The multiplication rule is simply matrix multiplication. In general, an element of any compact group such as $SU(2)$ can be written as

$$u = e^{-iT_i\theta_i}, \tag{22.2}$$

where T_i are the generators of the group and θ_i are real parameters. The number of generators is determined by the conditions satisfied by u. In this case, $u^\dagger u = 1$ implies that $T_i^\dagger = T_i$, i.e. the T_is are hermitian 2×2 matrices, and $\det u = 1$ implies that $\text{Tr}(T_i) = 0$, i.e. T_is are traceless. There are only three possible traceless, Hermitian, 2×2 matrices. Therefore, the index $i = 1, 2, 3$. In addition, we can always normalize the matrices via the condition, $\text{Tr}(T_iT_j) = \frac{1}{2}\delta_{ij}$. Finally, we need to choose a set of matrices with these conditions. The choice defines a particular convention. In the case of $SU(2)$, this convention is given in terms of the Pauli matrices, i.e. $T_i = \frac{\tau_i}{2}$. This two-dimensional complex vector space that the u matrices act on is the defining representation of the group. The generators satisfy the commutation relations, or Lie algebra,

$$[T_i, \, T_j] = i\epsilon_{ijk}T_k. \tag{22.3}$$

Finally, following Cartan, we define a set of raising and lowering generators,

$$T_\pm = T_1 \pm iT_2, \tag{22.4}$$

and the diagonal generator, T_3. They satisfy the commutation relations,

$$[T_3, \, T_\pm] = \pm T_\pm, \tag{22.5}$$
$$[T_+, \, T_-] = 2T_3.$$

We can now follow Cartan to determine all the finite-dimensional irreducible representations of the group $SU(2)$. The generators $\{T_3, \vec{T}^2\}$ are a complete set of commuting generators of $SU(2)$. We then have the eigenstates given by

$$\vec{T}^2|I,\ t_3\rangle = I(I+1)|I,\ t_3\rangle, \tag{22.6}$$

$$T_3|I,\ t_3\rangle = t_3|I,\ t_3\rangle.$$

Irreducible representations are defined in terms of the so-called highest weight state, where t_3 is a weight, and the highest weight is given by t_3^{MAX}. By definition,

$$T_+|I,t_3^{MAX}\rangle \equiv 0. \tag{22.7}$$

However, we have

$$T_3T_\pm|I,\ t_3\rangle \equiv ([T_3,\ T_\pm]+T_\pm T_3)|I,\ t_3\rangle$$

$$= (\pm 1 + t_3)T_\pm|I,\ t_3\rangle.$$

We therefore conclude that

$$T_\pm|I,\ t_3\rangle \equiv N_\pm|I,\ t_3\pm 1\rangle, \tag{22.8}$$

where $N_\pm$ is a normalization factor. Define the normalization of states via

$$\langle I,\ t_3'|I,\ t_3\rangle \equiv \delta_{t_3 t_3'}. \tag{22.9}$$

Now evaluate $N_\pm$. We find

$$N_\pm^2 = \langle I,\ t_3|T_\mp T_\pm|I,\ t_3\rangle. \tag{22.10}$$

Using the identities

$$\vec{T}^2 \equiv \frac{1}{2}(T_+T_- + T_-T_+) + T_3^2 \tag{22.11}$$

and $[T_+,\ T_-] = 2T_3$, we have

$$N_\pm^2 = \langle I,\ t_3|\frac{1}{2}\{T_\mp,\ T_\pm\} + \frac{1}{2}[T_\mp,\ T_\pm]|I,\ t_3\rangle$$

$$= \langle I,\ t_3|(\vec{T}^2 - T_3^2 \mp T_3)|I,\ t_3\rangle$$

$$= I(I+1) - t_3(t_3 \pm 1)$$

or, or up to an arbitrary phase, we have

$$N_\pm(t_3) = \sqrt{I(I+1) - t_3(t_3 \pm 1)}. \tag{22.12}$$

Now, by definition, $N_+(t_3^{MAX}) \equiv 0$. Therefore, $t_3^{MAX} = I$. All other states in the irreducible representation can then by obtained by acting with T_-, $2t_3^{MAX} = 2I$ times, until we reach the minimum value of $t_3 = t_3^{MIN}$, which satisfies $T_-|I,t_3^{MIN}\rangle = 0$, where $t_3^{MIN} \equiv -I$. Finally, the quantity $2I+1$ (which counts the number of states in the irreducible representation) is an integer and thus all finite irreducible representations of $SU(2)$ have value, $I = 0,\ \frac{1}{2},\ 1,\ \frac{3}{2},\ \dots$.

Consider the generators of $SU(2)$ in the $(2I+1)\times(2I+1)$ matrix representation. In general, we have

$$(T_+)_{t_3 t_3'}^I \equiv \langle I,\ t_3|T_+|I,\ t_3'\rangle$$

$$(T_3)_{t_3 t_3'}^I \equiv \langle I,\ t_3|T_3|I,\ t_3'\rangle = t_3\ \delta_{t_3 t_3'}.$$

For example, let $I = 1$, then we have

$$T_3 = \begin{pmatrix} 1 & 0 & 0 \\ 0 & 0 & 0 \\ 0 & 0 & -1 \end{pmatrix}$$

$$T_+ = \begin{pmatrix} 0 & \sqrt{2} & 0 \\ 0 & 0 & \sqrt{2} \\ 0 & 0 & 0 \end{pmatrix}$$

$$T_- = T_+^\dagger = \begin{pmatrix} 0 & 0 & 0 \\ \sqrt{2} & 0 & 0 \\ 0 & \sqrt{2} & 0 \end{pmatrix}.$$

Finally, it is straightforward to show that

$$[T_3, \ T_\pm] = \pm T_\pm$$
$$[T_+, \ T_-] = 2T_3,$$

i.e. the commutation relations in every irreducible representation are the same as those in the defining representation.

22.1 Exercises

22.1 Find the generators of $SU(2)$ in the $I = 3/2$ representation, as we discussed with

$$(T_i)_{t_3 t_3'} \equiv \langle t_3 | T_i | t_3' \rangle \tag{22.13}$$

for $i = 1, 2, 3$, i.e. write out the three 4×4 matrices.

Show (without actually commuting the matrices) that these matrices satisfy the Lie algebra

$$[T_i, \ T_j] = i\epsilon_{ijk} T_k. \tag{22.14}$$

23 $SU(3)$ Symmetry

In Figs. 21.1–21.4 we see the strong interacting states which were found experimentally and then organized by Gell-Mann into two-dimensional plots in terms of strong hypercharge, $Y = S + B$, and the third component of isospin, I_3. This led to the discovery of the flavor group $SU(3)$ by Gell-Mann (the Eightfold Way) and Ne'eman (Gell-Mann, 1961, 1962; Ne'eman, 1961) and, subsequently, the quark model by Gell-Mann and Zweig (Gell-Mann, 1964; Zweig, 1964a,b).[1] Experimental verification of the quark model came quickly with the discovery of the Ω^- at Brookhaven. But we will come to this in due time. First, let's discuss the representations of the symmetry group $SU(3)$. A nice reference for $SU(3)$ can be found in Carruthers (1966).

By definition, we have

$$SU(3) = \{u \mid u = 3 \times 3 \text{ complex matrices}; \ u^\dagger u = 1, \ \det u = 1\}, \tag{23.1}$$

with the group product given by matrix multiplication. The matrices u can be written in the form $u = e^{-iT_a\theta_a}$, where T_a are the generators of the group and θ_a are real parameters. The constraint $u^\dagger u = 1$ implies that $T_a = T_a^\dagger$, and $\det u = 1$ implies that $\text{Tr}(T_a) = 0$, i.e. they are traceless, Hermitian 3×3 matrices. There are only eight such matrices. Therefore, the index $a = 1, \ldots, 8$. We then normalize the matrices via the condition, $\text{Tr}(T_a T_b) = \frac{1}{2}\delta_{ab}$. We thus obtain the generators of $SU(3)$ in the defining (or fundamental) representation given by

$$T_a \equiv \frac{\lambda_a}{2} \qquad \lambda_a - \quad \text{Gell-Mann matrices}$$

$$\lambda_1 = \begin{pmatrix} 0 & 1 & 0 \\ 1 & 0 & 0 \\ 0 & 0 & 0 \end{pmatrix} \qquad \lambda_2 = \begin{pmatrix} 0 & -i & 0 \\ i & 0 & 0 \\ 0 & 0 & 0 \end{pmatrix} \qquad \lambda_3 = \begin{pmatrix} 1 & 0 & 0 \\ 0 & -1 & 0 \\ 0 & 0 & 0 \end{pmatrix}$$

$$\lambda_4 = \begin{pmatrix} 0 & 0 & 1 \\ 0 & 0 & 0 \\ 1 & 0 & 0 \end{pmatrix} \qquad \lambda_5 = \begin{pmatrix} 0 & 0 & -i \\ 0 & 0 & 0 \\ i & 0 & 0 \end{pmatrix}$$

$$\lambda_6 = \begin{pmatrix} 0 & 0 & 0 \\ 0 & 0 & 1 \\ 0 & 1 & 0 \end{pmatrix} \qquad \lambda_7 = \begin{pmatrix} 0 & 0 & 0 \\ 0 & 0 & -i \\ 0 & i & 0 \end{pmatrix} \qquad \lambda_8 = \frac{1}{\sqrt{3}}\begin{pmatrix} 1 & 0 & 0 \\ 0 & 1 & 0 \\ 0 & 0 & -2 \end{pmatrix}.$$

---

[1] According to Gell-Mann, he took the currents entering the weak decays of states in the baryon octet and commuted them to see if they closed on an algebra. He then learned that the algebra was that of the group $SU(3)$.

Table 23.1 The values for the structure constants, f_{abc}

a	b	c	f_{abc}
1	2	3	1
1	4	7	$\frac{1}{2}$
1	5	6	$-\frac{1}{2}$
2	4	6	$\frac{1}{2}$
2	5	7	$\frac{1}{2}$
3	4	5	$\frac{1}{2}$
3	6	7	$-\frac{1}{2}$
4	5	8	$\frac{\sqrt{3}}{2}$
6	7	8	$\frac{\sqrt{3}}{2}$

Strong hypercharge is then defined to be $Y \equiv \frac{2}{\sqrt{3}} T_8$. These matrices satisfy the commutation relations (or Lie algebra)

23.1 Lie Algebra of $SU(3)$

Given the Gell-Mann matrices, we can now evaluate the Lie algebra for $SU(3)$. We have $[T_a, T_b] = i f_{abc} T_c$, where f_{abc} are the structure constants of the group (see Table 23.1). However, a more useful form of the Lie algebra was defined by Cartan as follows. We shall define a set of raising and lowering operators, as we did previously for the group $SU(2)$. We have

$$T_\pm \equiv T_1 \pm i T_2,$$
$$V_\pm \equiv T_4 \pm i T_5,$$
$$U_\pm \equiv T_6 \pm i T_7.$$

It turns out that the group $SU(3)$ has three distinct $SU(2)$ subgroups embedded within it. Of course, there is the isospin subgroup, $SU(2)^I$, defined by the commutation relations

$$[T_3, T_\pm] = \pm T_\pm,$$
$$[T_+, T_-] = 2 T_3.$$

There is also U-spin, $SU(2)^U$, defined by

$$[U_+, U_-] = \frac{3}{2} Y - T_3 \equiv 2 U_3,$$
$$[U_3, U_\pm] = \pm U_\pm;$$

and there is V-spin, $SU(2)^V$, defined by

$$[V_+, \ V_-] = \frac{3}{2}Y + T_3 \equiv 2V_3.$$
$$[V_3, \ V_\pm] = \pm V_\pm.$$

We also can show that

$$[T_3, \ U_\pm] = \mp \frac{1}{2} \, U_\pm,$$
$$[T_3, \ V_\pm] = \pm \frac{1}{2} \, V_\pm.$$

Finally, to complete the commutation relations, we have

$$[Y, \ T_\pm] = 0 = [Y, \ T_3], \quad [Y, \ U_\pm] = \pm U_\pm, \quad [Y, \ V_\pm] = \pm V_\pm, \qquad (23.2)$$
$$[T_+, \ V_+] = [T_+, \ U_-] = [U_+, \ V_+] = 0,$$
$$[T_+, \ V_-] = -U_-, \quad [U_+, \ V_-] = T_-, \quad [T_+, \ U_+] = V_+.$$

23.2 Complete Set of Commuting Operators

A complete set of commuting operators can be given by

$$\left\{ T_3, \ Y, \ \sum_{a=1}^{8}(T_a)^2 \equiv C_2(R) \ \text{ quadratic Casimir + a cubic Casimir operator} \right\}.$$

$$(23.3)$$

We can then label the states by the eigenvalues of the operators T_3, Y. We have $|t_3, \ y\rangle$ such that

$$T_3|t_3, \ y\rangle = t_3|t_3, \ y\rangle,$$
$$Y|t_3, \ y\rangle = y|t_3, \ y\rangle.$$

We know from isospin that $T_\pm$ raises and lowers t_3 by ± 1. However, $T_\pm$ leaves the value y unchanged, since $YT_\pm|t_3, \ y\rangle = T_\pm Y|t_3, \ y\rangle = yT_\pm|t_3, \ y\rangle$.
  Similarly,

- $V_\pm$ raises and lowers t_3 by $\pm\frac{1}{2}$,
- $U_\pm$ lowers and raises t_3 by $\mp\frac{1}{2}$,
- $U_\pm$, $V_\pm$ raise and lower y by ± 1.

  Consider the y, t_3 plane, where we describe the action of the raising and lowering operators on states $|t_3, \ y\rangle$. The space of states of the group is called weight space, and the diagram, Fig. 23.1, which describes the action of the generators on the weights, is called the root diagram. In comparison, the root diagram for $SU(2)$ is given in Fig. 23.2.
  Irreducible representations of $SU(3)$ (generalizations of the three-dimensional defining representation) are obtained by starting with the highest weight state,

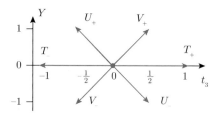

Fig. 23.1 y, t_3 plane and action of the $SU(3)$ generators on states.

Fig. 23.2 t_3 line and action of the $SU(2)$ generators on states.

$|t_3^{MAX}, y\rangle$ such that $T_+ |t_3^{MAX}, y\rangle \equiv 0$. For example, let us start with the highest weight state, we call $|\Sigma^+\rangle \equiv |1, 0\rangle$ such that $T_+|\Sigma^+\rangle = 0$.[2] We can obtain all the other states in this irreducible representation by using raising and lowering operators. Note, however, that

$$V_+|\Sigma^+\rangle = U_-|\Sigma^+\rangle \equiv 0, \qquad (23.4)$$

since they raise t_3.

Now let's evaluate $U_3|\Sigma^+\rangle$. We have $U_3|\Sigma^+\rangle = (\frac{3}{4}Y - \frac{1}{2}T_3)|1, 0\rangle = -\frac{1}{2}|\Sigma^+\rangle$. Given that $U_-|\Sigma^+\rangle \equiv 0$, we conclude that $U_-|\Sigma^+\rangle$ is the lowest weight state of a U-spin doublet. Then, define the state $|p\rangle \equiv U_+|\Sigma^+\rangle \equiv |\frac{1}{2}, 1\rangle$. We also have $U_+|p\rangle \equiv 0$, since $|p\rangle$ is the highest weight state of a U-spin doublet.

Consider $V_3|\Sigma^+\rangle = (\frac{3}{4}Y + \frac{1}{2}T_3)|1, 0\rangle = \frac{1}{2}|\Sigma^+\rangle$. Also, since $V_+|\Sigma^+\rangle \equiv 0$, we see that the state $|\Sigma^+\rangle$ is the highest weight state of a V-spin doublet. Define $V_-|\Sigma^+\rangle \equiv |\Xi^0\rangle \equiv |\frac{1}{2}, -1\rangle$, which is the lowest weight state of the V-spin doublet.

In Fig. 23.3 we show the action of the lowering operator T_- on the states Σ^+, p, Ξ^0.

Note, $T_3|p\rangle = \frac{1}{2}|p\rangle$ and $T_+|p\rangle \equiv T_+U_+|\Sigma^+\rangle \equiv [T_+, U_+]|\Sigma^+\rangle \equiv V_+|\Sigma^+\rangle = 0$. Therefore, $|p\rangle$ is the highest weight state of a T-spin doublet. Define $T_-|p\rangle \equiv |n\rangle$. Similarly, $T_-|\Xi^0\rangle \equiv |\Xi^-\rangle$. Also, we have $U_+|\Xi^-\rangle = V_-|n\rangle \equiv |\Sigma^-\rangle$ is unique.

What about $T_-|\Sigma^+\rangle =$? We know that $|\Sigma^+\rangle$ is the highest weight state and the third component of an isospin triplet. Therefore, we have $T_-|\Sigma^+\rangle = \sqrt{2}|\Sigma^0\rangle$ and acting once more with T_- we obtain $|\Sigma^-\rangle$. However, the question is, is the center state unique? If yes, then we should get to the same state from all directions. But we will show now that this is not possible. Consider the two states given by

$$(A) \quad V_-|p\rangle \equiv a|\Sigma^0\rangle + b|\Lambda\rangle, \qquad (23.5)$$
$$(B) \quad U_+|\Xi^0\rangle \equiv a'|\Sigma^0\rangle + b'|\Lambda\rangle, \qquad (23.6)$$

---

[2] This does not necessarily have to be identified with the baryons. It could just as well be identified with the π^+ or ρ^+ with the same quantum numbers.

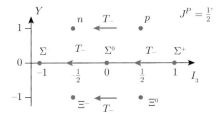

Fig. 23.3   Action of the lowering operator T_- on the states Σ^+, p, Ξ^0.

where, if the center was unique, the constants b and b' would be zero. Define

$$|\Lambda\rangle \equiv |0,\,0\rangle, \tag{23.7}$$

which is an isosinglet, i.e. $T_a|\Lambda\rangle \equiv 0$, $a = 1, 2, 3$, since it has no partners. Let's evaluate the constants a, b, a', b'. To do this we calculate the norms of the states (A) and (B). For (A) we have[3]

$$\langle p|\underbrace{V_+V_-}_{[V_+,\,V_-]\,=\,2V_3}|p\rangle = a^2\underbrace{\langle\Sigma^0|\Sigma^0\rangle}_{1} + b^2\underbrace{\langle\Lambda|\Lambda\rangle}_{1}.$$

We thus obtain $a^2 + b^2 = 2$. Similarly, from (B) we find the norm of

$$\langle\Xi^0|U_-U_+|\Xi^0\rangle,$$

which gives $a'^2 + b'^2 = 2$. So far, we only have two equations for four unknowns. Therefore, we need two more equations. We obtain these by using the commutator $[U_+,\,V_-] \equiv T_-$. We then have

$$(U_+V_- - V_-U_+ - T_-)|\Sigma^+\rangle \equiv 0,$$
$$U_+\,|\Xi^0\rangle - V_-|p\rangle - \sqrt{2}|\Sigma^0\rangle = 0,$$
$$(a' - a - \sqrt{2})|\Sigma^0\rangle + (b' - b)|\Lambda\rangle = 0.$$

We now have our two additional equations, i.e. $b' = b$ and $a' = a + \sqrt{2}$. The solution is given by $a' = \frac{1}{\sqrt{2}}$, $a = -\frac{1}{\sqrt{2}}$, $b' = b = \sqrt{\frac{3}{2}}$.

To summarize, we have constructed an irreducible representation of $SU(3)$: **8**, the octet (see the baryon octet in Fig. 21.1). Isospin $\otimes$ hypercharge are subgroups of the full flavor group, i.e. $SU(3)_{flavor} \supset SU(2)^I \otimes U(1)^Y$, with generators $\{T_a a = 1, \ldots, 8\} \supset \{T_i i = 1, 2, 3 \oplus Y\}$. Vector bosons and pseudo-scalars form an octet + singlet under $SU(3)$, i.e. **8** + **1** (see the vector-meson octet, Fig 21.4, and pseudo-scalar octet, Fig. 21.3). The additional vector-boson and pseudo-scalar $SU(3)$ singlets are given by ϕ and η', respectively. Last, but not least, for homework you will construct the decuplet representation by starting with the highest weight

---

[3] We use the fact that

$$V_+|p\rangle \equiv V_+U_+|\Sigma^+\rangle = [V_+,\,U_+]|\Sigma^+\rangle \equiv 0, \tag{23.8}$$

using the commutation relations, Eqn. 23.2.

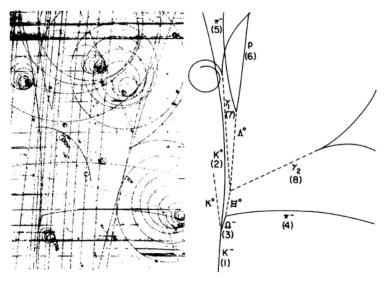

$K^- + p \rightarrow K^+ + K^0 + \Omega^-$ with the subsequent decay of the $\Omega^- \rightarrow \Xi^0 + \pi^-$, then $\Xi^0 \rightarrow \Lambda^0 + \pi^0$ and finally, $\Lambda^0 \rightarrow \pi^- + p$ and $\pi^0 \rightarrow \gamma_1 + \gamma_2$. Reprinted with permission from Barnes et al. (1964). Copyright (1964) by the American Physical Society.

state, $|\Delta^{++}\rangle \equiv |\frac{3}{2}, 1\rangle$. It is a remarkable fact that the electric charge of all these states is given by the simple formula,

$$Q = T_3 + \frac{Y}{2}. \tag{23.9}$$

Note, at the time the flavor group $SU(3)$ was proposed by Gell-Mann and Ne'eman, there was one state missing in the decuplet, i.e. the $|\Omega^-\rangle = |0, -2\rangle$. It has baryon number, $B = 1$, and strangeness, $S = -3$. It was discovered experimentally in a bubble-chamber experiment with $K^- + p \rightarrow K^+ + K^0 + \Omega^-$ (Barnes et al., 1964) (see Fig. 23.4), soon after the prediction of the Eightfold Way and the Gell-Mann–Okubo mass formula (Gell-Mann, 1962; Okubo, 1962) for the masses of the baryon octet and decuplet.[4]

23.3 The Quark Model

In this section we describe the fundamental (or defining) representation of $SU(3)$. Define the triplet state (**3**), $|\psi_i\rangle$, $i = 1, 2, 3$, and its transformation under $SU(3)$ given by

$$\mathcal{U}|\psi_i\rangle \equiv |\psi_i\rangle' = |\psi_j\rangle u^j{}_i, \tag{23.10}$$

---

[4] We will derive the Gell-Mann–Okubo mass formula in the next section.

where $\mathcal{U} \equiv e^{-i\theta_a \hat{T}_a}$ and $\hat{T}_a \equiv \int d^3\vec{x} j^0_a(x)$ is an operator on Fock space. To set notation, the matrix u has the form $u^j{}_i \equiv \begin{pmatrix} u^1{}_1 & u^1{}_2 & u^1{}_3 \\ u^2{}_1 & u^2{}_2 & u^2{}_3 \\ u^3{}_1 & u^3{}_2 & u^3{}_3 \end{pmatrix} \equiv (e^{-i\theta_a T_a})^j{}_i$, satisfying $u^\dagger u = 1$, $\det u = 1$. Finally, $(T_a)^j{}_i \equiv (\frac{\lambda_a}{2})_{ji}$. Under an infinitesimal transformation, we have $\delta_a |\psi_i\rangle = |\psi_i\rangle' - |\psi_i\rangle$, which implies that

$$\hat{T}_a |\psi_i\rangle = |\psi_j\rangle (T_a)^j{}_i. \tag{23.11}$$

Again, to set notation, we have $(u^*)^j{}_i \equiv \begin{pmatrix} u^{1*}{}_1 & u^{1*}{}_2 & u^{1*}{}_3 \\ u^{2*}{}_1 & u^{2*}{}_2 & u^{2*}{}_3 \\ u^{3*}{}_1 & u^{3*}{}_2 & u^{3*}{}_3 \end{pmatrix} \equiv (e^{i\theta_a T^*_a})^j{}_i$, and $(u^\dagger)^j{}_i \equiv (u^*)^i{}_j$, with $(u^\dagger)^j{}_i (u)^i{}_k \equiv \delta^j{}_k$. Note, $(T^*_a)^j{}_i \equiv (T^T_a)^j{}_i$ by the hermiticity of T_a. We now define $(\bar{T}_a)_j{}^i \equiv -(T^T_a)^j{}_i \equiv -(\frac{\lambda^T_a}{2})_{ji}$. With this notation, we have $(u^*)^j{}_i \equiv (e^{-i\theta_a \bar{T}_a})_j{}^i$.

The conjugate representation is defined by $(\bar{3})$, $|\psi^i\rangle$, $i = 1, 2, 3$, and transforms as follows

$$|\psi^i\rangle' \equiv |\psi^j\rangle (u^*)^j{}_i \equiv (u^\dagger)^i{}_j |\psi^j\rangle \equiv \mathcal{U} |\psi^i\rangle. \tag{23.12}$$

Then, an infinitesimal transformation of $(\bar{3})$ is given by

$$\hat{T}_a |\psi^i\rangle = -|\psi^j\rangle (T^T_a)^j{}_i = |\psi^j\rangle (\bar{T}_a)_j{}^i. \tag{23.13}$$

It is called the conjugate representation since $\langle \psi_i | \equiv |\psi_i\rangle^*$ transforms as a $(\bar{3})$, i.e. $|\psi_i\rangle' = |\psi_j\rangle u^j{}_i$, which implies that $|\psi_i\rangle'^* = |\psi_j\rangle^* u^{*j}{}_i$. Note, if $[T_a, T_b] = i f_{abc} T_c$, then

$$[\bar{T}_a, \bar{T}_b] = i f_{abc} \bar{T}_c. \tag{23.14}$$

What are the quantum numbers, $\{t_3, y\}$ of the $\mathbf{3}$? We use the triplet representation given by

$$T_3 \equiv \frac{\lambda_3}{2} = \begin{pmatrix} \frac{1}{2} & 0 & 0 \\ 0 & -\frac{1}{2} & 0 \\ 0 & 0 & 0 \end{pmatrix}$$

$$Y \equiv \frac{2}{\sqrt{3}} T_8 = \begin{pmatrix} \frac{1}{3} & 0 & 0 \\ 0 & \frac{1}{3} & 0 \\ 0 & 0 & -\frac{2}{3} \end{pmatrix}.$$

Since $\hat{T}_a |\psi_i\rangle = |\psi_j\rangle (T_a)^j{}_i$, we have

$$|\psi_1\rangle = |\frac{1}{2}, \frac{1}{3}\rangle,$$

$$|\psi_2\rangle = |-\frac{1}{2}, \frac{1}{3}\rangle,$$

$$|\psi_3\rangle = |0, -\frac{2}{3}\rangle.$$

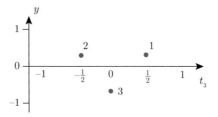

Fig. 23.5 The triplet representation on the weight diagram.

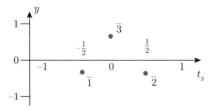

Fig. 23.6 The anti-triplet representation on the weight diagram.

On the weight diagram, the triplet representation is given in Fig. 23.5.

Similarly, the quantum numbers of the anti-triplet ($\bar{\mathbf{3}}$) is given by

$$|\psi^1\rangle = |-\frac{1}{2},\ -\frac{1}{3}\rangle,$$
$$|\psi^2\rangle = |\frac{1}{2},\ -\frac{1}{3}\rangle,$$
$$|\psi^3\rangle = |\ 0,\ \frac{2}{3}\rangle,$$

(see Fig. 23.6).

If we use the formula for the electric charge of the hadronic states, Eqn. 23.9, we can now identify electric charges of the triplet and anti-triplet states. We have

$$\psi_i = \begin{pmatrix} u \\ d \\ s \end{pmatrix} \quad \text{with charge} \quad \begin{matrix} \frac{2}{3} \\ -\frac{1}{3} \\ -\frac{1}{3} \end{matrix}, \tag{23.15}$$

i.e. the up. down and strange quarks, and

$$\psi^i = \begin{pmatrix} u^c \\ d^c \\ s^c \end{pmatrix} \quad \text{with charge} \quad \begin{matrix} -\frac{2}{3} \\ \frac{1}{3} \\ \frac{1}{3} \end{matrix}, \tag{23.16}$$

the charge-conjugate anti-quarks. At the moment, these are just names given to the fundamental representation states. What do they have to do with the observed hadrons?

23.4 Direct (or Tensor) Product States

Recall for the rotation group, $SU(2)$, we can form direct products of two irreducible representations. For example, $\frac{1}{2} \otimes 1 = \frac{1}{2} \oplus \frac{3}{2}$, which is given by the $SU(2)$ Clebsch–Gordan decomposition. The direct product state is written as $|I^1, I_3^1\rangle \otimes |I^2, I_3^2\rangle$ and the generators are given by $T_i \equiv T_i^1 + T_i^2$, i.e. they are a direct sum of generators, where T^1 acts only on $|I^1, I_3^1\rangle$ and T^2 acts only on $|I^2, I_3^2\rangle$, such that

$$T_i(|I^1, I_3^1\rangle \otimes |I^2, I_3^2\rangle) \equiv (T_i^1 |I^1, I_3^1\rangle) \otimes |I^2, I_3^2\rangle + |I^1, I_3^1\rangle \otimes (T_i^2 |I^2, I_3^2\rangle).$$

Consider for $SU(3)$, we now derive the Clebsch–Gordan decomposition,

$$\mathbf{3 \otimes \bar{3} = 1 + 8}.$$

For example, consider the tensor product (using a short-hand notation)

$$|_1^2\rangle \equiv |\psi_1\rangle \otimes |\psi^2\rangle. \tag{23.17}$$

In general, an infinitesimal transformation on such a state is given by

$$\hat{T}_a |_j^i\rangle = |_k^i\rangle (T_a)^k{}_j - |_j^k\rangle (T_a)^i{}_k. \tag{23.18}$$

We can thus calculate the quantum numbers of the product state. For example,

$$\hat{T}_3 |_1^2\rangle = \frac{1}{2}|_1^2\rangle - (-\frac{1}{2})|_1^2\rangle$$
$$\equiv |_1^2\rangle \equiv (t_3(1) + t_3(\bar{2}))|_1^2\rangle,$$
$$\hat{Y} |_1^2\rangle = \frac{1}{3} |_1^2\rangle - (\frac{1}{3})|_1^2\rangle$$
$$\equiv 0 \equiv (y(1) + y(\bar{2})) |_1^2\rangle.$$

Thus, in a tensor product, the quantum numbers add as vectors in weight space. For the general case, we can use a graphical description (adding vectors in weight space). Consider Fig. 23.7. We see that the state $|_1^2\rangle \equiv |1, 0\rangle$ is the highest weight state of the octet representation, $\mathbf{8}$, and is the product of the two highest weights

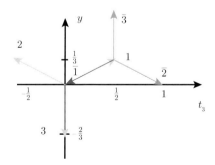

The highest weight state of the octet representation is obtained by the direct product state $|_1^2\rangle$.

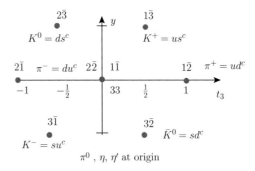

Fig. 23.8 The meson octet in terms of quarks.

of the triplet and anti-triplet. Thus, we obtain the full octet representation in an obvious notation, as in Fig. 23.8. In the quark model, the vector and pseudo-scalar octets have the quantum numbers of quark anti-quark bound states. Quarks have baryon number, $B_q = \frac{1}{3}$. Note, also, if the triplet states have baryon number, $\frac{1}{3}$, and $S = Y - B$, then u, d have strangeness, 0, and s has strangeness, -1.

In order to completely specify the three states at the origin in Fig. 23.8, we start with the highest weight state and use the isospin lowering operator to identify the π^0. We have

$$\hat{T}_- |{}^2_1\rangle (= |\pi^+\rangle) = -\sqrt{2}|\pi^0\rangle$$
$$= |{}^2_k\rangle (T_-)^k{}_1 - |{}^k_1\rangle (T_-)^2{}_k$$
$$= |{}^2_2\rangle - |{}^1_1\rangle,$$

where the minus sign on the first step is convention and we used

$$T_- = \begin{pmatrix} 0 & 0 & 0 \\ 1 & 0 & 0 \\ 0 & 0 & 0 \end{pmatrix}.$$

Hence, we find

$$|\pi^0\rangle = |0,\ 0\rangle = \frac{1}{\sqrt{2}} \left(uu^c - dd^c \right).$$

Similarly, we can find

$$|\eta\rangle = \frac{1}{\sqrt{6}} \left(|{}^1_1\rangle + |{}^2_2\rangle - 2|{}^3_3\rangle \right) = \frac{1}{\sqrt{6}} \left(uu^c + dd^c - 2ss^c \right).$$

The η is an isospin singlet, but a member of an $SU(3)$ octet. Finally, the η' is an $SU(3)$ singlet, given by

$$|\eta'\rangle = \frac{1}{\sqrt{3}} \left(|{}^1_1\rangle + |{}^2_2\rangle + |{}^3_3\rangle \right) = \frac{1}{\sqrt{3}} \left(uu^c + dd^c + ss^c \right).$$

Using the orthogonality relation,

$$\langle {}^{i'}_{j'} | {}^i_j \rangle \equiv \delta^i_{i'}\, \delta^{j'}_j,$$

the π^0, η, η' are orthogonal.

Some other direct product states of interest are

$$\mathbf{3} \otimes \mathbf{3} = \bar{\mathbf{3}} \oplus \mathbf{6},$$

$$\mathbf{3} \otimes \mathbf{6} = \mathbf{8} \oplus \mathbf{10},$$

$$\mathbf{3} \otimes \mathbf{3} \otimes \mathbf{3} = [\bar{\mathbf{3}} \oplus \mathbf{6}] \otimes \mathbf{3} = \mathbf{1} \oplus \mathbf{8}_1 \oplus \mathbf{8}_2 \oplus \mathbf{10}.$$

23.5 Exercises

23.1 Given the Gell-Mann matrices, $\{\lambda_a\}$, The generators of $SU(3)$ in the triplet representation, $T_a \equiv \frac{\lambda_a}{2}$, satisfy the Lie algebra of $SU(3)$, i.e.

$$[T_a, \ T_b] = i f_{abc} T_c, \tag{23.19}$$

where f_{abc} are the structure constants given in Table 23.1.

(a) Show that the generators,

$$\bar{T}_a \equiv -T_a^T, \tag{23.20}$$

satisfy the same algebra. This is the conjugate representation, $\bar{\mathbf{3}}$.

(b) Use the Jacobi identity,

$$[[T_a, \ T_b], \ T_c] + [[T_b, \ T_c], \ T_a] + [[T_c, \ T_a], \ T_b] \equiv 0, \tag{23.21}$$

to show that the eight 8×8 matrices,

$$(F_a)_{bc} \equiv -i f_{abc}, \quad a, b, c = 1 \ldots 8, \tag{23.22}$$

satisfy the Lie algebra of $SU(3)$. This is the octet (or adjoint) representation.

(c) (i) Given the highest weight state, $|t_3, y\rangle$ with $t_3 = 3/2$, $y = 1$, use raising and lowering operators to complete the multiplet.

(ii) Draw the weight diagram for this representation. Assuming $Q = T_3 + \frac{Y}{2}$, identify the charge of each state.

24 Tests of $SU(3)$ Symmetry

We now want to give two examples of the consequences of $SU(3)$ flavor symmetry. These are:

1. Coleman–Glashow relation (1961) (Coleman and Glashow, 1961) – electromagnetic corrections to the baryon octet;
2. Gell-Mann – Okubo relation (Gell-Mann, 1962; Okubo, 1962) – the largest symmetry-breaking correction to the baryon octet and decuplet.

24.1 Coleman–Glashow Relation

Recall the $SU(3)$ commutators

$$[U_\pm, T_3] = \pm \frac{1}{2} U_\pm,$$

$$[U_\pm, Y] = \mp U_\pm$$

and the definition

$$U_3 \equiv \frac{3}{4} Y - \frac{1}{2} T_3.$$

Using the empirical formula for the electric charge of hadrons,

$$Q \equiv T_3 + \frac{Y}{2},$$

we have

$$[U_\pm, Q] = 0 = [U_3, Q].$$

Now consider the baryon octet, Fig. 21.1. We see that all the states in a U-spin multiplet have the same electric charge. We also have

$$[U_i, H'_{EM}] = 0. \tag{24.1}$$

We want to calculate the electromagnetic corrections to the baryon masses, valid to first order in

$$H'_{EM} = -e^2 \int d^3\vec{x}\, d^3\vec{y}\, j^{EM}_\mu(x)\, j^\mu_{EM}(y)\, \Delta^\gamma(x - y), \tag{24.2}$$

i.e. radiative corrections to baryon masses to order e^2 in perturbation theory. Note, order e corrections vanish! Using time-independent perturbation theory, we have

$$\delta m_p \simeq \langle p | H' | p \rangle. \tag{24.3}$$

Since U-spin commutes with H'_{EM}, all elements of a U-spin multiplet receive the same correction. Therefore, we have

$$\delta m_p = \delta m_{\Sigma^+}, \quad \delta m_{\Sigma^-} = \delta m_{\Xi^-}, \quad \delta m_n = \delta m_{\Xi^0}. \tag{24.4}$$

In the limit that isospin is a good symmetry, i.e. broken only by electromagnetic interactions and the up–down quark mass difference, $\Delta \equiv m_u - m_d,$[1] we have the relations

$$m_p = m_N + \delta m_p + \Delta/2$$
$$m_n = m_N + \delta m_n - \Delta/2$$
$$m_{\Sigma^+} = m_\Sigma + \delta m_{\Sigma^+} + \Delta$$
$$m_{\Sigma^-} = m_\Sigma + \delta m_{\Sigma^-} - \Delta$$
$$m_{\Xi^0} = m_\Xi + \delta m_{\Xi^0} + \Delta/2$$
$$m_{\Xi^-} = m_\Xi + \delta m_{\Xi^-} - \Delta/2.$$

$$m_n - m_p = \delta m_n - \delta m_p - \Delta$$
$$m_{\Xi^-} - m_{\Xi^0} = \delta m_{\Xi^-} - \delta m_{\Xi^0} - \Delta$$
$$m_{\Sigma^-} - m_{\Sigma^+} = \delta m_{\Sigma^-} - \delta m_{\Sigma^+} - 2\Delta.$$

Thus,

$$m_n - m_p + m_{\Xi^-} - m_{\Xi^0} = \delta m_n - \delta m_p + \delta m_{\Xi^-} - \delta m_{\Xi^0} - 2\Delta$$
$$m_{\Sigma^-} - m_{\Sigma^+} = \delta m_{\Sigma^-} - \delta m_{\Sigma^+} - 2\Delta.$$

Using Eqn. 24.4, we obtain the Coleman–Glashow relation

$$m_n - m_p + m_{\Xi^-} - m_{\Xi^0} = m_{\Sigma^-} - m_{\Sigma^+}. \tag{24.5}$$

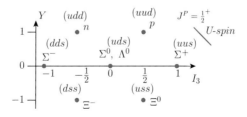

Fig. 24.1   Quark model composition of the baryon octet.

[1] See Fig. 24.1 and note, that the original derivation of Coleman–Glashow did not assume the quark model. Instead, they used tensor analysis. We will discuss tensor analysis in the next section.

Let's now compare this relation with data. We have from the PDG 2018:

$$(m_n - m_p) \, [\text{MeV}] = 1.29333205 \pm 0.00000051,$$
$$(m_{\Xi^-} - m_{\Xi^0}) \, [\text{MeV}] = 6.85 \pm 0.21,$$
$$(m_{\Sigma^-} - m_{\Sigma^+}) \, [\text{MeV}] = 8.08 \pm 0.08.$$

Thus we have excellent agreement with Eqn. 24.5, i.e.

$$[\text{LHS}] \ 8.14 \pm 0.21 = [\text{RHS}] \ 8.08 \pm 0.08. \tag{24.6}$$

24.2 Tensor Methods

Before considering the Gell-Mann–Okubo mass relations, we should learn a useful formalism known as tensor methods.[2] Consider a tensor product state made of n **3**s and m $\overline{\bf 3}$s given by $|^{i_1...i_m}_{j_1...j_n}\rangle$. The transformation under $SU(3)$ is given by

$$|^{i_1...i_m}_{j_1...j_n}\rangle' \equiv |^{k_1...k_m}_{\ell_1...\ell_n}\rangle (u^*)^{k_1}{}_{i_1} \cdots (u^*)^{k_m}{}_{i_m} u^{\ell_1}{}_{j_1} \cdots u^{\ell_n}{}_{j_n}, \tag{24.7}$$

with infinitesimal transformation given by

$$\hat{T}_a \, |^{i_1...i_m}_{j_1...j_n}\rangle = \sum_{\ell=1}^{n} |^{i_1...i_m}_{j_1...k...j_n}\rangle (T_a)^k{}_{j_\ell} - \sum_{\ell=1}^{m} |^{i_1...k...i_m}_{j_1...j_n}\rangle (T_a)^{i_\ell}{}_k. \tag{24.8}$$

The inner product in this space is given by

$$\langle^{i'_1...i'_m}_{j'_1...j'_n}|^{i_1...i_m}_{j_1...j_n}\rangle \equiv \delta^{i_1}{}_{i'_1} \cdots \delta^{i_m}{}_{i'_m} \, \delta^{j'_1}{}_{j_1} \cdots \delta^{j'_n}{}_{j_n}. \tag{24.9}$$

Let $|V\rangle$ be a particular state in this tensor product space. We have

$$|V\rangle \equiv |^{i_1...i_m}_{j_1...j_n}\rangle V^{j_1...j_n}_{i_1...i_m}, \tag{24.10}$$

where $V^{j_1...j_n}_{i_1...i_m}$ are the tensor components. For example, the baryon state, $|\Sigma^+\rangle = |^2_1\rangle = |^i_j\rangle \, V^j_i$ implies that the tensor component $V^1_2 = 1$ and all others vanish. Let's evaluate the action of an $SU(3)$ transformation on the tensor components. We have

$$\mathcal{U}|V\rangle \equiv |V\rangle' \equiv |^{i_1...i_m}_{j_1...j_n}\rangle' V^{j_1...j_n}_{i_1...i_m}$$
$$= |^{i'_1...i'_m}_{j'_1...j'_n}\rangle (u^*)^{i'_1}{}_{i_1} \cdots (u^*)^{i'_m}{}_{i_m} u^{j'_1}{}_{j_1} \cdots u^{j'_n}{}_{j_n} \times V^{j_1...j_n}_{i_1...i_m}$$
$$= |V'\rangle \equiv |^{i'_1...i'_m}_{j'_1...j'_n}\rangle V'^{j'_1...j'_n}_{i'_1...i'_m}.$$

Thus, we find

$$V'^{j'_1...j'_n}_{i'_1...i'_m} = (u^*)^{i'_1}{}_{i_1} \cdots (u^*)^{i'_m}{}_{i_m} u^{j'_1}{}_{j_1} \cdots u^{j'_n}{}_{j_n} V^{j_1...j_n}_{i_1...i_m}. \tag{24.11}$$

Finally, an infinitesimal transformation on $|V\rangle$ is given by

$$\hat{T}_a|V\rangle \equiv |T_a V\rangle = \sum_{\ell=1}^{n} |^{i_1...i_m}_{j_1...k...j_n}\rangle (T_a)^k{}_{j_\ell} V^{j_1...j_\ell...j_n}_{i_1...i_m} - \sum_{\ell=1}^{m} |^{i_1...k...i_m}_{j_1...j_n}\rangle (T_a)^{i_\ell}{}_k V^{j_1...j_n}_{i_1...i_\ell...i_m}$$
$$\equiv |^{i'_1...i'_m}_{j'_1...j'_n}\rangle (T_a V)^{j'_1...j'_n}_{i'_1...i'_m}.$$

[2] This discussion follows closely the book by Georgi (Georgi, 1982).

Thus, we have

$$(T_a V)_{i_1\ldots i_m}^{j_1\ldots j_n} = \sum_{\ell=1}^{n} (T_a)^{j_l}{}_k V_{i_1\ldots i_m}^{j_1\ldots k\ldots j_n} - \sum_{\ell=1}^{m} (T_a)^{k}{}_{i_\ell} V_{i_1\ldots k\ldots i_m}^{j_1\ldots j_n}. \tag{24.12}$$

Now, it is important to identify so-called invariant tensor states and invariant tensors. There are only three kinds.

(1)
$$|\delta\rangle \equiv |{}^{i}_{j}\rangle \delta_i^j, \tag{24.13}$$

$$|\delta\rangle' \equiv |{}^{i'}_{j'}\rangle \underbrace{u^{*i'}{}_i}_{(u^\dagger)^i{}_{i'}} u^{j'}{}_j \delta_i^j$$

$$\equiv |{}^{i'}_{j'}\rangle \delta_{i'}^{j'} \equiv |\delta\rangle,$$

since $(u^\dagger)^i{}_{i'} u^{j'}{}_j \delta_i^j \equiv \delta_{i'}^{j'}$. Thus, $|\delta\rangle$ is an invariant state and δ_i^j is an invariant tensor, i.e. $\delta' = \delta$.

(2)
$$|\epsilon\rangle \equiv |_{ijk}\rangle \, \epsilon^{ijk}, \tag{24.14}$$

$$|\epsilon\rangle' \equiv |_{i'j'k'}\rangle \, u^{i'}{}_i \, u^{j'}{}_j \, u^{k'}{}_k \, \epsilon^{ijk}.$$

But $u^{i'}{}_i u^{j'}{}_j u^{k'}{}_k \epsilon^{ijk} \equiv \underbrace{\det u}_{1} \epsilon^{i'j'k'}$. Thus,

$$|\epsilon\rangle' = |\epsilon\rangle \tag{24.15}$$

and ϵ^{ijk} is an invariant tensor.

(3) ϵ_{ijk} is also an invariant tensor.

24.3 Tensor Analysis and the Clebsch–Gordan Decomposition

We can use tensor analysis to construct irreducible representations of $SU(3)$ and the Clebsch–Gordan decomposition. In general, the vector components, $V_{i_1\ldots i_m}^{j_1\ldots j_n}$, transform as a reducible representation of $SU(3)$. Note that the symmetry of the tensor under interchange of the upper (or lower) indices is invariant under an $SU(3)$ transformation. Since the states $|\psi_1\rangle$ and $|\psi^2\rangle$ are highest weight states, it implies that the highest weight state with m upper indices and n lower indices is given by $|V_H\rangle \equiv |{}^{222\ldots}_{111\ldots}\rangle$, and tensor components, $V_{H\,i_1\ldots i_m}^{j_1\ldots j_n} \equiv \delta_1^{j_1} \cdots \delta_1^{j_n} \delta_{i_1}^2 \cdots \delta_{i_m}^2$.

This defines the largest irreducible representation in this tensor product. All other states in this irreducible representation are obtained using raising or lowering operators (or via a finite $SU(3)$ rotation). Note: $V_{H\,i_1\ldots i_m}^{j_1\ldots j_n}$ is

- symmetric under interchange of $j_1\ldots j_n$ or $i_1\ldots i_m$,
- $\sum_{i_\ell j_{\ell'}} \delta_{j_{\ell'}}^{i_\ell} V_{H\,i_1\ldots i_\ell\ldots i_m}^{j_1\ldots j_{\ell'}\ldots j_n} \equiv 0$.

Both of these properties are preserved by an $SU(3)$ transformation ([2] is preserved since δ is an invariant tensor).

Now consider the procedure for constructing all irreducible representations under $SU(3)$ in the direct product of m upper and n lower indices. Start with the general tensor $V^{j_1 \dots j_n}_{i_1 \dots i_m}$.

(A)   Then start with the highest weight state. It satisfies the following:

(1) the trace of any upper index with any lower index vanishes,
(2) it is totally symmetric under interchange of upper or lower indices.

(B)   Construct all states in the same representation as the highest weight state by $SU(3)$ rotations. They can be written in terms of the traceless tensors

$$V^{(j_1 \dots j_n)}_{(i_1 \dots i_m)},$$

which also satisfies (1) and (2), where parentheses indicate symmetrization of indices.

Convention: Label states by the number of upper and lower indices of the highest weight state, where[3]

$$\mathbf{3} \equiv (1,0), \quad \bar{\mathbf{3}} \equiv (0,1) \quad \text{and} \quad V^{(j_1 \dots j_n)}_{(i_1 \dots i_m)} \equiv (n,m).$$

The dimension of the representation[4] (n,m) is given by the formula

$$\frac{(n+m+2)(n+1)(m+1)}{2}$$
$$= \frac{3 \cdot 4 \cdots n+2}{n!} \times \frac{3 \cdot 4 \cdots m+2}{m!} - \frac{3 \cdot 4 \cdots n+1}{(n-1)!} \times \frac{3 \cdot 4 \cdots m+1}{(m-1)!}.$$

(C)   Now, use invariant tensors to project out tensor product states of lower dimension, i.e.

$$\delta^{i_1}_{j_1} V^{j_1 \dots j_n}_{i_1 \dots i_m}, \quad \text{or} \quad \epsilon^{i_1 i_2 i_3} V^{j_1 \dots j_n}_{i_1 \dots i_m}.$$

These states are, in general, reducible. The highest weight state in this reduced set is again totally symmetric in upper and lower indices, and traceless. It is either the state $(n-1, m-1)$ or $(n, m-3)$ or $(n-3, m)$.

(D)   Continue with this process until the resulting state is irreducible.

Let's consider three examples of the above process.

Example 1:

$\mathbf{3} \otimes \bar{\mathbf{3}} = \mathbf{1} \oplus \mathbf{8}$ – We have the tensor state $v^i u_j |^j_i\rangle$ where v^i is the tensor of particle 1 and u_j is the tensor of particle 2. We now decompose the general tensor $v^i u_j$ in terms of irreducible representations by

---

[3] These are known as Dynkin weights.
[4] The dimension of the symmetric n index tensor is given in terms of, for example, the number of
    ways of placing n indistinguishable balls in k distinguishable boxes (with $k = 3$), or $\frac{(n+k-1)!}{n!(k-1)!}$.

$$v^i u_j = \left(v^i u_j - \frac{1}{3}\, \delta^i_j v^k u_k \right) \quad (1,1) \quad dim\ 8 \tag{24.16}$$

$$+ \frac{1}{3}\, \delta^i_j v^k u_k \quad (0,0) \quad dim\ 1.$$

Consider the Clebsch–Gordan coefficients for the state $|^1_1\rangle$ with $v^i = \delta^i_1,\ u_j = \delta^1_j$. We have

$$|^1_1\rangle = v^i u_j |^j_i\rangle \tag{24.17}$$

$$= \left(|^1_1\rangle - \frac{1}{3}\, \delta^i_j \right) |^j_i\rangle + \frac{1}{3}\, \delta^i_j |^j_i\rangle$$

$$= \left(\frac{2}{3}\, |^1_1\rangle - \frac{1}{3}\, |^2_2\rangle - \frac{1}{3}\, |^3_3\rangle \right)$$

$$+ \frac{1}{3}\, \left(|^1_1\rangle + |^2_2\rangle + |^3_3\rangle \right)$$

$$= \sqrt{\frac{2}{3}} \left(\sqrt{\frac{2}{3}}\, |^1_1\rangle - \frac{1}{\sqrt{6}}\, |^2_2\rangle - \frac{1}{\sqrt{6}}\, |^3_3\rangle \right) \quad \text{octet}$$

$$+ \frac{1}{\sqrt{3}} \left(\frac{|^1_1\rangle + |^2_2\rangle + |^3_3\rangle}{\sqrt{3}} \right) \quad \text{singlet}.$$

We're not quite finished. Recall that there are two states at the center of the octet. Given $|\pi^0\rangle = \frac{1}{\sqrt{2}}(|^1_1\rangle - |^2_2\rangle)$, $|\eta\rangle = \frac{1}{\sqrt{6}} \left(|^1_1\rangle + |^2_2\rangle - 2|^3_3\rangle \right)$ and the singlet $|\eta'\rangle = \frac{1}{\sqrt{3}} \left(|^1_1\rangle + |^2_2\rangle + |^3_3\rangle \right)$. We have

$$|^1_1\rangle = a|\pi^0\rangle + b|\eta\rangle + c|\eta'\rangle. \tag{24.18}$$

Comparing Eqns. 24.17 and 24.18 we find the Clebsch–Gordan coefficients,

$$a = \frac{1}{\sqrt{2}}, \quad b = \frac{1}{\sqrt{6}}, \quad c = \frac{1}{\sqrt{3}}. \tag{24.19}$$

Example 2:

$\mathbf{3} \otimes \mathbf{3}$ is given in terms of $v^i u^j |_{ij}\rangle$. The irreducible tensor components are given by

$$v^i u^j = \frac{1}{2}(v^i u^j + v^j u^i) \quad (2,0) \quad dim\ 6$$

$$+ \frac{1}{2}(v^i u^j - v^j u^i) \equiv \frac{1}{2}\epsilon^{ijk}\epsilon_{k\ell m} v^\ell u^m \quad (0,1) \quad dim\ 3.$$

Hence we find

$$\mathbf{3} \otimes \mathbf{3} = \mathbf{6} \oplus \bar{\mathbf{3}}. \tag{24.20}$$

We have also obtained the Clebsch–Gordan coefficients with this method. Consider the state $|_{12}\rangle$. It is a linear combination of a $\mathbf{6}$ and a $\bar{\mathbf{3}}$. We have $v^i = \delta^i_1,\ u^j = \delta^j_2$. Then

$$|_{12}\rangle = \frac{1}{2}(|_{12}\rangle + |_{21}\rangle)$$

$$+ \frac{1}{2}(|_{12}\rangle - |_{21}\rangle).$$

Then, using the normalization of states given by $\langle {}_{i'j'} | {}_{ij} \rangle = \delta^{i'}_i \delta^{j'}_j$, we obtain

$$|_{12}\rangle = \frac{1}{\sqrt{2}} \left(\frac{|12\rangle + |21\rangle}{\sqrt{2}} \right) \rightarrow \mathbf{6} \tag{24.21}$$

$$+ \frac{1}{\sqrt{2}} \left(\frac{|12\rangle - |21\rangle}{\sqrt{2}} \right) \rightarrow \mathbf{\bar{3}},$$

where the Clebsch–Gordan coefficients are $\frac{1}{\sqrt{2}}$ in both cases.

Example 3:

We would now like to define the tensor components of the baryon $(B^i{}_j)$ and meson $(\pi^i{}_j)$ octets. Recall, earlier we had defined the state $|\Sigma^+\rangle \equiv |^2_1\rangle$. Now we define its tensor component by the relation $B^1{}_2 \equiv \Sigma^+$ where Σ^+ is just the numerical value for this tensor component. In a similar fashion we now define

$$|p\rangle \equiv |^3_1\rangle \Rightarrow B^1{}_3 \equiv p,$$

$$|\Sigma^0\rangle \equiv \frac{1}{\sqrt{2}}(|^1_1\rangle - |^2_2\rangle) \Rightarrow B^1{}_1 = -B^2{}_2 = \frac{1}{\sqrt{2}} \Sigma^0,$$

$$|\Lambda\rangle \equiv \frac{1}{\sqrt{6}}(|^1_1\rangle + |^2_2\rangle - 2|^3_3\rangle) \Rightarrow B^1{}_1 = B^2{}_2 = -\frac{1}{2} B^3{}_3 = \frac{1}{\sqrt{6}} \Lambda.$$

In general, we have

$$|B\rangle \equiv B^i{}_j |^j_i\rangle, \tag{24.22}$$

where, for example, $|\Sigma^+\rangle = \Sigma^+ |^2_1\rangle$. We then find the complete set of tensor components for the baryon octet given by

$$B^i{}_j \equiv \begin{pmatrix} \frac{\Sigma^0}{\sqrt{2}} + \frac{\Lambda}{\sqrt{6}} & \Sigma^+ & p \\ \Sigma^- & -\frac{\Sigma^0}{\sqrt{2}} + \frac{\Lambda}{\sqrt{6}} & n \\ \Xi^- & \Xi^0 & -\frac{2\Lambda}{\sqrt{6}} \end{pmatrix}. \tag{24.23}$$

Similarly, for the meson octet, we have

$$\pi^i{}_j \equiv \begin{pmatrix} \frac{\pi^0}{\sqrt{2}} + \frac{\eta}{\sqrt{6}} & \pi^+ & K^+ \\ \pi^- & -\frac{\pi^0}{\sqrt{2}} + \frac{\eta}{\sqrt{6}} & K^0 \\ K^- & \bar{K}^0 & -\frac{2\eta}{\sqrt{6}} \end{pmatrix}. \tag{24.24}$$

The octet components transform as follows under $SU(3)$. We have

$$B'^i{}_j \equiv B^{i'}{}_{j'} u^{*i'}{}_i u^{j'}{}_j$$

$$= (u^\dagger)^i{}_{i'} B^{i'}{}_{j'} u^{j'}{}_j,$$

or

$$B' \equiv u^\dagger B u. \tag{24.25}$$

Similarly, we have

$$\pi' \equiv u^\dagger \pi u. \tag{24.26}$$

Consider conjugate tensor states

$$\langle V| \equiv V^{j_1 \ldots j_n}{}_{i_1 \ldots i_m}{}^* \langle^{i_1 \ldots i_m}_{j_1 \ldots j_n}| \tag{24.27}$$
$$\equiv \bar{V}^{i_1 \ldots i_m}_{j_1 \ldots j_n} \langle^{i_1 \ldots i_m}_{j_1 \ldots j_n}|.$$

The second line explicitly takes into account the fact that conjugate states transform in the conjugate representation. In particular, we have the $SU(3)$ transformation given by

$$\langle V|' = \bar{V}^{i_1 \ldots i_m}_{j_1 \ldots j_n} \langle^{i'_1 \ldots i'_m}_{j'_1 \ldots j'_n}| (u^*)^{j'_1}{}_{j_1} \cdots (u^*)^{j'_n}{}_{j_n} u^{i'_1}{}_{i_1} \cdots u^{i'_m}{}_{i_m} \tag{24.28}$$
$$= (\bar{V}')^{i'_1 \ldots i'_m}_{j'_1 \ldots j'_n} \langle^{i'_1 \ldots i'_m}_{j'_1 \ldots j'_n}| \equiv \langle V'|,$$

where

$$(\bar{V}')^{i'_1 \ldots i'_m}_{j'_1 \ldots j'_n} \equiv u^{i'_1}{}_{i_1} \cdots u^{i'_m}{}_{i_m} (u^*)^{j'_1}{}_{j_1} \cdots (u^*)^{j'_n}{}_{j_n} \bar{V}^{i_1 \ldots i_m}_{j_1 \ldots j_n}. \tag{24.29}$$

Theorem 24.1 *Given two tensor states of the same rank, $|V_1\rangle$, $|V_2\rangle$, the inner product is a pure number and is (and must be) an $SU(3)$ invariant.*

Proof

$$\# \equiv \langle V_2|V_1\rangle = \langle V_2|\mathcal{U}^\dagger \mathcal{U}|V_1\rangle \equiv \langle V_2'|V_1'\rangle. \tag{24.30}$$

In terms of the tensor components we have on the left side of the equation

$$\langle V_2|V_1\rangle = (\bar{V}_2)^{i'_1 \ldots i'_m}_{j'_1 \ldots j'_n} (V_1)^{j_1 \ldots j_n}_{i_1 \ldots i_m} \langle^{i'_1 \ldots i'_m}_{j'_1 \ldots j'_n}|^{i_1 \ldots i_m}_{j_1 \ldots j_n}\rangle.$$

Finally, using the inner product for tensor states, Eqn. 24.9, we find

$$(\bar{V}_2)^{i_1 \ldots i_m}_{j_1 \ldots j_n} V_1{}^{j_1 \ldots j_n}_{i_1 \ldots i_m} = (\bar{V}_2')^{i_1 \ldots i_m}_{j_1 \ldots j_n} V_1'{}^{j_1 \ldots j_n}_{i_1 \ldots i_m}. \tag{24.31}$$

Thus, when all indices of the two tensors, V_1, $\bar{V}_2$, are contracted, the result is an $SU(3)$-invariant, pure number. □

Let's now define *tensor operators* on Fock space, $O^{i_1 \ldots i_m}_{j_1 \ldots j_n}$, with the transformation law

$$O' \equiv \mathcal{U} O \mathcal{U}^\dagger \tag{24.32}$$
$$O'^{i_1 \ldots i_m}_{j_1 \ldots j_n} = O^{i'_1 \ldots i'_m}_{j'_1 \ldots j'_n} u^{j'_1}{}_{j_1} \cdots u^{j'_n}{}_{j_n} (u^*)^{i'_1}{}_{i_1} \cdots (u^*)^{i'_m}{}_{i_m},$$

i.e. it transforms like a tensor state, such that (for example) $O|0\rangle$ creates a tensor state, where $\mathcal{U}|0\rangle = |0\rangle$ is an $SU(3)$-invariant vacuum state. An arbitrary tensor operator is of the form

$$W \equiv \omega^{j_1 \ldots j_n}_{i_1 \ldots i_m} O^{i_1 \ldots i_m}_{j_1 \ldots j_n}, \tag{24.33}$$

where $\omega^{j_1 \ldots j_n}_{i_1 \ldots i_m}$ are tensor components, transforming as a tensor under $SU(3)$.

24.4 Gell-Mann–Okubo Mass Formula

We now derive the mass formula for the baryon octet and decuplet, which takes into account $SU(3)$ flavor symmetry breaking to first order. The strong-interaction Hamiltonian is assumed to be given by

$$H_{strong} \equiv H_0 + H_8, \tag{24.34}$$

where H_0 is an $SU(3)$ singlet tensor operator and H_8 is the tensor operator (in the Y direction) which gives the largest mass splitting among states (ignoring isospin breaking). The baryon masses are given by

$$M_{B'B} = \langle B'|H_{st}|B\rangle, \tag{24.35}$$

where $M_{B'B}$ is the baryon mass matrix and $|B\rangle \equiv |B(\vec{p} = 0)\rangle$ is the baryon state at rest. We define $H_0 \equiv O$ (a scalar operator) and $H_8 \equiv (T_8)^i{}_j O^j{}_i$, where $O^j{}_i$ is an octet tensor operator, satisfying $Tr O^j{}_i = 0$ and $T_8 \equiv \frac{1}{\sqrt{12}} \begin{pmatrix} 1 & 0 & 0 \\ 0 & 1 & 0 \\ 0 & 0 & -2 \end{pmatrix}$ are the tensor components.

The generators of $SU(3)$, $\hat{T}_a$, have the following commutation relations with tensor operators, O and $O^j{}_i$. We have

$$\left[\hat{T}_a,\, O\right] = 0$$

$$\left[\hat{T}_a,\, O^j{}_i\right] = O^j{}_k (T_a)^k{}_i - O^k{}_i (T_a)^j{}_k \equiv [O,\, T_a]^j{}_i.$$

The last term is a matrix commutator.

The baryon octet states are given in Eqn. 24.22, and the conjugate state is given by

$$\langle B| \equiv B^{*i}{}_j |^j_i\rangle^* \equiv \langle^j_i|(\bar{B})^j{}_i \tag{24.36}$$

where $(\bar{B})^j{}_i \equiv B^{*i}{}_j \equiv (B^\dagger)^j{}_i$.

Theorem 24.2 *Consider the matrix elements of tensor operators in the octet state,*

$$\langle^{j'}_{i'}|O^k{}_\ell|^j_i\rangle.$$

This is an $SU(3)$ tensor. Moreover, it is an invariant $SU(3)$ tensor.

Proof

$$\langle^{j'}_{i'}|O^k{}_\ell|^j_i\rangle' \equiv \langle^{j'}_{i'}|\mathcal{U}^\dagger \mathcal{U} O^k{}_\ell \mathcal{U}^\dagger \mathcal{U}|^j_i\rangle \tag{24.37}$$

$$= \langle^{m'}_{n'}|O^{k'}{}_{\ell'}|^m_n\rangle u^{m'}{}_{j'} u^{*n'}{}_{i'} u^{*k'}{}_k u^{\ell'}{}_\ell u^{*m}{}_j u^n{}_i$$

$$\equiv \langle^{j'}_{i'}|O^k{}_\ell|^j_i\rangle.$$

Hence, we conclude that $\langle^{j'}_{i'}|O^k{}_\ell|^j_i\rangle$ is an invariant tensor. $\qquad\square$

Similarly, it is easy to see that the tensor given by

$$\langle^{j'}_{i'}|O|^j_i\rangle \tag{24.38}$$

is an invariant tensor.

Given these preliminaries, we are now ready to derive the Gell-Mann–Okubo mass formula. We have

$$\langle B|H_0|B\rangle \equiv \langle^{j'}_{i'}|O|^j_i\rangle(B^\dagger)^{j'}{}_{i'}B^i{}_j$$

and

$$\langle^{j'}_{i'}|O|^j_i\rangle \equiv \langle 8||O||8\rangle \delta^j_{j'}\delta^{i'}_i ,$$

where $\langle 8||O||8\rangle$ is the Wigner–Eckart reduced matrix element. Thus for this term we find

$$\langle B|H_0|B\rangle = \langle 8||O||8\rangle\mathrm{Tr}(B^\dagger B) \equiv M_0\mathrm{Tr}(B^\dagger B). \tag{24.39}$$

Then we have

$$\langle B|H_8|B\rangle \equiv \langle^{j'}_{i'}|O^k{}_\ell|^j_i\rangle(B^\dagger)^{j'}{}_{i'}(T_8)^\ell{}_k B^i{}_j.$$

In this case we should expect two different reduced matrix elements, since the tensor $\langle^{j'}_{i'}|O^k{}_\ell|^j_i\rangle$ is of the form $\langle 8|8\otimes 8\rangle = \langle 8|8_1\oplus 8_2+\cdots\rangle$, i.e. there are two different octets in the direct product $\mathbf{8}\otimes\mathbf{8}=\mathbf{1}\oplus\mathbf{8_1}\oplus\mathbf{8_2}\oplus\mathbf{10}\oplus\mathbf{\overline{10}}\oplus\mathbf{27}$. Moreover, only the octets on the right-hand side can contribute to the reduced matrix elements. It is then easy to see that

$$\langle^{j'}_{i'}|O^k{}_\ell|^j_i\rangle \equiv X\delta^{i'}_i\,\delta^j_\ell\delta^k_{j'}+Y\delta^{i'}_\ell\,\delta^k_i\delta^j_{j'} \tag{24.40}$$

where X and Y are the two reduced matrix elements. Note, we have used the fact that for any octet state $\delta^i_j|^j_i\rangle \equiv 0$. Note, also that $H_8\neq\hat{T}_8$. Since the baryon states are eigenstates of $\hat{T}_8$, there is only one possible reduced matrix element with $X=-Y$. Finally, we obtain

$$\langle B|H_8|B\rangle = X\mathrm{Tr}(B^\dagger BT_8)+Y\mathrm{Tr}(B^\dagger T_8 B). \tag{24.41}$$

We can now write the results in terms of the baryon tensor components. We have

$$\mathrm{Tr}(B^\dagger B) = (|N|^2+|\Sigma|^2+|\Lambda|^2+|\Xi|^2),$$
$$\mathrm{Tr}(B^\dagger BT_8) = \frac{1}{12}\left(|\Sigma|^2+|\Xi|^2-|\Lambda|^2-2|N|^2\right)$$
$$\mathrm{Tr}(BB^\dagger T_8) = \frac{1}{12}\left(|\Sigma|^2+|N|^2-|\Lambda|^2-2|\Xi|^2\right)$$

where $|N|^2\equiv|p|^2+|n|^2$ and similarly for the other isospin multiplets.

Putting it all together, we find the baryon masses for isospin multiplets

$$M_N = M_0 - \frac{X}{6} + \frac{Y}{12},$$
$$M_\Lambda = M_0 - \frac{X}{12} - \frac{Y}{12},$$
$$M_\Sigma = M_0 + \frac{X}{12} + \frac{Y}{12},$$
$$M_\Xi = M_0 + \frac{X}{12} - \frac{Y}{6}.$$

There are four masses given in terms of three arbitrary parameters, so we obtain one relation:

$$2(M_N + M_\Xi) = 3M_\Lambda + M_\Sigma. \tag{24.42}$$

Let's now compare it with experiment. The baryon masses are approximately given in MeV by $M_N \sim 940$, $M_\Lambda \sim 1115$, $M_\Sigma \sim 1190$, $M_\Xi \sim 1320$. Thus, the left-hand side is $\sim 2 \times 2260$ and the right-hand side is $\sim 3345 + 1190 = 4535$. The Gell-Mann–Okubo mass relation seems to be satisfied much too well. It is only a first-order correction with approximately 30% splittings.

What about the baryon decuplet, Fig. 21.2? We now need to find the matrix element of H_{strong} in the 10-dimensional representation of $SU(3)$, i.e.

$$\langle 10|H_0 + H_8|10\rangle.$$

This is not as difficult as it might seem. The first term, $\langle 10|H_0|10\rangle$ gives a common mass to all states. The second term is of the form

$$\langle 10|H_8|10\rangle = \langle 10|8 \otimes 10\rangle,$$

where the tensor product $\mathbf{8} \otimes \mathbf{10} = \mathbf{10} \oplus \mathbf{8} \oplus \mathbf{27} \oplus \mathbf{35}$. Thus, there is a unique $\mathbf{10}$, which implies that there can only be one reduced matrix element. This also implies that all octet operators must have the same reduced matrix element. This means that, up to normalization, the matrix element of H_8 is equal to the matrix element of $\hat{T}_8$. But $\hat{T}_8 \propto \hat{Y}$. The different isospin multiplets in the $\mathbf{10}$ are all separated by one unit of hypercharge. Thus, we obtain the mass formula for the decuplet,

$$m_{\Sigma^*} - m_\Delta = m_{\Xi^*} - m_{\Sigma^*} = m_{\Omega^-} - m_{\Xi^*}, \tag{24.43}$$

or, putting in the numbers in MeV with $m_\Delta \sim 1232$, $m_{\Sigma^*} \sim 1385$, $m_{\Xi^*} \sim 1530$, $m_{\Omega^-} \sim 1672$, we find $153 \sim 145 \sim 142$. Again, the agreement with data is better than might be expected. And the discovery of the Ω^- just required one event in the bubble chamber, Fig. 23.4. Note the splitting in the hypercharge direction is easily understood in the quark model. It is just due to the mass difference between the strange quark and the up, down quarks.

To summarize, we have used the flavor group $SU(3)$ to describe the low-lying hadronic states, both mesons and baryons. We have also used this symmetry to understand symmetry-breaking effects, i.e. both electromagnetism and strong hypercharge breaking. These results described the mass splitting of the baryons quite well. One can also use $SU(3)$ symmetry to compare strong-interaction cross-sections between states in $SU(3)$ multiplets. $SU(3)$ is a continuous global symmetry which is explicitly broken by small corrections; either due to electromagnetic and weak interaction processes or, as we now understand, by mass splittings between the up, down and strange quarks. The non-relativistic quark model of Gell-Mann and Zweig had many successes. Nevertheless, as Gell-Mann would often remark, quarks are just fictitious objects (or mathematical constructs) which are used to do calculations, but they don't really exist in nature. This idea lasted for more than 10 years, before there was a major paradigm shift, which came with the discovery of the J/ψ in 1976 and the charm quark. Only then did quarks become "real."

PART VI

SPONTANEOUS SYMMETRY BREAKING

25 Spontaneous Symmetry Breaking

Up until now we have studied *explicit symmetry breaking*, i.e. the Hamiltonian is given by $H = H_0 + H_I$ where H_0 is symmetric and H_I breaks the symmetry. Now we want to discuss the phenomenon of *spontaneous symmetry breaking*. In this case, the Hamiltonian is symmetric, but the vacuum is not.

There is a simple statistical mechanics analog which illustrates the distinction between the two different breaking mechanisms. Consider the Heisenberg ferromagnet, a system of classical spins $\vec{S}_i$ (transforming as a vector under rotations) located at spatial lattice sites labeled by the index i, with $\vec{S}_i^{\,2} \equiv 1$. The Hamiltonian for the system is given by

$$H_0 = -J \sum_{\langle ij \rangle} \vec{S}_i \cdot \vec{S}_j,$$

where $\sum_{\langle ij \rangle}$ means sum over all nearest-neighbor spins i and j. The constant $J > 0$ is such that the energy of the system is lower for parallel spins, all pointing in the same direction. The partition function for a canonical ensemble of such spins is given by

$$Z_T = \sum_{\{\vec{S}_i\}} e^{-\beta H_0},$$

where $\beta \equiv \frac{1}{kT}$, T is temperature and all values of each $\vec{S}_i$ are summed over. Given an arbitrary function of the spins, $A \equiv A(\vec{S}_i)$, the thermal average of A is given by

$$\langle A \rangle_T \equiv \frac{\sum_{\{\vec{S}_i\}} e^{-\beta H_0} A(\vec{S}_i)}{Z_T}.$$

This system has a continuous global symmetry defined by $\vec{S}_i{}' \equiv R\vec{S}_i$ or

$$S'_{i\alpha} \equiv R_{\alpha\beta} S_{i\beta}, \quad \alpha, \beta = 1, 2, 3$$

such that $R^T R = 1$. This is an $SO(3)$ transformation, i.e. the rotation group. Since R is independent of i, it is a global symmetry operation and $H'_0(\vec{S}_i) \equiv H_0(\vec{S}_i{}') = H_0(\vec{S}_i)$, i.e. H_0 is invariant. For temperatures, T large, the spins are pointing in random directions.

Consider *explicit symmetry breaking* where $H = H_0 + H_I^z$, i.e. add to H_0 a symmetry-breaking interaction, $H_I^z \equiv \mu \sum_i S_i^z$, with $\mu > 0$. Such a term could be due to an external magnetic field in the z direction. This term is minimized if

$S_i^z = -1$ for all i. Moreover, it breaks the rotational symmetry, since

$$H_I^{z'} \equiv \mu \sum_i S_{i3}' = R_{3\alpha}\mu \sum_i S_{i\alpha}.$$

In this case, a special direction has been chosen. As a consequence of explicit symmetry breaking,

- there is an induced magnetic field in the system,

$$\frac{1}{N}\langle\sum_{i=1}^{N} S_{i3}\rangle_T \neq 0;$$

- the degeneracy of magnetic quantum numbers (spin up vs. spin down) is lifted in the presence of the external magnetic field, and we have the Zeeman effect.

Consider now *spontaneous symmetry breaking*. The Hamiltonian is $H = H_0$ and is rotation invariant, i.e. $\mu = 0$. Define the magnetization

$$\vec{M} \equiv \frac{1}{N}\langle\sum_{i=1}^{N} \vec{S}_i\rangle_T. \tag{25.1}$$

For temperatures greater than some critical temperature, i.e. $T > T_c$, we have $\vec{M} = 0$. (For $\mu \neq 0$, we have $\vec{M} \propto -\mu\hat{z}$.) For $T < T_c$, we have spontaneous magnetization with $\vec{M} \neq 0$. The spins tend to align, choosing some direction randomly. In this case ($\mu = 0$), the Hamiltonian is rotationally invariant, but the ground state is not, i.e.

$$M_\alpha' = \frac{1}{N}\langle\sum_{i=1}^{N} S_{i\alpha}'\rangle_T \equiv \frac{1}{N}\langle\sum_{i=1}^{N} R_{\alpha\beta}S_{i\beta}\rangle_T = R_{\alpha\beta}M_\beta. \tag{25.2}$$

The ground state transforms as a vector.

Theorem 25.1   *As a consequence of spontaneously breaking a continuous global symmetry, there exist spin waves with NO energy gap.*

Proof   Here we give a heuristic proof. In general, the dispersion relation of a spin excitation (spin wave) is given by

$$E_{\vec{k}} \simeq \vec{k}^{\,2} + \Delta,$$

where Δ is the energy gap and $k \sim \frac{1}{\lambda}$ is the wave number. This is valid for $T > T_c$. However, for $T < T_c$, consider a spin wave with wavelength λ (see Fig. 25.1). Note, as $\lambda \to \infty$, $\vec{k} \to 0$, the excitation approaches a global rotation of the system. This costs no energy, thus $E_{\vec{k}} \to 0$ as $\vec{k} \to 0$. $\square$

The spin wave of statistical mechanics has a particle physics analog. If a continuous global symmetry is spontaneously broken, there exists massless excitations; so-called Nambu–Goldstone bosons.

Fig. 25.1 Spin wave with wavelength λ.

25.1 Spontaneously Breaking a Discrete Global Symmetry

Before leaving statistical mechanics, consider a system with a discrete global symmetry, i.e. the Ising model with classical spins, $S_i = \pm 1$. The Hamiltonian is given by

$$H_0 = -J \sum_{\langle ij \rangle} S_i S_j.$$

In this case, for $T < T_c$, we have[1]

$$M \equiv \frac{1}{N} \langle \sum_{i=1}^{N} S_i \rangle \neq 0,$$

i.e. spontaneous magnetization, and for $T > T_c$, $M = 0$. There are no gapless spin waves due to spontaneously breaking a discrete symmetry, since you cannot go continuously from spin up to spin down. A consequence of discrete symmetry breaking is domain walls. Consider for $T < T_c$, Fig. 25.2. There is a finite amount of energy/area localized in the wall connecting the spin-up and spin-down phases.

What is the analog of the Ising system in a relativistic field theory? Instead of $S_i = \pm 1$ for all i, we have a real scalar field $\phi(x)$, which takes on values $-\infty < \phi(x) < \infty$ for all x. We have a Lagrangian density given by

$$\mathcal{L}(\phi) = \frac{1}{2}(\partial_\mu \phi)^2 - V(\phi),$$

with

$$V(\phi) = \frac{\lambda}{4}\left(\phi^2 - \frac{m^2}{\lambda}\right)^2 \equiv \frac{\lambda \phi^4}{4} - \frac{m^2 \phi^2}{2} + const.$$

(see Fig. 25.3). The two vacuum states with $\langle \phi \rangle_0 \approx \pm \frac{m}{\sqrt{\lambda}}$ are degenerate. The

Domain wall

Fig. 25.2 Domain wall located at the boundary of two spin domains.

[1] In one dimension $T_c = 0$.

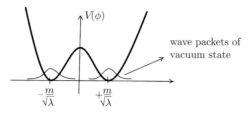

Fig. 25.3 The two vacuum states with $\langle \phi \rangle \approx \pm \frac{m}{\sqrt{\lambda}}$ are degenerate.

Lagrangian has a discrete global symmetry, $\phi \to -\phi$, which can be generated by the Fock-space operator, $\mathcal{P}$, such that

$$\phi' \equiv \mathcal{P}\phi\mathcal{P}^\dagger = -\phi, \quad \text{with} \ \ \mathcal{L}(\phi') = \mathcal{L}(\phi).$$

The two vacuum states are related by $\mathcal{P}|\Omega_+\rangle = |\Omega_-\rangle$ with $\langle\Omega_+|\phi|\Omega_+\rangle = \frac{m}{\sqrt{\lambda}}$ and $\langle\Omega_-|\phi|\Omega_-\rangle = -\frac{m}{\sqrt{\lambda}}$. Finally, we have

$$\frac{m}{\sqrt{\lambda}} \equiv \langle\Omega_+|\phi|\Omega_+\rangle$$
$$= \langle\Omega_+|\mathcal{P}^\dagger\mathcal{P}\phi\mathcal{P}^\dagger\mathcal{P}|\Omega_+\rangle = -\langle\Omega_-|\phi|\Omega_-\rangle.$$

Note, possible classical vacua are given by space–time-independent solutions to the equations of motion:

$$\partial^\mu \frac{\partial\mathcal{L}}{\partial\partial^\mu\phi} - \frac{\partial\mathcal{L}}{\partial\phi} = \Box\phi + V'(\phi) = 0,$$

with

$$V'(\phi_0) \equiv \frac{\partial V}{\partial\phi}(\phi_0) = \lambda\left(\phi_0^2 - \frac{m^2}{\lambda}\right)\phi_0 = 0,$$
$$\partial_\mu\phi_0 = 0.$$

The second condition is required by translation invariance of the vacuum. Note: $\phi_0 \equiv \langle\phi\rangle_0$ at the classical level and $i\partial_\mu\phi_0 \equiv \langle\left[\hat{P}_\mu, \phi\right]\rangle_0 = 0$ since $\hat{P}_\mu|\Omega\rangle = 0$.

Their are three solutions to the classical equations of motion. They are $\phi_0 = 0, \pm\frac{m}{\sqrt{\lambda}}$. The solution $\phi_0 = 0$ is fine classically (it's like a pencil standing on its point), but quantum mechanically it is unstable. Thus, the only stable vacua are $\phi_0 = \pm\frac{m}{\sqrt{\lambda}}$, which are degenerate because of the discrete symmetry. Consider another indication of instability. We have $V''(\phi_0) \equiv \lambda(3\phi_0^2 - \frac{m^2}{\lambda}) \equiv m_\phi^2$, where m_ϕ is the mass of the ϕ. Then, $V''(\phi_0 = 0) = -m^2 < 0$ is tachyonic (the signal for an unstable vacuum). On the other hand, $V''(\phi_0 = \pm\frac{m}{\sqrt{\lambda}}) = 2m^2 > 0$ is an indication of a stable vacuum.

In order to do perturbation theory, we want the free scalar field to be written in terms of creation and annihilation operators, and thus have zero expectation value in the Fock vacuum. We thus define a shifted scalar field,

$$\phi' = \phi - \frac{m}{\sqrt{\lambda}},$$

where we've chosen to perturb around the vacuum with $\phi_0 = \frac{m}{\sqrt{\lambda}}$. Now we have $\langle \phi' \rangle_0 = \phi_0 - \frac{m}{\sqrt{\lambda}} \equiv 0$. Thus,

$$\phi(x) = \frac{m}{\sqrt{\lambda}} + \phi'(x) = \phi_0 + \int \frac{d^3\vec{p}}{(2\pi)^3 2E_p} [a(\vec{p})e^{-ip\cdot x} + h.c.],$$

with the perturbative Fock vacuum satisfying $a(\vec{p})|0\rangle \equiv 0$.

The Lagrangian for the ϕ' quantum states is given by

$$\mathcal{L}(\phi') = \frac{1}{2}(\partial_\mu \phi')^2 - V(\phi' + \frac{m}{\sqrt{\lambda}}),$$

with

$$\tilde{V}(\phi') \equiv V(\phi' + \frac{m}{\sqrt{\lambda}}) = \frac{\lambda}{4}(\phi'^2 + \frac{2m}{\sqrt{\lambda}}\phi')^2$$

$$= \frac{\lambda}{4}\phi'^4 + m\sqrt{\lambda}\,\phi'^3 + m^2\phi'^2.$$

Note, in terms of the ϕ' field, there is no apparent symmetry, $\phi \to -\phi$.

Let's compare our field theory with two degenerate vacua with a quantum-mechanical system with one degree of freedom. We consider the Hamiltonian,

$$H = \frac{1}{2}p^2 + V(x),$$

with the potential $V(x)$ given in Fig. 25.4. In this case, there is NO degenerate ground state. Since the energy barrier is finite, there is tunneling and the two states centered at $x = \pm x_0$ mix. The ground state is therefore unique, given by a symmetric superposition of order $\psi_+(x) \sim N(e^{-\frac{1}{2}(x-x_0)^2} + e^{-\frac{1}{2}(x+x_0)^2})$. In quantum field theory, however, the Hamiltonian is given by

$$H = \int d^3\vec{x} \left(\frac{1}{2}\Pi_\phi^2(\vec{x}) + \frac{1}{2}(\vec{\nabla}\phi)^2 + V(\phi) \right).$$

In this case, the energy barrier is infinite $\propto \int d^3\vec{x}(\Delta V)$, where ΔV is the height of the barrier. There is no tunneling and the two vacua are degenerate and orthogonal, i.e. $\langle \Omega_+ | \Omega_- \rangle \equiv 0$. No local process can take the system from one vacuum to the other.

In the early universe, such theories with discrete symmetries might be dangerous. At high temperatures, the free energy would have a unique vacuum at the origin. But as the universe cools there can be a second-order phase transition with two degenerate vacua. Parts of the universe in thermal equilibrium may end up in the $|\Omega_+\rangle$ vacuum and other parts may end up in the $|\Omega_-\rangle$ vacuum. When these two regions of the universe come in causal contact there will be a domain wall

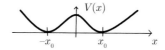

The quantum-mechanical potential with *apparently* degenerate ground states at $\pm x_0$.

between them. This domain wall has energy per unit area and will contribute to the expansion of the universe, perhaps with disastrous consequences. Thus, theories with discrete global symmetries are typically avoided.

25.2 Spontaneously Breaking a Continuous Global Symmetry

In this section we discuss Goldstone's theorem, which follows from the spontaneous breaking of a continuous global symmetry.

Spontaneous symmetry breaking in the context of particle physics was first considered by both Nambu and Goldstone and their co-workers Nambu (1960), Nambu and Jona-Lasinio (1961a), Nambu and Jona-Lasinio (1961b), Goldstone (1961) and Goldstone et al. (1962). The first version of spontaneous symmetry breaking of continuous global symmetries is due to Nambu. It is commonly referred to as dynamical symmetry breaking. In the following we present a theorem due to Goldstone.

Theorem 25.2   *There exists a massless boson for every symmetry generator Q_a such that $Q_a|\Omega\rangle \neq 0$. Note, Q_a is a bosonic charge satisfying commutation relations.*[2]

Proof   Consider N real scalar fields, ϕ_j, $j = 1, \ldots, N$, with the Lagrangian density given by

$$\mathcal{L}(\phi) = \frac{1}{2}(\partial_\mu \phi)^2 - V(\phi).$$

We want the Lagrangian to be invariant under a global symmetry. Consider the dot product $\phi \cdot \phi \equiv \sum_{i=1}^{N} (\phi_i)^2$. This is invariant under the transformation,

$$\delta\phi = i\epsilon_a T_a \phi$$

$$\text{or} \quad \delta\phi_i \equiv i\epsilon_a (T_a)_{ij} \phi_j$$

for some matrices, T_a. We have

$$\delta(\phi \cdot \phi) = 0 \equiv \delta\phi \cdot \phi + \phi \cdot \delta\phi$$
$$= i\epsilon_a (T_a)_{ij} \; \phi_j \phi_i + \phi_i (i\epsilon_a (T_a)_{ij} \phi_j)$$
$$= i\epsilon_a \phi_i [(T_a^T)_{ij} + (T_a)_{ij}] \phi_j \equiv 0$$

implies that

$$T_a = -T_a^T. \tag{25.3}$$

---

[2] If Q_a is a fermionic charge, satisfying anti-commutation relations, then Q_a creates massless fermions.

In addition, iT_a is real, since ϕ are real. Therefore, the symmetry group is given by

$$SO(N) = \{R \,|R = e^{i\epsilon_a T_a}, \quad N \times N \text{ real orthogonal matrices}; R^T R = 1, \det R = 1\}. \tag{25.4}$$

Clearly, $\partial_\mu \phi \cdot \partial^\mu \phi$ and $V(\phi) \equiv f(\phi \cdot \phi)$ are invariant under $SO(N)$. Therefore, the Lagrangian density,

$$\mathcal{L}(\phi) = \frac{1}{2} \partial_\mu \phi \cdot \partial^\mu \phi - V(\phi), \tag{25.5}$$

is $SO(N)$ invariant. As an example, the potential for a $U(1)$ global symmetry operation is plotted in Fig. 25.5.

Define the quantity $V_{j_1 \ldots j_n}(\phi) \equiv \frac{\partial^n}{\partial \phi_{j_1} \cdots \phi_{j_n}} V(\phi)$. Classical vacua satisfy

$$\phi_0 \equiv \langle \Omega | \phi | \Omega \rangle \equiv \langle \phi \rangle_0,$$
$$\partial_\mu \phi_0 \equiv 0,$$
$$V_j(\phi_0) \equiv 0.$$

In addition, for stability, we require a positive mass-squared condition, $V_{jk}(\phi_0) \geq 0$, i.e. local minimum of the potential. Now expand about the classical vacuum, $\tilde{\phi} \equiv \phi - \phi_0$, such that $\langle \tilde{\phi} \rangle_0 \equiv 0$. We have

$$\mathcal{L}(\tilde{\phi}) = \frac{1}{2}(\partial_\mu \tilde{\phi})^2 - V(\phi_0) - \tilde{\phi}_j V_j(\phi_0) - \frac{1}{2}\tilde{\phi}_j \tilde{\phi}_k V_{jk}(\phi_0) - V_I(\tilde{\phi}), \tag{25.6}$$

where the last term contains the interactions. The mass squared for $\tilde{\phi}$ is given by

$$m_{jk}^2 \equiv V_{jk}(\phi_0) \geq 0, \tag{25.7}$$

which implies that there are no tachyons.

Consider the action of the group $SO(N)$ in terms of a unitary operation on Fock space. We have

$$\phi'(x) \equiv e^{-i\epsilon_a Q_a} \phi(x) e^{i\epsilon_a Q_a} = e^{i\epsilon_a T_a} \phi(x), \tag{25.8}$$

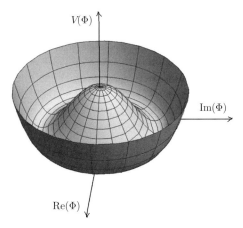

Fig. 25.5   The potential, roughly describing a theory with a $U(1)$ symmetry (using Mathematica).

where Q_a are conserved charges. Infinitesimally, we have

$$\phi'(x) = \phi(x) - i\epsilon_a \left[Q_a, \ \phi(x)\right] + \ldots$$
$$= \phi(x) + i\epsilon_a T_a \phi(x) + \ldots,$$

which implies that

$$[Q_a, \ \phi(x)] = -T_a \phi(x). \tag{25.9}$$

Consider the rotation of the vacuum state. We have

$$\langle \phi' \rangle_0 \equiv \langle \Omega | e^{-i\epsilon_a Q_a} \phi(x) e^{i\epsilon_a Q_a} | \Omega \rangle. \tag{25.10}$$

If $Q_a |\Omega\rangle \equiv 0$, then $\langle \phi' \rangle_0 = \langle \phi \rangle_0$. In general we have

$$\delta\langle\phi\rangle \equiv (\langle\phi'\rangle_0 - \langle\phi\rangle_0)$$
$$= -i\epsilon_a \langle\Omega|[Q_a, \ \phi(x)]|\Omega\rangle$$
$$= i\epsilon_a T_a \langle\Omega|\phi(x)|\Omega\rangle \equiv i\epsilon_a T_a \phi_0.$$

If

$$Q_a |\Omega\rangle \equiv 0 \tag{25.11}$$

for all a, then the vacuum is invariant under the symmetry operation and thus

$$T_a \phi_0 \equiv 0 \tag{25.12}$$

for all a, i.e. there is no symmetry breaking. On the other hand, if

$$T_a \phi_0 \neq 0 \tag{25.13}$$

for some a, then there is spontaneous symmetry breaking, and likewise,

$$Q_a |\Omega\rangle \neq 0 \tag{25.14}$$

for the same a.

Note: if there exists a subset of T_as in the group G which annihilate the vacuum, then this subset forms an unbroken symmetry subgroup group H. The proof is simple. If $T_a \phi_0 = 0$ and $T_b \phi_0 = 0$, then $[T_a, \ T_b] \phi_0 = 0$. But $[T_a, \ T_b] = i f_{abc} T_c$ and $T_c \in H$ (i.e. $T_c \phi_0 = 0$). Therefore, H is a subgroup of G, since its Lie algebra closes. For example, in the present case, the field vector $\vec{\phi}_0 \neq 0$ can always be rotated into the direction $(\phi_0)_N \neq 0$ with $(\phi_0)_i = 0$, $i = 1, \cdots N-1$, using an $SO(N)$ rotation. Then the subgroup $SO(N-1)$ remains unbroken.

We will now look at the mass matrix and show that for each charge that doesn't annihilate the vacuum, i.e. $Q_a |\Omega\rangle \neq 0$, there exists a massless boson. Since V is invariant under $\delta\phi$ (i.e. an $SO(N)$ rotation), we have

$$0 \equiv V(\phi + \delta\phi) - V(\phi) = i V_k(\phi)\epsilon_a (T_a)_{k\ell} \phi_\ell + O(\delta\phi^2). \tag{25.15}$$

Differentiating with respect to ϕ_j, we find to order ϵ_a,

$$V_{jk}(\phi)(T_a)_{k\ell}\phi_\ell + V_k(\phi)(T_a)_{kj} \equiv 0. \tag{25.16}$$

Now set $\phi = \phi_0$ and use $V_k(\phi_0) = 0$. We find

$$V_{jk}(\phi_0)(T_a)_{k\ell}(\phi_0)_\ell = 0$$

or equivalently,

$$m_{jk}^2(T_a\phi_0)_k = 0. \tag{25.17}$$

There are now two possibilities.

1. If $T_a\phi_0 = 0$, then Eqn. 25.17 is true trivially,
2. If $T_a\phi_0 \neq 0$ for $a = 1, \ldots, M$, then there are M eigenvectors of m^2 with zero eigenvalue., i.e. there are M massless Nambu–Goldstone bosons (one for each broken generator). Let $V_k{}^\ell \equiv (T_\ell\phi_0)_k$, $\ell = 1, \ldots, M$, then $\phi_k V_k{}^\ell \equiv \phi_G^\ell$ are the M linearly independent Nambu–Goldstone bosons.

$\square$

26 Spontaneous Symmetry Breaking in Hadronic Physics

Why is the pion so light, compared to the rho (ρ) and omega (ω) mesons? In this chapter we will begin to address this issue. We will discuss the σ model of Gell-Mann and Levy (Gell-Mann and Levy, 1960). Other references include Adler and Dashen (1968), Treiman et al. (1972), Georgi (1984) and Donoghue et al. (1992). By the end of the chapter, we will emphasize the significance of the $SU(2)_L \otimes SU(2)_R$ chiral algebra and the fact that spontaneous chiral symmetry breaking makes many predictions which have been tested experimentally.

26.1 Chiral Symmetry

In an earlier chapter we introduced the nucleon doublet, Eqn. 19.5, $\psi_N = \begin{pmatrix} p \\ n \end{pmatrix}$ where p, n are four-component Dirac spinors. ψ_N transforms as a doublet under isospin, such that

$$\psi'_N(x) \equiv e^{i\epsilon_a T_a}\psi_N(x),$$
$$\delta\psi_N \equiv i\epsilon_a T_a \ \psi_N(x),$$

with $T_a \equiv \frac{\tau_a}{2}$. Consider the free Lagrangian

$$\mathcal{L}(x) \equiv \bar{\psi}i\partial\!\!\!/\psi - \hat{m}\bar{\psi}\psi - \frac{\Delta m}{2}\bar{\psi}\tau_3\psi \tag{26.1}$$

(I have dropped the subscript N).

$$\bar{\psi}\psi \equiv \bar{p}p + \bar{n}n \tag{26.2}$$
$$\bar{\psi}\tau_3\psi \equiv \bar{p}p - \bar{n}n.$$

Therefore, the mass of the proton and neutron are given by

$$m_p = \hat{m} + \frac{\Delta m}{2} \tag{26.3}$$
$$m_n = \hat{m} - \frac{\Delta m}{2}$$

and

$$\Delta m = m_p - m_n$$
$$\hat{m} = \frac{m_p + m_n}{2}.$$

The term with Δm explicitly breaks $SU(2)$ invariance.

What is the continuous global symmetry of the kinetic term in the Lagrangian?[1] Consider, Eqn. 10.29, $\psi \equiv \psi_L + \psi_R$ with $\psi_L \equiv P_L\psi$, $\psi_R \equiv P_R\psi$ and $P_L \equiv \frac{1-\gamma_5}{2}$, $P_R \equiv \frac{1+\gamma_5}{2}$. We have

$$\bar{\psi}i\slashed{\partial}\psi = \bar{\psi}_L i\slashed{\partial}\psi_L + \bar{\psi}_R i\slashed{\partial}\psi_R. \tag{26.4}$$

Now consider the transformations

$$\delta\psi_L \equiv i\epsilon_a^L T_a \psi_L$$
$$\text{and} \quad L \to R$$
$$\delta\psi_R \equiv i\epsilon_a^R T_a \psi_R.$$

In the massless limit, these are symmetries of the theory, and we have conserved currents and charges given by

$$j_a^\mu(x) \equiv \bar{\psi}\gamma^\mu T_a \psi \quad Q_a \equiv \int d^3\vec{x} j_a^0(x), \tag{26.5}$$

$$j_{aL}^\mu(x) \equiv \bar{\psi}\gamma^\mu P_L T_a \psi \quad Q_{aL} \equiv \int d^3\vec{x} j_{aL}^0(x),$$

$$j_{aR}^\mu(x) \equiv \bar{\psi}\gamma^\mu P_R T_a \psi \quad Q_{aR} \equiv \int d^3\vec{x} j_{aR}^0(x),$$

which are the generators of $SU(2)_I$ isospin, $SU(2)_L$, $SU(2)_R$, respectively. The kinetic term is invariant under Q_{aL}, Q_{aR} transformations, independently.

The Lie algebra of $SU(2)_L \otimes SU(2)_R$ is given by

$$[Q_{aL}, Q_{bL}] = i\epsilon_{abc} Q_{cL}, \tag{26.6}$$
$$[Q_{aR}, Q_{bR}] = i\epsilon_{abc} Q_{cR},$$
$$[Q_{aL}, Q_{bR}] = 0,$$

and

$$Q_a \equiv Q_{aL} + Q_{aR} \tag{26.7}$$

are the vector charges, which define the diagonal subgroup, $SU(2)_I \subset SU(2)_L \otimes SU(2)_R$. The other linear combinations,

$$Q_{a5} \equiv -(Q_{aL} - Q_{aR}), \tag{26.8}$$

are axial-vector charges.

---

[1] In fact, the largest continuous global symmetry is $U(2)_L \otimes U(2)_R$.

Theorem 26.1 *Any mass term explicitly breaks the axial symmetry.*

Proof   We have

$$\bar{\psi}\psi \equiv \bar{\psi}_L \psi_R + \bar{\psi}_R \psi_L. \tag{26.9}$$

Now let's define an infinitesimal chiral transformation given by

$$\delta\psi \equiv (\epsilon_a^L P_L + \epsilon_a^R P_R)T_a\psi \tag{26.10}$$

and define

$$\epsilon_a \equiv \frac{\epsilon_a^L + \epsilon_a^R}{2} \tag{26.11}$$

$$\epsilon_{a5} \equiv \frac{\epsilon_a^L - \epsilon_a^R}{2}.$$

The transformation ϵ_a (a pure isospin rotation) acts on L, R fields with the same phase, while ϵ_{a5} (a pure chiral rotation) acts on L, R fields with opposite phases. Using the inverse relations,

$$\epsilon_a^L \equiv \epsilon_a + \epsilon_{a5}, \tag{26.12}$$

$$\epsilon_a^R \equiv \epsilon_a - \epsilon_{a5},$$

we can rewrite the general chiral transformation by

$$\delta\psi \equiv i(\epsilon_a - \epsilon_{a5}\gamma_5)T_a\psi. \tag{26.13}$$

We then have

$$\begin{aligned}
\delta(\bar{\psi}\psi) &\equiv (\delta\bar{\psi})\psi + \bar{\psi}(\delta\psi) \tag{26.14}\\
&= -i\bar{\psi}T_a(\epsilon_a + \epsilon_{a5}\gamma_5)\psi + i\bar{\psi}T_a(\epsilon_a - \epsilon_{a5}\gamma_5)\psi\\
&= -2i\epsilon_{a5}\,\bar{\psi}T_a\gamma_5\psi \neq 0.
\end{aligned}$$

$\square$

26.2 Gell-Mann–Levy Model

Gell-Mann and Levy (1960) showed that we can consider a theory of pions and nucleons in which chiral symmetry is spontaneously broken. Then the consequences are:

- pions are massless Nambu–Goldstone bosons;
- nucleon mass is generated spontaneously.

Note,

$$\frac{m_\pi^2}{m_N^2} \simeq \frac{1}{50}. \tag{26.15}$$

So this is not a bad first step; but then pions must obtain a small mass. We will then generate this small pion mass by explicitly breaking the chiral symmetry by a small amount.

Let the field $\Sigma(x)$ be a 2×2 matrix, including pion fields. We have

$$\Sigma(x) \equiv \sigma(x) + i\tau_a \pi_a(x), \quad a = 1, 2, 3. \tag{26.16}$$

$\sigma(x)$, $\pi_a(x)$ are real scalar fields. Finite $SU(2)_L \otimes SU(2)_R$ chiral transformations on the nucleon doublet are given by

$$\psi'_L \equiv L\psi_L \quad L \equiv e^{i\epsilon_a^L T_a}, \quad L^\dagger L = 1, \quad \det L = 1 \tag{26.17}$$
$$\psi'_R \equiv R\psi_R \quad R \equiv e^{i\epsilon_a^R T_a}, \quad R^\dagger R = 1, \quad \det R = 1.$$

We then let Σ transform as follows

$$\Sigma' \equiv L\Sigma R^\dagger. \tag{26.18}$$

We now construct a Lagrangian with pions and nucleons, which is invariant under a global $SU(2)_L \otimes SU(2)_R$ symmetry. We have

$$\mathcal{L}(\Sigma, \psi) \equiv \bar{\psi} i \partial\!\!\!/ \psi - g\bar{\psi}_L \Sigma \psi_R - g\bar{\psi}_R \Sigma^\dagger \psi_L + \mathcal{L}(\Sigma). \tag{26.19}$$

We want $\mathcal{L}(\Sigma', \psi') \equiv \mathcal{L}(\Sigma, \psi)$. The kinetic term is chiral invariant. The next two terms are also invariant, since

$$\bar{\psi}'_L \Sigma' \psi'_R \equiv (\bar{\psi}_L L^\dagger)(L\Sigma R^\dagger)(R\psi_R) = \bar{\psi}_L \Sigma \psi_R. \tag{26.20}$$

Therefore, we just need the last term to be invariant. But $\text{Tr}(\Sigma^\dagger \Sigma)$ and $\text{Tr}(\partial_\mu \Sigma^\dagger \partial^\mu \Sigma)$ are invariant. Hence,

$$\mathcal{L}(\Sigma) \equiv \frac{1}{4} \text{Tr}(\partial_\mu \Sigma^\dagger \partial^\mu \Sigma) - V(\text{Tr}(\Sigma^\dagger \Sigma)) \tag{26.21}$$

is a perfectly good invariant Lagrangian for the Σ field.

Before we analyze this chiral Lagrangian, let's first show how the fields π_a, σ transform under isospin and chiral symmetry. We have

$$\delta\Sigma \sim \Sigma' - \Sigma = i(\epsilon_{aL} T_a \Sigma - \epsilon_{aR} \Sigma T_a) \tag{26.22}$$

or

$$\delta(\sigma + i\tau_a \pi_a) = i\epsilon_a [T_a, \Sigma] + i\epsilon_{a5} \{T_a, \Sigma\}$$
$$= i\epsilon_a \left[\frac{\tau_a}{2}, \sigma + i\tau_b \pi_b\right] + i\epsilon_{a5} \left\{\frac{\tau_a}{2}, \sigma + i\tau_b \pi_b\right\}. \tag{26.23}$$

Using the relations $[\tau_a, \tau_b] = 2i\epsilon_{abc}\tau_c$, $\{\tau_a, \tau_b\} = 2\delta_{ab}$, we have

$$\delta\sigma = -\epsilon_{a5} \pi_a, \tag{26.24}$$
$$\delta\pi_a = \epsilon_{a5}\sigma - \epsilon_b \epsilon_{abc} \pi_c.$$

In terms of operators on Fock space, we have

$$\psi' = e^{-i\epsilon Q} \psi e^{i\epsilon Q} = e^{i\epsilon T} \psi, \tag{26.25}$$
$$\Sigma' = e^{-i\epsilon Q} \Sigma e^{i\epsilon Q}$$

where the operator Q is either Q_L or Q_R and the parameter ϵ is ϵ^L or ϵ^R, respectively, or, in general, we have $\epsilon Q = \epsilon_a Q_a - \epsilon_{a5} Q_{a5}$. We then have

$$\delta \psi = -i\epsilon \left[Q, \ \psi\right], \tag{26.26}$$
$$\delta \Sigma = -i\epsilon \left[Q, \ \Sigma\right].$$

Putting it all together, we find

$$\delta \sigma = -\epsilon_{a5} \pi_a = i\epsilon_{a5} \left[Q_{a5}, \ \sigma\right], \tag{26.27}$$
$$\delta \pi_a = \epsilon_{a5} \sigma - \epsilon_b \epsilon_{abc} \pi_c = i\epsilon_{b5} \left[Q_{b5}, \ \pi_a\right] - i\epsilon_b \left[Q_b, \ \pi_a\right].$$

Finally, we obtain

$$[Q_{a5}, \ \sigma] = i\pi_a, \tag{26.28}$$
$$[Q_a, \ \sigma] = 0,$$
$$[Q_{a5}, \ \pi_b] = -i\delta_{ab}\sigma,$$
$$[Q_a, \ \pi_b] = i\epsilon_{abc}\pi_c.$$

We find that the field σ is an iso-scalar and π_a is an iso-vector (consistent with pion isospin). Under axial transformations, the pions and sigma transform into each other.

Another way of understanding these results is to once again consider the transformation $\Sigma' = L\Sigma R^\dagger$ under $SU(2)_L \otimes SU(2)_R$. We see that Σ is in the $(\mathbf{2}, \bar{\mathbf{2}})$ representation. Moreover, the group $SU(2)_L \otimes SU(2)_R \approx SO(4)$, i.e. they are homomorphic. Under $SO(4)$, the fields (σ, π_a) transforms as a four-vector.

26.3 Spontaneously Breaking $SU(2)_L \otimes SU(2)_R$

We have $\Sigma^\dagger \equiv \sigma - i\tau_a \pi_a$ and, thus,

$$\Sigma^\dagger \Sigma = \mathbb{I}\sigma^2 + \tau_a \pi_a \tau_b \pi_b \equiv \mathbb{I}(\sigma^2 + \sum_a \pi_a^2).$$

Hence,

$$\frac{1}{2}\text{Tr}(\Sigma^\dagger \Sigma) \equiv \sigma^2 + \vec{\pi}^{\,2}. \tag{26.29}$$

Therefore,

$$\mathcal{L}(\Sigma) = \frac{1}{2}(\partial_\mu \sigma)^2 + \frac{1}{2}(\partial_\mu \vec{\pi})^2 - V(\sigma^2 + \vec{\pi}^{\,2}), \tag{26.30}$$
$$\mathcal{L}(\Sigma, \psi) = \bar{\psi} i \partial\!\!\!/ \psi - g\sigma \underbrace{\bar{\psi}_L \psi_R + \bar{\psi}_R \psi_L}_{\bar{\psi}\psi} - ig\pi_a \underbrace{\bar{\psi}_L \tau_a \psi_R - \bar{\psi}_R \tau_a \psi_L}_{\bar{\psi}\tau_a \gamma_5 \psi}.$$

Since $\bar{\psi}\psi$, $\bar{\psi}\tau_a \gamma_5 \psi$ is (even, odd) under parity, and the strong interactions conserve parity, we conclude that σ, π have $J^P = 0^+$, 0^-, respectively.

Consider the scalar potential

$$V(\sigma^2 + \vec{\pi}^{\,2}) \equiv \frac{\lambda}{4}\left[(\sigma^2 + \vec{\pi}^{\,2}) - F_\pi^2\right]^2. \tag{26.31}$$

At the minimum of V, we have $\langle \Sigma \rangle_0 \neq 0$. Since the theory is invariant under global $SU(2)_L \otimes SU(2)_R \approx SO(4)$, we can use the symmetry to rotate the vacuum expectation value (VEV) such that $\langle \sigma \rangle_0 \neq 0$, $\langle \pi_a \rangle_0 = 0$. We then have

$$\frac{\partial V}{\partial \sigma} = \lambda(\sigma^2 - F_\pi^2)\sigma = 0, \tag{26.32}$$

which implies the possible vacua given by

$$\sigma_0 = 0, \pm F_\pi. \tag{26.33}$$

$\sigma_0 = 0$ is a local maximum of the potential and thus not a stable vacuum state. We can also use the symmetry to rotate $-\sigma_0 \to \sigma_0$. Therefore, without loss of generality, we choose to expand our theory around the vacuum with

$$\sigma_0 = \langle \sigma \rangle_0 = F_\pi. \tag{26.34}$$

As a consequence, chiral symmetry is spontaneously broken. Why? Consider

$$\langle \delta \pi_a \rangle \equiv \epsilon_{a5}\langle \sigma \rangle_0 \equiv \epsilon_{a5}F_\pi \neq 0. \tag{26.35}$$

On the other hand, the left-hand side of the equation is also given by

$$i\epsilon_{a5}\langle [Q_{a5},\ \pi_a] \rangle_0 \tag{26.36}$$

with no sum over a. Therefore we find

$$Q_{a5}|\Omega\rangle \neq 0 \tag{26.37}$$

for all a. Moreover, π_a are massless, since $Q_{a5} \equiv \int d^3\vec{x}\, j_{a5}^0(x)$ creates zero-energy states (it commutes with the Hamiltonian) with zero momentum out of the vacuum. Note, also,

$$Q_a|\Omega\rangle = 0, \tag{26.38}$$

since

$$\langle \delta \pi_a \rangle = -\epsilon_b \epsilon_{abc}\langle \pi_c \rangle = 0. \tag{26.39}$$

Therefore, isospin is not broken.

Let's now explicitly calculate the particle spectrum in the vacuum with $\sigma_0 = F_\pi$. We define the perturbative σ field by $\tilde{\sigma} = \sigma - F_\pi$ such that $\langle \tilde{\sigma} \rangle_0 = 0$. Then, we have

$$\mathcal{L}(\psi, \tilde{\sigma}, \pi_a) = \bar{\psi}i\partial\!\!\!/\psi - gF_\pi \bar{\psi}\psi - g\tilde{\sigma}\bar{\psi}\psi - ig\pi_a \bar{\psi}\tau_a \gamma_5 \psi \tag{26.40}$$
$$+\frac{1}{2}(\partial_\mu \tilde{\sigma})^2 + \frac{1}{2}(\partial_\mu \vec{\pi})^2 - \frac{\lambda}{4}(\tilde{\sigma}^2 + \vec{\pi}^{\,2} + 2\tilde{\sigma}F_\pi)^2.$$

The masses are determined by the quadratic terms in the Lagrangian. We find

$$m_N = gF_\pi,\ m_{\tilde{\sigma}} = 2\lambda F_\pi^2,\ m_\pi = 0. \tag{26.41}$$

To summarize,

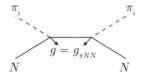

Fig. 26.1 We define the pion–nucleon coupling, $g = g_{\pi NN}$.

- the pions are massless Nambu–Goldstone bosons,
- the nucleon obtains a dynamical mass.

We have some additional very interesting consequences of the theory.

(1) The first is known as the Goldberger–Treiman relation.

Consider Fig. 26.1. $g_{\pi NN}$ can be measured in $\pi{-}N$ scattering using dispersion relations. Goldberger and Treiman find $\frac{g_{\pi NN}^2}{4\pi} \simeq 14.4 \pm 0.4$. The constant F_π (the pion decay constant) can be measured in pion decay. In weak-interaction decays of pions, the vector and axial-vector currents couple to W bosons. The vector and axial-vector currents are given by

$$j_a^\mu \equiv \bar{\psi}\gamma^\mu T_a \psi + \frac{1}{2}\epsilon_{abc}\pi_b \overleftrightarrow{\partial^\mu} \pi_c \qquad (26.42)$$

$$j_{a5}^\mu \equiv \bar{\psi}\gamma^\mu \gamma_5 T_a \psi + \sigma \overleftrightarrow{\partial^\mu} \pi_a.$$

Therefore, after shifting the σ field, we obtain

$$j_{a5}^\mu \supset F_\pi \partial^\mu \pi_a + \text{terms quadratic in the fields.} \qquad (26.43)$$

We thus obtain a weak-interaction decay diagram schematically of the form in Fig. 26.2.

The relation between the nucleon mass, the pion decay constant and the pion–nucleon coupling, Eqn. 26.41, is a first attempt in a derivation of the Goldberger–Treiman relation (Goldberger and Treiman, 1958),

$$m_N g_A = g_{\pi NN} F_\pi, \qquad (26.44)$$

with $g_{\pi NN} \sim 13.45$, $F_\pi \sim 93$ MeV and the constant, $g_A \sim 1.26$. We shall discuss a more complete derivation later, at which time we shall define g_A. In the meantime, we notice that this equation predicts the proton mass, which fits to within 6%.

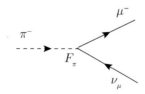

Fig. 26.2 The pion decay constant, F_π, is measured in π decay.

(2) **Adler zeros**

The pion–nucleon coupling is given in terms of the interaction term

$$\mathcal{L} \supset \pi_a \bar{\psi} \tau_a \gamma_5 \psi. \tag{26.45}$$

The matrix element between nucleons is given by

$$\langle N | \bar{\psi} \tau_a \gamma_5 \psi(0) | N \rangle. \tag{26.46}$$

However, consider the following identity,

$$- i \partial_\mu \bar{\psi} \gamma^\mu \gamma_5 \tau_a \psi \equiv -i \bar{\psi} \overleftarrow{\slashed{\partial}} \gamma_5 \tau_a \psi + i \bar{\psi} \gamma_5 \tau_a \slashed{\partial} \psi \tag{26.47}$$
$$\doteq 2 m_N \bar{\psi} \gamma_5 \tau_a \psi,$$

where the last equality follows by using the equations of motion (on shell). Using this result, we can now evaluate the pion–nucleon coupling, see Fig. 26.3. We have

$$\langle N(p') | \bar{\psi} \tau_a \gamma_5 \psi(0) | N(p) \rangle = \frac{-i}{2 m_N} \langle N(p') | \partial_\mu \bar{\psi} \gamma^\mu \gamma_5 \tau_a \psi(0) | N(p) \rangle \tag{26.48}$$
$$\equiv \frac{1}{2 m_N} \langle N(p') | \left[\hat{P}_\mu, \ \bar{\psi} \gamma^\mu \gamma_5 \tau_a \psi(0) \right] | N(p) \rangle$$
$$\equiv \frac{q_\mu}{2 m_N} \langle N(p') | \bar{\psi} \gamma^\mu \gamma_5 \tau_a \psi(0) | N(p) \rangle \overset{q_\mu \to 0}{\longrightarrow} 0.$$

This is known as the Adler zero, i.e. the pion vertex vanishes as the pion four-momentum is taken to zero. A consequence of this phenomenon for massless Nambu–Goldstone bosons is that their exchange does not produce a long range, $\left(\frac{1}{r} \right)$, potential, even if $m_\pi = 0$.

(3) **Pion mass** $\neq 0$

The small mass of the pion implies that chiral symmetry is explicitly broken by a small term in $\mathcal{L}$. Consider adding the term to Eqn. 26.19,

$$\delta \mathcal{L}(\Sigma) = \frac{m_\pi^2}{2} (\sigma^2 + \vec{\pi}^{\,2} - 2 F_\pi \sigma + F_\pi^2), \tag{26.49}$$

where the term linear in F_π explicitly breaks the $SU(2)_L \otimes SU(2)_R$ symmetry. Now shift $\sigma = \tilde{\sigma} + F_\pi$. We have

$$\delta \mathcal{L}(\Sigma) = \frac{m_\pi^2}{2} (\tilde{\sigma}^2 + 2 F_\pi \tilde{\sigma} + F_\pi^2 + \vec{\pi}^{\,2} - 2 F_\pi \tilde{\sigma} - 2 F_\pi^2 + F_\pi^2) \equiv \frac{m_\pi^2}{2} (\tilde{\sigma}^2 + \vec{\pi}^{\,2}). \tag{26.50}$$

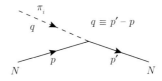

Fig. 26.3   Pion–nucleon coupling.

As a consequence, we now have the mass relations,

$$m_{\tilde{\sigma}}^2 = 2\lambda F_\pi^2 + m_\pi^2,$$
(26.51)

$$m_{\pi_a}^2 = m_\pi^2.$$

(4) Partially conserved axial current

Consider the vector and axial-vector currents for $SU(2)_L \otimes SU(2)_R$ given in terms of an up and down quark doublet,

$$q(x) \equiv \begin{pmatrix} u(x) \\ d(x) \end{pmatrix}.$$
(26.52)

We have

$$j_{a5}^\mu(x) \equiv \bar{q}(x)\gamma^\mu\gamma_5 T_a q(x),$$
(26.53)

$$j_a^\mu(x) \equiv \bar{q}(x)\gamma^\mu T_a q(x),$$

with $T_a \equiv \frac{\tau_a}{2}$. The corresponding vector and axial charges are given by

$$Q_{a5} \equiv \int d^3\vec{x}\, j_{a5}^0(x),$$
(26.54)

$$Q_a \equiv \int d^3\vec{x}\, j_a^0(x).$$

Spontaneous symmetry breaking is then given by

$$Q_{a5}|\Omega\rangle \neq 0.$$
(26.55)

Finally, Goldstone's theorem reduces to the equation

$$\langle [Q_{a5},\, \bar{q}(0)\gamma_5 T_b q(0)]\rangle_0 = \frac{1}{2}\delta_{ab}\langle \bar{q}(0)q(0)\rangle_0 \neq 0.$$
(26.56)

Note, the quark condensate, $\langle\bar{q}(0)q(0)\rangle_0$, is an order parameter for spontaneous symmetry breaking. Eqn. 26.56 implies that the operator, Q_{a5}, creates a pion state out of the vacuum, which is degenerate with the vacuum, assuming the symmetry is exact. The pion state is a pseudo-scalar, also created by $\bar{q}(0)\gamma_5 T_a q(0)$. Let's now define the matrix element

$$\langle\Omega|j_{b5}^\mu(0)|\pi_a(q)\rangle = iF_\pi q^\mu \delta_{ab},$$
(26.57)

where we assume that the leading contribution to the axial-vector current at low energy is given by $j_{b5}^\mu(0) \sim -F_\pi \partial^\mu \pi_b(0) = -iF_\pi \left[\hat{P}^\mu,\, \pi_b(0)\right]$ or

$$\partial_\mu j_{b5}^\mu(x) \sim -F_\pi \,\Box\, \pi_b(x) \doteq F_\pi \, m_\pi^2 \, \pi_b(x).$$
(26.58)

Thus, we have

$$\langle\Omega|\underbrace{\partial_\mu j_b^\mu(0)}_{i\left[\hat{P}_\mu,\, j_{b5}^\mu(0)\right]}|\pi_a(q)\rangle = -iq_\mu \,\langle\Omega|j_{b5}^\mu(0)|\pi_a(q)\rangle = F_\pi m_\pi^2 \delta_{ab}.$$
(26.59)

Note,

$$\partial_\mu j_{b5}^\mu \equiv 0 \quad \Rightarrow \quad m_\pi^2 = 0. \tag{26.60}$$

Thus, the axial-vector current is not conserved. However, the symmetry breaking is small; hence, there is a partially conserved axial current (PCAC).

26.4 General Derivation of the Goldberger–Treiman Relation

The axial-vector form factor between a neutron and proton state is given by

$$\langle p(k')|j_5^{\mu+}(0)|n(k)\rangle \equiv \bar{u}_p(k')\left[\gamma^\mu\gamma_5 g_A(q^2) + q^\mu\gamma_5 h_A(q^2)\right]u_n(k) \tag{26.61}$$

with $q \equiv k - k'$ and $j_5^{\mu+}(x) \equiv j_5^{\mu 1}(x) + ij_5^{\mu 2}(x)$, where $1, 2$ are $SU(2)$ indices. $g_A(q^2)$, $h_A(q^2)$ are form factors. This vertex appears in neutron β decay. In addition, the matrix element of the axial-vector current between the vacuum and a single pion state is given, using PCAC, by

$$\langle \Omega | j_5^{\mu+}(0)| \frac{\pi_1 - i\pi_2}{\sqrt{2}}\rangle \equiv i\sqrt{2}F_\pi q^\mu. \tag{26.62}$$

Note, $\pi^-(x) \equiv (\pi^+(x))^\dagger \equiv \frac{\pi_1(x)+i\pi_2(x)}{\sqrt{2}}$, such that the state

$$|\pi^-\rangle \equiv \frac{1}{\sqrt{2}}|\pi_1 - i\pi_2\rangle \equiv \pi^+|\Omega\rangle \equiv (\pi^-)^\dagger|\Omega\rangle. \tag{26.63}$$

We also have the identities, $\langle\Omega|\pi_a(0)|\pi_b\rangle \equiv \delta_{ab}$, $\langle\Omega|\pi^-(0)|\pi^-\rangle \equiv 1$.

We now have

$$\langle p(k')|\partial_\mu j_5^{\mu+}(0)|n(k)\rangle = -iq_\mu\langle p(k')|j_5^{\mu+}(0)|n(k)\rangle \tag{26.64}$$
$$= \bar{u}_p(k')\left[-i\slashed{q}\gamma_5 g_A(q^2) - iq^2\gamma_5 h_A(q^2)\right]u_n(k).$$

Consider the left-hand side (LHS) of the equation, we have

$$\partial_\mu j_5^{\mu+}(0) \simeq \sqrt{2}F_\pi m_\pi^2(\pi^+(0))^\dagger + \ldots. \tag{26.65}$$

Thus on the left-hand side of Eqn. 26.64, we have[2]

$$\text{LHS} \simeq \sqrt{2}F_\pi m_\pi^2\langle p(k')|(\pi^+(0))^\dagger|n(k)\rangle. \tag{26.66}$$

Moreover by using the Lehmann–Symanzik–Zimmerman (LSZ) formalism, we have the relation which defines the pion–nucleon coupling, i.e.

$$\langle p(k')|\pi^+(-q)n(k)\rangle \equiv -\sqrt{2}g_{\pi NN}(q^2)\bar{u}_p(k')\gamma_5 u_n(k) \tag{26.67}$$
$$= i(-q^2 + m_\pi^2)\langle p(k')|(\pi^+(0))^\dagger|n(k)\rangle. \tag{26.68}$$

---

[2] Note, we expect the term linear in the pion field dominates near $q^2 \approx 0$.

Hence, we have

$$\text{LHS} \simeq \frac{2F_\pi m_\pi^2 g_{\pi NN}(q^2)}{-q^2 + m_\pi^2} i\bar{u}_p(k')\gamma_5 u_n(k). \tag{26.69}$$

Note, the physical pion–nucleon coupling, given by $g_{\pi NN} = g_{\pi NN}(q^2 = m_\pi^2)$ is measured in $\pi - N$ scattering using dispersion relations (see, for example, Gasiorowicz, 1966, chs. 21 and 22).

Now, relating the left-hand side to the right-hand side of Eqn. 26.64, we have

$$\frac{2F_\pi m_\pi^2 g_{\pi NN}(q^2)}{-q^2 + m_\pi^2} i\bar{u}_p(k')\gamma_5 u_n(k) = \bar{u}_p(k') \left[i(\not{k}'\gamma_5 + \gamma_5\not{k})g_A(q^2) - iq^2\gamma_5 h_A(q^2) \right] u_n(k)$$

or, finally,

$$\frac{2F_\pi m_\pi^2 g_{\pi NN}(q^2)}{-q^2 + m_\pi^2} \equiv 2m_N g_A(q^2) - q^2 h_A(q^2). \tag{26.70}$$

In the limit $q^2 \to 0$ we obtain the Goldberger–Treiman relation,

$$m_N g_A(0) = F_\pi g_{\pi NN}(0) \approx F_\pi g_{\pi NN}. \tag{26.71}$$

Note, we can now use the Goldberger–Treiman relation to show that the pion pole only appears in the form factor, $h_A(q^2)$. We have

$$- q^2 h_A(q^2) \equiv 2F_\pi g_{\pi NN}(q^2) \left[\frac{m_\pi^2}{-q^2 + m_\pi^2} \right] - 2m_N g_A(q^2) \tag{26.72}$$

$$\simeq 2F_\pi g_{\pi NN} \left[\frac{m_\pi^2}{-q^2 + m_\pi^2} - 1 \right]$$

or

$$h_A(q^2)_{q^2 \simeq m_\pi^2} \simeq \frac{2F_\pi g_{\pi NN}}{q^2 - m_\pi^2}. \tag{26.73}$$

This is consistent with the Adler zeros and the derivative coupling of Nambu–Goldstone bosons. Hence $g_A(q^2)$ is finite at $q^2 = m_\pi^2$.

26.5 Non-Linear Sigma Model

In the Sigma model of Gell-Mann and Levy, the Σ field transformed linearly under $SU(2)_L \otimes SU(2)_R$. Consider instead a new $\tilde{\Sigma}$ field defined by the expression

$$\tilde{\Sigma} \equiv F_\pi \left(e^{i\frac{\tau_a \pi_a}{F_\pi}} \right), \tag{26.74}$$

where $a = 1, 2, 3$ and π_a is the pion iso-triplet. By definition, $\tilde{\Sigma}$ transforms under $SU(2)_L \otimes SU(2)_R$ by

$$\tilde{\Sigma}' = L\tilde{\Sigma}R^\dagger. \tag{26.75}$$

π_a transforms linearly as a vector under isospin, but non-linearly under the axial symmetry generated by Q_{a5}. This defines the non-linear Sigma model. Note also that the VEV of $\tilde{\Sigma}$ is given by

$$\langle \tilde{\Sigma} \rangle = F_\pi, \tag{26.76}$$

i.e. the symmetry is spontaneously broken to $SU(2)$ isospin.

Now define the field $\Sigma \equiv \tilde{\Sigma}/F_\pi$. We can then define an $SU(2)_L \otimes SU(2)_R$ invariant Lagrangian for this non-linear Sigma model given by

$$\mathcal{L} = F_\pi^2 \, \mathrm{Tr}(\partial^\mu \Sigma^\dagger \partial_\mu \Sigma). \tag{26.77}$$

Chiral Lagrangians have also been generalized to $SU(3)_L \otimes SU(3)_R$ symmetry. Consider the Lagrangian defined by

$$\mathcal{L} = \frac{f^2}{4} \, \mathrm{Tr}(\partial^\mu U^\dagger \partial_\mu U), \tag{26.78}$$

with

$$U \equiv e^{2iT_a \pi_a / f}, \tag{26.79}$$

where $f = 2\sqrt{2} F_\pi$ and

$$T_a \pi_a \equiv \frac{1}{\sqrt{2}} \begin{pmatrix} \frac{\pi^0}{\sqrt{2}} + \frac{\eta}{\sqrt{6}} & \pi^+ & K^+ \\ \pi^- & -\frac{\pi^0}{\sqrt{2}} + \frac{\eta}{\sqrt{6}} & K^0 \\ K^- & \bar{K}^0 & -\frac{2\eta}{\sqrt{6}} \end{pmatrix}. \tag{26.80}$$

In summary, we have discussed $SU(2)_L \otimes SU(2)_R$ chiral symmetry as an approximate symmetry of the strong-interaction Lagrangian. Spontaneous breaking of the symmetry leads to an explanation of why pseudo-scalar bosons are so light compared to either vector mesons or baryons, i.e. they are so-called pseudo-Nambu–Goldstone bosons, which obtain small masses due to explicit chiral symmetry breaking. Non-linear chiral Lagrangians describe the interactions of pseudo-scalar mesons without the σ field, which in some sense was just introduced in order to define a scalar potential that spontaneously breaks the chiral symmetry. In the next chapter we shall use the chiral algebra to predict the value of the axial-vector coupling.

Current Algebra and the Adler–Weisberger Relation

27.1 Adler–Weisberger Relation

In this chapter we shall derive the Adler–Weisberger relation and the π–N scattering amplitude (Adler, 1965; Weisberger, 1965). The current algebra given by

$$[Q_a,\ Q_b] = i\epsilon_{abc}\ Q_c, \tag{27.1}$$
$$[Q_a,\ Q_{b5}] = i\epsilon_{abc}\ Q_{c5},$$
$$[Q_{a5},\ Q_{b5}] = i\epsilon_{abc}\ Q_c,$$

is a non-linear relation between the axial charges and isospin. Isospin conservation, $\partial_\mu j_a^\mu \equiv 0$, and $Q_a|\Omega\rangle \equiv 0$ implies that states are eigenstates of isospin and the non-linear algebra has solutions given by the eigenvalues of $Q_3 \equiv T_3$ and Clebsch–Gordan relations. This is the so-called conserved vector current hypothesis (CVC) (Feynman and Gell-Mann, 1958). The matrix element of the isospin current between a proton and neutron is given by

$$\langle p(k')|j^{\mu+}(0)|n(k)\rangle = \bar{u}_p(k')\left[\gamma^\mu f_1(q^2) - i\Sigma^{\mu\nu}q_\nu f_2(q^2)\right]u_n(k). \tag{27.2}$$

The form factors f_1, f_2 satisfy $f_1(0) \equiv 1$, as guaranteed by isospin invariance and charge conservation, and $f_2(0)$ is related to the anomalous magnetic moment.

If the axial-vector current were also conserved, $\partial_\mu j_{a5}^\mu = 0$ and $Q_{a5}|\Omega\rangle = 0$, then we would also have $g_A(0) \equiv 1$. But $Q_{a5}|\Omega\rangle \neq 0$ and $g_A(0) \neq 1$. The axial charge is renormalized by strong interactions. The question posed by Gell-Mann was, can the non-linear relation, Eqn. 27.1, be used to fix the value of g_A? This is the essence of the Adler–Weisberger relation. We start with the commutator

$$\left[Q_5^+(t),\ Q_5^-(t)\right] = 2T_3, \tag{27.3}$$

which then implies that

$$\langle p(p_2)|\left[Q_5^+(t),\ Q_5^-(t)\right]|p(p_1)\rangle \equiv (2\pi)^3 2E_{p_1}\delta^3(\vec{p}_1 - \vec{p}_2). \tag{27.4}$$

Using partially conserved axial current (PCAC) and $F_\pi g_{\pi NN} = m_n g_A$, we have

$$\frac{dQ_5^+(t)}{dt} \sim \int d^3\vec{x}\,\frac{m_\pi^2 m_N g_A}{g_{\pi NN}}(\pi^+(x))^\dagger. \tag{27.5}$$

Then, inserting a complete set of states into the commutator, Eqn. 27.4, we obtain the Adler–Weisberger relation,

$$1 - \frac{1}{g_A^2} = -\frac{2m_N^2}{\pi g_{\pi NN}^2} \int_{\nu_0}^{\infty} d\nu \frac{[\sigma_{tot}^{\pi^- p}(\nu) - \sigma_{tot}^{\pi^+ p}(\nu)]}{\nu}. \tag{27.6}$$

Theoretically, plugging in the data for the total cross-sections, it is found that $g_A = 1.24$, while experimentally, as measured in neutron beta decay, $g_A = 1.259 \pm 0.017$. Of course, the derivation of the Adler–Weisberger relation requires a great deal of ingenuity and general knowledge of the properties of scattering amplitudes, such as Lorentz invariance, analyticity and crossing symmetry. In the next section, we will present the derivation, since it is quite illuminating.

27.2 Derivation of the Adler–Weisberger Relation

Let's first define the relevant scattering amplitude. We have

$$\langle \pi^a(q_2)N(p_2)|\pi^b(q_1)N(p_1)\rangle \equiv i(2\pi)^4\delta^4(p_1 + q_1 - p_2 - q_2)T_{\pi n}^{ab}. \tag{27.7}$$

Theorem 27.1   *In general, the scattering amplitude can be written in the form*

$$T_{\pi N}^{ab} = \bar{u}(p_2)\underbrace{\left[A^{ab} + \gamma \cdot \frac{(q_1 + q_2)}{2} B^{ab}\right]}_{\Gamma} u(p_1), \tag{27.8}$$

where A, B are functions of the Mandelstam variables, s, t, u.

Proof   The most general form for the matrix Γ is given by

$$\Gamma = A + B^\mu \gamma_\mu + C^{\mu\nu}[\gamma_\mu, \ \gamma_\nu] + D^\mu \gamma_\mu \gamma_5 + E\gamma_5. \tag{27.9}$$

Now we have several constraints.

(a) Parity invariance of the strong interactions implies that $D = E = 0$.
(b)

$$C^{\mu\nu} = C_1 p_1^\mu p_2^\nu + C_2 p_1^\mu q_1^\nu + \dots. \tag{27.10}$$

But using

$$(\not{p}_1 - m_N)u(p_1) = 0, \tag{27.11}$$
$$\bar{u}(p_2)(\not{p}_2 - m_N) = 0,$$
$$p_1 + q_1 = p_2 + q_2,$$

we find that one of the vector indices can always be eliminated.

Thus, only A and B survive. Now, $s \equiv (p_1 + q_1)^2$, $t \equiv (p_2 - p_1)^2$, $u \equiv (p_1 - q_2)^2$, with $s + t + u \equiv 2m_\pi^2 + 2m_N^2$. However, we can also use the Lorentz-invariant variables,

$$\nu \equiv q_1 \cdot \frac{(p_1 + p_2)}{2} = q_2 \cdot \frac{(p_1 + p_2)}{2}, \tag{27.12}$$

$$\nu_B \equiv -\frac{q_1 \cdot q_2}{2}.$$

$\square$

A^{ab}, B^{ab} are 2×2 matrices in isospin space, which, assuming isospin invariance, transform as

$$A^{ab} = \delta^{ab} A^{(+)} + \frac{1}{2}[\tau^a, \ \tau^b]A^{(-)}, \tag{27.13}$$

$$B^{ab} = \delta^{ab} B^{(+)} + \frac{1}{2}[\tau^a, \ \tau^b]B^{(-)},$$

such that an isospin rotation of the nucleons can be compensated by an isospin rotation of the pions. For example, the operator, Eqn. 26.45, is invariant under isospin.

The Adler–Weisberger relation is given in terms of the total π–N cross-section, which, via the optical theorem, is determined by the forward scattering amplitude. We have $q_1 = q_2$, $p_1 = p_2 \equiv p$. In this limit we have

$$\bar{u}(p_2, s_2)\gamma^\mu u(p_1, s_1) = \frac{(p_2 + p_1)^\mu}{2m_N}\bar{u}(p_2, s_2)u(p_1, s_1) = 2p^\mu \delta_{s_1 s_2}, \tag{27.14}$$

where we've used $\bar{u}(p, s_2)u(p, s_1) = 2m_N\delta_{s_1 s_2}$. Finally, the forward elastic scattering amplitude has the form

$$T_{\pi N}^{\pm}\left(\nu, \ \nu_B\left(=-\frac{m_\pi^2}{2}\right), \ q_1^2(=m_\pi^2), \ q_2^2(=m_\pi^2)\right) = \left(A^{\pm}(\nu) + \frac{\nu}{m_N}B^{\pm}(\nu)\right)2m_N. \tag{27.15}$$

Let's now use the optical theorem. We have

$$2\mathrm{Im}T_{ii} \equiv 4qs^{1/2}\,\sigma_{tot}(i). \tag{27.16}$$

The kinematics, in the forward direction in the center of momentum system (CMS) we have $p_1 = p_2$, $\vec{p}_1 + \vec{q}_1 = 0$ or $\vec{p}_1 = -\vec{q}_1 \equiv -\vec{q}$ and $q \equiv |\vec{q}|$. In addition, we have $s = (q_1 + p_1)^2 = m_\pi^2 + m_N^2 + 2q_1 \cdot p_1(\equiv 2\nu)$. Thus, $\nu = \frac{s - m_\pi^2 - m_N^2}{2}$ and

$$2qs^{1/2} = \sqrt{\lambda(s, m_\pi^2, m_N^2)} \overset{m_\pi \approx 0}{\to} s - m_N^2 \simeq 2\nu. \tag{27.17}$$

Hence,

$$2\mathrm{Im}T_{ii} \sim 4\nu\sigma_{tot}(i) \tag{27.18}$$

or, finally,

$$\mathrm{Im}T_{\pi N}^{-}\left(\nu, -\frac{m_\pi^2}{2}, m_\pi^2, m_\pi^2\right) \simeq 2\nu\sigma_{tot}^{-}(\nu). \tag{27.19}$$

Note, later we shall discuss $T_{\pi N}^{+}$.

Of course, the next question is, what is $\sigma_{tot}^-(\nu)$ in terms of $\pi^- p$ or $\pi^+ p$ scattering? From the definition, Eqn. 27.15, we have

$$\langle \pi^a(q_2)N(p_2)|\pi^b(q_1)N(p_1)\rangle \equiv i\epsilon^{abc}\langle N|\tau_c|N\rangle T_{\pi N}^- \quad \text{for} \quad a \neq b \tag{27.20}$$

$$\langle \pi^a(q_2)N(p_2)|\pi^b(q_1)N(p_1)\rangle = \underbrace{\langle N|\mathbb{I}|N\rangle}_{1} T_{\pi N}^+, \quad a,b = 1,2,3, \quad \text{for} \quad a = b.$$

Thus,

$$\langle \pi^1(q_2)N(p_2)|\pi^2(q_1)N(p_1)\rangle = i\langle N|\tau_3|N\rangle T_{\pi N}^- \tag{27.21}$$

$$\langle \pi^2(q_2)N(p_2)|\pi^2(q_1)N(p_1)\rangle = \langle N|N\rangle T_{\pi N}^+.$$

Consider the amplitudes for $\pi^- p$ and $\pi^+ p$ scattering. We have

$$\langle \pi^- p|\pi^- p\rangle = \langle \frac{\pi^1 + i\pi^2}{\sqrt{2}}p|\frac{\pi^1 - i\pi^2}{\sqrt{2}}p\rangle \equiv T_{\pi N}^+ + T_{\pi N}^-, \tag{27.22}$$

$$\langle \pi^+ p|\pi^+ p\rangle \equiv T_{\pi N}^+ - T_{\pi N}^-$$

and therefore, we have

$$T_{\pi N}^- \equiv \frac{1}{2}(\langle \pi^- p|\pi^- p\rangle - \langle \pi^+ p|\pi^+ p\rangle). \tag{27.23}$$

Thus, using the optical theorem, we have

$$\text{Im}T_{\pi N}^- \equiv \frac{1}{2}(\text{Im}T_{\pi^- p}^{elastic} - \text{Im}T_{\pi^+ p}^{elastic}) = \nu(\sigma_T(\pi^- p) - \sigma_T(\pi^+ p)). \tag{27.24}$$

Hence,

$$\sigma_{tot}^-(\nu) \equiv \frac{1}{2}(\sigma_T(\pi^- p) - \sigma_T(\pi^+ p)). \tag{27.25}$$

Now, recall crossing symmetry, Section 13.5. The top diagram in Fig. 27.1 represents the amplitude, $T_{\pi N}^-(\nu, \dots)$ with $\nu = q_1 \cdot \frac{(p_1 + p_2)}{2}$. The bottom diagram is the crossed process. It is easy to see that, under crossing, $\nu \to -\nu$ and $a \leftrightarrow b$. Hence

$$T_{\pi N}^-(\nu, \dots) = -T_{\pi N}^-(-\nu, \dots). \tag{27.26}$$

Consider now the analyticity assumption, i.e.

$$\frac{T_{\pi N}^-(\nu, \dots)}{\nu} \tag{27.27}$$

is an analytic function of ν, except at physical thresholds where there are either poles or cuts. For a reference to the analyticity assumption of scattering amplitudes, see Itzykson and Zuber (1980, pp. 245–255). This property is valid to all finite orders in perturbation theory or it can be proven if one assumes that matrix elements of fields are polynomial bounded. In addition, we have by crossing symmetry,

$$\frac{T_{\pi N}^-(\nu, \dots)}{\nu} \tag{27.28}$$

is a symmetric function of ν. In Fig. 27.2 we show the complex ν plane with the position of the poles and cuts on the real axis. The poles occur at $\nu = \pm\frac{m_\pi^2}{2}$ and the

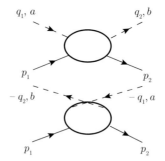

Fig. 27.1 Crossing symmetry for πN scattering. The top diagram represents the amplitude, $T_{\pi N}^{-}(\nu, \dots)$ with $\nu = q_1 \cdot \frac{(p_1 + p_2)}{2}$. The bottom diagram is the crossed process.

Fig. 27.2 Complex ν plane.

cuts begin at threshold $\nu_0 = m_\pi m_N$. Note, in Fig. 27.3 we display two Feynman diagrams which illustrate the origin of the poles and cuts.

We can now use Cauchy's theorem to define a closed contour, Γ, in the complex ν plane which avoids the poles and cuts (see Fig. 27.4). Using Cauchy's theorem we find

$$\frac{T_{\pi N}^{-}(\nu, \dots)}{\nu} = \frac{1}{2\pi i} \oint_\Gamma d\nu' \frac{T_{\pi N}^{-}(\nu', \dots)}{(\nu' - \nu)\nu'}. \tag{27.29}$$

We shall use this formula to relate $T_{\pi N}^{-}$ to the integral over $\mathrm{Im} T_{\pi N}^{-}$, which is known from experiment. We will later use chiral Ward identities to prove that

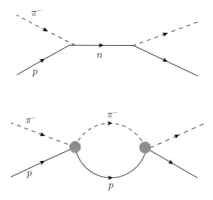

Fig. 27.3 These Feynman diagrams illustrate the origin for both the pole and cut contributions to the πN amplitude.

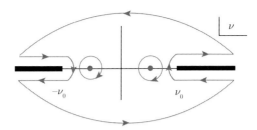

Fig. 27.4 We define a closed contour, Γ, in the complex ν plane, which avoids the poles and cuts.

$$\lim_{\nu \to 0} \frac{T^-_{\pi N}(\nu, -\frac{m^2_\pi}{2}, m^2_\pi, m^2_\pi)}{\nu} \approx \lim_{\nu \to 0} \frac{T^-_{\pi N}(\nu, 0, 0, 0)}{\nu} = \frac{1 - g^2_A}{F^2_\pi}. \tag{27.30}$$

Hence, the amplitude is finite in the limit $\nu \to 0$.

Let's now evaluate the right-hand side (RHS) of Eqn. 27.29. We have

$$\frac{T^-_{\pi N}(\nu, \dots)}{\nu} = \frac{1}{2\pi i} \int_{\nu_0}^{\infty} d\nu' \frac{T^-_{\pi N}(\nu' + i\epsilon)}{(\nu' + i\epsilon - \nu)(\nu' + i\epsilon)} \tag{27.31}$$

$$+ \frac{1}{2\pi i} \int_{\infty}^{\nu_0} d\nu' \frac{T^-_{\pi N}(\nu' - i\epsilon)}{(\nu' - i\epsilon - \nu)(\nu' - i\epsilon)}$$

$$+ \frac{1}{2\pi i} \int_{-\infty}^{-\nu_0} d\nu' \frac{T^-_{\pi N}(\nu' + i\epsilon)}{(\nu' + i\epsilon - \nu)(\nu' + i\epsilon)}$$

$$+ \frac{1}{2\pi i} \int_{-\nu_0}^{-\infty} d\nu' \frac{T^-_{\pi N}(\nu' - i\epsilon)}{(\nu' - i\epsilon - \nu)(\nu' - i\epsilon)}$$

$$+ \text{contour at } \infty + \text{single particle poles.}$$

Combining terms we find (assuming that the integral over the contour at infinity vanishes)

$$\frac{T^-_{\pi N}(\nu, \dots)}{\nu} = \lim_{\epsilon \to 0} \frac{1}{2\pi i} \int_{\nu_0}^{\infty} d\nu' \frac{(T^-_{\pi N}(\nu' + i\epsilon) - T^-_{\pi N}(\nu' - i\epsilon))}{(\nu' - \nu)\nu'} \tag{27.32}$$

$$+ \lim_{\epsilon \to 0} \frac{1}{2\pi i} \int_{\nu_0}^{\infty} d\nu' \frac{(T^-_{\pi N}(\nu' + i\epsilon) - T^-_{\pi N}(\nu' - i\epsilon))}{(\nu' + \nu)\nu'}$$

$$+ \text{single particle poles.}$$

We now want to show that

$$2i \text{Im} T^-_{\pi N}(\nu') = \lim_{\epsilon \to 0}(T^-_{\pi N}(\nu' + i\epsilon) - T^-_{\pi N}(\nu' - i\epsilon)). \tag{27.33}$$

ASIDE on analytic functions: Assume that the function,

$$f(z) \equiv u(z) + iv(z), \text{with } z \equiv x + iy, \tag{27.34}$$

is an analytic function in a connected region, Ω. Then $f(z)$ satisfies

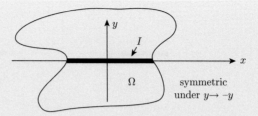

Fig. 27.5   The connected region, Ω, in the complex z plane.

$$\frac{\partial u}{\partial x} = \frac{\partial v}{\partial y},\tag{27.35}$$
$$\frac{\partial u}{\partial y} = -\frac{\partial v}{\partial x}.$$

As a consequence, we have

$$\left(\frac{\partial^2}{\partial x^2} + \frac{\partial^2}{\partial y^2}\right) u \equiv \nabla u = 0,\tag{27.36}$$
$$\nabla v = 0,$$

i.e. u and v are harmonic functions. Now define $\bar{z} \equiv x - iy$. We then have

$$\frac{1}{4}\nabla u \equiv \frac{\partial^2}{\partial z \partial \bar{z}} u,\tag{27.37}$$

which implies

$$\frac{\partial u}{\partial \bar{z}} = \frac{\partial v}{\partial \bar{z}} = 0.\tag{27.38}$$

Now define the conjugate function,

$$\overline{f(\bar{z})} \equiv u(\bar{z}) - iv(\bar{z}).\tag{27.39}$$

We have that $\overline{f(\bar{z})}$ is an analytic function in the connected region, $\bar{z} \in \Omega^*$. Consider a symmetric region, such that $\Omega = \Omega^*$ (see Fig. 27.5). Since $\Omega = \Omega^*$ and Ω is connected, this implies that Ω intersects the real axis along an open interval, I, as shown in Fig. 27.5. Now assume that $f(z)$ is analytic in Ω and real on at least one interval, I, of the real axis, i.e. $f(z) - \overline{f(\bar{z})}$ vanishes on I. As a result, we shall now show that

$$f(z) = \overline{f(\bar{z})} \in \Omega.\tag{27.40}$$

Theorem 27.2   *If an analytic function*

$$f(z) = \overline{f(\bar{z})}\tag{27.41}$$

along an open interval, I, of the real axis, then

$$f(z) = \overline{f(\bar{z})} \quad \in \Omega.\tag{27.42}$$

Proof   Assume

$$f(z) = \overline{f(\bar{z})}, \quad \text{for } z \in \Omega, \tag{27.43}$$

then we have

$$u(z) = u(\bar{z}), \quad v(z) = -v(\bar{z}), \tag{27.44}$$

which implies that

$$u(x, y) = u(x, -y), \quad v(x, y) = -v(x, -y). \tag{27.45}$$

Now for $x \in I$, we have

$$u(x, 0) \neq 0, \quad v(x, 0) = 0. \tag{27.46}$$

Consider expanding $u(x, y)$ in a Taylor series expansion about $y = 0$. The first few terms in this Taylor series expansion are given (for $x \in I$) by

$$\frac{\partial u}{\partial y}\Big|_{y=0} = -\frac{\partial v}{\partial x}\Big|_{y=0} = 0, \tag{27.47}$$

$$\frac{\partial^2 u}{\partial y^2}\Big|_{y=0} = -\frac{\partial^2 u}{\partial x^2}\Big|_{y=0} \neq 0,$$

$$\frac{\partial^3 u}{\partial y^3}\Big|_{y=0} = +\frac{\partial^3 v}{\partial x^3}\Big|_{y=0} = 0.$$

These are the first few terms in the Taylor series expansion, which converges in Ω. There are only terms which are even under $y \to -y$. Hence we find

$$u(x, y) = u(x, -y) \in \Omega. \tag{27.48}$$

Similarly, for $v(x, y)$ (and $x \in I$) we have

$$\frac{\partial v}{\partial y}\Big|_{y=0} = \frac{\partial u}{\partial x}\Big|_{y=0} \neq 0, \tag{27.49}$$

$$\frac{\partial^2 v}{\partial y^2}\Big|_{y=0} = -\frac{\partial^2 v}{\partial x^2}\Big|_{y=0} = 0,$$

$$\frac{\partial^3 v}{\partial y^3}\Big|_{y=0} = -\frac{\partial^3 u}{\partial x^3}\Big|_{y=0} \neq 0.$$

All terms even under $y \to -y$ vanish. Hence,

$$v(x, y) = -v(x, -y) \in \Omega. \tag{27.50}$$

Thus, we have

$$f(z) \left[\equiv u(x + iy) + iv(x + iy)\right] = \overline{f(\bar{z})} \left[\equiv u(x - iy) - iv(x - iy)\right] \in \Omega, \tag{27.51}$$

since

$$u(x + iy) = u(x - iy), \quad v(x + iy) = -v(x - iy) \in \Omega. \tag{27.52}$$

$\square$

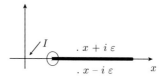

Fig. 27.6   The discontinuity across the cut, $\text{Disc}[f(x)] \equiv f(x + i\epsilon) - f(x - i\epsilon)$.

Now, as stated earlier, if $f(z) = \overline{f(\bar{z})} \in \Omega$, then we have

$$u(z) = u(\bar{z}) \tag{27.53}$$
$$v(z) = -v(\bar{z}).$$

Therefore, we have

$$f(x + i\epsilon) = \bar{f}(x - i\epsilon), \tag{27.54}$$
$$f(x - i\epsilon) = \bar{f}(x + i\epsilon).$$

Define the discontinuity in the function $f(x)$, Fig. 27.6,

$$\text{Disc}[f(x)] \equiv f(x + i\epsilon) - f(x - i\epsilon). \tag{27.55}$$

Then we have

$$\begin{aligned}
\text{Disc}[f(x)] &\equiv f(x + i\epsilon) - f(x - i\epsilon) \tag{27.56}\\
&= f(x + i\epsilon) - \bar{f}(x + i\epsilon)\\
&\equiv 2i\text{Im}f(x + i\epsilon).
\end{aligned}$$

Hence, going back to our original problem, Eqn. 27.33, we have

$$2i\text{Im}T^-_{\pi N}(\nu) = \lim_{\epsilon \to 0}(T^-_{\pi N}(\nu + i\epsilon) - T^-_{\pi N}(\nu - i\epsilon)). \tag{27.57}$$

Thus, we find

$$\frac{T^-_{\pi N}(\nu, \ldots)}{\nu} = \frac{2}{\pi} \int_{\nu_0}^{\infty} d\nu' \frac{\text{Im}T^-_{\pi N}(\nu')}{(\nu'^2 - \nu^2)} + \text{single particle poles} \tag{27.58}$$

or, using the optical theorem, Eqn. 27.24,

$$\lim_{\nu \to 0}\left(\frac{T^-_{\pi N}(\nu, \ldots)}{\nu}\right) = \frac{2}{\pi} \int_{\nu_0}^{\infty} \frac{d\nu'}{\nu'}(\sigma_T(\pi^- p) - \sigma_T(\pi^+ p)) + \text{single particle poles}. \tag{27.59}$$

We are almost there. Now we need to consider the left-hand side of Eqn. 27.59 using low energy theorems, and also evaluate the pole terms. Using the Lehmann–Symanzik–Zimmerman (LSZ) formalism, it can be seen that

$$T^{ab}_{\pi N} \equiv i \int d^4x e^{iq_1 \cdot x}(q_1^2 - m_\pi^2)(q_2^2 - m_\pi^2)\langle N(p_2)|T(\pi^a(x)\pi^b(0))|N(p_1)\rangle \tag{27.60}$$

$$= i\frac{(q_1^2 - m_\pi^2)(q_2^2 - m_\pi^2)}{m_\pi^4 F_\pi^2} \int d^4x e^{iq_1 \cdot x}\langle N(p_2)|T(\partial^\mu j^a_{\mu 5}(x)\partial^\nu j^b_{\nu 5}(0))|N(p_1)\rangle.$$

Note, the second line is a consequence of PCAC.

A brief digression:

Define

$$T_{\mu\nu}^{ab} \, (2\pi)^4 \delta^4(p_1 + q_1 - p_2 - q_2) \tag{27.61}$$

$$\equiv \int d^4x \, d^4y \, e^{iq_1 \cdot x} e^{-iq_2 \cdot y} \langle N(p_2)|T(j_{\mu 5}^a(x) j_{\nu 5}^b(y))|N(p_1)\rangle.$$

These two amplitudes, Eqns. 27.60 and 27.61, as we shall now show, are related by axial-vector Ward identities. Consider

$$\partial_x^\mu \partial_y^\nu T(j_{\mu 5}^a(x) j_{\nu 5}^b(y))$$

$$\equiv \partial_x^\mu \partial_y^\nu [\theta(x^0 - y^0) j_{\mu 5}^a(x) j_{\nu 5}^b(y) + \theta(y^0 - x^0) j_{\nu 5}^b(y) j_{\mu 5}^a(x)]$$

$$= \partial_x^\mu \left[\theta(x^0 - y^0) j_{\mu 5}^a(x) \partial_y^\nu \, j_{\nu 5}^b(y) + \theta(y^0 - x^0) \partial_y^\nu j_{\nu 5}^b(y) j_{\mu 5}^a(x) \right.$$

$$\left. - \delta(x^0 - y^0) j_{\mu 5}^a(x) j_{05}^b(y) + \delta(y^0 - x^0) j_{05}^b(y) j_{\mu 5}^a(x) \right]$$

$$\partial_x^\mu \partial_y^\nu \, T(j_{\mu 5}^a(x) \, j_{\nu 5}^b(y)) \tag{27.62}$$

$$= T(\partial^\mu j_{\mu 5}^a(x) \partial^\nu j_{\nu 5}^b(y)) \qquad (1)$$

$$+ \, \delta(x^0 - y^0)[j_{05}^a(x), \, \partial^\nu j_{\nu 5}^b(y)] \qquad (2)$$

$$- \, \partial_x^\mu (\delta(x^0 - y^0)[j_{\mu 5}^a(x), \, j_{05}^b(y)]) \qquad (3).$$

Now, calculate the expectation value of Eqn. 27.62, i.e.

$$\int d^4x \, d^4y \, e^{iq_1 \cdot x} e^{-iq_2 \cdot y} \langle N(p_2)|[\text{Eqn. 27.62}]|N(p_1)\rangle \tag{27.63}$$

and neglect the overall factor of $(2\pi)^4 \delta^4(q_1 + p_1 - q_2 - p_2)$, which is obtained by using translation invariance inside the matrix element and integrating over y. We have

$$q_1^\mu q_2^\nu \int d^4x \, e^{iq_1 \cdot x} \langle N(p_2)|T(j_{\mu 5}^a(x) j_{\nu 5}^b(0))|N(p_1)\rangle \tag{27.64}$$

$$\equiv \int d^4x \, e^{iq_1 \cdot x} \left\{ \langle N(p_2)|T(\partial^\mu j_{\mu 5}^a(x) \partial^\nu j_{\nu 5}^b(0))|N(p_1)\rangle \quad (1) \right.$$

$$+ \langle N(p_2)|\delta(x^0)[j_{05}^a(x), \partial^\nu j_{\nu 5}^b(0)]|N(p_1)\rangle \qquad (2)$$

$$\left. - iq_1^\mu \, \langle N(p_2)|\delta(x^0)[j_{05}^b(0), \, j_{\mu 5}^a(x)]|N(p_1)\rangle \right\} \qquad (3).$$

Now we are interested in the limit $q_1 = q_2$, $p_1 = p_2 \equiv p$, i.e. forward scattering. The first term on the right-hand side, using PCAC, is related to $T_{\pi N}^{ab}$. Consider term (2) in the limit $q \to 0$. Define

$$\lim_{q \to 0} \sigma_N^{ab}(p, q) \equiv \sigma_N^{ab} \tag{27.65}$$

$$\equiv i \int d^4x \, \delta(x^0) \langle N(p)|[j_{05}^a(x), \partial^\nu j_{\nu 5}^b(x^0, \vec{0})]|N(p)\rangle.$$

σ_N^{ab} is known as the πN σ term. We now want to show that it is actually symmetric under the interchange, $a \leftrightarrow b$, and thus does not contribute to $T_{\pi N}^-$.

Proof

$$\sigma_N^{ab} = i \int d^3\vec{x} \langle N(p)|[j_{05}^a(0,\vec{x}), \partial^0 j_{05}^b(0,\vec{0})]|N(p)\rangle, \tag{27.66}$$

where the spatial component of the partial derivative vanishes upon integrating over all space. We then have

$$\sigma_N^{ab} = \lim_{x^0 \to 0} \left\{ i\partial^0 \int d^3\vec{x} \langle N(p)|[j_{05}^a(x^0,\vec{x}), j_{05}^b(x^0,\vec{0})]|N(p)\rangle \right. \tag{27.67}$$

$$\left. - i \int d^3\vec{x} \langle N(p)|[\partial^0 j_{05}^a(x^0,\vec{x}), j_{05}^b(x^0,\vec{0})]|N(p)\rangle (\equiv \sigma_N^{ba}) \right\}. \tag{27.68}$$

Hence, we have

$$\sigma_N^{ab} - \sigma_N^{ba} = \lim_{x^0 \to 0} i\partial^0 \int d^3\vec{x} \langle N(p)|[j_{05}^a(x^0,\vec{x}), j_{05}^b(x^0,\vec{0})]|N(p)\rangle. \tag{27.69}$$

But

$$[j_{05}^a(x), j_{05}^b(x^0,\vec{0})] \equiv i\epsilon^{abc} j_0^c(x)\delta^3(\vec{x}) \tag{27.70}$$

and therefore

$$\partial^0 \int d^3\vec{x}[j_{05}^a(x), j_{05}^b(x^0,\vec{0})] = i\epsilon^{abc}\partial^0 \int d^3\vec{x} j_0^c(x)\delta^3(\vec{x}) \tag{27.71}$$

$$= \epsilon^{abc} \int d^3\vec{x}[\hat{P}^0, \ j_0^c(x)]\delta^3(\vec{x}) \tag{27.72}$$

vanishes on shell. Hence, we conclude that

$$\sigma_N^{ab} = \sigma_N^{ba} \equiv \sigma_N \delta^{ab}. \tag{27.73}$$

$\square$

We now have

$$q^\mu q^\nu T_{\mu\nu}^{ab} = -i(q^2 - m_\pi^2)^{-2} m_\pi^4 F_\pi^2 T_{\pi N}^{ab} \qquad (1) \tag{27.74}$$

$$-i\sigma_N^{ab}(p,q) \qquad (2)$$

$$-iq^\mu \int d^4x e^{iq\cdot x}\delta(x^0)\langle N(p)|[j_{05}^b(0), \ j_{\mu 5}^a(x)]|N(p)\rangle \qquad (3).$$

Using current algebra,

$$\delta(x^0 - y^0)[Q_5^b(y^0), \ j_{\mu 5}^a(x)] \equiv \delta(x^0 - y^0) \int d^3\vec{y} \ [j_{05}^b(y^0,\vec{y}), \ j_{\mu 5}^a(x)]$$

$$\equiv i\epsilon^{bac} j_\mu^c(x)\delta(x^0 - y^0), \tag{27.75}$$

which implies that

$$\delta(x^0)[j_{05}^b(0), \ j_{\mu 5}^a(x)] = -i\epsilon^{abc} j_\mu^c(x)\delta^4(x). \tag{27.76}$$

Moreover, we have

$$\langle N(p)|j_\mu^c(x)|N(p)\rangle \equiv \bar{u}(p)\frac{\tau_c}{2}\gamma_\mu u(p). \tag{27.77}$$

The normalization is fixed by the fact that it is a conserved current and the generator of isospin transformations.

Thus, we now have

$$q^\mu q^\nu T_{\mu\nu}^{ab} = -i(q^2 - m_\pi^2)^{-2} m_\pi^4 F_\pi^2 T_{\pi N}^{ab} \tag{27.78}$$

$$-i\sigma_N^{ab}(p,q) - q^\mu \epsilon^{abc}\bar{u}(p)\gamma_\mu \frac{\tau^c}{2}u(p)$$

$$= -i(q^2 - m_\pi^2)^{-2} m_\pi^4 F_\pi^2 T_{\pi N}^{ab} - i\sigma_N^{ab}(p,q) + i\nu\,\frac{[\tau^a,\,\tau^b]}{2}.$$

Note, we used the relations

$$\bar{u}(p)\gamma_\mu u(p) = 2p_\mu \delta_{s_1 s_2} \tag{27.79}$$

and

$$-2q\cdot p\epsilon^{abc}\,\frac{\tau^c}{2} \equiv i\nu\,\frac{[\tau^a,\,\tau^b]}{2}. \tag{27.80}$$

Soft-Pion Limit

We are interested in the limit that $q^2 \to 0$ first and then $q^\mu \to 0$ (i.e. the soft-pion limit). Since $\nu \equiv \frac{q_1\cdot(p_1+p_2)}{2} = q\cdot p$, this implies that as $q^\mu \to 0$, then $\nu \to 0$. Thus we are interested in the limit

$$\lim_{\nu\to 0}\left(\frac{T_{\pi N}^-(\nu,\dots)}{\nu}\right). \tag{27.81}$$

Now we need to evaluate the left-hand side of Eqn. 27.78. We have

$$q^\mu q^\nu T_{\mu\nu}^{ab} = \text{pole terms } + \dots. \tag{27.82}$$

See Fig. 27.7 for the Feynman diagrams contributing to the pole terms.

Pole Terms

$$g_A^2 iq^\mu q^\nu \bar{u}(p)\left\{\gamma_\mu\gamma_5\frac{\tau^a}{2}\frac{\not{p}+\not{q}+m_N}{(p+q)^2-m_N^2}\gamma_\nu\gamma_5\frac{\tau^b}{2} + \gamma_\nu\gamma_5\frac{\tau^b}{2}\frac{\not{p}-\not{q}+m_N}{(p-q)^2-m_N^2}\gamma_\mu\gamma_5\frac{\tau^a}{2}\right\}u(p).$$
$$\tag{27.83}$$

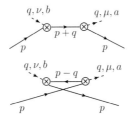

Feynman diagrams contributing to the pole terms in Eqn. 27.78.

Using $(p+q)^2 - m_N^2 = 2\nu + q^2$ and $(p-q)^2 - m_N^2 = q^2 - 2\nu$, we have

$$ig_A^2 \bar{u}(p) \left\{ \slashed{q}\gamma_5 \frac{\slashed{p}+\slashed{q}+m_N}{q^2+2\nu} \slashed{q}\gamma_5 \frac{\tau^a}{2}\frac{\tau^b}{2} + \slashed{q}\gamma_5 \frac{\slashed{p}-\slashed{q}+m_N}{q^2-2\nu} \slashed{q}\gamma_5 \frac{\tau^b}{2}\frac{\tau^a}{2} \right\} u(p)$$

$$= ig_A^2 \bar{u}(p) \left\{ \slashed{q}\frac{\slashed{p}+\slashed{q}-m_N}{q^2+2\nu} \slashed{q} \frac{\tau^a}{2}\frac{\tau^b}{2} + \slashed{q}\frac{\slashed{p}-\slashed{q}-m_N}{q^2-2\nu} \slashed{q}\frac{\tau^b}{2}\frac{\tau^a}{2} \right\} u(p). \tag{27.84}$$

We can use the following identities to simply further. We have

$$\slashed{q}\slashed{q} = q^\mu q^\nu \gamma_\mu \gamma_\nu = q^\mu q^\nu \frac{\{\gamma_\mu \cdot \gamma_\nu\}}{2} = q^2, \tag{27.85}$$

$$\slashed{q}\slashed{p}\slashed{q} = q^\mu q^\nu \gamma_\mu \slashed{p}\gamma_\nu = q^\mu q^\nu (-\gamma_\mu \gamma_\nu \slashed{p} + 2\gamma_\mu p_\nu) = -q^2\slashed{p} + 2\slashed{q}\nu, \tag{27.86}$$

and $\slashed{p}u(p) = m_N u(p)$. We obtain

$$ig_A^2 \bar{u}(p) \left\{ \frac{-2q^2 m_N + \slashed{q}(q^2+2\nu)}{q^2+2\nu} \frac{\tau^a}{2}\frac{\tau^b}{2} + \frac{-2q^2 m_N - \slashed{q}(q^2-2\nu)}{q^2-2\nu} \frac{\tau^b}{2}\frac{\tau^a}{2} \right\} u(p). \tag{27.87}$$

Next, using

$$\bar{u}(p)u(p) = 2m_N \delta_{s_1 s_2}, \quad \bar{u}(p)\slashed{q}u(p) = 2\nu \delta_{s_1 s_2}, \tag{27.88}$$

we have

$$(q^\mu q^\nu T_{\mu\nu}^{ab})|_{\text{pole term}} = \frac{2ig_A^2}{q^4 - 4\nu^2} \left\{ (q^2 - 2\nu)(2\nu^2 + q^2(\nu - 2m_N^2)) \frac{\tau^a}{2}\frac{\tau^b}{2} \right.$$

$$\left. + (q^2 + 2\nu)(2\nu^2 - q^2(\nu + 2m_N^2)) \frac{\tau^b}{2}\frac{\tau^a}{2} \right\} \delta_{s_1 s_2}. \tag{27.89}$$

Expanding the parentheses we finally find

$$(q^\mu q^\nu T_{\mu\nu}^{ab})|_{\text{pole term}} = \frac{2ig_A^2}{q^4 - 4\nu^2} \left\{ (q^4 - 4\nu^2 + 4q^2 m_N^2)\nu \left[\frac{\tau^a}{2}, \frac{\tau^b}{2}\right] \right.$$

$$\left. + 2q^2(\nu^2 - q^2 m_N^2)\left\{\frac{\tau^a}{2}, \frac{\tau^b}{2}\right\} \right\}$$

$$= \lim_{\substack{q^2 \to 0 \\ \nu \text{ fixed}}} ig_A^2 \frac{[\tau^a, \tau^b]}{2} \nu. \tag{27.90}$$

Let's now show that the pole term dominates near $q^2 \to 0$, since otherwise we have propagators of the form (with $M \geq m_\pi + m_N$)

$$\frac{1}{(p+q)^2 - M^2} = \frac{1}{m_N^2 - M^2 + q^2 + 2\nu} \tag{27.91}$$

$$= \lim_{q^2 \to 0} \frac{1}{2\nu + m_N^2 - M^2} < \frac{1}{2\nu - (2m_\pi m_N + m_\pi^2)} \ll \frac{1}{2\nu} \quad \text{for} \quad \nu \to 0.$$

Putting it all together, Eqns. 27.78, 27.90, we obtain at $q^2 \to 0$, ν fixed,

$$ig_A^2 \frac{[\tau^a, \, \tau^b]}{2} \nu = -iF_\pi^2 T_{\pi N}^{ab} - i\sigma_N^{ab}(p, q) + i\frac{[\tau^a, \, \tau^b]}{2} \nu. \tag{27.92}$$

After these many steps we find

$$\lim_{\nu \to 0} \left(\frac{T_{\pi N}^{ab}(\nu, 0, 0, 0)}{\nu} \right) = \left(\frac{1 - g_A^2}{F_\pi^2} \right) \frac{[\tau^a, \, \tau^b]}{2} - \frac{\sigma_N \delta^{ab}}{F_\pi^2 \nu}. \tag{27.93}$$

Finally, we now obtain the left-hand side of Eqn. 27.59. We have

$$\lim_{\nu \to 0} \left(\frac{T_{\pi N}^-(\nu, 0, 0, 0)}{\nu} \right) = \frac{1 - g_A^2}{F_\pi^2} \tag{27.94}$$

and, similarly, we have

$$\lim_{\nu \to 0} \left(T_{\pi N}^+(\nu, 0, 0, 0) \right) = -\frac{\sigma_N}{F_\pi^2}. \tag{27.95}$$

Therefore, Eqn. 27.59 is now given by

$$\frac{1 - g_A^2}{F_\pi^2} = \frac{2}{\pi} \int_{\nu_0}^\infty \frac{d\nu'}{\nu'} (\sigma_T(\pi^- p) - \sigma_T(\pi^+ p)) + \text{single particle poles}. \tag{27.96}$$

However, we can now show that these pole terms, Fig. 27.8, do not contribute to the amplitude $T_{\pi N}^-$ in the soft pion limit, $q^2 \to 0$ first, followed by $\nu \to 0$.

The Feynman amplitude is given by

$$g_{\pi NN}^2 \bar{u}(p) \left\{ \gamma_5 \frac{\slashed{p} + \slashed{q} + m_N}{(p + q)^2 - m_N^2} \gamma_5 \frac{\tau^a}{2} \frac{\tau^b}{2} + \gamma_5 \frac{\slashed{p} - \slashed{q} + m_N}{(p - q)^2 - m_N^2} \gamma_5 \frac{\tau^b}{2} \frac{\tau^a}{2} \right\} u(p) \tag{27.97}$$

$$= g_{\pi NN}^2 \left\{ \frac{-2\nu}{q^2 + 2\nu} \frac{\tau^a}{2} \frac{\tau^b}{2} + \frac{2\nu}{q^2 - 2\nu} \frac{\tau^b}{2} \frac{\tau^a}{2} \right\} \delta_{s_1 s_2}$$

$$= -\frac{2g_{\pi NN}^2}{(q^4 - 4\nu^2)} \left\{ (q^2 - 2\nu) \frac{\tau^a}{2} \frac{\tau^b}{2} \nu - (q^2 + 2\nu) \frac{\tau^b}{2} \frac{\tau^a}{2} \nu \right\}$$

$$= -\frac{2g_{\pi NN}^2}{(q^4 - 4\nu^2)} \left\{ q^2 \left[\frac{\tau^a}{2}, \frac{\tau^b}{2} \right] \nu - 2\nu^2 \left\{ \frac{\tau^a}{2}, \frac{\tau^b}{2} \right\} \right\}$$

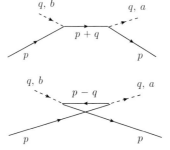

q, b   q, a
p + q
p   p

q, b   q, a
p − q
p   p

Fig. 27.8 Feynman diagrams contributing to the pole terms in Eqn. 27.96.

$$\xrightarrow{q^2 \to 0} -g^2_{\pi NN} \left\{ \frac{\tau^a}{2}, \frac{\tau^b}{2} \right\}.$$

Therefore, there is no contribution of these pole terms to $T^-_{\pi N}$ in the soft-pion limit.

As the last step, using the Goldberger–Treiman relation, $F^2_\pi / g^2_A = m^2_N / g^2_{\pi NN}$, Eqn. 27.96, becomes

$$\frac{1}{g^2_A} = 1 + \frac{2m^2_N}{\pi g^2_{\pi NN}} \int^\infty_{\nu_0} \frac{d\nu}{\nu} [\sigma^{\pi^- p}_T (\nu) - \sigma^{\pi^+ p}_T (\nu)], \tag{27.98}$$

i.e. the Adler–Weisberger relation!

PART VII

ROAD TO THE STANDARD MODEL: QUANTUM CHROMODYNAMICS

28 Quantum Chromodynamics

28.1 Five Puzzles of the Quark Model

In 1971, Harald Fritzsch and Murray Gell-Mann defined the current algebra of the up, down and strange quarks, $SU(3)_L \otimes SU(3)_R$, to discuss deep inelastic scattering (a subject which we shall discuss shortly). In their analysis they assumed that quark currents could be used to derive experimental consequences. They also noted that there were some consequences of three spin-1/2 quarks which did not agree with data. To fix this problem it was suggested that quarks come in three colors, and that gauge particles, called gluons, held the quarks together to form color singlet bound states (Fritzsch and Gell-Mann, 1971; Bardeen et al., 1972; Fritzsch et al., 1973). It is these paradoxes of the non-relativistic quark model which we discuss in this chapter.

Consider the quark model with the quark field $q(x)$ given by

$$q(x) = \begin{pmatrix} u(x) \\ d(x) \\ s(x) \end{pmatrix}. \tag{28.1}$$

The free-field Lagrangian is then

$$\mathcal{L} = \bar{q}(x) i \not{\partial} q(x) = \bar{q}_L(x) i \not{\partial} q_L(x) + \bar{q}_R(x) i \not{\partial} q_R(x). \tag{28.2}$$

The Lagrangian is invariant under the chiral symmetry, $SU(3)_L \otimes SU(3)_R$. However, it was thought that this symmetry is effectively spontaneously broken by a condensate of the form

$$\langle \bar{u}u \rangle = \langle \bar{d}d \rangle = \langle \bar{s}s \rangle \neq 0, \tag{28.3}$$

which breaks

$$SU(3)_L \otimes SU(3)_R \rightarrow SU(3)_{flavor}. \tag{28.4}$$

As a consequence, there is an octet of Nambu–Goldstone bosons:

$$\{\pi(140), K(494), \eta(550)\}. \tag{28.5}$$

Puzzle 1: The actual symmetry of the Lagrangian, Eqn. 28.2, is

$$U(3)_L \otimes U(3)_R = SU(3)_L \otimes SU(3)_R \otimes U(1)_B \otimes U(1)_{axial}, \tag{28.6}$$

where $U(1)_B$ is the baryon number symmetry. $U(1)_{axial}$ is also spontaneously broken and therefore there should be a ninth Nambu–Goldstone boson. But the state $\eta'(958)$ is much heavier than the pseudo-scalar octets. The question is, why?

Puzzle 2: The spin, parity and charge conjugation quantum numbers, J^{PC}, of the pseudo-scalar and vector bosons satisfy

$$\{\pi,\ K,\ \eta,\ \eta'\}\quad 0^{-+}\quad S = 0 \tag{28.7}$$
$$\{\rho,\ K^*,\ \omega,\ \phi\}\quad 1^{--}\quad S = 1.$$

These quantum numbers can successfully be described in terms of the non-relativistic quark model as $\bar{q}q$ bound states with orbital angular momentum, $L = 0$. We then have

$$P = (-1)^{L+1},\quad C = (-1)^{L+S} \tag{28.8}$$

(see Chapter 16, Section 16.4).

These results can also be described in terms of the non-relativistic symmetry group,

$$SU(6) \equiv SU(3)_{flavor} \otimes SU(2)_{spin}. \tag{28.9}$$

The quark field then transforms as

$$q(x) = \left(\mathbf{3}, \pm\frac{1}{2}\right). \tag{28.10}$$

The 35 generators of $SU(6)$ in the **6** representation are given by

$$\left(\frac{\lambda_a}{2} \otimes \mathbb{I}_2\right),\ \left(\mathbb{I}_3 \otimes \frac{\tau_i}{2}\right),\ \left(\frac{\lambda_a}{2} \otimes \frac{\tau_i}{2}\right). \tag{28.11}$$

In the language of non-relativistic $SU(6)$, the bosons are in the tensor-product representation,

$$\mathbf{6} \otimes \bar{\mathbf{6}} = \mathbf{1} \oplus \mathbf{35}, \tag{28.12}$$

where the **35** contains the eight pseudo-scalar bosons and the 27 states of the nine vector bosons times their three spins. On the other hand, the baryons are in the symmetric tensor product representation

$$(\mathbf{6} \otimes \mathbf{6} \otimes \mathbf{6})_S = \mathbf{56}. \tag{28.13}$$

This includes the baryon decuplet with four spin states and the octet with two spin states, for a total of 56 states. Great! But the problem is that quarks are spin-1/2 fermions and therefore the bound-state wave function should be totally anti-symmetric. However, it seems to be symmetric in space, i.e. $L = 0$, and symmetric under flavor and spin, i.e. under $SU(6)$; therefore, it is totally symmetric. So what is it anti-symmetric under??

Puzzle 3: The decay rate for $\pi^0 \to 2\gamma$ can be calculated with the triangle graphs, Fig. 28.1. It is found that

$$\Gamma^{theory}_{\pi^0 \to 2\gamma} = \frac{1}{9}\Gamma^{exp.}_{\pi^0 \to 2\gamma}. \tag{28.14}$$

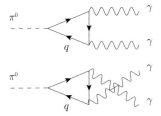

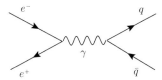

Fig. 28.1 Feynman diagrams contributing to π^0 decay to two photons.

Fig. 28.2 Feynman diagrams contributing to $e^+e^- \to$ hadrons.

Puzzle 4: Define the ratio

$$R \equiv \frac{\sigma(e^+e^- \to q\bar{q}\{= hadrons\})}{\sigma(e^+e^- \to \mu^+\mu^-)}. \tag{28.15}$$

The Feynman graph for the process $e^+e^- \to$ hadrons is given in Fig. 28.2. Then it is found that

$$R^{theory} = \frac{1}{3}R^{exp.}. \tag{28.16}$$

Puzzle 5: What is the strong force that binds quarks and anti-quarks to make hadrons, and binds hadrons together to make nuclei? The solution is a new color force, quantum chromodynamics (QCD). The Lagrangian for QCD is given by

$$\mathcal{L} = \sum_f \bar{q}_f(x)(i\not{D} - m)q_f(x) - \frac{1}{4}G^2_{\mu\nu}, \tag{28.17}$$

where f is a flavor index and the quark fields now also have a color index, $q^a_f(x)$, with $a = 1, 2, 3$. Baryons and mesons are color singlets given by

$$q^a_{f_1} q^b_{f_2} q^c_{f_3} \epsilon_{abc}, \quad \bar{q}_{af_1} q^a_{f_2}. \tag{28.18}$$

This resolves the statistics problem of Puzzle 2. But it doesn't really resolve the issue of why only finite-energy color singlet states exist. For the answer to this question we need to wait until Chapter 30.

28.2 Local Non-Abelian Symmetry

Consider a specific example of quarks defined by the Dirac spinor field $q_f(x)$, $f = 1, \ldots, n_f$. f is a flavor quantum number, for example, $q_1 = u$, $q_2 = d$, $q_3 = s$, etc. For simplicity of the discussion, let's now take $n_f = 1$ and put aside the flavor quantum number, so that we can focus on color. For a general non-Abelian gauge theory we have

$$q^a(x), \ a = 1, \ldots, N \qquad (28.19)$$

where a is the color quantum number. For QCD, $N = 3$.[1]

Now consider the Lagrangian

$$\mathcal{L} = \bar{q}_a(i\slashed{\partial} - m)q^a. \qquad (28.20)$$

It is invariant under a global $SU(N)$ symmetry, where

$$SU(N) = \{u|N \times N \ \text{complex matrices}; u^\dagger u = \mathbb{I}_N, \det u = 1\}. \qquad (28.21)$$

For a global symmetry, the matrices u are independent of space–time. The transformations are given by

$$q'(x) = uq(x), \ \text{or} \ q^{a'}(x) = u^a{}_b q^b(x), \qquad (28.22)$$

$$\bar{q}'(x) = \bar{q}(x)u^\dagger$$

where

$$u^a{}_b = (e^{iT_A\theta_A})^a{}_b. \qquad (28.23)$$

The conditions

$$u^\dagger u = \mathbb{I}_N \Rightarrow T_A = T_A^\dagger, \quad \det u = 1 \Rightarrow \text{Tr } T_A = 0. \qquad (28.24)$$

Thus, the color index takes on values, $A = 1, \ldots, N^2 - 1$. Given a convention for the $N \times N$ matrices, $(T_A)^a{}_b$, and the normalization $\text{Tr}(T_A T_B) = \frac{1}{2}\delta_{AB}$, we obtain the commutation relations

$$[T_A, \ T_B] = if_{ABC}T_C. \qquad (28.25)$$

Now consider the local $SU(N)$ transformation given by

$$u^a{}_b(x) = (e^{iT_A\theta_A(x)})^a{}_b. \qquad (28.26)$$

We shall then, in analogy with quantum electrodynamics (QED), define a covariant derivative

$$D_\mu q(x) \equiv (\partial_\mu + ig\tilde{A}_\mu)q(x), \qquad (28.27)$$

where

$$\tilde{A}_\mu \equiv A_{\mu B}(x)(T_B)^a{}_b \qquad (28.28)$$

are $N \times N$ traceless, Hermitian matrices and $A_{\mu B}$ are $N^2 - 1$ gauge fields. The covariant derivative is defined such that it satisfies the equation

$$D'_\mu q'(x) \equiv uD_\mu q(x), \qquad (28.29)$$

$$(\partial_\mu + ig\tilde{A}'_\mu)uq(x) \equiv u(\partial_\mu + ig\tilde{A}_\mu)q(x).$$

---

[1] Non-Abelian gauge theory or Yang–Mills theory (Yang and Mills, 1954) was first discussed as a possible local isospin symmetry. It didn't quite work since the rho meson was massive.

Now we can solve for the transformation law of the gauge field, which satisfies this relation. We have

$$(\partial_\mu u)q + u\partial_\mu q + ig\tilde{A}'_\mu uq = u\partial_\mu q + igu\tilde{A}_\mu q$$
$$\Rightarrow \quad ig\tilde{A}'_\mu u = igu\tilde{A}_\mu - \partial_\mu u \tag{28.30}$$

(since it must be true for any q) or

$$\tilde{A}'_\mu = u\tilde{A}_\mu u^\dagger + \frac{i}{g}(\partial_\mu u)u^\dagger. \tag{28.31}$$

With this definition of the covariant derivative, it is easy to see that the Lagrangian

$$\mathcal{L}_{q,A} = \bar{q}(x)(i\slashed{D} - m)q(x) \tag{28.32}$$

is gauge invariant.

Now we need a kinetic term for the gauge field. Define

$$\tilde{F}_{\mu\nu} \equiv -\frac{i}{g}\left[D_\mu,\ D_\nu\right], \tag{28.33}$$

where the covariant derivatives are operators acting on an N-dimensional vector space. We then have

$$\tilde{F}_{\mu\nu} \equiv -\frac{i}{g}\left[(\partial_\mu + ig\tilde{A}_\mu),\ (\partial_\nu + ig\tilde{A}_\nu)\right]$$

or

$$\tilde{F}_{\mu\nu} \equiv \partial_\mu \tilde{A}_\nu - \partial_\nu \tilde{A}_\mu + ig\left[\tilde{A}_\mu,\ \tilde{A}_\nu\right]. \tag{28.34}$$

This is clearly a generalization of the Abelian field strength.

But how does $\tilde{F}_{\mu\nu}$ transform under the local $SU(N)$ gauge transformation? We have

$$\tilde{F}'_{\mu\nu}q' \equiv -\frac{i}{g}\left[D'_\mu,\ D'_\nu\right]uq \tag{28.35}$$
$$\Rightarrow \quad \tilde{F}'_{\mu\nu}uq \equiv -\frac{i}{g}u\left[D_\mu,\ D_\nu\right]q \equiv u\tilde{F}_{\mu\nu}q.$$

The latter equation follows directly from the property of the covariant derivative, Eqn. 28.29. We therefore obtain the transformation law,

$$\tilde{F}'_{\mu\nu} = u\tilde{F}_{\mu\nu}u^\dagger. \tag{28.36}$$

Finally, a gauge-invariant Lagrangian for the gauge field is given by

$$\mathcal{L}_{gauge} = -\frac{1}{2}\text{Tr}(\tilde{F}^{\mu\nu}\tilde{F}_{\mu\nu}). \tag{28.37}$$

It will be useful for later to work out the form of the field strength and transformation laws for the field, and field strength in component form. The components of the gauge fields are given by

$$A_A^\mu = 2\text{Tr}(T_A\tilde{A}^\mu) = 2\text{Tr}(T_AT_B)A_B^\mu. \tag{28.38}$$

$$F_A^{\mu\nu} = 2\mathrm{Tr}(T_A \tilde{F}^{\mu\nu}) = 2\mathrm{Tr}\left(T_A(\partial^\mu \tilde{A}^\nu - \partial^\nu \tilde{A}^\mu + ig\left[\tilde{A}^\mu,\ \tilde{A}^\nu\right])\right) \tag{28.39}$$

or

$$F_A^{\mu\nu} = (\partial^\mu A_A^\nu - \partial^\nu A_A^\mu - g f_{ABC} A_B^\mu A_C^\nu), \tag{28.40}$$

where we used

$$\left[\tilde{A}^\mu,\ \tilde{A}^\nu\right] = [T_B,\ T_C]\, A_B^\mu A_C^\nu = i f_{BCD} T_D A_B^\mu A_C^\nu \tag{28.41}$$

and

$$f_{ABC} \equiv -2i\mathrm{Tr}(T_A\,[T_B,\ T_C]) = f_{BCA} = -f_{BAC}. \tag{28.42}$$

The latter two identities, cyclical symmetry and anti-symmetry under the interchange of two adjacent indices, follow from the symmetry properties of the trace.

Consider now infinitesimal gauge transformations. Let $u = e^{i\tilde{\theta}}$. We have

$$\delta\tilde{F}^{\mu\nu} = i\left[\tilde{\theta},\ \tilde{F}^{\mu\nu}\right] \tag{28.43}$$

or

$$\delta F_A^{\mu\nu} = -f_{ABC}\theta_B F_C^{\mu\nu}. \tag{28.44}$$

$$\delta\tilde{A}^\mu = i\left[\tilde{\theta},\ \tilde{A}^\mu\right] - \frac{1}{g}\,\partial^\mu\theta_A \equiv -\frac{1}{g}\left[D^\mu,\tilde{\theta}\right] \tag{28.45}$$

or

$$\delta A_A^\mu = -f_{ABC}\theta_B A_C^\mu - \frac{1}{g}\partial^\mu\,\theta_A \equiv -\frac{1}{g}(D^\mu\theta)_A, \tag{28.46}$$

where

$$(D^\mu\theta)_A \equiv \partial^\mu\theta_A - g f_{ABC} A_B^\mu\theta_C \equiv \left[\delta_{AC}\partial^\mu + ig(T_B^{adj}A_B^\mu)_{AC}\right]\theta_C \tag{28.47}$$

and the $SU(N)$ generators in the adjoint representation are given by

$$(T_B^{adj})_{AC} \equiv i f_{ABC}. \tag{28.48}$$

Finally,

$$\delta q^a = i(\tilde{\theta}\,q)^a \equiv i\theta_A(T_A)^a{}_b q^b. \tag{28.49}$$

For QCD, we have $N = 3$ with the symmetry $SU(3)_C$ and Lagrangian,

$$\mathcal{L}_{QCD} = \bar{q}(i\slashed{D} - m)q - \frac{1}{2}\mathrm{Tr}(\tilde{G}_{\mu\nu}\tilde{G}^{\mu\nu}), \tag{28.50}$$

with eight gluons.

QCD gives a resolution to Puzzle 1. The $U(1)_{axial}$ current is given by $j_{\mu 5} = \bar{q}\gamma_\mu\gamma_5 q$ and axial charge $Q_5 = \int d^3\vec{x} j_5^0(x)$. We also have the relation

$$\langle[Q_5,\ \bar{q}\gamma_5 q]\rangle_0 = -2\langle\bar{q}\,q\rangle_0 \neq 0. \tag{28.51}$$

Thus, there should be a ninth pseudo-scalar Nambu–Goldstone boson. And the problem is even more disturbing than the fact that the η' is heavy. Weinberg (1975)

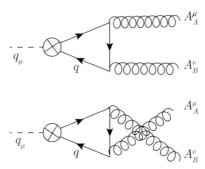

Fig. 28.3 The triangle diagram with the $U(1)_{axial}$ current, with momentum q_μ, at one vertex and gluons at the other two vertices. Up and down quarks are in the loop.

has shown, using chiral Lagrangian analysis, that $m_{\eta'} < \sqrt{3}m_{\pi^0}$. The solution, from QCD, is that the ninth axial current is not conserved, i.e.

$$\partial^\mu j_{\mu 5} = \frac{g^2}{32\pi^2} \sum_A G_A^{\mu\nu} G_A^{\rho\sigma} \epsilon_{\mu\nu\rho\sigma}. \tag{28.52}$$

There is a QCD anomaly due to the triangle diagram in Fig. 28.3 (see Appendix E).

The resolution to Puzzle 3 can be found in the Feynman diagram for π^0 decay, Fig. 28.1, there are now three colors for each quark in the loop. This increases the amplitude by a factor of three and thus the theoretical rate increases by a factor of nine.

To resolve Puzzle 4, observe that in the calculation of the rate for $e^+e^- \to \bar{q}\,q$, there is now an additional overall factor of three in the cross-section coming from the sum over three colors for each quark produced.

The understanding of the resolution of Puzzle 5 will be discussed in Chapters 29 and 30.

29 Quantizing Non-Abelian Gauge Theory

In order to find the Feynman rules for a gauge theory, we need to include gauge fixing, just as was necessary for quantum electrodynamics (QED). How to gauge fix a non-Abelian gauge theory and at the same time preserve unitarity was a problem since Yang and Mills introduced the $SU(2)$ gauge theory in 1954. That theory was invented to quantize a theory of ρ mesons; it had one major problem. While ρ mesons are massive, the local $SU(2)$ gauge theory only described massless gauge bosons. Nevertheless, it was an interesting theory to quantize, since it provided a non-Abelian gauge theory that was in fact much simpler than general relativity, which is the local gauge theory of Poicaré invariance. However, it was known that at one loop, the non-Abelian gauge theory was not unitary. Unphysical gauge states entered the quantum loops. A solution to this problem, which worked at one loop, was given by Feynman (1963). He suggested adding a "dopey" particle in the loop, which canceled the contribution of the unphysical states going in the loop. But he wasn't able to extend the proof to higher orders in perturbation theory.

The solution to the problem of unitarity of non-Abelian gauge theories, to all orders in perturbation theory, was provided by Faddeev and Popov (1967) using path-integral methods. This was then followed by a series of papers ('t Hooft, 1971b,a; Lee, 1972, 1974) on the method applied to theories of massless, as well as massive, gauge bosons, and then a new regularization scheme and proof of renormalizability of these theories ('t Hooft and Veltman, 1972) (see also Abers and Lee (1973)). We will not attempt to discuss the details of this story. Instead, we give the final result for the gauge fixed Lagrangian for non-Abelian gauge theories. This is the so-called quantum Lagrangian and it includes the so-called Faddeev–Popov ghost degrees of freedom, i.e. the "dopey" particles discussed by Feynman needed to make the one-loop corrections unitary. The quantum Lagrangian has a new global symmetry known as BRST invariance (Becchi et al., 1976; Tyutin, 1975), which leads to Ward identities for Green's functions and assists in the renormalization procedure.

Here we are interested in the quantum Lagrangian because it leads directly to the Feynman rules for non-Abelian gauge theories. The classical Lagrangian for the gauge fields is given in Eqn. 28.37. Then, with a particular choice of gauge fixing term, we obtain the quantum Lagrangian given by

$$\mathcal{L}_q = \mathcal{L}_{cl} + \mathcal{L}_{gauge\ fixing} + \mathcal{L}_{ghost} \tag{29.1}$$
$$= -\frac{1}{4}F^{A\mu\nu}F^A_{\mu\nu} - \frac{1}{2\xi}(\partial^\mu A^A_\mu)^2 - \bar{\eta}^A(\partial^\mu D_\mu)^{AB}\eta_B.$$

The fields $\eta_A(x)$, $\bar{\eta}_A(x)$ are anti-commuting scalar fields, i.e. the Faddeev–Popov ghosts. $\mathcal{L}_q$ is no longer invariant under the local gauge symmetry. It is, however invariant under the global BRST symmetry defined by

$$\delta A_\mu^A(x) = -\omega D_\mu^{AB}\eta_B(x), \tag{29.2}$$

$$\delta\eta_A(x) = -\frac{g\,\omega}{2} f_{ABC}\eta_B(x)\eta_C(x),$$

$$\delta\bar{\eta}_A = \frac{\omega}{\xi}\,\partial_\mu A_A^\mu(x),$$

where ω is an anti-commuting, space–time-constant, Grassmann parameter.

Proof

$$\delta\mathcal{L}_{cl} = 0, \tag{29.3}$$

since

$$\delta A_\mu^A(x) = -D_\mu^{AB}(\omega\eta_B(x)), \tag{29.4}$$

where we just replaced the local gauge transformation parameter, $\theta_B(x)$, by the field, $\omega\eta_B(x)$. Then,

$$\delta\mathcal{L}_{g-f} = \frac{\omega}{\xi}(\partial^\mu A_\mu^A(x))(\partial^\nu D_\nu^{AB}\eta_B(x)). \tag{29.5}$$

And

$$\delta\mathcal{L}_{gh} = -\delta\bar{\eta}_A(\partial^\mu D_\mu\eta)_A - \bar{\eta}_A\partial^\mu\delta(D_\mu\eta)_A \tag{29.6}$$

$$= -\frac{\omega}{\xi}(\partial^\mu A_\mu^A(x))(\partial^\nu D_\nu^{AB}\eta_B(x)) - \bar{\eta}_A\partial^\mu\delta(D_\mu\eta)_A. \tag{29.7}$$

Therefore, to complete the proof that $\delta\mathcal{L}_q = 0$, we just need to show that $\delta(D_\mu\eta)_A = 0$.

$$\delta(D_\mu\eta)_A = \delta(\partial_\mu\eta_A - gf_{ABC}A_\mu^B\eta_C) \tag{29.8}$$

$$= \partial_\mu\left(-\frac{g\,\omega}{2}\,f_{ABC}\eta_B\eta_C\right)$$

$$-gf_{ABC}\left(-\omega D_\mu^{BD}\eta_D\right)\eta_C$$

$$-gf_{ABC}A_\mu^B\left(-\frac{g\,\omega}{2}\,f_{CDE}\eta_D\eta_E\right)$$

$$= -\frac{g\omega}{2}\,f_{ABC}\partial_\mu(\eta_B\eta_C)$$

$$+g\omega f_{ABC}(\partial_\mu\eta_B)\eta_C$$

$$+g^2\omega f_{ABC}(f_{EBD}A_\mu^E\eta_D)\eta_C$$

$$+\frac{g^2\omega}{2}\,f_{ABC}A_\mu^B f_{CDE}\eta_D\eta_E.$$

Now check that it vanishes to order g. We have

$$-\frac{g\omega}{2}\,f_{ABC}(\eta_B(\partial_\mu\eta_C) - (\partial_\mu\eta_B)\eta_C) \equiv 0, \tag{29.9}$$

since the term in parentheses is symmetric under the interchange $B \leftrightarrow C$.

Now to order g^2 we have

$$\frac{g^2\omega}{2}A_\mu^B\eta_D\eta_E\ (2f_{ACE}f_{BCD}+f_{ABC}f_{CDE}) \tag{29.10}$$

$$=\ldots(f_{ACE}f_{BCD}-f_{ACD}f_{BCE}+f_{ABC}f_{CDE}). \tag{29.11}$$

It turns out that this also vanishes, since the $SU(N)$ generators in the adjoint representation are given by

$$(F_A)_{BC}=-if_{ABC}, \tag{29.12}$$

with commutation relations

$$(F_A)_{EC}(F_B)_{CD}-(F_B)_{EC}(F_A)_{CD}=if_{ABC}(F_C)_{ED} \tag{29.13}$$

$$-f_{AEC}f_{BCD}+f_{BEC}f_{ACD}=f_{ABC}f_{CED}$$

or

$$f_{ACE}f_{BCD}-f_{ACD}f_{BCE}+f_{ABC}f_{CDE}\equiv0. \tag{29.14}$$

$$\square$$

29.1 Perturbation Theory for Non-Abelian Gauge Theories – R_ξ Gauge

Consider first the pure gauge sector. The free Lagrangian is given by setting $g=0$ in Eqn. 29.1, i.e.

$$\mathcal{L}_0=-\frac{1}{4}(\partial_\mu A_\nu^A-\partial_\nu A_\mu^A)^2-\frac{1}{2\xi}(\partial^\mu A_\mu^A)^2-\bar{\eta}_A\Box\eta_A. \tag{29.15}$$

The interacting Lagrangian is then given by

$$\mathcal{L}_I=\frac{g}{2}(\partial_\mu A_\nu^A-\partial_\nu A_\mu^A)f_{ABC}A_B^\mu A_C^\nu \tag{29.16}$$

$$-\frac{g^2}{4}\,f_{ABC}f_{AB'C'}A_{\mu B}A_{\nu C}A_{B'}^\mu A_{C'}^\nu$$

$$+gf_{ABC}\bar{\eta}_A\partial^\mu(A_\mu^B\eta_C).$$

In order to derive the Green's functions we add external sources to the free action, S_0. We have

$$S_{free}=S_0+\int_x(J_A^\mu A_\mu^A+\bar{\eta}_A\bar{s}_A+\eta_As_A) \tag{29.17}$$

$$=\int_x\left(A_\mu^A\frac{1}{2}K^{\mu\nu}A_\nu^A+J_A^\mu A_\mu^A\right)$$

$$+\int_x(\bar{\eta}_AK\eta_A+\bar{\eta}_A\bar{s}_A+\eta_As_A),$$

where

$$K^{\mu\nu} \equiv g^{\mu\nu} \,\Box - \partial^\mu \partial^\nu \left(1 - \frac{1}{\xi}\right) \tag{29.18}$$

and

$$K \equiv -\Box. \tag{29.19}$$

The gauge propagators/Green's functions are defined as follows:

$$K_x^{\mu\nu} \Delta_{\nu\gamma}^{\xi}(x-y) = \delta^\mu{}_\gamma \delta^4(x-y) \quad \text{gauge propagator} \tag{29.20}$$
$$K_x \Delta(x-y) = \delta^4(x-y) \quad \text{ghost propagator}.$$

The solutions are given by

$$\Delta_{\mu\nu}^{\xi}(x-y) = \int \frac{d^4p}{(2\pi)^4} \, e^{-ip\cdot x} \, \frac{-g_{\mu\nu} + (1-\xi)\frac{p_\mu p_\nu}{p^2}}{p^2 + i\epsilon}, \tag{29.21}$$
$$\Delta(x-y) = \int \frac{d^4p}{(2\pi)^4} \frac{e^{-ip\cdot x}}{p^2 + i\epsilon}.$$

Note, ξ is an arbitrary parameter. The Feynman or Landau gauges are given by $\xi = 1, \, 0$, respectively.

29.2 Feynman Rules

Given the propagators and vertices, the Feynman rules follow essentially from those given in Chapter 13, Section 13.2, with one exception. Since the ghosts satisfy anti-commutation relations, there is a factor of -1 for every closed ghost loop and there are never any external ghost lines. The propagators are given by, for gauge bosons in the renormalizable R_ξ gauge (Fig. 29.1),

$$\langle 0|T(A_\mu^A(x)A_\nu^B(0))|0\rangle = i\Delta_{\mu\nu}^\xi(x)\delta^{AB}. \tag{29.22}$$

And for the ghosts (Fig. 29.2),

$$\langle 0|T(\eta_A(x)\bar\eta_B(0))|0\rangle = i\Delta(x)\delta_{AB}. \tag{29.23}$$

For the gauge-boson self interactions, there is both a three- and four-point vertex given by

$$gf_{ABC}[(r_\mu - q_\mu)g_{\nu\rho} + (p_\nu - r_\nu)g_{\mu\rho} + (q_\rho - p_\rho)g_{\mu\nu}], \tag{29.24}$$

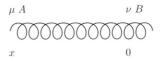

$\mu\,A$ \qquad\qquad $\nu\,B$

x \qquad\qquad\qquad 0

Fig. 29.1 Gauge-boson propagator in the R_ξ gauge.

$$
\begin{array}{ccc}
A & & B \\
\cdots\cdots\cdots\cdots\blacktriangleleft\cdots\cdots\cdots\cdots & & \\
x & & 0
\end{array}
$$

Fig. 29.2   Ghost propagator.

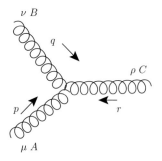

Fig. 29.3   Three-point gauge vertex with all momenta pointing inward.

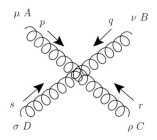

Fig. 29.4   Four-point gauge vertex with all momenta pointing inward.

for the three-point vertex (Fig. 29.3) and

$$- g^2 [f^{ABE} f^{CDE} (g_{\mu\rho} g_{\nu\sigma} - g_{\mu\sigma} g_{\nu\rho}) \tag{29.25}$$

$$+ f^{ACE} f^{BED} (g_{\mu\sigma} g_{\rho\nu} - g_{\mu\nu} g_{\rho\sigma})$$

$$+ f^{ADE} f^{BCE} (g_{\mu\nu} g_{\rho\sigma} - g_{\mu\rho} g_{\sigma\nu})]$$

for the four-point vertex (Fig. 29.4). The gauge–ghost vertex is given by (Fig. 29.5)

$$g f^{ABC} p_\mu. \tag{29.26}$$

Now we can add the matter Lagrangian,

$$\mathcal{L}_{matter} = \bar{q}(i\slashed{D} - m)q. \tag{29.27}$$

The free Lagrangian is then given by

$$\mathcal{L}_{0\,matter} = \bar{q}(i\slashed{\partial} - m)q \tag{29.28}$$

and the interacting term given by

$$\mathcal{L}_{I\,matter} = -g A_\mu^A \bar{q} \gamma^\mu T_A q \tag{29.29}$$

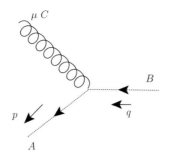

Fig. 29.5   Gauge–ghost vertex.

Fig. 29.6   Fermion propagator.

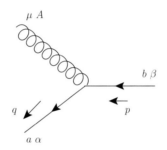

Fig. 29.7   Gauge-fermion vertex.

with the Fermion propagator (Fig. 29.6) given by

$$\langle 0|T(q_\alpha^a(x)\bar{q}_{b\beta}(0))|0\rangle = i(S_F(x))_{\alpha\beta}\ \delta^a{}_b \tag{29.30}$$

and the gauge-fermion vertex (Fig. 29.7) given by

$$-ig(\gamma^\mu)_{\alpha\beta}(T_A)^a{}_b \tag{29.31}$$

30 Renormalization

In this book we focused on so-called tree diagrams, i.e. Feynman diagrams with no loops. In a course on quantum field theory you will learn that loop diagrams are divergent. In order to make sense of them, these divergent integrals must be made finite by some regularization procedure. Once finite, we can subtract the divergent part and define finite n-point Green's functions. These are the so-called renormalized Green's functions. For a non-renormalizable field theory, every n-point Green's function is divergent and in order to define the theory, the value of each renormalized n-point Green's function must be determined by experiment. As a consequence, such theories have no predictions. On the other hand, a renormalizable theory is one in which there are only a finite number of primitive divergent Green's functions. Hence, once the values of these primitively divergent n-point Green's functions are determined by experiment, then the rest can be predicted.

Green's functions depend on the external momenta and thus their finite values are fixed at some particular momentum or energy scale. Since this scale is arbitrary, no physical observable should depend on this arbitrary renormalization scale. This fact leads to the so-called renormalization group (RG) equations, or Callan–Symanzik equations, which tell us how coupling constants change with the renormalization scale, such that physical observables are invariant (Callan, 1970; Symanzik, 1970a,b, 1971; 't Hooft, 1973; Gross and Wilczek, 1973a); for a pedagogical review of broken scale invariance and the Callan–Symanzik equations, see Coleman's Erice lectures (Coleman, 1985). This procedure defines what are called running coupling constants. For example, consider quantum electrodynamics (QED). The electromagnetic fine-structure constant, α, as measured in low-energy (or equivalently, at large distances) $\gamma-e$ scattering (Compton scattering), or in atomic physics experiments, has a value of $\alpha^{-1} \sim 137$. But at higher energies (or shorter distances) the value of α increases, such that at the weak scale $\alpha^{-1} \sim 128$.

In QED this result can be obtained at one loop in the photon propagator (Fig. 30.1), see Appendices C and D. The renormalized theory defines the electromagnetic coupling constant at some scale, $\alpha(\mu)$. Then at the scale $\mu = m_e$, we find $\alpha^{-1}(m_e) \sim 137$. However, the one-loop graph contains a virtual electron–positron pair, which acts like a dielectric that partially screens the charge on the electron. Thus, at higher energies (or shorter distances) the effective charge on the electron increases, since there is less screening. The running coupling constant satisfies the following RG equation

$$\frac{de}{d\ln(\mu/\mu_0)} \equiv \beta(e) = \frac{e^3}{12\pi^2} + \dots, \tag{30.1}$$

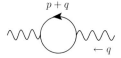

Fig. 30.1 The one-loop correction to the photon propagator where p is the loop momentum.

Fig. 30.2 The one-loop beta function for QED with just an electron in the loop. The arrow shows the direction α goes with increasing values of t, i.e. the ultra-violet (UV) limit. The opposite direction is the infra-red (IR) limit.

where the ... corresponds to the contribution of higher loops. We can also rewrite this equation in terms of $\alpha \equiv \frac{e^2}{4\pi}$. We have

$$\frac{d\alpha}{d\ln(\mu^2/\mu_0^2)} \equiv \beta(\alpha) = \frac{e}{4\pi}\beta(e) = \frac{\alpha^2}{3\pi} + O(\alpha^3). \tag{30.2}$$

In the literature, one typically defines the dimensionless scale parameter $t \equiv \ln(\mu^2/\mu_0^2)$. Then the RG equation becomes

$$\frac{d\alpha}{dt} = \beta(\alpha) = \frac{\alpha^2}{3\pi} + \ldots, \tag{30.3}$$

where $\alpha(t)$ is the running coupling. We plot the so-called one-loop beta function for QED (in Fig. 30.2) with just an electron in the loop. The arrow shows the direction α goes with increasing values of t, i.e. the ultraviolet limit. The opposite direction is the infrared limit. We see that $\alpha(t)$ increases as we come closer and closer to the bare charge.

The one-loop RG equation can be solved exactly. We have

$$\int_{\alpha_0}^{\alpha} \frac{d\alpha'}{(\alpha')^2} = -\frac{1}{\alpha'}\big|_{\alpha_0}^{\alpha} = \frac{t}{3\pi}$$

or

$$\frac{1}{\alpha(\mu)} = \frac{1}{\alpha(\mu_0)} - \frac{t}{3\pi}$$

and, finally,

$$\alpha(\mu) = \frac{\alpha(\mu_0)}{1 - \frac{\alpha(\mu_0)}{3\pi}\ln(\mu^2/\mu_0^2)}. \tag{30.4}$$

We take $\mu_0 = m_e$ and $\alpha(\mu_0)^{-1} \sim 137\ldots$, then it is clear that at some value of $\mu = \Lambda \gg \mu_0$, $\alpha(\Lambda) = \infty$. This is known as the Landau pole and clearly, long before one reaches energies of order Λ, when $\alpha(\mu) \sim 1$, the perturbation expansion is no longer valid. The fact that perturbation theory breaks down suggests that QED

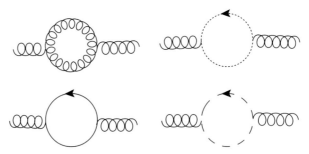

Fig. 30.3 The one-loop contributions to wave-function renormalization.

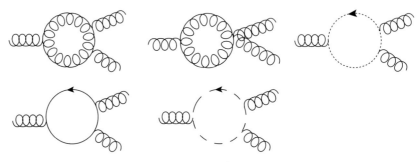

Fig. 30.4 The one-loop contributions to vertex renormalization.

is an incomplete theory and should be incorporated into a more complete theory below the energy scale Λ. In QED at one loop we have

$$\Lambda = m_e e^{\frac{3\pi}{2\alpha(\mu_0)}} \gg M_{Pl} = 2.4 \times 10^{18} \text{ GeV}. \tag{30.5}$$

For a general non-Abelian gauge theory with gauge symmetry group G, the beta function has been calculated to three loops. For two-loop results, see for example Jones (1974), Caswell (1974) and Machacek and Vaughn (1983). They include the one-loop contributions to wave function (Fig. 30.3) and vertex renormalization (Fig. 30.4). We have

$$\frac{d\alpha}{d\ln(\mu^2/\mu_0^2)} \equiv \beta(\alpha) = -\beta_0 \frac{\alpha^2}{4\pi} - \beta_1 \frac{\alpha^3}{16\pi^2}, \tag{30.6}$$

with

$$\beta_0 = \frac{11}{3} C_2(G) - \frac{2}{3} T(R) N_f - \frac{1}{3} T(R) N_s \tag{30.7}$$

$$\beta_1 = \frac{34}{3} C_2(G)^2 - \left[\frac{10}{3} C_2(G) + 2 C_2(R)\right] T(R) N_f - \left[\frac{2}{3} C_2(G) + 4 C_2(R)\right] T(R) N_s,$$

where

$$C_2(R) \mathbb{I}_{d(R)} \equiv \sum_A (T_A^{(R)} T_A^{(R)}), \tag{30.8}$$

$$T(R) \, \delta_{AB} \equiv Tr(T_A^{(R)} T_B^{(R)}), \tag{30.9}$$

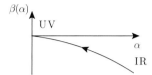

Fig. 30.5 The one-loop beta function for QCD with $n_f < 6$. The arrow shows the direction α goes with increasing values of t, i.e. the ultra-violet limit. The opposite direction is the infra-red limit.

$T^{(R)}$ are the generators of the group in the representation R and G is the adjoint representation. (N_f, N_s) are the number of Weyl fermions and complex scalars, respectively. Note, we also have the identity

$$d(G)T(R) \equiv d(R)C_2(R), \tag{30.10}$$

where $d(G)$, $d(R)$ is the dimension of the adjoint and R representation, respectively.

For $G = SU(N)$, we have $C_2(SU(N)) = N$, $T(N) = \frac{1}{2}$ and $C_2(N) = \frac{N^2-1}{2N}$. Therefore, the one-loop beta function for quantum chromodynamics (QCD) with $N = 3$ is given by (Gross and Wilczek, 1973a,b, 1974; Politzer, 1973):

$$\beta(\alpha) = - \left[11 - \frac{2}{3} n_{flavors} \right] \frac{\alpha^2}{4\pi}, \tag{30.11}$$

where $N_f \equiv 2n_{flavors}$. Note, for $n_f < 33/2$, $\beta(\alpha) < 0$ (see Fig. 30.5). We see that as $\mu \to \infty$, $\alpha \to 0$. This is called asymptotic freedom. The solution to the RG equation at one loop is given by

$$\alpha(\mu) = \frac{\alpha_0}{1 + \frac{\alpha_0 \beta_0}{4\pi} \ln(\mu^2/\mu_0^2)}, \tag{30.12}$$

where $\alpha_0 \equiv \alpha(\mu_0)$. If we define the scale, Λ, by

$$\frac{\alpha_0 \beta_0}{4\pi} \ln(\mu_0^2/\Lambda^2) \equiv 1, \tag{30.13}$$

then we obtain the simple relation

$$\alpha(\mu) = \frac{4\pi}{\beta_0 \ln(\mu^2/\Lambda^2)}. \tag{30.14}$$

Λ is called the QCD scale or Λ_{QCD}. As $\mu \to \Lambda$, $\alpha(\mu)$ diverges. Experimentally, we have $\alpha(M_Z) \sim 0.118$ and perturbation expansions provide a good approximation to data (see Fig. 30.6). At energies below about 2 GeV, QCD becomes non-perturbative. Note, also, $\Lambda_{QCD} \sim 300$ MeV.

To summarize, in QED, fermion loops cause screening of electric charge. In fact, in QCD, matter loops (fermion or scalar) also contribute to the screening of color charge. However, gauge boson loops contribute to anti-screening. In QED, photons are electrically neutral. On the other hand, gluons have color. They seem to spread the color charge over a finite distance. Thus, as one goes to smaller distance (or larger energies), one sees less color charge.

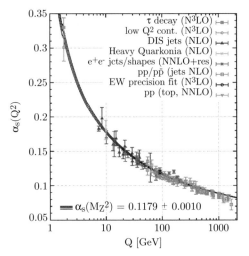

Fig. 30.6 Experimental measurements of the strong-interaction coupling constant $\alpha_s(Q^2)$ as compared to the theory (solid line) (Tanabashi et al., 2018).

30.1 Four-Dimensional Lattice Gauge Theory

In standard lattice gauge theory calculations, as defined by Wilson (1976), the path integral which defines the Standard Model quark and gluon interactions is placed on a four-dimensional Euclidean lattice, with quarks sitting at the lattice sites and gluons sitting on the links (see Fig. 30.7). The continuum gauge kinetic action is given by

$$S_{gauge}^{cont.} = -\int d^4x \, \frac{1}{4g_{latt}^2} \, \text{Tr}[\tilde{F}_{\mu\nu}(x)^2] \tag{30.15}$$

where

$$\tilde{F}_{\mu\nu} = \partial_\mu \tilde{A}_\nu - \partial_\nu \tilde{A}_\mu + i[\tilde{A}_\mu, \, \tilde{A}_\nu] \tag{30.16}$$

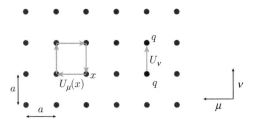

Fig. 30.7 Sketch of a two-dimensional slice through the $\mu-\nu$ plane of a lattice, showing gluon fields lying on links and forming either the plaquette product appearing in the gauge action or a component of the covariant derivative connecting quark and anti-quark fields. Taken from Tanabashi et al. (2018).

and the lattice equivalent is given by

$$S_{gauge}^{latt.} = \beta \sum_{x,\mu,\nu} [1 - \frac{1}{3} \text{ReTr}[U_\mu(x)U_\nu(x+a\hat{\mu})U_\mu^\dagger(x+a\hat{\nu})U_\nu^\dagger(x)]], \qquad (30.17)$$

with $\beta = 6/g_{latt.}^2$ and the link operator $U_\mu(x)$ is given by

$$U_\mu(x) = e^{ia\tilde{A}_\mu(x)}. \qquad (30.18)$$

Similarly, the action for quarks in the continuum is given by

$$S_q^{cont.} = \int d^4x \bar{q}(i\gamma_\mu D_\mu + m_q)q, \qquad (30.19)$$

with $D_\mu = \partial_\mu + i\tilde{A}_\mu$. In the lattice equivalent, the continuum derivative is replaced by the finite difference,

$$D_\mu q(x) \to \frac{1}{2a}[U_\mu(x)q(x+a\hat{\mu}) - U_\mu(x-a\hat{\mu})^\dagger q(x-a\hat{\mu})]. \qquad (30.20)$$

This defines the naive fermion action, which actually describes 2^D continuum fermions. Moreover, chiral symmetry is explicitly broken on the lattice. These problems are dealt with in several different ways which we will not discuss further here (for more details, see Tanabashi et al., 2018). The theory retains local gauge invariance, while the finite lattice turns a continuous infinity of states into a discrete infinity and at the same time regularizes the integrals, since all UV divergences in field theory occur at infinite momentum or equivalently at short distances. The cut-off is thus given by $1/a$. The path integrals, including fermions, are then used to calculate hadronic masses and quark matrix elements necessary for comparison to experiment. As an example of some of the successes of lattice calculations, see Fig. 30.8.

Finally, one can study phase transitions in QCD using lattice techniques. One defines QCD on a Euclidean space–time lattice where one assumes periodic boundary conditions in the "time" direction with length given by $\beta \equiv 1/kT$, where in this case T = temperature. One of the order parameters one studies is the Wilson loop, defined by the exponent of the product of gauge links that is defined on the circumference of a disk with radius R. If the result of the path integral decreases exponentially with the area of the disk, then this signifies the confining phase. While if the result only decreases with the circumference of the disk, then this indicates the Coulomb phase. At zero temperature, QCD is in the confining phase. When the disk is rotated into the time direction, the Wilson loop describes a quark and anti-quark separated by a distance R and the area law corresponds to the energy between the quark–anti-quark pair, which grows linearly with the separation, R, integrated over a time, T. A perimeter law then corresponds to just the Coulomb potential between the quarks. One can also study the matrix element of $\bar{q}\,q$ pairs in the vacuum. This is the order parameter for chiral symmetry breaking. One finds $\langle \bar{q}\,q \rangle_T \neq 0$ for $\leq T_C$. It turns out that the critical temperature for both the confinement and chiral phase transitions occur at approximately the same critical temperature, $T_C \sim \Lambda_{QCD}$.

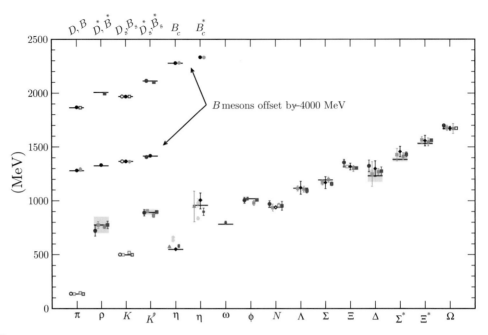

Fig. 30.8 Hadron spectrum from lattice QCD. Comprehensive results for mesons and baryons are from the MILC (Aubin et al., 2004; Bazavov et al., 2010), PACS-CS (Aoki et al., 2009), BMW (Durr et al., 2008), QCDSF (Bietenholz et al., 2011) and ETM (Alexandrou et al., 2014) collaborations. Results for η and η' are from the RBC & UKQCD (Christ et al., 2010), Hadron Spectrum (Dudek et al., 2011) (also the only ω mass) and UKQCD collaborations (Gregory et al., 2012; Michael et al., 2013). Results for heavy–light hadrons from the Fermilab–MILC (Bernard et al., 2011) and HPQCD collaborations (Gregory et al., 2011; Mohler and Woloshyn, 2011; Dowdall et al., 2012). Circles, squares, diamonds and triangles stand for staggered, Wilson twisted-mass Wilson and chiral sea quarks, respectively. Asterisks represent anisotropic lattices. Open symbols denote the masses used to fix parameters. Filled symbols (and asterisks) denote results. Note, different points represent different number of ensembles (i.e. lattice spacing and sea-quark mass). Black symbols stand for results with 2+1+1 flavors of sea quarks. Horizontal bars (gray boxes) denote experimentally measured masses (widths). b-flavored meson masses are offset by −4000 MeV. Taken from Tanabashi et al. (2018).

30.2 Strong Coupling Limit of QCD on a Three-Dimensional Lattice

Consider the Lagrangian for the gauge sector of QCD. We have

$$\mathcal{L}_{gauge} = -\frac{1}{4} G^A_{\mu\nu}\, G_A^{\mu\nu} \equiv \frac{1}{2}\, (\vec{E}_A^{\,2} - \vec{B}_A^{\,2}). \tag{30.21}$$

Color flux

Fig. 30.9 A quark and anti-quark separated by a distance R has a color-flux tube stretched between them.

Let's rescale

$$A_A^\mu \to A_A^\mu / g. \tag{30.22}$$

Then the Lagrangian becomes

$$\mathcal{L}_{gauge} = \frac{1}{2g^2} \, (\vec{E}_A^{\,2} - \vec{B}_A^{\,2}). \tag{30.23}$$

We can then find the Hamiltonian density. In the gauge $A_A^0 = 0$, we have

$$\Pi_i^A \equiv \frac{1}{g^2} \, E_i^A \tag{30.24}$$

and

$$\mathcal{H} = \frac{1}{2} \, g^2 \, \vec{\Pi}_A^2 + \frac{1}{2g^2} \, \vec{B}_A^{\,2}, \tag{30.25}$$

with the constraint that

$$\vec{\nabla} \cdot \vec{\Pi}_A = \rho_A, \tag{30.26}$$

where $\rho_A = J_A^0$ is the color charge density. Note, Π_i^A is the canonical momentum to the gauge field, A_i^A.

In the strong coupling limit, $g \gg 1$, color flux costs energy and the lowest energy state (i.e. the QCD vacuum) has no color electric flux (Wilson, 1976; Banks et al., 1977). Moreover, if one now places a quark and anti-quark in the vacuum, separated by a distance R, there is an energy cost given by

$$E = \frac{g^2}{2} \, R \tag{30.27}$$

(see Fig. 30.9). Gauge invariance requires color electric flux to extend from the quark to the anti-quark. Moreover, it doesn't spread out like electric flux, since the energy cost would be too large. Instead it forms a narrow flux tube with energy growing linearly with R. This is the phenomenon of infra-red slavery in QCD. Only bound color singlet states have finite energy. In Banks et al. (1977) the hadronic spectrum was calculated on a spatial lattice with lattice spacing, a, as a perturbation series in $1/g^2$. Then the continuum limit was taken such that $g^2 \to 0$, as $a \to 0$ using Pade approximation.[1] Reasonable results were found.

---

[1] Note, this is similar to condensed-matter theory in which one uses a high-temperature expansion of the partition function to then understand critical behavior for small T.

This phenomenon is similar but dual to the Meissner effect in a superconductor where magnetic flux is excluded. The lowest energy configuration with magnetic flux confines the magnetic flux to vortex tubes. In the QED vacuum in a superconductor, $\vec{B} = 0$ and both electric charge Q and $\vec{E}$ are undefined. On the other hand in the QCD vacuum, both color charge, Q_A, and color electric field, $\vec{E}_A$, are zero. Then the color magnetic field, $\vec{B}_A$, is undefined.

31 Deep Inelastic Electron–Nucleon Scattering

The deep inelastic scattering experiment of electrons on nucleons is an analog of Rutherford scattering of alpha particles on gold foil. Since the nucleon is not a point-like particle, with a size of order a fermi, it is expected that there would only be a small-angle scattering. However, this is not what occurred! Large-angle scattering was the evidence for point-like constituents of the nucleon. These experiments were responsible for the discovery of quarks.

31.1 The Cross-Section

In these experiments the nucleon was shattered, so the only variables that were measured experimentally were the energy and momentum of the outgoing electron. The nucleons are in their rest frame with momentum $p = (M, \vec{0})$ and the incident electrons have momentum $k = (E, \vec{k})$, while the outgoing electron has momentum $k' = (E', \vec{k}\,')$ (see Figs. 31.1 and 31.2). The momentum $q \equiv k - k'$ and we define the parameters $\nu \equiv \frac{p \cdot q}{M}$ and $W^2 \equiv p_n^2 \equiv (q + p)^2$. The kinematic variable ν is the energy loss of the electron and takes on values

$$0 \leq \nu = E - E'. \tag{31.1}$$

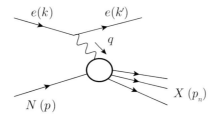

Fig. 31.1   Phenomenological diagram for deep inelastic scattering.

Fig. 31.2   Defining the scattering angle θ in the laboratory frame.

The photon momentum satisfies $q^2 = (k - k')^2 = 2m_e^2 - 2k \cdot k' \simeq -2EE'(1 - \cos\theta)$ when we neglect the electron mass for $m_e \ll E$. We therefore have

$$q^2 = -4EE' \sin^2\frac{\theta}{2} \leq 0. \tag{31.2}$$

We then define the quantity $Q^2 \equiv -q^2 > 0$ and the scaling parameter, Feynman x, with

$$x \equiv \frac{Q^2}{2M\nu} \geq 0. \tag{31.3}$$

Since $W^2 \equiv q^2 + 2q \cdot p + M^2 \geq M^2$ or $q^2 + 2M\nu \geq 0$, we find

$$x \leq 1. \tag{31.4}$$

Thus,

$$0 \leq x \leq 1. \tag{31.5}$$

Now the scattering amplitude is given by the expression

$$T_n = e^2 \bar{u}(k', s')\gamma^\mu u(k, s)\frac{1}{q^2}\langle n|j_\mu^{EM}(0)|p, \sigma\rangle \tag{31.6}$$

and the unpolarized differential cross-section is

$$d\sigma_n = \frac{1}{|\vec{v}|}\frac{1}{2M}\frac{1}{2E}\frac{d^3\vec{k}\,'}{(2\pi)^3 2k_0'}\prod_{i=1}^{n}\frac{d^3\vec{p}_i}{(2\pi)^3 2p_0^i}\frac{1}{4}\sum_{\sigma,s,s'}|T_n|^2(2\pi)^4\delta^4(k+p-k'-p_n), \tag{31.7}$$

where p_i are the momenta of the outgoing particles, with $p_n \equiv \sum_{i=1}^{n} p_i$.

Summing over all possible final states $\{n\}$, we obtain

$$\frac{d^2\sigma}{d\Omega dE'} = \frac{\alpha^2}{q^4}\left(\frac{E'}{E}\right)\ell^{\mu\nu}W_{\mu\nu}, \tag{31.8}$$

where

$$\ell_{\mu\nu} \equiv \frac{1}{2}\mathrm{Tr}(\slashed{k}'\gamma_\mu\slashed{k}\gamma_\nu) = 2(k_\mu' k_\nu + k_\nu' k_\mu - g_{\mu\nu}\underbrace{k \cdot k'}_{-q^2/2}) \tag{31.9}$$

and

$$W_{\mu\nu}(p, q) \equiv \frac{1}{4M}\sum_{\sigma}\sum_{n}\int\prod_{i=1}^{n}\left[\frac{d^3\vec{p}_i}{(2\pi)^3 2p_0^i}\right]\langle p, \sigma|j_\mu^{EM}(0)|n\rangle\langle n|j_\nu^{EM}(0)|p, \sigma\rangle$$

$$\times \underbrace{(2\pi)^3\delta^4(p_n - p - q)}_{\frac{1}{2\pi}\int d^4x\, e^{i(p+q-p_n)\cdot x}}$$

$$\equiv \frac{1}{4M}\sum_{\sigma}\int\frac{d^4x}{2\pi}e^{iq\cdot x}\langle p, \sigma|j_\mu^{EM}(x)\, j_\nu^{EM}(0)|p, \sigma\rangle, \tag{31.10}$$

where we used

$$\mathbb{I} \equiv \sum_n \int \prod_{i=1}^{n} \left[\frac{d^3 \vec{p}_i}{(2\pi)^3 2p_0^i} \right] |n\rangle\langle n|. \tag{31.11}$$

Now let's prove the following lemma:

Proof

$$0 \equiv \int \frac{d^4 x}{2\pi} e^{iq\cdot x} \langle p, \sigma | j_\nu^{EM}(0) \, j_\mu^{EM}(x) | p, \sigma \rangle \tag{31.12}$$

$$= \sum_n \int \frac{d^4 x}{2\pi} e^{i(p_n - p + q)\cdot x} \langle p, \sigma | j_\nu^{EM}(0) | n \rangle \langle n | j_\mu^{EM}(0) | p, \sigma \rangle$$

$$= \sum_n (2\pi)^3 \delta^4(p_n - p + q) \langle p, \sigma | j_\nu^{EM}(0) | n \rangle \langle n | j_\mu^{EM}(0) | p, \sigma \rangle.$$

But in the laboratory frame, $q_0 = \nu > 0$ and the delta function implies that $E_n = p_0 - q_0 \equiv M - \nu \leq M$. However, there is no such state $|n\rangle$ with $E_n \leq M$. $\square$

Using the above lemma, we have

$$W_{\mu\nu}(p, q) = \frac{1}{4M} \sum_\sigma \int \frac{d^4 x}{2\pi} e^{iq\cdot x} \langle p, \sigma | \left[j_\mu^{EM}(x), \, j_\nu^{EM}(0) \right] | p, \sigma \rangle. \tag{31.13}$$

Note, $q^\mu W_{\mu\nu} \equiv 0$, since $\partial^\mu j_\mu^{EM} \equiv 0$. The most general form for $W_{\mu\nu}$ consistent with **C, P**, Lorentz invariance and current conservation is given by

$$W_{\mu\nu}(p, q) = \left[-W_1 \left(g_{\mu\nu} - \frac{q_\mu q_\nu}{q^2} \right) + \frac{W_2}{M^2} \left(p_\mu - \frac{p \cdot q}{q^2} q_\mu \right) \left(p_\nu - \frac{p \cdot q}{q^2} q_\nu \right) \right], \tag{31.14}$$

where $W_{1,2}$ are Lorentz invariant functions of q^2, ν, called *structure functions*. We then have

$$\frac{d^2 \sigma}{d\Omega dE'} = \frac{\alpha^2}{4E^2 \sin^4 \frac{\theta}{2}} \left(2W_1 \sin^2 \frac{\theta}{2} + W_2 \cos^2 \frac{\theta}{2} \right). \tag{31.15}$$

In order to gain some insight into the problem, consider the contribution of one nucleon to $W_{\mu\nu}$, i.e. elastic electron–nucleon scattering. Using the form factors obtained in Chapter 15, Section 15.2,

$$\langle N(p') | j_\mu^{EM}(0) | N(p) \rangle = \bar{u}(\vec{p}\,', s') \left[\gamma_\mu F_1(q^2) + i \frac{\Sigma_{\mu\nu}}{M} q^\nu \kappa F_2(q^2) \right] u(\vec{p}, s), \tag{31.16}$$

we find

$$W_1^{elastic}(q^2, \nu) = \delta(q^2 + 2M\nu) \frac{q^2}{2M} \, G_M^2(q^2), \tag{31.17}$$

$$W_2^{elastic}(q^2, \nu) = \delta(q^2 + 2M\nu) \frac{2M}{(1 - \frac{q^2}{4M^2})} \left[G_E^2(q^2) - \frac{q^2}{4M^2} G_M^2(q^2) \right].$$

Note, the form factors G_E, G_M are given in terms of F_1, F_2 in Eqn. 15.41. Recall, these form factors have a dipole momentum dependence at large q^2 with $G_E^p(q^2) \sim \frac{1}{(1 - \frac{q^2}{.7 GeV^2})^2}$. These form factors were the indication that the nucleon has structure, since it decreases at large $|q^2|$. We thus expect the inelastic cross-section to fall off just as fast at high $|q^2|$, just like one might have expected in the example of the Thomson model of the atom.

We now perform a conventional change of variables. Instead of (q^2, ν), we switch to the variables (x, y) given by Feynman

$$x = \frac{Q^2}{2M\nu}, \tag{31.18}$$

which satisfies

$$0 \le x \le 1, \tag{31.19}$$

and for elastic scattering,

$$x = 1 \tag{31.20}$$

and

$$y \equiv \frac{\nu}{E} = 1 - \frac{E'}{E}. \tag{31.21}$$

y is thus the fraction of the electron energy transferred to the hadron. Since $0 \le E' \le E$, we have

$$0 \le y \le 1. \tag{31.22}$$

We then have

$$dxdy = Jd\theta dE', \tag{31.23}$$

where the Jacobian is obtained using $q^2 = -4EE' \sin^2 \frac{\theta}{2}$, $\nu = E - E'$ and

$$J \equiv \begin{vmatrix} \frac{\partial x}{\partial \theta} & \frac{\partial x}{\partial E'} \\ \frac{\partial y}{\partial \theta} & \frac{\partial y}{\partial E'} \end{vmatrix} = \frac{E' \sin \theta}{EyM}. \tag{31.24}$$

We then obtain the new expression for the differential cross-section given by

$$\frac{d^2\sigma}{dxdy} = \frac{E}{E'} 2\pi yM \frac{d^2\sigma}{d\Omega dE'} \tag{31.25}$$

$$= \frac{2\pi\alpha^2}{MEx^2y^2} \left[xy^2 MW_1 + (1 - y - \frac{M}{2E}xy)\nu W_2 \right].$$

31.2 Bjorken Scaling

In 1969, Bjorken (Bjorken, 1969) suggested using new structure functions defined in terms of dimensionless, Lorentz scalar functions of p, q.[1] We have

---

[1] Bjorken used the scaling parameter $\omega = 1/x$.

$$G_1\left(x, \frac{q^2}{M^2}\right) \equiv MW_1(q^2, \nu), \tag{31.26}$$

$$G_2\left(x, \frac{q^2}{M^2}\right) \equiv \nu W_2(q^2, \nu). \tag{31.27}$$

In the limit that $-q^2/M^2 \gg 1$, with x fixed, he defined the scaling functions

$$G_i(x, q^2/M^2) \to F_i(x); \tag{31.28}$$

functions of x only. This is the scaling limit. Note, if the nucleon was a soft ball with form factors vanishing at large $|q^2|$, then we would expect

$$F_i(x) \equiv 0. \tag{31.29}$$

In the scaling limit, Eqn. 31.25 becomes

$$\frac{d^2\sigma}{dxdy} = \frac{4\pi\alpha^2 s}{Q^4}\left[xy^2 F_1(x) + (1-y)F_2(x)\right], \tag{31.30}$$

where we've used the relations $xy \equiv \frac{Q^2}{2ME}$, $s = 2ME$ and, since $E \gg M$, we neglected the term proportional to M/E in Eqn. 31.25.

Bjorken scaling was discovered experimentally for $Q^2 \geq 2 \text{ GeV}^2$ at Stanford Linear Accelerator Center (SLAC) (Breidenbach et al., 1969), i.e. the structure functions depend on Q^2 and ν only as a function of the dimensionless variable x.

31.3 Parton Model

Soon after Bjorken scaling was discovered experimentally, a description of the proton in terms of point-like constituents, or partons, was derived (Feynman, 1969a,b; Bjorken and Paschos, 1969; Drell and Yan, 1971). Consider the nucleon in the infinite momentum frame with momentum $p_z \gg M$, $|\vec{p}_\perp|$. Then the nucleon four-vector

$$p^\mu = (E, \vec{p}) \simeq \left(p_z + \frac{m_\perp^2}{2p_z}, \vec{p}_\perp, p_z\right), \tag{31.31}$$

where $m_\perp^2 \equiv M^2 + \vec{p}_\perp^2$. If the nucleon, in the rest frame, is a spherically symmetric ball, then in the infinite-momentum frame, the ball is contracted to a thin pancake in the z direction, Fig. 31.3. Consider the nucleon made of N free fundamental

rest frame   infinite-momentum frame

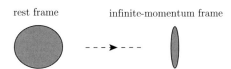

Fig. 31.3   Nucleon in the infinite-momentum frame.

point-like partons, each carrying a fraction x_i of the nucleon momentum and q_i of its charge, with $i = 1, \ldots, N$ and

$$\sum_{i=1}^{N} x_i = 1, \quad \sum_{i=1}^{N} q_i = Q. \tag{31.32}$$

Neglecting the transverse momentum, we have

$$p_i^{\mu} \simeq x_i p^{\mu} \tag{31.33}$$

and

$$\sum_{i=1}^{N} p_i^{\mu} = p^{\mu}. \tag{31.34}$$

If the interactions between the individual partons can be neglected, then we can calculate the contribution to νW_2 and $M W_1$ due to the scattering of a photon on a single parton. When would this be feasible? The interaction timescale for scattering,

$$\Delta t_{int} \simeq \frac{1}{|q^2|^{1/2}}, \tag{31.35}$$

whereas the time-scale for partons to rearrange in the nucleon,

$$\Delta t_{re} \simeq \frac{1}{M}. \tag{31.36}$$

For large $|q^2|$, we have

$$\Delta t_{int} \ll \Delta t_{re} \tag{31.37}$$

and we can use the impulse approximation when kicking the parton out of the nucleon, i.e. neglect the rescattering of the parton with the final-state nucleon minus the parton.

For a point-like parton with charge 1 we can use the results Eqn. 31.17. We take

$$F_1(q^2) \equiv 1, \quad F_2(q^2) \equiv 0, \tag{31.38}$$

thus

$$G_E(q^2) = G_M(q^2) = 1. \tag{31.39}$$

We then find

$$\nu W_2^{elastic}(q^2, \nu)|_{\text{ point-like}} = \delta(q^2 + 2M\nu)2M\nu = \delta\left(1 + \frac{q^2}{2M\nu}\right) \equiv \delta(1 - x). \tag{31.40}$$

Now, for scattering on an individual parton with momentum p_i and charge q_i as depicted in Fig. 31.4, we have

$$\nu_i = \frac{q \cdot p_i}{M} \equiv x_i \nu. \tag{31.41}$$

We then have

$$\nu W_2^{elastic}(q^2, \nu, x_i) \equiv q_i^2 \delta(q^2 + 2M\nu_i)2M\nu_i = q_i^2 \delta\left(1 - \frac{x}{x_i}\right). \tag{31.42}$$

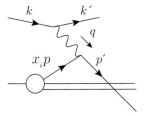

Fig. 31.4 Electron scattering off of a single parton in the nucleon.

This gives part of the scattering cross-section off a single parton. The full νW_2 part of the scattering cross-sections requires us to sum over all possible partons, i, and integrate over the fraction of momentum carried by each parton, where $f_i(x_i)$ is the probability of finding the ith parton with momentum fraction x_i. The result is given by

$$\nu W_2(Q^2, \nu) = \sum_i \int_0^1 dx_i f_i(x_i) \nu W_2^{elastic}(Q^2, \nu, x_i)$$

$$= \sum_i \int_0^1 dx_i f_i(x_i) q_i^2 \delta(1 - \frac{x}{x_i}),$$

$$\nu W_2(Q^2, \nu) = \sum_i q_i^2 f_i(x) x \equiv F_2(x). \tag{31.43}$$

Note, we've used the identity $\delta(1 - \frac{x}{x_i}) = x\delta(x_i - x)$.

Similarly, for a point-like particle with charge 1, we have

$$MW_1^{elastic}(q^2, \nu)|_{\text{point-like}} = \delta(q^2 + 2M\nu)\frac{q^2}{2} \tag{31.44}$$

and thus for a parton we have

$$MW_1^{elastic}(q^2, \nu, x_i) = \frac{q_i^2}{x_i} \delta(q^2 + 2M\nu_i)\frac{q^2}{2} = \frac{q_i^2}{2x_i} \delta\left(1 - \frac{x_i}{x}\right), \tag{31.45}$$

where the factor of $1/x_i$ is the result of a flux factor, which shall become evident in our more careful derivation which follows. We then find the full contribution to MW_1 given by

$$MW_1(Q^2, \nu) = \sum_i \int_0^1 dx_i f_i(x_i) MW_1^{elastic}(q^2, \nu, x_i)$$

$$= \sum_i \int_0^1 dx_i f_i(x_i) \frac{q_i^2}{2x_i} \delta\left(1 - \frac{x_i}{x}\right),$$

$$MW_1(Q^2, \nu) = \sum_i \frac{q_i^2}{2} f_i(x) \equiv F_1(x). \tag{31.46}$$

Now let's derive the contribution of partons to deep inelastic scattering from first principles. Consider the kinematics in Fig. 31.4. Then, following the analysis of Eqn. 31.10, except that now we are scattering off partons, we have

$$\omega^i_{\mu\nu} = \frac{1}{4x_i M} \sum_{spins} \int \left[\frac{d^3\vec{p}\,'}{(2\pi)^3 2p'_0} \right] \langle x_i p, \sigma | j^{EM}_\mu(0) | p', \sigma' \rangle \langle p', \sigma' | j^{EM}_\nu(0) | x_i p, \sigma \rangle$$

$$\times (2\pi)^3 \delta^4(p' - x_i p - q) \tag{31.47}$$

$$= \frac{q_i^2}{4x_i M} \sum_{spins} \bar{u}(x_i p) \gamma_\mu u(p') \bar{u}(p') \gamma_\nu u(x_i p) \times \delta(p'_0 - x_i p_0 - q_0)/2p'_0.$$

Note, the factor of $1/x_i$ is part of the flux factor.

Let's now consider the two factors in Eqn. 31.47 separately. For the first factor we have

$$\frac{1}{2} \sum_{spins} \bar{u}(x_i p) \gamma_\mu u(x_i p + q) \bar{u}(x_i p + q) \gamma_\nu u(x_i p) \simeq \frac{1}{2} \text{Tr}[x_i(\not{p} + M)\gamma_\mu (x_i(\not{p} + M) + \not{q})\gamma_\nu],$$
$$\tag{31.48}$$

where we've used $p' = x_i p + q$. We then have

$$\frac{1}{2} \text{Tr}[x_i(\not{p} + M)\gamma_\mu (x_i(\not{p} + M) + \not{q})\gamma_\nu] = 2x_i[p_\mu(x_i p_\nu + q_\nu) + p_\nu(x_i p_\mu + q_\mu) \tag{31.49}$$

$$-p \cdot (x_i p + q)g_{\mu\nu}] + 2x_i^2 M^2 g_{\mu\nu}$$

$$= 4x_i^2 M^2 \left(\frac{p_\mu p_\nu}{M^2} \right) + 2x_i(p_\mu q_\nu + p_\nu q_\mu)$$

$$-2x_i M \nu g_{\mu\nu}.$$

The second factor gives

$$\delta(p'_0 - x_i p_0 - q_0)/2p'_0 = \theta(p'_0)\delta(p_0'^2 - (x_i p_0 + q_0)^2) \tag{31.50}$$

$$= \theta(p'_0)\delta(p'^2 - (x_i p + q)^2) \quad \text{since} \quad \vec{p}\,' = x_i \vec{p} + \vec{q}$$

$$= \theta(x_i p_0 + q_0)\delta(p'^2 - (x_i^2 p^2 + 2x_i p \cdot q + q^2)) \quad \text{but on shell}$$
$$(x_i p)^2 = p'^2$$

$$= \theta(x_i p_0 + q_0)\delta(2M\nu x_i + q^2) = \theta(x_i p_0 + q_0)\delta(x_i - x)/2M\nu.$$

Putting it all together, we find

$$\omega^i_{\mu\nu} = \frac{q_i^2}{2x_i M} \left(4x_i^2 M^2 \left(\frac{p_\mu p_\nu}{M^2} \right) + 2x_i(p_\mu q_\nu + p_\nu q_\mu) - 2x_i M\nu \; g_{\mu\nu} \right) \delta(x_i - x)/2M\nu$$

$$= q_i^2 \delta(x_i - x) \left[\frac{x_i}{\nu} \left(\frac{p_\mu p_\nu}{M^2} \right) + \frac{1}{2M^2\nu}(p_\mu q_\nu + p_\nu q_\mu) - \frac{1}{2M} \; g_{\mu\nu} \right]$$

$$\equiv \left[-\omega_1^i \left(g_{\mu\nu} - \frac{q_\mu q_\nu}{q^2} \right) + \frac{\omega_2^i}{M^2} \left(p_\mu - \frac{p \cdot q}{q^2} q_\mu \right) \left(p_\nu - \frac{p \cdot q}{q^2} q_\nu \right) \right]. \tag{31.51}$$

Using $x_i = x = \frac{-q^2}{2M\nu}$, $\nu = \frac{q \cdot p}{M}$, we find

$$\nu \omega_2^i \equiv q_i^2 x_i \delta(x_i - x) \equiv \nu W_2(q^2, \nu, x_i), \tag{31.52}$$

$$M\omega_1^i \equiv \frac{q_i^2}{2}\delta(x_i - x) \equiv MW_1(q^2, \nu, x_i).$$

This can then be compared with the results of Eqns. 31.42 and 31.45.

Then upon summing over all partons, i, times the probability for finding parton i with fraction of momentum, x_i, we have

$$W_{\mu\nu} \equiv \sum_i \int_0^1 dx_i f_i(x_i) \omega_{\mu\nu}^i. \tag{31.53}$$

Once again we find the scaling functions,

$$F_1(x) \equiv MW_1 \equiv \sum_i \int_0^1 dx_i f_i(x_i) M\omega_1^i = \sum_i \frac{q_i^2}{2} f_i(x) \tag{31.54}$$

and

$$F_2(x) \equiv \nu W_2 \equiv \sum_i \int_0^1 dx_i f_i(x_i)\nu\omega_2^i = \sum_i q_i^2 x f_i(x). \tag{31.55}$$

The two structure functions satisfy the relation

$$F_2(x) = 2x F_1(x), \tag{31.56}$$

which is known as the Callan–Gross relation (Callan and Gross, 1969). It was derived assuming that partons are fermions. If instead partons were bosons with spin 0, then we would find

$$F_1(x) \equiv 0. \tag{31.57}$$

Experimental results are consistent with scaling at low Q^2 (Breidenbach et al., 1969; Friedman et al., 1972), see Fig. 31.5. In addition, the Callan–Gross relation is satisfied (Bodek et al., 1979), Fig. 31.6; thus, charged partons are fermions! In fact, they are quarks!! At higher values of Q^2, scaling violations occur due to quantum chromodynamics (QCD) corrections (Bodek et al., 1974; Tanabashi et al., 2018). We shall discuss these in more detail shortly.

Let's now consider deep inelastic electron–nucleon scattering and the specific dependence on the fact that the charged partons are quarks. The electromagnetic current is given by

$$j_\mu^{EM} = \frac{2}{3}\bar{u}\gamma_\mu u - \frac{1}{3}\bar{d}\gamma_\mu d - \frac{1}{3}\bar{s}\gamma_\mu s + \dots. \tag{31.58}$$

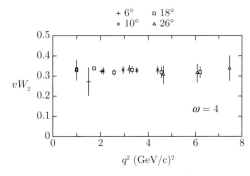

Fig. 31.5   Measurement of νW_2 at fixed value of $1/x = \omega = 4$. Figure from Bloom et al. (1970).

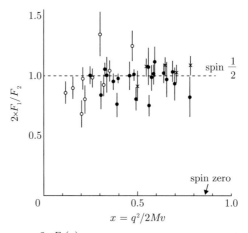

Fig. 31.6 The Callan–Gross relation, $\frac{2 \times F_1(x)}{F_2(x)}$. Figure taken from Perkins (1981). This is evidence that the partons are spin-1/2 particles.

Since

$$F_2(x)/x = \sum_i q_i^2 f_i(x), \tag{31.59}$$

we have for the proton

$$F_2^P(x)/x = \frac{4}{9}(u^P(x) + \bar{u}^P(x)) + \frac{1}{9}(d^P(x) + \bar{d}^P(x)) + \frac{1}{9}(s^P(x) + \bar{s}^P(x)) + \ldots, \tag{31.60}$$

where

$$u^P(x) \equiv f_u^P(x) \tag{31.61}$$

is the probability of the u parton with fraction of momentum x in the proton, i.e. these are the so-called *parton distribution functions* (PDFs).

Flavor $SU(3)$ symmetry implies the following sum rules. In the non-relativistic quark model, the proton is made up of only two up, one down and zero strange quarks. Thus we have

$$N_u^P = \int_0^1 dx(u^P(x) - \bar{u}^P(x)) = 2, \tag{31.62}$$

$$N_d^P = \int_0^1 dx(d^P(x) - \bar{d}^P(x)) = 1,$$

$$N_s^P = \int_0^1 dx(s^P(x) - \bar{s}^P(x)) = 0.$$

In QCD, however, the proton is made up of two up, one down and zero strange quarks, these are the so-called *valence quarks*, and then an infinite number of quark–anti-quark pairs (the so-called *sea quarks*) and gluons, Fig. 31.7. We can then rewrite the PDFs,

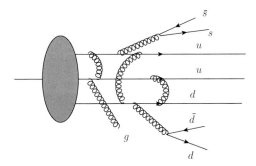

Fig. 31.7 The proton state in QCD.

$$u^p(x) \equiv u_V^p(x) + u_S^p(x), \tag{31.63}$$
$$d^p(x) \equiv d_V^p(x) + d_S^p(x),$$
$$\bar{u}_V^p(x) = \bar{d}_V^p(x) = 0 = s_V^p(x) = \bar{s}_V^p(x).$$

It is then reasonable to assume that the sea-quark distributions are identical for up and down quarks; however, since strange quarks are heavier, their distributions can be different. We then have

$$c(x) \equiv u_S(x) = \bar{u}_S(x) = d_S(x) = \bar{d}_S(x), \tag{31.64}$$
$$s(x) \equiv s_S(x) = \bar{s}_S(x).$$

Isospin symmetry then implies the following relations:

$$u_V^p(x) = d_V^n(x) \equiv 2a(x), \tag{31.65}$$
$$d_V^p(x) = u_V^n(x) \equiv b(x).$$

Given these definitions, we can now write the structure functions $F(x)/x$ for protons and neutrons in terms of these PDFs. We have

$$F_2^p(x)/x = \frac{1}{9}\left(8\,a(x) + b(x) + 10c(x) + 2s(x)\right) \tag{31.66}$$

$$F_2^n(x)/x = \frac{1}{9}\left(2\,a(x) + 4\,b(x) + 10c(x) + 2s(x)\right).$$

Note: the structure functions, $F_2^p(x), F_1^p(x), F_2^n(x), F_1^n(x)$ were initially determined experimentally in deep inelastic $e-p$ and $e-d$ scattering (Friedman et al., 1972). In addition, the PDFs are positive semi-definite. Using this fact, and Eqn. 31.66, Nachtmann (1972a,b) derived some inequalities which are very useful in understanding the experimental results. He found

$$\frac{1}{4} \leq \frac{F_2^n(x)}{F_2^p(x)} \leq 4. \tag{31.67}$$

Consider the following limits of the ratio $F_2^n(x)/F_2^p(x)$:

$$(1) \qquad c,\, s,\, a \to 0 \Longrightarrow 4 \tag{31.68}$$

$$(2) \qquad c,\, s,\, b \to 0 \Longrightarrow 1/4$$

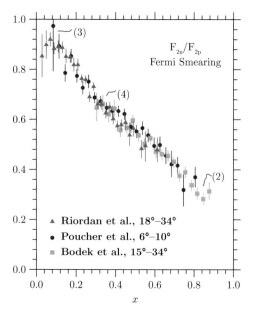

Fig. 31.8 The ratio $F^n(x)/F^p(x)$ (Petratos et al., 2000) using the results from Bodek et al. (1973), Poucher et al. (1974) and Riordan et al. (1974). Note: the numbers correspond to the regions described in Eqn. 31.68

$$(3) \qquad c,\, s \gg a,\, b \Longrightarrow 1$$
$$(4) \quad c,\, s \simeq 0,\, a = b \Longrightarrow 2/3.$$

The experimental results on the ratio were obtained early on (Bodek et al., 1973; Poucher et al., 1974; Riordan et al., 1974). A nice compilation of the results is given in Fig. 31.8. At low values of x the PDFs are dominated by sea quarks. At $x \sim 1/3$, valence quarks dominate, with each quark carrying one-third of the nucleon's momentum. While at $x \sim 1$, the up (down) valence quark dominates for the proton (neutron), respectively. In both cases, at $x \sim 1$, there is a ud pair near $x \to 0$. This is consistent with lattice gauge calculations of the nucleon wave function, which shows it to be more like a two-body state with the leading valence quark at one end and a ud pair at the other end.

Define P_q as the fraction of momentum carried by quarks in the nucleon. Given the nucleon momentum p with $p^2 = M^2$, we have

$$P_q = \int_0^1 dx (xp)[2\, a(x) + b(x) + 4\, c(x) + 2\, s(x)]. \tag{31.69}$$

Note: using Eqn. 31.66, we have

$$\frac{1}{x}\, (F_2^p(x) + F_2^n(x)) = \frac{5}{9} \left(2\, a(x) + b(x) + 4\, c(x) + \underbrace{\frac{4}{5}\, s(x)}_{2s(x) - \frac{6}{5} s(x)} \right). \tag{31.70}$$

Hence,

$$P_q = \frac{9}{5} \, p \int_0^1 dx (F_2^p(x) + F_2^n(x) + \frac{6}{5} \, xs(x)) \tag{31.71}$$

$$\geq \frac{9}{5} \, p \int_0^1 dx (F_2^p(x) + F_2^n(x))$$

and we expect the lower bound to be a good estimate, since sea quarks dominate at small x implying that they carry a small fraction of the nucleon momentum. Experimentally, we have

$$P_q \simeq \frac{9}{5}(0.28 \pm 0.03) \sim 0.5 \tag{31.72}$$

as extracted from Miller et al. (1972) Bodek et al. (1973) and Altarelli (1974). Therefore, only $\sim$ half of the momentum of the nucleon is carried by quarks!

The parton model predicts scaling. However, QCD predicts scaling violations due to the running coupling constant, which generates a logarithmic dependence on Q^2. Corrections to scaling are obtained using an operator product expansion beginning with the representation of $W_{\mu\nu}$, Eqn. 36.9, or the Altarelli–Parisi equations (Kogut and Susskind, 1974; Altarelli and Parisi, 1977). As a result, the PDFs become

$$F_i(x) \Longrightarrow F_i(x, \, Q^2). \tag{31.73}$$

As a result of scaling violations, as Q^2 increases, the PDFs at small x get larger, which is compensated by the large x values getting smaller, see Fig. 31.9. As Q^2 increases, the experiments resolve the nucleon at shorter distances. As a result it is seen that there are more and more quark–anti-quark pairs and gluons visible in the nucleon wave function. The Altarelli–Parisi equations quantify this behavior using splitting functions which describe the probability of a parton splitting into two others.

31.4 Jets

We have learned that, in deep inelastic scattering, the photon scatters off a point-like parton, knocking it out of the nucleon at large scattering angles. But we never see free quarks. As the energetic quark leaves the nucleon, a flux tube connecting it to the rest of the nucleon is formed. This flux tube costs energy. Pretty soon, when the energy in the flux tube exceeds the mass of a quark–anti-quark pair, the pair is created out of the vacuum, thus forming a color singlet jet carrying the momentum of the ejected quark. This was nicely described in a very simple $1 + 1$ dimensional quantum electrodynamics (QED) model (Casher et al., 1973) and then generalized QCD (Field and Feynman, 1977, 1978). By definition, a jet is a collection of color singlet hadrons which carry the momentum of the ejected quark. The description of this process of *hadronization* is accomplished phenomenologically with the definition of *fragmentation functions*, which give the probability for a particular parton

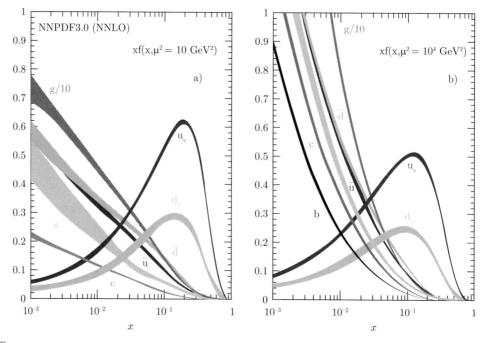

Fig. 31.9 The bands are x times the unpolarized parton distributions $f(x)$ (where $f = u_v, d_v, \bar{u}, \bar{d}, s \simeq \bar{s}, c \simeq \bar{c}, b \simeq \bar{b}, g$) obtained in NNLO NNPDF3.0 global analysis (Rojo et al., 2015) at scales $\mu^2 = 10$ GeV^2 (a) and $\mu^2 = 10^4$ GeV^2 (b), with $\alpha_s(M_Z^2) = 0.118$. Note the PDFs at small x increase as $\mu^2 \sim Q^2$ increases (Tanabashi et al., 2018). For a color version of this figure, please see the color plate section.

$(q, \bar{q}, g)$ to form particular hadrons (π, K, N, $\bar{N}$) carrying a fraction of the momentum of the parton. These jets were first observed in the data in e^+e^- collisions at the SPEAR storage ring at SLAC at a center-of-mass (CM) energy $E_{CM} = 7.4$ GeV (Hanson et al., 1975). The angular distribution for the two jet events was identical to that for $e^+e^- \to \mu^+\mu^-$ as expected by the theory for spin-1/2 quarks.

31.5 Discovery of the Gluon

In high-energy e^+e^- collisions at the PETRA collider in the Deutsches Elektronen Synchrotron (DESY), four different detectors, (TASSO, MARK J, PLUTO and JADE) were able to clearly see two- and three-jet events (Brandelik et al., 1979; Barber et al., 1979a; Berger et al., 1979; Bartel et al., 1980). The collider operated at CM energies $E_{CM} = 12 \to 32$ GeV. At these higher energies several important confirmations of QCD were made. The ratio

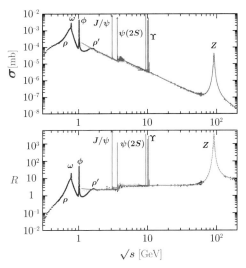

World data on the ratio $R(s) = \sigma(e^+e^- \to hadrons, s)/\sigma(e^+e^- \to \mu^+\mu^-, s)$ where $\sigma(e^+e^- \to hadrons, s)$ is the experimental cross-section corrected for initial state radiation and electron–positron vertex loops, and $\sigma(e^+e^- \to \mu^+\mu^-, s) = 4\pi\alpha^2(s)/3s$. Data errors are total below 2 GeV and statistical above 2 GeV. The curves are an educative guide: the broken one is a naive quark–parton model prediction, and the solid one is a three-loop perturbative QCD prediction. Breit–Wigner parameterizations of J/ψ, $\psi(2S)$, and $\Upsilon(nS)$, $n = 1, 2, 3, 4$ are also shown (Tanabashi et al., 2018).

$$R \equiv \frac{\sigma(e^+e^- \to q\bar{q} \ \{= hadrons\})}{\sigma(e^+e^- \to \mu^+\mu^-)} \qquad (31.74)$$

was measured to be consistent with data, *as long as there were three colors of quarks produced*. In addition, the value increased incrementally as the threshold for the heavier b quark was crossed (see Fig. 31.10). Three-jet events, predicted in Ellis et al. (1976) and DeGrand et al. (1977), were identified as gluon Bremsstrahlung from a quark. The detailed energy flow in the three-jet events is well described by QCD (Barber et al., 1979b). The experimental measure of the cross-over from two to three jets is given in terms of the parameter, *thrust*, defined by

$$T = max \sum_i |\vec{E}^i \cdot \vec{e}_1|/E_{vis} \qquad (31.75)$$

where $\vec{E}^i$ is the energy flow whose direction is given by the direction in the detector and the magnitude given by the deposited energy, E_{vis} is the total visible energy and the unit vector $\vec{e}_1$ is the thrust axis which, by definition, is the direction that maximizes T. $T = 1$ for two-jet events and, for $T < 1$, three-jet events appear. They are, by definition, co-planar events (see Fig. 31.11 [Barber et al., 1979b]). Two additional axes are defined, the major axis $\vec{e}_2 \perp \vec{e}_1$ given by

$$F_{major} = max \sum_i |\vec{E}^i \cdot \vec{e}_2|/E_{vis} \qquad (31.76)$$

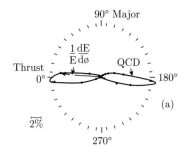

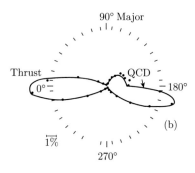

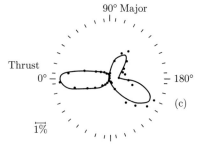

Fig. 31.11 Jet structure as a function of the observable thrust, T (Barber et al., 1979b). The angular energy flow diagram $E^{-1}dE/d\varphi$ is projected onto the thrust-major plane. Each point represents the energy contained in a $10°$ angular bin. (a) Events with $T > 0.9$. (b) Events with $0.8 < T < 0.9$ and $O_b > 0.1$. (c) Events with $T < 0.8$ and $O_b > 0.1$.

and $\vec{e}_3 \perp \vec{e}_{1,2}$ with

$$F_{minor} = \sum_i |\vec{E}^i \cdot \vec{e}_3|/E_{vis}. \tag{31.77}$$

Finally, the observable *oblateness*, O, is defined as

$$O = F_{major} - F_{minor}. \tag{31.78}$$

For states with three well-separated jets coming from $q\,\bar{q}\,g$, $O \approx 2\langle p_t \rangle_{gluon}/\sqrt{s}$. One example of a three-jet event from the TASSO collaboration in given in Fig. 31.12.

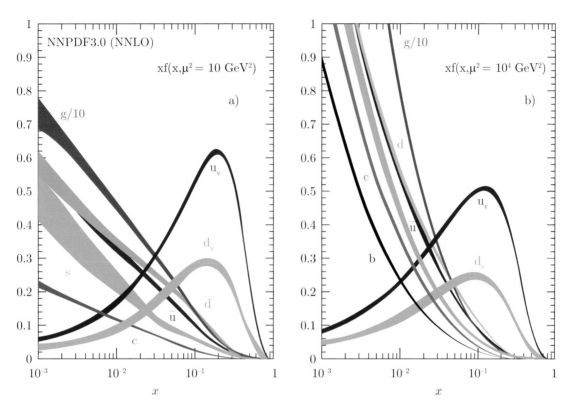

Figure 31.9 The bands are x times the unpolarized parton distributions $f(x)$ (where $f = u_v, d_v, \bar{u}, \bar{d}, s \simeq \bar{s}, c \simeq \bar{c}, b \simeq \bar{b}, g$) obtained in NNLO NNPDF3.0 global analysis (Rojo et al., 2015) at scales $\mu^2 = 10$ GeV^2 (a) and $\mu^2 = 10^4$ GeV^2 (b), with $\alpha_s(M_Z^2) = 0.118$. Note the PDFs at small x increase as $\mu^2 \sim Q^2$ increases (Tanabashi et al., 2018).

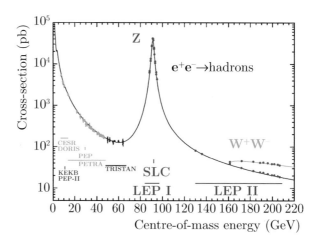

Figure 33.11 The Z peak in this data is from the analysis of $e^+e^- \to \mu^+\mu^-$ scattering with the Aleph detector (Schael et al., 2006) at the LEP. The lower-energy data come from experiments from many other colliders. Note, Tristan was built with the goal of finding the top quark, but its peak center-of-mass energy was below the top-quark threshold.

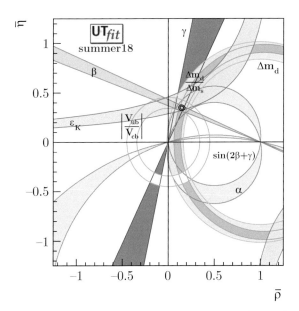

Figure 40.4 Experimental constraints on the CKM matrix obtained by UTfitter. Figure taken from UTfit Collaboration (n.d.).

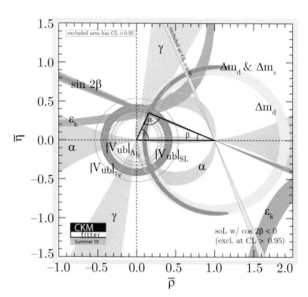

Figure 40.5 Experimental constraints on the CKM matrix obtained by CKMfitter. Figure taken from CKMfitter Collaboration (n.d.).

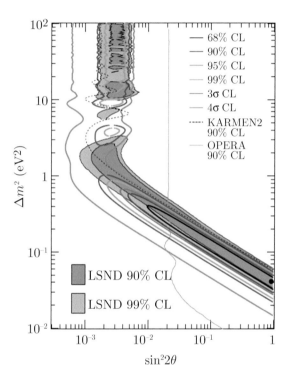

Figure 45.1 The combined fit to LSND and Mini-BooNE data for ν_e and $\bar{\nu}_e$ appearance (Aguilar-Arevalo et al., 2018). The shaded areas show the 90% and 99% confidence level (CL) LSND $\bar{\nu}_\mu \rightarrow \bar{\nu}_e$ allowed regions. The black point shows the Mini-BooNE best-fit point. Also shown are 90% CL limits from the KARMEN and OPERA experiments. Published under the CC BY 4.0 license.

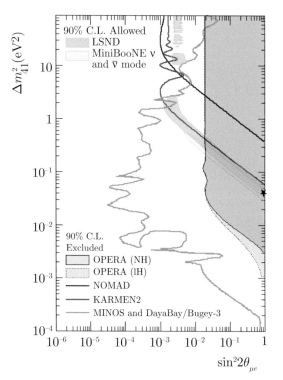

Figure 45.2 Note, the 90% CL allowed regions from LSND and Mini-BooNE, in contrast to the excluded regions found by the NOMAD experiments (Astier et al., 2003), KARMEN2 (Armbruster et al., 2002), MINOS (Adamson et al., 2019) and DayaBay/Bugey3 (Adamson et al., 2016) shown as 90% CL exclusion regions. Taken from the OPERA collaboration. (Agafonova et al., 2019). Published under the CC BY 4.0 licensed.

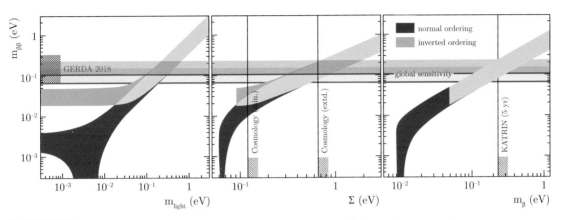

Figure 45.5 Constraints are shown, left to right, as function of the lightest neutrino mass m_{light}, the sum of neutrino masses Σ, and the effective neutrino mass m_β. Contours follow from a scan of the Majorana phases with the central oscillation parameters from NuFIT 4.0 (Esteban et al., 2019). The dark horizontal band shows the upper limits on $m_{\beta\beta} \equiv m_{ee}$ obtained by GERDA; the lighter horizonal band shows those from combining sensitivities of all leading experiments in the field (see Table 45.2). The vertical lines denote $\Sigma = 0.12$ eV and $\Sigma = 0.66$ eV, a stringent limit from cosmology (Aghanim et al., 2020) and a cosmological model with 12 free parameters, i.e. extended as opposed to the minimal model with 7 free parameters (Tanabashi et al., 2018), as well as $m_\beta = 0.23$ eV, the 5-year sensitivity of the KATRIN Tritium beta decay experiment (Angrik et al., 2005). Hatching denotes the excluded parameter space. Figure from Agostini et al. (2019). Reprinted with permission from AAAS.

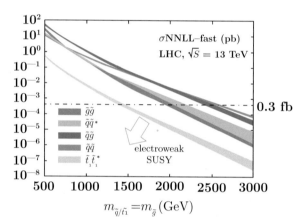

Figure 52.4 Supersymmetric (SUSY)-particle production cross-section at the LHC (Masciovecchio, 2019).

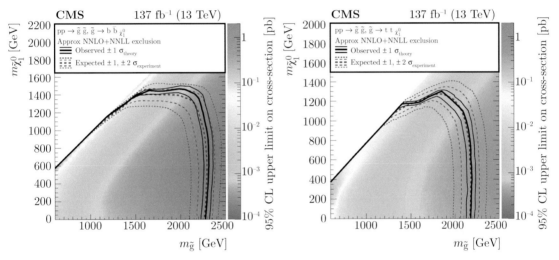

Figure 52.7 (left) Simplified model with $BR(\tilde{g} \to b\bar{b}\tilde{\chi}_1^0) = 1$, and (right) simplified model with $BR(\tilde{g} \to t\bar{t}\tilde{\chi}_1^0) = 1$ (Sirunyan et al., 2019). Published under the CC BY 4.0 license.

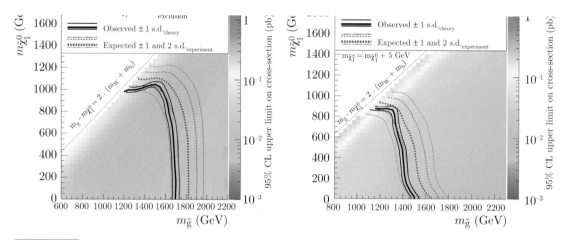

Figure 52.8 Gluino bound with simplified models in the search for jets missing E_T and leptons. (left) $\tilde{g} \to t\bar{t}\tilde{\chi}_1^0$ and (right) $\tilde{g} \to tb\tilde{\chi}_1^\pm$, 100% of the time (Sirunyan et al., 2020).

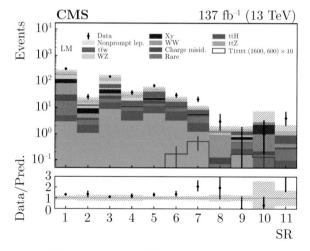

Figure 52.9 The LM signal regions (Sirunyan et al., 2020).

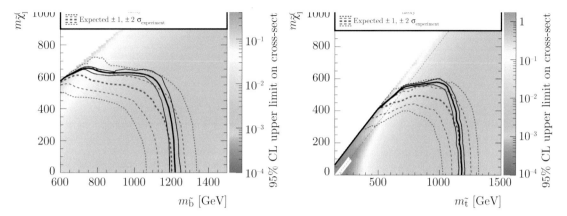

Figure 52.11 (left) Simplified model with $BR(\tilde{b} \to b\tilde{\chi}_1^0) = 1$, and (right) simplified model with $BR(\tilde{t} \to t\tilde{\chi}_1^0) = 1$ (Sirunyan et al., 2019). Published under the CC BY 4.0 license.

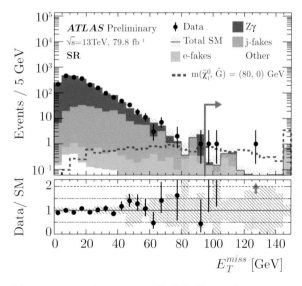

Figure 52.15 ATLAS data for Higgs to two photons via GMSB. Note, the excess events in the signal region (to the right of the arrow) (ATLAS Collaboration, 2018).

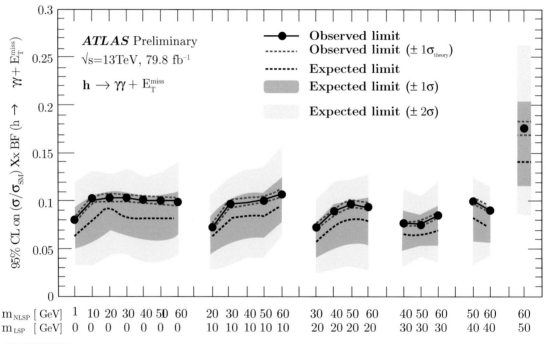

Figure 52.16 Bounds on the branching fraction of Higgs boson decays to two photons plus E_T^{miss} depending on the NLSP and LSP masses (ATLAS Collaboration, 2018).

Table 31.1 Measurement of α_s at 30 GeV using all PETRA experiments (Schopper, 1981)			
Experiment	α_s	Statistical error	Systematic error
JADE	0.18	±0.03	±0.03
MARK J	0.19	±0.02	±0.04
PLUTO	0.16	±0.03	±0.03
TASSO	0.17	±0.02	±0.03

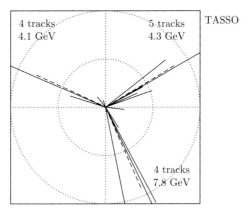

Fig. 31.12 A three-jet event projected into the event plane, TASSO collaboration (Wolf et al., 1979).

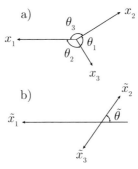

Fig. 31.13 (a) The jet angles, θ_i, defined in the CM frame (Brandelik et al., 1980). (b) $\tilde{\theta}$ is defined in the rest frame of partons 2 and 3.

The ratio of the three-jet to two-jet cross-section is a measure of the strong-interaction coupling constant (Schopper, 1981) (see Table 31.1).

The spin of the gluon was measured by TASSO (Brandelik et al., 1980) using the theoretical analysis (Ellis and Karliner, 1979). See also the nice reviews of Soding and Wolf (1981) and Ellis (2014)). The three-jet events can be described in terms of the observables $x_i = E^i/\sqrt{s}$, the energy fraction carried by each jet and the angles, θ_i, defined in Fig. 31.13. The x_i satisfy $x_1 + x_2 + x_3 = 1$ and TASSO defines

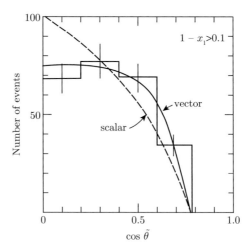

Fig. 31.14 Evidence for a spin-1 gluon using the TASSO detector at PETRA (Brandelik et al., 1980).

$x_3 \ll x_2 \ll x_1$, so x_1 is the thrust of the $q\,\bar{q}\,g$ system. For partons with mass much less than the CM energy, the angles and energy fractions, x_i, are related by

$$x_i = \frac{\sin\theta_i}{\sin\theta_1 + \sin\theta_2 + \sin\theta_3}. \tag{31.79}$$

Then, the angle $\tilde{\theta}$ is defined as in Fig. 31.13 in the rest frame of partons 2 and 3. As can be seen in Fig. 31.14, a spin-1 gluon fits the data as opposed to that of a putative spin-0 gluon.

Finally the three-gluon decay of the Upsilon is consistent with gluons with spin 1 – see the review by Soding and Wolf (1981)).

LHC Physics and Parton Distribution Functions

The Large Hadron Collider (LHC) at CERN is a proton–proton collider with center-of-mass (CM) energy starting out at $E_{CM} = 7$ or 8 TeV and up to 13 or 14 TeV. It has a luminosity ranging from $\mathcal{L}_{low} \sim 10^{33}$ cm^{-2} s^{-1} to $\mathcal{L}_{high} \sim 10^{34}$ cm^{-2} s^{-1}. The number of events observed per unit time in a detector is roughly given by the number of events per unit time $= \sigma \times \mathcal{L}$, and estimating the cross-section for new physics as roughly $\sigma_{new} \sim \alpha_s^2/\text{TeV}^2 \sim 10^{-36}$ cm$^2 = 1$ pb, for $\alpha_s \sim 0.1$ and 1 TeV$^{-1} \sim 10^{-17}$ cm, we have an event rate $\sigma_{new} \times \mathcal{L}_{low} \sim 10^{-3}/$s $\sim$ 1 event/15 minutes.

This should be compared to the background. The strong-interaction cross-section, $\sigma_{strong} \sim 30$ mb $= 3 \times 10^{-26}$ cm^2, corresponding an event rate, $\sigma_{strong} \times \mathcal{L}_{low} \sim 10^7$ events/s. This is 10 orders of magnitude larger than the expected signals. The cross-sections for top, W and Z production are

$$\sigma^{WW} \sim 100 \text{ pb} \Longrightarrow \text{event rate} \sim 0.1/\text{s} \tag{32.1}$$
$$\sigma^{t\bar{t}} \sim 1000 \text{ pb} \Longrightarrow \text{event rate} \sim 1/\text{s}$$
$$\sigma^{W,Z} \sim 10^5 \text{ pb} \Longrightarrow \text{event rate} \sim 100/\text{s},$$

which are also $100-10^5$ times larger than the signal.

Clearly, in order to find the signal in this immense background, we must focus on the highest-energy events using detectors with 4π coverage. These multi-purpose detectors have the following onion-like features. A central silicon tracker close to the beam pipe, then an electromagnetic calorimeter layer, followed by a hadronic calorimeter and, finally, muon chambers (see Fig. 32.1). The detector is built in this way in order to afford particle identification (see Table 32.1). The central tracker sees charged particles and sometimes even secondary vertices due to b, c or τ decays. The EM calorimeter registers energy loss due to photons, electrons and positrons. The hadronic calorimeter registers hadrons and, finally, the muon chamber sees any particle which is minimally ionizing and makes it all the way to the chambers. In the Standard Model this, to a good approximation, only includes muons.

In order to predict both Standard Model and new physics processes in proton–proton collisions, let's discuss a bit of the theory. We know that protons are made of partons which carry a fraction x of the proton momentum. Hence, the hard-scattering processes in a proton–proton collision can be represented as in Fig. 32.2. We define the proton momenta using light-cone coordinates with

$$p^{\pm} \equiv p^0 \pm p^z \tag{32.2}$$

Table 32.1 LHC detector particle identification				
	Tracker	EM calorimeter	Hadronic calorimeter	Muon chambers
γ	–	✓	–	–
$e^{\pm}$	✓	✓	–	–
$\pi^{\pm},\ K^{\pm},\ p$	✓	–	✓	–
$\pi^{0},\ K^{0},\ \bar{K}^{0},\ n$	–	–	✓	–
$\mu^{\pm}$	✓	–	–	✓
$b,\ c,\ \tau$	✓	–	–	–

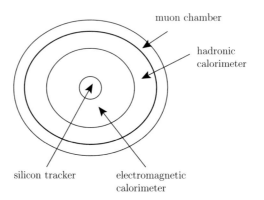

Fig. 32.1   Onion-like structure of an LHC detector.

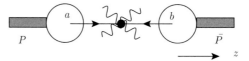

Fig. 32.2   Parton a–parton b hard process in a proton–proton collision.

and transverse momentum as

$$\vec{p}_T. \tag{32.3}$$

Note, the four-vector product

$$A \cdot B = \frac{1}{2}(A^{+}B^{-} + A^{-}B^{+}) - \vec{A}_T \cdot \vec{B}_T. \tag{32.4}$$

We choose

$$P \equiv (p^{+}, p^{-}, \vec{p}_T) = \sqrt{s}(1,0,0,0) \tag{32.5}$$

and

$$\bar{P} \equiv (\bar{p}^{+}, \bar{p}^{-}, \vec{p}_T) = \sqrt{s}(0,1,0,0), \tag{32.6}$$

where $p^{0} = p^{z} = \sqrt{s}/2, \bar{p}^{0} = -\bar{p}^{z} = \sqrt{s}/2$. We have chosen to work in light-cone coordinates since they transform simply under boosts in the z direction. Define the transverse energy

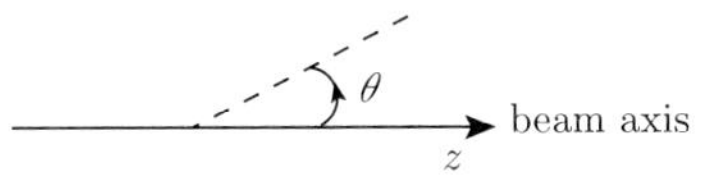

Fig. 32.3 Rapidity is related to the scattering angle θ.

$$E_T = \sqrt{\vec{p}_T^2 + m^2}. \tag{32.7}$$

Then

$$p^0 \equiv E_T \cosh y, \quad p^z \equiv E_T \sinh y \quad \text{and} \quad p^\pm \equiv E_T e^{\pm y}, \tag{32.8}$$

where y is known as *rapidity* and describes a boost in the z direction. Clearly, under an additional boost, using $\cosh(\eta+y) = \cosh(\eta)\cosh(y) + \sinh(\eta)\sinh(y)$ and $\sinh(\eta + y) = \cosh(\eta)\sinh(y) + \sinh(\eta)\cosh(y)$,

$$y \to y + \eta \tag{32.9}$$

and

$$p^\pm \to e^{\pm\eta} p^\pm. \tag{32.10}$$

There is a one-to-one correspondence between rapidity y and the scattering angle θ, Fig. 32.3.

$$\theta = 90° \implies y = 0 \tag{32.11}$$
$$\theta = 0° \implies y = \infty$$
$$\theta = 180° \implies y = -\infty$$
$$\theta = 10° \implies y \sim 3.5.$$

We can now write Lorentz-invariant phase space in terms of the light-cone coordinates. We have

$$(2\pi)^3 dLips(p) = d^4 p \delta(p^2 - m^2)\theta(p^0) = \frac{1}{2} dp^+ \wedge dp^- d^2\vec{p}_T \delta(p^+ p^- - E_T^2), \tag{32.12}$$

where we define $p^\pm \equiv \rho e^{\pm y}$ and

$$dp^+ \wedge dp^- = (d\rho e^{+y} + \rho e^{+y} dy) \wedge (d\rho e^{-y} - \rho e^{-y} dy) \tag{32.13}$$
$$= 2\rho \, dy \wedge d\rho. \tag{32.14}$$

Thus we have

$$(2\pi)^3 dLips(p) = \frac{1}{2} d\rho^2 dy \, d^2\vec{p}_T \delta(\rho^2 - E_T^2) = \frac{1}{2} dy \, d^2\vec{p}_T. \tag{32.15}$$

$dLips$ is flat in the coordinates, y, $\vec{p}_T$.

In the hard-scattering process, the two partons have four-momentum given by

$$x_a P = \sqrt{s}(x_a, 0, 0, 0), \quad x_b \bar{P} = \sqrt{s}(0, x_b, 0, 0), \tag{32.16}$$

while the outgoing particles have energy given by

$$p_j = (E_j e^{+y_j}, E_j e^{-y_j}, \vec{p}_{Tj}). \tag{32.17}$$

Momentum conservation then requires

$$\sqrt{s}x_a \equiv \sum_j E_j e^{y_j}, \tag{32.18}$$

$$\sqrt{s}x_b \equiv \sum_j E_j e^{-y_j},$$

$$\vec{0} \equiv \sum_j \vec{p}_{Tj}.$$

The parton CM energy is given by

$$s_{ab} = (x_a P + x_b \bar{P})^2 = 2x_a x_b \, P \cdot \bar{P} \equiv x_a x_b s. \tag{32.19}$$

In the parton model, the probability of finding a with momentum fraction between x_a and $x_a + dx_a$ is given by the parton distribution function (PDF),

$$f_a(x_a)dx_a. \tag{32.20}$$

We defined the PDFs in deep inelastic scattering. However, it was shown by Collins, Soper and Sterman in their *factorization theorem* (Collins et al., 1985) that the same PDFs can be used in hadron–hadron collisions. In fact, this is how the PDFs in Fig. 31.9 are obtained, which is implicit in Table 36.1. Note, in Fig. 31.9, at small x the PDFs are power-law distributions in x. Moreover, gluons dominate at small x. The differential cross-section for proton–proton collisions is given by

$$d\sigma = \int_0^1 dx_a dx_b f_a(x_a) f_b(x_b) \left(\frac{|T|^2}{2s_{ab}}\right)(2\pi)^4 \delta^4(x_a P + x_b \bar{P} - \sum_{j=1}^n p_j)\prod_{j=1}^n \frac{dy_j d\vec{p}_{Tj}}{2(2\pi)^3}$$

$$= f_a(x_a) f_b(x_b)\left(\frac{|T|^2}{2x_a x_b s}\right)(2\pi)^4 \prod_{j=1}^n \frac{dy_j d\vec{p}_{Tj}}{2(2\pi)^3}\, \delta^2(\sum_{j=1}^n \vec{p}_{Tj}). \tag{32.21}$$

Note, the term

$$\left(\frac{|T|^2}{2x_a x_b s}\right) = d\sigma_{ab} \tag{32.22}$$

is the hard-scattering cross-section and the values of x_a, x_b are fixed by momentum conservation.

For the case $n = 2$, Fig. 32.4, we have

$$d\sigma \sim f_a(x_a) f_b(x_b) \left(\frac{|T(ab \to 12)|^2}{2x_a x_b s}\right) dy_1 dy_2 d^2\vec{p}_{T1}, \tag{32.23}$$

(where we have ignored constant factors) and

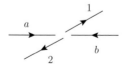

Fig. 32.4   Hard-scattering process.

$$\sqrt{s}x_a = E_1 e^{y_1} + E_2 e^{y_2},$$ (32.24)

$$\sqrt{s}x_b = E_1 e^{-y_1} + E_2 e^{-y_2}.$$

The Mandelstam invariants for the hard-scattering process are given by

$$s_{ab} \equiv s x_a x_b = E_1^2 + E_2^2 + 2E_1 E_2 \cosh(y_2 - y_1)$$ (32.25)

and

$$t_{ab} \equiv (x_a P - p_1)^2 = -(E_1 E_2 e^{(y_2-y_1)} + (\vec{p}_{T1})^2).$$ (32.26)

Note, s_{ab} and t_{ab} only depend on $(y_2 - y_1)$, since they are Lorentz invariant. Also $\sum_{spins} |T(ab \to 12)|^2$ only depends on s_{ab}, t_{ab}.

Let's change variables. We have

$$dy_1 dy_2 d^2\vec{p}_T = d\bar{y} d\Delta y d^2\vec{p}_T,$$ (32.27)

where

$$\bar{y} \equiv \frac{y_1 + y_2}{2}, \quad \Delta y \equiv y_2 - y_1.$$ (32.28)

Then, we change variables from $(\Delta y, \vec{p}_T)$ to $(s_{ab}, \vec{p}_T)$. We have

$$d\bar{y} d\Delta y d^2\vec{p}_T \equiv J d\bar{y} ds_{ab} d^2\vec{p}_T = \frac{d\bar{y} ds_{ab} d^2\vec{p}_T}{\sqrt{s_{ab}^2 - 2s_{ab}(E_1^2 + E_2^2) + (E_1^2 - E_2^2)^2}}.$$ (32.29)

Consider a simple example. Let $M_1 = M_2$ and $E_1 = E_2 \equiv E_T$. Then,

$$\sqrt{s}x_a = \sqrt{s_{ab}} e^{\bar{y}}$$ (32.30)

$$\sqrt{s}x_b = \sqrt{s_{ab}} e^{-\bar{y}}.$$

The differential cross-section is of order

$$d\sigma \sim f_a(\sqrt{x_a x_b} e^{\bar{y}}) f_b(\sqrt{x_a x_b} e^{-\bar{y}}) \frac{\sum_{spins} |T_{(ab \to 12)}(s_{ab}, \vec{p}_T)|^2}{s_{ab}} J d\bar{y} ds_{ab} d^2\vec{p}_T,$$ (32.31)

with the Jacobian

$$J = \frac{1}{\sqrt{s_{ab}(s_{ab} - 4E_T^2)}}.$$ (32.32)

We can then integrate over $\bar{y}$ and obtain

$$d\sigma \sim \frac{\sum_{spins} |T_{(ab \to 12)}(s_{ab}, \vec{p}_T)|^2}{s_{ab}} J \rho_{ab}(s_{ab}) ds_{ab} d^2\vec{p}_T,$$ (32.33)

where

$$\rho_{ab}(s_{ab}) \equiv \int d\bar{y} f_a(\sqrt{x_a x_b}\ e^{\bar{y}}) f_b(\sqrt{x_a x_b}\ e^{-\bar{y}})$$ (32.34)

is the *parton luminosity*. In the small-x regime, with x_a, $x_b \leq 0.1$, the PDFs scale as a power law in x, which is approximately the same for gluon–gloun, gluon–quark, etc., and therefore

$$\rho_{ab} \sim (x_a x_b)^{-2}$$ (32.35)

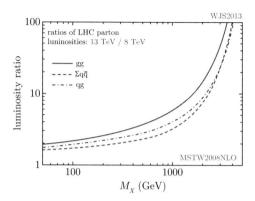

Fig. 32.5   13/8 TeV LHC luminosity ratios for $s_{ab} \sim M_X^2$ (Stirling, n.d.).

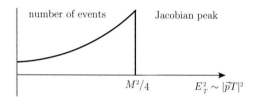

Fig. 32.6   Hard-scattering process.

for $\sqrt{s_{ab}} = M_X \sim 1$ TeV (see Fig. 32.5; Stirling, n.d.).[1]

Consider two examples of new physics processes.

(1) The first example is the production cross-section for a heavy resonance with mass M. We have $ab \to$ resonance $\to 12$ where 1, 2 are massless Standard Model particles. In this case, the hard-scattering amplitude goes like

$$|T|^2 \sim \frac{1}{(s_{ab} - M^2)^2 + M^2 \Gamma^2} \sim constant \times \delta(s_{ab} - M^2) \tag{32.36}$$

in the narrow-resonance approximation. The production cross-section then has the form

$$d\sigma \sim \frac{J}{M^2} f_a \left(\frac{M}{E_{CM}} e^{\bar{y}} \right) f_b \left(\frac{M}{E_{CM}} e^{-\bar{y}} \right) d\bar{y} d^2 \vec{p}_T \tag{32.37}$$

$$= \frac{1}{\sqrt{1 - \frac{4E_T^2}{M^2}}} f_a \left(\frac{M}{E_{CM}} e^{\bar{y}} \right) / M f_b \left(\frac{M}{E_{CM}} e^{-\bar{y}} \right) / M d\bar{y} d^2 (\vec{p}_T / M).$$

This gives a Jacobian peak at $E_T = M/2$ as given in Fig. 32.6.

Note, the rapidity distribution is approximately flat for $\frac{M}{E_{CM}} e^{\bar{y}} \leq 0.1$. The total cross-section for the production of a heavy resonance is of the form

---

[1] I have assumed $x_a \sim x_b \sim M_X/E_{CM}$ and $\rho(E_{CM}) \propto (x_a x_b)^{-a}$.

$$\sigma_{tot} \sim \frac{1}{M^2} \int d\bar{y} f_a\left(\frac{M}{E_{CM}} e^{\bar{y}}\right) f_b\left(\frac{M}{E_{CM}} e^{-\bar{y}}\right) \tag{32.38}$$

$$= \frac{1}{M^2} \rho_{ab}(M^2) \sim \begin{cases} \frac{E_{CM}^4}{M^6} & \text{gluon}-\text{gluon} \\ \frac{E_{CM}^{2.8}}{M^{4.8}} & u, \bar{u}, \, d, \bar{d} \end{cases}$$

Note, the significant enhancement of the total cross-section for $E_{CM} \gg M$, which is due to the dominance of sea quarks and gluons at small x (see Eqn. 32.35).

(2) For the second example, consider the cross-section $ab \to 12$ for producing new particles, beyond the Standard Model, with equal mass, M. We have

$$d\sigma \sim \frac{\sum_{spins} |T_{(ab \to 12)}(s_{ab}, \vec{p}_T)|^2}{s_{ab}} \frac{1}{\sqrt{s_{ab}(s_{ab} - 4E_T^2)}} \tag{32.39}$$

$$\times f_a(\sqrt{x_a x_b} \, e^{\bar{y}}) f_b(\sqrt{x_a x_b} \, e^{-\bar{y}}) d\bar{y} \, ds_{ab} d^2 \vec{p}_T$$

$$= \frac{\sum_{spins} |T_{(ab \to 12)}(s_{ab}, \vec{p}_T)|^2}{s_{ab}} \rho_{ab} \frac{1}{\sqrt{s_{ab}(s_{ab} - 4E_T^2)}} ds_{ab} d^2 \vec{p}_T.$$

The first two factors decrease as s_{ab} increases. Thus, the cross-section is dominated by small $s_{ab} = x_a x_b s$. Hence, heavy particles are predominantly produced at rest with $s_{ab} \sim 4M^2$, since there are more partons at small x. This means that at a proton–proton collider one has a significant advantage over an $e^+ e^-$ collider for the production of heavy particles. The cross-section to produce heavy particles with mass, M, in an $e^+ e^-$ collider is of order

$$\sigma_{(e^+ e^- \to \chi \bar{\chi})} \sim \frac{\alpha^2}{4M^2}, \tag{32.40}$$

while in a proton–proton collider with $E_{CM} \gg 2M$, the cross-section for producing particles with color roughly goes as

$$\sigma(pp \to \chi \bar{\chi}) \sim \frac{\alpha_s^2}{4M^2} \rho_{ab} \sim \frac{\alpha_s^2}{4M^2} \begin{cases} \left(\frac{E_{CM}}{M}\right)^4 & \text{gluon}-\text{gluon} \\ \left(\frac{E_{CM}}{M}\right)^{2.8} & u, \bar{u}, \, d, \bar{d} \end{cases}, \tag{32.41}$$

where x_a, $x_b \sim M/E_{CM}$, (see Eqn. 32.35). Hence, there is a significant increase in the production cross-section in a hadronic collider with an increase of energy. For example, for the LHC going from $E_{CM} = 8$ TeV to 13 TeV, there was approximately a factor of seven increase in the production cross-section for new physics. Unfortunately, if nothing new is discovered after the initial data collection, then the experiment becomes background limited and new physics can only slowly be discovered with time.

PART VIII

ROAD TO THE STANDARD MODEL: ELECTROWEAK THEORY

33 The Electroweak Theory

The Standard Model includes the strong color forces for quarks and gluons and, in addition, the electroweak interactions for all fermions. The full spectrum of fermions includes three families of quarks and leptons as given in Table 33.1. The electroweak interactions are chiral and therefore are most simply described in terms of Weyl spinors as discussed in Sections 10.2 and 10.3.

33.1 Brief Review of Spinors

The fields and Lagrangian for a free massive electron and neutrino are given by Eqns. 10.62 and 10.63. We have

$$\psi_e = \left(\begin{array}{c} \chi_e \\ i\sigma_2\chi_{\bar{e}}^* \end{array} \right), \qquad \psi_\nu = \left(\begin{array}{c} \chi_\nu \\ 0 \end{array} \right). \tag{33.1}$$

The Lagrangian is given by

$$\mathcal{L} = \chi_e^\dagger(i\bar{\sigma}^\mu\partial_\mu)\chi_e + \chi_{\bar{e}}^\dagger(i\bar{\sigma}^\mu\partial_\mu)\chi_{\bar{e}} + m_e(\chi_{\bar{e}\alpha}\epsilon^{\alpha\beta}\chi_{e\beta} + h.c.) \tag{33.2}$$
$$+ \chi_\nu^\dagger(i\bar{\sigma}^\mu\partial_\mu)\chi_\nu + m_\nu(\chi_{\nu\alpha}\epsilon^{\alpha\beta}\chi_{\nu\beta} + h.c.).$$

The discrete symmetries, $(\mathbf{C}, \mathbf{P}, \mathrm{T})$, for fermions were discussed in Chapters 16–18. Rewriting the results there in terms of Weyl fermions, we have for Dirac fermions

$$\psi \equiv \left(\begin{array}{c} \chi_L \\ \chi_R \end{array} \right) \equiv \left(\begin{array}{c} \chi_1 \\ i\sigma_2\chi_2^* \end{array} \right). \tag{33.3}$$

C:

$$\psi^C \equiv i\gamma_2\psi^* = i \left(\begin{array}{cc} 0 & -\sigma_2 \\ \sigma_2 & 0 \end{array} \right) \left(\begin{array}{c} \chi_L^* \\ \chi_R^* \end{array} \right) \tag{33.4}$$
$$= i \left(\begin{array}{c} -\sigma_2(i\sigma_2\chi_2) \\ \sigma_2\chi_1^* \end{array} \right) = \left(\begin{array}{c} \chi_2 \\ i\sigma_2\chi_1^* \end{array} \right).$$

Thus, under **C**:

$$\chi_1 \leftrightarrow \chi_2. \tag{33.5}$$

P:

$$\psi^P \equiv \gamma_0\psi \equiv \left(\begin{array}{c} \chi_R \\ \chi_L \end{array} \right). \tag{33.6}$$

Table 33.1 Three families of quarks and leptons in the Standard Model. The charge Q is in units of $e = 1.6 \times 10^{-19}$ C.

Q			
0	ν_e	ν_μ	ν_τ
-1	e	μ	τ
$2/3$	u	c	t
$-1/3$	d	s	b

Hence, under **P**:

$$\chi_L \leftrightarrow \chi_R. \tag{33.7}$$

CP:

$$\psi^{CP} \equiv \gamma_0 \psi^C \equiv \begin{pmatrix} i\sigma_2 \chi_1^* \\ \chi_2 \end{pmatrix}. \tag{33.8}$$

Hence, under **CP**:

$$\chi_1 \to i\sigma_2 \chi_1^*, \quad \chi_2 \to -i\sigma_2 \chi_2^*. \tag{33.9}$$

Thus, **CP** can be defined on a single Weyl spinor, which contains a left-handed particle with its own right-handed anti-particle.

33.2 Electroweak Theory of Leptons

The electroweak theory of leptons was first discussed by Glashow (1961) in which the $SU(2) \otimes U(1)_Y$ gauge symmetry was introduced and the weak mixing angle was defined. In order to obtain any phenomenological results from this theory, mass terms must be added in an ad hoc manner. The use of the Higgs mechanism (Higgs, 1964; Englert and Brout, 1964; Guralnik et al., 1964; Higgs, 1966) to resolve the problem of mass was done by Weinberg and Salam (Weinberg, 1967; Salam, 1968). This theory was more predictive than Glashow's theory, since it predicted a new particle, the BEH boson, and it related the W and Z masses. Let's first consider the electroweak theory for leptons introduced by Glashow.

We consider the electroweak theory for one family of leptons. The fields include the three Weyl spinors for one lepton family, i.e.

$$\nu_e \equiv \chi_\nu, \quad e \equiv \chi_e, \quad \bar{e} \equiv \chi_{\bar{e}}. \tag{33.10}$$

The gauge group is

$$SU(2) \otimes U(1)_Y. \tag{33.11}$$

Table 33.2 The three lepton fields and their transformation under the electroweak symmetry group

Fields	$SU(2) \otimes U(1)_Y$
$\ell \equiv \begin{pmatrix} \nu_e \\ e \end{pmatrix}$	$(2,\ -1)$
$\bar{e}$	$(1,\ 2)$

And the three fields are in the following representation of the gauge symmmetry given in Table 33.2. The $U(1)_Y$ symmetry is known as weak hypercharge and electric charge is given by the relation

$$Q \equiv T_3 + \frac{Y}{2}. \tag{33.12}$$

The generators of $SU(2) \otimes U(1)_Y$ are given by

$$\{T_a,\ a = 1, 2, 3;\ Y\}, \tag{33.13}$$

satisfying the commutation relations

$$[T_a,\ T_b] = i\epsilon_{abc}T_c \quad SU(2), \tag{33.14}$$
$$[T_a,\ Y] \equiv 0.$$

The action of the generators on the fields is given by

$$T_a\ell \equiv \frac{\tau_a}{2}\,\ell \qquad T_a\bar{e} \equiv 0 \tag{33.15}$$
$$Y\ell \equiv -\ell \qquad Y\bar{e} \equiv 2\bar{e}\ .$$

The interaction of the fermions with the gauge bosons are completely determined by their quantum numbers and the covariant derivative. We have

$$D^\mu \equiv \partial^\mu + igW_a^\mu T_a + i\frac{g'}{2}B^\mu Y, \tag{33.16}$$

where g, g' are independent gauge coupling constants and W_a^μ, $a = 1, 2, 3$; B^μ are the independent gauge bosons for the $SU(2)$; $U(1)_Y$ gauge groups. An invariant Lagrangian is then given by[1]

$$\mathcal{L} = \left[\ell^\dagger i\bar{\sigma}_\mu D^\mu\ell + \bar{e}^\dagger i\bar{\sigma}_\mu D^\mu\bar{e} - \frac{1}{4}W_{\mu\nu a}W_a^{\mu\nu} - \frac{1}{4}B_{\mu\nu}B^{\mu\nu}\right]. \tag{33.17}$$

The field strengths, $W_a^{\mu\nu}$, $B^{\mu\nu}$, are given by

$$W_a^{\mu\nu} \equiv \partial^\mu W_a^\nu - \partial^\nu W_a^\mu - g\epsilon_{abc}W_b^\mu W_c^\nu, \tag{33.18}$$
$$B^{\mu\nu} \equiv \partial^\mu B^\nu - \partial^\nu B^\mu.$$

[1] In a footnote in Glashow (1961), Glashow thanks Gell-Mann for saying that the theory looks much simpler in terms of Weyl spinors.

We have not included any mass terms in the Lagrangian because, as we discuss later, they break the gauge symmetry. However, before discussing mass terms, it is useful to rewrite the Lagrangian, Eqn. 33.17, in Dirac notation, so that we can compare this simple equation with the standard notation.

33.3 Converting to Dirac Notation

The free fermionic Lagrangian is given by (recall $\gamma^\mu = \begin{pmatrix} 0 & \sigma^\mu \\ \bar{\sigma}^\mu & 0 \end{pmatrix}$)

$$\mathcal{L}^f_{free} = \bar{\psi}_{\nu L} i \gamma^\mu \partial_\mu \psi_{\nu L} + \bar{\psi}_e i \gamma^\mu \partial_\mu \psi_e \tag{33.19}$$

where, as before,

$$\psi_e \equiv \begin{pmatrix} e \\ i\sigma_2 \bar{e}^* \end{pmatrix}, \quad \psi_{\nu L} \equiv \begin{pmatrix} \nu_e \\ 0 \end{pmatrix} = P_L \psi_\nu. \tag{33.20}$$

The interaction Lagrangian is then

$$\mathcal{L}^f_{int} = -g W^\mu_a \ell^* \bar{\sigma}_\mu T_a \ell - \frac{g'}{2} B^\mu \ell^* \bar{\sigma}_\mu Y \ell \tag{33.21}$$

$$- g W^\mu_a \bar{e}^* \bar{\sigma}_\mu T_a \bar{e} - \frac{g'}{2} B^\mu \bar{e}^* \bar{\sigma}_\mu Y \bar{e}$$

$$= -\frac{g}{2} W^\mu_a \ell^* \bar{\sigma}_\mu \tau_a \ell + \frac{g'}{2} B^\mu \ell^* \bar{\sigma}_\mu \ell$$

$$- g' B^\mu \bar{e}^* \bar{\sigma}_\mu \bar{e}.$$

Now let's define the raising and lowering operators,

$$\tau^\pm \equiv \frac{\tau_1 \pm i\tau_2}{2}, \quad \tau^+ = \begin{pmatrix} 0 & 1 \\ 0 & 0 \end{pmatrix}, \quad \tau^- = \begin{pmatrix} 0 & 0 \\ 1 & 0 \end{pmatrix}. \tag{33.22}$$

Then the interaction Lagrangian separates into the off-diagonal and diagonal terms given, by

$$\mathcal{L}^f_{off-diag} = -\frac{g}{2} \left[(W^\mu_1 + iW^\mu_2) \ell^* \bar{\sigma}_\mu \tau^- \ell + (W^\mu_1 - iW^\mu_2) \ell^* \bar{\sigma}_\mu \tau^+ \ell \right]$$

$$\mathcal{L}^f_{diag} = -\frac{g}{2} W^\mu_3 \ell^* \bar{\sigma}_\mu \tau_3 \ell + \frac{g'}{2} B^\mu \ell^* \sigma_\mu \ell - g' B^\mu \bar{e}^* \sigma_\mu \bar{e}. \tag{33.23}$$

Now define $W^\pm$ by

$$W^\mu_\mp \equiv \frac{W^\mu_1 \pm iW^\mu_2}{\sqrt{2}}. \tag{33.24}$$

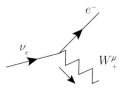

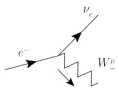

Fig. 33.1 Vertices associated with the charged current interactions.

Then we have

$$\mathcal{L}^f_{off\text{-}diag} = -\frac{g}{\sqrt{2}} \left[W^\mu_- \left(\nu^*_e \ e^* \right) \bar{\sigma}_\mu \begin{pmatrix} 0 & 0 \\ 1 & 0 \end{pmatrix} \begin{pmatrix} \nu_e \\ e \end{pmatrix} \right]$$
$$+ W^\mu_+ \left(\nu^*_e \ e^* \right) \bar{\sigma}_\mu \begin{pmatrix} 0 & 1 \\ 0 & 0 \end{pmatrix} \begin{pmatrix} \nu_e \\ e \end{pmatrix} \right]$$
$$= -\frac{g}{\sqrt{2}} \left[W^\mu_- \ e^* \bar{\sigma}_\mu \nu_e + W^\mu_+ \ \nu^*_e \bar{\sigma}_\mu e \right]. \qquad (33.25)$$

Note, W^μ_- annihilates a negatively charged vector boson and creates a positively charged vector boson. Also, $W^\mu_+ \equiv (W^\mu_-)^\dagger$. These are the so-called *charged current interactions* described by the Feynman diagrams, Fig. 33.1. Also note, charge is conserved at each vertex.

In Dirac notation, we have

$$\mathcal{L}^f_{off\text{-}diag} = -\frac{g}{\sqrt{2}} \left[W^\mu_- \ \bar{\psi}_e \gamma_\mu P_L \psi_{\nu_e} + W^\mu_+ \ \bar{\psi}_{\nu_e} \gamma_\mu P_L \psi_e \right]. \qquad (33.26)$$

Define

$$J^{weak}_\mu \equiv \bar{\psi}_e \gamma_\mu (1 - \gamma_5) \psi_{\nu_e}. \qquad (33.27)$$

Then

$$\mathcal{L}^f_{off\text{-}diag} = -\frac{g}{2\sqrt{2}} \left[W^\mu_- J^{weak}_\mu + W^\mu_+ J^{weak\dagger}_\mu \right]. \qquad (33.28)$$

Thus, by looking at the low-energy behavior of the tree diagram for a charged, massive W exchange (using the propagator given in Eqn. 11.42 with $m = M_W$), we obtain the Fermi theory of weak interactions given by (see Fig. 33.2)

$$\mathcal{L}_{eff} \equiv -\frac{G_F}{\sqrt{2}} \ J^{weak}_\mu J^{\mu \ weak\dagger} \qquad (33.29)$$

with

$$\frac{G_F}{\sqrt{2}} \equiv \frac{g^2}{8M^2_W}. \qquad (33.30)$$

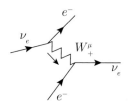

Fig. 33.2 Four-Fermi weak-interaction vertex.

Now consider the diagonal weak interaction, Eqn 33.23. Writing it in a more symmetric way, we have

$$\mathcal{L}_{diag}^{f} = -\ell^* \bar{\sigma}_\mu \left[g W_3^\mu T_3 + \frac{g'}{2} B^\mu Y \right] \ell - \bar{e}^* \bar{\sigma}_\mu \left[g W_3^\mu T_3 + \frac{g'}{2} B^\mu Y \right] \bar{e}. \qquad (33.31)$$

We have two neutral gauge bosons and one should be the photon which couples to electric charge, Eqn 33.12. Therefore, we express the term in brackets as

$$g W_3^\mu T_3 + \frac{g'}{2} B^\mu Y \equiv e A^\mu Q + Z^\mu Q_Z. \qquad (33.32)$$

The quantity Q_Z is defined by this equation, A^μ, Z^μ are orthogonal fields, e is electric charge and A^μ is the photon. Let

$$A^\mu \equiv \sin\theta \ W_3^\mu + \cos\theta \ B^\mu \qquad (33.33)$$
$$Z^\mu \equiv \cos\theta \ W_3^\mu - \sin\theta \ B^\mu, \text{i.e.}$$

$$\begin{pmatrix} A^\mu \\ Z^\mu \end{pmatrix} \equiv \mathcal{O} \begin{pmatrix} B^\mu \\ W_3^\mu \end{pmatrix}, \qquad (33.34)$$

where

$$\mathcal{O} = \begin{pmatrix} \cos\theta & \sin\theta \\ -\sin\theta & \cos\theta \end{pmatrix} \qquad (33.35)$$

is an orthogonal transformation. We can then solve for the values of $\sin\theta$, Q_Z in terms of g, g', e. Note: θ is known as the *weak mixing angle*, sometimes written as θ_W. Use the inverse transformation

$$\begin{pmatrix} B^\mu \\ W_3^\mu \end{pmatrix} = \begin{pmatrix} \cos\theta_W & -\sin\theta_W \\ \sin\theta_W & \cos\theta_W \end{pmatrix} \begin{pmatrix} A^\mu \\ Z^\mu \end{pmatrix} \qquad (33.36)$$

and substitute into Eqn. 33.32. We have

$$(g\sin\theta_W A^\mu + g\cos\theta_W Z^\mu)T_3 + (\tfrac{g'}{2}\cos\theta_W A^\mu - \tfrac{g'}{2}\sin\theta_W Z^\mu)Y$$
$$= A^\mu(g\sin\theta_W T_3 + \tfrac{g'}{2}\cos\theta_W Y) + Z^\mu(g\cos\theta_W T_3 - \tfrac{g'}{2}\sin\theta_W Y)$$
$$\equiv e A^\mu (T_3 + \tfrac{Y}{2}) + Z^\mu Q_Z. \qquad (33.37)$$

Equating the second and third lines, we find

$$e \equiv g \sin \theta_W = g' \cos \theta_W, \quad \tan \theta_W \equiv \frac{g'}{g} \tag{33.38}$$

and

$$Q_Z \equiv g \cos \theta_W (T_3 - \tan^2 \theta_W \underbrace{\frac{Y}{2}}_{Q-T_3}) = g \cos \theta_W (T_3 \underbrace{(1 + \tan^2 \theta_W)}_{1/\cos^2 \theta_W} - \tan^2 \theta_W Q).$$

Hence

$$Q_Z \equiv \frac{g}{\cos \theta_W} (T_3 - \sin^2 \theta_W Q). \tag{33.39}$$

We now have

$$\mathcal{L}_{diag}^f = -\ell^* \bar{\sigma}_\mu [e A^\mu Q + Z^\mu Q_Z] \ell - \bar{e}^* \bar{\sigma}_\mu [e A^\mu Q + Z^\mu Q_Z] \bar{e}. \tag{33.40}$$

Let's consider the photon and Z terms separately. For the photon we have

$$e A^\mu [e^* \bar{\sigma}_\mu e - \bar{e}^* \bar{\sigma}_\mu \bar{e}] \tag{33.41}$$

and using

$$\bar{e}^* \bar{\sigma}_\mu \bar{e} = (-i e_R \sigma_2) \bar{\sigma}_\mu (i \sigma_2 e_R^*) = e_R \sigma_\mu^T e_R^* = -e_R^* \sigma_\mu e_R \tag{33.42}$$

where R refers to right-handed (see Eqn. 10.36), we find

$$e A^\mu \bar{\psi}_e \gamma_\mu \psi_e. \tag{33.43}$$

For Z we have

$$-\frac{g}{\cos \theta_W} Z^\mu \left[\ell^* \bar{\sigma}_\mu (T_3 - \sin^2 \theta_W Q) \ell + \bar{e}^* \sigma_\mu (T_3 - \sin^2 \theta_W Q) \bar{e} \right]$$

$$= -\frac{g}{\cos \theta_W} Z^\mu \left[\frac{1}{2} \nu_e^* \bar{\sigma}_\mu \nu_e + \left(-\frac{1}{2} + \sin^2 \theta_W \right) e^* \bar{\sigma}_\mu e - \sin^2 \theta_W \underbrace{\bar{e}^* \bar{\sigma}_\mu \bar{e}}_{-e_R^* \sigma_\mu e_R} \right]$$

$$= -\frac{g}{\cos \theta_W} Z^\mu \left[\frac{1}{2} \bar{\psi}_{\nu_e} \gamma_\mu P_L \psi_{\nu_e} + \left(-\frac{1}{2} + \sin^2 \theta_W \right) \bar{\psi}_e \gamma_\mu P_L \psi_e + \sin^2 \theta_W \bar{\psi}_e \gamma_\mu P_R \psi_e \right]$$

$$\equiv -\frac{g}{2 \cos \theta_W} Z^\mu \left[\bar{\psi}_{\nu_e} \gamma_\mu (g_V^{\nu_e} - \gamma_5 g_A^{\nu_e}) \psi_{\nu_e} + \bar{\psi}_e \gamma_\mu (g_V^e - \gamma_5 g_A^e) \psi_e \right], \tag{33.44}$$

where g_V, g_A are defined by this equation. We have

$$g_V^{\nu_e} = g_A^{\nu_e} = \frac{1}{2}, \tag{33.45}$$

$$g_V^e = \left(-\frac{1}{2} + 2 \sin^2 \theta_W \right) = -\frac{1}{2} (1 - 4 \sin^2 \theta_W),$$

$$g_A^e = -\frac{1}{2}.$$

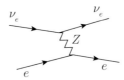

Fig. 33.3 Elastic $\nu_e\,e$ scattering.

We can re-express the relation,

$$\frac{G_F}{\sqrt{2}} \equiv \frac{g^2}{8M_W^2} = \frac{e^2}{8M_W^2\sin^2\theta_W} \tag{33.46}$$

as a formula for the W mass,

$$M_W^2 = \frac{\sqrt{2}e^2}{8G_F\sin^2\theta_W} = \frac{\pi\alpha}{\sqrt{2}G_F\sin^2\theta_W}. \tag{33.47}$$

G_F, α are measured experimentally. However, $\sin\theta_W$ is arbitrary, but can also be measured experimentally using neutral current processes mediated by the Z boson.[2] Consider the elastic $\nu_e\,e$ scattering process of Fig. 33.3. The effective four-fermi interaction, obtained at low energy, is given by

$$\frac{g^2}{4\cos^2\theta_W M_Z^2}J_{\mu Z}J_Z^\mu. \tag{33.48}$$

Note: with the Higgs mechanism we obtain one more prediction, i.e.

$$M_W = M_Z\cos\theta_W. \tag{33.49}$$

Thus, the W mass and the magnitude of the neutral current was fixed once $\sin\theta_W$ was measured.

The first neutral-current interaction was observed in 1973 at CERN using the Gargamelle detector (Hasert et al., 1973a,b). A total of 375,000 ν_μ pictures and 360,000 $\bar\nu_\mu$ pictures were taken and only one event was found (see Fig. 33.4). A high-energy proton beam was incident on a hadronic target. All of the hadronic material produced in the interaction was stopped in a steel barrier (except for neutrinos produced in the decay $\pi^- \to \mu^-\bar\nu_\mu$, $\pi^+ \to \mu^+\nu_\mu$). The neutrinos and anti-neutrinos went through. A single electron was visible in the detector, getting its energy from the neutral-current interaction. The results were consistent with the electroweak theory. Note, since the Z couplings depend on $\sin\theta_W$, the weak mixing angle and M_Z were determined.

---

[2] Searches for a massive W boson with mass of order 2 GeV and above were performed over the years. However, once $\sin\theta_W$ was measured there was now a prediction for the W mass.

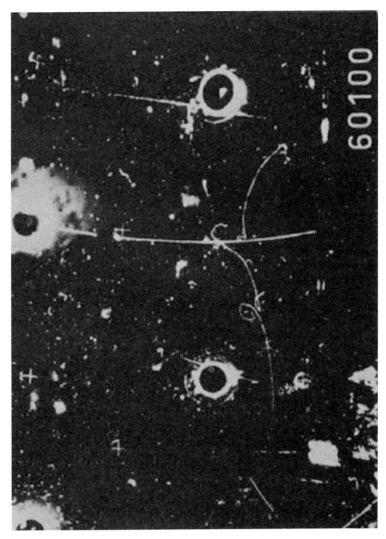

Fig. 33.4 Elastic $\bar{\nu}_\mu\ e$ scattering observed by the Gargamelle collaboration (Hasert et al., 1973b).

In summary, the electroweak interactions, given by the simple Lagrangian of Eqn. 33.17 for one family of leptons, is given in Dirac notation by the Lagrangian

$$\mathcal{L} = \bar{\psi}_{\nu L} i\gamma^\mu \partial_\mu \psi_{\nu L} + \bar{\psi}_e i\gamma^\mu \partial_\mu \psi_e \tag{33.50}$$

$$-\frac{g}{\sqrt{2}}\left[W_-^\mu\ \bar{\psi}_e \gamma_\mu P_L \psi_{\nu_e} + W_+^\mu\ \bar{\psi}_{\nu_e}\gamma_\mu P_L \psi_e\right]$$

$$+eA^\mu \bar{\psi}_e \gamma_\mu \psi_e$$

$$-\frac{g}{2\cos\theta_W}Z^\mu\left[\bar{\psi}_{\nu_e}\gamma_\mu(g_V^{\nu_e} - \gamma_5 g_A^{\nu_e})\psi_{\nu_e} + \bar{\psi}_e\gamma_\mu(g_V^e - \gamma_5 g_A^e)\psi_e\right]$$

$$-\frac{1}{4}W_{\mu\nu a}W_a^{\mu\nu} - \frac{1}{4}B_{\mu\nu}B^{\mu\nu},$$

with g_V, g_A in Eqn. 33.45. We also have

$$e = g \sin \theta_W, \quad \tan \theta_W = \frac{g'}{g}, \tag{33.51}$$

$$M_W = \left(\frac{\alpha \pi}{\sqrt{2} G_F} \right)^{1/2} \frac{1}{\sin \theta_W} = \frac{37.3 \ GeV}{\sin \theta_W}.$$

Note the Lagrangian has no mass terms. This is because any mass term violates the $SU(2) \otimes U(1)_Y$ gauge symmetry. In terms of Weyl spinors, any mass term is a bilinear in the fields. The terms of the form $\ell\ell, \ell\bar{e}$ and $\bar{e}\bar{e}$ have $U(1)_Y$ charges $-2, +1, +4$, respectively. In addition, for one family of leptons, the operator

$$\ell_{i\alpha} \ell_{j\beta} \epsilon^{\alpha\beta} \epsilon^{ij} \equiv 0, \tag{33.52}$$

is $SU(2)$ invariant, where i, j are $SU(2)$ indices and α, β are spin indices. However, it vanishes identically due to the anti-commutation relations of the spinors. The operator

$$\ell_{i\alpha} \epsilon^{ij} (T_a)_j{}^k \ell_{k\beta} \epsilon^{\alpha\beta} \tag{33.53}$$

is an $SU(2)$ vector and

$$\ell_i \bar{e} \tag{33.54}$$

is an $SU(2)$ doublet. Therefore, $SU(2) \otimes U(1)_Y$ forbids all possible mass terms. In addition, the possible mass terms for the gauge bosons,

$$M_W^2 W_{\mu+} W_-^\mu + \frac{1}{2} M_Z^2 Z_\mu Z^\mu \tag{33.55}$$

violate the gauge symmetries

$$\delta W_a^\mu = -\frac{1}{g} \partial^\mu \theta_a - \epsilon_{abc} \theta_b W_c^\mu, \tag{33.56}$$

$$\delta B^\mu = -\frac{1}{g'} \partial^\mu \theta.$$

33.4 Phenomenological Lagrangian

We can add mass terms to the Lagrangian of Eqn. 33.50 and pursue some phenomenology at the tree level. We shall see that our results are perfectly consistent with tree-level calculations in the full theory with a Higgs boson. The difference is that this theory is not renormalizable. In the next chapter we will complete the theory including the Higgs mechanism. The Lagrangian is given by Eqn. 33.50 with the addition of the mass terms

$$\mathcal{L}_{mass} = -m_e \bar{\psi}_e \psi_e + M_W^2 W_{\mu+} W_-^\mu + \frac{1}{2} M_Z^2 Z_\mu Z^\mu. \tag{33.57}$$

Let's consider the Feynman rules for this theory. We divide the Lagrangian up into a free and interacting piece. We have

$$\mathcal{L}_0 = \bar{\psi}_{\nu L} i\slashed{\partial} \psi_{\nu L} + \bar{\psi}_e (i\slashed{\partial} - m_e)\psi_e \tag{33.58}$$
$$- \frac{1}{2} W^0_{\mu\nu+} W^{\mu\nu0}_- + M^2_W W_{\mu+} W^\mu_-$$
$$- \frac{1}{4} W^{0Z}_{\mu\nu} W^{\mu\nu0Z} + \frac{1}{2} M^2_Z Z_\mu Z^\mu - \frac{1}{4} F_{\mu\nu} F^{\mu\nu}.$$

Note:

$$W^{\mu\nu0}_+ \equiv \partial^\mu W^\nu_+ - \partial^\nu W^\mu_+, \quad W^\mu_\pm \equiv \frac{W^\mu_1 \mp i W^\mu_2}{\sqrt{2}} \tag{33.59}$$

and

$$W^\mu_+ W_{\mu-} = \frac{(W^\mu_1)^2 + (W^\mu_2)^2}{2}. \tag{33.60}$$

In addition,

$$- \frac{1}{2} W^0_{\mu\nu+} W^{\mu\nu0}_- \equiv - \frac{1}{4} \left[(W^{\mu\nu0}_1)^2 + (W^{\mu\nu0}_2)^2 \right]. \tag{33.61}$$

Finally,

$$W^{\mu\nu0}_Z \equiv \partial^\mu Z^\nu - \partial^\nu Z^\mu \tag{33.62}$$

and

$$- \frac{1}{4} \left[(W^{\mu\nu0}_3)^2 + (B^{\mu\nu})^2 \right] = - \frac{1}{4} \left[(W^{\mu\nu0}_Z)^2 + (F^{\mu\nu})^2 \right]. \tag{33.63}$$

The last equation follows from the definition of A^μ, Z^μ in terms of an orthogonal transformation of W^μ_3, B^μ.

Propagators and Interactions

For the neutrino propagator we have[3]

$$i S^\nu_F = \frac{i P_L}{\slashed{p}} \equiv \text{FT} \langle T(\psi^\nu_L(x) \bar{\psi}^\nu_L(0)) \rangle; \tag{33.64}$$

since

$$P_L \psi^\nu_L \equiv \psi^\nu_L \tag{33.65}$$

it must be true that

$$P_L S^\nu_F = S^\nu_F. \tag{33.66}$$

For the Z and $W^\pm$ propagators we have (see Eqn. 11.42)

$$G_{\mu\nu}(p) = \frac{(-g_{\mu\nu} + \frac{p_\mu p_\nu}{m^2})}{p^2 - m^2 + i\epsilon}, \tag{33.67}$$

with $m = M_Z$, M_W, respectively. Note, both $Z_\mu, W^\pm_\mu$ have three helicity states, whereas a massless gauge boson has only two helicity states. We have to wonder, where did the extra three degrees of freedom come from? Nevertheless, we shall see

---

[3] FT stands for Fourier transform.

that the propagators for this phenomenological Lagrangian are exactly the same as obtained in the so-called "unitary gauge" of the full electroweak theory with Higgs bosons.

Let's first consider processes due to the charged-current interactions given in Eqn. 33.50 with vertex

$$-i\frac{g}{\sqrt{2}}(\gamma_\mu P_L)_{\alpha\beta}, \tag{33.68}$$

as shown in Fig. 33.5.

33.5 Muon Beta Decay

Given these vertices and its generalization for muons, we can calculate the muon decay amplitude

$$T(\mu^- \rightarrow e^-\bar{\nu}_e\nu_\mu) \tag{33.69}$$

given by the Feynman diagram, Fig. 33.6. We have

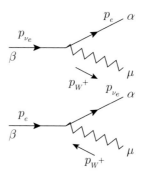

Fig. 33.5   Charged-current vertices.

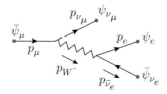

Fig. 33.6   Feynman diagram for μ^- decay. The fields defining the external legs in position space are displayed.

$$T(\mu^- \to e^- \bar{\nu}_e \nu_\mu) = \left[-\frac{ig}{\sqrt{2}} \bar{u}(p_{\nu_\mu}) \gamma_\mu P_L u(p_\mu) \right] \tag{33.70}$$

$$\times \, i \frac{\left[-g^{\mu\nu} + \frac{p_{W^-}^\mu p_{W^-}^\nu}{M_W^2} \right]}{p_W^2 - M_W^2 + i\epsilon}$$

$$\times \left[-\frac{ig}{\sqrt{2}} \bar{u}(p_e) \gamma_\nu P_L v(p_{\bar{\nu}_e}) \right],$$

where $p_{W^-} = p_e + p_{\bar{\nu}_e} = p_\mu - p_{\nu_\mu}$, $\not{p}_e u_e = m_e u_e$, $\not{p}_{\bar{\nu}_e} v_{\nu_e} = 0$ and $p_W^2 = m_\mu^2 - 2p_\mu \cdot p_{\nu_\mu} \simeq m_\mu^2 \ll M_W^2$.

Let's first check terms proportional to $p_{W^-}^\nu$ and show that they are negligible. We have

$$\frac{p_{W^-}^\nu}{M_W} \, \bar{u}(p_e) \gamma_\nu P_L v(p_{\bar{\nu}_e}) = \bar{u}(p_e)(\not{p}_e P_L + \not{p}_{\bar{\nu}_e} P_L) v(p_{\bar{\nu}_e})/M_W \tag{33.71}$$

$$= \frac{m_e}{M_W} \, \bar{u}(p_e) P_L v(p_{\bar{\nu}_e}).$$

We can neglect this term, and the one on the other vertex, since together they give a contribution of order $\left(\frac{m_e m_\mu}{M_W^2} \right) \simeq 10^{-7}$. Thus, we now have

$$T \simeq \frac{-ig^2}{8M_W^2} \left[\bar{u}_{\nu_\mu} (\gamma_\mu - \gamma_\mu \gamma_5) u_\mu \right] \left[\bar{u}_e (\gamma^\mu - \gamma^\mu \gamma_5) v_{\bar{\nu}_e} \right] \tag{33.72}$$

$$\equiv \langle e \bar{\nu}_e \nu_\mu | i\mathcal{L}_{eff} | \mu \rangle,$$

where the effective low-energy theory is given by

$$\mathcal{H}_{eff} = -\mathcal{L}_{eff} = \frac{G_F}{\sqrt{2}} \, J_\mu(e) J^\mu(\mu)^\dagger, \tag{33.73}$$

as in Eqn. 33.29.

The electron distribution in μ decay can be evaluated. Define the variable

$$x \equiv \frac{2E_e}{m_\mu}, \tag{33.74}$$

with $0 \le x \le 1$ and the angle θ defined in Fig. 33.7. Then the electron distribution

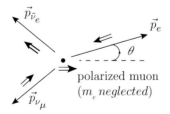

Fig. 33.7   The angle θ for μ decay is defined in terms of the electron momentum with respect to the muon spin. The double arrows are the helicity of the electron and neutrinos, neglecting their mass.

is given by

$$\frac{d^2 N_e}{dx\, d\cos\theta} \equiv \frac{1}{\Gamma_\mu}\frac{d^2\Gamma_\mu}{dx\, d\cos\theta} \tag{33.75}$$
$$= x^2[3 - 2x \mp \cos\theta(2x - 1)],$$

with the $-(+)$ sign for $\mu^-(\mu^+)$ decay. Integrating over the scattering angle θ, we obtain

$$\frac{dN_e}{dx} = \frac{1}{\Gamma_\mu}\frac{d\Gamma_\mu}{dx} = 6x^2\left[1 - \frac{2}{3}x\right]. \tag{33.76}$$

The muon decay rate (at tree level) is given by

$$\Gamma_\mu \equiv \int_0^1 dx \int_{-1}^1 d\cos\theta\, \frac{d^2\Gamma_\mu}{dx\, d\cos\theta} = \frac{G_F^2 m_\mu^5}{192\pi^3}\left(1 - \frac{8m_e^2}{m_\mu^2}\right). \tag{33.77}$$

The Fermi constant, G_F, is measured in muon decay. Given the muon lifetime $\tau_\mu = (2.19698110.0000022)\times 10^{-6}$ s (and including small radiative corrections which we shall discuss later) one obtains

$$G_F = 1.1663787(6) \times 10^{-5} \text{ GeV}^{-2}. \tag{33.78}$$

Sometimes the Fermi constant measured in muon decay is denoted in the literature by G_μ.

Our calculation used the Standard Model $V - A$ interaction, which was not initially understood to be the case. In 1950, Michel assumed, for muon decay, the two-component neutrino theory, (ν_μ, ν_e) with $m_\nu = 0$, and lepton number conservation (Michel, 1950). He then wrote down the most general four-Fermi interaction with no derivative couplings and showed that it leads to the result

$$\frac{dN_e}{dx} \equiv \frac{1}{\Gamma_\mu}\frac{d\Gamma_\mu}{dx} = 6x^2\left[\left(2 - \frac{4}{3}\rho\right) - \left(2 - \frac{16}{9}\rho\right)x\right], \tag{33.79}$$

where the Michel ρ parameter takes on values $0 \le \rho \le 1$, see Fig. 33.8. Moreover, for the $V - A$ interaction, $\rho = 3/4$. Therefore, using the spectrum of the charged lepton observable in the decay, we can test whether $\rho = 0.75$. The Michel ρ parameter was measured over the years. A plot of ρ_μ as a function of year (see Lee, 1981) is given in Fig. 33.9. It is amusing that the central value slowly increased over the years, with each subsequent measurement within the errors of the previous measurement. The measured value (Tanabashi et al., 2018) is now

$$\rho_\mu = 0.74979 \pm 0.00026. \tag{33.80}$$

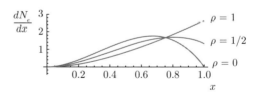

Fig. 33.8   The electron spectrum as a function of the Michel ρ parameter (using Mathematica).

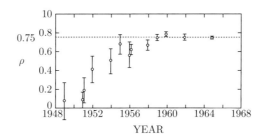

Fig. 33.9   The measurement of the Michel ρ parameter as a function of year. Republished with permission of Taylor & Francis Informa UK Ltd – Books, from "Particle Physics and Introduction to Field Theory," T.D. Lee, 01/01/1981 (Lee, 1981); permission conveyed through Copyright Clearance Center, Inc.

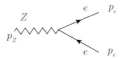

Fig. 33.10   Feynman diagram for Z decay.

For τ decay, $\tau \to \mu \bar{\nu}_\mu \nu_\tau,\ e \bar{\nu}_e \nu_\tau$, the measured value is

$$\rho_\tau(e \text{ or } \mu) = 0.745 \pm 0.008. \tag{33.81}$$

Thus, tau decay is also $V - A$.

Consider now the decay rate for $Z \to e^+ e^-$, Fig. 33.10, with $p_Z = p_e + p_{\bar{e}}$ and vertex

$$\frac{-ig}{2\cos\theta_W}\gamma_\mu(g_V^e - \gamma_5 g_A^e). \tag{33.82}$$

The decay amplitude is given by

$$T(Z \to e^+ e^-) = \frac{-ig}{2\cos\theta_W}\bar{u}(p_e)\gamma_\mu(g_V^e - \gamma_5 g_A^e)v(p_{\bar{e}})\epsilon^\mu(\vec{p}_Z, \lambda). \tag{33.83}$$

And the differential decay rate is given by

$$d\Gamma(Z \to e^+ e^-) = \frac{(2\pi)^4 \delta^4(p_e + p_{\bar{e}} - p_Z)}{2E_Z}\left(\frac{1}{3}\sum_{spins}|T|^2\right)\frac{d^3\vec{p}_e}{(2\pi)^3 2E_e}\frac{d^3\vec{p}_{\bar{e}}}{(2\pi)^3 2E_{\bar{e}}}. \tag{33.84}$$

We have

$$\begin{aligned}
|T|^2 &= \frac{g^2}{4\cos^2\theta_W}\bar{u}(p_e)\gamma_\mu(g_V^e - \gamma_5 g_A^e)v(p_{\bar{e}})\epsilon^\mu(\vec{p}_Z, \lambda)v^\dagger(p_{\bar{e}})(g_V^e - \gamma_5 g_A^e) \\
&\quad \times \gamma_\nu^\dagger \gamma_0^\dagger u(p_e)\epsilon^{\nu*}(\vec{p}_Z, \lambda) \\
&= \frac{g^2}{4\cos^2\theta_W}\bar{u}(p_e)\gamma_\mu(g_V^e - \gamma_5 g_A^e)v(p_{\bar{e}})\epsilon^\mu(\vec{p}_Z, \lambda)\ \bar{v}(p_{\bar{e}})\gamma_\nu(g_V^e - \gamma_5 g_A^e) \\
&\quad \times u(p_e)\epsilon^{\nu*}(\vec{p}_Z, \lambda).
\end{aligned} \tag{33.85}$$

Then summing over spins and averaging over the initial spin we find

$$\left(\frac{1}{3}\sum_{spins}|T|^2\right) = \frac{g^2}{12\cos^2\theta_W}\mathrm{Tr}\left[\gamma_\mu(g_V^e - \gamma_5 g_A^e)(\not{p}_{\bar{e}} - m_e)\gamma_\nu(g_V^e - \gamma_5 g_A^e)(\not{p}_e + m_e)\right]$$

$$\times\left(-g^{\mu\nu} + \frac{p_Z^\mu p_Z^\nu}{M_Z^2}\right). \tag{33.86}$$

Once again the terms containing $\frac{p_Z^\mu}{M_Z}$ can be neglected and we find

$$\left(\frac{1}{3}\sum_{spins}|T|^2\right) \simeq \frac{-g^2}{12\cos^2\theta_W}\mathrm{Tr}\left[\gamma_\mu(g_V^e - \gamma_5 g_A^e)(\not{p}_{\bar{e}} - m_e)\gamma^\mu(g_V^e - \gamma_5 g_A^e)(\not{p}_e + m_e)\right]$$

$$= \frac{-g^2}{12\cos^2\theta_W}\left\{\mathrm{Tr}\left[\gamma_\mu(\not{p}_{\bar{e}} - m_e)\gamma^\mu(\not{p}_e + m_e)\right](g_V^e)^2\right.$$

$$- \mathrm{Tr}\left[\gamma_\mu\gamma_5(\not{p}_{\bar{e}} - m_e)\gamma^\mu(\not{p}_e + m_e)\right](g_A^e g_V^e)$$

$$- \mathrm{Tr}\left[\gamma_\mu(\not{p}_{\bar{e}} - m_e)\gamma^\mu\gamma_5(\not{p}_e + m_e)\right](g_A^e g_V^e)$$

$$\left.+ \mathrm{Tr}\left[\gamma_\mu\gamma_5(\not{p}_{\bar{e}} - m_e)\gamma^\mu\gamma_5(\not{p}_e + m_e)\right](g_A^e)^2\right\}. \tag{33.87}$$

The first and fourth term inside the brackets give

$$- 16\left(\frac{p_e \cdot p_{\bar{e}}}{2} + m_e^2\right) \tag{33.88}$$

and the second and third terms vanish, since the only non-zero term with γ_5 is given by

$$\mathrm{Tr}(\gamma_\mu\gamma_\alpha\gamma_\nu\gamma_\beta\gamma_5) = 4i\epsilon_{\mu\alpha\nu\beta}. \tag{33.89}$$

Thus,

$$\mathrm{Tr}(\gamma_\mu\gamma_\alpha\gamma^\mu\gamma_\beta\gamma_5) \equiv 0. \tag{33.90}$$

We therefore find

$$\left(\frac{1}{3}\sum_{spins}|T|^2\right) = \frac{8g^2}{12\cos^2\theta_W}(p_e \cdot p_{\bar{e}} + 2m_e^2)[(g_V^e)^2 + (g_A^e)^2]. \tag{33.91}$$

Using $M_Z^2 = p_Z^2 = (p_e + p_{\bar{e}})^2 = 2m_e^2 + 2p_e \cdot p_{\bar{e}}$, we have $p_e \cdot p_{\bar{e}} = \frac{M_Z^2 - 2m_e^2}{2} \simeq \frac{M_Z^2}{2}$. Hence, we have

$$\left(\frac{1}{3}\sum_{spins}|T|^2\right) - \frac{g^2 M_Z^2}{3\cos^2\theta_W}[(g_V^e)^2 + (g_A^e)^2] \tag{33.92}$$

and

$$\Gamma(Z \to e^+ e^-)|_{rest} = \frac{\left(\frac{1}{3}\sum_{spins}|T|^2\right)}{16\pi M_Z} = \frac{g^2 M_Z}{48\pi\cos^2\theta_W}[(g_V^e)^2 + (g_A^e)^2]. \tag{33.93}$$

Table 33.3 The quarks and leptons transform under the Standard Model gauge group	
Fields	$SU(3) \otimes SU(2) \otimes U(1)_Y$
$\ell \equiv \begin{pmatrix} \nu_e \\ e \end{pmatrix}$	$(1, 2, -1)$
$\bar{e}$	$(1, 1, 2)$
$q \equiv \begin{pmatrix} u \\ d \end{pmatrix}$	$(3, 2, \frac{1}{3})$
$\bar{u}$	$(\bar{3}, 1, -\frac{4}{3})$
$\bar{d}$	$(\bar{3}, 1, \frac{2}{3})$

Using the result of Eqn. 34.66, we can rewrite the coefficient

$$\frac{g^2 M_Z}{48\pi \cos^2 \theta_W} = \frac{G_F M_Z^3}{6\pi\sqrt{2}} \tag{33.94}$$

and we obtain the final result

$$\Gamma(Z \to e^+ e^-)|_{rest} = \frac{G_F M_Z^3}{6\pi\sqrt{2}}[(g_V^e)^2 + (g_A^e)^2]. \tag{33.95}$$

Given G_F (Eqn. 33.78), $M_Z \sim 91.1$ GeV, and $\sin^2 \theta_W \sim 0.23$, we find

$$\Gamma(Z \to e^+ e^-) \sim 84 \text{ MeV}. \tag{33.96}$$

Note: the formulae for Z decay into any fermion is easily obtained. We have

$$\Gamma(Z \to \nu_e \bar{\nu}_e)|_{rest} = \frac{G_F M_Z^3}{6\pi\sqrt{2}}[(g_V^\nu)^2 + (g_A^\nu)^2] \tag{33.97}$$

and

$$\Gamma(Z \to q\bar{q})|_{rest} = \frac{3 G_F M_Z^3}{6\pi\sqrt{2}}[(g_V^q)^2 + (g_A^q)^2]. \tag{33.98}$$

Under the Standard Model gauge symmetry $SU(3) \otimes SU(2) \otimes U(1)_Y$, the quarks and leptons transform as given in Table 33.3. It is straightforward to evaluate the vector and axial-vector couplings of quarks to the Z. We have

$$g_V^u = \frac{1}{2} - \frac{4}{3}\sin^2 \theta_W, \tag{33.99}$$

$$g_A^u = \frac{1}{2},$$

$$g_V^d = -\frac{1}{2} + \frac{2}{3}\sin^2 \theta_W,$$

$$g_A^d = -\frac{1}{2}.$$

The Dirac spinors are defined by

$$\psi_u \equiv \begin{pmatrix} u \\ i\sigma_2 \bar{u}^* \end{pmatrix} \tag{33.100}$$

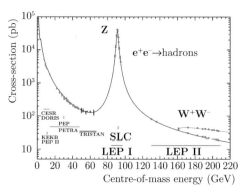

Fig. 33.11 The Z peak in this data is from the analysis of $e^+e^- \to \mu^+\mu^-$ scattering with the Aleph detector (Schael et al., 2006) at the LEP. The lower-energy data come from experiments from many other colliders. Note, Tristan was built with the goal of finding the top quark, but its peak center-of-mass energy was below the top-quark threshold. For a color version of this figure, please see the color plate section.

and

$$\psi_d \equiv \begin{pmatrix} d \\ i\sigma_2 \bar{d}^* \end{pmatrix}. \tag{33.101}$$

33.6 Discovery of the $W^\pm$ and Z^0 Bosons

Beginning in 1983 at the CERN $\bar{p}p$ collider, the UA1 (Arnison et al., 1983a,b,c, 1984a,b) and UA2 collaborations (Banner et al., 1983; Bagnaia et al., 1983, 1984) reported observations of high-energy electrons and muons due to the production and decay of the W boson and high-energy electron and muon pairs due to the production and decay of the Z boson. The W and Z mass are now measured to be 80.379 ± 0.012 GeV and 91.1876 ± 0.0021 GeV, respectively (Tanabashi et al., 2018). The Z mass was measured much better at the CERN e^+e^- colliders, the Large Electron–Positron (LEP) and LEPII. The total width of the Z boson is 2.4952 ± 0.0023 GeV. In addition, all the visible partial decay modes of the Z boson have been measured leading to an invisible width of the Z boson, 499.0 ± 1.5 MeV which is consistent with just three light neutrinos coupled to the Z. In Fig. 33.11, we show the Z peak from the Aleph detector at the LEP (Schael et al., 2006).

34 Electroweak Symmetry Breaking

In Section 25.2, we discussed the spontaneous symmetry breaking of a continuous, global symmetry. If you recall, we found that there exists a massless Nambu–Goldstone boson for each broken symmetry generator. In a particular Gell-Mann–Levy model which was constructed with an $SU(2)_L \otimes SU(2)_R$ chiral symmetry for nucleons and the Σ field, we saw that the nucleon was massless at tree level. However, after spontaneously breaking $SU(2)_L \otimes SU(2)_R \rightarrow SU(2)_I$, i.e. isospin, the nucleons obtain mass $m_N = gF_\pi$ and there were three massless Nambu–Goldstone bosons identified as the pion isospin triplet.

In this chapter, we study the spontaneous breaking of a continuous, local symmetry and the resultant Higgs mechanism. As a start, consider the simple case of an ordinary metal (conductor). There exists Debye screening, since we have

$$-\vec{\nabla}^2 A_0 = \langle e j_0 e^{-\frac{\int d^3 \vec{y} e j_0 A_0}{T}} \rangle_T + e j_0^{ext} \tag{34.1}$$
$$\simeq -\frac{n_e e^2}{T} A_0 + e j_0^{ext}.$$

The Debye screening length is given by

$$\lambda_{Debye}^{-1} = \left(\frac{n_e e^2}{T} \right)^{1/2}. \tag{34.2}$$

As a consequence, the time component of the gauge field obtains an effective mass and there are no electric fields interior to the conductor. On the other hand, a magnetic field does enter an ordinary conductor and the photon is massless, i.e. the transverse degrees of freedom are not affected by Debye screening.

The ground state of a conductor $|g\rangle$ satisfies,

$$Q|g\rangle = 0, \tag{34.3}$$

where $Q = \int d^3\vec{x} \; j^0(x)$ is the charge operator. Hence, the net charge in the conductor is zero. Now consider a superconductor. We have

$$(\partial_t^2 - \vec{\nabla}^2) A_i = e j_i. \tag{34.4}$$

Define the Cooper pair operator

$$C \equiv e^\dagger(\vec{k}_F) e^\dagger(-\vec{k}_F), \tag{34.5}$$

where

$$[Q, \; C] = -2C. \tag{34.6}$$

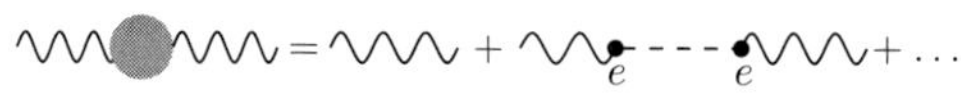

Fig. 34.1 The full photon propagator iG_{ij} is obtained by summing over all charge-density insertions.

In a conductor we have

$$\langle g|C|g\rangle \equiv 0. \tag{34.7}$$

In a superconductor, for $T < T_C$ we have

$$\Delta \equiv \langle g|C|g\rangle \neq 0. \tag{34.8}$$

As a consequence

$$Q|g\rangle \neq 0 \tag{34.9}$$

and $U(1)_{QED}$ is spontaneously broken. Define the amplitude

$$\langle \vec{k}|j^0(0)|g\rangle = if\omega, \tag{34.10}$$

where the charge density creates the state $|\vec{k}\rangle$, i.e. a charge-density wave. If we set the electric charge e to zero, then the charge-density wave is a massless Nambu–Goldstone boson due to spontaneously breaking the global $U(1)$ symmetry. Now turn on e and the charge-density wave couples to the photon and gives a contribution to the photon polarization tensor. We have

$$i\Pi^{ij} = -i(\vec{k}^2\delta^{ij} - k^ik^j)\Pi(\vec{k}^2) \equiv -i\vec{k}^2\Delta^{ij}\Pi(\vec{k}^2) \tag{34.11}$$
$$= FT\langle g|T(j^i(x)j^j(0))|g\rangle.$$

Graphically, the full photon propagator is given by the geometric sum of all charge-density wave insertions, as in Fig. 34.1. In general,

$$\langle \vec{k}|j^\mu(0)|g\rangle = ifk^\mu, \tag{34.12}$$

with $k^\mu = (\omega, \vec{k})$ and gauge invariance, i.e. $\partial_\mu j^\mu(0) = 0$ requires $k^2 = 0$. Therefore the contribution of the charge-density wave to the polarization tensor is given by

$$\Pi(k^2) = -f^2/k^2 = \frac{f^2}{\vec{k}^2} \quad \text{for} \quad \omega = 0. \tag{34.13}$$

We then have

$$iG_{ij} = -i\frac{\Delta_{ij}}{\vec{k}^2} + \left(-i\frac{\Delta_{im}}{\vec{k}^2}\right)\left[-i\vec{k}^2\Delta_{mn}e^2\Pi(\vec{k}^2)\right]\left(-i\frac{\Delta_{nj}}{\vec{k}^2}\right) + \dots . \tag{34.14}$$

Using the identity

$$\Delta_{km}\Delta_{mj} = \Delta_{kj} \tag{34.15}$$

we have

$$iG_{ij} = \frac{-i\Delta_{ij}}{\vec{k}^{\,2}} \left[1 + \sum_{n=1}^{\infty} (-e^2\Pi(\vec{k}^{\,2}))^n \right] \tag{34.16}$$

$$= \frac{-i\Delta_{ij}}{\vec{k}^{\,2}(1 + e^2\Pi(\vec{k}^{\,2}))} = \frac{-i\Delta_{ij}}{\vec{k}^{\,2} + m_\gamma^2}, \tag{34.17}$$

where

$$m_\gamma^2 = e^2 f^2, \tag{34.18}$$

i.e. the photon obtains a mass !! This is the Higgs mechanism for a superconductor and as a consequence we have the Meissner effect. Given the equation of motion for the vector potential, we have

$$-\vec{\nabla}^2 A_i = -m_\gamma^2 A_i + e j_i^{ext}, \tag{34.19}$$

with the solution

$$A_i(\vec{x}) = -e \int d^3\vec{y} \frac{e^{-m_\gamma|\vec{x}-\vec{y}|}}{|\vec{x}-\vec{y}|} j_i(\vec{y})^{ext}. \tag{34.20}$$

Hence, magnetic fields don't penetrate into the superconductor and there is a skin depth given by

$$\lambda = \frac{\hbar}{m_\gamma c}. \tag{34.21}$$

34.1 The Higgs Mechanism of Electroweak Symmetry Breaking

Consider a scalar field ϕ transforming under $SU(2) \otimes U(1)_Y$ as $(2, 1)$, i.e. a doublet under $SU(2)$ with hypercharge, $Y = 1$. We then have

$$\phi = \begin{pmatrix} \phi^+ \\ \phi^0 \end{pmatrix}. \tag{34.22}$$

The generators acting on the field satisfy

$$T_a\phi = \frac{\tau_a}{2}\phi, \quad Y\phi = \phi \tag{34.23}$$

and the electric charge is determined by $Q = T_3 + \frac{Y}{2}$ as usual. Therefore, $Q\phi^+ = \phi^+$, $Q\phi^0 = 0$.

We can now write a gauge-invariant coupling of ϕ to the electron and neutrino. We have

$$\mathcal{L}_e = \lambda_e \bar{e}\phi^\dagger \ell + h.c. \tag{34.24}$$

We can check gauge invariance, given the transformation of the fields under $SU(2) \otimes U(1)_Y$ by

$$\ell' = U\ell, \tag{34.25}$$

$$\phi' = U\phi,$$

$$\bar{e}' = U\bar{e},$$

with

$$U = e^{i(T_a \theta_a + \frac{Y}{2}\theta)} \tag{34.26}$$

and the charges of the fields, Table 33.2. Multiplying out the terms, we get

$$\mathcal{L}_e = \lambda_e \bar{e} \, (\phi^{+\dagger} \phi^{0\dagger}) \begin{pmatrix} \nu_e \\ e \end{pmatrix} + h.c \tag{34.27}$$

$$= \lambda_e \bar{e} \, (\phi^{+\dagger} \nu_e + \phi^{0\dagger} e) + h.c$$

Note, if

$$\langle \phi^0 \rangle = \frac{v}{\sqrt{2}}, \quad \langle \phi^+ \rangle = 0, \tag{34.28}$$

then $SU(2) \otimes U(1)_Y$ is spontaneously broken, leaving $U(1)_{QED}$ unbroken. As a consequence, we would expect three massless Nambu–Goldstone bosons (assuming $g = g' = 0$). In addition, we have a massive electron with

$$m_e = \frac{\lambda_e v}{\sqrt{2}} \tag{34.29}$$

and a massless neutrino.

Now consider the full Lagrangian,

$$\mathcal{L} = \mathcal{L}_{gauge} + \mathcal{L}_{scalar} + \mathcal{L}_{fermion}. \tag{34.30}$$

We have

$$\mathcal{L}_{fermion} = \ell^\dagger (i\bar{\sigma}_\mu D^\mu)\ell + \bar{e}^\dagger (i\bar{\sigma}_\mu D^\mu)\bar{e} + \mathcal{L}_e. \tag{34.31}$$

For the scalar Lagrangian we take

$$\mathcal{L}_{scalar} = (D_\mu \phi)^\dagger (D^\mu \phi) - V(\phi), \tag{34.32}$$

with

$$D^\mu \phi = (\partial^\mu + igW_a^\mu T_a + i\frac{g'}{2}B^\mu Y)\phi \tag{34.33}$$

and

$$V(\phi) = \frac{\lambda}{2} \left(\phi^\dagger \phi - \frac{v^2}{2} \right)^2. \tag{34.34}$$

V is renormalizable and gauge invariant. At the minimum of the potential, the scalar field ϕ necessarily has a non-zero vacuum expectation value (VEV).

We now show that we can always choose the expectation value for ϕ to be in the ϕ^0 direction and real. Note, we are free to use global $SU(2) \otimes U(1)_Y$ rotations without any loss of generality.

Proof Let

$$\langle\phi\rangle = \begin{pmatrix} a \\ b \end{pmatrix}, \tag{34.35}$$

where a, b are arbitrary complex constants given by $a \equiv e^{i\theta_1} a_r$, $b \equiv e^{i\theta_2} b_r$. Now we can use a global $SU(2)$ rotation to obtain

$$\langle\phi'\rangle = \begin{pmatrix} e^{i\frac{\omega_3}{2}} a \\ e^{-i\frac{\omega_3}{2}} b \end{pmatrix} \tag{34.36}$$

and choose

$$\frac{\omega_3}{2} + \theta_1 = \theta_2 - \frac{\omega_3}{2} \equiv \alpha. \tag{34.37}$$

We have

$$\omega_3 = \theta_2 - \theta_1, \quad \alpha = \frac{\theta_1 + \theta_2}{2} \tag{34.38}$$

and

$$\langle\phi'\rangle = e^{i\alpha} \begin{pmatrix} a_r \\ b_r \end{pmatrix}. \tag{34.39}$$

Then, with a second $SU(2)$ rotation given by

$$e^{i\omega_2 \frac{\tau_2}{2}} \equiv \cos\frac{\omega_2}{2} + i\tau_2 \sin\frac{\omega_2}{2}, \tag{34.40}$$

we have

$$\begin{aligned}
\langle\phi''\rangle &= \begin{pmatrix} \cos\frac{\omega_2}{2} & \sin\frac{\omega_2}{2} \\ -\sin\frac{\omega_2}{2} & \cos\frac{\omega_2}{2} \end{pmatrix} e^{i\alpha} \begin{pmatrix} a_r \\ b_r \end{pmatrix} \\
&= e^{i\alpha} \begin{pmatrix} 0 \\ \frac{v}{\sqrt{2}} \end{pmatrix},
\end{aligned} \tag{34.41}$$

with v real. In this case, the rotation angle ω_2 satisfies

$$\tan\frac{\omega_2}{2} = -\frac{a_r}{b_r} \tag{34.42}$$

and

$$\frac{v}{\sqrt{2}} = (a_r^2 + b_r^2)^{1/2}. \tag{34.43}$$

Finally, with a third $SU(2)$ rotation we have

$$\langle\phi'''\rangle = e^{i\alpha} \begin{pmatrix} 0 \\ e^{-i\frac{\omega_3'}{2}} \frac{v}{\sqrt{2}} \end{pmatrix} \equiv \begin{pmatrix} 0 \\ \frac{v}{\sqrt{2}} \end{pmatrix}, \tag{34.44}$$

with $\omega_3' = 2\alpha$. $\qquad\qquad\square$

Now let's consider the spectrum of scalar states (assuming for the moment that $g = g' = 0$). Since the scalar doublet is given in terms of two complex scalar fields, we can always define the four real scalar fields by

$$\phi^0 \equiv \frac{\varphi_1 + i\varphi_2}{\sqrt{2}}, \quad \phi^+ \equiv \frac{\varphi_3 + i\varphi_4}{\sqrt{2}}. \tag{34.45}$$

In terms of the four real scalar fields, the scalar potential is now given by

$$V(\phi) = \frac{\lambda}{8} \left(\sum_{i=1}^{4} \varphi_i^2 - v^2 \right)^2. \tag{34.46}$$

And clearly at the minimum of the potential we can choose the expectation values

$$\langle \varphi_1 \rangle \equiv v, \quad \langle \varphi_i \rangle \equiv 0, \ i = 2, 3, 4. \tag{34.47}$$

In order to perturb around the minimum we define the shifted scalar fields,

$$\varphi_1 = \tilde{\varphi}_1 + v, \quad \varphi_i = \tilde{\varphi}_i, \ i = 2, 3, 4. \tag{34.48}$$

We then have

$$V(\tilde{\varphi}) = \frac{\lambda}{8} \left((\tilde{\varphi}_1 + v)^2 + \sum_{i=2}^{4} (\tilde{\varphi}_i)^2 - v^2 \right)^2 \tag{34.49}$$

$$= \frac{\lambda}{8} \left(\sum_{i=1}^{4} \tilde{\varphi}_i^2 + 2v\tilde{\varphi}_1 \right)^2$$

$$= \frac{\lambda}{8} \left[\left(\sum_{i=1}^{4} \tilde{\varphi}_i^2 \right)^2 + 4v\tilde{\varphi}_1 \sum_{i=1}^{4} \tilde{\varphi}_i^2 + 4v^2\tilde{\varphi}_1^2 \right].$$

Clearly,

$$m_{\tilde{\varphi}_1}^2 = \lambda v^2 \tag{34.50}$$

and we will identify

$$\tilde{\varphi}_1 \equiv h \tag{34.51}$$

as the physical massive Higgs field. For all the others, we have

$$m_{\tilde{\varphi}_i} = 0, \quad i = 2, 3, 4, \tag{34.52}$$

i.e. the three Nambu–Goldstone bosons as expected.

Now turn on the gauge couplings, g, g' and evaluate the kinetic term for the scalar field. Define the VEV

$$\phi_V \equiv \begin{pmatrix} 0 \\ \frac{v}{\sqrt{2}} \end{pmatrix} \tag{34.53}$$

and the shifted doublet

$$\phi = \phi_V + \tilde{\phi}. \tag{34.54}$$

Then the kinetic term is given by

$$(D_\mu\phi)^\dagger(D^\mu\phi) = (D_\mu\phi_V + D_\mu\tilde\phi)^\dagger(D^\mu\phi_V + D^\mu\tilde\phi) \tag{34.55}$$
$$= (D_\mu\phi_V)^\dagger(D^\mu\phi_V) \tag{1}$$
$$+ \left[(D_\mu\tilde\phi)^\dagger(D^\mu\phi_V) + h.c.\right] \tag{2}$$
$$+ (D_\mu\tilde\phi)^\dagger(D^\mu\tilde\phi). \tag{3}$$

Consider the first term. We have

$$(D_\mu\phi_V)^\dagger(D^\mu\phi_V) \equiv (\dots)\left(igW_a^\mu\frac{\tau_a}{2} + i\frac{g'}{2}B^\mu\right)\phi_V \tag{34.56}$$
$$= \phi_V^\dagger\left[\left(gW_{\mu a}\frac{\tau_a}{2} + \frac{g'}{2}B_\mu\right)\left(gW_b^\mu\frac{\tau_b}{2} + \frac{g'}{2}B^\mu\right)\right]\phi_V$$
$$= \frac{1}{2}W_{\mu a}W_b^\mu\phi_V^\dagger\left(\frac{g^2}{4}\{\tau_a,\ \tau_b\}\right)\phi_V$$
$$+ W_{\mu a}B^\mu\phi_V^\dagger\left(\frac{2gg'}{4}\tau_a\right)\phi_V \tag{34.57}$$
$$+ B_\mu B^\mu\phi_V^\dagger\left(\frac{g'^2}{4}\mathbb{I}\right)\phi_V.$$

Now using the identities

$$\{\tau_a,\ \tau_b\} = 2\delta_{ab}\mathbb{I}, \tag{34.58}$$
$$\tau^\pm \equiv \frac{\tau_1 \pm i\tau_2}{2},$$
$$W_\mu^\pm \equiv \frac{W_\mu^1 \mp iW_\mu^2}{\sqrt{2}},$$
$$W_{\mu a}\tau_a \equiv \sqrt{2}(W_\mu^+\tau^+ + W_\mu^-\tau^-) + W_{\mu 3}\tau_3,$$

we obtain

$$(D_\mu\phi_V)^\dagger(D^\mu\phi_V) = \frac{1}{2}W_{\mu a}W_a^\mu\left(\frac{g^2v^2}{4}\right) + W_{\mu 3}B^\mu\left(-\frac{gg'v^2}{4}\right) + B_\mu B^\mu\left(\frac{g'^2v^2}{8}\right). \tag{34.59}$$

We can identify the charged W mass, since

$$\frac{1}{2}\left(W_{\mu 1}W_1^\mu + W_{\mu 2}W_2^\mu\right) \equiv W_{\mu+}W_-^\mu \tag{34.60}$$

and thus

$$M_W = \frac{gv}{2}. \tag{34.61}$$

For the neutral gauge boson, we have the mass term

$$(W_{\mu 3}\ B_\mu)\begin{pmatrix} g^2 & -gg' \\ -gg' & g'^2 \end{pmatrix}\begin{pmatrix} W_3^\mu \\ B^\mu \end{pmatrix}\frac{v^2}{8}. \tag{34.62}$$

Using $\tan\theta_W = g'/g$ and

$$\sin\theta_W \equiv \frac{g'}{\sqrt{g^2 + g'^2}}, \quad \cos\theta_W \equiv \frac{g}{\sqrt{g^2 + g'^2}} \tag{34.63}$$

we have

$$(W_{\mu 3}B_\mu)\begin{pmatrix} \cos^2\theta_W & -\cos\theta_W\sin\theta_W \\ -\cos\theta_W\sin\theta_W & \sin^2\theta_W \end{pmatrix}\begin{pmatrix} W_3^\mu \\ B^\mu \end{pmatrix}(g^2+g'^2)\frac{v^2}{8} \equiv \frac{1}{2}M_Z^2 Z_\mu Z^\mu \tag{34.64}$$

with

$$M_Z \equiv \frac{\sqrt{g^2 + g'^2}\; v}{2}. \tag{34.65}$$

In the last step we used the defining equation for the Z boson, Eqn. 33.33, with $Z^\mu = \cos\theta_W W_3^\mu - \sin\theta_W B^\mu$. Note, the following important relation:

$$M_W \equiv M_Z \cos\theta_W. \tag{34.66}$$

In addition, the photon remains massless!!

To understand this last fact a bit better, let's derive the neutral mass term again, but now using the relation, Eqn. 33.32. Then, for the neutral bosons, W_3^μ, B^μ, Eqn. 34.56 becomes

$$\phi_V^\dagger\left(eA_\mu Q + Z_\mu Q_Z\right)\left(eA^\mu Q + Z^\mu Q_Z\right)\phi_V. \tag{34.67}$$

Using the fact that

$$Q\phi_V \equiv 0, \tag{34.68}$$

this reduces to

$$\phi_V^\dagger Q_Z^2 \phi_V Z_\mu Z^\mu, \tag{34.69}$$

with $Q_Z = \frac{g}{\cos\theta_W}(T_3 - \sin^2\theta_W Q)$ and the photon remains massless because $U(1)_{QED}$ is unbroken. We thus obtain the mass term

$$\frac{g^2}{4\cos^2\theta_W}\,\phi_V^\dagger \tau_3^2 \phi_V\, Z_\mu Z^\mu = \frac{g^2 v^2}{8\cos^2\theta_W} Z_\mu Z^\mu \equiv \frac{1}{2}M_Z^2 Z_\mu Z^\mu \tag{34.70}$$

and the relation,

$$M_Z = \frac{gv}{2\cos\theta_W} \equiv M_W/\cos\theta_W. \tag{34.71}$$

Now consider the last two terms of Eqn. 34.55,

$$\left[(D_\mu\tilde\phi)^\dagger(D^\mu\phi_V) + h.c.\right] + (D_\mu\tilde\phi)^\dagger(D^\mu\tilde\phi) \tag{34.72}$$

$$= i(\partial_\mu\tilde\phi)^\dagger(gW_a^\mu T_a + \tfrac{g'}{2}B^\mu Y)\phi_V + h.c. + (\partial_\mu\tilde\phi)^\dagger(\partial^\mu\tilde\phi) + \text{interactions}.$$

The first term corresponds to a mixing between the Nambu–Goldstone boson and gauge boson fields. In more detail, we have

$$i(\partial_\mu \tilde{\phi})^\dagger \left[\frac{g}{\sqrt{2}}(W_+^\mu \tau^+ + W_-^\mu \tau^-) + (eA^\mu Q + Z^\mu Q_Z) \right] \phi_V + h.c.$$

$$= i(\partial_\mu \tilde{\phi}^+)^\dagger W_+^\mu (\tfrac{gv}{2}) - i(\partial_\mu \tilde{\phi}^0)^\dagger \frac{Z^\mu}{\sqrt{2}} \left(\frac{gv}{2\cos\theta_W} \right) + h.c. \qquad (34.73)$$

$$= i(\partial_\mu \tilde{\phi}^+)^\dagger W_+^\mu M_W + i(\partial_\mu \tilde{\phi}^-)^\dagger W_-^\mu M_W + i\partial_\mu \underbrace{\left(\frac{\tilde{\phi}^0 - \tilde{\phi}^{0\,\dagger}}{\sqrt{2}} \right)}_{i\tilde{\varphi}_2} Z^\mu M_Z.$$

Note, $\tilde{\varphi}_1 \equiv h$ does not mix at all.

34.2 "Unitary" Gauge

In this section we shall define the "unitary" gauge. In the same way that we showed that, given the most general VEV, $\langle\phi\rangle = \begin{pmatrix} a \\ b \end{pmatrix}$ can be rotated into the direction $\langle\phi\rangle = \begin{pmatrix} 0 \\ \frac{v}{\sqrt{2}} \end{pmatrix}$ using global $SU(2) \otimes U(1)_Y$ transformations, we can now use local gauge transformations to rotate $\tilde{\varphi}_i$, $i = 2,3,4$ to zero. This becomes clearer if instead of writing

$$\phi(x) = \tilde{\phi}(x) + \phi_V \qquad (34.74)$$

we write

$$\phi(x) = e^{i\frac{\tau_a}{2}\chi_a(x)} \begin{pmatrix} 0 \\ \frac{v+h(x)}{\sqrt{2}} \end{pmatrix}, \qquad (34.75)$$

where χ_a, $a = 1,2,3$ are re-definitions of the fields, $\tilde{\varphi}_i$, $i = 2,3,4$. We can then use local $SU(2)$ rotations,

$$\phi'(x) = e^{-i\frac{\tau_a}{2}\chi_a(x)}\phi(x), \qquad (34.76)$$

where $\theta_a(x) \equiv -\chi_a(x)$ (see Eqn. 34.26) are local gauge parameters such that

$$\phi'(x) = \begin{pmatrix} 0 \\ \frac{v+h(x)}{\sqrt{2}} \end{pmatrix}. \qquad (34.77)$$

In the "unitary" gauge the Lagrangian takes the form

$$\mathcal{L} = \mathcal{L}_{gauge} + (D_\mu\phi)^\dagger(D^\mu\phi) - V(\phi) + \mathcal{L}_{fermion}. \qquad (34.78)$$

Then, with the transformations to the unitary gauge given by

$$\ell'(x) = U(x)\ell(x) \equiv e^{-iT_a\chi_a(x)}\ell(x), \qquad (34.79)$$

$$\bar{e}'(x) = \bar{e}(x),$$

$$\tilde{W}'_\mu(x) = U(x)W_\mu(x)U^\dagger(x) + \frac{i}{g}(\partial_\mu U(x))U^\dagger(x),$$

we obtain

$$\mathcal{L} = \mathcal{L}_{gauge} \tag{34.80}$$

$$+ \left(D_\mu \begin{pmatrix} 0 \\ \frac{v+h(x)}{\sqrt{2}} \end{pmatrix}\right)^\dagger \left(D^\mu \begin{pmatrix} 0 \\ \frac{v+h(x)}{\sqrt{2}} \end{pmatrix}\right) - \frac{\lambda}{8}\left((v+h(x))^2 - v^2\right)^2$$

$$+ \mathcal{L}_{fermion} - \frac{\lambda_e(v+h(x))}{\sqrt{2}}\bar{\psi}_e\psi_e$$

$$= \mathcal{L}_{gauge} + M_W^2\, W_\mu^+ W^{\mu-} + \frac{1}{2}M_Z^2 Z_\mu Z^\mu$$

$$+ \frac{1}{2}\left(D_\mu \begin{pmatrix} 0 \\ h(x) \end{pmatrix}\right)^\dagger\left(D^\mu \begin{pmatrix} 0 \\ h(x) \end{pmatrix}\right) + \frac{1}{\sqrt{2}}\left(D_\mu \begin{pmatrix} 0 \\ h(x) \end{pmatrix}\right)^\dagger(D^\mu\phi_V) + h.c.$$

$$- \frac{\lambda}{8}\left((v+h(x))^2 - v^2\right)^2$$

$$+ \mathcal{L}_{fermion} - \frac{\lambda_e(v+h(x))}{\sqrt{2}}\bar{\psi}_e\psi_e.$$

The second line gives

$$\frac{1}{2}(\partial_\mu h)^2 + \frac{1}{2}(0 \;\; \partial_\mu h)(igT_a W_a^\mu + i\frac{g'}{2}B^\mu Y)\begin{pmatrix} 0 \\ h \end{pmatrix} \tag{34.81}$$

$$- \frac{1}{2}(0 \;\; h)(igT_a W_{\mu a} + i\frac{g'}{2}B_\mu Y)\begin{pmatrix} 0 \\ \partial^\mu h \end{pmatrix}$$

$$+ \frac{1}{2}(0 \;\; h)(gT_a W_{\mu a} + \frac{g'}{2}B_\mu Y)^2\begin{pmatrix} 0 \\ h \end{pmatrix}$$

$$+ \frac{1}{2}(0 \;\; h)(gT_a W_{\mu a} + \frac{g'}{2}B_\mu Y)^2\begin{pmatrix} 0 \\ v \end{pmatrix}.$$

Putting it all together we have

$$\mathcal{L} = \mathcal{L}_{gauge} + M_W^2\left(1 + \frac{h}{v}\right)^2 W_\mu^+ W^{\mu-} + \frac{1}{2}M_Z^2\left(1 + \frac{h}{v}\right)^2 Z_\mu Z^\mu$$

$$+ \frac{1}{2}(\partial_\mu h)^2 - \frac{1}{2}m_h^2 h^2 - \frac{\lambda}{8}(h^4 + 4vh^3) \tag{34.82}$$

$$+ \bar{\psi}_{\nu L}i\gamma^\mu\partial_\mu\psi_{\nu L} + \bar{\psi}_e i\gamma^\mu\partial_\mu\psi_e - m_e\left(1 + \frac{h}{v}\right)\bar{\psi}_e\psi_e$$

$$- \frac{g}{\sqrt{2}}\left[W_-^\mu\, \bar{\psi}_e\gamma_\mu P_L\psi_{\nu_e} + W_+^\mu\, \bar{\psi}_{\nu_e}\gamma_\mu P_L\psi_e\right]$$

$$+ eA^\mu\bar{\psi}_e\gamma_\mu\psi_e$$

$$- \frac{g}{2\cos\theta}Z^\mu\left[\bar{\psi}_{\nu_e}\gamma_\mu(g_V^{\nu_e} - \gamma_5 g_A^{\nu_e})\psi_{\nu_e} + \bar{\psi}_e\gamma_\mu(g_V^e - \gamma_5 g_A^e)\psi_e\right].$$

Clearly, the Higgs boson couples to matter proportional to mass. Note, in the "unitary" gauge, the Feynman rules for the gauge and fermion sector are identical to those studied in Chapter 33, Section 33.4. The novel physics with the Higgs mechanism is the relation between the W and Z mass, Eqn. 34.66, and the Higgs boson. In addition, the theory is now renormalizable. However, this is not easy to see in the "unitary" gauge, although the simple argument is that the theory is renormalizable by power counting prior to spontaneous symmetry breaking. Moreover, spontaneous symmetry breaking only affects the infrared behavior of Green's functions. Hence, the theory should still be renormalizable. Nevertheless, most higher-loop calculations in the electroweak theory are carried out in the so-called 't Hooft–R_ξ gauge, which we discuss now.

34.3 't Hooft–R_ξ Gauge

Recall the first term in Eqn. 34.72. We have

$$
\left[(D_\mu\tilde{\phi})^\dagger(D^\mu\phi_V)+h.c.\right] = i\left[(\partial_\mu+igW_{\mu a}\frac{\tau_a}{2}+i\frac{g'}{2}B_\mu)\tilde{\phi}\right]^\dagger\left(gW_b^\mu\frac{\tau_b}{2}+\frac{g'}{2}B^\mu\right)\phi_V+h.c.
$$

$$
= i(\partial_\mu\tilde{\phi})^\dagger\left(gW_b^\mu\frac{\tau_b}{2}+\frac{g'}{2}B^\mu\right)\phi_V+h.c.+\ldots
$$

upon integration by parts $= -i\tilde{\phi}^\dagger\left(g\partial_\mu W_b^\mu\frac{\tau_b}{2}+\frac{g'}{2}\partial_\mu B^\mu\right)\phi_V+h.c.+\ldots. \quad (34.83)$

The 't Hooft gauge fixing term is then given by

$$
\mathcal{L}_{gf} \equiv -\frac{1}{2\xi}\left(f_a(W_\mu,\tilde{\phi})\right)^2 - \frac{1}{2\xi}\left(f(B_\mu,\tilde{\phi})\right)^2, \quad (34.84)
$$

with

$$
f_a = \partial_\mu W_a^\mu - ig\xi\left(\tilde{\phi}^\dagger\frac{\tau_a}{2}\phi_V - \phi_V^\dagger\frac{\tau_a}{2}\tilde{\phi}\right), \quad (34.85)
$$

$$
f = \partial_\mu B^\mu - ig'\xi\left(\tilde{\phi}^\dagger\frac{1}{2}\phi_V - \phi_V^\dagger\frac{1}{2}\tilde{\phi}\right).
$$

Thus,

$$
\quad (34.86)
$$

$$
\mathcal{L}_{gf} = -\frac{1}{2\xi}(\partial_\mu W_a^\mu)^2 + \left[ig\partial_\mu W_a^\mu\left(\tilde{\phi}^\dagger\frac{\tau_a}{2}\phi_V\right)+h.c.\right] + \frac{g^2\xi}{2}\left(\tilde{\phi}^\dagger\frac{\tau_a}{2}\phi_V - \phi_V^\dagger\frac{\tau_a}{2}\tilde{\phi}\right)^2
$$

$$
-\frac{1}{2\xi}(\partial_\mu B^\mu)^2 + \left[i\frac{g'}{2}\partial_\mu B^\mu(\tilde{\phi}^\dagger\phi_V)+h.c.\right] + \frac{g'^2\xi}{4}(\tilde{\phi}^\dagger\phi_V - \phi_V^\dagger\tilde{\phi})^2.
$$

The terms in square brackets cancel the Nambu–Goldstone boson–gauge boson mixing terms in Eqn. 34.72 and 34.83.

Once we have the gauge fixing term, which eliminates the crossed terms between Nambu–Goldstone bosons and gauge bosons, we must then determine the Fadeev–Popov ghost Lagrangian. Without going into the Fadeev–Popov procedure here, I will just present the ghost Lagrangian that results. It can be shown that

$$
\mathcal{L}_{ghost} = -\int d^4x d^4y(\omega_a^\dagger(x)\chi^\dagger(x))\mathcal{M}(x,y)\begin{pmatrix}\omega_b(y)\\\chi(y)\end{pmatrix}, \quad (34.87)
$$

where ω_a, $a = 1, 2, 3$; χ are the Fadeev–Popov ghosts. The 4×4 matrix $\mathcal{M}$ is given by

$$\mathcal{M} = \begin{pmatrix} M_f(x, y)_{ab} & M_f(x, y)_a \\ M_f(x, y)_b & M_f(x, y) \end{pmatrix}, \tag{34.88}$$

with

$$M_f(x, y)_{ab} = \left\{ \partial_\mu^y [\delta_{ab} \partial_y^\mu + g\epsilon_{abc} W_c^\mu] + g^2 \xi [|\phi_V|^2 \frac{\delta_{ab}}{2} + \tilde{\phi}^\dagger \frac{\tau_b \tau_a}{4} \phi_V + \phi_V^\dagger \frac{\tau_a \tau_b}{4} \tilde{\phi}] \right\}$$
$$\times \delta^4(x - y)$$

$$M_f(x, y)_a = \frac{gg'}{2} \xi [\phi_V^\dagger \tau_a \phi_V + \tilde{\phi}^\dagger \frac{\tau_a}{2} \phi_V + \phi_V^\dagger \frac{\tau_a}{2} \tilde{\phi}] \delta^4(x - y)$$

$$M_f(x, y) = \left\{ \Box + \frac{g'^2 \xi}{4} [2|\phi_V|^2 + \tilde{\phi}^\dagger \phi_V + \phi_V^\dagger \tilde{\phi}] \right\} \delta^4(x - y). \tag{34.89}$$

Recall, the Higgs doublets has components defined by

$$\tilde{\phi} = \begin{pmatrix} \frac{\varphi_3 + i\tilde{\varphi}_4}{\sqrt{2}} \equiv \tilde{\phi}^+ \\ \frac{h + i\tilde{\varphi}_2}{\sqrt{2}} \end{pmatrix} \tag{34.90}$$

and, similarly, we define the Fadeev–Popov ghosts by

$$\omega^\pm \equiv \frac{1}{\sqrt{2}} (\omega_1 \mp i\omega_2), \quad \omega_Z \equiv \cos\theta_W \omega_3 - \sin\theta_W \chi, \quad \omega_\gamma \equiv \sin\theta_W \omega_3 + \cos\theta_W \chi. \tag{34.91}$$

Given the full Lagrangian, $\mathcal{L} = \mathcal{L}_{gauge} + \mathcal{L}_{gf} + \mathcal{L}_{ghost}$, we can now determine the propagators and vertices. Here, we just present the propagators for the theory. We have the propagators in momentum space given by

$$W \Longrightarrow \frac{i \left[-g_{\mu\nu} + \frac{(1-\xi)k_\mu k_\nu}{k^2 - \xi M_W^2} \right]}{k^2 - M_W^2 + i\epsilon}, \tag{34.92}$$

$$Z \Longrightarrow \frac{i \left[-g_{\mu\nu} + \frac{(1-\xi)k_\mu k_\nu}{k^2 - \xi M_Z^2} \right]}{k^2 - M_Z^2 + i\epsilon}, \tag{34.93}$$

$$\tilde{\phi}^\pm \Longrightarrow \frac{i}{k^2 - \xi M_W^2 + i\epsilon}, \tag{34.94}$$

$$\tilde{\phi}_2 \Longrightarrow \frac{i}{k^2 - \xi M_Z^2 + i\epsilon}, \tag{34.95}$$

$$\tilde{\phi}_1 = h \Longrightarrow \frac{i}{k^2 - m_h^2 + i\epsilon}, \tag{34.96}$$

$$\omega^\pm \Longrightarrow \frac{i}{k^2 - \xi M_W^2 + i\epsilon}, \tag{34.97}$$

$$\omega_Z \Longrightarrow \frac{i}{k^2 - \xi M_Z^2 + i\epsilon}, \tag{34.98}$$

$$\omega_\gamma \Longrightarrow \frac{i}{k^2 + i\epsilon}, \tag{34.99}$$

with $\xi = 1$, 't Hooft–Feynman gauge; $\xi = 0$, Landau gauge; and $\xi \to \infty$, unitary gauge.

35

Electroweak Phenomena

The electroweak parameter $\rho = \frac{M_W^2}{M_Z^2 \cos^2 \theta_W}$ is experimentally very close to 1. It is important theoretically to understand this fact. Since it is a tree-level result, perhaps it is just an accident and radiative corrections might give large corrections. We shall show that, in fact, this result is due to an approximate symmetry of the electroweak theory when spontaneously broken by a scalar doublet. This fact is then used to constrain possible theories beyond the Standard Model. As we shall see later, the largest correction to the ρ parameter comes from the top-quark mass.

35.1 Custodial $SU(2)$

Consider the effective field theory below M_W. The only gauge symmetry below M_W is $U(1)_{EM}$. There is a massless photon with renormalizable interactions to quarks and leptons and four-Fermi interactions describing weak interactions. For the latter, we have

$$\mathcal{L}_{eff} = -i\frac{G_F}{\sqrt{2}} J_\mu^{ch} J^{\mu\ ch\dagger} - i\frac{G_F \rho}{\sqrt{2}} (J_{\mu 3} - 2\sin^2 \theta_W J_\mu^{EM})^2, \tag{35.1}$$

where $\frac{G_F}{\sqrt{2}} = \frac{g^2}{8M_W^2} \equiv \frac{1}{2v^2}$ and

$$\frac{G_F \ \rho}{\sqrt{2}} \equiv \frac{g^2}{8M_Z^2 \cos^2 \theta_W}, \tag{35.2}$$

with

$$\rho \equiv \frac{M_W^2}{M_Z^2 \cos^2 \theta_W} = 1 \tag{35.3}$$

at tree level. By definition we have

$$J_{\mu a} \equiv \bar{\ell}\gamma_\mu (1 - \gamma_5) T_a \ell. \tag{35.4}$$

The question that we want to address is, why is $\rho = 1$?

The current

$$\begin{aligned}
J_\mu^{ch} &\equiv 2\bar{\psi}_e \gamma_\mu P_L \psi_{\nu_e} + \dots \tag{35.5} \\
&= \bar{\psi}_e \gamma_\mu (1 - \gamma_5) \psi_{\nu_e} + \dots \\
&= \bar{\ell}\gamma_\mu (1 - \gamma_5) \tau^- \ell + \dots
\end{aligned}$$

and

$$J_\mu^{ch} J^{\mu\ ch\dagger} = \bar{\ell}\gamma_\mu(1-\gamma_5)\tau^-\ell\ \bar{\ell}\gamma^\mu(1-\gamma_5)\tau^+\ell \tag{35.6}$$

$$= \bar{\ell}\gamma_\mu(1-\gamma_5)\frac{\tau_1}{2}\ell\ \bar{\ell}\gamma^\mu(1-\gamma_5)\frac{\tau_1}{2}\ell$$

$$+ \bar{\ell}\gamma_\mu(1-\gamma_5)\frac{\tau_2}{2}\ell\ \bar{\ell}\gamma^\mu(1-\gamma_5)\frac{\tau_2}{2}\ell$$

$$\equiv J_{\mu 1}J_1^\mu + J_{\mu 2}J_2^\mu.$$

We now show that in the limit $g' \to 0$, $\theta_W \to 0$, the theory has a larger symmetry. Moreover,

$$\mathcal{L}_{eff} \to -i\frac{G_F}{\sqrt{2}}\ J_{\mu a}J_a^\mu,\ a = 1, 2, 3 \tag{35.7}$$

and the $SU(2)$ symmetry of $\mathcal{L}_{eff}$ is no accident! Consider the Higgs doublet

$$\phi = \begin{pmatrix} \phi^+ \\ \phi^0 \end{pmatrix} \tag{35.8}$$

and define the following 2×2 matrix of fields

$$\begin{pmatrix} \phi^{0*} & \phi^+ \\ -\phi^{+*} & \phi^0 \end{pmatrix} \equiv \frac{1}{\sqrt{2}}\begin{pmatrix} \varphi_1 - i\varphi_2 & \varphi_3 + i\varphi_4 \\ -\varphi_3 + i\varphi_4 & \varphi_1 + i\varphi_2 \end{pmatrix} \tag{35.9}$$

$$\equiv \frac{1}{\sqrt{2}}(\sigma + i\tau_a\pi_a) \equiv \frac{1}{\sqrt{2}}\Sigma,$$

where the new fields are defined by this equation and we have

$$\sigma \equiv \varphi_1,\quad \pi_1 \equiv \varphi_4,\quad \pi_2 \equiv \varphi_3,\quad \pi_3 \equiv -\varphi_2. \tag{35.10}$$

The similarity with the notation used in Section 26.2 is intentional. The field Σ transforms under a global $SU(2)_L \otimes SU(2)_R$ symmetry defined by

$$\Sigma' = L\Sigma R^\dagger, \tag{35.11}$$

where in fact,

$$L \in SU(2)_L \equiv SU(2)_{weak} \tag{35.12}$$

and

$$R \equiv e^{i\frac{\tau_a}{2}\omega_a} \in SU(2)_R. \tag{35.13}$$

The transformation of Σ under the group $SU(2)_R$ is given by

$$\Sigma' = \Sigma e^{-i\frac{\tau_a}{2}\omega_a} \tag{35.14}$$

or, infinitesimally,

$$\delta(\phi^{0*}\ \phi^+) \simeq -i(\phi^{0*}\ \phi^+)\frac{\tau_a}{2}\,\omega_a \tag{35.15}$$

$$\delta\begin{pmatrix} \phi^{0*} \\ \phi^+ \end{pmatrix} = +i\left(\frac{-\tau_a^T}{2}\right)\omega_a\begin{pmatrix} \phi^{0*} \\ \phi^+ \end{pmatrix}$$

$$= i(T_{aR})\omega_a\begin{pmatrix} \phi^{0*} \\ \phi^+ \end{pmatrix}.$$

Hence,

$$T_{3R} \begin{pmatrix} \phi^{0*} \\ \phi^+ \end{pmatrix} = \begin{pmatrix} \frac{-\tau_3^T}{2} \end{pmatrix} \begin{pmatrix} \phi^{0*} \\ \phi^+ \end{pmatrix} = \frac{1}{2} \begin{pmatrix} -\phi^{0*} \\ \phi^+ \end{pmatrix}. \tag{35.16}$$

But note,

$$Y\phi = \phi \tag{35.17}$$

and thus

$$Y\phi^+ = \phi^+, \quad Y\phi^{0*} = -\phi^{0*}. \tag{35.18}$$

Hence, we can identify

$$\frac{Y}{2} \equiv T_{3R}, \tag{35.19}$$

i.e. weak hypercharge is the $U(1)$ subgroup of $SU(2)_R$.

Note:

$$V(\phi) \equiv \frac{\lambda}{8} \left(\frac{1}{2} \text{Tr}(\Sigma^\dagger \Sigma) - v^2 \right)^2. \tag{35.20}$$

Thus, $V(\phi)$ is invariant under the global symmetry $SU(2)_L \otimes SU(2)_R$. However, only the subgroup,

$$SU(2)_L \otimes U(1)_R \tag{35.21}$$

is gauged. Thus, the gauge symmetry (with $g' \neq 0$) breaks the global $SU(2)_R$. However, when we take the limit $g' \to 0$, we retain $SU(2)_L \otimes SU(2)_R$ as a good symmetry of $\mathcal{L}_{scalar}$, since $(D_\mu\phi)^\dagger D^\mu\phi$ is invariant under the global $SU(2)_L \otimes SU(2)_R$. $\mathcal{L}_{gauge}$ is also invariant under global $SU(2)_L \otimes SU(2)_R$ when $g' \to 0$, since the field B^μ is an $SU(2)_L \otimes SU(2)_R$ singlet $(1,\ 1)$ and $W_{\mu a}$ transforms under $SU(2)_L \otimes SU(2)_R$ as $(3,\ 1)$. Only $\mathcal{L}_{fermion}$ breaks this symmetry either via mass terms or via the absence of the field $\bar{\nu}_e$.

Consider a theory with both a left-handed lepton doublet, $\ell = \begin{pmatrix} \nu_e \\ e \end{pmatrix}$, and a left-handed anti-lepton doublet,

$$\bar{\ell} = \begin{pmatrix} \bar{\nu}_e \\ \bar{e} \end{pmatrix}, \tag{35.22}$$

transforming under $SU(2)_L \otimes SU(2)_R$ as $(2,\ 1)$, $(1,\ 2)$, respectively, with

$$T_{aR}\ \bar{\ell} = \begin{pmatrix} \frac{-\tau_a^T}{2} \end{pmatrix} \bar{\ell}. \tag{35.23}$$

Note, in this case we have

$$\frac{Y}{2} = T_{3R} + \frac{1}{2}(B - L), \tag{35.24}$$

which actually generalizes to quarks and

$$Q = T_{3L} + T_{3R} + \frac{1}{2}(B - L). \tag{35.25}$$

The mass term for this hypothesized theory is given by

$$\mathcal{L}_{mass} = -\lambda_e \bar{\ell} \Sigma^\dagger \ell + h.c.$$

(35.26)

This term is invariant under global $SU(2)_L \otimes SU(2)_R$ with

$$\ell' = L\ell, \quad \bar{\ell}' = \bar{\ell}R^\dagger, \quad \Sigma' = L\Sigma R^\dagger, \quad \Sigma'^\dagger = R\Sigma^\dagger L^\dagger.$$

(35.27)

Given the Higgs doublet, ϕ, we can define the $SU(2)$ doublet field,

$$\tilde{\phi} \equiv i\tau_2 \phi^* = \begin{pmatrix} \phi^{0*} \\ -\phi^{+*} \end{pmatrix}.$$

(35.28)

Then we have

$$\Sigma = (\tilde{\phi} \ \ \phi)$$

(35.29)

and

$$\Sigma^\dagger = \begin{pmatrix} \tilde{\phi}^\dagger \\ \phi^\dagger \end{pmatrix}.$$

(35.30)

And the mass term becomes

$$\mathcal{L}_{mass} = \lambda_e(\bar{e}\phi^\dagger \ell + \bar{\nu}_e \tilde{\phi}^\dagger \ell).$$

(35.31)

The first term gives mass to the electron, while the second term would give mass to the electron neutrino, IF $\bar{\nu}_e$ existed. Moreover, they would be degenerate.

Therefore, we have shown that global $SU(2)_L \otimes SU(2)_R$ is an approximate symmetry of the electroweak interactions, broken only by the small coupling g' and the electron mass, since the electron neutrino is not degenerate with the electron. When we consider quarks, it will become apparent that the largest symmetry breaking is due to the top-quark mass. Now consider the vacuum expectation value,

$$\langle \Sigma \rangle = \frac{1}{\sqrt{2}} \begin{pmatrix} v & 0 \\ 0 & v \end{pmatrix},$$

(35.32)

which spontaneously breaks $SU(2)_L \otimes SU(2)_R \rightarrow SU(2)_{vector}$, i.e. $L = R$. As a consequence of the $SU(2)_{vector}$ we have

$$\mathcal{L}_{eff} \rightarrow -i\frac{G_F}{\sqrt{2}} J_{\mu a}J_a^\mu, \ a = 1, 2, 3,$$

(35.33)

i.e. $\rho = 1$. The $SU(2)_{vector}$ is called "custodial $SU(2)$" (Susskind, 1979; Sikivie et al., 1980). Note, we would not have the global $SU(2)_L \otimes SU(2)_R$ symmetry, if the Higgs was part of an $SU(2)_L$ triplet, instead of a doublet. Since the global custodial $SU(2)$ symmetry is explicitly broken, the electroweak ρ parameter gets radiative corrections. The dominant correction comes from top-quark loops. We shall discuss this later.

35.2 Elastic Neutrino–Electron Scattering

In Eqn. 35.1, we have the effective weak Lagrangian for low-energy lepton scattering. For processes such as elastic $\nu_\mu - e$ scattering, only the second term contributes and we have

$$\mathcal{L}_{eff} \simeq -i \frac{G_F}{\sqrt{2}} [\bar{\nu}_\mu \gamma_\mu (g_V^\nu - g_A^\nu \gamma_5) \nu_\mu] [\bar{e} \gamma^\mu (g_V^e - g_A^e \gamma_5) e], \qquad (35.34)$$

where $g_V^\nu = g_A^\nu = \frac{1}{2}$; $g_V^e = -\frac{1}{2}(1 - 4\sin^2\theta_W)$, $g_A^e = -\frac{1}{2}$. However, for elastic $\nu_e - e$ scattering, both terms contribute and we have

$$\mathcal{L}_{eff} \simeq -i \frac{G_F}{\sqrt{2}} [\bar{\nu}_e \gamma_\mu (g_V^\nu - g_A^\nu \gamma_5) \nu_e] [\bar{e} \gamma^\mu (\tilde{g}_V^e - \tilde{g}_A^e \gamma_5) e], \qquad (35.35)$$

where $\tilde{g}_{V,A}^e = 1 + g_{V,A}^e$. The additional contribution comes from a Fierz rearrangement of the term Eqn. 35.6 to give

$$\bar{e} \gamma_\mu (1 - \gamma_5) \nu_e \bar{\nu}_e \gamma^\mu (1 - \gamma_5) e = \bar{\nu}_e (1 - \gamma_5) \nu_e \bar{e} \gamma^\mu (1 - \gamma_5) e. \qquad (35.36)$$

Then, for the scattering process, $\nu(p_1) e(p_2) \to \nu(p_3) e(p_4)$ or $\bar{\nu}(p_1) e(p_2) \to \bar{\nu}(p_4) e(p_3)$, we obtain

$$\frac{1}{2} \sum_{s_1, s_2, s_3, s_4} |T|^2 = 16 G_F^2 \left[(\tilde{g}_V^e + \tilde{g}_A^e)^2 p_1 \cdot p_2 p_3 \cdot p_4 \right.$$
$$\left. + (\tilde{g}_V^e - \tilde{g}_A^e)^2 p_1 \cdot p_4 p_2 \cdot p_3 - ((\tilde{g}_V^e)^2 - (\tilde{g}_A^e)^2) m_e^2 p_1 \cdot p_3 \right]. \qquad (35.37)$$

We then obtain

$$\frac{d\sigma_{\nu e}}{dy} = \frac{G_F^2 m_e E_\nu}{2\pi} \left[(\tilde{g}_V^e + \tilde{g}_A^e)^2 + (\tilde{g}_V^e - \tilde{g}_A^e)^2 (1 - y^2) - ((\tilde{g}_V^e)^2 - (\tilde{g}_A^e)^2) \frac{m_e}{E_\nu} y \right] \qquad (35.38)$$

and

$$\frac{d\sigma_{\bar{\nu} e}}{dy} = \frac{G_F^2 m_e E_\nu}{2\pi} \left[(\tilde{g}_V^e - \tilde{g}_A^e)^2 + (\tilde{g}_V^e + \tilde{g}_A^e)^2 (1 - y^2) - ((\tilde{g}_V^e)^2 - (\tilde{g}_A^e)^2) \frac{m_e}{E_\nu} y \right], \qquad (35.39)$$

where $y \equiv \frac{p_1 \cdot p_3}{p_1 \cdot p_2}$. In the laboratory frame where $p_1 = (E_\nu, 0, 0, E_\nu)$, $p_2 = (m_e, \vec{0})$, $p_3 = (E'_\nu, \vec{p}_3)$, $p_4 = (E_e, \vec{p}_4)$, we have

$$y = \frac{E_e - m_e}{E_\nu}, \quad 0 \leq y \leq \left(1 + \frac{m_e}{2E_\nu} \right)^{-1}. \qquad (35.40)$$

Note, y_{max} is obtained when the scattered neutrino goes in the opposite direction of the incoming neutrino.

35.3 Forward–Backward Asymmetry in $e^+e^- \to \mu^+\mu^-$ Scattering

In this section, we consider the process $e^+e^- \to \mu^+\mu^-$ scattering with γ, Z interference (Fig. 35.1) which was measured at the Large Electron–Positron collider (LEP) at CERN. The amplitude takes the rough form

$$T \sim (V_e - A_e)(V_\mu - A_\mu) \tag{35.41}$$

and therefore the total cross-section contains terms of the form

$$\sigma_{tot} \sim \sum_{spins} |T|^2 \tag{35.42}$$

$$\sim \sum_{spins} \{VVVV + VVVA + V_\mu V_e A_\mu A_e + V_e V_e A_\mu A_\mu + AAAA + \dots \}.$$

The second term only contributes to helicity-dependent cross-sections. Let's focus on the third term,

$$V_\mu V_e A_\mu A_e. \tag{35.43}$$

It changes sign under the interchange $\mu^+ \leftrightarrow \mu^-$, since the vector current is charge-conjugation odd, while the axial-vector current is charge-conjugation even. The

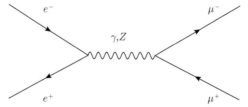

Fig. 35.1 $e^+e^- \to \mu^+\mu^-$ scattering with γ, Z interference.

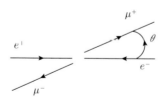

Fig. 35.2 $e^+e^- \to \mu^+\mu^-$ scattering angle, θ, defined in the CMS.

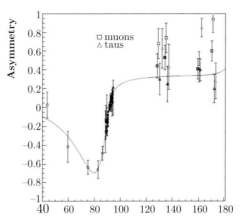

Fig. 35.3 Forward–backward asymmetry in $e^+e^- \to \mu^+\mu^-$ and $e^+e^- \to \tau^+\tau^-$ as a function of energy from the DELPHI experiment at the LEP. The interference of the γ and Z contributions gives the asymmetry variation with energy, as indicated by the Standard Model curve. Reprinted with permission from Gaillard et al. (1999). Copyright (1999) by the American Physical Society.

other terms do not change sign. Define the scattering angle θ in the center of momentum system (CMS), see Fig. 35.2. Forward (backward) scattering is defined for

$$0 \le \theta \le \frac{\pi}{2} \quad \left(\frac{\pi}{2} \le \theta \le \pi \right) \tag{35.44}$$

and the forward (backward) scattering cross-section is given by

$$\sigma_F \equiv \int_0^{\frac{\pi}{2}} d\theta \frac{d\sigma}{d\theta} \quad \left(\sigma_B \equiv \int_{\frac{\pi}{2}}^{\pi} d\theta \frac{d\sigma}{d\theta} \right). \tag{35.45}$$

Finally, the forward–backward asymmetry is defined by

$$A_{FB} \equiv \frac{\sigma_F - \sigma_B}{\sigma_F + \sigma_B} \ne 0. \tag{35.46}$$

At the Z pole we have

$$A_{FB} \simeq \frac{3 g_V^e g_V^\mu g_A^e g_A^\mu}{[(g_V^e)^2 + (g_A^e)^2][(g_V^\mu)^2 + (g_A^\mu)^2]}. \tag{35.47}$$

The forward–backward asymmetry for muons and taus has been measured at the LEP and the γ, Z interference is consistent with Standard Model expectations (see Fig. 35.3 using DELPHI data; Abreu et al., 2000; Gaillard et al., 1999).

Combining data from forward–backward scattering, $e^+e^- \to \mu^+\mu^-$, $\tau^+\tau^-$, $c\bar{c}$, $b\bar{b}$ at LEP, from $p\bar{p} \to e^+e^-$, $\mu^+\mu^-$ at the Tevatron and the Large Hadron Collider

Table 35.1 The different definitions of the weak mixing angle. The on-shell definition $s_W^2 \equiv 1 - M_W^2/M_Z^2$ is usually reported for neutrino scattering experiments. Table from Tanabashi et al. (2018).

Scheme	Notation	Value	Parametric uncertainty
On-shell	s_W^2	0.22343	±0.00007
$\overline{MS}$	$\hat{s}_Z^2$	0.23122	±0.00003
$\overline{MS}$ ND	$\hat{s}_{ND}^2$	0 23142	±0.00003
$\overline{MS}$	$\hat{s}_0^2$	0.23857	±0.00005
Effective angle	$\bar{s}_\ell^2$	0.23154	±0.00003

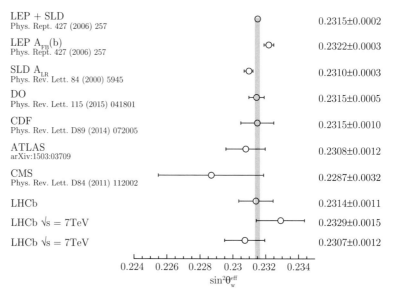

Fig. 35.4 A comparison of the $\sin^2 \theta_W^{eff}$ measurement at the LHCb and other experiments. The combined LEP and SLD collaboration measurement is indicated by the vertical grey band (from (Aaij et al., 2015)). Published under the CC BY 4.0 license.

(LHC) data (see Fig. 35.4 taken from; Schael et al., 2006; Aaij et al., 2015) gives a combined value for

$$\sin^2 \theta_W^{eff} = 0.23142 \pm 0.00073 \pm 0.00052 \pm 0.00056, \qquad (35.48)$$

where the uncertainties are statistical, systematic and theoretical, respectively. Note, there are several definitions of the weak mixing angle, which differ by radiative corrections. The different definitions are listed in Table 35.1. Additional data on $\sin^2 \theta_W$ comes from neutrino scattering experiments (see Chapter 36).

36

Deep Inelastic Scattering Revisited

In Chapter 31 we considered deep inelastic electron nucleon scattering in the context of quantum electrodynamics (QED). Deep inelastic scattering in the context of the electroweak theory adds some new ingredients to the analysis. In the case of $ep \to eX$, i.e. neutral-current scattering, we can now have both photons and Z bosons exchanged. In addition, we now have charged-current processes, such as $ep \to \nu X$ and $\bar{\nu}p \to e^+ X$.

In these experiments the nucleon was shattered, so the only variables that were measured experimentally were the energy and momentum of the outgoing electron. The nucleons are in their rest frame with momentum $p = (M, \vec{0})$ and the incident electrons have momentum $k = (E, \vec{k})$, while the outgoing electron has momentum $k' = (E', \vec{k}\,')$ (see Figs. 36.1 and 36.2). The momentum $q \equiv k - k'$ and we define the parameters $\nu \equiv \frac{p \cdot q}{M}$ and $W^2 \equiv p_n^2 \equiv (q + p)^2$. The kinematic variable ν is the energy loss of the electron and takes on values

$$0 \le \nu = E - E'. \tag{36.1}$$

The photon momentum satisfies $q^2 = (k - k')^2 = 2m_e^2 - 2k \cdot k' \simeq -2EE'\,(1 - \cos\theta)$ when we neglect the electron mass for $m_e \ll E$. We therefore have

$$q^2 = -4EE'\,\sin^2\frac{\theta}{2} \le 0. \tag{36.2}$$

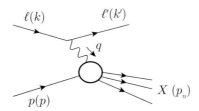

Fig. 36.1 Phenomenological diagram for deep inelastic scattering.

Fig. 36.2 Defining the scattering angle θ in the laboratory frame.

We then define the quantity $Q^2 \equiv -q^2 > 0$ and the scaling parameter, Feynman x with

$$x \equiv \frac{Q^2}{2M\nu} \geq 0. \tag{36.3}$$

Since $W^2 \equiv q^2 + 2q \cdot p + M^2 \geq M^2$ or $q^2 + 2M\nu \geq 0$, we find

$$x \leq 1. \tag{36.4}$$

Thus,

$$0 \leq x \leq 1. \tag{36.5}$$

We also define $y = \nu/E$ satisfying $0 \leq y \leq 1$.

Now the scattering amplitude for neutral-current scattering is a sum of both photon and Z exchanges, while the scattering amplitude for charged-current exchanges just includes the W boson (Anselmino et al., 1994; Forte et al., 2001; Tanabashi et al., 2018). Squaring the scattering amplitudes and summing over all possible final states $X = \{n\}$, we obtain

$$\frac{d^2\sigma}{dxdy} = \frac{2\pi M\nu}{E'} \frac{d^2\sigma}{d\Omega dE'} = \frac{2\pi y\alpha^2}{Q^4} \sum_j \eta_j L_j^{\mu\nu} \, W_{\mu\nu}^j. \tag{36.6}$$

For neutral-current processes, the summation is over $j = \gamma, Z$ and γZ, where the latter represents the interference between the photon and Z exchange. For charged currents, $j = W$. The lepton tensor $L_{\mu\nu}$ depends on the exchanged boson. For incoming leptons of charge $e = \pm 1$ and helicity $\lambda = \pm 1/2$,

$$L_{\mu\nu}^\gamma = [\bar{u}(k')\gamma_\mu u(k)]^* [\bar{u}(k')\gamma_\nu u(k)]$$
$$= 2(k'_\mu k_\nu + k'_\nu k_\mu - g_{\mu\nu} \, (k \cdot k' - m_\ell^2) - 2i\lambda\epsilon_{\mu\nu\alpha\beta}k^\alpha k'^\beta), \tag{36.7}$$
$$L_{\mu\nu}^{\gamma Z} = [\bar{u}(k')\gamma_\mu(g_V - g_A\gamma_5)u(k)]^* [\bar{u}(k')\gamma_\nu u(k)]$$
$$= (g_V + 2e\lambda g_A)L_{\mu\nu}^\gamma,$$
$$L_{\mu\nu}^Z = [\bar{u}(k')\gamma_\mu(g_V - g_A\gamma_5)u(k)]^* [\bar{u}(k')\gamma_\nu(g_V - g_A\gamma_5)u(k)]$$
$$= (g_V + 2e\lambda g_A)^2 L_{\mu\nu}^\gamma,$$
$$L_{\mu\nu}^W = [\bar{u}(k')\gamma_\mu(1 - \gamma_5)u(k)]^* [\bar{u}(k')\gamma_\nu(1 - \gamma_5)u(k)]$$
$$= (1 + 2e\lambda)^2 L_{\mu\nu}^\gamma,$$

where $g_V^\ell = -\frac{1}{2}(1 - 4\sin^2\theta_W)$, $g_A^\ell = -\frac{1}{2}$ with $\ell = e$, μ. The factors η_j give the ratios of the corresponding propagator and couplings squared

$$\eta_\gamma = 1; \quad \eta_{\gamma Z} = \left(\frac{G_F M_Z^2}{2\sqrt{2}\pi\alpha}\right)\left(\frac{Q^2}{Q^2 + M_Z^2}\right); \tag{36.8}$$
$$\eta_Z = \eta_{\gamma Z}^2; \quad \eta_W = \frac{1}{2}\left(\frac{G_F M_W^2}{4\pi\alpha} \frac{Q^2}{Q^2 + M_W^2}\right)^2.$$

For unpolarized targets, we have

$$W_{\mu\nu}(p,q) = \frac{1}{4\pi} \sum_\sigma \int d^4x \, e^{iq\cdot x} \langle p,\sigma| \left[j_\mu^\dagger(x), \, j_\nu(0) \right] |p,\sigma\rangle, \qquad (36.9)$$

where j_μ is the appropriate electroweak current. Note, $q^\mu W_{\mu\nu} \equiv 0$, since $\partial^\mu j_\mu \equiv 0$. The most general form for $W_{\mu\nu}$ consistent with Lorentz invariance and current conservation is given by

$$W_{\mu\nu}(p,q) = \left[(-g_{\mu\nu} + \frac{q_\mu q_\nu}{q^2})F_1(x,Q^2) + \frac{(p_\mu - \frac{p\cdot q}{q^2}q_\mu)(p_\nu - \frac{p\cdot q}{q^2}q_\nu)}{p\cdot q} F_2(x,Q^2) \right.$$
$$\left. - i\epsilon_{\mu\nu\alpha\beta}\frac{q^\alpha p^\beta}{2p\cdot q} F_3(x,Q^2) \right], \qquad (36.10)$$

where $F_{1,2,3}$ are the *structure functions*. We then have for unpolarized nucleons

$$\frac{d^2\sigma}{dxdy} = \frac{4\pi\alpha^2}{xyQ^2}\eta_i \left\{ y^2 x F_1^i + \left(1 - y - \frac{x^2y^2M^2}{Q^2}\right) F_2^i \mp \left(y - \frac{y^2}{2}\right) x F_3^i \right\}, \qquad (36.11)$$

where $i = NC, CC$ corresponds to neutral-current ($eN \to eX$) or charged-current ($eN \to \nu X$) or ($\nu N \to eX$) processes, respectively. For incoming neutrinos, $L_{\mu\nu}^W$ in Eqn. 36.7, e,λ correspond to the outgoing charged lepton. In the last term in Eqn. 36.11, the $-$ sign is for an incoming e^+ or $\bar\nu$ and the $+$ sign is for an incoming e^- or ν. The factor $\eta^{NC} = 1$ for unpolarized $e^\pm$ beams, while $\eta^{CC} = (1 \pm 2\lambda)^2 \eta_W$ with $\pm$ for $\ell^\pm$, where λ is the helicity of the incoming lepton, and $\eta^{CC} = 4\eta_W$ for incoming neutrinos.

The neutral-current structure functions $F_2^\gamma, F_2^{\gamma Z}, F_2^Z$ for $e^\pm N \to e^\pm X$ are given by

$$F_2^{NC} = F_2^\gamma - (g_V^e \pm 2\lambda g_A^e)\eta_{\gamma Z} F_2^{\gamma Z} + (g_V^{e\,2} + g_A^{e\,2} \pm 4\lambda g_V^e g_A^e)\eta_Z F_2^Z. \qquad (36.12)$$

Similarly, for F_1^{NC}, and

$$x F_3^{NC} = -(g_A^e \pm 2\lambda g_V^e)\eta_{\gamma Z} x F_3^{\gamma Z} + [2g_V^e g_A^e \pm 2\lambda(g_V^{e\,2} + g_A^{e\,2})]\eta_Z x F_3^Z. \qquad (36.13)$$

The neutral-current structure functions $F_2^\gamma, F_2^{\gamma Z}, F_2^Z$ for $\nu(\bar\nu)N \to \nu(\bar\nu)X$ are given by

$$F_2^{NC} = \eta_Z F_2^Z. \qquad (36.14)$$

Similarly for F_1^{NC}, and

$$x F_3^{NC} = \eta_Z x F_3^Z. \qquad (36.15)$$

The charged-current structure functions are given by $F_1^{CC} = F_1^W, F_2^{CC} = F_2^W, F_3^{CC} = F_3^W$.

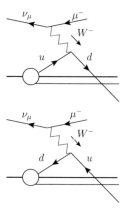

Fig. 36.3 $\bar{\nu}_\mu p$ charged-current inelastic scattering.

For the neutral-current process $ep \to eX$, the structure functions given in terms of quark distribution functions, $q = \{u, d, s, c\}$, are

$$[F_2^\gamma, F_2^{\gamma Z}, F_2^Z] = x \sum_q [e_q^2, 2e_q g_V^q, g_V^{q\,2} + g_A^{q\,2}](q + \bar{q}), \tag{36.16}$$

$$[F_3^\gamma, F_3^{\gamma Z}, F_3^Z] = \sum_q [0, 2e_q g_A^q, 2g_V^q g_A^q](q - \bar{q}),$$

with

$$F_2(x) = 2xF_1(x). \tag{36.17}$$

For the neutral-current process $\nu p \to \nu X$, the formula (Eqn. 36.11) simplifies and we have

$$\begin{aligned}
\frac{d^2\sigma_{NC}(\nu p)}{dxdy} &\approx \frac{4\pi\alpha^2}{xyQ^2}\left\{y^2 xF_1 + (1 - y)F_2 + \left(y - \frac{y^2}{2}\right)xF_3\right\} \\
&= \frac{4\pi\alpha^2}{xyQ^2}2x\eta_Z\{[g_L^{q\,2} + g_R^{q\,2}(1 - y)^2]q + [g_R^{\bar{q}\,2} + g_L^{\bar{q}\,2}(1 - y)^2]\bar{q}\} \\
&= \frac{2G_F^2 ME}{\pi}\{[g_L^{q\,2} + g_R^{q\,2}(1 - y)^2]xq(x) + [g_R^{\bar{q}\,2} + g_L^{\bar{q}\,2}(1 - y)^2]x\bar{q}(x)\},
\end{aligned} \tag{36.18}$$

where $g_{L,R} = \frac{1}{2}(g_V \pm g_A)$ and for $\bar{\nu}p \to \bar{\nu}X$ we interchange $g_L \leftrightarrow g_R$.

For the charge-current process $\bar{\nu}_\mu p \to \mu^+ X$ (and similarly for the crossed-process with $\mu \to e$, i.e. $e^- p \to \nu_e X$), we have (see Fig. 36.3)

$$\begin{aligned}
F_2^{W^-} &= 2x(u + \bar{d} + \bar{s} + c \dots), \\
F_3^{W^-} &= 2(u - \bar{d} - \bar{s} + c \dots).
\end{aligned} \tag{36.19}$$

Thus, we have

$$\frac{d^2\sigma_{CC}(\bar{\nu}_\mu p)}{dxdy} \approx \frac{2G_F^2 ME}{\pi}x\{[u(x) + c(x)](1 - y)^2 + [\bar{d}(x) + \bar{s}(x)]\}. \tag{36.20}$$

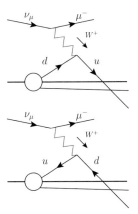

Fig. 36.4 $\nu_\mu p$ charged-current inelastic scattering.

Note, for $\nu_\mu p \to \mu^- X$ and $e^+ p \to \bar{\nu}_e X$,

$$F^{W^+}(u, d, s, c) = F^{W^-}(d, u, c, s) \tag{36.21}$$

(see Fig. 36.4), and

$$\frac{d^2\sigma_{CC}(\nu_\mu p)}{dxdy} \approx \frac{2G_F^2 ME}{\pi} x\{[d(x) + s(x)] + [\bar{u}(x) + \bar{c}(x)](1 - y)^2\}. \tag{36.22}$$

Charge-current scattering on neutrons, using the notation

$$u_n(x) = d_p(x) \equiv d(x); \quad d_n(x) = u_p(x) \equiv u(x), \tag{36.23}$$
$$s_n(x) = s_p(x) \equiv s(x); \quad c_n(x) = c_p(x) \equiv c(x),$$

is given by

$$\frac{d^2\sigma_{CC}(\nu_\mu n)}{dxdy} \approx \frac{2G_F^2 ME}{\pi} x\{[u(x) + s(x)] + [\bar{d}(x) + \bar{c}(x)](1 - y)^2\} \tag{36.24}$$

and

$$\frac{d^2\sigma_{CC}(\bar{\nu}_\mu n)}{dxdy} \approx \frac{2G_F^2 ME}{\pi} x\{[d(x) + c(x)](1 - y)^2 + [\bar{u}(x) + \bar{s}(x)]\}. \tag{36.25}$$

Muon-neutrino and muon-anti-neutrino neutral-current and charged-current scattering experiments were performed by the FMMF (Bogert et al., 1985), CCFR (Reutens et al., 1985), CDHS (Abramowicz et al., 1986), CHARM (Allaby et al., 1987), CHARM II (Vilain et al., 1994), CCFR (McFarland et al., 1998) and NuTev (McFarland et al., 2002) collaborations on isoscalar targets, i.e. targets with equal numbers of protons and neutrons. The muon-neutrino beams were produced in the decays of pions and kaons. Charge exchange processes were identified via the outgoing muon or anti-muon with a vertex in common with the hadronic shower. Events with an hadronic shower without an accompanying muon or anti-muon were identified as neutral-current events. Finally, in order to avoid elastic scattering events, the hadronic energy, W, was required to satisfy $W > 4$ GeV (Allaby et al., 1987).

Let's consider the cross-section predictions for the neutral-current and charged-current processes on an isoscalar target and ignoring the c and s content of the nucleon, i.e. $q = \{u, d\}$. For neutral currents we have

$$\frac{d^2\sigma_{NC}(\nu_\mu N)}{dxdy} \approx \frac{2G_F^2 ME}{2\pi}\{[g_L^{q\,2} + g_R^{q\,2}(1-y)^2]xq(x) + [g_R^{q\,2} + g_L^{q\,2}(1-y)^2]x\bar{q}(x)\}$$
(36.26)

and

$$\frac{d^2\sigma_{NC}(\bar{\nu}_\mu N)}{dxdy} \approx \frac{2G_F^2 ME}{2\pi}\{[g_R^{q\,2} + g_L^{q\,2}(1-y)^2]xq(x) + [g_L^{q\,2} + g_R^{q\,2}(1-y)^2]x\bar{q}(x)\}.$$
(36.27)

For charged currents we have

$$\frac{d^2\sigma_{CC}(\nu_\mu N)}{dxdy} \approx \frac{2G_F^2 ME}{2\pi}x\{q(x) + \bar{q}(x)(1-y)^2\}$$
(36.28)

and

$$\frac{d^2\sigma_{CC}(\bar{\nu}_\mu N)}{dxdy} \approx \frac{2G_F^2 ME}{2\pi}x\{q(x)(1-y)^2 + \bar{q}(x)\}.$$
(36.29)

For the case of neutral currents, the values of x, y cannot be directly measured. Therefore we consider the total inelastic cross-sections defined by

$$\sigma = \int_0^1 \int_0^1 dxdy \frac{d\sigma}{dxdy}.$$
(36.30)

We have

(36.31)

$$\sigma_{NC}(\nu_\mu N) = \frac{G_F^2 ME}{\pi} \int_0^1 dx \left\{ g_L^2 \left[xq^p(x) + \frac{1}{3}x\bar{q}^p(x) \right] + g_R^2 \left[\frac{1}{3}xq^p(x) + x\bar{q}^p(x) \right] \right\},$$

$$\sigma_{NC}(\bar{\nu}_\mu N) = \frac{G_F^2 ME}{\pi} \int_0^1 dx \left\{ g_L^2 \left[\frac{1}{3}xq^p(x) + x\bar{q}^p(x) \right] + g_R^2 \left[\frac{1}{3}x\bar{q}^p(x) + xq^p(x) \right] \right\},$$

$$\sigma_{CC}(\nu_\mu N) = \frac{G_F^2 ME}{\pi} \int_0^1 dx \left\{ xq^p(x) + \frac{1}{3}x\bar{q}^p(x) \right\},$$

$$\sigma_{CC}(\bar{\nu}_\mu N) = \frac{G_F^2 ME}{\pi} \int_0^1 dx \left\{ \frac{1}{3}xq^p(x) + x\bar{q}^p(x) \right\},$$

where $q^p = u + d$, $\bar{q}^p = \bar{u} + \bar{d}$ (as defined in Eqn. 36.23), $g_L^2 \equiv \sum_{q=u,d} g_L^{q\,2} = \frac{1}{2} - \sin^2\theta_W + \frac{5}{9}\sin^4\theta_W$ and $g_R^2 = \sum_{q=u,d} g_R^{q\,2} = \frac{5}{9}\sin^4\theta_W$.

We now define the ratios (the Llewelyn Smith relations; Llewellyn Smith, 1983)

$$R_\nu \equiv \frac{\sigma_{NC}(\nu_\mu N)}{\sigma_{CC}(\nu_\mu N)} = g_L^2 + rg_R^2$$
(36.32)

and

$$R_{\bar{\nu}} \equiv \frac{\sigma_{NC}(\bar{\nu}_\mu N)}{\sigma_{CC}(\bar{\nu}_\mu N)} = g_L^2 + \frac{1}{r}g_R^2,$$
(36.33)

Table 36.1 The main processes relevant to global parton distribution function (PDF) analyses, ordered in three groups: fixed-target experiments, HERA and the $p\bar{p}$ Tevatron/pp LHC. For each process we give an indication of their dominant partonic subprocesses, the primary partons which are probed and the approximate range of x constrained by the data. Table taken from (Tanabashi et al., 2018).

Process	Subprocess	Partons	x range
$l^\pm \{p, n\} \to l^\pm X$	$\gamma^* q \to q$	$q,\bar{q},g$	$x \gtrsim 0.01$
$l^\pm n/p \to l^\pm X$	$\gamma^* d/u \to d/u$	d/u	$x \gtrsim 0.01$
$pp \to \mu^+\mu^- X$	$u\bar{u}, d\bar{d} \to \gamma^*$	$\bar{q}$	$0.015 \lesssim x \lesssim 0.35$
$pn/pp \to \mu^+\mu^- X$	$(u\bar{d}) / (\bar{u}u) \to \gamma^*$	$\bar{d}/\bar{u}$	$0.015 \lesssim x \lesssim 0.35$
$\bar{v}(v)\, N \to \mu^-(\mu^+)X$	$W^* q \to q\varphi$	$q,\bar{q}$	$0.01 \lesssim x \lesssim 0.5$
$vN \to \mu^-\mu^+ X$	$W^* s \to c\varphi$	s	$0.01 \lesssim x \lesssim 0.2$
$\bar{v}\, N \to \mu^+(\mu^-)X$	$W^* \bar{s} \to \bar{c}$	$\bar{s}$	$0.01 \lesssim x \lesssim 0.2$
$e^\pm p \to e^\pm X$	$\gamma^* q \to q$	$g,q,\bar{q}$	$10^{-4} \lesssim x \lesssim 0.1$
$e^\pm\, p \to e^- X$	$W^+ - d,s'' \to -u,c''$	d,s	$x \gtrsim 0.01$
$e^\pm\, p \to e^\pm c\bar{c}X,\, e^\pm b\bar{b}X$	$\gamma^* c \to c, \gamma^* g \to c\bar{c}$	c,b,g	$10^{-4} \lesssim x \lesssim 0.01$
$e^\pm\, p \to \text{jet} + X$	$\gamma^* g \to q\bar{q}$	g	$0.01 \lesssim x \lesssim 0.1$
$\bar{p}p, pp \to \text{jet} + X$	$gg, qg, qq \to 2j$	g,q	$0.00005 \lesssim x \lesssim 0.5$
$p\bar{p} \to (W^\pm \to l^\pm v)X$	$ud \to W^+, \bar{u}\bar{d} \to W^-$	$u,d,\bar{u},\bar{d}$	$x \gtrsim 0.05$
$pp \to (W^\pm \to l^\pm v)X$	$u\bar{d} \to W^+, \bar{d}u \to W^-$	$u,d,\bar{u},\bar{d},g$	$x \gtrsim 0.001$
$p\bar{p}(pp) \to (Z^\pm \to l^+l^-)X$	$uu, dd,..(\bar{u}\bar{u},..) \to Z$	$u,d,..(g)$	$x \gtrsim 0.001$
$pp \to W^-c, W^+\bar{c}$	$gs \to W^-c$	$s,\bar{s}$	$x \sim 0.01$
$pp \to (\gamma^* \to l^+l^-)X$	$u\bar{u}, d\bar{d},.. \to \gamma^*$	$\bar{q},g$	$x \gtrsim 10^{-5}$
$pp \to (\gamma^* \to l^+l^-)X$	$u\gamma, d\gamma,.. \to \gamma^*$	γ	$x \gtrsim 10^{-2}$
$pp \to b\bar{b}X,\, t\bar{t}X$	$gg \to b\bar{b}, t\bar{t}$	g	$x \gtrsim 10^{-5}, 10^{-2}$
$pp \to \text{exclusive } J/\psi,\, \Upsilon$	$\gamma^*(gg) \to J/\psi, \Upsilon$	g	$x \gtrsim 10^{-5}, 10^{-4}$
$pp \to \Upsilon X$	$gq \to \gamma q, g\bar{q} \to \gamma\bar{q}$	g	$x \gtrsim 0.005$

with

$$r \equiv \frac{\sigma_{CC}(\bar{\nu}_\mu N)}{\sigma_{CC}(\nu_\mu N)}. \tag{36.34}$$

The ratio

$$r \equiv \frac{\frac{1}{3} + \epsilon}{1 + \frac{\epsilon}{3}}, \quad \text{with } \epsilon \equiv \frac{\int_0^1 x\bar{q}^p(x)}{\int_0^1 xq^p(x)}. \tag{36.35}$$

In the naive-quark model, $\epsilon = 0$ and $r = \frac{1}{3}$. These tree-level relations must be corrected to include the heavier charm and strange quarks, the Cabibbo–Kobayashi–Maskawa (CKM) mixing angles and radiative corrections. These were included by the CHARM Collaboration, which determined $r_{avg} = 0.44$ and $\sin^2\theta_W = 0.236 \pm 0.012(m_c - 1.5) \pm 0.005(exp) \pm 0.003(theory)$. They also obtain an experimental value for $\rho = 1.030 \pm 0.020$. Combining results from similar experiments by the FMMF (Bogert et al., 1985), CCFR (Reutens et al., 1985) and CDHS (Abramowicz et al., 1986) collaborations, CHARM obtains $\sin^2 \theta_W^{avg} = 0.233 \pm 0.004(exp) \pm 0.005(theory)$. The present value for $\sin^2 \theta_W(on\text{-}shell) = 0.22343 \pm 0.00007$ (Tanabashi et al., 2018).

There is also the Paschos–Wolfenstein ratio (Paschos and Wolfenstein, 1973),

$$R^- = \frac{\sigma_{NC}(\nu_\mu N) - \sigma_{NC}(\bar{\nu}_\mu N)}{\sigma_{CC}(\nu_\mu N) \quad \sigma_{CC}(\bar{\nu}_\mu N)} = g_L^2 - g_R^2 = \frac{1}{2} - \sin^2 \theta_W, \tag{36.36}$$

which was used to determine $\sin^2 \theta_W$. Since it has larger uncertainties due to taking a difference, it required a more accurate experiment. The NuTev Collaboration (McFarland et al., 2002) determined $\sin^2 \theta_W$ using the Paschos–Wolfenstein relation. They found $\sin^2 \theta_W = 0.22773 \pm 0.00135(stat) \pm 0.00093(syst)$ plus small corrections due the top quark and Higgs loops. This is 3σ above the present result from precision electroweak data and may be due to several possible unaccounted-for effects.

Let me end this section with an update on the structure of the nucleon in quantum chromodynamics. There are more than just up-, down-, strange-quark and gluon PDFs. At higher energies, charm- and bottom-quark content can be relevant. In addition to the spin-independent structure functions, F_1, F_2 and F_3, there are spin-dependent ones as well. Finally, in Table 36.1 (Tanabashi et al., 2018), we show the data which are used to obtain all the PDFs. Moreover, there are several international collaborations who are in the business of evaluating the PDFs.

37 Weak Interactions of Quarks

Up until now we have basically ignored quarks. Beta decay of hadrons is now understood at the quark level. Consider the following processes for $n \to p + e^- + \bar{\nu}_e$, Fig. 37.1, and for $\Lambda \to p + e^- + \bar{\nu}_e$, Fig. 37.2. The latter process is suppressed with a branching ratio, $BR(\Lambda \to p + e^- + \bar{\nu}_e) \sim 10^{-3}$. In fact, for Λ decay, the processes $\Lambda \to p\pi^-$, $n\pi^0$ dominate, Figs. 37.3 and 37.4, with $BR(\Lambda \to p\pi^-, n\pi^0) \sim 1$.

Historically, in the first attempts to incorporate quarks and describe the above processes, one defined the weak eigenstates

$$q \equiv \begin{pmatrix} u \\ d' \end{pmatrix}, \quad \bar{u}, \; \bar{d}, \; \bar{s} \tag{37.1}$$

with

$$d' \equiv d \cos\theta_C + s \sin\theta_C. \tag{37.2}$$

θ_C is the Cabibbo angle (Cabibbo, 1963; Cabibbo et al., 2003, 2004) with

$$\sin\theta_C \sim 0.22. \tag{37.3}$$

The weak couplings to $W_\pm$ are given by

$$\mathcal{L} \sim -\frac{g}{2\sqrt{2}} (W_-^\mu \, J_\mu^h + h.c.), \tag{37.4}$$

where the weak hadronic current is given by

$$J_\mu^h \equiv (\bar{\psi}_d \cos\theta_C + \bar{\psi}_s \sin\theta_C)\gamma_\mu(1 - \gamma_5)\psi_u, \tag{37.5}$$

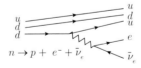

Fig. 37.1 Neutron beta decay.

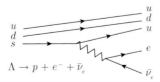

Fig. 37.2 Lambda beta decay.

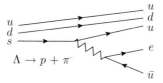

Fig. 37.3   Hadronic decay of lambda to $p\pi^-$.

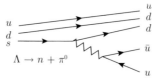

Fig. 37.4   Hadronic decay of lambda to $n\pi^0$.

with the Dirac fields given by

$$\psi_u \equiv \left(\begin{array}{c} u \\ i\sigma_2 \bar{u}^* \end{array} \right), \tag{37.6}$$

$$\psi_d \equiv \left(\begin{array}{c} d \\ i\sigma_2 \bar{d}^* \end{array} \right),$$

$$\psi_s \equiv \left(\begin{array}{c} s \\ i\sigma_2 \bar{s}^* \end{array} \right).$$

37.1 Some Classic Weak Decays

Neutron Beta Decay

The neutron beta decay amplitude, $n \to pe\bar{\nu}_e$, is given by W exchange and since the neutron–proton mass difference is so small (of order 1.3 MeV), we can use the effective low-energy theory. We have

$$T \propto \frac{G_F}{\sqrt{2}} \, \cos(\theta_C)\bar{u}_p\gamma^\mu(1 - g_A\gamma_5)u_n\bar{u}_e\gamma_\mu(1 - \gamma_5)v_{\nu_e}. \tag{37.7}$$

This gives the decay rate

$$\Gamma_n = \frac{G_F^2 \cos(\theta_C)^2}{2\pi^3}(1 + 3g_A^2) \int_{m_e}^{\Delta} p_e E_e(\Delta - E_e)^2 dE_e \approx \frac{G_F^2 \cos(\theta_C)^2(1 + 3g_A^2)\Delta^5}{192\pi^3}, \tag{37.8}$$

with $\Delta = m_n - m_p$. It is this process which allows for the measurement of g_A, the axial-vector constant that was discussed in Section 26.4.

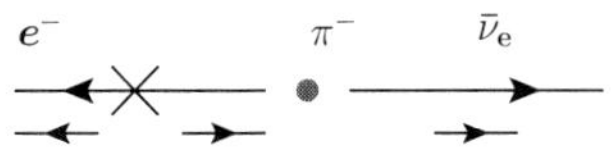

Fig. 37.5 Decay of $\pi^- \to e^- + \bar{\nu}_e$.

Pion Decay

In the case of pseudo-scalar boson decay, for example $\pi^- \to e^- \bar{\nu}_e$, we have the amplitude given by

$$T \propto \frac{G_F}{\sqrt{2}} \cos(\theta_C) \sqrt{2} F_\pi q^\mu \bar{u}_e \gamma_\mu (1 - \gamma_5) v_{\nu_e} \tag{37.9}$$
$$= G_F \cos(\theta_C) F_\pi m_e \bar{u}_e (1 - \gamma_5) v_{\nu_e}.$$

This gives a pion partial decay rate into electrons and electron neutrinos as

$$\Gamma(\pi^- \to e^- \bar{\nu}_e) = m_\pi \frac{(G_F \cos(\theta_C) F_\pi m_e)^2}{4\pi} \left(1 - \frac{m_e^2}{m_\pi^2}\right)^2. \tag{37.10}$$

By measuring the $\pi^\pm$ lifetime one obtains the pion decay constant $F_\pi \sim 93$ MeV. For $\pi^- \to \mu^- \bar{\nu}_\mu$ we just replace m_e with m_μ. The factor of the electron or muon mass in the pion decay rate is a consequence of an approximate chiral symmetry. In Fig. 37.5 we describe the decay of a $\pi^- \to e^- \bar{\nu}_e$. The arrows under the lines correspond to chirality, which in the massless limit is equivalent to helicity. Chirality is conserved at the vertex. But a spin-0 pion cannot decay to two massless fermions, since their helicities add up to one. The **X** on the electron line corresponds to a mass insertion which flips the chirality and allows the process to occur! Hence, as a consequence of approximate chirality conservation, the ratio of partial widths,

$$R \equiv \frac{\Gamma(\pi^- \to e^- \bar{\nu}_e)}{\Gamma(\pi^- \to \mu^- \bar{\nu}_\mu)} = \frac{m_e^2}{m_\mu^2} \left(\frac{m_\pi^2 - m_e^2}{m_\pi^2 - m_\mu^2}\right)^2 = 1.28 \times 10^{-4}. \tag{37.11}$$

In order to make neutrino beams one crashes a beam of protons onto a fixed target, which produces a host of pions, kaons and other hadrons. The pions are then channeled via magnets into a long horn where they decay, predominantly into muon neutrinos or anti-neutrinos, depending on whether they were π^- or π^+, respectively.

37.2 Flavor-Changing Neutral Currents

However, there is an immediate problem with this approach when one includes neutral currents. The problem is due to flavor-changing neutral currents (FCNCs).

Consider the weak coupling to the Z given by

$$\mathcal{L} \sim -\frac{g}{\cos\theta_W}Z^\mu\left[q^*\bar{\sigma}_\mu(T_3 - \sin^2\theta_W Q)q + \bar{u}^*\bar{\sigma}_\mu(T_3 - \sin^2\theta_W Q)\bar{u}\right. \tag{37.12}$$
$$\left. + \bar{d}^*\bar{\sigma}_\mu(T_3 - \sin^2\theta_W Q)\bar{d} + \bar{s}^*\bar{\sigma}_\mu(T_3 - \sin^2\theta_W Q)\bar{s}\right]$$
$$\equiv -\frac{g}{\cos\theta_W}Z^\mu\left[u^*\,\bar{\sigma}_\mu\left(\frac{1}{2} - \frac{2}{3}\sin^2\theta_W\right)u + d'^*\bar{\sigma}_\mu\left(-\frac{1}{2} + \frac{1}{3}\sin^2\theta_W\right)d'\right.$$
$$\left. + \bar{u}^*\bar{\sigma}_\mu(\frac{2}{3}\sin^2\theta_W)\bar{u} + \bar{d}^*\bar{\sigma}_\mu(-\frac{1}{3}\sin^2\theta_W)\bar{d} + \bar{s}^*\bar{\sigma}_\mu\left(\frac{1}{3}\sin^2\theta_W\right)\bar{s}\right].$$

The term with d' is the problem. It gives a term in the Lagrangian at tree level with the FCNC,

$$\mathcal{L} \propto \cos\theta_C \sin\theta_C(d^*\bar{\sigma}_\mu\, s + s^*\bar{\sigma}_\mu\, d)\left(-\frac{1}{2} + \frac{1}{3}\sin^2\theta_W\right). \tag{37.13}$$

This current changes strangeness by one unit, i.e. $\Delta S = 1$. It contributes to processes such as

$$K^0 \to \mu^+\mu^- \tag{37.14}$$

as in Fig. 37.6. We can estimate the predicted rate for this process. It should be comparable to the rate for the process $K^+ \to \mu^+\nu_\mu$, Fig. 37.7, i.e. we expect

$$\Gamma(K^0 \to \mu^+\mu^-) \approx \Gamma(K^+ \to \mu^+\nu_\mu). \tag{37.15}$$

However, experimentally we have

$$\Gamma(K^0 \to \mu^+\mu^-) \sim 10^{-8}\Gamma(K^+ \to \mu^+\nu_\mu). \tag{37.16}$$

This is the problem! This $\Delta S = 1$ neutral-current process is suppressed. In addition, $\Delta S = 2$ neutral-current processes are also suppressed, for example in the process

$$K^0 \to \bar{K}^0 \tag{37.17}$$

in Fig. 37.8. The amplitude is of order G_F. It contributes to the mass splitting. However, experimentally we find

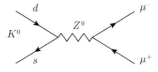

Fig. 37.6 The decay $K^0 \to \mu^+\mu^-$.

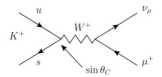

Fig. 37.7 The decay $K^+ \to \mu^+\nu_\mu$.

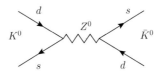

Fig. 37.8 $K^0 - \bar{K}^0$ mixing.

$$\frac{\Delta m_{K^0 - \bar{K}^0}}{m_K} \sim 10^{-14}, \tag{37.18}$$

which is of order $(G_F \Lambda^2_{QCD})^2$.

The solution to the FCNC problem was discovered by Glashow, Iliopoulos and Maiani (Glashow et al., 1970) and is known as the GIM mechanism. They introduced a fourth quark into the theory, i.e. the charm quark, c, such that (c, s) form a second quark family given by

$$\begin{pmatrix} c \\ s' \end{pmatrix}, \; \bar{c}, \; \bar{s} \tag{37.19}$$

where

$$s' \equiv s \cos \theta_C - d \sin \theta_C. \tag{37.20}$$

We then have the weak eigenstate fields d', s' given by a 2×2 orthogonal transformation, V_C, in terms of the mass eigenstate fields d, s.

$$\begin{pmatrix} d' \\ s' \end{pmatrix} = \begin{pmatrix} \cos \theta_C & \sin \theta_C \\ -\sin \theta_C & \cos \theta_C \end{pmatrix} \begin{pmatrix} d \\ s \end{pmatrix} \equiv V_C \begin{pmatrix} d \\ s \end{pmatrix}. \tag{37.21}$$

Now the contribution to the $\Delta S = 1$ neutral current becomes

$$\mathcal{L} \sim -\frac{g}{\cos \theta_W} Z^\mu \left[(d'^* \, \bar{\sigma}_\mu \, d' + s'^* \, \bar{\sigma}_\mu \, s') \left(-\frac{1}{2} + \frac{1}{3} \sin^2 \theta_W \right) \right]. \tag{37.22}$$

But

$$(d'^* \bar{\sigma}_\mu d' + s'^* \bar{\sigma}_\mu \, s') \equiv (d' \; s')^* \bar{\sigma}_\mu \begin{pmatrix} d' \\ s' \end{pmatrix} = (d \; s)^* \underbrace{V_C^\dagger V_C}_{\equiv \mathbb{I}} \, \bar{\sigma}_\mu \begin{pmatrix} d \\ s \end{pmatrix}$$

$$= (d^* \bar{\sigma}_\mu d + s^* \bar{\sigma}_\mu s), \tag{37.23}$$

i.e. there are no FCNC effects at order G_F!

Note, FCNCs are not eliminated completely. They reappear at higher orders. For example, consider $K^0 \to \mu^+ \mu^-$, Fig. 37.9. The weak hadronic current is now of the form

$$J^h_\mu \equiv (\bar{\psi}_d, \; \bar{\psi}_s) V_C^\dagger \gamma_\mu (1 - \gamma_5) \begin{pmatrix} \psi_u \\ \psi_c \end{pmatrix}. \tag{37.24}$$

The result is of order

$$(G_F^2 V_{is}^* V_{id} m_i^2), \quad i = u, c. \tag{37.25}$$

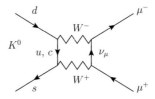

Fig. 37.9 $K^0 \to \mu^+\mu^-$ at one loop.

Hence, because of the GIM mechanism, the amplitude vanishes if $m_u = m_c$. We then have the result of Gaillard and Lee (1974),

$$\frac{\Gamma(K^0 \to \mu^+\mu^-)}{\Gamma(K^+ \to \mu^+\nu_\mu)} \simeq \frac{\sin^2\theta_C (m_c^2 G_F^2)^2}{\sin^2\theta_C G_F^2} = (m_c^2 G_F)^2 \simeq (10^{-5} \times (1.5)^2)^2 \left(\frac{m_c}{1.5\ GeV}\right)^4,$$

(37.26)

which was used to place an upper limit on the charm-quark mass. Given the experimental results

$$\Gamma(K_L^0 \to \mu^+\mu^-) = \frac{\mathrm{BR}(K^0 \to \mu^+\mu^-)}{\tau_{K_L^0}} = \frac{6.8 \times 10^{-9}}{5 \times 10^{-8}\ \mathrm{s}}$$

(37.27)

$$\Gamma(K^+ \to \mu^+\nu_\mu) = \frac{0.63}{1.24 \times 10^{-8}\ \mathrm{s}}$$

implies

$$\frac{\Gamma(K^0 \to \mu^+\mu^-)}{\Gamma(K^+ \to \mu^+\nu_\mu)} \sim 2.7 \times 10^{-9}$$

(37.28)

and

$$m_c \le 2.2\ \mathrm{GeV}.$$

(37.29)

The charm-quark mass is about 1.5 GeV, so this result was pretty good. However, we will see that, in fact, the top quark actually dominates the result.

PART IX

THE STANDARD MODEL

Three-Family Model

38.1 Prologue

The charm quark was discovered in 1974 by two groups. They discovered the $c\bar{c}$ charmonium bound state, known as the J/ψ. A group led by B. Richter discovered the $\psi_{1S}(3.1)$ and $\psi_{2S}(3.7)$ at Stanford Linear Accelerator Center (SLAC) (Augustin et al., 1974; Abrams et al., 1974) as a resonance in $e^+e^- \to \mu^+\mu^-$ interactions. The ψ_{2S} decays to $\psi_{1S} + \pi^+\pi^-$. At the same time, S. Ting at Brookhaven National Laboratory (BNL) discovered the $J(3.1)$ as a narrow e^+e^- resonance in the reaction, $p + Be \to e^+e^-X$.

The charmonium bound states are well described by a non-relativistic Schrödinger equation with both a linear and Coulomb potential (Eichten et al., 1975, 1978). The success of this simple description for all the charmonium states, including vector mesons, pseudo-scalars, etc., and their decays, led to a paradigm shift where quarks were now regarded as real particles, and not just fictitious mathematical tools for calculating the hadron spectrum and deep inelastic scattering. Hadrons containing a single charm particle and a light anti-quark are known as $\bar{D}$ mesons.

A search for heavy leptons between the years 1972 and 1977 by M. Perl at SLAC led to the discovery of the tau lepton in the process $e^+e^- \to \tau^+\tau^- \to e^+\mu^- + X$ (Perl et al., 1975; Perl, 1976), where the decay $\tau^- \to e^- + \bar{\nu}_e + \nu_\tau$ or $\tau \to \mu^- + \bar{\nu}_\mu + \nu_\tau$ was assumed (Tsai, 1971). There were no direct predictions for this third family of charged leptons; however, the signature for the production of these heavy leptons had little background.[1] Nevertheless, there was a prediction for at least a third family of quarks which we discuss in this chapter.

A narrow resonance in the $\mu^+\mu^-$ channel with a mass of 9.5 GeV was discovered at Fermilab in the process $p + Pt \to \mu^+\mu^- + X$ (Herb et al., 1977). It was the Upsilon resonance $\Upsilon(9.5)$ or a $b\bar{b}$ bound state. The $\Upsilon_{1S}(9.5)$, $\Upsilon_{2S}(10)$, $\Upsilon_{3S}(10.36)$ are below threshold for decays into $\bar{B}$ mesons made of a bottom quark and a light anti-quark. On the other hand, the $\Upsilon_{4S}(10.6)$ decays directly to $\bar{B}B$. These processes were studied for many years with the e^+e^- storage ring, CESR, at Cornell (Loomis, 1981). As might be expected, the bottom quark in B mesons has a weak decay into up or charm quarks. It was thus part of an electroweak doublet. However, it took many more years to find the top quark, which was only discovered in 1995 in $\bar{p}p$ collisions at Fermilab (Abe et al., 1995).

---

[1] We'll discuss the discovery of the tau neutrino later.

38.2 Six-Quark Model

The six-quark model was introduced by Kobayashi and Maskawa to incorporate **CP** violation. The complete Standard Model now has three families of quarks and leptons with the local gauge symmetry $SU(3) \otimes SU(2)_L \otimes U(1)_Y$. The quark and lepton fields are given in terms of the left-handed Weyl spinors:

$$q = \begin{pmatrix} u \\ d \end{pmatrix}_i, \ u_i, \ \bar{d}_i, \quad \ell_i = \begin{pmatrix} \nu \\ e \end{pmatrix}_i, \ \bar{e}_i, \ \bar{\nu}_i, \quad i - 1, 2, 3, \quad (38.1)$$

with Standard Model charges given by

$$\mathcal{T}_A\, q = \frac{1}{2}\lambda_A q, \ \mathcal{T}_A\bar{u} = -\frac{1}{2}\lambda_A^T \bar{u}, \ \mathcal{T}_A\bar{d} = -\frac{1}{2}\lambda_A^T\bar{d},$$
$$\mathcal{T}_A\ell = \mathcal{T}_A\bar{e} = \mathcal{T}_A\bar{\nu} = 0 \quad (38.2)$$

(where λ_A, $A = 1, \cdots, 8$ are the 3×3 Gell-Mann matrices).

$$T_a q = \frac{1}{2}\tau_a q, \ T_a\ell = \frac{1}{2}\tau_a\ell,$$
$$T_a\bar{u} = T_a\bar{d} = T_a\bar{e} = T_a\bar{\nu} = 0 \quad (38.3)$$

(where τ_a, $a = 1, 2, 3$ are the 2×2 Pauli matrices). We use the convention $Tr(\mathcal{T}_A\, \mathcal{T}_B) = \frac{1}{2}\delta_{AB}$ and $Tr(T_a\, T_b) = \frac{1}{2}\delta_{ab}$.

$$Y q = \frac{1}{3}\, q, \ Y\bar{u} = -\frac{4}{3}\bar{u}, \ Y\, \bar{d} = \frac{2}{3}\bar{d}, \ Y\, \ell = -\ell, \ Y\bar{e} = +2\bar{e}, \ Y\bar{\nu} = 0. \quad (38.4)$$

All gauge interactions are given by the covariant derivative

$$D_\mu \equiv \left(\partial_\mu + ig T_a W_{\mu a} + ig'\frac{Y}{2}B_\mu + ig_s \mathcal{T}_A G_{\mu A} \right), \quad (38.5)$$

where $\mathcal{T}_A$, T_a, Y are the generators of $SU(3)_{color}$, $SU(2)_L$, $U(1)_Y$, respectively. Note, we also have the Higgs doublet

$$\phi = \begin{pmatrix} \phi^+ \\ \phi^0 \end{pmatrix}, \quad (38.6)$$

with

$$\mathcal{T}_A\phi = 0, \ T_a\phi = \frac{1}{2}\tau_a\phi, \ Y\phi = \phi, \quad (38.7)$$

and since neutrinos have mass we have allowed for the possibility of adding the left-handed anti-neutrino fields, $\bar{\nu}_i$. They have no gauge quantum numbers. Finally, this is defined as the *weak eigenstate basis* for the fields, as opposed to the *mass eigenstate basis*. Let's now discuss mass.

38.3 Yukawa Couplings

Consider the most general Yukawa couplings.

Leptons:

$$\lambda^e_{jk} \bar{e}_j \phi^\dagger \ell_k + h.c. \tag{38.8}$$

and given

$$\langle \phi \rangle = \begin{pmatrix} 0 \\ \frac{v}{\sqrt{2}} \end{pmatrix} \tag{38.9}$$

we obtain the charged-lepton mass term,

$$\bar{e}^T m^e e + h.c., \tag{38.10}$$

with

$$m^e_{jk} \equiv \lambda^e_{jk} v / \sqrt{2}. \tag{38.11}$$

Down quarks:

$$\lambda^d_{jk} \bar{d}_j \phi^\dagger q_k + h.c. \tag{38.12}$$

and we obtain the down-quark mass term,

$$\bar{d}^T \, m^d \, d + h.c., \tag{38.13}$$

with

$$m^d_{jk} \equiv \lambda^d_{jk} v / \sqrt{2}. \tag{38.14}$$

Up quarks:

$$\lambda^u_{jk} \bar{u}_j \tilde{\phi}^\dagger q_k + h.c. \tag{38.15}$$

and we obtain the up-quark mass term,

$$\bar{u}^T m^u u + h.c., \tag{38.16}$$

with

$$m^u_{jk} \equiv \lambda^u_{jk} v / \sqrt{2}. \tag{38.17}$$

Note, as discussed earlier, the Higgs field

$$\tilde{\phi} \equiv i\tau_2 \phi^* = \begin{pmatrix} \phi^{0*} \\ -\phi^{+*} \end{pmatrix} \tag{38.18}$$

transforms as an $SU(2)_L$ doublet with $Y = -1$.

Neutrinos: with the addition of left-handed anti-neutrinos we can have a Yukawa term of the form

$$\lambda^\nu_{jk} \bar{\nu}_j \tilde{\phi}^\dagger \ell_k + h.c. \tag{38.19}$$

and we obtain the neutrino mass term,

$$\bar{\nu}^T m^\nu \nu + h.c., \tag{38.20}$$

with

$$m^\nu_{jk} \equiv \lambda^\nu_{jk} v/\sqrt{2}. \tag{38.21}$$

We will see later that neutrinos have mass, but $m_\nu \ll m_e$, and thus the Yukawa couplings must be exceedingly small. We might also have an effective neutrino mass term of the form

$$\frac{g_{ij}}{M} (\tilde{\phi}^\dagger \ell_i)(\tilde{\phi}^\dagger \ell_j). \tag{38.22}$$

This is the so-called Weinberg operator, which has dimensions of mass to the power of five and must be considered as an effective interaction due to new physics at a scale M, with $M \gg v$. It gives a Majorana mass matrix,

$$m^\nu_{ij} = \frac{g_{ij} v^2}{M}. \tag{38.23}$$

For the moment we will assume that $m_\nu = 0$.

38.4 Mass Matrices and Mass Eigenstate Basis

Consider the charged-fermion Yukawa matrices,

$$\lambda^e_{jk}, \ \lambda^u_{jk}, \ \lambda^d_{jk}, \tag{38.24}$$

which may be a consequence of some more fundamental theory. We don't know now what that theory is, but we can use low-energy data to discover some properties of that theory. For the sake of this discussion, let's assume that the Yukawa matrices are, in general, arbitrary complex 3×3 matrices, which must be diagonalized to find the mass eigenstates. The fermion kinetic energy terms have a large global symmetry that can be used to redefine the fields in order to diagonalize the mass matrices. Given the Lagrangian kinetic terms,

$$\mathcal{L}_{kin} = \ell^\dagger i \not{D} \ell + \bar{e}^\dagger i \not{D} \bar{e} + q^\dagger i \not{D} q + \bar{u}^\dagger i \not{D} \bar{u} + \bar{d}^\dagger i \not{D} \bar{d}, \tag{38.25}$$

we find the global symmetry,

$$U^\ell_3 \otimes U^{\bar{e}}_3 \otimes U^q_3 \otimes U^{\bar{u}}_3 \otimes U^{\bar{d}}_3. \tag{38.26}$$

Consider for example charged leptons, where the mass term in the Lagrangian is

$$\mathcal{L}_{mass} \supset \bar{e}^T_0 m^e e_0 + h.c., \tag{38.27}$$

where now we define the fields with the 0 subscript to be fields in the weak eigenstate basis and the new fields,

$$\bar{e}_0^T \equiv \bar{e}^T U_{\bar{e}}^\dagger, \quad e_0 \equiv U_\ell e, \tag{38.28}$$

to be mass eigenstates, since the unitary matrices U_ℓ and $U_{\bar{e}}$ are defined such that

$$U_{\bar{e}}^\dagger m^e U_\ell \equiv m_{diag}^e = \begin{pmatrix} m_e & 0 & 0 \\ 0 & m_\mu & 0 \\ 0 & 0 & m_\tau \end{pmatrix}. \tag{38.29}$$

As a result we now have

$$\mathcal{L}_{mass} \supset \bar{e}^T m_{\text{diag}}^e e + h.c.. \tag{38.30}$$

Theorem 38.1 *A complex matrix M can be diagonalized by two unitary transformations.*

Proof First this can be made plausible by the fact that $MM^\dagger$ and $M^\dagger M$ are both Hermitian matrices, which can be diagonalized by two unitary transformations with positive eigenvalues (assuming no zero eigenvalues). We have

$$U^\dagger M^\dagger M U = M_D^2 \tag{38.31}$$

or

$$[U^\dagger M^\dagger M U]_{\alpha\gamma} = \delta_{\alpha\gamma}(M_D^2)_\alpha > 0 \tag{38.32}$$

and

$$(M_D^2)_\alpha = \sum_\beta (U^\dagger M^\dagger)_{\alpha\beta}(MU)_{\beta\alpha} = \sum_\beta (MU)_{\beta\alpha}^*(MU)_{\beta\alpha} \equiv \sum_\beta |(MU)_{\beta\alpha}|^2. \tag{38.33}$$

Define M_D as the positive square root of the matrix M_D^2 and define

$$A \equiv UM_D U^\dagger. \tag{38.34}$$

Then we have

$$A^2 \equiv UM_D^2 U^\dagger \equiv M^\dagger M. \tag{38.35}$$

Define

$$V \equiv MA^{-1} \tag{38.36}$$

such that

$$V^\dagger V = A^{-1\dagger} M^\dagger M A^{-1} \tag{38.37}$$

but

$$A^{-1\dagger} \equiv A^{-1} = UM_D^{-1}U^\dagger \tag{38.38}$$

and therefore we have

$$V^\dagger V = A^{-1}M^\dagger M A^{-1} = U M_D^{-1}(U^\dagger M^\dagger M U) M_D^{-1} U^\dagger = U U^\dagger \equiv \mathbb{I}, \qquad (38.39)$$

i.e. V is unitary. Now,

$$M = VA = (VU)M_D U^\dagger \equiv \bar{U} M_D U^\dagger \qquad (38.40)$$

and, finally,

$$\bar{U}^\dagger M U = M_D, \qquad (38.41)$$

where both U and $\bar{U}$ are unitary and

$$\bar{U}^\dagger M M^\dagger \bar{U} \equiv M_D^2. \qquad (38.42)$$

$\square$

For up quarks we have

$$U_{\bar{u}}^\dagger m^u U_q = m_D^u \equiv \begin{pmatrix} m_u & 0 & 0 \\ 0 & m_c & 0 \\ 0 & 0 & m_t \end{pmatrix}. \qquad (38.43)$$

And for down quarks,

$$U_{\bar{d}}^\dagger m^d U_d = m_D^d \equiv \begin{pmatrix} m_d & 0 & 0 \\ 0 & m_s & 0 \\ 0 & 0 & m_b \end{pmatrix}. \qquad (38.44)$$

Note, in general,

$$U_d \neq U_q. \qquad (38.45)$$

This corresponds to the possibility of up–down symmetry breaking for the states in quark doublets.

Now consider the weak currents. Previously, the weak currents were written in terms of the weak eigenstates, but these are not necessarily the mass eigenstates. Again, we shall now explicitly use the subscript, 0, to distinguish the weak eigenstate basis. The $W_\pm$ vertices are given by

$$W_+^\mu \, (q_0^\dagger \tau_+ \bar{\sigma}_\mu q_0 + \ell_0^\dagger \tau_+ \bar{\sigma}_\mu \ell_0) + h.c. \qquad (38.46)$$
$$= W_+^\mu \, (u_0^\dagger \bar{\sigma}_\mu d_0 + \nu_0^\dagger \bar{\sigma}_\mu e_0) + h.c. \quad .$$

Now let's change to the mass eigenstate basis. We have

$$e_0 = U_e e, \quad \nu_0 = U_\nu \nu, \qquad (38.47)$$
$$u_0 = U_u u, \quad d_0 = U_d d, \quad \text{etc},$$

where $U_u \equiv U_q$ and $U_e \equiv U_\ell$. In terms of the mass eigenstates, the W vertices become

$$W_+^\mu (u^\dagger \bar{\sigma}_\mu (U_u^\dagger U_d) d + \nu^\dagger \bar{\sigma}_\mu (U_\nu^\dagger U_e) e) + h.c. \qquad (38.48)$$

Note, for neutrinos, we will choose

$$U_\nu \equiv U_e, \tag{38.49}$$

which defines the neutrino flavor basis, i.e. a W_+ changes an electron into, by definition, an electron neutrino, etc. In addition, for massless neutrinos, we are free to make this choice and there would be no leptonic mixing angles, and individual lepton numbers, L_e, L_μ, L_τ, are conserved. However, we now know, due to neutrino oscillations, that separate lepton numbers are not conserved. It is still possible, however, that the sum, $L = L_e + L_\mu + L_\tau$, is still conserved.

For quarks, however, we define the Cabibbo–Kobayashi–Maskawa (CKM) mixing matrix, given by

$$V_{CKM} \equiv U_u^\dagger U_d, \tag{38.50}$$

and now we have

$$W_+^\mu (u^\dagger \bar{\sigma}_\mu V_{CKM} d + \nu^\dagger \bar{\sigma}_\mu e) + h.c. \tag{38.51}$$

We simplify notation with $V \equiv V_{CKM}$ with $V^\dagger V \equiv \mathbb{I}$, a 3×3 unitary matrix. Note, since neutral currents are diagonal in flavor space, no mixing angles appear in couplings to the photon or Z boson.

The CKM matrix has nine real parameters, but an orthogonal 3×3 matrix has only three arbitrary angles. Thus, V contains three real angles and six phases. However, not all the phases in V are observable. After we diagonalize all mass terms in the Lagrangian, we still have $U(1)^{u,d}$ vector-like phase transformations of all fermions of the form

$$U(1)^{u,\,d} = \begin{pmatrix} e^{i\alpha_{u,d}} & 0 & 0 \\ 0 & e^{i\beta_{u,d}} & 0 \\ 0 & 0 & e^{i\gamma_{u,d}} \end{pmatrix}. \tag{38.52}$$

We can use these global phase redefinitions of the fields to put the complex 3×3 matrix, V, in the form

$$V \equiv \begin{pmatrix} a & b & c \\ d & A_{22} & A_{23} \\ e & A_{32} & A_{33} \end{pmatrix}, \tag{38.53}$$

where a, b, c, d, e are real. Thus, the complex $V = V_{CKM}$ has at most 13 real observable parameters. But, once again, $V^\dagger V = \mathbb{I}$ implies nine constraints. Hence, there are only four observable real parameters which correspond to three angles and one phase.

38.5 Kobayashi–Maskawa Model and Three Families

In general, for n generations of quarks, let's count the number of parameters in V, where V is an $n \times n$ unitary matrix. We have

$$n^2 - \underbrace{2n - 1}_{\text{phase redefinitions}} = n^2 - 2n + 1. \tag{38.54}$$

For $n = 2$, there is only one real angle,

$$V_C = \begin{pmatrix} \cos\theta_C & \sin\theta_C \\ -\sin\theta_C & \cos\theta_C \end{pmatrix}, \tag{38.55}$$

and V_C is real. However, for $n = 3$, there are four parameters, three real angles and one phase. It was therefore concluded in Kobayashi and Maskawa (1973) that in order for V_{CKM} to be complex, we must have the number of generations $n \geq 3$.

But what is significant about a complex V_{CKM}? We now show that a complex CKM matrix breaks **CP** invariance. Consider the charged-current Lagrangian given by

$$\mathcal{L} = g W_+^\mu(x) u^\dagger(x) \bar{\sigma}_\mu V d(x) + g W_-^\mu(x) d^\dagger(x) \bar{\sigma}_\mu V^\dagger u(x) \tag{38.56}$$

and the **CP** transformations given by

$$u_{CP}(x') = i\sigma_2 u^*(x), \quad d_{CP}(x') = i\sigma_2 d^*(x), \tag{38.57}$$
$$u_{CP}^\dagger(x') = -i u^T(x)\sigma_2 \quad \text{with } x^{0\prime} = x^0, \ \vec{x}\,' = -\vec{x}$$
$$W_{+CP}^i(x') = -\eta_W W_-^i(x), \quad W_{+CP}^0(x') = \eta_W W_-^0(x).$$

Then, the **CP** transform of the Lagrangian is given by

$$\begin{aligned}
\mathcal{L}_{CP} &= g\eta_W W_-^\mu(x)(-i u^T(x)\sigma_2)\sigma_\mu V(i\sigma_2 d^*(x)) \\
&\quad + g\eta_W W_+^\mu(x)(-i d^T(x)\sigma_2)\sigma_\mu V^\dagger(i\sigma_2 u^*(x)) \\
&= -g\eta_W W_-^\mu(x) d^\dagger(x)(\sigma_2\sigma_\mu\sigma_2)^T V^T u(x) \\
&\quad - g\eta_W W_+^\mu(x) u^\dagger(x)(\sigma_2\sigma_\mu\sigma_2)^T V^* d(x).
\end{aligned} \tag{38.58}$$

Note, the minus sign in the second equality is due to the anti-commutation of the fermion fields. Also, σ_μ replaced $\bar{\sigma}_\mu$ to take into account the different transformation of the spatial and time component of the gauge field. Now, using the identity

$$(\sigma_2\sigma_\mu\sigma_2)^T = (\bar{\sigma}_\mu^*)^T \equiv \bar{\sigma}_\mu \tag{38.59}$$

and choosing $\eta_W = -1$, we find

$$\mathcal{L}_{CP} = g W_+^\mu(x) u^\dagger(x) \bar{\sigma}_\mu V^* d(x) + g W_-^\mu(x) d^\dagger(x) \bar{\sigma}_\mu V^T u(x) \equiv \mathcal{L} \tag{38.60}$$

if and only if

$$V \equiv V^*. \tag{38.61}$$

Thus **CP** invariance requires $V = V^*$. Note:

$$\mathcal{L} = \mathcal{L}_{gauge} + \mathcal{L}_{scalar} + \mathcal{L}_{fermion}, \tag{38.62}$$

with

$$\mathcal{L}_{fermion} = \mathcal{L}_{kinetic} + \mathcal{L}_{Yukawa} \tag{38.63}$$

and all other sectors of the theory, other than the coupling to the W, are explicitly **CP** invariant.

For two families, the theory is automatically **CP** invariant. For three families, Kobayashi and Maskawa noted that there can be one **CP**-violating phase. As far as we know from experiment, all CP violation thus far observed in the quark sector can be described by this one phase in the CKM matrix.

39 Determining V_{CKM} and Quark Masses

In this chapter we review the experimental determination of the mass and mixing parameters in the Standard Model. This 20-year experimental program confirmed the three-family model of Kobayashi–Maskawa and the origin of CP violation in the quark sector, as the single phase in the Cabibbo–Kobayashi-Maskawa (CKM) matrix. It is these parameters which will hopefully be explained by some new physics beyond the Standard Model.

We have

$$V \equiv V_{CKM} = \begin{pmatrix} V_{ud} & V_{us} & V_{ub} \\ V_{cd} & V_{cs} & V_{cb} \\ V_{td} & V_{ts} & V_{tb} \end{pmatrix}. \tag{39.1}$$

A particularly useful and suggestive form of the CKM matrix is known as the Wolfenstein form (Wolfenstein, 1983; Buras et al., 1994), which is explicitly unitary to order λ^3. It is given by

$$V \simeq \tag{39.2}$$
$$\begin{pmatrix} 1 - \frac{\lambda^2}{2} - \frac{\lambda^4}{8} & \lambda & A\lambda^3(\rho - i\eta) \\ -\lambda + \frac{1}{2}A^2\lambda^5[1 - 2(\rho + i\eta)] & 1 - \frac{\lambda^2}{2} - \frac{1}{8}\lambda^4(1 + 4A^2) & A\lambda^2 \\ A\lambda^3[1 - (1 - \frac{\lambda^2}{2})(\rho + i\eta)] & -A\lambda^2 + \frac{1}{2}A\lambda^4[1 - 2(\rho + i\eta)] & 1 - \frac{1}{2}A^2\lambda^4 \end{pmatrix} + O(\lambda^6),$$

which satisfies

$$V^\dagger V = \mathbb{I} + O(\lambda^6). \tag{39.3}$$

This parametrization has two nice features

1. It contains three real parameters, A, λ, ρ, and one **CP**-violating phase, η, where $A \simeq 1$.
2. It shows the hierarchical structure of the CKM matrix with dominant mixing between the first two families and less mixing between the first and third or second and third families. The parameter $\lambda \approx \sin\theta_C$.

Approximately 20 or more years of high-energy experiment have gone into verifying this basic flavor structure of the Standard Model. There are many consistency relations which must be satisfied. In this chapter we want to summarize some of these results. Note, unless specified otherwise, these results are from the Particle Data Book (Tanabashi et al., 2018).

Nuclear beta decay determines the quantity

$$|V_{ud}| = 0.97420 \pm 0.00021. \tag{39.4}$$

Note, the value of $G_F \equiv G_\mu$ is determined by the process $\mu \to e\nu_\mu\bar{\nu}_e$ and

$$\Gamma(n \to pe^-\bar{\nu}_e) \sim G_F^2 |V_{ud}|^2. \tag{39.5}$$

The determination of V_{us} comes from many sources, including $K \to \pi\mu\nu_\mu$ or $K \to \pi e\nu_e$ for $K = \{K_L, K_S, K^\pm\}$ and the appropriate final states conserving charge and lepton number, and also semi-leptonic hyperon decays. We then have

$$|V_{us}| = 0.2243 \pm 0.0005. \tag{39.6}$$

The light-quark masses can be obtained using chiral Lagrangian analysis. The most precise relation for light-quark masses is given by the Kaplan, Manohar, Leutwyler ellipse parameter (Gasser and Leutwyler, 1985; Kaplan and Manohar, 1986; Leutwyler, 1990, 1996). We have

$$Q^2 = \frac{m_s^2 - \hat{m}^2}{m_d^2 - m_u^2} \approx \frac{M_K^2}{M_\pi^2} \frac{M_K^2 - M_\pi^2}{M_{K^0}^2 - M_{K^+}^2}(1 + O(m^2)) \tag{39.7}$$

or

$$1 = \frac{m_u^2}{m_d^2} + \frac{1}{Q^2}\left(\frac{m_s^2}{m_d^2} - \frac{1}{4}(1 + \frac{m_u}{m_d})^2\right) \simeq \frac{m_u^2}{m_d^2} + \frac{1}{Q^2}\frac{m_s^2}{m_d^2}. \tag{39.8}$$

(Leutwyler, 1996) obtains

$$Q = 22.7 \pm 0.8. \tag{39.9}$$

Kaplan and Manohar show that this uncertainty in the precise value of the up-quark mass is due to instanton effects of quantum chromodynamics (QCD) (Kaplan and Manohar, 1986). Then we have the next best understood mass ratio,

$$\frac{m_d}{m_s} = \frac{M_{K^0}^2 + M_{\pi^+}^2 - M_{K^+}^2}{M_{K^0}^2 - M_{\pi^+}^2 + M_{K^+}^2} \simeq 0.05 \pm 0.015, \tag{39.10}$$

and finally, from lattice gauge theory, we have (Aoki et al., 2014)

$$m_s(2 \text{ GeV}) = 93.5 \pm 2.0 \text{ MeV}. \tag{39.11}$$

Note, the notation, $m(\mu)$, defines the running quark mass at the scale μ. For the light quarks, the running masses are all defined at the scale $\mu = 2$ GeV.

The CKM parameter $|V_{cd}|$ can be obtained from deep inelastic $\nu_\mu N$ scattering on isoscalar targets by analyzing di-muon events, i.e. $\nu_\mu + d(s) \to c + \mu^-$, respectively, followed by the decay $c \to s(d) + \mu^+ + \nu_\mu$. Comparing this process to single-muon events produced via $\nu_\mu + d \to u + \mu^-$, one can infer $|V_{cd}|$ with knowledge of the parton distribution functions (PDFs). $|V_{cd}|$ and $|V_{cs}|$ can also be obtained via the decay $D^+(c\bar{d}) \to \pi\ell\nu$ or $D^+(c\bar{d}) \to \mu^+\nu_\mu$ and $D(\bar{c}u) \to K\ell\nu$ or $D_s^+(c\bar{s}) \to \mu^+\nu_\mu$ or $D_s^+(c\bar{s}) \to \tau^+\nu_\tau$. For example, $D^+ \to \bar{K}^0 e^+ \nu_e$ corresponds to $c\bar{d} \to s\bar{d} + e^+ + \nu_e$. The analyses require knowledge of the vector form factor or f_D, the D decay constant, coming from lattice calculations. The results are

$$|V_{cd}| = 0.218 \pm 0.004, \quad |V_{cs}| = 0.997 \pm 0.017. \tag{39.12}$$

The parameter $|V_{cb}|$ can be determined by either exclusive decays, such as $B \to D^* \ell \bar{\nu}_\ell$ using heavy-quark effective theory or via inclusive processes, such as $B \to X_c \ell \bar{\nu}$. The result for exclusive decays is

$$|V_{cb}| = (41.9 \pm 2.0) \times 10^{-3} \tag{39.13}$$

and, for inclusive processes,

$$|V_{cb}| = (42.2 \pm 0.8) \times 10^{-3}. \tag{39.14}$$

The result is dominated by the inclusive measurements.

For the parameter $|V_{ub}|$, the results are similarly obtained from both exclusive, such as $B^0 \to \pi^- \ell^+ \nu_\ell$ or $B^0 \to \rho^- \ell^+ \nu_\ell$, and inclusive, such as $B \to X_u \ell \bar{\nu}_\ell$, processes. The results from inclusive data give

$$|V_{ub}| = (4.49 \pm 0.28) \times 10^{-3} \tag{39.15}$$

and, from exclusive data,

$$|V_{ub}| = (3.70 \pm 0.16) \times 10^{-3}, \tag{39.16}$$

where the errors have been summed in quadrature. These results disagree by more than 2σ. Note, both the theoretical and the experimental errors are independent in the two different measurements. Finally, $|V_{ub}|$ can be independently determined via the decay $B \to \tau \bar{\nu}$ with the result

$$|V_{ub}| = (4.01 \pm 0.37) \times 10^{-3}, \tag{39.17}$$

which is consistent with the two other determinations. In addition, the CERN LHCb experiment measures the ratio

$$|V_{ub}/V_{cb}| = 0.083 \pm 0.006. \tag{39.18}$$

There are international collaborations [CKMfitter, UTfit] which use the data to determine the four Wolfenstein parameters. The results (Tanabashi et al., 2018) are

$$\lambda = 0.22453 \pm 0.00044 \approx |V_{us}|, \tag{39.19}$$

$$\left(\frac{|V_{cb}|}{\lambda^2} \right) = A = 0.836 \pm 0.015,$$

$$|V_{ub}/V_{cb}|/\lambda = \frac{A\lambda^3(\rho^2 + \eta^2)^{1/2}}{A\lambda^3} = (\rho^2 + \eta^2)^{1/2} = 0.38 \pm 0.04.$$

Note, **CP** violation implies that $\eta \neq 0$, assuming that the only source of **CP** violation is contained in V_{CKM}.

We also have the heavy-quark masses given by

$$m_c(m_c) = (1.28 \pm 0.025) \text{ GeV}, \tag{39.20}$$
$$m_b(m_b) = (4.18 \pm 0.03) \text{ GeV},$$
$$M_b - M_c = (3.4 \pm 0.2) \text{ GeV},$$
$$M_t = (173.5 \pm 0.6 \pm 0.8) \text{ GeV}.$$

Note, M corresponds to the quark pole mass. Finally, $M_b - M_c$ is determined using heavy-quark effective theory.

> ASIDE: The inverse propagator has the form (see Appendix C)
>
> $$\not{p} - m - \Sigma(p), \tag{39.21}$$
>
> where $\Sigma(p)$ is obtained by loop corrections to the fermion propagator and, in general,
>
> $$\Sigma(p) = A(p^2)\not{p} + B(p^2). \tag{39.22}$$
>
> The pole mass is then give by
>
> $$\not{p} = m + mA(m^2) + B(m^2), \tag{39.23}$$
>
> where $m \equiv m(\mu)$ is by definition the running mass. At one loop, in QCD, A and B are of order g_s^2. Given a one-loop calculation, we find
>
> $$M = m(m)\left(1 + \frac{4\alpha_s(m)}{3\pi}\right). \tag{39.24}$$

In summary, the magnitude of the CKM matrix elements are given by

$$
V_{CKM}
$$
$$
= \begin{pmatrix}
0.97446 \pm 0.00010 & 0.22452 \pm 0.00044 & 0.00365 \pm 0.00012 \\
0.22438 \pm 0.00044 & 0.97359 + 0.00010 - 0.00011 & 0.04214 \pm 0.00076 \\
0.00896 + 0.00024 - 0.00023 & 0.04133 \pm 0.00074 & 0.999105 \pm 0.000032
\end{pmatrix}.
$$
$$
\tag{39.25}
$$

39.1 CP Violation

In 1954, Dalitz (Dalitz, 1954, 1994) and others wondered about the two particles, θ^+, τ^+, with identical mass and lifetimes, but with different parity final states, i.e.

$$\tau^+ \to \pi^+\pi^+\pi^- \qquad J^P = 0^-, \tag{39.26}$$
$$\theta^+ \to \pi^+\pi^0 \qquad J^P = 0^+.$$

The resolution of the $\theta - \tau$ puzzle was that both **C** and **P** are violated in weak interactions due to the $V - A$ currents and that the states

$$\theta^+ = \tau^+ \equiv K^+. \tag{39.27}$$

This became apparent after Lee and Yang (Lee and Yang, 1956b) questioned whether parity was in fact conserved in weak interactions and then the discovery of parity violation in Co^{60} decay in 1957 by C.S. Wu et al. (Wu et al., 1957).

Now let's assume that **CP** is an exact symmetry and we define the **CP** eigenstates

$$|K_1^0\rangle \equiv \frac{1}{\sqrt{2}}(|K^0\rangle - |\bar{K}^0\rangle) \tag{39.28}$$

$$|K_2^0\rangle \equiv \frac{1}{\sqrt{2}}(|K^0\rangle + |\bar{K}^0\rangle)$$

where under **CP** we have

$$\mathbf{CP}|K^0\rangle \equiv -|\bar{K}^0\rangle \tag{39.29}$$

and thus

$$\mathbf{CP}|K_1^0\rangle \equiv |\bar{K}_1^0\rangle \tag{39.30}$$
$$\mathbf{CP}|K_2^0\rangle \equiv -|\bar{K}_2^0\rangle.$$

Then, assuming **CP** invariance, we expect the decay modes

$$\begin{array}{ll} K_1^0 \to \pi^+\pi^-, \ \pi^0\pi^0 & \mathbf{CP} \text{ even final state,} \\ K_2^0 \to \pi^+\pi^-\pi^0, \ 3\pi^0 & \mathbf{CP} \text{ odd final state.} \end{array} \tag{39.31}$$

Since three-body phase space is smaller than two-body phase space, K_2^0 has a longer lifetime than K_1^0. We thus define

$$K_L \equiv K_2^0, \tag{39.32}$$

with

$$\tau_L = (5.170 \pm 0.040) \times 10^{-8} \text{ s}, \tag{39.33}$$

and

$$K_S \equiv K_1^0 \tag{39.34}$$

with

$$\tau_S = (0.8934 \pm 0.0008) \times 10^{-10} \text{ s}. \tag{39.35}$$

In 1964, Christenson et al. (1964) measured the decay

$$K_L \to 2\pi, \tag{39.36}$$

which was evidence for **CP** violation. They determined the two **CP**-violating observables,

$$\eta_{+-} \equiv \frac{T(K_L^0 \to \pi^+\pi^-)}{T(K_S^0 \to \pi^+\pi^-)}, \tag{39.37}$$

and, similarly, η_{00}. They are now determined to have values (Tanabashi et al., 2018)

$$|\eta_{+-}| = (2.232 \pm 0.011) \times 10^{-3}, \quad |\eta_{00}| = (2.220 \pm 0.011) \times 10^{-3}. \tag{39.38}$$

39.2 $K^0 - \bar{K}^0$ Mixing

K_L, K_S are **CP** eigenstates, but **CP** is violated. K^0, $\bar{K}^0$ are strong-interaction eigenstates, but not electroweak eigenstates, i.e. they can mix. In this section, we begin the discussion of the phenomenon of $K^0 - \bar{K}^0$ oscillations. We define the wave function

$$|\psi(\tau)\rangle = a_1(\tau)|K^0\rangle + a_2(\tau)|\bar{K}^0\rangle, \tag{39.39}$$

which describes a coherent beam of K^0, $\bar{K}^0$ as a function of proper time, τ. Equivalently we can write

$$\psi(\tau) = \begin{pmatrix} a_1(\tau) \\ a_2(\tau) \end{pmatrix}. \tag{39.40}$$

Under $\Theta = $ **CPT**, we have

$$\langle \bar{K}^0| = \Theta|K^0\rangle. \tag{39.41}$$

The Hamiltonian

$$H = H_{strong} + H_{EM} + H_W \tag{39.42}$$

is **CPT** invariant. K^0 is an eigenstate of $H_{strong} + H_{EM}$ with strangeness $+1$ and, by **CPT**, $\bar{K}^0$ is an eigenstate of $H_{strong} + H_{EM}$ with strangeness -1. The quark content is given by

$$K^0(\bar{s}d), \quad \bar{K}^0(s\bar{d}). \tag{39.43}$$

In addition, we have

$$(H_{strong} + H_{EM})|K^0\rangle = m_K|K^0\rangle, \tag{39.44}$$
$$(H_{strong} + H_{EM})|\bar{K}^0\rangle = m_K|\bar{K}^0\rangle.$$

Under weak interactions we have

$$\langle \pi^+\pi^-|H_W|K^0\rangle \neq 0. \tag{39.45}$$
$$\langle \pi^+ e^- \bar{\nu}_e|H_W|K^0\rangle \neq 0, \quad \text{etc.},$$

and similarly for $\bar{K}^0$. Thus K^0, $\bar{K}^0$ can mix and decay. We can describe the time evolution of $\psi(\tau)$ by the Schrödinger equation (Weisskopf and Wigner, 1930b,a)

$$i\frac{d}{d\tau}\psi(\tau) = \left(M - i\frac{\Gamma}{2}\right)\psi(\tau), \tag{39.46}$$

where Γ takes into account the probability of leakage into the continuum. Without loss of generality we can take $M = M^\dagger$, $\Gamma = \Gamma^\dagger$, since given

$$M = \begin{pmatrix} M_{11} & M_{12} \\ M_{12}^* & M_{22} \end{pmatrix}, \tag{39.47}$$

$$\Gamma = \begin{pmatrix} \Gamma_{11} & \Gamma_{12} \\ \Gamma_{12}^* & \Gamma_{22} \end{pmatrix},$$

we have

$$M - i\frac{\Gamma}{2} = \begin{pmatrix} M_{11} - i\frac{\Gamma_{11}}{2} & M_{12} - i\frac{\Gamma_{12}}{2} \\ M_{12}^* - i\frac{\Gamma_{12}^*}{2} & M_{22} - i\frac{\Gamma_{22}}{2} \end{pmatrix}. \tag{39.48}$$

Note, this is an arbitrary complex matrix. We also see that

$$\frac{d}{d\tau}|\psi|^2 = -\psi^\dagger \Gamma \psi \leq 0. \tag{39.49}$$

Theorem 39.1 Γ *is a positive matrix with*

$$\Gamma_{11} \geq 0, \quad \Gamma_{22} \geq 0, \quad \det \Gamma \geq 0. \tag{39.50}$$

Proof

$$\psi^\dagger \Gamma \psi \geq 0 \text{ for arbitrary } \psi. \tag{39.51}$$

Let

$$\psi = \begin{pmatrix} a_1 \\ 0 \end{pmatrix}, \text{ implies } \Gamma_{11} \geq 0. \tag{39.52}$$

Similarly, $\Gamma_{22} \geq 0$. Eqn. 39.51 also implies that all eigenvalues of Γ are ≥ 0. Hence, $\det \Gamma \geq 0$. $\square$

39.2.1 Perturbation Expansion in H_W

Let $H = (H_{strong} + H_{EM}) + H_W \equiv H_0 + H_W$, then

$$(M - i\frac{\Gamma}{2})_{\alpha\beta} = \langle\alpha|H|\beta\rangle + \sum_n \frac{\langle\alpha|H_W|n\rangle\langle n|H_W|\beta\rangle}{m_K - (m_n - i\epsilon)} + O(H_W^3), \tag{39.53}$$

where $\epsilon = 0^+$, α, β can be K^0 or $\bar{K}^0$ and $|n\rangle$ is any eigenstate of H_0 (except for $|K^0\rangle$ or $|\bar{K}^0\rangle$) with eigenvalues m_n. Now we use the identity

$$\int_{-a}^{a} dz \frac{f(z)}{z \pm i\epsilon} = \mathcal{P}\int_{-a}^{a} dz \frac{f(z)}{z} \mp i\pi f(0), \tag{39.54}$$

with

$$\mathcal{P}\int_{-a}^{a} dz \frac{f(z)}{z} = \lim_{\epsilon \to 0}\left[\int_{-a}^{-\epsilon} dz \frac{f(z)}{z} + \int_{\epsilon}^{a} dz \frac{f(z)}{z}\right]. \tag{39.55}$$

$\mathcal{P}$ is the principal value. Hence

$$\frac{1}{m_K - (m_n - i\epsilon)} = \mathcal{P}\frac{1}{m_K - m_n} - i\pi\delta(m_n - m_K) \tag{39.56}$$

and we have

$$M_{\alpha\beta} = \langle\alpha|H|\beta\rangle + \sum_n \mathcal{P}\frac{\langle\alpha|H_W|n\rangle\langle n|H_W|\beta\rangle}{m_K - m_n} \tag{39.57}$$

and

$$\Gamma_{\alpha\beta} = 2\pi \sum_n \langle \alpha | H_W | n \rangle \langle n | H_W | \beta \rangle \delta(m_K - m_n), \qquad (39.58)$$

neglecting terms of order H_W^3.

Theorem 39.2 *Assuming* **CP$\mathcal{T}$** *invariance, independent of $\mathcal{T}$ invariance, we have $M_{11} = M_{22}$ and $\Gamma_{11} = \Gamma_{22}$.*

Proof

$$
\begin{aligned}
M_{11} &= \langle K^0 | \Theta^\dagger \Theta H \Theta^\dagger \Theta | K^0 \rangle + \sum_n \mathcal{P} \frac{\langle K^0 | \Theta^\dagger \Theta H_W \Theta^\dagger \Theta | n \rangle \langle n | \Theta^\dagger \Theta H_W \Theta^\dagger \Theta | K^0 \rangle}{m_K - m_n} \\
&= \langle \bar{K}^0 | \Theta H \Theta^\dagger | \bar{K}^0 \rangle + \sum_{\bar{n}} \mathcal{P} \frac{\langle \bar{K}^0 | \Theta H_W \Theta^\dagger | \bar{n} \rangle \langle \bar{n} | \Theta H_W \Theta^\dagger | \bar{K}^0 \rangle}{m_K - m_n} \qquad (39.59) \\
&= \langle \bar{K}^0 | H | \bar{K}^0 \rangle + \sum_{\bar{n}} \mathcal{P} \frac{\langle \bar{K}^0 | H_W | \bar{n} \rangle \langle \bar{n} | H_W | \bar{K}^0 \rangle}{m_{\bar{K}} - m_{\bar{n}}} \\
&\equiv M_{22}.
\end{aligned}
$$

We used

$$\underbrace{\Theta H_0 \Theta^\dagger \Theta | n \rangle}_{m_n \langle \bar{n} |} = \underbrace{\langle \bar{n} | H_0}_{m_{\bar{n}} \langle \bar{n} |} \qquad (39.60)$$

and $M_{22} = M_{22}^*$. Similarly, $\Gamma_{11} = \Gamma_{22}$. $\qquad \square$

Theorem 39.3 *If $\mathcal{T}$ invariance holds, then independent of* **CP$\mathcal{T}$** *we have*

$$\frac{\Gamma_{12}^*}{\Gamma_{12}} = \frac{M_{12}^*}{M_{12}}. \qquad (39.61)$$

Proof

$$
\begin{aligned}
\Gamma_{12} &= 2\pi \sum_n \langle K^0 | H_W | n \rangle \langle n | H_W | \bar{K}^0 \rangle \delta(m_K - m_n) \qquad (39.62) \\
&= 2\pi \sum_n \langle K^0 | \mathcal{T}^\dagger \mathcal{T} H_W \mathcal{T}^\dagger \mathcal{T} | n \rangle \langle n | \mathcal{T}^\dagger \mathcal{T} H_W \mathcal{T}^\dagger \mathcal{T} | \bar{K}^0 \rangle \delta(m_K - m_n) \\
&= 2\pi \sum_{n'} e^{-i\bar{\alpha}} \langle \bar{K}^0 | \mathcal{T} H_W \mathcal{T}^\dagger | n' \rangle \langle n' | \mathcal{T} H_W \mathcal{T}^\dagger | K^0 \rangle e^{i\alpha} \delta(m_K - m_{n'}),
\end{aligned}
$$

where we've used

$$
\begin{aligned}
\mathcal{T} | K^0 \rangle &= e^{-i\alpha} \langle K^0 | \qquad (39.63) \\
\mathcal{T} | \bar{K}^0 \rangle &= e^{-i\bar{\alpha}} \langle \bar{K}^0 |.
\end{aligned}
$$

Then assume

$$\mathcal{T} H_W \mathcal{T}^\dagger = H_W \tag{39.64}$$

and we have

$$\Gamma_{12} \equiv e^{-i(\bar{\alpha}-\alpha)} \Gamma_{21}. \tag{39.65}$$

But

$$\Gamma_{21} \equiv \Gamma_{12}^*. \tag{39.66}$$

Hence,

$$\Gamma_{12}^* = e^{i(\bar{\alpha}-\alpha)} \Gamma_{12}. \tag{39.67}$$

Similarly,

$$M_{12}^* = e^{i(\bar{\alpha}-\alpha)} M_{12}. \tag{39.68}$$

$\square$

39.2.2   Eigenvalues

Since $M - i\frac{\Gamma}{2}$ is a 2×2 matrix, it has two eigenvectors $|L\rangle$, $|S\rangle$, which satisfy

$$\left(M - i\frac{\Gamma}{2} \right) |j\rangle = \left(m_j - i\frac{\gamma_j}{2} \right) |j\rangle, \tag{39.69}$$

with $j = L$, S. m_j are the masses and $\tau_j = \gamma_j^{-1}$ are the lifetimes (see Eqns. 39.33 and 39.35). The mass difference

$$\Delta m \equiv m_L - m_S = (3.489 \pm 0.009) \times 10^{-6} \text{ eV} \tag{39.70}$$

is known very precisely, considering that

$$\frac{\Delta m}{m_K} \simeq 7.08 \times 10^{-15} \sim \left(\frac{H_W}{H_{strong}} \right)^2 \tag{39.71}$$

with

$$\frac{H_W(\Delta S \pm 1)}{H_{st}} \simeq \frac{G_F m_p^2}{4\pi} \sin \theta_C \simeq 10^{-7}. \tag{39.72}$$

Let's now summarize what we have learned. We have

$$M - i\frac{\Gamma}{2} \equiv \begin{pmatrix} A & B \\ C & A \end{pmatrix}, \tag{39.73}$$

with

$$A \equiv M_{11} - i\frac{\Gamma_{11}}{2} = M_{22} - i\frac{\Gamma_{22}}{2} \equiv \bar{M} - i\frac{\bar{\Gamma}}{2} \tag{39.74}$$

and

$$B = M_{12} - i\frac{\Gamma_{12}}{2}, \quad C = M_{12}^* - i\frac{\Gamma_{12}^*}{2}. \tag{39.75}$$

Then we have

$$\frac{1}{2} Tr \left(M - i\frac{\Gamma}{2} \right) = \bar{M} - i\frac{\bar{\Gamma}}{2} \tag{39.76}$$

$$\det \left(M - i\frac{\Gamma}{2} \right) = A^2 - BC.$$

Then the eigenvalues can be written in the form

$$\bar{M} - i\frac{\bar{\Gamma}}{2} \pm \frac{1}{2} \left(\Delta m - i\frac{\Delta\Gamma}{2} \right) \tag{39.77}$$

and

$$\det \left(M - i\frac{\Gamma}{2} \right) = A^2 - BC = A^2 - \frac{1}{4} \left(\Delta m - i\frac{\Delta\Gamma}{2} \right)^2 \tag{39.78}$$

or

$$\Delta m - i\frac{\Delta\Gamma}{2} = 2\sqrt{BC}. \tag{39.79}$$

We shall look for eigenstates of the form

$$\begin{pmatrix} p \\ q \end{pmatrix}, \quad \begin{pmatrix} p \\ -q \end{pmatrix}. \tag{39.80}$$

Then

$$\begin{pmatrix} A & B \\ C & A \end{pmatrix} \begin{pmatrix} p \\ q \end{pmatrix} = \begin{pmatrix} Ap + Bq \\ Cp + Aq \end{pmatrix} = (A + \sqrt{BC}) \begin{pmatrix} p \\ q \end{pmatrix} \tag{39.81}$$

implies

$$Bq = \sqrt{BC}p, \quad Cp = \sqrt{BC}q \tag{39.82}$$

or

$$\frac{p}{q} = \sqrt{\frac{B}{C}}. \tag{39.83}$$

Similarly,

$$\begin{pmatrix} A & B \\ C & A \end{pmatrix} \begin{pmatrix} p \\ -q \end{pmatrix} = \begin{pmatrix} Ap - Bq \\ Cp - Aq \end{pmatrix} = (A - \sqrt{BC}) \begin{pmatrix} p \\ -q \end{pmatrix} \tag{39.84}$$

implies

$$Bq = \sqrt{BC}p, \quad Cp = \sqrt{BC}q \tag{39.85}$$

or

$$\frac{p}{q} = \sqrt{\frac{B}{C}}. \tag{39.86}$$

p, q are normalized, such that

$$|p|^2 + |q|^2 = 1. \tag{39.87}$$

We then define

$$K_{\substack{L \\ S}} = p|K^0\rangle \pm q|\bar{K}^0\rangle. \tag{39.88}$$

Theorem 39.4   *Note,*

$$\xi \equiv \langle K_S | K_L \rangle \tag{39.89}$$

is non-zero if and only if **CP** *is violated. We have*

$$\xi \equiv (\ \langle K^0 | p^* - \langle \bar{K}^0 | q^* \ )(\ p | K^0 \rangle + q | \bar{K}^0 \rangle\ ) = |p|^2 - |q|^2. \tag{39.90}$$

Proof

$$\frac{|p|^2}{|q|^2} = \sqrt{\frac{|B|^2}{|C|^2}}. \tag{39.91}$$

But

$$|B|^2 = \left(M_{12}^* + i\frac{\Gamma_{12}^*}{2} \right) \left(M_{12} - i\frac{\Gamma_{12}}{2} \right) \tag{39.92}$$

$$= |M_{12}|^2 + \frac{|\Gamma_{12}|^2}{4} + \frac{i}{2}(\Gamma_{12}^* M_{12} - M_{12}^* \Gamma_{12})$$

and

$$|C|^2 = \left(M_{12} + i\frac{\Gamma_{12}}{2} \right) \left(M_{12}^* - i\frac{\Gamma_{12}^*}{2} \right) \tag{39.93}$$

$$= |M_{12}|^2 + \frac{|\Gamma_{12}|^2}{4} - \frac{i}{2}(\Gamma_{12}^* M_{12} - M_{12}^* \Gamma_{12}).$$

Finally,

$$(\Gamma_{12}^* M_{12} - M_{12}^* \Gamma_{12}) \equiv M_{12} \Gamma_{12} \left(\frac{\Gamma_{12}^*}{\Gamma_{12}} - \frac{M_{12}^*}{M_{12}} \right) = 0 \tag{39.94}$$

if $\mathcal{T}$ is invariant. Hence $|p|^2 - |q|^2 \neq 0$ if and only if **CP** is violated, since **CP$\mathcal{T}$** is conserved. $\qquad\qquad\qquad\square$

39.3 $K^0 - \bar{K}^0$ Oscillations: Measuring Δm_K

The states K^0, $\bar{K}^0$ are strong and electromagnetic eigenstates, but not electroweak eigenstates. Thus there can be vacuum oscillations that can be measured. However, the most accurate measurement of Δm_K comes using the regenerator method. Simply stated, in the regenerator method a neutral kaon beam is created via strong interactions. It then propagates as a superposition of K_L and K_S. After a sufficient amount of time, the K_S decay out of the beam and one is left with a K_L beam. Then one sends this beam through a thin target. In the target the K^0 and $\bar{K}^0$ components of the K_L scatter with different phase shifts. As a result the beam then once more has a K_S component. For more details see Geweniger et al. (1974), who found

$$\Delta m_K = (0.534 \pm 0.003) \times 10^{10} \text{ s}^{-1}. \tag{39.95}$$

40   CP-Violating Parameters ϵ_K and ϵ'_K

We defined the **CP**-violating observables η_{+-}, η_{00} in Eqn. 39.37. Now we define the theoretically motivated **CP**-violating observables, ϵ_K and ϵ'_K, via the expressions

$$\eta_{+-} \equiv \epsilon_K + \epsilon'_K \tag{40.1}$$

and

$$\eta_{00} \equiv \epsilon_K - 2\epsilon'_K. \tag{40.2}$$

Note, in terms of isospin amplitudes, we have

$$|\pi^+ \pi^-\rangle = \frac{1}{\sqrt{6}}|\pi\pi, 2\rangle + \frac{1}{\sqrt{3}}|\pi\pi, 0\rangle \tag{40.3}$$

$$|\pi^0 \pi^0\rangle = \sqrt{\frac{2}{3}}|\pi\pi, 2\rangle - \frac{1}{\sqrt{3}}|\pi\pi, 0\rangle$$

for $I = 2, 0$ and the isospin 1 contribution is not allowed by Bose statistics. We then have

$$\eta_{+-} \equiv \frac{\frac{1}{\sqrt{6}}(\langle\pi\pi, 2| + \sqrt{2}\langle\pi\pi, 0|)H_W(p|K^0\rangle + q|\bar{K}^0\rangle)}{\frac{1}{\sqrt{6}}(\langle\pi\pi, 2| + \sqrt{2}\langle\pi\pi, 0|)H_W(p|K^0\rangle - q|\bar{K}^0\rangle)}, \tag{40.4}$$

$$\eta_{00} \equiv \frac{\frac{1}{\sqrt{3}}(\langle\pi\pi, 2|\sqrt{2} - \langle\pi\pi, 0|)H_W(p|K^0\rangle + q|\bar{K}^0\rangle)}{\frac{1}{\sqrt{3}}(\langle\pi\pi, 2|\sqrt{2} - \langle\pi\pi, 0|)H_W(p|K^0\rangle - q|\bar{K}^0\rangle)}.$$

Define

$$L_I = \langle\pi\pi, I|H_W|K_L\rangle, \tag{40.5}$$

$$S_I = \langle\pi\pi, I|H_W|K_S\rangle.$$

Then

$$\eta_{+-} \equiv \frac{L_2 + \sqrt{2}L_0}{S_2 + \sqrt{2}S_0} = \epsilon + \epsilon', \tag{40.6}$$

$$\eta_{00} \equiv \frac{\sqrt{2}L_2 - L_0}{\sqrt{2}S_2 - S_0} \equiv \epsilon - 2\epsilon'$$

and thus

$$L_2 + \sqrt{2}L_0 = (\epsilon + \epsilon')(S_2 + \sqrt{2}S_0), \tag{40.7}$$

$$\sqrt{2}L_2 - L_0 = (\epsilon - 2\epsilon')(\sqrt{2}S_2 - S_0).$$

Then, in matrix form, we have

$$\begin{pmatrix} \sqrt{2} & 1 \\ -1 & \sqrt{2} \end{pmatrix} \begin{pmatrix} L_0 \\ L_2 \end{pmatrix} = \begin{pmatrix} \sqrt{2}(\epsilon + \epsilon') & \epsilon + \epsilon' \\ -(\epsilon - 2\epsilon') & \sqrt{2}(\epsilon - 2\epsilon') \end{pmatrix} \begin{pmatrix} S_0 \\ S_2 \end{pmatrix} \tag{40.8}$$

or, multiplying by the inverse matrix,

$$\frac{1}{3} \begin{pmatrix} \sqrt{2} & -1 \\ 1 & \sqrt{2} \end{pmatrix}, \tag{40.9}$$

we have

$$\begin{pmatrix} L_0 \\ L_2 \end{pmatrix} = \begin{pmatrix} \epsilon & \sqrt{2}\epsilon' \\ \sqrt{2}\epsilon' & \epsilon - \epsilon' \end{pmatrix} \begin{pmatrix} S_0 \\ S_2 \end{pmatrix}. \tag{40.10}$$

We then obtain

$$L_0 = \epsilon S_0 + \sqrt{2} S_2 \epsilon' \tag{40.11}$$

or

$$\epsilon' = \frac{L_0 - \epsilon S_0}{\sqrt{2} S_2}. \tag{40.12}$$

Similarly, we have

$$L_2 = \epsilon S_2 + \epsilon'(\sqrt{2} S_0 - S_2) \tag{40.13}$$

$$= \epsilon S_2 + (\sqrt{2} S_0 - S_2)\frac{(L_0 - \epsilon S_0)}{\sqrt{2} S_2}.$$

Solving for ϵ we find

$$\epsilon = \frac{L_0 - \frac{L_2 S_2}{(S_0 - \frac{S_2}{\sqrt{2}})}}{S_0 \left(1 - \frac{S_2^2}{(S_0^2 - \frac{S_2 S_0}{\sqrt{2}})}\right)} \tag{40.14}$$

or

$$\epsilon \simeq \frac{L_0}{S_0}\left(1 + \frac{S_2^2}{S_0^2}\right) - \frac{L_2 S_2}{S_0^2} + O\left(\frac{L_2 S_2^2}{S_0^3}, \frac{L_0}{S_0}\frac{S_2^3}{S_0^3}\right). \tag{40.15}$$

Finally, inserting Eqn. 40.15 into Eqn. 40.12, and neglecting higher-order terms in S_2/S_0, we find

$$\epsilon' \simeq \frac{L_2}{\sqrt{2} S_0} - \epsilon \frac{S_2}{\sqrt{2} S_0}. \tag{40.16}$$

Before we continue, let's explain why we can neglect higher-order terms in (S_2/S_0). Define

$$H_W \equiv H_1 + H_1^\dagger. \tag{40.17}$$

Then

$$\langle \pi\pi, I | H_1 | K^0 \rangle \equiv A_I e^{i\delta_I}, \tag{40.18}$$

$$\langle \pi\pi, I | H_1^\dagger | \bar{K}^0 \rangle \equiv A_I^* e^{i\delta_I}.$$

δ_I is the strong-interaction $\pi\pi$ phase shift.

Theorem 40.1   **CP**$\mathcal{T}$ *invariance implies given*

$$\langle \pi\pi, I | H_1 | K^0 \rangle \equiv A_I e^{i\delta_I},$$

then

$$\langle \pi\pi, I | H_1^\dagger | \bar{K}^0 \rangle \equiv A_I^* e^{i\delta_I}.$$

δ_I *is the strong-interaction* $\pi\pi$ *phase shift.*

Proof

$$
\begin{aligned}
A_I e^{i\delta_I} &\equiv \langle \pi\pi, I | H_1 | K^0 \rangle = \langle \pi\pi, I(free) | U(\infty, 0) H_W | K^0 \rangle &&(40.19)\\
&= \langle \pi\pi, I(free) | \Theta^\dagger \Theta U(\infty, 0) H_W \Theta^\dagger \Theta | K^0 \rangle \\
&= \langle \bar{K}^0 | H_W \underbrace{\Theta U(\infty, 0) \Theta^\dagger}_{U^\dagger(-\infty, 0)} | \pi\pi, I(free) \rangle \\
&= \langle \pi\pi, I(free) | U(-\infty, 0) H_W | \bar{K}^0 \rangle^* \\
&= \langle \pi\pi, I(free) | S^\dagger U(\infty, 0) H_W | \bar{K}^0 \rangle^* &&(40.20)\\
&\equiv e^{2i\delta_I} \langle \pi\pi, I(free) | U(\infty, 0) H_W | \bar{K}^0 \rangle^* &&(40.21)\\
&= e^{2i\delta_I} \langle \pi\pi, I | H_W | \bar{K}^0 \rangle^*,
\end{aligned}
$$

where Eqn. 40.20 uses the identity $U(\infty, 0) \equiv S U^\dagger(0, -\infty) \equiv S U(-\infty, 0)$, with $S \equiv U(\infty, 0) U(0, -\infty)$, and Eqn. 40.21 uses the identity for the strong-interaction phase shift

$$e^{-2i\delta_I} \equiv \langle \pi\pi, I(free) | S | \pi\pi, I(free) \rangle. \qquad (40.22)$$

The bottom line is

$$\langle \pi\pi, I | H_1 | \bar{K}^0 \rangle \equiv A_I^* e^{i\delta_I} \qquad (40.23)$$

and

$$\frac{\langle \pi\pi, I | H_1 | K^0 \rangle}{\langle \pi\pi, I | H_1 | \bar{K}^0 \rangle^*} \equiv e^{2i\delta_I}. \qquad (40.24)$$

$\square$

Experimentally, we have

$$\left| \frac{S_2}{S_0} \right|^2 \simeq \left| \frac{A_2}{A_0} \right|^2 \simeq \frac{1}{400} = 0.25\%, \qquad (40.25)$$

i.e. the $\Delta I = 1/2$ rule. Thus

$$\epsilon \simeq \frac{L_0}{S_0} + O(0.25\%) \qquad (40.26)$$

$$\epsilon' \simeq \frac{L_2}{\sqrt{2} S_0} - \epsilon \times O(5\%).$$

Now let's continue our derivation of the final formulae for ϵ and ϵ'.

$$L_0 = \langle \pi\pi, 0 | H_W | (p|K^0\rangle + q|\bar{K}^0\rangle) \tag{40.27}$$

and

$$S_0 = \langle \pi\pi, 0 | H_W | (p|K^0\rangle - q|\bar{K}^0\rangle). \tag{40.28}$$

Hence,

$$\epsilon = \frac{pA_0 + qA_0^*}{pA_0 - qA_0^*} \tag{40.29}$$
$$= \frac{-\frac{p}{q}e^{i\omega_0} + 1}{-\frac{p}{q}e^{i\omega_0} - 1},$$

where

$$-e^{i\omega_0} \equiv \frac{A_0}{A_0^*}, \quad \frac{p}{q} = \sqrt{\frac{B}{C}}. \tag{40.30}$$

We shall return to this expression later.

Define

$$\bar{B} \equiv e^{i\omega_0} B, \ \bar{C} \equiv e^{-i\omega_0} C. \tag{40.31}$$

Then

$$\frac{p}{q} = \sqrt{\frac{\bar{B}}{\bar{C}}} e^{-i\omega_0} \tag{40.32}$$

and

$$\epsilon = \frac{\sqrt{\frac{\bar{B}}{\bar{C}}} - 1}{\sqrt{\frac{\bar{B}}{\bar{C}}} + 1} = \frac{\sqrt{\bar{B}} - \sqrt{\bar{C}}}{\sqrt{\bar{B}} + \sqrt{\bar{C}}} = \frac{\bar{B} - \bar{C}}{\bar{B} + \bar{C} + 2\sqrt{\bar{B}\bar{C}}}. \tag{40.33}$$

Now, using Eqn. 39.75, we have

$$\bar{B} - \bar{C} = 2i\mathrm{Im}\bar{M}_{12} + \mathrm{Im}\bar{\Gamma}_{12}, \tag{40.34}$$
$$\bar{B} + \bar{C} = 2\mathrm{Re}\bar{M}_{12} - i\mathrm{Re}\bar{\Gamma}_{12},$$
$$\simeq 2\sqrt{\bar{B}\bar{C}} = 2\sqrt{BC} \equiv \Delta m - i\frac{\Delta\Gamma}{2},$$

where

$$\Delta m \simeq 2\mathrm{Re}\bar{M}_{12}, \quad \Delta\Gamma \simeq 2\mathrm{Re}\bar{\Gamma}_{12}. \tag{40.35}$$

Hence, finally we have

$$\epsilon = \frac{i\mathrm{Im}\bar{M}_{12} + \frac{1}{2}\mathrm{Im}\bar{\Gamma}_{12}}{\Delta m - \frac{i}{2}\Delta\Gamma}. \tag{40.36}$$

Now let's obtain the final formula for

$$\epsilon' \simeq \frac{L_2}{\sqrt{2}S_0} = \frac{(pA_2 + qA_2^*)e^{i\delta_2}}{\sqrt{2}(pA_0 - qA_0^*)e^{i\delta_0}} \tag{40.37}$$

$$= \frac{1}{\sqrt{2}} \left(\frac{\frac{p}{q}A_2 + A_2^*}{\frac{p}{q}A_0 - A_0^*} \right) e^{i(\delta_2 - \delta_0)}.$$

We have neglected the term proportional to ϵ, since $\epsilon \ll 1$. Recall, using Eqn. 40.29 we can solve for p/q in terms of ϵ. We have

$$\frac{p}{q} = \left(\frac{1+\epsilon}{1-\epsilon} \right) e^{-i\omega_0}. \tag{40.38}$$

Thus, we find

$$\epsilon' = \frac{1}{\sqrt{2}} \left(\frac{(\frac{1+\epsilon}{1-\epsilon})e^{-i\omega_0} A_2 + A_2^*}{(\frac{1+\epsilon}{1-\epsilon})e^{-i\omega_0} A_0 - A_0^*} \right) e^{i(\delta_2 - \delta_0)} \tag{40.39}$$

$$= \frac{1}{\sqrt{2}} \left(\frac{(\frac{1+\epsilon}{1-\epsilon})\frac{A_2}{A_0} - \frac{A_2^*}{A_0^*}}{(\frac{1+\epsilon}{1-\epsilon}) + 1} \right) e^{i(\delta_2 - \delta_0)},$$

where we've used the identity,

$$\frac{A_0^*}{A_0} = -e^{-i\omega_0}. \tag{40.40}$$

Then, expanding in ϵ, we have

$$\epsilon' \simeq \frac{i}{\sqrt{2}} \mathrm{Im} \left(\frac{A_2}{A_0} \right) e^{i(\delta_2 - \delta_0)} + O(\epsilon \frac{A_2}{A_0}). \tag{40.41}$$

40.1 Phase Conventions

Let's start with the equations,

$$e^{i\omega_0} \frac{p}{q} \equiv \frac{1+\epsilon}{1-\epsilon} \tag{40.42}$$

and

$$e^{i\omega_0} = -\frac{A_0}{A_0^*} \equiv -\frac{\langle \pi\pi, 0|H_1|K^0 \rangle}{\langle \pi\pi, 0|H_1^\dagger|\bar{K}^0 \rangle}. \tag{40.43}$$

Then recall that the states

$$|K^0\rangle = |\bar{s}d\rangle, \quad |\bar{K}^0\rangle = |s\bar{d}\rangle \tag{40.44}$$

transform under phase redefinitions of the quarks by

$$|K^0\rangle \to e^{i(\theta_d - \theta_s)}|K^0\rangle, \quad |\bar{K}^0\rangle \to e^{-i(\theta_d - \theta_s)}|\bar{K}^0\rangle. \tag{40.45}$$

Hence, the phase $e^{i\omega_0}$ transforms as

$$e^{i\omega_0} \rightarrow e^{i(\omega_0 + 2(\theta_d - \theta_s))}. \tag{40.46}$$

However, the states

$$K_{\substack{L \\ S}} = p|K^0\rangle \pm q|\bar{K}^0\rangle \tag{40.47}$$

are unchanged by an arbitrary phase redefinition. Thus, we have

$$p \rightarrow p e^{-i(\theta_d - \theta_s)}, \quad q \rightarrow q e^{i(\theta_d - \theta_s)} \tag{40.48}$$

and

$$\frac{p}{q} \rightarrow \frac{p}{q} e^{-2i(\theta_d - \theta_s)}. \tag{40.49}$$

Therefore,

$$\frac{p}{q} e^{i\omega_0} \rightarrow \frac{p}{q} e^{i\omega_0} \tag{40.50}$$

is phase invariant. We can always choose a phase convention such that

$$\omega_0 \equiv 0. \tag{40.51}$$

With this phase convention we now have

$$\frac{p}{q} = \frac{1 + \epsilon}{1 - \epsilon} \tag{40.52}$$

and $|p|^2 + |q|^2 = 1$. Hence,

$$K_{\substack{L \\ S}} = \frac{1}{2(1 + |\epsilon|^2)} \left((1 + \epsilon)|K^0\rangle \pm (1 - \epsilon)|\bar{K}^0\rangle \right). \tag{40.53}$$

Therefore, $\epsilon \equiv \epsilon_K$ is a measure of **CP** violation in the state vector, i.e. in the mixing of K_L, K_S. Whereas $\epsilon' \equiv \epsilon'_K$ is a measure of **CP** violation in the $K \rightarrow 2\pi$ decay amplitude, i.e. $\text{Im}\left(\frac{A_2}{A_0}\right)$.

Experimentally, we have (Tanabashi et al., 2018)

$$|\epsilon_K| = (2.228 \pm 0.011) \times 10^{-3}, \quad \text{Re}(\epsilon'_K/\epsilon_K) = (1.65 \pm 0.26) \times 10^{-3}. \tag{40.54}$$

40.2 Theoretical Calculation of ϵ_K

Given the formula for ϵ_K, Eqn. 40.36, we can use experimental results to simplify it even further. We have

$$\Gamma_{12} = 2\pi \sum_n \int d\phi_n \langle K^0|H_W|n\rangle \langle n|H_W|\bar{K}^0\rangle \delta(m_n - m_K) \tag{40.55}$$

$$\simeq \phi_0 A_0^* A_0^* + \phi_2 A_2^* A_2^* + \phi_{\pi\pi} A_3^* A_3^*,$$

where $d\phi_n$ is a phase-space measure with ϕ_I, $I = 0, 2$ representing the phase space for two-body final states with definite isospin. Note, the first term is approximately 400 times larger than all the other terms as a result of the $\Delta I = 1/2$ rule. Thus, we have

$$\bar{\Gamma}_{12} \equiv \Gamma_{12} e^{i\omega_0} = -\phi_0 A_0 A_0^* \times (1 + O(0.25\%)) \qquad (40.56)$$

and consequently

$$\mathrm{Im}\bar{\Gamma}_{12} = 0 \qquad (40.57)$$

plus terms suppressed by a factor of $1/400$. Note, also experimentally, we have

$$\Delta m = m_L - m_S = (3.489 \pm 0.009) \times 10^{-6} \text{ eV} \qquad (40.58)$$

and

$$\begin{aligned}
\Delta\Gamma = \gamma_L - \gamma_S &\simeq -\gamma_S \\
&= -\frac{\hbar}{(0.8934 \pm 0.0008) \times 10^{-10} \text{ s}} \\
&= -\frac{6.582 \times 10^{-16} \text{ eV} \cdot \text{s}}{0.8934 \times 10^{-10} \text{ s}} = -7.3674 \times 10^{-6} \text{ eV}
\end{aligned}$$

or

$$\Delta\Gamma/2 = 3.6837 \times 10^{-6} \text{ eV}. \qquad (40.59)$$

Therefore, experimentally it is found that

$$\Delta m \simeq -\frac{\Delta\Gamma}{2} \times 0.947 \qquad (40.60)$$

or

$$\Delta m - i\frac{\Delta\Gamma}{2} \simeq \Delta m(1 + i) = \sqrt{2}\,\Delta m e^{i\pi/4}. \qquad (40.61)$$

As a result we have the final theoretical expression for ϵ_K given by

$$\epsilon_K \simeq \frac{i\mathrm{Im}\bar{M}_{12}e^{-i\pi/4}}{\sqrt{2}\Delta m}, \qquad (40.62)$$

which is valid to 5%. In the calculation of ϵ_K we use the experimentally measured value for Δm.

The leading-order contribution to ϵ_K comes from the one-loop diagram in Fig. 40.1. This box diagram produces an effective local operator given by

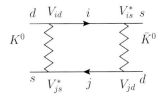

One-loop contribution to $\bar{M}_{12}$ with time going from left to right and $i, j = u, c, t.$

$$\mathcal{H}_2(0) = \frac{G_F^2 M_W^2}{16\pi^2} \lambda_i^* \lambda_j^* S_{ij} \eta_{ij} \times \bar{s}\gamma^\mu (1-\gamma_5)d\,\bar{s}\gamma_\mu(1-\gamma_5)d + h.c., \qquad (40.63)$$

where

$$\lambda_i^* \equiv V_{id}V_{is}^*, \qquad (40.64)$$

and the loop factors, the Inami–Lim functions (Inami and Lim, 1981), are

$$S_{ii} \equiv S(x_i) = \frac{x_i}{4}\left[1 + \frac{3-9x_i}{(1-x_i)^2} + \frac{6x_i^2 \ln x_i}{(x_i-1)^3}\right], \quad x_i \equiv \frac{m_i^2}{M_W^2}, \quad (40.65)$$

$$S_{ij}|_{i\neq j} \equiv S(x,y) = xy\left\{\left[\frac{1}{4} + \frac{3}{2(1-y)} - \frac{3}{4(1-y)^2}\right]\frac{\ln y}{y-x}\right.$$
$$\left. + (y \leftrightarrow x) - \frac{3}{4(1-x)(1-y)}\right\}.$$

The constants η_{ij} in Eqn. 40.63 are obtained via renormalization group running of the effective four-Fermi operators from the weak scale to the kaon mass. The charm and top quark loops dominate in the result, we have $\eta_c, \eta_t, \eta_{ct} = 0.85, 0.57, 0.\mathrm{f}36$. However, at one-loop level, $\eta_{ij} = 1$.

We then have

$$M_{21} \equiv \frac{\langle \bar{K}^0|H_2|K^0\rangle}{\langle \bar{K}^0|K^0\rangle} + \ldots \qquad (40.66)$$
$$= \frac{\langle \bar{K}^0|\mathcal{H}_2(0)|K^0\rangle}{2m_K} + \ldots,$$

where

$$H_2 \equiv \int d^3\vec{x}\, \mathcal{H}_2(\vec{x}). \qquad (40.67)$$

Plugging this into Eqn. 40.62, we find

$$|\epsilon_K| = \frac{|\mathrm{Im}\bar{M}_{21}|}{\sqrt{2}\Delta m} \qquad (40.68)$$
$$= \frac{G_F^2 M_W^2}{32\pi^2 m_K \sqrt{2}\Delta m}|\mathrm{Im}(\lambda_i\lambda_j)|S_{ij}\eta_{ij}\langle \bar{K}^0|(\bar{s}\gamma^\mu(1-\gamma_5)d)^2|K^0\rangle,$$

where we have ignored the phase $e^{i\omega_0}$. Later, we shall discuss a phase-invariant formalism. Note: **CP** violation in the K system vanishes if the up-type quarks were degenerate in mass.

The matrix element,

$$\langle \bar{K}^0|(\bar{s}\gamma^\mu(1-\gamma_5)d)^2|K^0\rangle \equiv \frac{8}{3}f_K^2 m_K^2 B_K, \qquad (40.69)$$

where B_K is called the quantum chromodynamics bag constant. The quantities B_K, f_K are determined by lattice gauge calculations. Now we have

$$|\epsilon_K| = \frac{G_F^2 M_W^2}{12\pi^2}\frac{m_K f_K^2 B_K}{\sqrt{2}\Delta m}|\mathrm{Im}(\lambda_i\lambda_j)|S_{ij}\eta_{ij}. \qquad (40.70)$$

One might wonder what is the origin of Eqn. 40.69. Recall that f_π and f_K are the pion and kaon decay constants defined via the partially conserved axial current (PCAC) relation,

$$\langle \bar{K}^0 | \bar{s}_a \gamma^\mu (1 - \gamma_5) d^b | \Omega \rangle \equiv \frac{\delta_a^{\ b}}{3} i f_K q^\mu, \qquad (40.71)$$

where a, b are color indices and $|\Omega\rangle$ is the strong-interaction vacuum state. In Eqn. 40.69 we have the relevant current squared between the K^0 and $\bar{K}^0$ states. As a first approximation, one can assume that the vacuum dominates in the intermediate state. Using the Fierz relation,

$$(\gamma^\mu P_L)_{\alpha\beta} (\gamma_\mu P_L)_{\delta\rho} \equiv -(\gamma^\mu P_L)_{\alpha\rho} (\gamma_\mu P_L)_{\delta\beta}, \qquad (40.72)$$

we find[1]

$$\begin{aligned}
&\langle \bar{K}^0 | \bar{s}_a \gamma^\mu (1 - \gamma_5) d^a | \Omega \rangle \langle \Omega | \bar{s}_b \gamma_\mu (1 - \gamma_5) d^b | K^0 \rangle \\
&+ \langle \bar{K}^0 | \bar{s}_a \gamma^\mu (1 - \gamma_5) d^b | \Omega \rangle \langle \Omega | \bar{s}_b \gamma_\mu (1 - \gamma_5) d^a | K^0 \rangle \\
&= f_K^2 m_K^2 [1 + \tfrac{1}{3}] \times 2 = \tfrac{8}{3} f_K^2 m_K^2 \qquad\qquad .
\end{aligned} \qquad (40.73)$$

The bag constant was defined originally by using the MIT bag model to calculate this strong-interaction deviation from simple vacuum insertion.

In the evaluation of ϵ_K the up-quark contribution is negligible since $S(x_u) \simeq \frac{m_u^2}{M_W^2} \ll 1$ and $S(x_u, x_i) \simeq x_u \ln x_u \ll 1$. Thus,

$$\begin{aligned}
|\epsilon_K| \simeq{}& \frac{G_F^2 M_W^2}{12\pi^2} \frac{m_K f_K^2 B_K}{\sqrt{2}\Delta m} [|\mathrm{Im}(\lambda_t^2)| S(x_t) \eta_t \\
&+ |\mathrm{Im}(\lambda_c^2)| S(x_c) \eta_c + 2|\mathrm{Im}(\lambda_c \lambda_t)| S(x_c, x_t) \eta_{ct}].
\end{aligned} \qquad (40.74)$$

However, up until now we have ignored $e^{i\omega_0} \equiv -\frac{A_0}{A_0^*}$. If we include this factor we can obtain a re-phase-invariant result.

40.2.1 Re-Phase-Invariant Formula

A re-phase-invariant formula for ϵ_K was obtained by Bjorken and Dunietz (1987). Recall $\bar{M}_{12} \equiv M_{12} e^{i\omega_0} \propto \lambda_i \lambda_j e^{i\omega_0}$, where $\lambda_i = V_{is} V_{id}^*$. Numerically, we know that $\epsilon_K' \leq 4 \times 10^{-6}$ and thus

$$\arg A_0 \simeq \arg A_2 \qquad (40.75)$$

to very high accuracy. In addition, we have $e^{i\omega_0} = -\frac{A_0}{A_0^*}$, which implies that $\omega_0 = 2 \arg A_0 \simeq 2 \arg A_2$. The leading contribution to A_2 is given by two Feynman diagrams, Figs. 40.2 and 40.3. The amplitude of Fig. 40.2 is proportional to $A_2 \sim b V_{us}^* V_{ud}$, while the Penguin amplitude is proportional to $\frac{\alpha G_F}{4\pi} \sum_i V_{is}^* V_{id} m_i^2 \ll A_2$. Therefore, neglecting the Penguin contribution and combining the above results, we find

$$e^{i\omega_0} \simeq -\frac{A_2}{A_2^*} \simeq -\frac{V_{us}^* V_{ud}}{V_{us} V_{ud}^*} \simeq -\frac{V_{us}^* V_{us}^* V_{ud} V_{ud}}{|V_{us}|^2 |V_{ud}|^2}. \qquad (40.76)$$

---

[1] The factor of two comes from the two possible orderings of the currents.

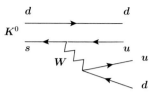

Fig. 40.2 $W^{\pm}$ tree contribution to $K^0 \to \pi^+\pi^-$.

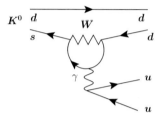

Fig. 40.3 Penguin graph contribution to $K^0 \to \pi^+\pi^-$.

Consider the quantity

$$\lambda_i V_{us}^* V_{ud} \equiv V_{is} V_{id}^* V_{us}^* V_{ud} \tag{40.77}$$
$$= V_{is} V_{us}^* V_{ud} V_{id}^* \equiv \Box_i \quad \text{for} \quad i = c, t$$
$$= |V_{us}|^2 |V_{ud}|^2 \quad \text{for} \quad i = u.$$

This result implies that the phase in ϵ_K must come from the internal c or t diagram. Under phase redefinitions of the quark mass eigenstates, the Cabibbo–Kobayashi–Maskawa (CKM) matrix elements transform as

$$V_{ij} \to e^{i(\theta_i^u - \theta_j^d)} V_{ij}. \tag{40.78}$$

As a consequence it is easy to see that $\Box_i$ is re-phase-invariant. In terms of these re-phase-invariant quantities, we have

$$\epsilon_K \simeq -e^{i\pi/4} \left(\frac{G_F^2 M_W^2}{12\pi^2} \frac{m_K f_K^2 B_K}{\sqrt{2}\Delta m} \right) \text{Im} \left\{ \sum_{ij=c,t} \frac{\Box_i \Box_j S_{ij} \eta_{ij}}{|V_{us}|^2 |V_{ud}|^2} \right\}. \tag{40.79}$$

We can also define

$$\text{Im}(\Box_c^2) \equiv 2\text{Re}(\Box_c)\text{Im}(\Box_c) \equiv -2\text{Re}(\Box_c)J, \tag{40.80}$$
$$\text{Im}(\Box_t^2) \equiv 2\text{Re}(\Box_t)\text{Im}(\Box_t) \equiv 2\text{Re}(\Box_t)J,$$

where J is the Jarlskog-invariant measure of **CP** violation.

$$J \equiv \text{Im}(\Box_t) = \text{Im}(V_{ts} V_{us}^* V_{ud} V_{td}^*) \tag{40.81}$$

or using the unitarity of the CKM matrices, i.e.

$$\sum_i V_{ti} V_{ui}^* = 0, \tag{40.82}$$

we have, equivalently,

$$J = -\text{Im}(V_{tb}V_{ub}^*V_{ud}V_{td}^*). \tag{40.83}$$

Note,

$$\sum_i \Box_i \equiv 0 \tag{40.84}$$

as a consequence of the unitarity of the CKM matrix. In addition, $\text{Im}\Box_u = 0$, therefore $\text{Im}(\Box_c) = -\text{Im}(\Box_t)$. Finally,

$$\text{Im}(\Box_c\Box_t) = \text{Re}(\Box_c)\text{Im}(\Box_t) + \text{Re}(\Box_t)\text{Im}(\Box_c) = -J(\text{Re}(\Box_t) - \text{Re}(\Box_c)). \tag{40.85}$$

Hence, in re-phase-invariant notation, we have

$$\epsilon_K \simeq -e^{i\pi/4}\left(\frac{G_F^2 M_W^2}{6\pi^2}\frac{m_K f_K^2 B_K}{\sqrt{2}\Delta m}\right) \tag{40.86}$$

$$\times \frac{J}{|V_{us}|^2|V_{ud}|^2}\left\{-\text{Re}(\Box_c)S_c\eta_c + \text{Re}(\Box_t)S_t\eta_t - (\text{Re}(\Box_t) - \text{Re}(\Box_c))S_{ct}\eta_{ct}\right\}.$$

Using

$$S_c \sim \frac{m_c^2}{M_W^2}, \quad S_t \simeq 0.6\frac{m_t^2}{M_W^2}, \quad S_{ct} \sim \frac{m_c^2}{m_W^2}\ln\frac{m_t^2}{m_c^2}, \tag{40.87}$$

and

$$\left(\frac{G_F^2 M_W^2 f_K^2}{\sqrt{2}\,6\pi^2}\frac{m_K}{\Delta m}\right) = 6.25\left(\frac{f_K}{165\,MeV}\right)^2, \tag{40.88}$$

we have

$$|\epsilon_K| \sim (6.25B_K)\left(\frac{f_K}{165\,\text{MeV}}\right)^2\frac{J}{|V_{ud}|^2|V_{us}|^2} \tag{40.89}$$

$$\times \left\{-\text{Re}(\Box_c)\frac{m_c^2}{M_W^2}\eta_c + \text{Re}(\Box_t)0.6\frac{m_t^2}{M_W^2}\eta_t - (\text{Re}(\Box_t)-\text{Re}(\Box_c))\frac{m_c^2}{m_W^2}\ln\left(\frac{m_t^2}{m_c^2}\right)\eta_{ct}\right\}.$$

Clearly, for large $m_t \gg m_c$ the middle term dominates and we have

$$|\epsilon_K| \simeq 6.25B_K\left(\frac{f_K}{165\,\text{MeV}}\right)^2\frac{J}{\lambda^2}\text{Re}(\Box_t)0.6\frac{m_t^2}{M_W^2}\eta_t. \tag{40.90}$$

For an evaluation of ϵ_K'/ϵ_K see, for example, Buchalla et al. (1990) and Kim et al. (1990).

40.3 ϵ_K in Terms of the Wolfenstein Parameters

Keeping only the top-quark contribution in Eqn. 40.74 and using Eqns. 40.87 and 40.88, we have

$$|\epsilon_K| \simeq 6.25B_K\left(\frac{f_K}{165\,\text{MeV}}\right)^2(A^2\lambda^5)^2\bar{\eta}(1-\bar{\rho})0.6\frac{m_t^2}{M_W^2}\eta_t, \tag{40.91}$$

where $\lambda_t \approx -A^2\lambda^5[1 - \bar\rho + i\bar\eta]$. Combining the measurement of ϵ_K and the theory, we obtain a curve in the $\bar\eta$, $\bar\rho$ plane. This result is presented in Figs. 40.4 and 40.5 as obtained by the international collaborations, UTfitter and CKMfitter. On these two figures, you can see the constraints coming from $\left|\frac{V_{ub}}{V_{cb}}\right|$ and ϵ_K plotted in the $\bar\eta$, $\bar\rho$ plane.

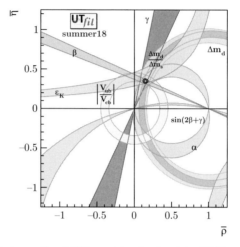

Fig. 40.4　Experimental constraints on the CKM matrix obtained by UTfitter. Figure taken from UTfit Collaboration (n.d.). For a color version of this figure, please see the color plate section.

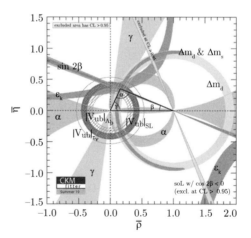

Fig. 40.5　Experimental constraints on the CKM matrix obtained by CKMfitter. Figure taken from CKMfitter Collaboration (n.d.). For a color version of this figure, please see the color plate section.

There are several additional constraints that we shall now discuss. Two of these come from the theory and measurements of B_d^0–$\bar{B}_d^0$ and B_s^0–$\bar{B}_s^0$ mixing. Experimentally, B_d^0–$\bar{B}_d^0$ mixing is obtained by, for example, $e^+e^- \to b\,\bar{b}$ at $\sqrt{s} = 10$ GeV by the BaBar collaboration at Stanford Linear Accelerator Center (SLAC) (Aubert et al., 2002; Beringer, 2002; Sakai, 2002). The $b,\bar{b}$ then hadronize and produce a $B,\bar{B}$ pair at rest. Then one observes the decays such as

$$\bar{B}_{b\bar{d}}^0 \to \bar{D}_{c\bar{d}}^+ + e^- + \bar{\nu}_e,$$
$$B_{b\bar{d}}^0 \to D_{\bar{c}d}^- + e^+ + \nu_e,$$
(40.92)

followed, prior to decaying, by an oscillation $B_d \to \bar{B}_d$ and

$$\bar{B}_{b\bar{d}}^0 \to \bar{D}_{c\bar{d}}^+ + e^- + \bar{\nu}_e.$$

With the observation of same-sign di-leptons, the latest result is given by (Tanabashi et al., 2018)

$$x_d \equiv \frac{\Delta m_{B_d}}{\Gamma_{B_d}} = 0.770 \pm 0.004.$$
(40.93)

Theoretically, B_d–$\bar{B}_d$ mixing is obtained by calculating the off-diagonal element in the B_d, $\bar{B}_d$ mass matrix. At one loop, the relevant Feynman diagram is given in Fig. 40.6. The amplitude is given in terms of the matrix element of an effective Hamiltonian, $\mathcal{H}_{eff}$, given by

$$\mathcal{H}_{eff} = \frac{G_F^2 M_W^2}{16\pi^2}\, V_{ib}^* V_{id} V_{jb}^* V_{jd} S_{ij} \eta_{ij}\, \bar{b}\gamma^\mu(1-\gamma_5)d\,\bar{b}\,\gamma_\mu(1-\gamma_5)d + h.c..$$
(40.94)

Then, the off-diagonal mass matrix element is given by

$$M_{12} \equiv \frac{\langle B_d^0|\mathcal{H}_{eff}(0)|\bar{B}_d^0\rangle}{2m_{B_d}} + \dots$$
(40.95)
$$= \frac{G_F^2 M_W^2}{12\pi^2} f_{B_d}^2 m_{B_d} B_{B_d} V_{ib}^* V_{id} V_{jb}^* V_{jd} S_{ij} \eta_{ij}$$

and

$$\Delta m_{B_d} \equiv \Delta m_d \approx 2|M_{12}|.$$
(40.96)

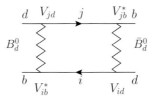

Fig. 40.6   One-loop Feynman diagram contributing to B_d^0–$\bar{B}_d^0$ mixing.

Experimentally, we have $\Gamma_{12} \ll M_{12}$. $\eta_t = 0.55$ and the term S_t dominates. The experimental observable,

$$x_d \equiv \frac{\Delta m_d}{\Gamma_{B_d}} = \frac{G_F^2}{6\pi^2 \Gamma_{B_d}} f_{B_d}^2 m_{B_d} B_{B_d} |V_{tb}|^2 |V_{td}|^2 m_t^2 \eta_t. \tag{40.97}$$

Lattice calculations give $f_{B_d} = 160 \pm 30$ MeV and $B_{B_d} = 1.30 \pm 0.15$. Finally, the dependence on the Wolfenstein parameters comes from

$$|V_{td}| \simeq A\lambda^3 [(1 - \bar{\rho})^2 + \bar{\eta}^2]. \tag{40.98}$$

Note, $|V_{tb}| \sim 1$.

Similarly, we can obtain $B_s^0 - \bar{B}_s^0$ mixing. Experimentally (Tanabashi et al., 2018),

$$\Delta m_{B_s} \equiv \Delta m_s = 1.1688 \pm 0.0014 \times 10^{-8} \text{ MeV}. \tag{40.99}$$

It was first measured by the Collider Detector at Fermilab (CDF). Theoretically, we obtain

$$\Delta m_{B_s} = \frac{G_F^2}{6\pi^2} f_{B_s}^2 m_{B_s} B_{B_s} |V_{tb}|^2 |V_{ts}|^2 m_t^2 \eta_t, \tag{40.100}$$

with

$$|V_{ts}| \simeq A\lambda^2. \tag{40.101}$$

40.4 Unitarity Triangle

Consider the sum

$$(V^\dagger V)_{bd} = V_{ud} V_{ub}^* + V_{cd} V_{cb}^* + V_{td} V_{tb}^* \equiv 0. \tag{40.102}$$

Each term is a complex number which defines a vector in a two-dimensional plane. Divide the sum by one of the terms and we obtain

$$\frac{V_{ud} V_{ub}^*}{V_{cd} V_{cb}^*} + \frac{V_{td} V_{tb}^*}{V_{cd} V_{cb}^*} + 1 = 0. \tag{40.103}$$

The sum of these three vectors in the two-dimensional plane form the "unitarity triangle," see Fig. 40.5. The upper vertex of the triangle is given by the end point of the vector

$$1 + \frac{V_{td} V_{tb}^*}{V_{cd} V_{cb}^*} = 1 + \frac{A\lambda^3 [1 - (1 - \frac{\lambda^2}{2})(\rho + i\eta)]}{(-\lambda)(A\lambda^2)} \tag{40.104}$$

$$= (\rho + i\eta)(1 - \lambda^2/2) + O(\lambda^4) \equiv (\bar{\rho} + i\bar{\eta}) + O(\lambda^4).$$

It is also given by

$$-\frac{V_{ud} V_{ub}^*}{V_{cd} V_{cb}^*} = -\frac{(1 - \lambda^2/2) A\lambda^3 (\rho + i\eta)}{(-\lambda) A\lambda^2} \tag{40.105}$$

$$= (\bar{\rho} + i\bar{\eta}) + O(\lambda^4).$$

The magnitude of the sides and the angles of the unitarity triangle, α, β, γ, can be measured independently (Beringer, 2002; Sakai, 2002). The **CP**-violating observable

$$a_{J/\psi K_S}^{CP} \equiv \frac{N(\bar{B}_d^0 \to J/\psi\ K_S) - N(B_d^0 \to J/\psi\ K_S)}{N(\bar{B}_d^0 \to J/\psi\ K_S) + N(B_d^0 \to J/\psi\ K_S)} \tag{40.106}$$

determines

$$\sin(2\beta) \tag{40.107}$$

and

$$a_{\pi^+\pi^-}^{CP} \equiv \frac{N(\bar{B}_d^0 \to \pi^+\pi^-) - N(B_d^0 \to \pi^+\pi^-)}{N(\bar{B}_d^0 \to \pi^+\pi^-) + N(B_d^0 \to \pi^+\pi^-)} \tag{40.108}$$

determines

$$\sin(2\alpha). \tag{40.109}$$

41 Effective Field Theories

In this chapter we discuss the concept of effective field theories which we shall apply to the Standard Model and physics beyond the Standard Model. We shall make use of the following references: Georgi et al. (1974), Appelquist and Carazzone (1975), Weinberg (1980) and Ovrut and Schnitzer (1981b), Ovrut and Schnitzer (1981a); and also a review article (Cohen, 1993). Consider an example of quantum electrodynamics with n_H heavy fermions with mass M and n_L light fermions with mass m, with $m \ll M$. For energies much less than M we intuitively expect that we can neglect heavy states and consider an effective theory with light fermions and photons only. This is the essence of the Appelquist–Carazzone (AC) theorem (Appelquist and Carazzone, 1975). The concept of effective field theories was defined by Weinberg (1980), but used earlier in the paper on grand unification by Georgi, Quinn and Weinberg (Georgi et al., 1974). In particular, our simplified discussion here will rely on the work of Ovrut and Schnitzer (1981b,a) who discuss the AC theorem in the context of the minimal subtraction $[\overline{MS}]$ renormalization scheme and dimensional regularization.

41.1 Vacuum Polarization

Consider the photon polarization tensor, Fig. 30.1, given by (see Appendix C)

$$\Pi_{\mu\nu}(q^2) \equiv (q_\mu q_\nu - g_{\mu\nu}q^2)\Pi(q^2). \tag{41.1}$$

For $q^2 < 0$ we have

$$\Pi(q^2) = -\frac{e^2}{2\pi^2}n_H \int_0^1 d\alpha\,\alpha(1-\alpha) \ln\left(\frac{-q^2\alpha(1-\alpha)+M^2}{\hat{\mu}^2}\right) \tag{41.2}$$
$$-\frac{e^2}{2\pi^2}n_L \int_0^1 d\alpha\,\alpha(1-\alpha) \ln\left(\frac{-q^2\alpha(1-\alpha)+m^2}{\hat{\mu}^2}\right),$$

where we've defined the scale $\hat{\mu}^2 \equiv 4\pi\mu^2 e^{-\gamma}$. Now considering the limit $-q^2 \ll M^2$, we have

$$\Pi(q^2) = -\left(\frac{e^2}{16\pi^2}\,\frac{4}{3}n_H\right)\ln\frac{M^2}{\hat{\mu}^2} + \Pi_L(q^2) + O(q^2/M^2). \tag{41.3}$$

The full photon propagator is given by (see Appendix C)

$$D_{\mu\nu}(q) = \frac{(1 + \Pi(q^2))^{-1}}{q^2}\left[-g_{\mu\nu} + [1 - \xi(1 + \Pi(q^2))]\frac{q_\mu q_\nu}{q^2}\right], \qquad (41.4)$$

given by the full Lagrangian

$$\mathcal{L} = -\frac{1}{4}(\partial_\mu A_\nu - \partial_\nu A_\mu)^2 - \frac{1}{2\xi}(\partial_\mu A^\mu)^2 \qquad (41.5)$$

$$+ \bar{\psi}_H i\partial\!\!\!/\psi_H - M\bar{\psi}_H\psi_H + e\bar{\psi}_H A\!\!\!/\psi_H$$

$$+ \bar{\psi}_L i\partial\!\!\!/\psi_L - m\bar{\psi}_L\psi_L + e\bar{\psi}_L A\!\!\!/\psi_L.$$

This is the high-energy effective theory which includes all particles. We can evaluate the $\beta(e)$ function and anomalous dimension $\gamma_M(e)$. They are given at one loop by (recall Eqn. 30.1)

$$\beta(e) = \frac{e^3}{16\pi^2}\frac{4}{3}(n_H + n_L), \quad \gamma_M(e) = -\frac{6e^2}{16\pi^2}. \qquad (41.6)$$

Now consider the effective field theory for $-q^2 \ll M^2$, which is obtained by integrating out heavy fields. For example, the heavy contribution to the polarization tensor is given by

$$\hat{\Pi}(q^2) = -\left(\frac{e^2}{16\pi^2}\frac{4}{3}n_H\right)\ln\frac{M^2}{\hat{\mu}^2} + O(q^2/M^2). \qquad (41.7)$$

To one loop order the photon propagator dressed with only heavy fermions in the loop is given by

$$\hat{D}_{\mu\nu}(q) = \frac{(1 + \hat{\Pi}(q^2))^{-1}}{q^2}\left[-g_{\mu\nu} + [1 - \xi(1 + \hat{\Pi}(q^2))]\frac{q_\mu q_\nu}{q^2}\right], \qquad (41.8)$$

neglecting terms of order $|q^2|/M^2 \ll 1$. Then

$$\hat{D}_{\mu\nu}(q) \overset{|q^2|\ll M^2}{\to} (1 + \hat{\Pi}(0))^{-1}D_{\mu\nu}(q)_e, \qquad (41.9)$$

where

$$D_{\mu\nu}(q)_e = \frac{1}{q^2}\left[-g_{\mu\nu} + [1 - \xi_e]\frac{q_\mu q_\nu}{q^2}\right] \qquad (41.10)$$

is the effective photon propagator, at tree level, with $\xi_e \equiv \xi(1 + \hat{\Pi}(0))$.

Now let's construct the effective low-energy theory with just light particles. The insertion of $\hat{D}_{\mu\nu}$ between two light-fermion lines (for $|q^2| \ll M^2$) is equivalent to the tree graph of effective light fields with the effective photon propagator $\hat{D}_{\mu\nu}(q)_e$ and effective coupling constant

$$e_e \equiv (1 + \hat{\Pi}(0))^{-1/2}e. \qquad (41.11)$$

Thus, the Lagrangian for the effective low-energy theory is given by

$$\mathcal{L}_{eff} = -\frac{1}{4}(\partial_\mu A_\nu^e - \partial_\nu A_\mu^e)^2 - \frac{1}{2\xi_e}(\partial^\mu A_\mu^e)^2 \tag{41.12}$$

$$+ \bar{\psi}_{Le}i\slashed{\partial}\psi_{Le} - m\bar{\psi}_{Le}\psi_{Le} + e_e\bar{\psi}_{Le}\slashed{A}_e\psi_{Le},$$

with

$$A_\mu^e \equiv (1 + \hat{\Pi}(0))^{1/2}A_\mu, \quad \psi_{Le} \equiv \psi_L. \tag{41.13}$$

Note,

$$e_e \equiv e\left[1 - \left(\frac{e^2}{16\pi^2}\frac{4}{3}n_H\right)\ln\frac{M'^2}{\hat{\mu}^2}\right]^{-1/2}$$

$$\simeq e + \frac{e^3}{16\pi^2}\frac{2}{3}n_H\,\ln\frac{M^2}{\hat{\mu}^2}. \tag{41.14}$$

Eqn. 41.14 is the threshold behavior relating couplings in the full theory to couplings in the effective low-energy theory.

Now consider the renormalization group (RG) behavior of the effective coupling (see Appendix D). In Eqn. 41.6 we gave the beta function for the full theory, above M. The beta function for the effective theory satisfies the Callan–Symanzik equation

$$\mu\frac{de_e}{d\mu} = \left(\mu\frac{\partial}{\partial\mu} + \beta(e)\frac{\partial}{\partial e} + \gamma_M M\frac{\partial}{\partial M}\right)e_e. \tag{41.15}$$

Plugging in the formulae in Eqn. 41.6, we have

$$\mu\frac{de_e}{d\mu} = \left(-\frac{e^3}{16\pi^2}\frac{4}{3}n_H + \frac{e^3}{16\pi^2}\frac{4}{3}(n_H + n_L) - \frac{6e^2}{16\pi^2}\left(\frac{e^3}{16\pi^2}\frac{4}{3}n_H\right)\right) \equiv \beta_e(e_e) + O(e_e^5). \tag{41.16}$$

Hence, we have

$$\beta_e(e_e) \equiv \frac{e_e^3}{16\pi^2}\frac{4}{3}n_L, \tag{41.17}$$

i.e. the RG equation for the effective coupling is given solely in terms of light fermion loops, calculated in the effective low-energy field theory. Heavy states only contribute to threshold effects (or matching conditions) relating the coupling in the full theory to the coupling in the effective low-energy theory at the scale $\hat{\mu} \sim M$. There are also higher-dimension operators, such as terms of order $|q^2|/M^2$ from $\hat{\Pi}(q^2)$, for example

$$\mathcal{L}_{eff} \supset \frac{1}{M^2}(\partial_\mu A_\nu^e - \partial_\nu A_\mu^e)\Box(\partial^\mu A^{\nu e} - \partial^\nu A^{\mu e}). \tag{41.18}$$

41.2 Effective Field Theories above and below M_Z

The Standard Model is, perhaps, an effective field theory valid below the Planck scale. The gauge group is $SU(3)_C \otimes SU(2)_L \otimes U(1)_Y$ with three families of quarks

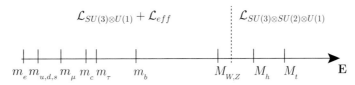

$$\mathcal{L}_{SU(3)\otimes U(1)} + \mathcal{L}_{eff} \qquad\qquad \mathcal{L}_{SU(3)\otimes SU(2)\otimes U(1)}$$

Fig. 41.1 The Standard Model described by two effective field theories, one above and one below M_Z.

and leptons. At the weak scale, the Higgs field obtains a non-zero vacuum expectation value, giving mass to the $W^\pm$, Z^0, h bosons, and quarks and leptons. Below the weak scale, the gauge group is $SU(3)_C \otimes U(1)_{EM}$. The top quark obtains mass of order 173 GeV, so that below the top-quark mass, there are only five quarks and three leptons in the effective low-energy theory. One can define a scale μ of order M_Z to divide the Standard Model into two effective field theories, above or below $\mu = M_Z$, as in Fig. 41.1. The low-energy theory, below M_Z, does not include the $W^\pm$, Z^0, the top quark or the Higgs boson. In principle, there can be a number of effective field theories. For example, for $\mu \geq m_t$, one has the complete Standard Model. Then for $M_Z \leq \mu \leq m_t$, one can integrate out the top quark, keeping all other states of the Standard Model. Finally for $m_b \leq \mu \ll M_Z$, one has all leptons, five quarks, the photon and gluons. Finally, for $\mu < m_b$ one has four quarks, and all leptons. Then there is one more threshold at the tau mass and another at the charm-quark mass. The up, down and strange quark masses are usually determined at 2 GeV.

Now consider some important threshold corrections. At the scale $\mu = M_Z$, one has the relation

$$M_W^2 = \frac{\alpha(M_Z)\pi}{\sqrt{2}G_F(M_Z)\sin^2\theta_W(M_Z)}. \tag{41.19}$$

Higher loop corrections to this result (evaluated at $\mu = M_Z$) are determined as an expansion in $\frac{\alpha_1}{4\pi} = \frac{\alpha}{4\pi\cos^2\theta_W}$, $\frac{\alpha_2}{4\pi} = \frac{\alpha}{4\pi\sin^2\theta_W} \leq 1\%$. These are small effects. The most important correction comes from the fact that $\alpha(M_Z) \neq \alpha(m_e) = \frac{1}{137...}$ and effective four-Fermi operators measured in neutral- and charged-current processes at lower energy must be renormalized from the scale $\mu \sim E_{exp}$ to $\mu = M_Z$. We use the RG equations, which sum an infinite series of logarithmic corrections obtained in $\mathcal{L}_{eff}(\mu < M_Z)$ due to α, α_s corrections of order $\frac{\alpha}{4\pi}\ln\frac{M_Z}{\mu}$ and $\frac{\alpha_s}{4\pi}\ln\frac{M_Z}{\mu}$.

Corrections to α

Using an approximation to Eqn. 30.3, we have

$$\alpha(\mu_2) \approx \alpha(\mu_1)\left[1 + \frac{2\alpha(\mu_1)}{3\pi}\sum_i Q_i^2\ln\frac{\mu_2}{\mu_1}\right], \tag{41.20}$$

where the sum is over all fermions in the effective theory between the scales μ_1 and μ_2. Therefore, integrating from m_e up to M_Z we obtain, at one loop,

$$\alpha(M_Z) \approx \alpha(m_e) \left[1 + \frac{2\alpha(m_e)}{3\pi} \sum_i Q_i^2 \ln \frac{M_Z}{m_i} \right] \simeq \frac{1}{128}, \qquad (41.21)$$

where m_i are the quark and lepton masses with $m_i < M_Z$. This happens to be the largest correction to Eqn. 41.19. We have

$$\frac{\alpha(M_Z) - \alpha(m_e)}{\alpha(M_Z)} \equiv \frac{\Delta\alpha}{\alpha(M_Z)} \approx \frac{9}{137} \sim 6\% \qquad (41.22)$$

increase in α. Now define

$$M_W^2 \equiv \frac{\alpha(m_e)\pi}{\sqrt{2}G_F(M_Z)\sin^2\theta_W(M_Z)} \left(\frac{1}{1 - \Delta r} \right) = \frac{(37.3 \text{ GeV})^2}{\sin^2\theta_W(M_Z)} \left(\frac{1}{1 - \Delta r} \right), \qquad (41.23)$$

where at low energies $G_F = G_\mu$. This equation can be rearranged to obtain an equation for $\sin^2\theta_W$ in terms of M_Z. We find

$$\sin^2\theta_W(M_Z) = \frac{1}{2} \left(1 - \sqrt{1 - \frac{4(37.3 \text{ GeV})^2}{M_Z^2(1 - \Delta r)}} \right). \qquad (41.24)$$

Δr gets contributions from $\Delta\alpha$, ΔG_F and radiative corrections from the top quark and Higgs boson. The largest contribution to $\Delta r \simeq 1 - \frac{\alpha(m_e)}{\alpha(M_Z)} = \frac{\Delta\alpha}{\alpha(M_Z)}$.

Corrections to G_F

First, consider just leptons. We have the effective four-Fermi operator

$$\frac{G_F}{\sqrt{2}} \bar{\nu}_e \gamma^\mu (1 - \gamma_5) e \bar{\mu} \gamma_\mu (1 - \gamma_5) \nu_\mu. \qquad (41.25)$$

In this case, $G_F \equiv G_\mu$. At one loop, there are two types of Feynman diagrams which contribute to the renormalization of this operator, given in Fig. 41.2. In the Landau gauge, the first term has no $\ln\mu$ dependence and the second diagram is finite. Thus, at one loop, $G_\mu(m_\mu) \simeq G_\mu(M_Z)$.

Consider now a hadronic–leptonic charged-current operator,

$$C(\mu)\bar{\mu}\gamma^\mu(1 - \gamma_5)\nu_\mu \bar{u}\gamma_\mu(1 - \gamma_5)d, \qquad (41.26)$$

as in Fig. 41.3. At one loop, the Wilson coefficient $C(\mu)$ is given by

$$C(\mu) = \frac{G_F}{\sqrt{2}} \left(1 + \frac{\alpha}{\pi} \ln \frac{M_Z}{\mu} \right). \qquad (41.27)$$

Thus,

$$G_F(M_Z) \simeq G_\mu \neq G_F(m_p), \qquad (41.28)$$

relevant for nuclear beta decay. The coefficient $C(\mu)$ satisfies an RG equation.

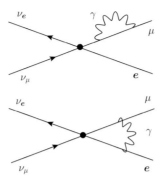

Fig. 41.2 Two types of one-loop diagrams (with $\xi = 0$) contributing to the RG running of this four-Fermi operator.

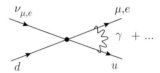

Fig. 41.3 One-loop diagrams contributing to the RG running of this four-Fermi operator.

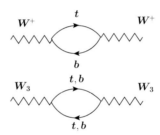

Fig. 41.4 One-loop corrections to the $W^+ - W^3$ mass difference.

Corrections to the Electroweak ρ Parameter

Radiative corrections to the electroweak ρ parameter come at one loop due to the $W^+ - W^3$ mass corrections, Fig. 41.4. The result is given by Einhorn et al. (1981)

$$M_{W^\pm}^2 - M_{W^3}^2 = \frac{3\alpha}{16\pi \sin^2 \theta_W} m_t^2, \tag{41.29}$$

where $m_t = \lambda_t \frac{v}{\sqrt{2}}$ and $e \equiv g_2 \sin \theta_W$. We then find

$$M_{W^3}^2 = M_W^2 (1 - \Delta) \quad \text{with} \quad \Delta \equiv \frac{M_{W^\pm}^2 - M_{W^3}^2}{M_{W^\pm}^2} = \frac{3\lambda_t^2}{32\pi^2} \tag{41.30}$$

and $M_W \equiv M_{W^\pm}$. Finally,

$$M_Z^2 \equiv \frac{M_{W^3}^2}{\cos^2 \theta_W} = \frac{M_W^2 (1 - \Delta)}{\cos^2 \theta_W}. \tag{41.31}$$

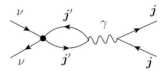

Fig. 41.5   One-loop diagram contributing to the correction to $\sin^2 \theta_W$.

Hence,

$$\rho \equiv \frac{M_W^2}{M_Z^2 \cos^2 \theta_W} = 1 + \Delta\rho, \qquad (41.32)$$

with

$$\Delta\rho \simeq \Delta = \frac{3\alpha}{16\pi \sin^2 \theta_W} \frac{m_t^2}{M_W^2} = \frac{3 G_F m_t^2}{8\sqrt{2}\pi^2}. \qquad (41.33)$$

Experimentally, $\rho = 1.01013 \pm 0.00005$ (Tanabashi et al., 2018), which places an upper bound on the top-quark mass of order 178.7 GeV. This is viewed as a prediction of the top-quark mass.

Corrections to Neutral-Current Processes

Consider, for example, four Fermi neutral-current neutrino scattering. At leading order, the effective operator is given by

$$\frac{G_F \rho}{\sqrt{2}} \bar{\nu}\gamma^\mu(1-\gamma_5)\nu \sum_j \bar{\psi}_j \gamma_\mu \left(T_{3j}(1-\gamma_5) - 2\sin^2 \theta_W (M_Z) Q_j\right)\psi_j. \qquad (41.34)$$

There is a radiative correction to $\sin^2 \theta_W$ due to the one-loop Feynman graph in Fig. 41.5. The vertex correction in Fig. 41.5 vanishes as $q^2 \to 0$ (where q is the photon momentum), since the neutrino has zero charge. Hence, the correction is of order

$$\sim e\bar{\nu}\gamma^\mu(1-\gamma_5)\nu \underbrace{\partial^\lambda F_{\lambda\mu}}_{e j_\mu^{EM}}. \qquad (41.35)$$

We thus obtain the corrected four-Fermi operator given by

$$\frac{G_F \rho}{\sqrt{2}} \bar{\nu}\gamma^\mu(1-\gamma_5)\nu \Big[\sum_j \bar{\psi}_j \gamma_\mu (T_{3j}(1-\gamma_5) - 2\sin^2 \theta_W (M_Z) Q_j)\psi_j \qquad (41.36)$$
$$- \sum_j \bar{\psi}_j \gamma_\mu Q_j \psi_j \sum_{j'} \frac{2\alpha}{3\pi} \ln \frac{M_Z}{m} Q_{j'}[T_{3j'} - 2\sin^2 \theta_W (M_Z) Q_{j'}]\Big],$$

where $m \equiv Max(\mu, m_{j'})$. Finally, we obtain the correction to the effective neutral-current operator at one loop given by (Dawson et al., 1981)

$$\frac{G_F \rho}{\sqrt{2}} \bar{\nu}\gamma^\mu(1-\gamma_5)\nu \left[\sum_j \bar{\psi}_j \gamma_\mu \left(T_{3j}(1-\gamma_5) - 2\sin^2 \theta_W (\mu) Q_j\right) \psi_j \right], \qquad (41.37)$$

where

$$\sin^2 \theta_W (\mu) = \sin^2 \theta_W (M_Z) \left[1 + \frac{\alpha}{3\pi} \sum_{j'} Q_{j'} \left(\frac{T_{3j'}}{\sin^2 \theta_W (M_Z)} - 2Q_{j'} \right) \ln \frac{M_Z}{m} \right]. $$
$$(41.38)$$

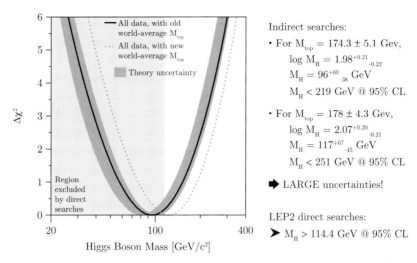

Indirect searches:

- For $M_{top} = 174.3 \pm 5.1$ Gev,

  $\log M_H = 1.98^{+0.21}_{-0.22}$

  $M_H = 96^{+60}_{-38}$ GeV

  $M_H < 219$ GeV @ 95% CL

- For $M_{top} = 178 \pm 4.3$ Gev,

  $\log M_H = 2.07^{+0.20}_{-0.21}$

  $M_H = 117^{+67}_{-45}$ GeV

  $M_H < 251$ GeV @ 95% CL

➡ LARGE uncertainties!

LEP2 direct searches:

➤ $M_H > 114.4$ GeV @ 95% CL

Fig. 41.6 Electroweak bounds on the Higgs mass, prior to its discovery in 2012, from Chuang's presentation at the Beach 04 meeting in 2004) (Chuang, 2004). The dotted line corresponds to the case of a heavier top quark.

This gives a few percent correction to $\sin^2 \theta_W$.

$\sin^2 \theta_W$ is measured in the processes of $\nu_e e$, $\bar{\nu}_e e$, $\nu_\mu e$, $\bar{\nu}_\mu e$ scattering, $\nu_\mu n$, $\nu_\mu p$ scattering, $A_{FB}(e^+ e^- \to \mu^+ \mu^-)$, Z decays and the Z mass. Each measurement is made at a different energy scale and probes different effective operators which have different radiative corrections. Prior to the discovery of the top quark and Higgs boson, the measurements of $\sin^2 \theta_W$ had a calculable dependence on m_t, m_h, which was used to fix the best-fit value of these masses. Once the top quark was discovered, the electroweak data provided a bound on the Higgs mass, see Fig. 41.6. LEP II data gave a lower bound on the Higgs mass of 114.4 GeV.

41.3 The Higgs Boson

The Higgs boson was discovered at the CERN LHC pp collider in 2012. The decay modes of the Higgs with the least amount of background are $h \to \gamma\gamma$ and $h \to ZZ^* \to \mu^+ \mu^- \mu^+ \mu^-$. These processes were seen by the center of momentum system (CMS) and ATLAS collaborations. See, for example, the observation in the $\gamma\gamma$ decay mode, Figs. 41.7 and 41.8.

A single Higgs boson is produced predominantly via gluon fusion, Fig. 41.9. In the two first observed modes, it decays via the inverse process, Fig. 41.10. The Higgs boson mass is measured to be 125.10 ± 0.25 GeV with $J^P = 0^+$ and, to within 10%, its couplings are consistent with the Standard Model predictions (Tanabashi et al., 2018).

Note, precision electroweak data show a $\sim 2\sigma$ tension between the W boson and top-quark mass measurements, see Fig. 41.11.

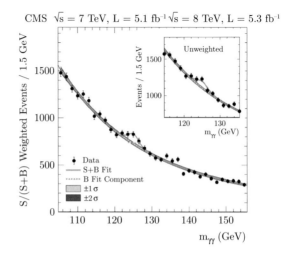

Higgs discovery in the 2γ channel by the CMS collaboration (Chatrchyan et al., 2012).

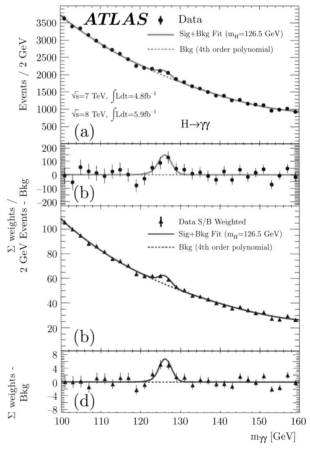

Higgs discovery in the 2γ channel by ATLAS collaboration (Aad et al., 2012).

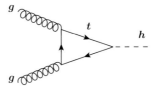

This is the gluon fusion process which provides the dominant production cross-section for the Higgs boson. Note, it goes through a triangle diagram with all fermions coupling to the Higgs in the loop. However, the top-quark contribution dominates.

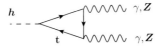

The Higgs decay mode to a pair of photons or Z proceeds via the same triangle diagram as in gluon fusion.

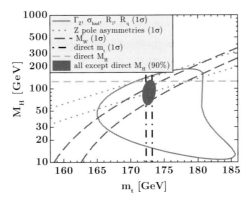

There is a clear tension between the W boson and top-quark mass measurements evident in this figure (Tanabashi et al., 2018). Removing the kinematic constraint on M_{Higgs} from the LHC gives the loop-level determination from the precision data, $M_{Higgs} = 90^{+17}_{-16}$ GeV, which is 1.9σ below the observed value.

42 Anomalies

This book has focused on the phenomenology of the Standard Model and just as much quantum field theory as is necessary to understand this phenomenology. We have made extensive use of symmetries to understand the Standard Model. These symmetries give us, via Noether's theorem, conserved currents and charges, which are the generators of those symmetries. All of this analysis has been completely classical. However, sometimes classical symmetries and the resultant conservation laws can be destroyed by quantum effects. This can be the case for either global or gauge currents whose divergence is anomalous.

We have already referred to this possibility in two instances.

1. The resolution of puzzle 3 in Section 28.1 was in fact due to an anomaly in the axial-vector current,

$$J_3^{\mu 5} = \bar{q}(x)\gamma_\mu\gamma_5 T_3 q(x). \tag{42.1}$$

The theory of a partially conserved axial current and the soft-pion limit would make $\Gamma(\pi^0 \to 2\gamma) \approx 0$. This is the Sutherland–Veltman theorem (Sutherland, 1967). However, due to triangle diagrams, it can be shown that (Adler, 1969; Bell and Jackiw, 1969)

$$\partial^\mu J_{\mu 5}^3 = \frac{e^2 D}{8\pi^2} F^{\mu\nu} \, {}^*F_{\mu\nu}, \tag{42.2}$$

where $F_{\mu\nu}$ is the electromagnetic field strength, ${}^*F_{\mu\nu} \equiv \frac{1}{2}\epsilon_{\mu\nu\lambda\rho}F^{\lambda\rho}$ and

$$D = \frac{3}{2}\mathrm{Tr}(\{Q,Q\}T_3), \tag{42.3}$$

where Q is the electric charge of the up, down and strange quarks with three colors each and T_3 is the isospin generator. We have

$$Q = \frac{1}{3}\begin{pmatrix} 2 & & \\ & -1 & \\ & & -1 \end{pmatrix}, \quad T_3 = \frac{1}{2}\begin{pmatrix} 1 & & \\ & -1 & \\ & & 0 \end{pmatrix}. \tag{42.4}$$

2. In the case of the η', the solution to the problem of where is the ninth pseudo-scalar Nambu–Goldstone boson is resolved due to the fact that the axial $U(1)$ current has a quantum chromodynamics (QCD) anomaly, as described in Eqn. 28.52, with

$$\partial^\mu j_{\mu 5} = \frac{g^2}{8\pi^2} \, \mathrm{Tr}(\tilde{G}^{\mu\nu} \, {}^*\tilde{G}_{\mu\nu}). \tag{42.5}$$

In addition, as a consequence of non-perturbative QCD effects, the η' obtains a large mass.

Anomalies were first calculated using the Feynman-diagram approach (Adler, 1969; Bell and Jackiw, 1969) (see Appendix E). In chiral gauge theories, such as the Standard Model, it is possible for anomalies to appear in gauge currents. This would be a disaster, making the theories non-renormalizable (Gross and Jackiw, 1972). Thus, they must be avoided at all costs. In this section, we will apply the analysis of Fujikawa (1979, 1980) which will accentuate the role of axial transformations and anomalies. Fujikawa uses the path-integral formulation of quantum field theory to analyze anomalies.

42.1 Anomalies via the Path-Integral Approach

Consider the general case of a non-Abelian gauge theory coupled to left-handed Dirac spinors. Note, as we have seen we can construct the Standard Model in terms of these fermion fields. The Lagrangian is given by

$$\mathcal{L} = \bar{\psi} i \slashed{D} P_L \psi - \frac{1}{4} \mathrm{Tr}(\tilde{F}_{\mu\nu} \tilde{F}^{\mu\nu}), \tag{42.6}$$

with

$$D_\mu = \partial_\mu + ig\tilde{A}_\mu, \quad \tilde{A}_\mu = T_A A_{\mu A}. \tag{42.7}$$

In the path-integral formalism, Green's functions are obtained via functional derivatives of the generating functional given by

$$Z[\eta, \bar{\eta}, J_\mu] = \int DA_\mu D\psi D\bar{\psi} e^{i\left[\int_x (\bar{\psi}_L i \slashed{D} \psi_L - \frac{1}{4} \mathrm{Tr}(\tilde{F}_{\mu\nu} \tilde{F}^{\mu\nu})) + sources\right]}. \tag{42.8}$$

Consider the generating functional

$$Z[0, 0, J_\mu] = \int DA_\mu e^{i(S_{gauge} + source)} \int D\psi D\bar{\psi} e^{i\left[\int_x (\bar{\psi}_L i \slashed{D} \psi_L)\right]} \tag{42.9}$$

and the chiral transformations

$$\psi_L(x) \to \psi'_L(x) = e^{i\alpha(x)\gamma^5} \psi_L(x) = e^{-i\alpha(x)} \psi_L(x), \tag{42.10}$$
$$\bar{\psi}_L(x) \to \bar{\psi}'_L(x) = \bar{\psi}_L(x) e^{i\alpha(x)\gamma^5} = \bar{\psi}_L(x) e^{i\alpha(x)}.$$

Then we have

$$\int_x \bar{\psi}'_L i \slashed{D} \psi'_L = \int_x \left[\bar{\psi}_L i \slashed{D} \psi_L + \bar{\psi}_L(-\alpha(x)\slashed{\partial})\psi_L + \bar{\psi}_L \slashed{\partial}(\alpha(x)\psi_L) + O(\alpha(x)^2) \right]$$

$$= \int_x \left[\bar{\psi}_L(x) i \slashed{D} \psi_L(x) + (\partial_\mu \alpha(x)) \bar{\psi}_L(x) \gamma^\mu \psi_L(x) \right] \tag{42.11}$$

$$= \int_x \left[\bar{\psi}_L(x) i \slashed{D} \psi_L(x) - \alpha(x) \partial_\mu (\bar{\psi}_L(x) \gamma^\mu \psi_L(x)) \right].$$

For $\alpha(x)$ constant, the action is invariant and $J^\mu \equiv -\bar{\psi}_L \gamma^\mu \psi_L$ is conserved, i.e. classically,

$$\frac{\delta S}{\delta \alpha(x)} = \partial_\mu J^\mu = 0. \tag{42.12}$$

However, since the path integral corresponds to an integral over all of field space, it should be invariant under a change of integration variables. Therefore, even for arbitrary $\alpha(x)$, the path integral should be invariant. As a consequence, the coefficient of $\alpha(x)$ must vanish.

Fujikawa then showed that the functional measure is *not* chiral invariant. Define the functional integral in terms of eigenvectors of $i\slashed{D}$.[1]

ASIDE :
We define

$$(i\slashed{D})\phi_m(x) = \lambda_m \phi_m(x) \quad \text{left eigenvectors}, \tag{42.13}$$
$$(-iD_\mu^\dagger \hat{\phi}_m(x))\gamma^\mu = \hat{\lambda}_m \hat{\phi}_m(x) \quad \text{right eigenvectors}.$$

Then we have

$$\psi(x) \equiv \sum_m a_m \phi_m(x), \quad \bar{\psi}(x) \propto \hat{\phi}_m(x). \tag{42.14}$$

Therefore,

$$i\slashed{D}\psi(x) = \sum_m a_m \lambda_m \phi_m(x) \tag{42.15}$$

and the fermion action is given by

$$\int_x \hat{\phi}_{m'}(x) i\slashed{D}\phi_m(x) = \lambda_m \int_x \hat{\phi}_{m'}(x)\phi_m(x), \tag{42.16}$$
$$\int_x -iD_\mu^\dagger \hat{\phi}_{m'}(x) \gamma^\mu \phi_m(x) = \hat{\lambda}_{m'} \int_x \hat{\phi}_{m'}(x)\phi_m(x).$$

Note, the left-hand sides of both equations are equal by Hermitian conjugation and thus the right-hand sides are also equal. Thus, for $m = m'$ we have $\lambda_m = \hat{\lambda}_m$ and for $m \neq m'$ we have $\int_x \hat{\phi}_{m'}(x)\phi_m(x) \equiv 0$. We choose to ortho-normalize the eigenfunctions by

$$\int_x \hat{\phi}_{m'}(x)\phi_m(x) \equiv \text{constant} \times \delta_{mm'}. \tag{42.17}$$

We can similarly argue that

$$\int_x \phi_{m'}^\dagger(x)\phi_m(x) = \delta_{m'm}. \tag{42.18}$$

We then choose to define

$$\bar{\psi}(x) = \sum_m \bar{b}_m \phi_m^\dagger(x). \tag{42.19}$$

---

[1] This is more precisely done in Euclidean space–time in a large box, such that all eigenvalues are discrete.

a_m, $\bar{b}_m$ are anti-commuting Grassmann coefficients.

Some examples of eigenvectors are given by

1. $A_\mu(x) \equiv 0$, then $\phi_m(x) = e^{-ik \cdot x} u(\vec{k}, s)$ and $i\partial\!\!\!/\phi_m(x) = e^{-ik \cdot x} k\!\!\!/ u(\vec{k}, s)$. Thus, $\lambda_m^2 = k^2$.

2. $A_\mu(x) \neq 0$ is a smooth function, and $A_\mu(x) \to 0$ as $|\vec{x}| \to \infty$ or it can be ignored as $\vec{k} \to \infty$, then $\phi_m(x) \to e^{-ik \cdot x} u(\vec{k}, s)$ as $|\vec{x}| \to \infty$ or large momenta.

For the chiral fields we have[2]

$$\phi_m^L(x) \equiv \left(\frac{1 - \gamma^5}{\sqrt{2}} \right) \phi_m(x) \quad \text{for} \quad \lambda_m > 0 \tag{42.20}$$

$$\equiv \left(\frac{1 - \gamma^5}{2} \right) \phi_m(x) \quad \text{for} \quad \lambda_m = 0,$$

$$\phi_m^R(x)^\dagger \equiv \phi_m^\dagger(x) \left(\frac{1 + \gamma^5}{\sqrt{2}} \right) \quad \text{for} \quad \lambda_m > 0 \tag{42.21}$$

$$\equiv \phi_m^\dagger(x) \left(\frac{1 + \gamma^5}{2} \right) \quad \text{for} \quad \lambda_m = 0.$$

Also,

$$\phi_m^{L\,\dagger}(x) \equiv \phi_m(x)^\dagger \left(\frac{1 - \gamma^5}{\sqrt{2}} \right) \quad \text{for} \quad \lambda_m > 0. \tag{42.22}$$

We then have

$$\int_x \phi_m^{L\,\dagger}(x) \phi_{m'}^L(x) = \delta_{mm'}, \tag{42.23}$$

$$\int_x \phi_m^{R\,\dagger}(x) \phi_{m'}^R(x) = \delta_{mm'}.$$

These basis vectors form a complete orthonormal set and we have

$$\psi_L(x) \equiv \sum_{\lambda_m \geq 0} a_m \phi_m^L(x), \tag{42.24}$$

$$\bar{\psi}_L(x) \equiv \sum_{\lambda_m \geq 0} \bar{b}_m \phi_m^{R\,\dagger}(x). \tag{42.25}$$

Under the chiral transformation we then have

$$\psi_L'(x) \approx (1 - i\alpha(x))\psi_L(x) = (1 - i\alpha(x)) \sum_n a_n \phi_n^L(x) \tag{42.26}$$

or

$$a_m' = \sum_n \int_x \phi_m^{L\,\dagger}(x)(1 - i\alpha(x))\phi_n^L(x) a_n \equiv \sum_n (\mathbb{I}_{m,n} - i C_{m,n}^L) a_n. \tag{42.27}$$

---

[2] Note, $\gamma_5 \phi_m^L(x)$ is an eigenvector of $D\!\!\!\!/$ with eigenvalue $-\lambda_m$, if ϕ_m^L has eigenvalue $+\lambda_m$. The reason for the normalization with $\sqrt{2}$ in the denominator is because we will sum, initially, over only positive values of λ_m.

The path-integral integration measure is given by[3]

$$d\mu \equiv D\psi_L D\bar{\psi}_L = \prod_m da_m \prod_m d\bar{b}_m \tag{42.28}$$

and under a chiral transformation we have

$$\prod_m da'_m = [\det(\mathbb{I}_{m,n} - iC^L_{m,n})]^{-1} \prod_n da_n. \tag{42.29}$$

Proof   Consider the Grassmann integral

$$\int d\theta f(\theta) \equiv N, \tag{42.30}$$

where N is a number and $\theta^2 \equiv 0$. This is by definition the integral over "all space." Therefore it must be invariant under a shift of integration variables, i.e. $\theta' = \theta + \epsilon$, where ϵ is a Grassmann constant. Then we have

$$N = \int d\theta' f(\theta') = \int d\theta'[f(0) + \theta' f'], \tag{42.31}$$

where $f' \equiv \frac{\partial}{\partial \theta'} f$. We also have, under the shift,

$$N = \int d\theta f(\theta + \epsilon) = \int d\theta[f(0) + (\theta + \epsilon)f'(0)]. \tag{42.32}$$

Thus, the solution is

$$\int d\theta \text{ constant} = 0 \quad \text{and} \quad \int d\theta'\theta' = \int d\theta\theta = 1, \tag{42.33}$$

where we've normalized the integral to one. As a consequence, a Grassmann integral satisfies

$$\int d\theta f \equiv \frac{\partial}{\partial \theta} f. \tag{42.34}$$

Finally, given a change of variables $\theta' = A\theta + \epsilon$ we find the Jacobian using the chain rule such that

$$\int d\theta' f(\theta') \equiv \frac{\partial}{\partial \theta'} f(A\theta + \epsilon) = \frac{\partial \theta}{\partial \theta'} \frac{\partial}{\partial \theta} f(A\theta + \epsilon) \tag{42.35}$$

$$= A^{-1} \int d\theta f(A\theta + \epsilon) = \int d\theta \left(\frac{\partial \theta'}{\partial \theta}\right)^{-1} f(\theta'(\theta)). \tag{42.36}$$

The generalization to n Grassmann variables is then given by

$$1 = \int da'_1 \cdots da'_n \, a'_n \cdots a'_1, \tag{42.37}$$

with the change of variables

$$a'_i = \sum_j A_{ij} \, a_j. \tag{42.38}$$

---

[3] Note, integration by a Grassmann variable is equivalent to differentiation, i.e. $da = \frac{d}{da}$.

The Jacobian of the transformation

$$\prod_i da'_i = \int (\det A)^{-1} \prod_i da_i. \tag{42.39}$$

This can be proven using the identity

$$a'_n \cdots a'_1 = (\det A)\, a_n \cdots a_1. \tag{42.40}$$

$\square$

Similarly, for $\bar{b}_n$ we have

$$\bar{\psi}'_L(x) \approx \bar{\psi}_L(x)(1 + i\alpha(x)) = \sum_n \bar{b}_n \phi_n^{R\dagger}(x)(1 + i\alpha(x)) \tag{42.41}$$

or

$$\bar{b}'_m = \sum_n \int_x \bar{b}_n \phi_n^{R\dagger}(x)(1 + i\alpha(x))\phi_m^R(x) \equiv \sum_n \bar{b}_n\, (\mathbb{I}_{n,m} + iC_{n,m}^R). \tag{42.42}$$

Hence,

$$\prod_m d\bar{b}'_m = [\det(\mathbb{I}_{n,m} + iC_{n,m}^R)]^{-1} \prod_n d\bar{b}_n. \tag{42.43}$$

Using the identity

$$\det(\mathbb{I}_{m,n} \pm iC_{m,n}) = \exp(\operatorname{Tr}\ln(\mathbb{I}_{m,n} \pm iC_{m,n})) \approx \exp(\pm i\operatorname{Tr} C_{m,n}), \tag{42.44}$$

we have the transformed integration measure given by

$$D\psi'_L D\bar{\psi}'_L = \prod_m da'_m \prod_m d\bar{b}'_m \tag{42.45}$$

$$= d\mu \exp\left\{ i \int_x \alpha(x) \sum_{\lambda_m \geq 0} [\phi_m^L(x)^\dagger \phi_m^L(x) - \phi_m^R(x)^\dagger \phi_m^R(x)] \right\}$$

$$= d\mu \exp\left\{ -i \int_x \alpha(x) \sum_{\text{all } \lambda_m} \phi_m(x)^\dagger \gamma_5 \phi_m(x) \right\}$$

$$\equiv d\mu \exp\left\{ -i \int_x \alpha(x) A(x) \right\}.$$

The coefficient of $\alpha(x)$ must still vanish. As a result, we now have

$$\frac{\delta S}{\delta \alpha(x)} = \partial_\mu J^\mu - A(x) \equiv 0. \tag{42.46}$$

There is an anomaly which we must evaluate. The definition for $A(X)$ is an ill-defined integral which we must regulate and, asymptotically at large x, change the basis vectors to plane waves.

$$A(x) \equiv \lim_{M \to \infty} \left(\sum_m \phi_m(x)^\dagger \gamma_5 e^{(\lambda_m/M)^2} \phi_m(x) \right)$$

$$= \lim_{M \to \infty} \left(\sum_m \phi_m^\dagger(x) \gamma_5 e^{(i\slashed{D}/M)^2} \phi_m(x) \right)$$

$$= \lim_{M \to \infty} \langle x | \mathrm{Tr}[\gamma^5 e^{(i\slashed{D}/M)^2}] | x \rangle.$$

The trace is over Dirac and gauge indices.

We use the identity

$$(i\slashed{D})^2 = iD_\mu \gamma^\mu i D_\nu \gamma^\nu \tag{42.47}$$

$$= (iD_\mu)(iD_\nu) \left(\frac{\{\gamma^\mu, \gamma^\nu\}}{2} + \frac{[\gamma^\mu, \gamma^\nu]}{2} \right)$$

$$= -D^2 - \frac{1}{2} D_\mu D_\nu [\gamma^\mu, \gamma^\nu]$$

$$= -D^2 - \frac{1}{2} [D_\mu, D_\nu] \gamma^\mu \gamma^\nu$$

$$= -D^2 - \frac{ig}{2} \tilde{F}_{\mu\nu} \gamma^\mu \gamma^\nu.$$

Then we have

$$A(x) = \lim_{M \to \infty} \langle x | \mathrm{Tr}[\gamma^5 e^{(-D^2 - \frac{ig}{2} \tilde{F}_{\mu\nu} \gamma^\mu \gamma^\nu)/M^2}] | x \rangle$$

$$= \lim_{M \to \infty} \mathrm{Tr} \left[\gamma^5 \frac{1}{2!} \left(\frac{ig}{2} \tilde{F}_{\mu\nu} \gamma^\mu \gamma^\nu)/M^2 \right)^2 \right] \langle x | e^{-\partial^2/M^2} | x \rangle,$$

where the first factor is the leading term with non-zero trace. In the second factor, in the limit $M \to \infty$, we can neglect the gauge contribution at large k^2.

The second factor is given by

$$\langle x | e^{-\partial^2/M^2} | x \rangle = \lim_{x \to y} \int \frac{d^4 k}{(2\pi)^4} e^{-ik \cdot (x-y)} e^{k^2/M^2} \tag{42.48}$$

$$= i \int \frac{d^4 k_E}{(2\pi)^4} e^{-k_E^2/M^2} = \frac{iM^4}{16\pi^4} (2\pi^2) \int_0^\infty u^3 du \, e^{-u^2}$$

$$= \frac{iM^4}{16\pi^2}.$$

Note, we have continued the momentum integral to Euclidean space with $k_0 \to ik_4$, $k^2 \to -k_E^2$ and $\int d^4 k \to i \int d^4 k_E$. Also, in the last integral we have defined $k_E \equiv Mu$ and $(2\pi^2)$ is a result of the angular integral in four dimensions.

The first factor becomes

$$-\frac{g^2}{8M^4} \mathrm{Tr} \left(\tilde{F}_{\mu\nu} \tilde{F}_{\lambda\rho} \right) \mathrm{Tr} \left[\gamma^5 \gamma^\mu \gamma^\nu \gamma^\lambda \gamma^\rho \right] = -\frac{g^2}{8M^4} \mathrm{Tr} \left(\tilde{F}_{\mu\nu} \tilde{F}_{\lambda\rho} \right) 4i\epsilon^{\mu\nu\lambda\rho} \tag{42.49}$$

$$= -i\frac{g^2}{2M^4} \mathrm{Tr} \left(\tilde{F}_{\mu\nu} \tilde{F}_{\lambda\rho} \right) \epsilon^{\mu\nu\lambda\rho}$$

$$= -i\frac{g^2}{M^4} \mathrm{Tr} \left({}^*\tilde{F}_{\mu\nu} \tilde{F}^{\mu\nu} \right)$$

where

$$*\tilde{F}^{\mu\nu} \equiv \frac{1}{2}\epsilon^{\mu\nu\lambda\rho}\tilde{F}_{\lambda\rho}. \tag{42.50}$$

Finally, we have

$$A(x) = \frac{g^2}{16\pi^2}\mathrm{Tr}\left(*\tilde{F}_{\mu\nu}\tilde{F}^{\mu\nu}(x)\right). \tag{42.51}$$

Thus

$$\partial_\mu J^\mu = \frac{g^2}{16\pi^2}\mathrm{Tr}\left(*\tilde{F}_{\mu\nu}\tilde{F}^{\mu\nu}(x)\right). \tag{42.52}$$

IF we now consider a general chiral transformation of the form

$$\psi'_L(x) = e^{i\alpha(x)\gamma_5 T^A}\psi_L(x) = e^{-i\alpha(x)T^A}\psi_L(x) \tag{42.53}$$

with T^A the generator of the gauge transformation, then the covariant derivative of the current

$$J^A_\mu(x) = -\bar{\psi}_L(x)\gamma_\mu T^A\psi_L(x) \tag{42.54}$$

obtains an anomaly given by the anomaly factor

$$A^A(x) = \sum_n \phi^\dagger_n(x)\gamma_5 T^A\phi_n(x) = \frac{g^2}{16\pi^2}\mathrm{Tr}\left(T^A \, *\tilde{F}_{\mu\nu}\tilde{F}^{\mu\nu}(x)\right). \tag{42.55}$$

Using the identity

$$d^{ABC} \equiv \mathrm{Tr}(T^A\{T^B, T^C\}) \tag{42.56}$$

we can also write this in the form

$$A^A(x) = \frac{g^2}{32\pi^2}d^{ABC} \, *F_{\mu\nu\,B}\,F^{\mu\nu}_C. \tag{42.57}$$

42.2 Gauge Anomalies in the Standard Model

In order for the Standard Model to be renormalizable, one needs that all gauge currents are conserved, i.e. there are no anomalies (Gross and Jackiw, 1972). Now let's check the anomalies of the SM gauge currents. In Table 33.3 we list all the fermions in the Standard Model. We shall consider the possible anomalies in turn.

- $SU(3)^3$: In this case we have quarks transforming under $T^A = \frac{\lambda^A}{2}$ and anti-quarks transforming under $\bar{T}^A = \frac{-\lambda^{A\,T}}{2}$. Thus, we have

$$6 \times \left[\mathrm{Tr}\left(\frac{\lambda^A}{2}\left\{\frac{\lambda^B}{2}, \frac{\lambda^C}{2}\right\}\right) + \mathrm{Tr}\left(\frac{-\lambda^{A\,T}}{2}\left\{\frac{-\lambda^{B\,T}}{2}, \frac{-\lambda^{C\,T}}{2}\right\}\right)\right] \equiv 0. \tag{42.58}$$

- $SU(2)^3$: In this case we have quark and lepton doublets. But $SU(2)$ is a real group with generators $T_a = \frac{\tau_a}{2}$ and $\bar{T}_a = -\frac{\tau_a^T}{2}$ related by a similarity transformation, $\bar{T}_a = C T_a C^{-1}$. As a consequence, $d^{abc} \equiv 0$ for all doublets.

- $U(1)_Y^3$: In this case we need just take the sum of all the charges. We have for one family of quarks and leptons

$$6\left(\frac{1}{27}\right) - 3\left(\frac{64}{27}\right) + 3\left(\frac{8}{27}\right) + 2(-1) + 8 \equiv 0 \qquad (42.59)$$

for q, $\bar{u}$, $\bar{d}$, ℓ, $\bar{e}$, respectively.

- $SU(3)^2 U(1)_Y$: We just need to consider quarks. The anomaly factor is proportional $\mathrm{Tr} T_A T_B Y$. We have for one family,

$$\frac{1}{2}\delta_{AB}\left(2\left(\frac{1}{3}\right) + \left(\frac{2}{3}\right) + \left(\frac{-4}{3}\right)\right) \equiv 0. \qquad (42.60)$$

- $SU(2)^2 U(1)_Y$: We just need to consider the quark and lepton doublets. The anomaly factor is proportional $\mathrm{Tr} T_a T_b Y$. We have for one family,

$$\frac{1}{2}\delta_{ab}\left(3\left(\frac{1}{3}\right) + (-1)\right) \equiv 0. \qquad (42.61)$$

Thus every gauge current in the Standard Model is anomaly free!

What about currents associated with global symmetries, such as baryon and lepton number? Since only left-handed quarks and leptons are in doublets, we have the $SU(2)^2$ contribution given by

$$\partial^\mu J_\mu^B = 3\left(\frac{1}{3}\right)\frac{g^2}{16\pi^2}\mathrm{Tr}\left(^*\tilde{W}_{\mu\nu}\tilde{W}^{\mu\nu}(x)\right), \qquad (42.62)$$

$$\partial^\mu J_\mu^L = \frac{g^2}{16\pi^2}\mathrm{Tr}\left(^*\tilde{W}_{\mu\nu}\tilde{W}^{\mu\nu}(x)\right).$$

Also, the $U(1)_Y^2$ contribution is given by

$$\partial^\mu J_\mu^B = 3\left(\frac{1}{3}\right)\frac{1}{4}\left[2\left(\frac{1}{3}\right)^2 - \left(\frac{2}{3}\right)^2 - \left(-\frac{4}{3}\right)^2\right]\frac{g'^2}{16\pi^2}\left(^*B_{\mu\nu}B^{\mu\nu}(x)\right) \quad (42.63)$$

$$\partial^\mu J_\mu^L = \frac{1}{4}[2(-1)^2 - (+2)^2]\frac{g'^2}{16\pi^2}\left(^*B_{\mu\nu}B^{\mu\nu}(x)\right)$$

Note,

$$\partial^\mu(J_\mu^B - J_\mu^L) = 0 \qquad (42.64)$$

in both cases. There is no $SU(3)^2$ contribution to $\partial^\mu J_\mu^B$ because there are an equal number of baryons and anti-baryons with opposite baryon number.

PART X

NEUTRINO OSCILLATIONS

43

Neutrino Oscillations: Atmospheric

Neutrino oscillations have been observed, which, we shall see, means that neutrinos have mass. The upper bound on the sum of their masses is constrained by cosmology to be less than about 1 eV. We don't know yet if neutrino masses are Majorana or Dirac. In the first case, the lepton number is violated and the process of neutrinoless double-beta decay might be observable. In this case a nucleus decay of the type $(A, Z) \to (A, Z+2) + 2e^-$ would be possible. However, in the latter case, the lepton number is conserved and this process is forbidden. There are many experiments in progress or under construction to search for this decay, with no observation to date.

In the next three chapters we shall focus on neutrino oscillations. We shall consider, in turn,

(1) atmospheric ν oscillations,
(2) solar ν oscillations,
(3) LSND/miniBoone data,

with brief discussions of other corroborating experiments.

43.1 The Missing Muon Neutrinos

Super-Kamiokande (Super-K) is a 50,000-ton water Cherenkov detector, using 13,000 photomultiplier tubes (PMTs) to detect light, see Fig. 43.1. It is located 1 km underground in Kamioka, Japan. Neutrinos interact in the water, scattering off nuclei or electrons, producing either electrons or muons (or even taus), which then produce Cherenkov light. Muons and electrons can be distinguished by their signals in the PMTs (Fig. 43.2). For example, a muon signal has sharp edges, as seen in Fig. 43.3, while an electron signal has fuzzier edges, due to scattering in the water, as shown in Fig. 43.4. Theoretically, atmospheric neutrinos are produced by cosmic rays hitting nuclei in the upper atmosphere (Gaisser and Honda, 2002). This produces pions and kaons, which decay, for example, via $\pi^+ \to \mu^+ + \nu_\mu$ followed by the decay $\mu^+ \to e^+ + \nu_e + \bar{\nu}_\mu$ and similarly for π^-, $K^\pm$. Thus, one expects two muon-type neutrinos for every electron-type neutrino. The flux of atmospheric neutrinos is expected to be approximately isotropic about the upper atmosphere,

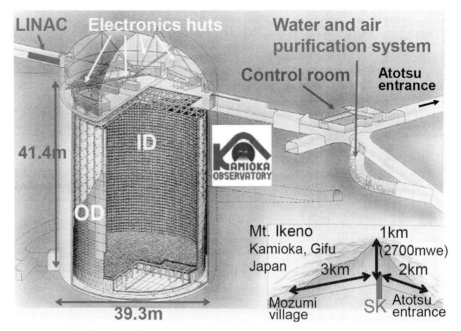

Fig. 43.1 An artist's rendition of the Super-K detector (Wilkes, 2011).

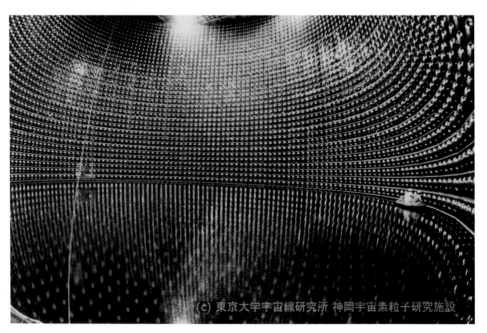

Fig. 43.2 Filling the Super-K detector with purified water when half filled, with a view of the PMTs and workers in a dinghy (Kamioka Observatory, ICRR (Institute for Cosmic Ray Research), The University of Tokyo, n.d.).

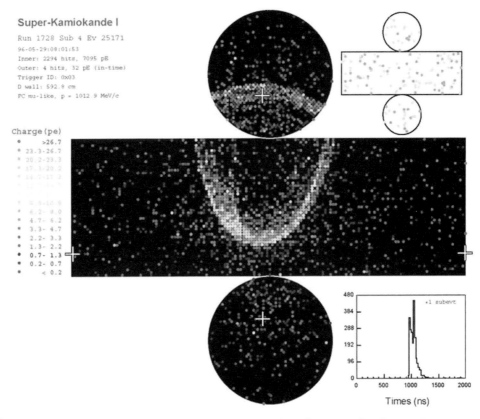

Fig. 43.3 Muon signal at Super-K (Kamioka Observatory, ICRR (Institute for Cosmic Ray Research), The University of Tokyo, n.d.).

i.e. equal in all directions. Super-K measures these neutrinos as a function of the zenith angle, θ, with neutrinos approaching from above, with $\cos\theta = 1$, or below, with $\cos\theta = -1$. Neutrinos coming from below travel about 10,000 km, while neutrinos coming from above travel only about 10 km (see Fig. 43.5). Hence, a measurement as a function of zenith angle is equivalent to a measurement as a function of the distance traveled. Note, the electron or muon, produced by the collision of a neutrino in water, essentially goes in the same direction as the neutrino was going. We give some results of Super-K (Fukuda et al., 1998), Fig. 43.6 (see also Fukuda et al., 1998, 2000; Ashie et al., 2005). There is a discrepancy of muon neutrinos coming from below, while there is no such discrepancy for electron neutrinos. With only three types of neutrinos, they conclude that the muon neutrinos oscillate into tau neutrinos. Moreover, the absence of tau neutrinos coming from below is excluded at 2.4σ (Abe et al., 2006). Finally there is no evidence for sterile neutrinos (Fukuda et al., 2000; Abe et al., 2006), which would be evident due to matter effects in the Earth, an effect we shall consider in the next chapter.

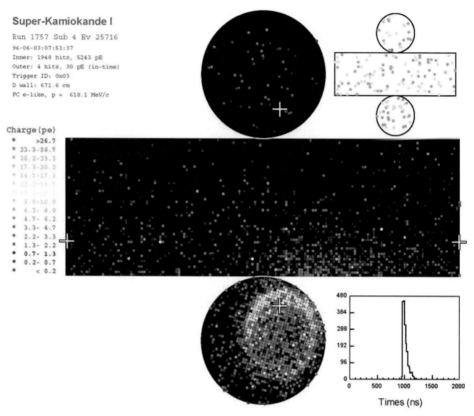

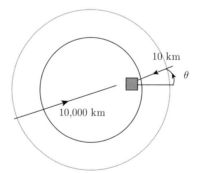

Fig. 43.5 Atmospheric neutrinos hit the upper atmosphere and produce neutrinos. The flux is measured as a function of the zenith angle, θ.

43.2 Neutrino Oscillations in Vacuum

Let's now consider the theory of neutrino oscillations (in vacuum). We define the neutrino flavor eigenstates, $|\nu_\alpha\rangle$, with $\alpha = \{e,\ \mu,\ \tau\}$ and mass eigenstates, $|\nu_i\rangle$,

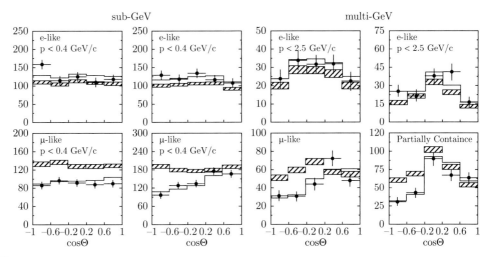

Fig. 43.6 The SuperKamiokande atmospheric neutrino results showing excellent agreement between the predicted and observed electron-like events, but a sharp depletion in the muon-like events for neutrinos coming from below, through the Earth. The results are fit very well by the assumption of $\nu_\mu \to \nu_\tau$ oscillations with maximal mixing. Reprinted with permission from Fukuda et al. (1998). Copyright (1998) by the American Physical Society.

with $i = 1, 2, 3$. Both form a complete set and they are related by a unitary transformation

$$|\nu_\alpha\rangle = U^*_{\alpha i} |\nu_i\rangle. \tag{43.1}$$

U is known as the PMNS Pontecorvo–Maki–Nakagawa–Sakata matrix (Maki et al., 1962; Pontecorvo, 1968). The weak-interaction Hamiltonian is given by

$$\mathcal{H}_W = \frac{g}{\sqrt{2}} W^-_\mu \sum_{\alpha=e,\mu,\tau} \bar{\ell}_\alpha \gamma^\mu \nu_{\alpha L} + h.c. \tag{43.2}$$

with

$$\nu_{\alpha L} = U_{\alpha i} \nu_{iL}. \tag{43.3}$$

We define the neutrino plane wave at position x and time t, given by

$$|\nu_\alpha, x\rangle = \sum_{j=1}^{3} U^*_{\alpha j} e^{-i(Et - p_j x)} |\nu_j\rangle, \tag{43.4}$$

where $p_j = \sqrt{E^2 - m_j^2} \simeq E - \frac{m_j^2}{2E}$ for $E \gg m_j$. Then we can rewrite this expression as

$$|\nu_\alpha, x\rangle = \sum_{j=1}^{3} U^*_{\alpha j} e^{-i(E(t - \frac{x}{c}) + \frac{m_j^2}{2E} x)} |\nu_j\rangle. \tag{43.5}$$

At $x = t = 0$, the atmospheric neutrinos are produced in the upper atmosphere by the cosmic protons hitting nuclei. We can then ask the question, what is the probability that a ν_α oscillates into a ν_β at a distance $L \cong ct$ from where it was produced? The probability is given by

$$P(\nu_\alpha \to \nu_\beta) = |\langle \nu_\beta, 0 | \nu_\alpha, L \rangle|^2 \tag{43.6}$$

$$= |\sum_{i=1}^{3} \langle \nu_i | U_{\beta i} \sum_{j=1}^{3} U_{\alpha j}^* e^{-i\frac{m_j^2}{2E}\frac{Lc^3}{\hbar}} |\nu_j \rangle|^2.$$

Using the orthogonality relation,

$$\langle \nu_i | \nu_j \rangle \equiv \delta_{ij}, \tag{43.7}$$

we have

$$P(\nu_\alpha \to \nu_\beta) = |\sum_{j=1}^{3} U_{\beta j} U_{\alpha j}^* e^{-i\frac{m_j^2 L}{2E}}|^2, \tag{43.8}$$

where we have again set $\hbar = c = 1$.

Let's enumerate some possibilities.

1. If all m_j^2 are equal, then $P(\nu_\alpha \to \nu_\beta) = \delta_{\alpha\beta}$, i.e. no oscillation.
2. If we want to study anti-neutrino oscillations, then we replace $U_{\alpha i}$ by $U_{\alpha i}^*$.
3. **CP$\mathcal{T}$** invariance requires $P(\nu_\alpha \to \nu_\beta) = P(\bar{\nu}_\beta \to \bar{\nu}_\alpha)$.
4. Oscillations occur as a function of L/E.
5. $P(\nu_\alpha \to \nu_\beta)$ is invariant under phase redefinitions of the type $U_{\alpha j} \to e^{i\phi_\alpha} U_{\alpha j} e^{i\phi_j}$.

Consider now, for simplicity, just two flavors of neutrinos. Given

$$U = \begin{pmatrix} \cos\theta & \sin\theta \\ -\sin\theta & \cos\theta \end{pmatrix}, \tag{43.9}$$

we have

$$P(\nu_{\alpha=1} \to \nu_{\beta=2}) = \left| -\sin\theta\cos\theta\, e^{-im_1^2 L/2E} + \cos\theta\sin\theta\, e^{-im_2^2 L/2E} \right|^2$$

$$= \cos^2\theta\sin^2\theta(2 - e^{-i(m_2^2 - m_1^2)L/2E} - h.c.) \tag{43.10}$$

$$= \sin^2 2\theta \frac{1}{2}\left(1 - \cos\frac{\Delta m^2 L}{2E}\right),$$

with $\Delta m^2 \equiv m_2^2 - m_1^2$. In addition, putting into the equation some relevant units, we define

$$\Delta \equiv \frac{\Delta m^2 L}{2E} = 2.54\,\Delta m^2 (\text{eV}^2)\,L(\text{km})/\,E(\text{GeV}) \tag{43.11}$$

and

$$\frac{1}{2}(1 - \cos\Delta) \equiv \sin^2\frac{\Delta}{2}. \tag{43.12}$$

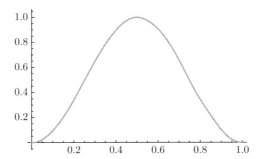

Fig. 43.7 $P(\nu_1 \to \nu_2)/\sin^2(2\theta)$ plotted as a function of L/L_{osc} (using Mathematica).

So finally, we have

$$P(\nu_1 \to \nu_2) = \sin^2 2\theta \, \sin^2\left(1.27\frac{\Delta m^2(\mathrm{eV}^2)\ L(\mathrm{km})}{E(\mathrm{GeV})}\right). \tag{43.13}$$

We can also define an oscillation length by the formula

$$\frac{\Delta m^2 L}{2E} \equiv 2\pi \, \frac{L}{L_{osc}}, \tag{43.14}$$

where

$$L_{osc} \equiv 2.48 \ \mathrm{km} \left(\frac{E}{\mathrm{GeV}}\right)\left(\frac{1 \ \mathrm{eV}^2}{\Delta m^2}\right). \tag{43.15}$$

So equivalently we have

$$P(\nu_1 \to \nu_2) = \frac{1}{2}\sin^2 2\theta \left(1 - \cos 2\pi L/L_{osc}\right). \tag{43.16}$$

In Fig. 43.7 we plot $P(\nu_1 \to \nu_2)$ as a function of L/L_{osc}, while in Fig. 43.8 it is plotted as a function of $\log(L/L_{osc})$. Note that at distances much larger than the oscillation length the probability varies rapidly between zero and one. In Fig. 43.9 we plot data points as a function of $\log(L/E)$, assuming no oscillation, taken from Super-K results (Kajita et al., 2016), with an overlay of the theory for the probability $P(\nu_\mu \to \nu_\mu)$ including oscillation parameters $\Delta m^2 = 0.0032 \ \mathrm{eV}^2$, $\sin 2\theta = 1$.

The K2K experiment used a beam of muon neutrinos with mean energy 1.3 GeV, produced with the proton synchotron at KEK, Japan, and detected at Super-K (the far detector), 250 km away (Ahn et al., 2003). There is also a near detector, only 300 m away. The experiment was designed to look for ν_μ disappearance. The signal can be distinguished from background by timing, since it had to come within a 1.5 μs window. The expected number of muon neutrino events, of the type $\nu_\mu + n \to \mu + p$, at Super-K was $80.1^{+6.2}_{-5.5}$, and only 56 events were observed. The results are consistent with that of Super-K atmospheric neutrino oscillations. The combined fit to both atmospheric neutrino results at Super-K and K2K determined

$$\Delta m^2 \sim 2.7 \times 10^{-3} \ \mathrm{eV}^2, \quad \sin^2 2\theta \sim 1. \tag{43.17}$$

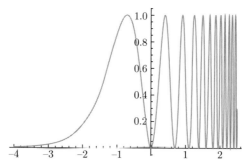

Fig. 43.8 $P(\nu_1 \to \nu_2)/\sin^2(2\theta)$ plotted as a function of $\log(L/L_{osc})$ (using Mathematica).

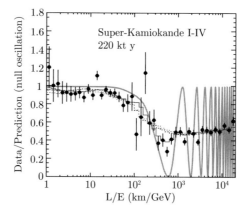

Fig. 43.9 $P(\nu_\mu \to \nu_\mu)$ plotted as a function of $\log(L/E_\nu)$ (using Mathematica). The data points are Super-K results as of 2016 (Kajita et al., 2016).

Neutrino Oscillations: Solar

44.1 The Solar-Neutrino Problem

Electron neutrinos are emitted by the Sun as a result of fusion reactions. The basic pp chain is described in Table 44.1. The spectrum of solar neutrinos is determined by a detailed model of the Sun. For example, the 2000 Bahcall–Pinsonneault model, Fig. 44.1, gives the solar-neutrino flux with respect to the processes in the pp chain. Note, the highest-energy neutrinos are those coming from ^{8}B. Their flux is extremely sensitive to the temperature at the solar core. However, the dominant neutrino flux is due to the low-energy pp neutrinos. Several experiments were performed to detect the solar neutrinos. Some of these are known as the chlorine, Kamiokande/Super-K, gallium, and Sudbury Neutrino Observatory (SNO) experiments. Note, the threshold for the gallium, chlorine and water Cherenkov (Super-K and SNO) experiments are also presented.

The first of these was the chlorine experiment of Ray Davis (Cleveland et al., 1998). This experiment was located in the Homestake Mine, South Dakota, at a depth of 4850 feet. The rock overburden is necessary to block out the cosmic-ray background. This experiment was the sole experiment looking for solar neutrinos during the period from 1968 to 1988. They used a vat of ^{37}Cl as the detector. When an electron neutrino scatters in the detector, we have the process, $\nu_e + {}^{37}\text{Cl} \to {}^{37}\text{Ar} + e^-$ with a reaction threshold of 0.814 MeV. Davis' results were consistently about one-third of that predicted by the solar model of Bahcall et al. (Bahcall, 2004, 2005). This is called the "solar neutrino problem."

The chlorine experiment was followed by several other experiments, for example, multi-purpose detectors such as Kamiokande and Super-K, using water Cherenkov detectors originally designed to look for nucleon decay, as well as astronomical neutrinos, such as supernova neutrinos, atmospheric neutrinos and solar neutrinos.[1] With a threshold of 9.3 MeV, they observed only about half of the expected number of solar neutrinos (Hirata et al., 1989). Later they reduced the threshold to 7.5 MeV, with similar results. They also conclusively determined that neutrinos were coming from the Sun (Hirata et al., 1991) (see Figs. 44.2 and 44.3). Super-K began operation in 1996, with a threshold of 6.5 MeV, and carried on the search for nucleon decay, supernovas, atmospheric and solar neutrinos

---

[1] In fact, Kamiokande II, along with the IMB experiment, observed neutrinos from supernova 1987A.

Table 44.1 The *pp* chain in the Sun. The average number of pp neutrinos produced per termination in the Sun is 1.85. For all other neutrino sources, the average number of neutrinos produced per termination is equal to the termination percentage/100 (Bahcall, 2004).

Reaction	Termination[a] Number	(%)	ν energy (MeV)
$p + p \rightarrow {}^2H + e^+ + \nu_e$	1a	100	≤ 0.42
or			
$p + e^- + p \rightarrow {}^2H + \nu_e$	1b *(pep)*	0.4	1.44
${}^2H + p \rightarrow {}^3He + \gamma$	2	100	
${}^3He + {}^4He \rightarrow \alpha + 2p$	3	85	
or			
${}^3He + {}^4He \rightarrow {}^7Be + \gamma$	4	15	
${}^7Be + e^- \rightarrow {}^7Li + \nu_e$	5	15	(90%) 0.86
${}^7Li + p \rightarrow 2\alpha$	6	15	(10%) 0.38
or			
${}^7Be + p \rightarrow {}^8B + \gamma$	7	0.02	
${}^8B \rightarrow {}^8Be^* + e^+ + \nu_e$	8	0.02	< 15
${}^8Be^* \rightarrow 2\alpha$	9	0.02	
or			
${}^3He + p \rightarrow {}^4He + e^+ + \nu_e$	10	0.00002	≤ 18.77

[a] The termination percentage is the fraction of terminations of the *pp* chain, $4p \rightarrow \alpha + 2e^+ 2\nu_e$, in which each reaction occurs. The results are averaged over the model of the current Sun. Since in essentially all terminations at least one *pp* neutrino is produced and in a few terminations one *pp* and one *pep* neutrino are created, the total of *pp* and *pep* terminations exceeds 100%.

(Hosaka et al., 2006; Cravens et al., 2008; Abe et al., 2011). Their results for solar neutrinos were consistent with those of Kamiokande. The chlorine, Kamiokande and Super-K experiments were only sensitive to the highest-energy solar neutrinos, the ${}^8B$ neutrinos.

The gallium experiments, GALLEX, SAGE and GNO, were designed to see the *pp* neutrinos (Abdurashitov et al., 2009; Altmann, 1998; Altmann et al., 2005). Like Super-K and Kamiokande earlier, they also measured approximately half of the expected number of neutrinos.

Then there were the SNO experiments, which were designed to clarify what was going on with the solar neutrino problem. In Fig. 44.4, the experimental measurement of solar neutrinos in the five different types of experiments are compared with the predicted flux using the 2004 Bahcall–Pinsonneault solar model (Bahcall, 2005).

Finally, there was the Borexino experiment, designed to see the ${}^7Be$ neutrinos (Bellini et al., 2010, 2011).

What were some of the proposed solutions to the solar neutrino problem?

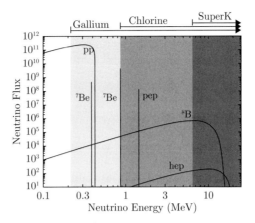

Fig. 44.1 The energy spectrum of neutrinos from the pp chain of interactions in the Sun, as predicted by the standard solar model. Neutrino fluxes from continuum sources (such as pp and ^{8}B) are given in the units of counts per cm^2 per second. The pp chain is responsible for more than 98% of the energy generation in the standard solar model. Neutrinos produced in the carbon-nitrogen-oxygen (CNO) chain are not important energetically and are difficult to detect experimentally. The arrows at the top of the figure indicate the energy thresholds for the relevant neutrino experiments. (Bahcall, 2000, with data from Bahcall et al., 1998).

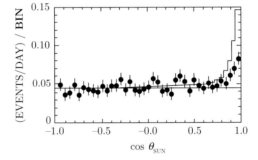

Fig. 44.2 Neutrinos observed in the Kamiokande II detector as a function of $\cos\theta_{Sun}$, where $\theta_{Sun} = 0$ is in the direction of the Sun. Reprinted with permission from Hirata et al. (1991). Copyright (1991) by the American Physical Society.

- There was the possibility that neutrino decay might solve the solar neutrino problem (Acker and Pakvasa, 1994).

- There was also the possibility of a combined solution to the dark-matter and solar-neutrino problem, known as the cosmion solution (Press and Spergel, 1985). The latter solution described a weakly interacting neutral particle with a mass in the range $5-60$ GeV and a cross-section on protons of 4×10^{-36} cm^2 (a WIMP), which would collect in the Sun and iso-thermalize the core. One only needed to reduce the core temperature, T_c, by a small amount to solve the ^{8}B neutrino

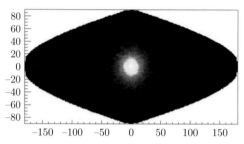

Fig. 44.3 More recent view of the Sun as seen via neutrinos in Super-K after 4500 days (Kamioka Observatory, ICRR (Institute for Cosmic Ray Research), The University of Tokyo, n.d.).

Total Rates: Standard Model vs. Experiment
Bahcall–Pinsonneault 2004

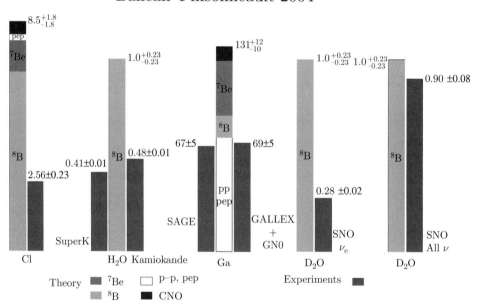

Fig. 44.4 The experimental measurements of solar neutrinos in the five different types of experiments (Bahcall, 2004). These are compared with the predicted flux using the 2004 Bahcall–Pinsonneault solar model.

problem, since the ^{8}B flux varies as T_c^{22}. In fact, the mass and cross-section of the cosmion was fixed by the requirements that the core temperature of the Sun is lowered by about 12% without affecting the temperature beyond $M \sim 0.05 M_{Sun}$, where most of the solar luminosity originates.

- There might be "just-so" vacuum oscillations (Glashow and Krauss, 1987) of electron neutrinos into some other flavor of neutrino, i.e. such that $\Delta m^2 = 5 -$

13×10^{-11} eV^2 and, on average, for the gallium experiments, 50% of the electron neutrinos oscillated into some other flavor.

- Finally, there was the suggestion of enhanced matter oscillations in the Sun (Rosen and Gelb, 1986; Langacker et al., 1987; Bahcall et al., 1987) using the so-called MSW effect (Mikheyev and Smirnov, 1985, 1986; Wolfenstein, 1978).

44.2 The SNO Experiment

It is now clear that the MSW effect plays the major role in understanding the solar-neutrino problem. We shall see that the different experiments were sensitive to different energy regimes and they observed different amounts of the solar-neutrino flux. This was an expected consequence of the MSW effect. However, it was the SNO experiment which in the end clinched the deal in favor of the MSW explanation (Ahmad et al., 2002). The SNO experiment was a water Cherenkov experiment like Super-K, with one major difference (Fig. 44.5). Instead of H_2O, they used heavy water, D_2O.

They measured three types of neutrino interactions, which were (1) the charged-current process, (CC) $\nu_e + D \rightarrow p + p + e^-$, and (2) the elastic-scattering process (ES), $\nu_x + e^- \rightarrow \nu_x + e^-$, which can proceed via neutral and charged currents for

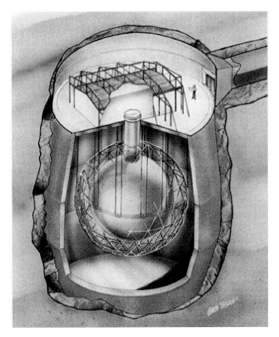

Fig. 44.5   The SNO detector. Reprinted with permission from McDonald (2016). Copyright (2016) by the American Physical Society.

Table 44.2 Two-neutrino flavor fit to SNO, all solar data and, lastly, including KamLAND. Reprinted with permission from Aharmim et al. (2013). Copyright (2013) by the American Physical Society.

Oscillation analysis	$\tan^2 \theta_{12}$	$\Delta m_{21}^2 (\mathrm{eV}^2)$	χ^2/NDF
SNO only (LMA)	$0.427^{+0.033}_{-0.029}$	$5.62^{+1.92}_{-1.36} \times 10^{-5}$	$1.39/3$
SNO only (LOW)	$0.427^{+0.043}_{-0.035}$	$1.35^{+0.35}_{-0.14} \times 10^{-7}$	$1.41/3$
Solar	$0.427^{+0.028}_{-0.028}$	$5.13^{+1.29}_{-0.96} \times 10^{-5}$	$108.07/129$
Solar + KamLAND	$0.427^{+0.027}_{-0.024}$	$7.46^{+0.20}_{-0.19} \times 10^{-5}$	

Fig. 44.6 The results of the SNO experiment. Reprinted with permission from Ahmad et al. (2002). Copyright (2002) by the American Physical Society.

$\nu_x = \nu_e$, or only via neutral currents for $x \neq e$. In the latter process, it was known that, theoretically, 15% of the results were due to $\nu_x = \nu_\mu$, ν_τ. Finally, (3) they measured neutral-current processes (NC), $\nu_x + D \rightarrow p + n + \nu_x$. They made use of the heavy water in three separate phases, which differed by the method used for the NC extraction. In the first phase, they detected the Cherenkov light from the 6.25 MeV gamma ray emitted when the neutron is captured on deuterium. In the second phase, about 2.5 tons of NaCl was added to the heavy water and the 8.6 MeV gamma ray, which was emitted when the neutron was captured on Cl, was detected. Finally, in phase three, the salt was removed and an array of ^{3}He proportional counters were installed to measure the neutrons directly.

The results of the SNO experiment after phase I is given in Fig. 44.6. One sees the bands associated with the CC, ES and NC measurements. In addition, one sees the expectation of the Standard Solar Model (SSM). It is apparent from the NC measurement that the total neutrino flux is consistent with that expected from the SSM.[2]

The two-flavor solution to the final SNO data (Aharmim et al., 2013) is given in Fig. 44.7. Note, there are two possible solutions, known as the LMA-MSW (large

---

[2] This is in fact a verification of the Standard Solar Model.

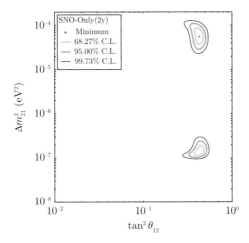

More results of the SNO experiment. Reprinted with permission from Aharmim et al. (2013). Copyright (2013) by the American Physical Society.

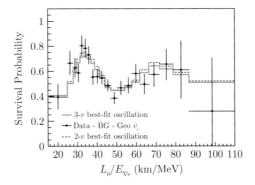

Ratio of the observed $\bar{\nu}_e$ spectrum to the expectation for nooscillation vs. L_0/E for the KamLAND data. $L_0 \sim 180$ km is the flux-weighted average reactor baseline. Reprinted with permission from (Gando et al., 2011). Copyright (2011) by the American Physical Society.

mixing angle) and LOW-MSW (low mixing angle) solutions, with a high and low value of Δm_{12}^2. A global fit combining all the solar neutrino data, chlorine, Super-K I,II,III, GALLEX, SAGE, GNO, and Borexino removes the LOW-MSW solution (Aharmim et al., 2013). Finally, also including KamLAND data (Gando et al., 2011), gives the last entry in Table 44.2.

The KamLAND detector contains 1 kton of liquid scintillator located in the Kamioka mine. It measures the $\bar{\nu}_e$ flux coming predominantly from 56 Japanese nuclear reactors. The flux-weighted average baseline to these reactors is ~ 180 km. They are looking for $\bar{\nu}_e$ disappearance, consistent with solar-neutrino oscillations. Their result is presented in Fig. 44.8.

44.3 The MSW Effect

Let's define the neutrino mass matrix, $m_{\alpha\beta}$, in the flavor basis. The term in the Lagrangian is given by

$$\frac{1}{2}\nu_\alpha m_{\alpha\beta}\nu_\beta \equiv \frac{1}{2}\nu^T m\nu \tag{44.1}$$

and

$$\nu_\alpha \equiv \sum_j U_{\alpha j}\hat{\nu}_j \equiv U\ddot{\nu}, \tag{44.2}$$

where by definition $\hat{\nu}$ are mass eigenstates, i.e.

$$\frac{1}{2}\nu^T m\nu = \frac{1}{2}\hat{\nu}^T U^T mU\hat{\nu} \equiv \frac{1}{2}\hat{\nu}^T \hat{m}\hat{\nu} \tag{44.3}$$

and

$$\hat{m} \equiv U^T mU = \begin{pmatrix} m_1 & & \\ & m_2 & \\ & & m_3 \end{pmatrix}. \tag{44.4}$$

Neutrino evolution in vacuum is described by the effective Hamiltonian in the mass eigenstate basis,

$$H_\nu^{vac} = \frac{1}{2E}\hat{\nu}^\dagger \hat{m}^2\hat{\nu}. \tag{44.5}$$

In the flavor basis we have

$$H_\nu^{vac} = \frac{1}{2E}\nu^\dagger m^\dagger m\nu \tag{44.6}$$

or

$$H_\nu^{vac} \equiv \frac{1}{2E}\nu^\dagger (U\hat{m}^2 U^\dagger)\nu. \tag{44.7}$$

Now consider matter effects.

$$H^{matter} = \frac{G_F}{\sqrt{2}}\bar{\nu}_\alpha\gamma_\mu(1-\gamma_5)\nu_\alpha \sum_f \bar{f}\gamma^\mu(g_V^f - \gamma_5 g_A^f)f \quad \text{neutral current}$$

$$+ \frac{G_F}{\sqrt{2}}\bar{e}_\alpha\gamma_\mu(1-\gamma_5)\nu_\alpha\, \bar{\nu}_\beta\gamma^\mu(1-\gamma_5)e_\beta \quad \text{charged current.} \tag{44.8}$$

In a thermal state with temperature T, we have the following thermal averages

$$\langle \bar{f}\gamma_\mu f\rangle_T = N_f\delta_{\mu 0}, \tag{44.9}$$

$$\langle \bar{f}\gamma_\mu\gamma_5 f\rangle_T \simeq 0 \tag{44.10}$$

where N_f is the number density of particles f with

$$N_e = N_p, \ N_n. \tag{44.11}$$

Then the thermal average of H^{matter} is given by

$$H^{matter} = \frac{2G_F}{\sqrt{2}} \nu_\alpha^\dagger P_L \nu_\alpha \sum_f N_f (\delta_{\alpha f} + g_V^f), \tag{44.12}$$

where the term with $\delta_{\alpha f}$ takes into account the Fierz transform of the charged-current term. Finally, we have

$$g_V^e = -\frac{1}{2} + 2\sin^2\theta_W, \tag{44.13}$$

$$g_V^p = 2g_V^u + g_V^d = \frac{1}{2} - 2\sin^2\theta_W,$$

$$g_V^n = g_V^u + 2g_V^d = -\frac{1}{2}.$$

Then we combine the vacuum and matter effective Hamiltonians to obtain

$$H_\nu^{matter} \equiv H_\nu^{vac} + H^{matter} \tag{44.14}$$

or

$$H_\nu^{matter} = \frac{1}{2E} U\hat{m}^2 U^\dagger$$
$$+ \frac{2G_F}{\sqrt{2}} \begin{pmatrix} N_e(1 + g_V^e) + N_p g_V^p + N_n g_V^n & 0 & 0 \\ 0 & N_e g_V^e + N_p g_V^p + N_n g_V^n & 0 \\ 0 & 0 & N_e g_V^e + N_p g_V^p + N_n g_V^n \end{pmatrix}$$
$$= \frac{1}{2E} U\hat{m}^2 U^\dagger + \frac{2G_F}{\sqrt{2}} \begin{pmatrix} N_e - \frac{1}{2}N_n & 0 & 0 \\ 0 & -\frac{1}{2}N_n & 0 \\ 0 & 0 & -\frac{1}{2}N_n \end{pmatrix}. \tag{44.15}$$

To simplify notation, consider two neutrino oscillations with the oscillation of electron neutrinos, $\nu_e \to \nu_X$, and the unitary matrix which diagonalizes this two-state mass matrix given by

$$U = \begin{pmatrix} \cos\theta & \sin\theta \\ -\sin\theta & \cos\theta \end{pmatrix}. \tag{44.16}$$

We then have

$$H_\nu^{matter} = \begin{pmatrix} \frac{m_1^2 \cos^2\theta + m_2^2 \sin^2\theta}{2E} + A_\alpha & \frac{\cos\theta\sin\theta\Delta m^2}{2E} \\ \frac{\cos\theta\sin\theta\Delta m^2}{2E} & \frac{m_1^2 \sin^2\theta + m_2^2 \cos^2\theta}{2E} + A_\beta \end{pmatrix}, \tag{44.17}$$

with

$$A_\alpha \equiv N_{\nu_\alpha} \sqrt{2} G_F, \tag{44.18}$$

$$N_{\nu_e} \equiv N_e - \frac{1}{2}N_n, \quad N_{\nu_\mu} = N_{\nu_\tau} \equiv -\frac{1}{2}N_n, \quad N_{\nu_s} \equiv 0 \tag{44.19}$$

and $\Delta m^2 = m_2^2 - m_1^2$. Subtracting a constant proportional to the identity matrix, we have

$$H_\nu^{matter} = \frac{1}{4E} \begin{pmatrix} A - \Delta m^2 \cos 2\theta & \Delta m^2 \sin 2\theta \\ \Delta m^2 \sin 2\theta & -A + \Delta m^2 \cos 2\theta \end{pmatrix} + c\mathbb{I}. \tag{44.20}$$

The term proportional to the identity is irrelevant when discussing oscillations. We have

$$A \equiv (A_\alpha - A_\beta)2E \equiv 2\sqrt{2}G_F E(N_{\nu_\alpha} - N_{\nu_\beta}). \tag{44.21}$$

The relevant Schrödinger equation for the problem is given by

$$i\frac{d}{dt}\nu_\alpha = (H_\nu^{matter})_{\alpha\beta}\nu_\beta. \tag{44.22}$$

There were detailed analyses of all possible solutions to this equation, depending on where neutrinos were produced in the Sun, on their energy, etc. The solution which actually fits the data is described by adiabatic evolution in the Sun. In the adiabatic approximation, the eigenvalues, of an effectively time-dependent Hamiltonian, change slowly with time. In this case, the potential seen by neutrinos changes with the density of electrons in the Sun. We have

$$H_\nu^{matter}(x)\psi_{mj}(x) = E_j(x)\psi_{mj}(x). \tag{44.23}$$

The eigenvalue equation is of the form

$$\begin{pmatrix} a & b \\ b & -a \end{pmatrix} \begin{pmatrix} \cos\theta_m \\ -\sin\theta_m \end{pmatrix} = E_- \begin{pmatrix} \cos\theta_m \\ -\sin\theta_m \end{pmatrix} \tag{44.24}$$

and

$$\begin{pmatrix} a & b \\ b & -a \end{pmatrix} \begin{pmatrix} \sin\theta_m \\ \cos\theta_m \end{pmatrix} = E_+ \begin{pmatrix} \sin\theta_m \\ \cos\theta_m \end{pmatrix}. \tag{44.25}$$

Also,

$$E_+ + E_- = 0 \quad \text{gives} \quad E_- = -E_+ \tag{44.26}$$

and

$$E_+E_- = -(a^2 + b^2), \tag{44.27}$$

thus,

$$E_+ = (a^2 + b^2)^{1/2}. \tag{44.28}$$

Given the two equations from Eqn. 44.24, we find

$$a\cos\theta_m\sin\theta_m - b\sin^2\theta_m = E_-\cos\theta_m\sin\theta_m \tag{44.29}$$
$$b\cos^2\theta_m + a\cos\theta_m\sin\theta_m = -E_-\cos\theta_m\sin\theta_m \tag{44.30}$$

and thus

$$b(\cos^2\theta_m - \sin^2\theta_m) = -2a\cos\theta_m\sin\theta_m \tag{44.31}$$

or

$$\tan 2\theta_m = -\frac{b}{a}. \tag{44.32}$$

Now using the fact that $a \equiv A - \Delta m^2\cos 2\theta$, $b \equiv \Delta m^2\sin 2\theta$, we have

$$\tan 2\theta_m = -\frac{\Delta m^2\sin 2\theta}{-\Delta m^2\cos 2\theta + A} = \frac{\tan 2\theta}{1 - \dfrac{A(x)}{\Delta m^2\cos 2\theta}}. \tag{44.33}$$

The two eigenfunctions are given by

$$\psi_- = \begin{pmatrix} \cos\theta_m \\ -\sin\theta_m \end{pmatrix}, \quad \psi_+ = \begin{pmatrix} \sin\theta_m \\ \cos\theta_m \end{pmatrix} \tag{44.34}$$

and

$$E_+^2 \equiv (A - \Delta m^2 \cos 2\theta)^2 + (\Delta m^2 \sin 2\theta)^2 = A^2 + (\Delta m^2)^2 - 2A\Delta m^2 \cos 2\theta \tag{44.35}$$

or

$$E_+ \equiv \Delta m^2 \left[1 + \frac{A^2}{(\Delta m^2)^2} - \frac{2A\cos 2\theta}{\Delta m^2} \right]^{1/2}. \tag{44.36}$$

Recall, for vacuum oscillations, we defined the oscillation length in Eqn. 43.14 and the probability for oscillation in Eqn. 43.16. We can define a similar quantity for oscillations in matter given by

$$\frac{E_+ L}{2E} \equiv 2\pi \frac{L}{L_m} \tag{44.37}$$

and the probabilities for oscillating given by

$$P(\nu_e \to \nu_x) = \sin^2 2\theta_m \sin^2 \pi \frac{L}{L_m} \tag{44.38}$$
$$P(\nu_e \to \nu_e) = 1 - P(\nu_e \to \nu_x).$$

We also have

$$\frac{L_m}{L_{osc}} \equiv \frac{\Delta m^2}{E_+} = \left[1 + \frac{A^2}{(\Delta m^2)^2} - \frac{2A\cos 2\theta}{\Delta m^2} \right]^{-1/2}. \tag{44.39}$$

Let's now analyze a very simplified example of matter oscillations in the Sun. We shall imagine that for the initial state we have an electron neutrino produced at the solar core, x_0. Define the quantity

$$\frac{A}{\Delta m^2} = \frac{2\sqrt{2}G_F E N_e}{\Delta m^2} \equiv \frac{L_{osc}}{L_0}. \tag{44.40}$$

Then we have (Eqn. 44.33)

$$\tan 2\theta_m(x_0) = \frac{\tan 2\theta}{\left(1 - \frac{L_{osc}}{L_0 \cos 2\theta} \right)}. \tag{44.41}$$

We plot $\tan 2\theta_m$ as a function of L_{osc}/L_0 in Fig. 44.9. The quantity L_{osc}/L_0 can be expressed as follows

$$\frac{L_{osc}}{L_0}(x_0) = 0.22 \left[\frac{E}{1 \text{ MeV}} \right] \left[\frac{\rho \, Y_e(x_0)}{100 \text{ g/cm}^3} \right] \left[\frac{7 \times 10^{-5} \text{ eV}^2}{\Delta m^2} \right] \tag{44.42}$$

where ρ is the mass density of the Sun and $Y_e(x) \equiv N_e(x)/(N_p(x) + N_n(x))$. Note, for $E > 10$ MeV, valid for the highest-energy ^{8}B neutrinos, we have $L_{osc}/L_0 > 2$ for $x_0 = 0$. From Fig. 44.9 we see that for $L_{osc}/L_0 \to +\infty$, we have $\tan 2\theta_m \to 0^-$ with $\theta_m \to \frac{\pi}{2}^-$ and $L_m/L_{osc} \to 0$. For our purposes, let's assume the limiting case,

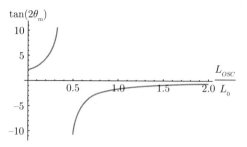

Fig. 44.9 A plot of $\tan 2\theta_m$ as a function of L_{osc}/L_0 (using Mathematica).

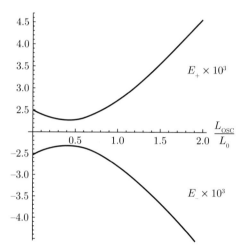

Fig. 44.10 The two eigenstates as a function of L_{osc}/L_0 (using Mathematica).

$\theta_m(x_0 = 0) \approx \frac{\pi}{2}^-$. Our initial state at the solar core is an electron neutrino which we define as

$$|\nu_e\rangle \equiv \begin{pmatrix} 1 \\ 0 \end{pmatrix}. \tag{44.43}$$

In general, the electron neutrino at the solar core is given by

$$|\nu_e\rangle_{core} \equiv \begin{pmatrix} 1 \\ 0 \end{pmatrix} \equiv \sin\theta_m(0)|\psi_+\rangle + \cos\theta_m(0)|\psi_-\rangle \approx |\psi_+\rangle. \tag{44.44}$$

As a solution to the adiabatic approximation the electron neutrino remains in the same energy eigenstate while traveling from the core to the surface. In Fig. 44.10 we show the eigenstates as a function of L_{osc}/L_0. At the surface,

$$\frac{L_{osc}}{L_0} \to 0^+, \quad \theta_m \to \theta. \tag{44.45}$$

Adiabatically, the angle $\theta_m(0)$ continuously goes to the vacuum angle θ.

$$\psi_- \to e^{i\alpha_-}(\cos\theta|\nu_e\rangle - \sin\theta|\nu_x\rangle), \tag{44.46}$$
$$\psi_+ \to e^{i\alpha_+}(\sin\theta|\nu_e\rangle + \cos\theta|\nu_x\rangle).$$

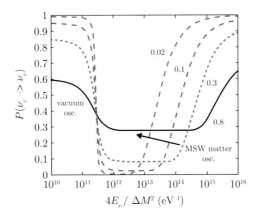

Fig. 44.11 Survival probability for ν_es created at the solar center as a function of the ratio $4E_\nu/\Delta m^2(\text{eV}^{-1})$. The curves are labeled by the value of $\sin^2 2\theta$ (McKeown and Vogel, 2004).

Thus,

$$|\nu_e\rangle_{surface} = e^{-i\alpha_+}\sin\theta|\psi_+\rangle + e^{-i\alpha_-}\cos\theta|\psi_-\rangle. \tag{44.47}$$

Given the solution for an electron neutrino at the core (Eqn. 44.44) and at the surface (Eqn. 44.47), we find the probability for electron neutrinos at the core to end up as electron neutrinos on the surface. We have

$$P(\nu_e^{core} \to \nu_e^{surface}) \equiv \left|\langle\nu_e^{surface}|\nu_e^{core}\rangle\right|^2 \tag{44.48}$$

$$= \left|e^{-i\alpha_+}\sin\theta\sin\theta_m(0) + e^{-i\alpha_-}\cos\theta\cos\theta_m(0)\right|^2$$

$$= \left(\cos^2\theta\cos^2\theta_m(0) + \sin^2\theta\sin^2\theta_m(0)\right) \tag{44.49}$$

$$\equiv \frac{1}{2}(1 + \cos 2\theta_m(0)\cos 2\theta),$$

where we've averaged over phases. If we assume $\theta_m(0) \simeq \frac{\pi}{2}$, we find

$$P(\nu_e^{core} \to \nu_e^{surface}) \approx \frac{1}{2}(1 - \cos 2\theta) \equiv \sin^2\theta. \tag{44.50}$$

Given the LMA-MSW solution, we have

$$\sin^2 2\theta \simeq 0.83, \Rightarrow \theta \simeq 32.8°, \quad \Delta m^2 \simeq 7 \times 10^{-5}\text{ eV}^2 > 0 \tag{44.51}$$

or $\sin\theta \simeq 0.54$, $\cos\theta \simeq 0.84$. Thus,

$$P(\nu_e^{core} \to \nu_e^{surface}) \approx 0.29. \tag{44.52}$$

Only one-third of the initial ν_es make it to the surface as ν_es. Stated otherwise, $\sim$ two-thirds of the initial ν_es oscillate into ν_xs at the surface. Of course, one must then consider the vacuum oscillations from the surface of the Sun to the detector.

The adiabatic approximation is valid when $\gamma \equiv \left|\frac{E\dot{\theta}_m}{E_+ - E_-}\right| \ll 1$. This is mostly valid for the LMA-MSW solution. However, there can be non-adiabatic corrections.

We then have

$$P(\nu_e^{core} \rightarrow \nu_e^{surface}) = \frac{1}{2} + \left(\frac{1}{2} - P_X\right) \cos 2\theta_m(0) \cos 2\theta \qquad (44.53)$$

$$\simeq \sin^2\theta + P_X \cos 2\theta \quad \text{for } \theta_m(0) = \frac{\pi}{2}$$

where P_X is the jump probability. $P_X \rightarrow 1$ as $E_\nu \rightarrow \infty$.

In a careful analysis of the MSW solution to the solar-neutrino problem, one needs to sum electron neutrinos created at different depths in the Sun and add up all the matter effects. As a simple way of understanding the effects of different experimental neutrino thresholds and the amount of neutrinos expected in the detector, consider Fig. 44.11 from McKeown and Vogel (2004). What is plotted is the survival probability for ν_es created at the solar center as a function of the ratio $4E_\nu/\Delta m^2(\text{eV}^{-1})$. Given the LMA-MSW solution, the ^{8}B and pp neutrinos produced at the solar core have,

$$(^8\text{B } \nu s) \quad \frac{4E}{\Delta m^2} > 2.55 \times 10^{11} \text{ eV}^{-1} \qquad (44.54)$$

$$(pp \ \nu s) \quad \frac{4E}{\Delta m^2} < 2.4 \times 10^{10} \text{ eV}^{-1}.$$

We see that about 60% of the pp neutrinos produced at the core survive to the surface, while only 40% of the ^{8}B neutrinos survive. Note, matter oscillations determine the sign of Δm^2 and the solution fitting the data has $\Delta m^2 > 0$. Also, we can see that the lowest mass state (see Eqn. 44.46) is predominantly an electron neutrino with $\cos^2\theta \sim 0.7$.

Neutrino Oscillations Cont'd: Neutrino Mass and Mixing Angles

45.1 LSND and Mini-BooNE

The Los Alamos experiment, known as LSND, observed $\bar{\nu}_\mu \to \bar{\nu}_e$ appearance via the decay $\pi^+ \to \mu^+ + \nu_\mu$ followed by the decay $\mu^+ \to \bar{\nu}_\mu + e^+ + \nu_e$. They fit the result with a value of

$$\Delta m^2 \geq 0.2 \text{ eV}^2 \quad \text{at} \quad 90\% \; CL. \tag{45.1}$$

With only three neutrinos, this would not be possible, since there can only be two different values of Δm^2. The Mini-BooNE experiment at FermiLab was designed to test this and both $\bar{\nu}_\mu \to \bar{\nu}_e$ and $\nu_\mu \to \nu_e$ were found to appear. The results from the Mini-BooNE collaboration (Aguilar-Arevalo et al., 2018) are given in Fig. 45.1. How would a sterile neutrino help explain this result? In the simplest explanation, U_{PMNS} would become a 4×4 matrix. Then $\nu_\mu \to \nu_e$ appearance would be given by the probability,

$$P(\nu_\mu \to \nu_e) = \sin^2 2\theta_{e\mu} \sin^2 \left(1.27 \frac{\Delta m^2(\text{eV}^2) \; L(\text{km})}{E(\text{GeV})} \right), \tag{45.2}$$

where

$$\sin^2 2\theta_{e\mu} \equiv 4|U_{e4}|^2 |U_{\mu 4}|^2. \tag{45.3}$$

Note, however, several other experiments, such as NOMAD, KARMEN, MINOS, DayaBay/Bugey and OPERA (Agafonova et al., 2019), have also searched for $\nu_\mu \to \nu_e$ appearance, excluding much of the allowed region (see Fig. 45.2). The Micro-BooNE experiment, now operating at FermiLab, was designed to test both the LSND and Mini-BooNE results. Verification of the LSND and Mini-BooNE results would require a new sterile neutrino to fit the data.

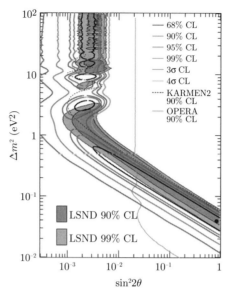

Fig. 45.1   The combined fit to LSND and Mini-BooNE data for ν_e and $\bar{\nu}_e$ appearance (Aguilar-Arevalo et al., 2018). The shaded areas show the 90% and 99% confidence level (CL) LSND $\bar{\nu}_\mu \to \bar{\nu}_e$ allowed regions. The black point shows the Mini-BooNE best-fit point. Also shown are 90% CL limits from the KARMEN and OPERA experiments. Published under the CC BY 4.0 license.

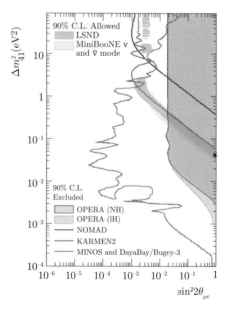

Fig. 45.2   Note, the 90% CL allowed regions from LSND and Mini-BooNE, in contrast to the excluded regions found by the NOMAD experiments (Astier et al., 2003), KARMEN2 (Armbruster et al., 2002), MINOS (Adamson et al., 2019) and DayaBay/Bugey3 (Adamson et al., 2016) shown as 90% CL exclusion regions. Taken from the OPERA collaboration (Agafonova et al., 2019). Published under the CC BY 4.0 license. For a color version of this figure, please see the color plate section.

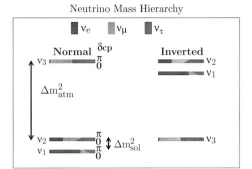

Fig. 45.3 The allowed hierarchy of neutrino masses consistent with solar and atmospheric neutrino oscillations (Qian and Vogel, 2015).

45.2 Summarizing Neutrino Results

The PMNS matrix has the general form

$$
U_{PMNS} = \begin{pmatrix} U_{e1} & U_{e2} & U_{e3} \\ U_{\mu1} & U_{\mu2} & U_{\mu3} \\ U_{\tau1} & U_{\tau2} & U_{\tau3} \end{pmatrix} \tag{45.4}
$$

$$
= \begin{pmatrix} 1 & 0 & 0 \\ 0 & c_{23} & s_{23} \\ 0 & -s_{23} & c_{23} \end{pmatrix} \begin{pmatrix} c_{13} & 0 & s_{13}e^{-i\delta} \\ 0 & 1 & 0 \\ -s_{13}e^{-i\delta} & 0 & c_{13} \end{pmatrix} \begin{pmatrix} c_{12} & s_{12} & 0 \\ -s_{12} & c_{12} & 0 \\ 0 & 0 & 1 \end{pmatrix} P_{12}
$$

or

$$
U_{PMNS} = \begin{pmatrix} c_{12}c_{13} & s_{12}c_{13} & s_{13}e^{-i\delta_{CP}} \\ -s_{12}c_{23} - c_{12}s_{13}s_{23}e^{i\delta_{CP}} & c_{12}c_{23} - s_{12}s_{13}s_{23}e^{i\delta_{CP}} & c_{13}s_{23} \\ s_{12}s_{23} - c_{12}s_{13}c_{23}e^{i\delta_{CP}} & -c_{12}s_{23} - s_{12}s_{13}c_{23}e^{i\delta_{CP}} & c_{13}c_{23} \end{pmatrix} P_{12}, \tag{45.5}
$$

where $c_{ij} \equiv \cos\theta_{ij}$, $s_{ij} \equiv \sin\theta_{ij}$ and

$$
P_{12} = \begin{pmatrix} e^{i\beta_1} & 0 & 0 \\ 0 & e^{i\beta_2} & 0 \\ 0 & 0 & 1 \end{pmatrix} \tag{45.6}
$$

contains two Majorana phases. Atmospheric neutrino oscillations give

$$
|U_{\mu3}| \approx |U_{\tau3}| \approx 1/\sqrt{2}, \quad \Delta m_{atm}^2 \equiv \pm\Delta m_{23}^2 \sim \pm 2 \times 10^{-3} \text{ eV}^2. \tag{45.7}
$$

The LMA-MSW solution to the solar neutrino problem give

$$
|U_{e2}| \approx |U_{\mu2}| \approx |U_{\tau2}| \approx 1/\sqrt{3}, \quad \Delta m_{sol}^2 \equiv \Delta m_{21}^2 \sim 7 \times 10^{-5} \text{ eV}^2 > 0. \tag{45.8}
$$

Assuming only three flavors of neutrinos, we have two possibilities, the normal (NH) or inverted (IH) mass hierarchy, given in Fig. 45.3.

Table 45.1 A global three-neutrino flavor analysis of neutrino oscillation data (Esteban et al., 2019). Published under the CC BY 4.0 license.

| | Normal ordering (best fit point, bfp) | | Inverted ordering ($\Delta\chi^2 = 9.3$) | |
	bft $\pm 1\sigma$	3σ range	bfp $\pm 1\sigma$	3σ range
$\sin^2\theta_{12}$	$0.310^{+0.013}_{-0.012}$	$0.275 \rightarrow 0.350$	$0.310^{+0.013}_{-0.012}$	$0.275 \rightarrow 0.350$
$\theta_{12}/°$	$33.82^{+0.78}_{-0.76}$	$31.61 \rightarrow 36.27$	$33.82^{+0.78}_{-0.75}$	$31.62 \rightarrow 36.27$
$\sin^2\theta_{23}$	$0.582^{+0.015}_{-0.019}$	$0.428 \rightarrow 0.624$	$0.582^{+0.015}_{-0.018}$	$0.433 \rightarrow 0.623$
$\theta_{23}/°$	$49.7^{+0.9}_{-1.1}$	$40.9 \rightarrow 52.2$	$49.7^{+0.9}_{-1.0}$	$41.2 \rightarrow 52.1$
$\sin^2\theta_{13}$	$0.02240^{+0.00065}_{-0.00066}$	$0.02044 \rightarrow 0.02437$	$0.02263^{+0.00065}_{-0.00066}$	$0.02067 \rightarrow 0.02461$
$\theta_{13}/°$	$8.61^{+0.12}_{-0.13}$	$8.22 \rightarrow 8.98$	$8.65^{+0.12}_{-0.13}$	$8.27 \rightarrow 9.03$
$\delta_{CP}/°$	217^{+40}_{-28}	$135 \rightarrow 366$	280^{+25}_{-28}	$196 \rightarrow 351$
$\dfrac{\Delta m^2_{21}}{10^{-5}\ \mathrm{eV}^2}$	$7.39^{+0.21}_{-0.20}$	$6.79 \rightarrow 8.01$	$7.39^{+0.21}_{-0.20}$	$6.79 \rightarrow 8.01$
$\dfrac{\Delta m^2_{3\ell}}{10^{-3}\ \mathrm{eV}^2}$	$+2.525^{+0.033}_{-0.031}$	$+2.431 \rightarrow 2.622$	$-2.512^{+0.034}_{-0.031}$	$-2.606 \rightarrow -2.413$

with SK-atm

Reactor neutrino experiments, Double Chooz (Abe et al., 2012), Daya Bay (An et al., 2012) and RENO (Ahn et al., 2012), have provided direct evidence for non-vanishing U_{e3}. A global three-neutrino flavor analysis of all neutrino oscillation data (excluding LSND and Mini-BooNE) is given in Esteban et al. (2019), Table 45.1. The **CP**-violating angle δ_{CP} is fit with large error bars.

A direct test of CP violation in neutrino oscillations is planned (Acciarri et al., 2015). The Long-Baseline Neutrino Facility (LBNF) located at FermiLab will produce a neutrino beam with a short base-line detector on site and a long-baseline detector, Deep Underground Neutrino Experiment (DUNE), located in the Sanford Underground Research Facility (SURF) 1300 km away in South Dakota. Of course, DUNE will be able to do many other measurements, such as testing for sterile neutrinos, nucleon decay and astrophysical neutrinos.

45.3 Neutrino Mass

Neutrinos may have Dirac masses as discussed earlier, Eqn. 38.21, or they may have Majorana masses, as given by the Weinberg operator, Eqn. 38.22. A small Majorana mass term is very natural. In fact, when we added the left-handed anti-neutrinos to generate a Dirac mass term, there was no symmetry preventing the Majorana mass term of the form

$$\frac{1}{2}\bar{\nu}_i M_{ij} \bar{\nu}_j, \tag{45.9}$$

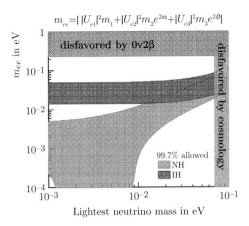

$$m_{ee}=|\,|U_{e1}|^2 m_1+|U_{e2}|^2 m_2 e^{2i\alpha}+|U_{e3}|^2 m_3 e^{2i\beta}|$$

Fig. 45.4 The mass parameter m_{ee} relevant for neutrinoless double-beta decay experiments (Qian and Vogel, 2015).

since the $\bar{\nu}_i$ have absolutely no Standard Model quantum numbers, i.e. they are *sterile* neutrinos. If the eigenvalues of M_{ij} are much larger than the weak scale, we can integrate these anti-neutrinos out of the low-energy theory. We then find an effective low-energy theory with three light Majorana neutrinos with mass given by

$$(m_\nu)_{ij} = (\lambda^{\nu T} M^{-1} \lambda^\nu)_{ij} v^2, \tag{45.10}$$

where λ^ν (Eqn. 38.19) is the neutrino Yukawa matrix. We don't need small coupling constants to generate small neutrino masses. What we need is a new scale in nature with $M \sim 10^{15}$ GeV to generate a neutrino mass $\sqrt{\Delta m_{atm}^2} \sim v^2/M \sim 0.05$ eV, such that $\Delta m_{atm}^2 \sim 3 \times 10^{-3}$ eV^2.

Finally, can a Majorana-neutrino mass be observed? The relevant mass parameter for neutrinoless double-beta decay experiments, where a nucleus (A, Z) decays to $(A, Z+2) + 2e^-$, is given by

$$m_{ee} = \left| |U_{e1}|^2 m_1 + |U_{e2}|^2 m_2 e^{2i\beta_1} + |U_{e3}|^2 m_3 e^{2i\beta_2} \right|. \tag{45.11}$$

In Fig. 45.4, the observable m_{ee} is evaluated for both the normal and inverted mass hierarchy.

More recently, the GERDA experiment in the Gran Sasso tunnel in Italy is pursuing the search for neutrinoless double-beta decay, see Agostini et al. (2019) and Fig. 45.5.

Experiments under construction, such as Majorana at SURF, may be able to measure $0\nu\beta\beta$ decay for the inverted mass hierarchy. Unfortunately, there are no experiments, that I know of, which can measure $0\nu\beta\beta$ decay for the normal mass hierarchy, UNLESS the three neutrinos are very degenerate with mass greater than $\sim 10^{-2}$ eV.

Table 45.2 Neutinoless double-beta decay experiments with sensitivities $S(T_{1/2})$ at 90% CL. The sensitivities have been converted to upper limits on the effective Majorana masses, m_{ee}. Table copied from Agostini et al. (2019)

Experiment	Isotope	$S(T_{1/2})(10^{25}\ \mathrm{yr})$	$m_{ee}(\mathrm{meV})$
KamLAND-Zen (Gando et al., 2016)	$^{136}\mathrm{Xe}$	5.6	76 to 234
CUORE (Alduino et al., 2018)	$^{130}\mathrm{Te}$	0.7	162 to 757
EXO-200 (Albert et al., 2018)	$^{136}\mathrm{Xe}$	3.7	93 to 287
CUPID-0 (Azzolini et al., 2018)	$^{82}\mathrm{Se}$	0.23	394 to 810
GERDA (Agostini et al., 2019)	$^{76}\mathrm{Ge}$	11	104 to 228
MAJORANA (Alvis et al., 2019)	$^{76}\mathrm{Ge}$	4.8	157 to 346
Combined			66 to 155

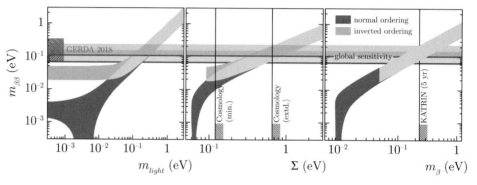

Fig. 45.5 Constraints are shown, left to right, as function of the lightest neutrino mass m_{light}, the sum of neutrino masses Σ, and the effective neutrino mass m_β. Contours follow from a scan of the Majorana phases with the central oscillation parameters from NuFIT 4.0 (Esteban et al., 2019). The dark horizontal band shows the upper limits on $m_{\beta\beta} \equiv m_{ee}$ obtained by GERDA; the lighter horizonal band shows those from combining sensitivities of all leading experiments in the field (see Table 45.2). The vertical lines denote $\Sigma = 0.12$ eV and $\Sigma = 0.66$ eV, a stringent limit from cosmology (Aghanim et al., 2020) and a cosmological model with 12 free parameters, i.e. extended as opposed to the minimal model with 7 free parameters (Tanabashi et al., 2018), as well as $m_\beta = 0.23$ eV, the 5-year sensitivity of the KATRIN Tritium beta decay experiment (Angrik et al., 2005). Hatching denotes the excluded parameter space. Figure from Agostini et al. (2019). Reprinted with permission from AAAS. For a color version of this figure, please see the color plate section.

45.4 Arbitrary Parameters of the Standard Model

The Standard Model of particle physics is very successful, explaining thousands of pieces of data with a few 2σ or 3σ discrepancies which are to be expected. On the other hand, there are some 24 arbitrary parameters in the Standard Model which must be fit to experiment. These parameters include ten quark masses and mixing angles, nine 'observable' lepton masses and mixing angles, three gauge coupling constants, the Higgs vacuum expectation value (VEV) and mass.[1]

However, there are some very enticing patterns. For some reason, quarks and leptons appear in three complete families. Their charges are consistent with the cancelation of all gauge anomalies. Neutrino masses, consistent with data, require either extremely small Yukawa coupling constants (for Dirac neutrino masses) or a very large mass scale, $M \sim 10^{15}$ GeV, for the right-handed neutrinos (for Majorana neutrino masses).

There is a hierarchy of quark and lepton masses and mixing angles. The third family is heavier than the second, and the second is heavier than the first. Moreover the hierarchy in the Cabibbo–Kobayashi–Maskawa (CKM) matrix is naturally explained by one parameter λ, ~ 0.22. In addition, the charge of the electron and that of the proton are equal and opposite. This fact is extremely important, since it allows for the existence of electrically neutral atoms. However, in the Standard Model, it is simply a consequence of the "crazy" charge assignments of quarks and leptons. Is this charge quantization an accident? Or is it determined by some fundamental physics? Finally, why is the strong coupling constant, $\alpha_s \sim 0.1$, larger than $\alpha_2 \sim 1/30$ (the $SU(2)$ fine-structure constant), which is larger than $\alpha' \sim 1/150$ at M_Z?

The Standard Model does not include the gravitational interactions described by the Planck scale, $M_{Pl} \sim 2.4 \times 10^{18}$ GeV. We expect new physics with new degrees of freedom to appear at this scale. Nevertheless, there is no explanation why the Higgs VEV, $v \sim 246$ GeV $\ll M$ or M_{Pl}. This is the gauge hierarchy problem. Fermion and gauge boson masses are protected by symmetry from large radiative corrections proportional to a possible new large scale of physics. On the other hand, scalar boson masses are not protected from radiative corrections and quite naturally tend to the largest scales of nature.

Last, but not least, it is by now well established that Standard Model particles apparently only make up 5% of the energy of the universe. Approximately 25% is dominated by some form of dark matter, while 70% is an approximately constant energy density known as dark energy. In addition, the universe is dominated by matter and not anti-matter. What is the origin of this asymmetry?

These are some of the questions that a theory beyond the Standard Model might be able to address. In the next few chapters we describe just one new idea for physics beyond the Standard Model.

---

[1] There is also, in principle, a strong CP-violating angle, θ, which must satisfy $\theta < 10^{-10}$.

PART XI

GRAND UNIFICATION

Grand Unification

46.1 Two Roads to Grand Unification

Grand unification of the strong and electroweak theories is a natural extension of the Standard Model. By 1973, the basic components of the Standard Model had been achieved. The theory was unitary and renormalizable. It included quarks and leptons as fundamental, point-like particles. It became natural to believe that quarks and leptons might be related by an $SU(4)_{color}$ symmetry, with lepton number as the fourth color (Pati and Salam, 1974).

One can first unify quarks and leptons into two irreducible representations of the group $SU(4)_C \otimes SU(2)_L \otimes SU(2)_R$, i.e. the so-called Pati–Salam (PS) group (Pati and Salam, 1974), where lepton number is the fourth color.

Then the PS fields,

$$Q = (q \, l), \quad \bar{Q} = (\bar{q} \, \bar{l}), \tag{46.1}$$

where

$$\bar{q} = \begin{pmatrix} \bar{u} \\ \bar{d} \end{pmatrix}, \, \bar{l} = \begin{pmatrix} \bar{\nu} \\ \bar{e} \end{pmatrix}, \tag{46.2}$$

transform as $(4, 2, 1) \oplus (\bar{4}, 1, \bar{2})$ under the PS symmetry. One can check that baryon number minus lepton number acting on a 4 of $SU(4)$ is given by

$$B - L = \begin{pmatrix} \frac{1}{3} & & & \\ & \frac{1}{3} & & \\ & & \frac{1}{3} & \\ & & & -1 \end{pmatrix} \tag{46.3}$$

and similarly electric charge is given by

$$Q = T_{3L} + T_{3R} + \frac{1}{2}(B - L). \tag{46.4}$$

Note, charge is quantized since it is embedded in a non-Abelian gauge group. One family is contained in two irreducible representations. Finally, if we require parity ($L \leftrightarrow R$) then there are two independent gauge couplings.

What about the Higgs? The two Higgs doublets H_u, H_d (e.g. $H_u = \phi$, $H_d = \tilde{\phi}$ as in Eqn. 35.29) are combined into one irreducible PS Higgs multiplet,

$$\mathcal{H} = (H_d \; H_u), \tag{46.5}$$

transforming as a $(1, 2, \bar{2})$ under PS symmetry. Thus, for one family, there is a unique renormalizable Yukawa coupling given by

$$\lambda \bar{\mathcal{Q}} \mathcal{H} \mathcal{Q}, \tag{46.6}$$

giving the Grand Unified Theory (GUT) relation

$$\lambda_t = \lambda_b = \lambda_\tau = \lambda_{\nu_\tau} \equiv \lambda. \tag{46.7}$$

Now the PS symmetry is *not* a grand-unified gauge group. However, since $SU(4) \approx SO(6)$ and $SU(2) \otimes SU(2) \approx SO(4)$ (where $\approx$ signifies a homomorphism), it is easy to see that the PS gauge group $\approx SO(6) \otimes SO(4) \subset SO(10)$ (Fritzsch and Minkowski, 1975; Georgi, 1975). In fact, one family of quarks and leptons is contained in the spinor representation of $SO(10)$, i.e.

$$SO(10) \to SU(4)_C \otimes SU(2)_L \otimes SU(2)_R$$
$$16 \to (4, 2, 1) \oplus (\bar{4}, 1, \bar{2}). \tag{46.8}$$

Hence by going to $SO(10)$ we have obtained quark–lepton unification (one family contained in one spinor representation) and gauge-coupling unification (one gauge group) (see Table 46.1).

But I should mention that there are several possible breaking patterns for $SO(10)$.

$$SO(10) \to SU(4)_C \otimes SU(2)_L \otimes SU(2)_R$$
$$\to SU(5) \otimes U(1)_X$$
$$\to SU(5)' \otimes U(1)_{X'}$$
$$\to SU(3)_C \otimes U(1)_{(B-L)} \otimes SU(2)_L \otimes SU(2)_R. \tag{46.9}$$

In order to preserve a prediction for gauge couplings we would require the breaking pattern

$$SO(10) \to SM \tag{46.10}$$

or

$$SO(10) \to SU(5) \to SM. \tag{46.11}$$

It will be convenient at times to work with the Georgi–Glashow GUT group $SU(5)$ (Georgi and Glashow, 1974). We have $16 \to 10 \oplus \bar{5} \oplus 1$.

46.2 Grand Unified Theory: $SU(5)$

In this chapter we will discuss the Georgi–Glashow GUT group $SU(5)$ (Georgi and Glashow, 1974). Let's identify the quarks and leptons of one family directly. We define the group $SU(5)$ by

$$SU(5) = \{U | U = 5 \times 5 \text{ complex matrix}; U^\dagger U = 1; \det U = 1\} \tag{46.12}$$

Table 46.1 Spinor representation of $SO(10)$ where this table explicitly represents the Cartan-Weyl weights for the states of one family of quarks and leptons. The double lines separate irreducible representations of $SU(5)$.

Grand Unification $-$ $SO(10)$

State	Y $= -\frac{2}{3}\Sigma(C) + \Sigma(W)$	Color C spins	Weak W spins
$\bar{\nu}$	0	$-\ -\ -$	$-\ -$
$\bar{e}$	2	$-\ -\ -$	$+\ +$
u_r	1/3	$+\ -\ -$	$-\ +$
d_r	1/3	$+\ -\ -$	$+\ -$
u_b	1/3	$-\ +\ -$	$-\ +$
d_b	1/3	$-\ +\ -$	$+\ -$
u_y	1/3	$-\ -\ +$	$-\ +$
d_y	1/3	$-\ -\ +$	$+\ -$
$\bar{u}_r$	$-4/3$	$-\ +\ +$	$-\ -$
$\bar{u}_b$	$-4/3$	$+\ -\ +$	$-\ -$
$\bar{u}_y$	$-4/3$	$+\ +\ -$	$-\ -$
$\bar{d}_r$	2/3	$-\ +\ +$	$+\ +$
$\bar{d}_b$	2/3	$+\ -\ +$	$+\ +$
$\bar{d}_y$	2/3	$+\ +\ -$	$+\ +$
ν	-1	$+\ +\ +$	$-\ +$
e	-1	$+\ +\ +$	$+\ -$

and the fundamental representation 5^α, $\alpha = 1,\ldots,5$ transforms as

$$5'^{\alpha} = U^{\alpha}{}_{\beta}\, 5^{\beta}. \tag{46.13}$$

We represent the unitary matrix U by

$$U = \exp(iT_A\, \omega_A), \tag{46.14}$$

where $\mathrm{Tr}(T_A) = 0$, $T_A^{\dagger} - T_A$, $A = 1,\ldots,24$ and $[T_A,\, T_B] = if_{ABC}T_C$, with f_{ABC} the structure constants of $SU(5)$. Under an infinitesimal transformation, we have

$$\delta_A 5^{\alpha} = i(T_A)^{\alpha}{}_{\beta}5^{\beta}\omega_A. \tag{46.15}$$

Let us now identify the $SU(3) \otimes SU(2) \otimes U(1)_Y$ subgroup of $SU(5)$. The $SU(3)$ subgroup is given by the generators

$$T_A = \begin{pmatrix} \frac{1}{2}\lambda_A & 0 \\ \hline 0 & 0 \end{pmatrix}, \quad A = 1, \ldots, 8. \tag{46.16}$$

And the $SU(2)$ subgroup is given by

$$T_A = \begin{pmatrix} 0 & 0 \\ \hline 0 & \frac{1}{2}\tau_{(A-20)} \end{pmatrix}, \quad A = 21, 22, 23. \tag{46.17}$$

The generators in $SU(5)/SU(3) \otimes SU(2) \otimes U(1)_Y$, i.e. those generators in $SU(5)$ but not in the Standard Model, are given by

$$T_A, \quad A = 9, \ldots, 20. \tag{46.18}$$

These are 12 generators of the form

$$\begin{pmatrix} 0 & \begin{matrix} 1 & 0 \\ 0 & 0 \\ 0 & 0 \end{matrix} \\ \hline \begin{matrix} 1 & 0 & 0 \\ 0 & 0 & 0 \end{matrix} & 0 \end{pmatrix}, \quad \begin{pmatrix} 0 & \begin{matrix} -i & 0 \\ 0 & 0 \\ 0 & 0 \end{matrix} \\ \hline \begin{matrix} i & 0 & 0 \\ 0 & 0 & 0 \end{matrix} & 0 \end{pmatrix}. \tag{46.19}$$

Let us now identify the hypercharge Y. The only remaining generator of $SU(5)$ commuting with the generators of $SU(3)$ and $SU(2)$ is given by

$$T_{24} = \sqrt{\frac{3}{5}} \begin{pmatrix} -1/3 & 0 & 0 & \\ 0 & -1/3 & 0 & 0 \\ 0 & 0 & -1/3 & \\ \hline & 0 & & \begin{matrix} 1/2 & 0 \\ 0 & 1/2 \end{matrix} \end{pmatrix} \equiv \sqrt{\frac{3}{5}}\frac{Y}{2}. \tag{46.20}$$

The overall normalization is chosen so that all the $SU(5)$ generators satisfy $\mathrm{Tr}(T_A T_B) = \frac{1}{2}\delta_{AB}$.

With these identifications, we see that the quantum numbers of a five-dimensional representation is given by

$$5^{\alpha} = \begin{pmatrix} d^a \\ \bar{\ell}^i \end{pmatrix}, \quad a = 1, 2, 3; i = 4, 5, \tag{46.21}$$

where d^a transforms as $(3, 1, -2/3)$ and $\bar{\ell}^i$ transforms as $(1, 2, +1)$ under the Standard Model. Note, electric charge is given by

$$Q = T_{23} + \sqrt{\frac{5}{3}}T_{24}. \tag{46.22}$$

Hence,

$$Q d^a = -\frac{1}{3}d^a, \tag{46.23}$$
$$Q \bar{\ell}^4 = \bar{\ell}^4 \equiv -\bar{e},$$
$$Q \bar{\ell}^5 = 0, \Rightarrow \bar{\ell}^5 \equiv \bar{\nu}.$$

Thus,

$$\bar{\ell} \equiv \begin{pmatrix} -\bar{e} \\ \bar{\nu} \end{pmatrix}, \quad Y_{\bar{\ell}} = +1. \tag{46.24}$$

Of course these are not the correct quantum numbers for any of the quarks and leptons, but the charge-conjugate states are just right. We have

$$\bar{5}'_\alpha = \bar{5}_\beta (U^\dagger)^\beta{}_\alpha \tag{46.25}$$

and

$$\delta_A \bar{5}_\alpha \equiv +i(-T_A^T)_\alpha{}^\beta \bar{5}_\beta \omega_A. \tag{46.26}$$

Hence,

$$\bar{5}_\alpha = \begin{pmatrix} \bar{d}_a \\ \ell'_i \end{pmatrix}, \quad a = 1, 2, 3; i = 4, 5, \tag{46.27}$$

with transformation properties

$$\bar{d} = (\bar{3}, 1, 2/3), \quad \ell' = (1, \bar{2}, -1) \tag{46.28}$$

and electric charge given by

$$Q\bar{d}_a = \frac{1}{3}\bar{d}_a, \tag{46.29}$$
$$Q\ell'_4 = -\ell'_4 \Rightarrow \ell'_4 \equiv -e,$$
$$Q\ell'_5 = 0, \quad \Rightarrow \ell'_5 \equiv \nu.$$

Thus,

$$\ell' \equiv \begin{pmatrix} -e \\ \nu \end{pmatrix}, \quad Y_{\ell'} = -1. \tag{46.30}$$

Once we have identified the states of the $\bar{5}$, we have no more freedom for the 10. The 10 transforms as an anti-symmetric tensor product of two 5s, i.e.

$$10^{\alpha\beta} = -10^{\beta\alpha} \propto 5_1^\alpha 5_2^\beta - 5_2^\alpha 5_1^\beta. \tag{46.31}$$

We have

$$10'^{\alpha\beta} = U^\alpha{}_\gamma U^\beta{}_\delta 10^{\gamma\delta} \tag{46.32}$$

and

$$\delta_A 10^{\alpha\beta} = i[(T_A)^\alpha{}_\gamma \delta^\beta{}_\delta + \delta^\alpha{}_\gamma (T_A)^\beta{}_\delta] 10^{\gamma\delta} \omega_A. \tag{46.33}$$

We find

$$10^{ab} \equiv \epsilon^{abc} (\bar{u})_c = (\bar{3}, 1, -4/3),$$
$$10^{ai} \equiv q^{ai} = (3, 2, 1/3),$$
$$10^{ij} \equiv \epsilon^{ij} \bar{e} = (1, 1, +2). \tag{46.34}$$

Note,

$$Y\bar{u} = -\frac{4}{3}\bar{u}, \tag{46.35}$$

$$Yq = \left(-\frac{2}{3} + 1\right)q \equiv \frac{1}{3}q,$$

$$Y\bar{e} = 2\bar{e}.$$

In addition, we have

$$Qq^{a4} = \frac{2}{3}q^{a4} \tag{46.36}$$

$$Qq^{a5} = -\frac{1}{3}q^{a5}$$

$$\Rightarrow q^{ai} \equiv \begin{pmatrix} u \\ d \end{pmatrix}.$$

To summarize we find

$$\bar{5}_\alpha = \begin{pmatrix} \bar{d}_1 \\ \bar{d}_2 \\ \bar{d}_3 \\ -e \\ \nu \end{pmatrix}, \quad 10^{\alpha\beta} = \frac{1}{\sqrt{2}} \left(\begin{array}{ccc|cc} 0 & \bar{u}_3 & -\bar{u}_2 & u^1 & d^1 \\ -\bar{u}_3 & 0 & \bar{u}_1 & u^2 & d^2 \\ \bar{u}_2 & -\bar{u}_1 & 0 & u^3 & d^3 \\ \hline -u^1 & -u^2 & -u^3 & 0 & \bar{e} \\ -d^1 & -d^2 & -d^3 & -\bar{e} & 0 \end{array} \right). \tag{46.37}$$

Note, it should be clear that, by embedding $SU(3) \otimes SU(2) \otimes U(1)_Y$ into a non-Abelian group, charge quantization is now understood.

We must also include the Higgs doublets in our $SU(5)$ model. The smallest non-trivial representation of $SU(5)$ is the 5-plet, so we shall place the Higgs doublets in a 5-plet,

$$H^\rho \equiv \begin{pmatrix} t^a \\ \begin{pmatrix} \phi^+ \\ \phi^0 \end{pmatrix} \end{pmatrix} \tag{46.38}$$

with $\phi \equiv \begin{pmatrix} \phi^+ \\ \phi^0 \end{pmatrix}$ and $Y\phi = \phi$.

Now that we have identified the states of one family in $SU(5)$, let us exhibit the Lagrangian (with gauge interactions). We have

$$\mathcal{L}_{fermion} = \bar{5}^\dagger_\alpha i(\bar{\sigma}_\mu D^\mu)_\alpha{}^\beta \bar{5}_\beta + 10^{\alpha\beta\dagger} i(\bar{\sigma}_\mu D^\mu)^{\alpha\beta}_{\gamma\delta} 10^{\gamma\delta} + |D_\mu H|^2, \tag{46.39}$$

where

$$D^\mu = \partial^\mu + ig_5 T_A A^\mu_A \tag{46.40}$$

and T_A is in the appropriate representation. We see that since there is only one gauge-coupling constant at the GUT scale we have

$$g_3 = g_2 = g_1 \equiv g_5, \tag{46.41}$$

where we have also identified the Standard Model gauge fields,

$$A_{24}^{\mu} \equiv B^{\mu}, \tag{46.42}$$

$$A_A^{\mu} \equiv g_A^{\mu}, \quad A = 1, \ldots, 8,$$

$$A_{20+a}^{\mu} \equiv W_a^{\mu}, \quad a = 1, 2, 3.$$

After weak-scale threshold corrections are included, we have in the effective field theory above M_Z,

$$g_3 \rightarrow g_s, \ g_2 \rightarrow g, \ g_1 \rightarrow \sqrt{\frac{5}{3}} \, g'. \tag{46.43}$$

The last equality follows from the fact that

$$g_5 T_{24} = g_1 \sqrt{\frac{3}{5}} \frac{Y}{2} \equiv g' \frac{Y}{2}. \tag{46.44}$$

We also have the relation

$$\sin^2 \theta_W (M_G) = \frac{(g')^2}{g^2 + (g')^2} = 3/8. \tag{46.45}$$

But these are tree-level relations which do not take into account threshold corrections at either the GUT or the weak scales nor the renormalization group (RG) running from the GUT scale to the weak scale. Consider first RG running. The one-loop RG equations are given by

$$\frac{d\alpha_i}{dt} = -\frac{b_i}{2\pi} \alpha_i^2, \tag{46.46}$$

where $\alpha_i = \frac{g_i^2}{4\pi}$, $i = 1, 2, 3$ and

$$b_i = \frac{11}{3} C_2(G_i) - \frac{2}{3} T_R N_F - \frac{1}{3} T_R N_S. \tag{46.47}$$

Note, $t = -\ln(\frac{M_G}{\mu})$. In addition, $\sum_A (T_A^2) = C_2(G_i)\mathbb{I}$, with T_A in the adjoint representation, defines the quadratic Casimir for the group G_i with $C_2(SU(N)) = N$ and $C_2(U(1)) = 0$. $\mathrm{Tr}(T_A T_B) = T_R \delta_{AB}$ for T_A in the representation R (for $U(1)_Y$, $T_R \equiv \frac{3}{5} \mathrm{Tr}(\frac{Y^2}{4})$) and $N_F(N_S)$ is the number of Weyl fermions (complex scalars) in representation R. The solution to the one-loop RG equations is given by

$$\alpha_i(M_Z)^{-1} = \alpha_G^{-1} - \frac{b_i}{2\pi} \ln\left(\frac{M_G}{M_Z}\right). \tag{46.48}$$

In what follows, we shall assume that the color-triplet Higgs states obtain mass at the GUT scale.

For the Standard Model we find

$$b_3 = 11 - \frac{2}{3} n_{fl} = 11 - \frac{4}{3} n_g, \tag{46.49}$$

$$b_2 = \frac{22}{3} - \frac{4}{3} n_g - \frac{1}{6} n_H,$$

$$b_1 = -\frac{2}{3} n_g \mathrm{Tr}_g \left(\frac{Y^2}{4}\right) \frac{3}{5} - \frac{1}{3} n_H \mathrm{Tr}_H \left(\frac{Y^2}{4}\right) \frac{3}{5}, \tag{46.50}$$

where n_{fl}, n_g, n_H are the number of flavors, generations and Higgs doublets, respectively.

$$\text{Tr}_g\left(\frac{Y^2}{4}\right) = \frac{1}{4}\left(6 \times \frac{1}{9} + 3 \times \frac{4}{9} + 3 \times \frac{16}{9} + 2 \times 1 + 4\right) = \frac{10}{3} \tag{46.51}$$

for $q, \bar{d}, \bar{u}, \ell, \bar{e}$, respectively. Hence,[1]

$$b_1 = -\frac{4}{3}n_g - \frac{1}{10}n_H. \tag{46.52}$$

The one-loop equations can be solved for the value of the GUT scale M_G and α_G in terms of the values of $\alpha_{EM}(M_Z)$ and $\sin^2\theta_W(M_Z)$ (Georgi et al., 1974). We have (without including weak-scale threshold corrections)

$$\alpha_2(M_Z) = \frac{\alpha_{EM}(M_Z)}{\sin^2\theta_W(M_Z)}, \quad \alpha_1(M_Z) = \frac{\frac{5}{3}\alpha_{EM}(M_Z)}{\cos^2\theta_W(M_Z)} \tag{46.53}$$

and we find

$$\alpha_1^{-1}(M_Z) - \alpha_2^{-1}(M_Z) = \left(\frac{3}{5} - \frac{8}{5}\sin^2\theta_W(M_Z)\right)\alpha_{EM}(M_Z)^{-1} = \left(\frac{b_2 - b_1}{2\pi}\right)\ln\left(\frac{M_G}{M_Z}\right), \tag{46.54}$$

which we use to solve for M_G.

Then we use

$$\alpha_G^{-1} = \sin^2\theta_W(M_Z)\alpha_{EM}(M_Z)^{-1} + \frac{b_2}{2\pi}\ln\left(\frac{M_G}{M_Z}\right) \tag{46.55}$$

to solve for α_G.

We can then predict the value for the strong coupling using

$$\alpha_3(M_Z)^{-1} = \alpha_G^{-1} - \frac{b_3}{2\pi}\ln\left(\frac{M_G}{M_Z}\right). \tag{46.56}$$

Given the experimental values $\sin^2\theta_W(M_Z) \approx 0.23$ and $\alpha_{EM}(M_Z)^{-1} \approx 128$ we find $M_G \approx 1.3 \times 10^{13}$ GeV with $n_H = 1, n_g = 3$ and $\alpha_G^{-1} \approx 42$ for the Standard Model with the one-loop prediction for $\alpha_3(M_Z) \approx 0.07$. How well does this agree with the data? According to the Particle Data Group the average value of $\alpha_s(M_Z) = 0.1181 \pm 0.0011$ (Tanabashi et al., 2018). So at one loop this $SU(5)$ GUT does not work well.

46.2.1 Yukawa Unification: $\lambda_b = \lambda_\tau$

The Yukawa couplings in $SU(5)$ for one family of quarks and leptons is given by

$$\frac{1}{4}\lambda 10^{\alpha\beta} 10^{\gamma\delta} H^\rho \epsilon_{\alpha\beta\gamma\delta\rho} + \lambda' 10^{\alpha\beta} \bar{5}_\alpha H^\dagger_\beta, \tag{46.57}$$

---

[1] Note, the terms in b_i, $i = 1, 2, 3$ depending on fermions are all the same. This is because the quarks and leptons come in complete $SU(5)$ families.

where

$$H^\rho \equiv \left(\begin{array}{c} t^a \\ \left(\begin{array}{c} \phi^+ \\ \phi^0 \end{array} \right) \end{array} \right),$$ (46.58)

with $\phi \equiv \left(\begin{array}{c} \phi^+ \\ \phi^0 \end{array} \right)$ and $Y\phi = \phi$, and

$$H^\dagger_\beta \equiv \left(\begin{array}{c} \bar{t}_a \\ \left(\begin{array}{c} \phi^- \\ \phi^{0\dagger} \end{array} \right) \end{array} \right),$$ (46.59)

with $\phi^\dagger \equiv \left(\begin{array}{c} \phi^- \\ \phi^{0\dagger} \end{array} \right)$ and $Y\phi^\dagger = -\phi^\dagger$. Note, the Standard Model Higgs doublet is now part of a 5-plet, which includes a color-triplet Higgs field, t^a. We shall see later that the color-triplet field must obtain mass near the GUT scale, in order to avoid rapid nucleon decay.

Consider the first term. We have

$$2\frac{1}{4}\lambda \underbrace{10^{ab}}_{\epsilon^{abc'}\bar{u}_{c'}} 10^{ci} H^j \epsilon_{abc}\epsilon_{ij}$$ (46.60)

$$= \lambda \bar{u}_c q^{ci} \phi^j \epsilon_{ij}.$$

When the Higgs doublet obtains a vacuum expectation value (VEV),

$$\langle \phi^j \rangle \equiv \frac{v}{\sqrt{2}} \, \delta^{j2},$$ (46.61)

we obtain the up-quark mass term,

$$m_u \bar{u}_c u^c, \quad m_u = \lambda \frac{v}{\sqrt{2}}.$$ (46.62)

Now consider the second term. We have

$$\lambda' 10^{\alpha\beta} \bar{5}_\alpha H^\dagger_\beta = \lambda' 10^{\alpha i} \bar{5}_\alpha H^\dagger_i$$ (46.63)
$$= \lambda' (10^{ai}\bar{5}_a + 10^{ji}\bar{5}_j) H^\dagger_i$$
$$= \lambda' (q^{ai}\bar{d}_a + \epsilon^{ji}\bar{e}\ell'_j)\phi^\dagger_i.$$

When the Higgs doublet gets an expectation value $\langle \phi^\dagger_i \rangle = \frac{v}{\sqrt{2}} \, \delta_{i2}$ we obtain two mass terms given by

$$m_d \bar{d}_a d^a + m_e \bar{e}e,$$ (46.64)

with

$$m_d = m_e = \lambda' \frac{v}{\sqrt{2}}.$$ (46.65)

Clearly this is a bad prediction for the first two families, since it would require $m_d = m_e$, $m_s = m_\mu$. Even though these are predictions at the GUT scale, we can define the approximate RG-invariant ratios of quark and lepton masses. For the light quarks and leptons, the contribution to the RG equations of their Yukawa couplings is negligible. The ratios are equal at the GUT scale, but clearly unequal at low energies. We have

$$20 \sim \frac{m_s}{m_d} = \frac{m_\mu}{m_e} \sim 200. \tag{46.66}$$

But what about the third family. We start with $\lambda_b(M_G) = \lambda_\tau(M_G)$ and use the RG equations to evaluate these running parameters at the weak scale. We have the RG equations

$$\frac{d}{dt}\left(\frac{\lambda_b(t)}{\lambda_\tau(t)}\right) = \frac{d}{dt}\left(\frac{m_b(t)}{m_\tau(t)}\right). \tag{46.67}$$

If we assume that strong interactions dominate in the RG equations, then

$$\frac{d}{dt}m_b = -\gamma m_b, \quad \frac{d}{dt}m_\tau = 0, \tag{46.68}$$

with γ at one loop given by

$$\gamma = \gamma_0 \frac{\alpha_3}{\pi}, \quad \gamma_0 = \frac{3}{2}\underbrace{C_2(R)}_{\frac{4}{3}} = 2. \tag{46.69}$$

Then the RG equation is

$$\frac{d}{dt}\ln\left(\frac{m_b(t)}{m_\tau(t)}\right) = -\gamma_0 \frac{\alpha_3}{\pi} = \frac{2\gamma_0}{b_3}\frac{d\ln\alpha_3}{dt}, \tag{46.70}$$

where we've used

$$\frac{d\ln\alpha_3}{dt} = -\frac{b_3}{2\pi}\alpha_3. \tag{46.71}$$

The solution is given by

$$\frac{m_b}{m_\tau}(M_Z) = \frac{m_b}{m_\tau}(M_G)\left[\frac{\alpha_3(M_Z)}{\alpha_3(M_G)}\right]^{2\gamma_0/b_3} \tag{46.72}$$

$$= 1 \times \left[\frac{\alpha_3(M_Z)}{\alpha_3(M_G)}\right]^{4/7} = (0.07 \times 42.4)^{4/7} = 1.86,$$

using $n_g = 3$, $b_3 = 7$. This can be compared to the experimental ratio, $\frac{m_b}{m_\tau}(M_Z) = 1.63$, which also does not work so well.

46.2.2 Spontaneously Breaking $SU(5) \rightarrow SU(3) \otimes SU(2) \otimes U(1)_Y$

Consider adding to our $SU(5)$ GUT, a scalar field, Σ transforming in the adjoint representation. We define

$$\Sigma^\alpha{}_\beta \equiv \sum_A \Sigma_A(T_A)^\alpha{}_\beta, \tag{46.73}$$

with $\text{Tr}\Sigma = 0$. The transformation law under $SU(5)$ is given by

$$\Sigma' = U\Sigma U^\dagger. \tag{46.74}$$

The Lagrangian for the Σ field is assumed to be

$$\text{Tr}\left[(D_\mu\Sigma)^\dagger(D_\mu\Sigma)\right] - \left[\text{Tr}(\Sigma^\dagger\Sigma) - \frac{V^2}{2}\right]^2. \tag{46.75}$$

Thus, at the minimum of the potential, Σ obtains a non-zero VEV with

$$\langle\Sigma\rangle = VT_{24}. \tag{46.76}$$

Note, there are several possible breaking patterns for $SU(5)$, but we have chosen one which breaks $SU(5) \to SU(3) \otimes SU(2) \otimes U(1)_Y$. Hence, the gauge bosons in $SU(5)/(SU(3) \otimes SU(2) \otimes U(1)_Y)$ obtain mass of order $g_G V$.

What about the Higgs bosons? We need to give the color-triplet component a large mass, while the Higgs doublet remains light. This can be accomplished by adding a term to the Lagrangian of the form[2]

$$\mathcal{L} \supset -M^2 H^\dagger H + \lambda_H H^\dagger \Sigma^2 H. \tag{46.77}$$

By choosing (or fine-tuning) $M = \sqrt{\frac{3\lambda_H}{20}}V$, one obtains a massless Higgs doublet and color triplets with mass, $M_T = V\sqrt{\frac{\lambda_H}{12}}$.

We can also make the Higgs doublet mass squared slightly negative, of order M_Z^2. However, in perturbation theory the Higgs doublet will obtain a one-loop correction to its mass squared of order $\delta m_h \sim \frac{\lambda_H^2}{16\pi^2}M_T^2$ which is much greater than M_Z^2. Hence, the fine-tuning required to keep the Higgs doublet light must be made at every order in perturbation theory. This is the "gauge hierarchy problem."

46.2.3 Nucleon Decay

Any GUT violates both baryon and lepton number (Gell-Mann et al., 1978). In Fig. 46.1, we give an example of the exchange of the heavy X boson, which demonstrates this fact. This diagram has two terms with the index $\beta = j, b$. We then have the

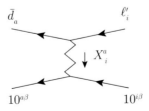

Fig. 46.1 X-boson exchange contributing to baryon- and lepton-number violation.

---

[2] We would also need a quartic Higgs coupling in the low energy theory $\propto (H^\dagger H)^2$.

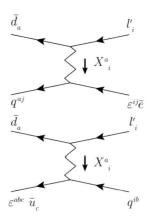

Fig. 46.2 X-boson exchange contributing to baryon- and lepton-number violation, where now the quarks and leptons in the 10-dimensional representation are made explicit.

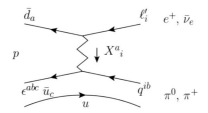

Fig. 46.3 X-boson exchange contributing to proton decay, $p \to e^+ + \pi^0$, $\bar{\nu}_e + \pi^+$. Similarly, by adding instead a spectator down quark we obtain the neutron decay processes, $n \to e^+ + \pi^-$, $\bar{\nu}_e + \pi^0$.

two diagrams in Fig. 46.2. In the diagram on top, $\Delta B = \Delta L = 0$. However, in the diagram on the bottom, we have $\Delta B = \Delta L = -1$. Hence, a conserved baryon number and lepton number cannot be defined, but $B - L$ is conserved.

In Fig. 46.3 we have added a spectator up quark and we see that we have the Feynman diagram for the proton decay processes, $p \to e^+ + \pi^0$, $\bar{\nu}_e + \pi^+$. Similarly, by adding instead a spectator down quark we obtain the neutron decay processes, $n \to e^+ + \pi^-$, $\bar{\nu}_e + \pi^0$. Theoretically, the nucleon lifetime is of order

$$\tau_N \sim \frac{M_G^4}{\alpha_G^2 m_N^5} \sim 1.4 \times 10^{24} yr \frac{(M_G/1.3 \times 10^{13} \text{ GeV})^4}{(\alpha_G \times 42)^2 (m_N/0.94 \text{ GeV})^5}. \tag{46.78}$$

This is a major problem, since Super-Kamiokande now places a lower bound on the proton lifetime, into the final state $e^+\pi^0$, of 1.6×10^{34} yr (Abe et al., 2017).

The color-triplet Higgs also contributes to proton decay via processes in Figs. 46.4 and 46.5, with the addition of a spectator quark. The vertices in this case are proportional to Yukawa couplings, which for the first family are much smaller than the GUT coupling constant. As a consequence, M_T can be as low as 10^{11} GeV without being excluded by Super-Kamiokande data.

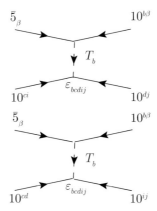

Color-triplet Higgs-boson exchange diagrams.

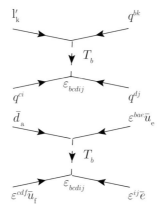

Color-triplet Higgs-boson exchange diagram with $\Delta B = \Delta L \neq 0$.

47 Supersymmetry

Supersymmetry (SUSY) is the largest possible extension of the Poincaré group which commutes with the S matrix. It takes bosons into fermions and vice versa. Global supersymmetry comes in four flavors, $N = 1, \ldots, 4$, where there are N supercharges, Q_i, $\bar{Q}_i$, $i = 1, \ldots, 4$. Note, $N = 8$ supersymmetry is the largest with all spins ≤ 2. An $N > 1$ supersymmetric gauge theory is necessarily non-chiral. Hence, since the Standard Model gauge theory is chiral, we shall only consider $N = 1$ SUSY here. Local Poincaré invariance implies Einstein's theory of general relativity and local SUSY implies supergravity. For further information on general SUSY theories, see Wess and Bagger (1992), Gates et al. (1983) and Ferrara (1987).

The simplest formalism for SUSY is the superfield formalism (Wess and Zumino, 1974a,b; Salam and Strathdee, 1974, 1975; Ferrara et al., 1974). We shall follow our chiral notation for fermions (Sections 10.2 and 10.3), with a little upgrade in notation.

47.1 Two-Component Spinors and the Lorentz Group

Define the group

$$SL(2, C) = \{M \mid 2 \times 2 \text{ complex matrices}, \det M = 1\}. \tag{47.1}$$

The group $SL(2, C) \approx SO(1, 3)$, i.e. the Lorentz group. The homomorphism is given by the following mapping. Given the four-vector momentum, $p_\mu = (E, -\vec{p})$, consider the matrix (recall σ^μ, Eqn. 10.39)

$$p_\mu \sigma^\mu \equiv \begin{pmatrix} E - p^3 & -(p^1 - ip^2) \\ -(p^1 + ip^2) & E + p^3 \end{pmatrix}. \tag{47.2}$$

We have

$$\det p_\mu \sigma^\mu = E^2 - \vec{p}^{\,2} \equiv p_\mu p^\mu. \tag{47.3}$$

The matrix $M \in SL(2, C)$ can be written in general as

$$M = \begin{pmatrix} \alpha & \beta \\ \gamma & \delta \end{pmatrix}, \tag{47.4}$$

with $\alpha\delta - \beta\gamma = 1$. Thus there are six parameters. A Lorentz transformation is given by

$$p'_\mu \sigma^\mu \equiv M p_\mu \sigma^\mu M^\dagger \tag{47.5}$$

such that $(p')^2 = \det p'_\mu \sigma^\mu = (\det M)^2 \det p_\mu \sigma^\mu = p^2$.

A left-handed Weyl fermion, ψ_α, transforms under $SL(2,C)$ as follows

$$\psi'_\alpha = M_\alpha{}^\beta \psi_\beta, \quad \psi'^\alpha = \psi^\beta (M^{-1})_\beta{}^\alpha \quad (0, 1/2). \tag{47.6}$$

There is an invariant tensor given by

$$\epsilon_{\alpha\beta}, \tag{47.7}$$

with

$$\epsilon_{\alpha\beta} = -\epsilon^{\alpha\beta} = \begin{pmatrix} 0 & -1 \\ 1 & 0 \end{pmatrix} \equiv (-i\sigma^2)_{\alpha\beta}. \tag{47.8}$$

Note, $\epsilon'_{\alpha\beta} = M_\alpha{}^\gamma M_\beta{}^\delta \epsilon_{\gamma\delta} \equiv \det M \; \epsilon_{\alpha\beta} = \epsilon_{\alpha\beta}$. ϵ can be used to raise and lower spinor indices. We have

$$\psi^\alpha = \epsilon^{\alpha\beta} \psi_\beta. \tag{47.9}$$

Then

$$(\xi\eta) \equiv \xi^\alpha \eta_\alpha \tag{47.10}$$

is invariant, since

$$\xi'^\alpha \eta'_\alpha = \xi^\beta (M^{-1})_\beta{}^\alpha M_\alpha{}^\gamma \eta_\gamma = \xi^\beta \eta_\beta. \tag{47.11}$$

Note, we also have

$$\epsilon^{\alpha\gamma} \epsilon_{\gamma\beta} \equiv \delta^\alpha{}_\beta, \tag{47.12}$$

thus

$$\psi_\alpha = \epsilon_{\alpha\beta} \psi^\beta. \tag{47.13}$$

We can also define the conjugate representation and its transformation rule, given by

$$\bar\psi'_{\dot\alpha} = M^*{}_{\dot\alpha}{}^{\dot\beta} \bar\psi_{\dot\beta}, \quad \bar\psi'^{\dot\alpha} = \bar\psi^{\dot\beta} (M^*)_{\dot\beta}^{-1\dot\alpha}, \quad (1/2, 0). \tag{47.14}$$

For the conjugate fields we have, similarly,

$$\bar\psi^{\dot\alpha} = \epsilon^{\dot\alpha\dot\beta} \bar\psi_{\dot\beta} \tag{47.15}$$

and

$$\bar\eta_{\dot\alpha} = \epsilon_{\dot\alpha\dot\beta} \bar\eta^{\dot\beta}, \tag{47.16}$$

with

$$\epsilon_{\dot\alpha\dot\beta} = -\epsilon^{\dot\alpha\dot\beta} = \begin{pmatrix} 0 & -1 \\ 1 & 0 \end{pmatrix}. \tag{47.17}$$

Note, however, in this case the invariant is given by

$$(\bar{\xi}'\bar{\eta}') \equiv \bar{\xi}'_{\dot\alpha}\bar{\eta}'^{\dot\alpha} = \bar{\xi}_{\dot\beta}(M^*)_{\dot\alpha}{}^{\dot\beta}(M^*)_{\dot\gamma}^{-1\,\dot\alpha}\bar{\eta}^{\dot\gamma} \equiv \bar{\xi}_{\dot\beta}\bar{\eta}^{\dot\beta}. \tag{47.18}$$

The wave equation for massless two-component spinors is given by

$$(\bar{\sigma}_\mu p^\mu)^{\dot\alpha\beta}\xi_\beta = 0 \ \ \text{or} \ \ (\sigma_\mu p^\mu)_{\alpha\dot\beta}\bar{\eta}^{\dot\beta} = 0, \tag{47.19}$$
$$(E + \vec{\sigma}\cdot\vec{p})\xi = 0 \ \ \text{or} \ \ (E - \vec{\sigma}\cdot\vec{p})\bar{\eta} = 0,$$

which gives

$$\frac{\vec{\sigma}\cdot\vec{p}}{E}\,\xi = -\xi \ \ \text{or} \ \ \frac{\vec{\sigma}\cdot\vec{p}}{E}\,\bar{\eta} = \bar{\eta}. \tag{47.20}$$

Hence, ξ is left-handed and $\bar{\eta}$ is right-handed. With this notation, the Lorentz generators are given by

$$(\sigma^{\mu\nu})_\alpha{}^\beta \equiv \frac{1}{4}(\sigma^\mu\bar{\sigma}^\nu - \sigma^\nu\bar{\sigma}^\mu)_\alpha{}^\beta, \tag{47.21}$$

$$(\bar{\sigma}^{\mu\nu})^{\dot\alpha}{}_{\dot\beta} \equiv \frac{1}{4}(\bar{\sigma}^\mu\sigma^\nu - \bar{\sigma}^\nu\sigma^\mu)^{\dot\alpha}{}_{\dot\beta}. \tag{47.22}$$

Finally, the Dirac four-component spinor, say for electrons, is given by

$$\psi_D = \begin{pmatrix} e_\alpha \\ \bar{e}^{\dot\alpha} \end{pmatrix}, \tag{47.23}$$

where

$$\bar{e}^{\dot\alpha} \equiv \epsilon^{\dot\alpha\dot\beta}\bar{e}_{\dot\beta} = (i\sigma^2)^{\alpha\beta}e^*_\beta. \tag{47.24}$$

The quantum electrodynamics Lagrangian for an electron is given by

$$\mathcal{L} = -\frac{1}{4}F^2_{\mu\nu} + e^\dagger(i\bar{\sigma}^\mu D_\mu)e + \bar{e}^\dagger(i\bar{\sigma}^\mu D_\mu)\bar{e} + (m_e(\bar{e}e) + h.c.). \tag{47.25}$$

47.2 $N = 1$ SUSY Algebra

SUSY is an extension of the Poincaré group, Chapter 2. Hence, we have the generators

$$\{\hat{P}_\mu,\ \hat{M}_{\mu\nu},\ \hat{Q}_\alpha,\ \hat{\bar{Q}}_{\dot\alpha},\ \hat{R}\}, \tag{47.26}$$

where $\hat{Q}_\alpha$, $\hat{\bar{Q}}_{\dot\alpha}$ are the supercharge operators and $\hat{R}$ is the generator of a $U(1)_R$ symmetry. The Lie algebra for the Poincaré generators was discussed earlier. The additional relations are given by

$$\left[\hat{M}_{\mu\nu}, \; \hat{Q}_\alpha\right] = -i(\sigma_{\mu\nu})_\alpha{}^\beta \hat{Q}_\beta, \tag{47.27}$$

$$\left[\hat{M}_{\mu\nu}, \; \hat{\bar{Q}}_{\dot\alpha}\right] = i\hat{\bar{Q}}_{\dot\beta}(\bar\sigma_{\mu\nu})^{\dot\beta}{}_{\dot\alpha},$$

$$\left[\hat{R}, \; \hat{Q}_\alpha\right] = \hat{Q}_\alpha,$$

$$\left[\hat{R}, \; \hat{\bar{Q}}_{\dot\alpha}\right] = -\hat{\bar{Q}}_{\dot\alpha},$$

$$\left[\hat{P}_\mu, \; \hat{Q}_\alpha\right] = 0,$$

$$\left\{\hat{Q}_\alpha, \; \hat{\bar{Q}}_{\dot\alpha}\right\} = 2\sigma^\mu_{\alpha\dot\alpha}\hat{P}_\mu,$$

$$\left\{\hat{Q}_\alpha, \; \hat{Q}_\beta\right\} = 0.$$

A direct consequence of global SUSY is that the vacuum energy is an order parameter for spontaneous SUSY breaking. We have

$$\left\{\hat{Q}_\alpha, \; \hat{\bar{Q}}_{\dot\alpha}\right\} = 2\sigma^\mu_{\alpha\dot\alpha}\hat{P}_\mu. \tag{47.28}$$

But

$$\left\{\hat{Q}_\alpha, \; \hat{\bar{Q}}_{\dot\alpha}\right\} \equiv \left\{\hat{Q}_\alpha, \; (\hat{Q}_\alpha)^\dagger\right\} \tag{47.29}$$

$$= \hat{Q}_\alpha \, (\hat{Q}_\alpha)^\dagger + (\hat{Q}_\alpha)^\dagger \, \hat{Q}_\alpha \equiv 2|\hat{Q}_\alpha|^2,$$

and thus

$$\sum_{\alpha=1}^{2} |\hat{Q}_\alpha|^2 = \mathrm{Tr}(\sigma_\mu \hat{P}^\mu) = 2\hat{P}^0 \geq 0. \tag{47.30}$$

Moreover the energy of the vacuum is given by

$$\langle\Omega| \left(\frac{1}{2}\sum_{\alpha=1}^{2} |\hat{Q}_\alpha|^2\right) |\Omega\rangle = \langle\Omega|\hat{P}^0|\Omega\rangle \equiv 0, \tag{47.31}$$

if and only if SUSY is not spontaneously broken.

47.3 Massless Representations of $N = 1$ SUSY

Consider the standard form of the four-momentum for massless particles, $\tilde{P}_\mu = (E, 0, 0, E)$ and single-particle states with helicity λ given by $|E, \lambda\rangle$. The Pauli–Lubanski spin vector is given by, Eqn. 3.7,

$$\hat{W}_\sigma \equiv -\frac{1}{2}\epsilon_{\mu\nu\rho\sigma}\hat{M}^{\mu\nu}\hat{P}^\rho \tag{47.32}$$

such that

$$\hat{W}_0|E, \lambda\rangle = \hat{\vec{J}} \cdot \hat{\vec{P}}|E, \lambda\rangle = \lambda E|E, \lambda\rangle. \tag{47.33}$$

Similarly, we have

$$\left\{\hat{Q}_\alpha, \ \hat{\bar{Q}}_{\dot{\alpha}}\right\} = 2\sigma^\mu_{\alpha\dot{\alpha}}\hat{P}_\mu \tag{47.34}$$

$$\doteq 2(\mathbb{I}_{2\times 2}E + \sigma^3_{\alpha\dot{\alpha}}E) \doteq 4E\begin{pmatrix} 1 & 0 \\ 0 & 0 \end{pmatrix},$$

where the symbol $\doteq$ means equal when acting on the state $|E, \ \lambda\rangle$. We also have

$$\left\{\hat{Q}_\alpha, \ \hat{Q}_\beta\right\} - 0. \tag{47.35}$$

Hence, we can define the creation and annihilation operators,

$$a^\dagger \equiv \frac{1}{2\sqrt{E}}\hat{Q}^\dagger_1, \quad a \equiv \frac{1}{2\sqrt{E}}\hat{Q}_1, \tag{47.36}$$

satisfying

$$\{a, \ a^\dagger\} \doteq 1, \quad \{a, \ a\} = 0. \tag{47.37}$$

Note,

$$\left\{\hat{Q}_2, \ \hat{Q}^\dagger_2\right\} \doteq 0 = \left\{\hat{Q}_2, \ \hat{Q}_2\right\} \tag{47.38}$$

means that $\hat{Q}_2 \doteq 0$.

Assume

$$a|E, \ \lambda\rangle = 0. \tag{47.39}$$

Now, what is $a^\dagger|E, \ \lambda\rangle$? We have

$$\hat{W}_0\hat{\bar{Q}}_{\dot{\alpha}}|E, \ \lambda\rangle = \left(\left[\hat{W}_0, \ \hat{\bar{Q}}_{\dot{\alpha}}\right] + \lambda E \ \hat{\bar{Q}}_{\dot{\alpha}}\right)|E, \ \lambda\rangle \tag{47.40}$$

$$= \left(iE(\bar{\sigma}_{12})\dot{\beta}_{\dot{\alpha}}\hat{\bar{Q}}_{\dot{\beta}} + \lambda E \ \hat{\bar{Q}}_{\dot{\alpha}}\right)|E, \ \lambda\rangle,$$

which follows from

$$\left[\hat{M}_{\mu\nu}, \ \hat{\bar{Q}}_{\dot{\alpha}}\right] = i\hat{\bar{Q}}_{\dot{\beta}}(\bar{\sigma}_{\mu\nu})^{\dot{\beta}}_{\ \dot{\alpha}}. \tag{47.41}$$

In addition, we have

$$i(\bar{\sigma}_{12})\dot{\beta}_{\dot{\alpha}} = \frac{i}{4}(\bar{\sigma}^1\sigma^2 - \bar{\sigma}^2\sigma^1) = -\frac{i}{4}(\sigma^1\sigma^2 - \sigma^2\sigma^1) \equiv \frac{1}{2}\sigma^3. \tag{47.42}$$

Thus,

$$\hat{W}_0\hat{\bar{Q}}_{\dot{\alpha}}|E, \ \lambda\rangle = \left(\frac{1}{2}(\sigma^3)\dot{\beta}_{\dot{\alpha}} + \lambda\delta\dot{\beta}_{\dot{\alpha}}\right)E\hat{\bar{Q}}_{\dot{\beta}}|E, \ \lambda\rangle$$

or

$$\hat{W}_0\hat{\bar{Q}}_1|E, \ \lambda\rangle = \left(\lambda + \frac{1}{2}\right)E\hat{\bar{Q}}_1|E, \ \lambda\rangle. \tag{47.43}$$

Thus, up to a normalization factor,

$$a^\dagger |E,\ \lambda\rangle = |E,\ \lambda + \frac{1}{2}\rangle. \qquad (47.44)$$

Hence, a minimal massless SUSY multiplet contains two states, i.e. with helicity λ and $\lambda + \frac{1}{2}$. For example, we shall define a chiral multiplet with helicities $(0,\ 1/2)$ and a gauge multiplet with helicities $(1/2,\ 1)$. In local SUSY, ... we can also have states with helicities $(3/2,\ 2)$.[1]

---

[1] The **CP**T conjugate states in the massless fields also include states with $\lambda \to -\lambda$.

In this chapter we introduce the concept of superfields residing in superspace. The formalism of superfields provides a very compact notation for describing the theory. Moreover, the superfield formalism allows for the construction of general supersymmetric Lagrangians in four dimensions where supersymmetry (SUSY) is verified without the necessity of using the equations of motion. This formalism is also applicable to theories of supergravity.

Superfields live in superspace

$$\{x_\mu, \ \theta_\alpha, \ \bar\theta_{\dot\alpha}\}, \tag{48.1}$$

where $\theta_\alpha, \bar\theta_{\dot\alpha}$ are classical Grassmann coordinates, satisfying

$$\{\theta_\alpha, \ \theta_\beta\} = \{\theta_\alpha, \ \bar\theta_{\dot\beta}\} = 0 \tag{48.2}$$

and

$$\bar\theta_{\dot\alpha} \equiv \theta_\alpha^\dagger. \tag{48.3}$$

Derivatives of these Grassmann coordinates are defined, such that

$$\partial_\alpha \equiv \frac{\partial}{\partial\theta^\alpha}, \quad \partial^\alpha \equiv \frac{\partial}{\partial\theta_\alpha} \tag{48.4}$$

and

$$\partial_\alpha\theta^\beta \equiv \delta_\alpha{}^\beta, \quad \partial^\alpha\theta_\beta \equiv \delta^\alpha{}_\beta \tag{48.5}$$
$$\partial^\alpha\theta^\beta = \epsilon^{\beta\alpha} \equiv -\epsilon^{\alpha\beta} \Rightarrow \partial^\alpha = -\epsilon^{\alpha\beta}\partial_\beta.$$

Similarly,

$$\bar\partial^{\dot\alpha} = \frac{\partial}{\partial\bar\theta_{\dot\alpha}}, \quad \bar\partial^{\dot\alpha}\bar\theta_{\dot\beta} \equiv \delta^{\dot\alpha}{}_{\dot\beta}. \tag{48.6}$$

Given the above, we can evaluate[1]

$$\partial_\alpha(\theta\theta) = 2\theta_\alpha \tag{48.7}$$

and defining, $\partial^2 \equiv \partial^\alpha\partial_\alpha$ and, similarly, $\bar\partial^2 \equiv \bar\partial_{\dot\alpha}\bar\partial^{\dot\alpha}$, we have

$$\partial^2(\theta\theta) = 4, \quad \bar\partial^2(\bar\theta\bar\theta) = 4. \tag{48.8}$$

We also define the Berezin integral for Grassmann variables by

$$\int d\theta_\alpha\theta^\beta \equiv \partial_\alpha\theta^\beta = \delta_\alpha{}^\beta, \tag{48.9}$$

---

[1] Note, $(\theta\theta) = 2\theta_2\theta_1 \neq 0$. We shall also use the notation, $\theta^2 \equiv (\theta\theta)$.

which satisfies the following requirement that the integral over all space is translation invariant and a pure number, i.e.

$$N = \int d\theta f(\theta) = \int d\theta f(\theta + \epsilon) = f'(0) \equiv \partial_\theta f(\theta), \tag{48.10}$$

where

$$f(\theta) \equiv f(0) + f'(0)\theta. \tag{48.11}$$

We shall normalize our Grassmann integrals such that

$$\int d^2\theta \equiv \frac{1}{4}\partial^2, \quad \int d^2\theta\, \theta^2 \equiv 1, \Rightarrow \quad \delta(\theta^2) \equiv \theta^2, \tag{48.12}$$

$$\int d^2\bar\theta \equiv \frac{1}{4}\bar\partial^2, \quad \int d^2\bar\theta\, \bar\theta^2 \equiv 1, \Rightarrow \quad \delta(\bar\theta^2) \equiv \bar\theta^2,$$

$$\int d^4\theta \equiv \int d^2\bar\theta \int d^2\theta \quad .$$

A general Lorentz scalar superfield given by

$$\begin{aligned}\Phi(x_\mu,\ \theta^\alpha,\ \bar\theta_{\dot\alpha}) &= f(x) + (\theta\phi(x)) + (\bar\theta\bar\chi(x)) \\ &\quad + \theta^2 m(x) + \bar\theta^2 n(x) + \theta\sigma^\mu\bar\theta v_\mu(x) \\ &\quad + \theta^2(\bar\theta\bar\lambda(x)) + \bar\theta^2(\theta\psi(x)) + \theta^2\bar\theta^2 d(x)\end{aligned} \tag{48.13}$$

has too many component fields. We need to define a SUSY-invariant constraint equation for Φ such that the field contains one Weyl fermion and one complex scalar, the so-called chiral multiplet.

A special SUSY transformation of the field Φ is given by

$$\Phi'(x_\mu,\ \theta,\ \bar\theta) = U\Phi(x_\mu,\ \theta,\ \bar\theta)U^\dagger = \Phi(x'_\mu,\ \theta',\ \bar\theta'), \tag{48.14}$$

with

$$U(\epsilon,\ \bar\epsilon) = e^{i(\epsilon\,\hat Q + \bar\epsilon\,\hat{\bar Q})} \tag{48.15}$$

and

$$\theta^{\alpha\prime} = \theta^\alpha - \epsilon^\alpha, \quad \bar\theta'_{\dot\alpha} = \bar\theta_{\dot\alpha} - \bar\epsilon'_{\dot\alpha}, \quad x^{\mu\prime} = x^\mu - i(\epsilon\sigma^\mu\bar\theta - \theta\sigma^\mu\bar\epsilon). \tag{48.16}$$

The SUSY algebra, like the Poincaré algebra, can be represented as derivative operators on functions, i.e. in this case functions on superspace. Recall, for example, $P^\mu \equiv i\partial^\mu$. Similarly, the supercharge operators can be represented as follows

$$Q_\alpha = i[\partial_\alpha + i\sigma^\mu_{\alpha\dot\beta}\bar\theta^{\dot\beta}\partial_\mu], \tag{48.17}$$

$$\bar Q_{\dot\alpha} = i[-\bar\partial_{\dot\alpha} - i\theta^\beta\sigma^\mu_{\beta\dot\alpha}\partial_\mu],$$

such that

$$\{Q_\alpha, \bar Q_{\dot\alpha}\} = 2\sigma^\mu_{\alpha\dot\alpha}P_\mu. \tag{48.18}$$

Under an infinitesimal transformation, we have

$$\delta\Phi(x_\mu,\ \theta,\ \bar\theta) = i\left[(\epsilon\ \hat Q + \bar\epsilon\ \hat{\bar Q}),\Phi(x_\mu,\ \theta,\ \bar\theta)\right] \tag{48.19}$$
$$\equiv i(\epsilon\ Q + \bar\epsilon\ \bar Q)\Phi(x_\mu,\ \theta,\ \bar\theta).$$

In order to obtain a supersymmetric constraint equation for Φ, we define invariant SUSY derivatives.

$$D_\alpha \equiv \partial_\alpha - i\sigma^\mu\ \bar\theta\ \partial_\mu, \tag{48.20}$$
$$\bar D_{\dot\alpha} \equiv -\partial_{\dot\alpha} + i\theta\ \sigma^\mu\ \partial_\mu,$$

satisfying

$$\left\{D_\alpha,\ \bar D_{\dot\alpha}\right\} = 2i\sigma^\mu_{\alpha\dot\alpha}\partial_\mu. \tag{48.21}$$

One can check that[2]

$$\left\{D_\alpha,\ \bar Q_{\dot\alpha}\right\} = 0 = \left\{Q_\alpha,\ \bar D_{\dot\alpha}\right\}. \tag{48.22}$$

We also have

$$\bar D_{\dot\alpha}y^\mu \equiv 0, \tag{48.23}$$

for

$$y^\mu \equiv x^\mu - i\theta\sigma^\mu\bar\theta. \tag{48.24}$$

Now define the constrained superfield by the equation

$$\bar D_{\dot\alpha}\Phi(x,\theta,\bar\theta) \equiv 0. \tag{48.25}$$

The solution is given by the general *left-chiral* superfield, i.e. since the spinor ψ is left chiral,

$$\Phi(x,\theta,\bar\theta) = \Phi(y,\ \theta) \equiv A(y) + \sqrt2\ (\theta\psi(y)) + \theta^2 F(y) \tag{48.26}$$
$$= A(x) + \sqrt2\ (\theta\psi(x)) + \theta^2 F(x)$$
$$- i\theta\sigma^\mu\bar\theta\partial_\mu A(x) + \frac{i}{\sqrt2}\theta^2(\partial_\mu\psi(x)\sigma^\mu\bar\theta) - \frac14\theta^2\bar\theta^2\Box A(x).$$

We can also define an anti-chiral superfield, such that $D_\alpha\bar\Phi = 0$, where $\bar\Phi = \bar\Phi(y^{\dagger\mu},\bar\theta)$. Note, if Φ is chiral, then $\bar\Phi \equiv \Phi^\dagger$ is anti-chiral.

Note, the mass dimension of $[\Phi] = 1$, since $[A] = 1$, $[\psi] = 3/2$. Therefore it is easy to see that the mass dimension of the SUSY coordinate $[\theta_\alpha] = -1/2$ and $[F] = 2$. Finally, the mass dimension $[d\theta] = 1/2$.

We can now evaluate the transformation properties of the component fields under the special SUSY transformation. Using Eqn. 48.19 we have

$$\delta A(x) = \sqrt2(\epsilon\psi(x)),$$
$$\delta\psi_\alpha(x) = \sqrt2 F(x)\epsilon_\alpha - i\sqrt2(\sigma^\mu\ \bar\epsilon)_\alpha\partial_\mu A(x), \tag{48.27}$$
$$\delta F(x) = i\sqrt2\partial_\mu\psi(x)\sigma^\mu\bar\epsilon.$$

---

[2] It is also easy to check that under the SUSY transformations, $D'_\alpha = D_\alpha$ and $\bar D'_{\dot\alpha} = \bar D_{\dot\alpha}$, i.e. they are invariant.

Note, a product of chiral superfields is a chiral superfield, i.e.

$$\bar{D}_{\dot{\alpha}}(\Phi_1\Phi_2) \equiv (\bar{D}_{\dot{\alpha}}\Phi_1)\Phi_2 + \Phi_1(\bar{D}_{\dot{\alpha}}\Phi_2) = 0, \tag{48.28}$$

if

$$(\bar{D}_{\dot{\alpha}}\Phi_1) = (\bar{D}_{\dot{\alpha}}\Phi_2) = 0. \tag{48.29}$$

Also, $\Phi^\dagger\Phi$ is NOT chiral. In fact, it is an effective real (Hermitian) superfield.

We can also define a real superfield given by

$$\begin{aligned}
V(x,\,\theta,\,\bar{\theta}) &= f(x) + (\theta\chi(x)) + (\bar{\theta}\bar{\chi}(x)) \\
&+ \theta^2 m(x) + \bar{\theta}^2 m^*(x) - \theta\sigma^\mu\bar{\theta}V_\mu(x) \\
&+ i\theta^2(\bar{\theta}\bar{\lambda}(x)) - i\bar{\theta}^2(\theta\lambda(x)) + \frac{1}{2}\theta^2\bar{\theta}^2 D(x),
\end{aligned} \tag{48.30}$$

with $f = f^*$, $V_\mu = V_\mu^*$, $D = D^*$. Under an infinitesimal special SUSY transformation, Eqn. 48.19, we have

$$\delta D(x) \propto \epsilon\sigma^\mu\partial_\mu\bar{\lambda}(x) + \partial_\mu\lambda(x)\sigma^\mu\bar{\epsilon}, \tag{48.31}$$

i.e. it is proportional to a total derivative.

Since the F term of the chiral superfield, and the D term of the real superfield, both transform as total space–time derivatives, we can construct an invariant supersymmetric action in terms of these components. We thus define the Wess–Zumino (WZ) SUSY-invariant action given by

$$S = \int d^4x d^4\theta \Phi^\dagger\Phi + \int d^4y d^2\theta W(\Phi) + h.c., \tag{48.32}$$

where

$$W(\Phi) = \frac{1}{2}m\Phi^2 + \frac{1}{3}\lambda\Phi^3 \tag{48.33}$$

is known as the superpotential.[3]

It takes some algebra to arrive at the WZ action in terms of the component fields. We find

$$\begin{aligned}
S = \int d^4x \Bigg\{ &|\partial_\mu A|^2 + \frac{i}{2}\bar{\psi}\bar{\sigma}^\mu\overleftrightarrow{\partial}_\mu\psi + |F|^2 \\
&+ F\frac{\partial W}{\partial\Phi}|_A - \frac{1}{2}\frac{\partial^2 W}{\partial\Phi^2}|_A(\psi\psi) + h.c. \Bigg\}.
\end{aligned} \tag{48.34}$$

The component field F is an auxiliary field which can be integrated out of the Lagrangian using the equations of motion. We then obtain

$$\begin{aligned}
S = \int d^4x \Bigg\{ &|\partial_\mu A|^2 + \frac{i}{2}\bar{\psi}\bar{\sigma}^\mu\overleftrightarrow{\partial}_\mu\psi \\
&- |mA + \lambda A^2|^2 - \frac{1}{2}(m + 2\lambda A)(\psi\psi) + h.c. \Bigg\}.
\end{aligned} \tag{48.35}$$

---

[3] Note, in the first term, the Grassmann integral $d^4\theta$ picks out the D term of the effective real superfield, while in the second term the integral $d^2\theta$ picks out the F term in the effective chiral superfield. Since each transforms as a total derivative under SUSY, the action is invariant.

Note, the scalar and fermion are degenerate, with mass m, and the scalar has both cubic and quartic self-interactions, as well as a Yukawa coupling to the Majorana fermion.

48.1 SUSY Gauge Theory

If the field $\Phi(y,\theta)$ also transforms under a non-Abelian gauge group, $\mathcal{G}$, via $\Phi' = e^{-i\Lambda}\Phi$, then Λ must also satisfy $\bar{D}_{\dot{\alpha}}\Lambda = 0$. Hence, the supersymmetric gauge parameter must also be chiral, such that Φ' is chiral, i.e. $\Lambda = \Lambda(y,\ \theta)$. There are many more SUSY gauge parameters than in a non-supersymmetric theory. However, given the real vector field in Eqn. 48.30, there are also many more fields than we might want, i.e. we only need the gauge field, $V_\mu(x)$, and the gaugino field, $\lambda(x)$. The rest are in fact, gauge artifacts. Under a supersymmetric gauge transformation the gauge superfield transforms by

$$e^{-2gV'} = e^{2i\Lambda^\dagger}e^{-2gV}e^{-2i\Lambda}. \tag{48.36}$$

There is a choice of gauge, called the WZ gauge, in which all of the extraneous component fields can be gauged away, using the freedom in $\Lambda(y,\ \theta)$. The real superfield is matrix valued, i.e. $V(x,\ \theta,\ \bar{\theta}) \equiv V_A(x,\ \theta,\ \bar{\theta})T_A$, where T_A are the generators of $\mathcal{G}$. In the WZ gauge, we have

$$V_{WZ}(x,\ \theta,\ \bar{\theta}) = -\theta\sigma^\mu\bar{\theta}V_\mu(x) + i\theta^2(\bar{\theta}\bar{\lambda}(x)) - i\bar{\theta}^2(\theta\lambda(x)) + \frac{1}{2}\theta^2\bar{\theta}^2 D(x). \tag{48.37}$$

The supersymmetric gauge field strength is given by

$$W_\alpha(y,\ \theta) = -\frac{1}{4}\bar{D}^2(e^{2gV}D_\alpha e^{-2gV}), \tag{48.38}$$

which, under a SUSY gauge transformation, transforms by

$$W'_\alpha = e^{2i\Lambda}W_\alpha e^{-2i\Lambda}. \tag{48.39}$$

After some algebra it can be shown that, in the WZ gauge, we have

$$\frac{1}{g}W^\alpha_{WZ} = i\lambda^\alpha(y) - \theta^\alpha D(y) + i(\sigma^{\mu\nu})^{\alpha\beta}\theta_\beta F_{\mu\nu}(y) + \theta^2\left(\sigma^\mu\left[\partial_\mu\bar{\lambda}(y) + ig[V_\mu,\ \bar{\lambda}(y)]\right]\right)^\alpha, \tag{48.40}$$

where

$$F_{\mu\nu} = \partial_\mu V_\nu - \partial_\nu V_\mu + ig\left[V_\mu,\ V_\nu\right]. \tag{48.41}$$

For example, consider SUSY quantum chromodynamics, with quarks in chiral superfields given by

$$Q^a(y,\ \theta) = \tilde{q}^a(y) + \sqrt{2}\ (\theta q^a(y)) + \theta^2 F_Q^a(y), \quad a = 1, 2, 3 \tag{48.42}$$
$$\bar{Q}_a(y,\ \theta) = \tilde{\bar{q}}_a(y) + \sqrt{2}\ (\theta \bar{q}_a(y)) + \theta^2 F_{\bar{Q}_a}(y).$$

The supersymmetric gauge transformation is given by

$$Q'(y,\ \theta) = e^{2i\Lambda(y,\ \theta)}Q(y,\ \theta), \quad \bar{Q}'(y,\ \theta) = e^{-2i\Lambda^T(y,\ \theta)}\bar{Q}(y,\ \theta), \tag{48.43}$$

with $\Lambda(y,\ \theta) = \Lambda_A(y,\ \theta)T_A$. The gauge and SUSY invariant action is given by

$$S = \frac{1}{2}\int d^4 y d^2\theta\ \frac{1}{4g^2}\ \mathrm{Tr}\left(W^\alpha(y,\ \theta)W_\alpha(y,\ \theta)\right) + h.c. \tag{48.44}$$
$$+ \int d^4 x d^4 \theta \left[Q^\dagger e^{-2gV} Q + \bar{Q}^\dagger e^{2gV^T}\bar{Q}\right]$$
$$+ \int d^4 y d^2 \theta\ [-m\bar{Q}(y,\theta)Q(y,\ \theta)] + h.c..$$

The second line above defines the so-called Kahler potential. In this case

$$K(\Phi_i, \Phi_i^\dagger, V) = \left[Q^\dagger e^{-2gV} Q + \bar{Q}^\dagger e^{2gV^T}\bar{Q}\right] \tag{48.45}$$

and the last line is the superpotential, with

$$W(Q,\ \bar{Q}) = -m\bar{Q}(y,\theta)Q(y,\ \theta). \tag{48.46}$$

The superpotential depends only on products of chiral superfields, i.e. it is a holomorphic function of the fields.

Upon integrating over the Grassmann coordinates we obtain the SUSY Lagrangian in terms of the component fields given by

$$\mathcal{L} = -\frac{1}{2}\mathrm{Tr}(F_{\mu\nu}^2) + 2i\mathrm{Tr}\left(\lambda^\dagger \bar{\sigma}^\mu D_\mu \lambda\right) + \mathrm{Tr}(D^2) \tag{48.47}$$
$$+ |D_\mu \tilde{q}|^2 + iq^\dagger \bar{\sigma}^\mu D_\mu q + |F_Q|^2 - \sqrt{2}ig\tilde{q}^*(\lambda q) + h.c - g\tilde{q}^* D\tilde{q}$$
$$+ |D_\mu \tilde{\bar{q}}|^2 + i\bar{q}^\dagger \bar{\sigma}^\mu D_\mu \bar{q} + |F_{\bar{Q}}|^2 + \sqrt{2}ig\tilde{\bar{q}}^*(\lambda^T \bar{q}) + h.c + g\tilde{\bar{q}}^* D^T \tilde{\bar{q}}$$
$$- m\left(F_{\bar{Q}}\tilde{q} + \tilde{\bar{q}}F_Q - (\bar{q}\ q)\right) + h.c..$$

The covariant derivatives are given by

$$D_\mu q \equiv (\partial_\mu + igV_\mu)q, \quad \text{and similarly for } \tilde{q} \tag{48.48}$$
$$D_\mu \bar{q} \equiv (\partial_\mu - igV_\mu^T)\bar{q},$$
$$D_\mu \lambda \equiv \partial_\mu \lambda + ig\left[V_\mu,\ \lambda\right].$$

The D and F auxiliary fields can be integrated out of the Lagrangian using their equations of motion. We then find

$$D_A = g \quad (\tilde{q}^* T_A \tilde{q} - \tilde{\bar{q}} T_A \tilde{\bar{q}}^*), \tag{48.49}$$

$$F_Q^* = m\tilde{\bar{q}}, \quad F_{\bar{Q}}^* = m\tilde{q}.$$

Plugging these back into the Lagrangian, we obtain the scalar potential given by

$$V(\tilde{q}, \tilde{\bar{q}}) = \frac{1}{2} \sum_A D_A^2 + |F_Q|^2 + |F_{\bar{Q}}|^2. \tag{48.50}$$

48.2 SUSY Quantum Electrodynamics

The simplest example of a supersymmetric gauge theory is SUSY quantum electrodynamics (QED) with two superfields,

$$E(y, \theta) = \tilde{e}(y) + \sqrt{2}(\theta e(y)) + \theta^2 F_e(y), \tag{48.51}$$
$$\bar{E}(y, \theta) = \tilde{\bar{e}}(y) + \sqrt{2}(\theta \bar{e}(y)) + \theta^2 F_{\bar{e}}(y),$$

and one real superfield,

$$V_{WZ}(x, \theta, \bar{\theta}) = -\theta \sigma^\mu \bar{\theta} A_\mu(x) + i\theta^2 (\bar{\theta}\bar{\lambda}(x)) - i\bar{\theta}^2 (\theta\lambda(x)) + \frac{1}{2}\theta^2\bar{\theta}^2 D(x). \tag{48.52}$$

The supersymmetric gauge field strength is given in the WZ gauge by

$$\frac{1}{e}W^\alpha(y, \theta) = i\lambda^\alpha(y) - D(y)\theta^\alpha + i(\sigma^{\mu\nu})^{\alpha\beta}\theta_\beta F_{\mu\nu}(y) + \theta^2(\sigma^\mu\partial_\mu\bar{\lambda}(y))_\alpha, \tag{48.53}$$

where

$$F_{\mu\nu} = \partial_\mu A_\nu - \partial_\nu A_\mu. \tag{48.54}$$

Then the supersymmetric generalization of QED (Eqn. 47.25) is given by

$$\mathcal{L} = \frac{1}{2} \int d^2\theta \frac{1}{4e^2} W^\alpha W_\alpha + h.c. \tag{48.55}$$

$$+ \int d^4\theta \left[E^\dagger e^{-2eV} E + \bar{E}^\dagger e^{+2eV} \bar{E} \right]$$

$$+ \int d^2\theta \left[-m_e \bar{E} E \right] + h.c..$$

Upon integrating θ and $\bar{\theta}$, we obtain SUSY QED in component form. We have

$$\mathcal{L} = -\frac{1}{4}F_{\mu\nu}^2 + i\bar{\lambda}^\dagger i\partial\!\!\!/\lambda + \frac{1}{2}D^2 \tag{48.56}$$

$$+ e^\dagger i\partial\!\!\!/e + \bar{e}^\dagger i\partial\!\!\!/\bar{e} + |D_\mu\tilde{e}|^2 + |D_\mu\tilde{\bar{e}}|^2 + |F_e|^2 + |F_{\bar{e}}|^2$$

$$+ eD\left(\tilde{e}^* Q\tilde{e} + \tilde{\bar{e}}^* Q\tilde{\bar{e}}\right) + i\sqrt{2}e\left(\tilde{e}^* Q(\lambda e) + \tilde{\bar{e}}^* Q(\lambda\bar{e}) - h.c.\right)$$

$$- m_e\left[\tilde{\bar{e}}F_e + F_{\bar{e}}\tilde{e} - (\bar{e}e)\right] + h.c.$$

The charge operator, Q, has eigenvalue -1 on e, $\tilde{e}$, eigenvalue $+1$ on $\bar{e}$, $\tilde{\bar{e}}$ and 0 on λ. The covariant derivative is given by $D_\mu = \partial_\mu + ieQA_\mu$.

The auxiliary field equations of motion give

$$F_e^* = m_e \tilde{\bar{e}}, \tag{48.57}$$

$$F_{\bar{e}}^* = m_e \tilde{e},$$

$$D = -e\,(\tilde{e}^* Q \tilde{e} + \tilde{\bar{e}}^* Q \tilde{\bar{e}}). \tag{48.58}$$

Plugging these back into the Lagrangian, we replace the auxiliary field terms with the scalar potential

$$V(\tilde{e},\,\tilde{\bar{e}}) = m_e^2(|\tilde{e}|^2 + |\tilde{\bar{e}}|^2) + \frac{1}{2}e^2(|\tilde{e}|^2 - |\tilde{\bar{e}}|^2)^2. \tag{48.59}$$

Consider what happens if we take $m_e = 0$. Then the scalar potential has a flat direction where $|\langle \tilde{e} \rangle| = |\langle \tilde{\bar{e}} \rangle| \neq 0$. This spontaneously breaks $U(1)_{QED}$ and we have a Higgs mechanism. The photon and photino obtain mass, but SUSY is unbroken.

49

SUSY $SU(5)$

In Chapter 46 we discussed non-supersymmetric $SU(5)$. This theory unified quarks and leptons into two irreducible representations. It solved the problem of charge quantization and also predicted nucleon decay. Unfortunately, once the three low-energy gauge couplings were measured precisely at the Large Electron–Positron Collider (LEP) collider, it became apparent that gauge coupling unification was a problem. In this chapter we show that supersymmetric $SU(5)$ solves two problems. Once superparticles are introduced into the renormalization group (RG) running of the gauge couplings, unification becomes highly successful, requiring only small threshold corrections at the Grand Unified Theory (GUT) scale. In addition, the GUT scale is increased by an order of magnitude compared to non-SUSY $SU(5)$ and thus nucleon decay, due to gauge-boson exchange, is significantly suppressed.

49.1 Superfields for SUSY $SU(5)$

In SUSY $SU(5)$ the matter fields transform as a $10 + \bar{5}$. These are left-handed Weyl spinors. To supersymmetrize this we just define chiral superfields with the same quantum numbers. We have three families of quarks and leptons in the superfields,

$$10(y,\ \theta) + \bar{5}(y,\ \theta), \tag{49.1}$$

along with their scalar partners. In addition, in non-supersymmetric $SU(5)$ we had one Higgs 5-plet. However, in SUSY $SU(5)$ we will require two Higgs multiplets given by the superfields,

$$H^\rho(y,\ \theta) + \bar{H}_\rho(y,\ \theta). \tag{49.2}$$

There are two reasons for this. In the first place, in order to give mass to down quarks and charged leptons, we used the conjugate Higgs field, $H^\dagger$. However, if H is a left-chiral superfield, then $H^\dagger$ is right chiral. But only products of left-chiral superfields can be in the superpotential. The second reason is that along with the Higgs doublet, we also have Higgs fermions. They have hypercharge $+1$. In order to cancel the $U(1)_Y SU(2)^2$ and $U(1)_Y^3$ anomalies, we need Higgs fermions with hypercharge -1.

Once again, assuming that $SU(5)$ is spontaneously broken at a GUT scale (to be determined) and that the only light states include those of the Minimal Supersymmetric Standard Model [MSSM], i.e. the $SU(3) \otimes SU(2) \otimes U(1)_Y$ gauge bosons and their fermionic partners ("gauginos"), the quarks and leptons of the Standard Model and their scalar partners ("squarks and sleptons") and the two doublets of Higgs bosons and their fermionic partners ("Higgsinos"), we can use the RG equations to determine α_G, M_G and $\alpha_3(M_Z)$. In Eqn. 46.47 we have the one-loop coefficient of the beta function for the RG running of gauge couplings. The generalization of the beta-function coefficient for the one-loop RG running in SUSY is given by (Dimopoulos et al., 1981; Dimopoulos and Georgi, 1981; Ibanez and Ross, 1981; Sakai, 1981)[1]

$$b_i = 3C_2(G_i) - T_R N_\chi, \tag{49.3}$$

where N_χ is the number of chiral fields in the representation R. For the MSSM we have

$$b_3^{MSSM} = 9 - 2N_g, \tag{49.4}$$
$$b_2^{MSSM} = 6 - 2N_g - N_{(H_u+H_d)},$$
$$b_1^{MSSM} = -2N_g - \frac{3}{5}N_{(H_u+H_d)},$$

where $N_{(H_u+H_d)}$ is the number of pairs of Higgs doublets. Thus for the MSSM (with $N_g = 3$, $N_{(H_u+H_d)} = 1$), we have

$$b_3^{MSSM} = 3, \tag{49.5}$$
$$b_2^{MSSM} = -1,$$
$$b_1^{MSSM} = -33/5.$$

Solving the one-loop RG equations for the MSSM, we find $M_G \approx 2.7 \times 10^{16}$ GeV, $\alpha_G^{-1} \approx 24$ and the predicted strong coupling $\alpha_3(M_Z) \approx 0.12$. How well does this agree with the data? Recall, $\alpha_s(M_Z) = 0.1181 \pm 0.0011$ (Tanabashi et al., 2018). So, at one loop, SUSY $SU(5)$ works quite well. In addition, the GUT scale has increased (Dimopoulos et al., 1981), so nucleon decay from gauge exchange is suppressed, compared to non-SUSY GUTs (see Fig. 49.1).

The superpotential necessary to give one family of quarks and leptons mass is given by

$$W = \frac{1}{4}\lambda 10^{\alpha\beta}10^{\gamma\delta}H^\rho \epsilon_{\alpha\beta\gamma\delta\rho} + \lambda' 10^{\alpha\beta}\bar{5}_\alpha \bar{H}_\beta. \tag{49.6}$$

Solving the RG equations for the top, bottom and tau Yukawa couplings one finds two solutions which fit low-energy data, i.e. for $\tan\beta \sim 1$ and $\tan\beta \sim 50$, as depicted in Fig. 49.2 taken from Barger et al. (1993). The latter solution is also valid for $\lambda_t = \lambda_b = \lambda_\tau$, i.e. top, bottom and tau Yukawa unification (see also Poh, 2017; Raby et al., 2017).[2]

---

[1] The beta function for SUSY at two loops can be found in Einhorn and Jones (1982) and Marciano and Senjanovic (1982).
[2] Note, the ratio $\frac{m_t}{m_b} = \frac{\lambda_t\langle H_u\rangle}{\lambda_b\langle H_d\rangle}$, where $H_u \subset H$ and $H_d \subset \bar{H}$. In order to fit the top–bottom mass ratio we require $\tan\beta \sim 50$, since at low energies we have $\lambda_t \sim \lambda_b$.

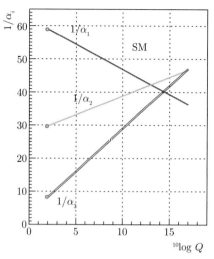 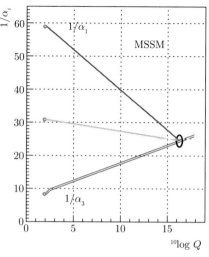

Fig. 49.1   Gauge-coupling unification in non-SUSY GUTs on the left vs. SUSY GUTs on the right using the LEP data as of 1991 (Amaldi et al., 1991; Ellis et al., 1991; Langacker and Luo, 1991). Note, the difference in the running for SUSY is the inclusion of supersymmetric partners of standard model (SM) particles at scales of order 1 TeV. Given the present accurate measurements of the three low-energy couplings, in particular $\alpha_s(M_Z)$, GUT scale threshold corrections are now needed to precisely fit the low-energy data. The dark blob in the plot on the right represents these model-dependent corrections. Figure taken from Kazakov (2000).

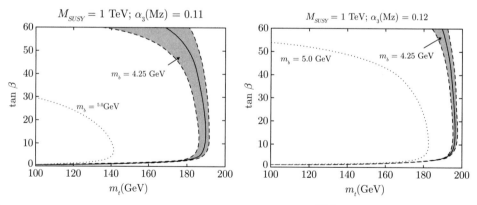

Fig. 49.2   The plot is $\tan\beta$ vs. the top-quark mass. Note, $\tan\beta \equiv \frac{\langle H_u \rangle}{\langle H_d \rangle}$ where $H_u \subset H$ and $H_d \subset \bar{H}$. The gray area is consistent with fitting the bottom quark and tau mass. Assuming $b-\tau$ Yukawa unification at M_G one finds two solutions fitting the top, bottom and tau masses. The latter solution is also valid for t, b, τ unification. Reprinted with permission from Barger et al. (1993). Copyright (1992) by the American Physical Society.

49.2 Nucleon Decay

In SUSY $SU(5)$, gauge bosons contribute to nucleon decay just as in non-SUSY $SU(5)$. The main difference is in the value of α_G and M_G. In SUSY GUTs, the GUT scale is of order 3×10^{16} GeV, as compared to the GUT scale in non-SUSY GUTs, which is of order 10^{15} GeV. Hence, the dimension-six baryon-number-violating operators are significantly suppressed in SUSY GUTs as noted by Dimopoulos, Raby and Wilczek and others (Dimopoulos et al., 1981; Dimopoulos and Georgi, 1981; Ibanez and Ross, 1981; Sakai, 1981) with $\tau_p \sim 10^{36 \pm 2}$ yr.

However, in SUSY GUTs there are additional sources for baryon-number violation – dimension-four and -five operators (Weinberg, 1982; Sakai and Yanagida, 1982). Although our notation does not change, when discussing SUSY GUTs all fields are implicitly bosonic superfields and the operators considered are the so-called F terms which contain two fermionic components and the rest scalars or products of scalars. Within the context of $SU(5)$ the dimension-four and -five operators have the form $(\mathbf{10}\ \bar{\mathbf{5}}\ \bar{\mathbf{5}}) \supset (\bar{U}\ \bar{D}\ \bar{D}) + (Q\ L\ \bar{D}) + (\bar{E}\ L\ L)$ and $(\mathbf{10}\ \mathbf{10}\ \mathbf{10}\ \bar{\mathbf{5}}) \supset (Q\ Q\ Q\ L) + (\bar{U}\ \bar{U}\ \bar{D}\ \bar{E}) + B$ and L conserving terms, respectively.[3] The dimension-four operators are renormalizable with dimensionless couplings; similar to Yukawa couplings. On the other hand, the dimension-five operators have a dimensionful coupling of order $(1/M_G)$.

The dimension-four operators violate baryon number or lepton number, respectively, but not both. The nucleon lifetime is extremely short if both types of dimension-four operators are present in the low-energy theory (see Fig. 49.3). In this case, if the coupling for the baryon-number-violating operator is λ and the lepton-number-violating operator is λ', then the product is constrained to satisfy $\lambda\ \lambda' < 10^{-27} \frac{\tilde{m}_d^2}{\text{TeV}^2}$ to be consistent with nucleon-decay bounds. However all

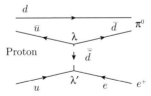

Fig. 49.3 The effective four-Fermi operator for proton decay obtained by having both baryon-(coupling λ) and lepton (coupling λ')-number-violating dimension-four operators.

---

[3] The dimension-four operators can have arbitrary couplings, λ_{ijk}, λ'_{ijk}, λ''_{ijk}, respectively. In a GUT we have $\lambda = \lambda' = \lambda''$ at M_G. There is also a dimension-three lepton-number-violating operator given by $H_u L_i$ with dimensionful coupling μ_i.

dimension-three and -four baryon- and lepton-violating operators can be eliminated by requiring R parity (Farrar and Fayet, 1978).[4]

In $SU(5)$, the Higgs doublets reside in a $\mathbf{5_H} \equiv \mathbf{H}$, $\mathbf{\bar{5}_H} \equiv \mathbf{\bar{H}}$ and R parity distinguishes the $\mathbf{\bar{5}}$ (quarks and leptons) from $\mathbf{\bar{5}_H}$ (Higgs). R parity (or its cousin, family reflection symmetry (see Dimopoulos and Georgi, 1981; Dimopoulos et al., 1982; Ellis et al., 1982) takes $F \to -F$, $H \to H$ with $F = \{\mathbf{10}, \mathbf{\bar{5}}\}$, $H = \{\mathbf{\bar{5}_H}, \mathbf{5_H}\}$. This forbids the dimension-four operator $(\mathbf{10}\,\mathbf{\bar{5}}\,\mathbf{\bar{5}})$, but allows the Yukawa couplings of the form $(\mathbf{10}\,\mathbf{\bar{5}}\,\mathbf{\bar{5}_H})$ and $(\mathbf{10}\,\mathbf{10}\,\mathbf{5_H})$. It also forbids the dimension-three, lepton-number-violating, operator $(\mathbf{\bar{5}}\,\mathbf{5_H}) \supset (L\,H_u)$ with a coefficient with dimensions of mass which, like the μ parameter, could be of order the weak scale and it forbids the dimension-five, baryon-number-violating, operator $(\mathbf{10}\,\mathbf{10}\,\mathbf{10}\,\mathbf{\bar{5}_H})$ $\supset (Q\,Q\,Q\,H_d) + \cdots$.

Note, in the MSSM, it is possible to retain R-parity-violating operators at low energy as long as they violate either baryon number or lepton number only, but not both. Such schemes are natural if one assumes a low-energy symmetry, such as lepton number, baryon number or a baryon parity (Ibanez and Ross, 1992a; Dreiner et al., 2006). However, these symmetries cannot be embedded in a GUT. Thus, in a SUSY GUT, only R parity can prevent unwanted dimension-four operators. Hence, by naturalness arguments, R parity must be a symmetry in the effective low-energy theory of any SUSY GUT.[5] In the theory with R parity, the lightest supersymmetric particle is stable. It is therefore a natural candidate for dark matter. In addition, SUSY particles can only be produced in pairs in the high-energy collisions of Standard Model particles.

Dimension-five baryon-number-violating operators may be forbidden at tree level by symmetries in $SU(5)$, etc. These symmetries are typically broken however by the vacuum expectation values responsible for the color-triplet Higgs masses. Consequently these dimension-five operators are generically generated via color-triplet higgsino exchange. Hence, the color-triplet partners of Higgs doublets must necessarily obtain mass of order the GUT scale. The dominant decay modes from dimension-five operators are $p \to K^+\bar{\nu}$ $(n \to K^0\bar{\nu})$. This is due to a simple symmetry argument; the operators $(Q_i\,Q_j\,Q_k\,L_l)$, $(\bar{U}_i\,\bar{U}_j\,\bar{D}_k\,\bar{E}_l)$ (where i, j, k, $l = 1, 2, 3$ are family indices and color and weak indices are implicit) must be invariant under $SU(3)_C$ and $SU(2)_L$. As a result, their color- and weak-doublet indices must be anti-symmetrized. However, since these operators are given by bosonic superfields, they must be totally symmetric under interchange of all indices. Thus, the first operator vanishes for $i = j = k$ and the second vanishes for $i = j$. Hence, a second- or third-generation particle must appear in the final state (Dimopoulos et al., 1982; Ellis et al., 1982).

---

[4] The particle charges under R parity is $(-1)^{3B+L+2s}$ such that all Standard Model particles are even and all new SUSY particles are odd. Thus, SUSY particles must necessarily be produced in pairs and the lightest SUSY particle is stable.

[5] SUSY GUTs in higher dimensions have other possibilities for eliminating both dimension-four and -five baryon- and lepton-number-violating operators; for a review, see Raby (2017).

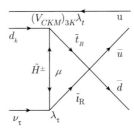

Fig. 49.4 The effective four-Fermi operator for proton decay obtained by integrating out sparticles at the weak scale (Lucas and Raby, 1997).

The dimension-five operator contribution to proton decay requires a sparticle loop at the SUSY scale to reproduce an effective dimension-six four-Fermi operator for proton decay (see Fig. 49.4). The loop factor (LF) is of the form[6]

$$(LF) \propto \frac{\lambda_t \, \lambda_\tau}{16\pi^2} \frac{\sqrt{\mu^2 + M_{1/2}^2}}{m_0^2}, \tag{49.7}$$

leading to a decay amplitude

$$A(p \to K^+ \bar{\nu}) \propto \frac{cc}{M_T^{eff}} \, (\text{LF}). \tag{49.8}$$

In any predictive SUSY GUT, the coefficients c are 3×3 matrices related to (but not identical to) Yukawa matrices. Thus, these tend to suppress the proton-decay amplitude. However, this is typically not sufficient to be consistent with the experimental bounds on the proton lifetime. Thus, it is also necessary to minimize the loop factor. This can be accomplished by taking $\mu, M_{1/2}$ *small* and m_0 *large*. Finally, the effective Higgs color-triplet mass M_T^{eff} must be MAXIMIZED. With these caveats, it is possible to obtain rough theoretical bounds on the proton lifetime given by Lucas and Raby (1996), Blazek et al. (1997), Altarelli et al. (2000), Babu et al. (2000) and Dermisek et al. (2001),

$$\tau_{p \to K^+ \bar{\nu}} \le \left(\frac{1}{3} - 3 \right) \times 10^{34} \text{ yr.} \tag{49.9}$$

Note, this is close to the Super-Kamiokande bound on this decay mode, 5.9×10^{33} years (Abe et al., 2014). In addition, the DUNE experiment is designed to be sensitive to this decay mode up to 10^{35} years after 10 years of running.

---

[6] $M_{1/2}$, m_0 are the universal gaugino mass and scalar masses, respectively, defined at the GUT scale.

PART XII

MINIMAL SUPERSYMMETRIC STANDARD MODEL

50

In this chapter we discuss the Minimal Supersymmetric Standard Model (MSSM). We describe the spectrum and interactions of the theory, both in terms of superfields and in terms of their component fields which may be discovered at colliders. Each Standard Model particle has a supersymmetric partner. Including R parity, all the coupling of the new supersymmetric particles have already been measured. The only unknown (and it is an enormous unknown) is the mass of all the superpartners. These are determined by how supersymmetry (SUSY) is spontaneously broken. Since there are many theoretical possibilities for the SUSY-breaking mechanism, it is necessary for experiment to resolve this ambiguity. Without this guidance we parameterize SUSY breaking in terms of soft SUSY breaking masses. However, since there are so many possibilities we limit the number of arbitrary parameters, arbitrarily. Once this is done we can discuss the Higgs potential and electroweak symmetry breaking. We then evaluate the sparticle spectrum in terms of these arbitrary soft-breaking parameters.

- The left-chiral superfields which make up the MSSM are as follows.

$$L = \begin{pmatrix} V \\ E \end{pmatrix} \bar{E} \bar{V}; \quad Q = \begin{pmatrix} U \\ D \end{pmatrix} \bar{U} \bar{D}, \tag{50.1}$$

with weak hypercharge $\{-1, \; +2, \; 0; \; \frac{1}{3}, \; -\frac{4}{3}, \; \frac{2}{3}\}$, respectively.

- There are two Higgs doublets which we label as

$$H_u = \begin{pmatrix} H^+ \\ H^0 \end{pmatrix}, \quad H_d = \begin{pmatrix} \bar{H}^0 \\ \bar{H}^- \end{pmatrix}, \tag{50.2}$$

with hypercharge $\{+1, \; -1\}$, respectively.

- The gauge interactions are fixed by the quantum numbers of the fields and the covariant derivative

$$D_\mu = \partial_\mu + i g_s T_A \mathcal{G}_{\mu A} + g T_a W_{\mu a} + g' \frac{Y}{2} B_\mu. \tag{50.3}$$

- The superpotential is given by

$$W = \lambda_e^{ij} \bar{E}_i L_j^a H_d^b \epsilon_{ba} + \lambda_d^{ij} \bar{D}_i Q_j^a H_d^b \epsilon_{ba} + \lambda_u^{ij} \bar{U}_i Q_j^a H_u^b \epsilon_{ab}$$
$$+ \lambda_\nu^{ij} \bar{V}_i L_j^a H_u^b \epsilon_{ab} - \frac{1}{2} M_{ij} \bar{V}_i \bar{V}_j + \mu H_u^a H_d^b \epsilon_{ab}. \tag{50.4}$$

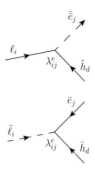

Fig. 50.1 An example of the new higgsino-type couplings derived from the superpotential.

Fig. 50.2 An example of the new gaugino-type couplings derived from the Kahler potential.

- Higgs vacuum expectation values (VEVs) are

$$\langle h^0 \rangle = \frac{v}{\sqrt{2}} \sin\beta, \quad \langle \bar{h}^0 \rangle = \frac{v}{\sqrt{2}} \cos\beta, \tag{50.5}$$

with the ratio, $\langle h^0 \rangle / \langle \bar{h}^0 \rangle \equiv \tan\beta$ arbitrary.

- We have assumed an R, parity where $H(y, \theta) \to H(y, -\theta)$, $F(y, \theta) \to -F(y, -\theta)$, such that all ordinary particles have R-charge $+1$, and all SUSY partners have R-charge -1. As a consequence, superpartners must be created in pairs at colliders and the lightest supersymmetric particle (LSP) is stable.

- The scalar potential is derived by using the equations of motion for the auxiliary fields.

The MSSM is the most general renormalizable Lagrangian consistent with SUSY, $SU(3) \otimes SU(2) \otimes U(1)_Y$, and R parity. For a review of SUSY and the MSSM, see Martin (1997).

An example of the new higgsino couplings is given in Fig. 50.1.[1] In Fig. 50.2 we represent an example of the new gaugino couplings. Finally, in Fig. 50.3 we give examples of the auxiliary couplings which produce the scalar potential.

---

[1] Our notation for the component fields: Standard Model particles are given with lower-case letters and the superpartners are given with a tilde. For example, the higgs boson in H_d is given by h_d and the higgsinos are given by $\tilde{h}_d$.

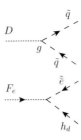

Fig. 50.3 An example of the F- and D-term couplings. When these fields are integrated out of the Lagrangian we obtain the scalar potential.

50.1 SUSY Non-Renormalization Theorems

- There are NO vertex corrections to the superpotential, W (not even finite corrections). Thus there is only wave-function renormalization.

- If SUSY is unbroken at tree level, then it is unbroken to all orders in perturbation theory. In Eqn. 48.27, we give the transformation of fermions under SUSY. Note, if the $\langle F \rangle \neq 0$, then SUSY is spontaneously broken and by taking the VEV of both sides we can prove the existence of a Nambu–Goldstone fermion, i.e. the goldstino. So $\langle F \rangle \neq 0$ is the order parameter for spontaneous SUSY breaking.[2] However,

$$F_i \equiv \frac{\partial W}{\partial \Phi_i}. \qquad (50.6)$$

Hence, if $F_i = 0$ at tree level, it will remain zero to all orders in perturbation theory. SUSY can be spontaneously broken by non-perturbative effects, i.e. the SUSY-breaking scale, Λ_{SUSY}, can be of order $e^{-\frac{8\pi^2}{g^2}} \mu$, where μ is some large scale, $\mu \sim M_{Pl}$. Therefore, it is natural to have a low-energy SUSY-breaking scale.

-

$$V(\Phi_i) = \sum_i |F_i|^2 + \frac{1}{2} \sum_A D_A^2. \qquad (50.7)$$

Thus, global SUSY implies that $V(\langle \Phi_i \rangle) = 0$.

- There are no quadratic divergences, since there is only wave-function renormalization.

---

[2] Note, $\langle D_A \rangle$ is also an order parameter for spontaneous SUSY breaking.

Global vs. Local SUSY

In global SUSY, the second derivative of the Kahler potential defines the Kahler metric, with

$$g^{I\bar{J}} \equiv \left(\frac{\partial^2 K}{\partial \Phi_I \partial \Phi_J^*} \right)^{-1}. \tag{50.8}$$

For a renormalizable theory, at tree level, we have $g^{I\bar{J}} = \delta^{I\bar{J}}$. However, for a non-renormalizable theory, the scalar potential is given by

$$V(\phi_I) = g^{I\bar{J}} \frac{\partial W}{\partial \phi_I} \frac{\partial W^*}{\partial \phi_J^*}. \tag{50.9}$$

In local SUSY (i.e. supergravity), the scalar potential becomes more complicated.

We have

$$V(\phi_I) = \left(g^{I\bar{J}}(D_I W)(D_J W)^* - \frac{3}{M_{Pl}^2} |W|^2 \right) e^{K/M_{Pl}^2}, \tag{50.10}$$

where the Kahler covariant derivative, D_I, is given by

$$D_I \equiv \frac{\partial}{\partial \phi_I} + \frac{K_I}{M_{Pl}^2}, \tag{50.11}$$

with $K_I \equiv \frac{\partial K}{\partial \phi_I}$. In local SUSY, if $D_I W \neq 0$, then SUSY is spontaneously broken. In this case the gravitino becomes massive by eating the goldstino. The gravitino mass is given by

$$m_{3/2} = e^{\frac{1}{2} K/M_{Pl}^2} \frac{W}{M_{Pl}^2}, \tag{50.12}$$

when we choose

$$V(\langle \phi_I \rangle) \equiv 0. \tag{50.13}$$

50.2 Soft SUSY Breaking

If SUSY is spontaneously broken at some scale, M, then in the effective low-energy theory, with $E \ll M$, there will only be soft-breaking terms in the effective Lagrangian (Dimopoulos and Georgi, 1981). Moreover, these SUSY soft-breaking terms preserve the property of NO quadratic divergences. A list of possible soft-breaking terms was given in Girardello and Grisaru (1982). For the MSSM, these are given as follows:

(1) scalar masses:

$$\tilde{\ell}_i^* \, (\tilde{m}_\ell^2)_{ij} \tilde{\ell}_j + \tilde{e}_i^* (\tilde{m}_{\bar{e}}^2)_{ij} \tilde{e}_j + \dots; \tag{50.14}$$

(2) A terms:

$$(A_e)_{ij} (\lambda_e)_{ij} \tilde{e}_i \tilde{\ell}_j h_d + \dots; \tag{50.15}$$

(3) gaugino masses:

$$- \left[M_3(\tilde{g}\tilde{g}) + M_2(\tilde{\omega}\tilde{\omega}) + M_1(\tilde{b}\tilde{b}) \right];$$ (50.16)

(4) Higgs mass:

$$- \mu B h_u h_d.$$ (50.17)

50.3 The MSSM Spectrum

In this brief section we introduce the notation for the MSSM particle spectrum. We first give the Standard Model particles, with the addition of a second Higgs doublet:

$$q_i = \begin{pmatrix} u \\ d \end{pmatrix}_i, \quad \bar{u}_i, \; \bar{d}_i, \; \ell_i = \begin{pmatrix} \nu \\ e \end{pmatrix}_i, \quad \bar{e}_i, \; \bar{\nu}_i, \quad i = 1, 2, 3$$ (50.18)

$$g_\mu^A, \; W_\mu^a, \; B_\mu,$$

$$h_u = \begin{pmatrix} h^+ \\ h^0 \end{pmatrix}, \qquad h_d = \begin{pmatrix} \bar{h}^0 \\ \bar{h}^- \end{pmatrix};$$

and their SUSY partners:

$$\tilde{q}_i = \begin{pmatrix} \tilde{u} \\ \tilde{d} \end{pmatrix}_i, \quad \tilde{\bar{u}}_i, \; \tilde{\bar{d}}_i, \; \tilde{\ell}_i = \begin{pmatrix} \tilde{\nu} \\ \tilde{e} \end{pmatrix}_i, \quad \tilde{\bar{e}}_i, \; \tilde{\bar{\nu}}_i, \quad i = 1, 2, 3$$ (50.19)

$$\tilde{g}^A, \; \tilde{\omega}^a, \; \tilde{b},$$

$$\tilde{h}_u = \begin{pmatrix} \tilde{h}^+ \\ \tilde{h}^0 \end{pmatrix}, \qquad \tilde{h}_d = \begin{pmatrix} \tilde{\bar{h}}^0 \\ \tilde{\bar{h}}^- \end{pmatrix}.$$

50.4 Electroweak Symmetry Breaking

Here is the Higgs potential relevant for electroweak breaking at tree level. We have

$$V = m_1^2 |h_u|^2 + m_2^2 |h_d|^2 + \mu B h_u^a h_d^b \epsilon_{ab} + h.c.$$ (50.20)

$$+ \frac{g^2}{2} \left(h_u^* \frac{\vec{\tau}}{2} h_u + h_d^* \frac{\vec{\tau}}{2} h_d \right)^2$$

$$+ \frac{g'^2}{2} \left(\frac{1}{2} |h_u|^2 - \frac{1}{2} |h_d|^2 \right)^2,$$

where $m_1^2 = m_{h_u}^2 + |\mu|^2$, $m_2^2 = m_{h_d}^2 + |\mu|^2$. Note, if $\langle h^0 \rangle = \langle \bar{h}^0 \rangle \neq 0$, the quartic terms in V vanish.

• Therefore, in order for the potential to be bounded from below we require

$$m_1^2 + m_2^2 > 2\mu B.$$ (50.21)

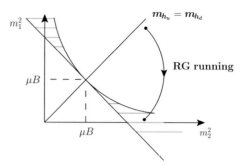

Fig. 50.4 The shaded area in the $m_1^2 - m_2^2$ plane is where V is (1) bounded from below and (2) electroweak symmetry is broken. The imagined RG flow goes from a point near where $m_{h_u} \sim m_{h_d}$ to a point in the allowed region of parameter space for electroweak symmetry breaking.

- Electroweak breaking requires

$$\det m_h^2 < 0. \tag{50.22}$$

Thus,

$$m_1^2 m_2^2 < \mu^2 B^2. \tag{50.23}$$

The shaded area in the $m_1^2 - m_2^2$ plane (see Fig. 50.4) is where V is (1) bounded from below and (2) electroweak symmetry is broken. The imagined renormalization group (RG) flow goes from a point near where $m_{h_u} \sim m_{h_d}$ to a point in the allowed region of parameter space for electroweak symmetry breaking. Assuming the RG flow takes us into the allowed region of V, we then obtain electroweak symmetry breaking with

$$\langle h_u \rangle = \begin{pmatrix} 0 \\ v_u \end{pmatrix}, \quad \langle h_d \rangle = \begin{pmatrix} v_d \\ 0 \end{pmatrix}, \tag{50.24}$$

with

$$v_u \equiv \frac{v}{\sqrt{2}} \sin \beta, \quad v_d \equiv \frac{v}{\sqrt{2}} \cos \beta. \tag{50.25}$$

We then have

$$M_W = \frac{gv}{2}, \quad M_Z^2 = \left(\frac{g'^2 + g^2}{4} \right) v^2 = \left| \frac{m_1^2 - m_2^2}{\cos 2\beta} \right| - m_1^2 - m_2^2 \tag{50.26}$$

and

$$\sin 2\beta = \frac{2\mu B}{m_1^2 + m_2^2}. \tag{50.27}$$

Once electroweak symmetry breaking occurs, the different gauge eigenstates mix to form mass eigenstates. In the next few pages we shall discuss this mixing. But now, we just state that the mass eigenstates are given by the following admixtures. We have

- Higgs:

$$h, \ H, \ A, \ H^{\pm};$$ (50.28)

- charginos:

$$\tilde{\omega}^{\pm}, \ \tilde{h}^{+}, \ \tilde{\bar{h}}^{-} \Rightarrow \tilde{\chi}^{\pm}_{1,2};$$ (50.29)

- neutralinos:

$$\tilde{\omega}_3, \ \tilde{b}, \ \tilde{h}^0, \ \tilde{\bar{h}}^0 \Rightarrow \tilde{\chi}^0_{1,2,3,4};$$ (50.30)

- gluino:

$$\tilde{g};$$ (50.31)

- squarks and sleptons: for example,

$$\tilde{u}, \ \tilde{\bar{u}} \Rightarrow \tilde{u}_{1,2};$$ (50.32)

- gravitino:

$$m_{3/2} \simeq \frac{\Lambda^2_{SUSY}}{M_{Pl}}.$$ (50.33)

In the MSSM with R parity, the only dimensionless couplings are those that already appear in the Standard Model. However, the new soft SUSY-breaking parameters, which determine the masses of all the SUSY partners, are completely undetermined. This makes it difficult to have a predictive theory. It also makes it difficult to say, with any precision, where one should search for the SUSY particles. In addition, there are many more sources of flavor and **CP** violation in supersymmetric theories which are severely constrained by experiment.

In order to minimize the number of arbitrary parameters, one defines the constrained MSSM [CMSSM] (Kane et al., 1994). The CMSSM assumes at the grand unification scale, M_G, the following boundary conditions for the soft SUSY-breaking parameters:

- gaugino masses:

$$M_i(M_G) = M_{1/2};$$ (50.34)

- universal scalar mass:

$$m_0;$$ (50.35)

- universal A parameter:

$$A_0;$$ (50.36)

- $\mu B \equiv m_3^2(M_G)$, and the SUSY parameter μ can be traded for the low-energy parameters,

$$M_Z, \ \tan\beta, \ \text{sign}(\mu).$$ (50.37)

Thus, there are only five parameters which define the low-energy theory. They are

$$M_{1/2}, \ m_0, \ A_0, \ \tan\beta, \ \text{sign}(\mu). \tag{50.38}$$

At energies, μ, below the Grand Unified Theory (GUT) scale, the parameters renormalize and we have

- gauginos:

$$M_3(\mu), \ M_2(\mu), \ M_1(\mu); \tag{50.39}$$

- scalar masses (3 × 3 Hermitian matrices for squarks and sleptons):

$$\tilde{m}_q^2(\mu), \ \tilde{m}_u^2(\mu), \ \tilde{m}_d^2(\mu), \ \tilde{m}_\ell^2(\mu), \ \tilde{m}_e^2(\mu), \ m_{h_u}^2(\mu), \ m_{h_d}^2(\mu); \tag{50.40}$$

- A parameters (3 × 3 complex parameters):

$$(A_u)(\mu)_{ij}, \ (A_d)(\mu)_{ij}, \ (A_e)(\mu)_{ij}. \tag{50.41}$$

50.5 Running the Gaugino and Higgs Masses

The RG equations for the gaugino masses at one loop is quite simple. Given

$$t = -\frac{1}{2\pi} \ln\left(\frac{M_G}{\mu}\right), \tag{50.42}$$

we have

$$\frac{d}{dt}\left(\frac{M_i(t)}{\alpha_i(t)}\right) = 0. \tag{50.43}$$

This is a result of the Feynman diagram of Fig. 50.5. Thus at one loop we have

$$M_3(M_Z) : M_2(M_Z) : M_1(M_Z) = \alpha_3(M_Z) : \alpha_2(M_Z) : \alpha_1(M_Z) \simeq 6 : 2 : 1. \tag{50.44}$$

Consider now the Higgs mass squared running. It obtains a one-loop gauge contribution to the running (Fig. 50.6). It also receives one-loop corrections proportional

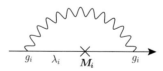

Fig. 50.5   The one-loop graph contributing to gaugino mass running.

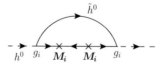

Fig. 50.6   The one-loop graph contributing to the Higgs mass squared running proportional to g_i^2.

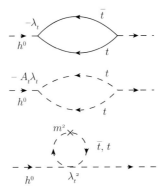

Fig. 50.7 The one-loop graph contributing to the Higgs mass squared running proportional to λ_t^2.

to the largest Yukawa coupling, λ_t, given in Fig. 50.7, due to the terms in the Lagrangian,

$$\mathcal{L} \supset \int d^2\theta \lambda_t \bar{T} Q_3 H_u - A_t \lambda_t \tilde{\bar{t}} \tilde{q}_3 h_u \qquad (50.45)$$

where $Q_3 = \begin{pmatrix} T \\ B \end{pmatrix}$. These one loop Feynman diagrams lead to the following RG equations.

$$8\pi^2 \frac{dm_{h_u}^2}{dt} = -4g_i^2 M_i^2 C_2^i + 3\lambda_t^2 (m_{h_u}^2 + \tilde{m}_t^2 + \tilde{m}_{\bar{t}}^2 + A_t^2), \qquad (50.46)$$

$$8\pi^2 \frac{d\tilde{m}_t^2}{dt} = -4g_i^2 M_i^2 C_2^i + \lambda_t^2 (m_{h_u}^2 + \tilde{m}_t^2 + \tilde{m}_{\bar{t}}^2 + A_t^2),$$

$$8\pi^2 \frac{d\tilde{m}_{\bar{t}}^2}{dt} = -4g_i^2 M_i^2 C_2^i + 2\lambda_t^2 (m_{h_u}^2 + \tilde{m}_t^2 + \tilde{m}_{\bar{t}}^2 + A_t^2),$$

where the factors of $(3, 1, 2)$ multiplying λ_t^2 are due to (three colors, a singlet, or a doublet), respectively, in the loop. The quadratic Casimir coefficients are

$$C_2^3(\tilde{t}) = C_2^3(\tilde{\bar{t}}) = 4/3, \quad C_2^3(h_u) \equiv 0, \qquad (50.47)$$

$$C_2^2(h_u) = C_2^2(\tilde{t}) = 3/4, \quad C_2^2(\tilde{\bar{t}}) \equiv 0,$$

$$C_2^1(h_u) = 1/4, \quad C_2^1(\tilde{t}) = 1/36, \quad C_2^1(\tilde{\bar{t}}) = 4/9, \qquad (50.48)$$

where $C_2^1 \equiv Y^2/4$ and $g_1^2 = \frac{5}{3}g'^2$. We see from Eqns. 50.42 and 50.46 that as μ decreases, t becomes more negative until we reach $\mu = M_Z$. Hence, gaugino terms make the scalars heavier, while Yukawa terms decrease the scalar masses at lower energies. Moreover, the Yukawa term for h_u has the largest coefficient. Thus, as depicted in Fig. 50.4, even though the Higgs mass squared at the GUT scale may be positive, it is driven negative at low energies near M_Z. Note, the stop mass squared remains positive, so only the electroweak symmetry is broken. This is known as radiative electroweak symmetry breaking (REWSB) (Ibanez and Ross,

1982, 1992b). Moreover, it was shown in Ibanez and Lopez (1983) that in order for REWSB to work, the top-quark mass must be greater than 50 GeV.

50.6 Mass Eigenstates after Electroweak Soft Breaking

Higgs Masses

At the low-energy point, the H_u mass squared is negative. In the allowed region of parameter space both h_u^0 and h_d^0 obtain VEVs. We define the shifted Higgs fields by

$$h_u^0 = \frac{1}{\sqrt{2}} \left(v \sin \beta + H \sin \alpha + h \cos \alpha + iA \cos \beta + i\mathcal{G}^0 \sin \beta \right), \qquad (50.49)$$

$$h_d^0 = \frac{1}{\sqrt{2}} \left(v \cos \beta + H \cos \alpha - h \sin \alpha + iA \sin \beta - i\mathcal{G}^0 \cos \beta \right)$$

for neutral scalars and

$$\begin{pmatrix} \bar{h}^{-*} \\ h^+ \end{pmatrix} = \begin{pmatrix} \sin \beta & -\cos \beta \\ \cos \beta & \sin \beta \end{pmatrix} \begin{pmatrix} H^+ \\ \mathcal{G}^+ \end{pmatrix} \qquad (50.50)$$

for charged scalars. Note the Nambu–Goldstone fields $\mathcal{G}^0$, $\mathcal{G}^\pm$ give mass to the Z^0, $W^\pm$, respectively. We have $v \sim 246$ GeV and $M_Z^2 = \frac{g^2 + g'^2}{4} v^2$. By minimizing the tree-level Higgs potential we find

$$-\frac{1}{8}(g^2 + g'^2)v^2(\sin^2 \beta - \cos^2 \beta) + m_{h_d}^2 + |\mu|^2 = m_3^2 \tan \beta, \qquad (50.51)$$

$$\frac{1}{8}(g^2 + g'^2)v^2(\sin^2 \beta - \cos^2 \beta) + m_{h_u}^2 + |\mu|^2 = m_3^2 \frac{1}{\tan \beta}.$$

The **CP** odd Higgs boson, A, has mass given by

$$m_A^2 = m_{h_u}^2 + m_{h_d}^2 + 2|\mu|^2 \equiv m_3^2 \left(\tan \beta + \frac{1}{\tan \beta} \right) > 2|m_3|^2, \qquad (50.52)$$

where the inequality is required by stability of the vacuum. The natural range for $\tan \beta$ is as follows:

- As $m_3 \to 0$, $\tan \beta \to \infty$, since λ_t drives $m_{h_u}^2 < 0$ and therefore $\langle h_u \rangle \neq 0$, but $\langle h_d \rangle = 0$.
- As m_3 increases, $\tan \beta$ decreases, since $m_3^2 = \mu B$ mixes the states h_u, h_d until $\tan \beta \to 1$.

There are two **CP**-even Higgs fields, h, H, defined at the tree level in terms of the mass term[3]

$$((h_d^0)_R^\dagger (h_u^0)_R^\dagger) \mathcal{M}^2 \begin{pmatrix} (h_d^0)_R \\ (h_u^0)_R \end{pmatrix} \qquad (50.53)$$

---

[3] The subscript R denotes the real part of the complex field.

with

$$\mathcal{M}^2 = \begin{pmatrix} \mathcal{M}_{11}^2 & \mathcal{M}_{12}^2 \\ \mathcal{M}_{12}^2 & \mathcal{M}_{22}^2 \end{pmatrix}. \tag{50.54}$$

The terms in the mass matrix are

$$\mathcal{M}_{11}^2 = m_A^2 \sin^2 \beta + M_Z^2 \cos^2 \beta, \tag{50.55}$$
$$\mathcal{M}_{22}^2 = m_A^2 \cos^2 \beta + M_Z^2 \sin^2 \beta,$$
$$\mathcal{M}_{12}^2 = -(m_A^2 + M_Z^2) \sin \beta \cos \beta.$$

The eigenvalues are then

$$\tag{50.56}$$
$$m_H^2 = \frac{1}{2}(m_A^2 + M_Z^2) + \frac{1}{2}\sqrt{(m_A^2 + M_Z^2)^2 - 4m_A^2 M_Z^2 (\cos^2 \beta - \sin^2 \beta)^2},$$
$$m_h^2 = \frac{1}{2}(m_A^2 + M_Z^2) - \frac{1}{2}\sqrt{(m_A^2 + M_Z^2)^2 - 4m_A^2 M_Z^2 (\cos^2 \beta - \sin^2 \beta)^2}.$$

or

$$m_h^2 = \frac{1}{2}(m_A^2 + M_Z^2) - \frac{1}{2}\sqrt{(m_A^2 - M_Z^2)^2 + (4m_A M_Z \cos \beta \sin \beta)^2} \leq M_Z^2$$
$$= \lim_{m_A^2 \gg M_Z^2} M_Z^2 \cos^2 2\beta. \tag{50.57}$$

Note, $m_h^2 = M_Z^2$ for $\cos \beta = 0$. The eigenstates are given by

$$H = \cos \alpha (h_d^0)_R + \sin \alpha (h_u^0)_R, \tag{50.58}$$
$$h = -\sin \alpha (h_d^0)_R + \cos \alpha (h_u^0)_R.$$

Finally, the charged Higgs field $H^\pm$ obtains mass

$$m_{H^\pm}^2 = m_A^2 + M_W^2. \tag{50.59}$$

We know since LEPII that the Higgs boson was heavier than 114.5 GeV. Thus, the tree-level mass is too small. Moreover, now we know that the Higgs boson has mass $m_h \sim 125$ GeV. Thus the Higgs boson must receive significant radiative corrections. At one loop we find

$$\Delta m_h^2 = \frac{3\lambda_t^2 m_t^2}{8\pi^2} \left(\ln \left(\frac{M_{SUSY}^2}{m_t^2} \right) + \frac{X_t^2}{M_{SUSY}^2} \left(1 - \frac{X_t^2}{12 M_{SUSY}^2} \right) \right), \tag{50.60}$$

where $X_t = A_t - \mu/\tan \beta$ and $M_{SUSY}^2 = m_{\tilde{t}_1} m_{\tilde{t}_2}$. For a SUSY-breaking scale, $M_{SUSY} \leq 1$ TeV, we have $m_h \leq 135$ GeV.

Squark and Slepton Masses

Now consider squark and slepton masses. The scalar mass terms are given by the 6×6 mass matrix,

$$-\mathcal{L} \supset (\tilde{u}^\dagger \ \tilde{u}^T) \begin{pmatrix} \tilde{m}_Q^2 + v_u^2 \lambda_u^\dagger \lambda_u + D_u & \lambda_u^\dagger (v_u A_u^\dagger - \mu v_d) \\ (v_u A_u - \mu v_d) \lambda_u & \tilde{m}_{\bar{u}}^2 + v_u^2 \lambda_u \lambda_u^\dagger + D_{\bar{u}} \end{pmatrix} \begin{pmatrix} \tilde{u} \\ \tilde{u}^* \end{pmatrix}, \tag{50.61}$$

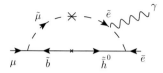

Fig. 50.8 A one-loop graph contributing to the process $\mu \to e\gamma$.

where $\tilde{u}$ and $\tilde{\bar{u}}^*$ are three-vectors (for three generations of up squarks) and

$$D_u = \frac{1}{4}(v_u^2 - v_d^2)\left(-g^2 + \frac{1}{3}g'^2\right)\mathbb{I}, \tag{50.62}$$

$$D_{\bar{u}} = \frac{1}{4}(v_u^2 - v_d^2)\left(-\frac{4}{3}g'^2\right)\mathbb{I}.$$

Recall the up-quark mass matrix is given by

$$m_u = \lambda_u v_u = \lambda_u \frac{v}{\sqrt{2}}\sin\beta, \tag{50.63}$$

which comes from the superpotential term

$$W \supset \bar{U}\lambda_u Q^a H_u^b \epsilon_{ab} + \mu H_u^a H_d^b \epsilon_{ab}. \tag{50.64}$$

The Higgs up auxiliary field equation of motion is then

$$- F^*_{h_u^0} = \tilde{\bar{u}}\lambda_u \tilde{u} - \mu \bar{h}^0, \tag{50.65}$$

which gives the scalar mass term

$$- \mathcal{L}_{SUSY} \supset |F_{h_u^0}|^2 \supset -\mu\tilde{\bar{u}}\lambda_u \tilde{u}\bar{h}^{0*} + h.c. \tag{50.66}$$

In addition, we have the soft SUSY breaking A term, which contributes

$$- \mathcal{L}_{SB} \supset \tilde{\bar{u}}A_u\lambda_u \tilde{u}h_u^0. \tag{50.67}$$

Putting them together we obtain the off-diagonal terms in the squark mass matrix given by

$$- \mathcal{L} \supset \tilde{\bar{u}}(v_u A_u - \mu v_d)\lambda_u \tilde{u}. \tag{50.68}$$

The SUSY Flavor Problem

SUSY scalars, in general, add more flavor violation into the low-energy theory, since the squark and slepton mass matrices are not necessarily diagonalized in the same basis as their fermionic partners. Consider the so-called SUSY flavor basis, where squark and slepton mass matrices are diagonalized in the same basis as the mass eigenstate basis for quarks and leptons. In this basis, consider the Feynman diagram for the process $\mu \to e\gamma$ in Fig. 50.8. The X on the scalar line is a $\Delta\tilde{m}^2_{\mu e}$ flavor-violating mass insertion. The amplitude for this process is given by

$$\Gamma(\mu \to e\gamma) \sim \frac{e^2}{16\pi}m_\mu^5\left(\frac{\alpha_Y}{8\pi}M_1\left(A_\mu/\lambda_\mu + \mu\tan\beta\right)\right)\frac{\Delta\tilde{m}^2_{\mu e}}{\tilde{m}^6}, \tag{50.69}$$

where $\alpha_Y = \frac{g'^2}{4\pi}$.

We now show that the CMSSM is minimal flavor violating (MFV). Let's transform the squark and slepton mass matrices to the SUSY flavor basis. Consider, for example, up quarks. The up-quark mass matrix is diagonalized by two unitary transformations, such that

$$m_u^{diag} = U_{\bar{u}}^\dagger \lambda_u U_u v_u, \tag{50.70}$$

where

$$U_u^\dagger \lambda_u^\dagger \lambda_u U_u = U_{\bar{u}}^\dagger \lambda_u \lambda_u^\dagger U_{\bar{u}} \tag{50.71}$$

are diagonal matrices. We have redefined the quark fields, such that the gauge eigenstates, u^0, are given in terms of the mass eigenstates, u, via

$$u_0 \equiv U_u u, \quad \bar{u}_0^T = \bar{u}^T U_{\bar{u}}^\dagger. \tag{50.72}$$

In the SUSY flavor basis we redefine the squarks and sleptons by the same unitary transformations. Thus we have

$$-\mathcal{L} \supset (\tilde{u}_0^\dagger \tilde{\bar{u}}_0^T) \tilde{m}_u^2 \begin{pmatrix} \tilde{u}_0 \\ \tilde{\bar{u}}_0^* \end{pmatrix} \tag{50.73}$$

$$= (\tilde{u}^\dagger U_u^\dagger \tilde{\bar{u}}^T U_{\bar{u}}^\dagger) \tilde{m}_u^2 \begin{pmatrix} U_u \tilde{u} \\ U_{\bar{u}} \tilde{\bar{u}}^* \end{pmatrix}$$

$$= (\tilde{u}^\dagger \tilde{\bar{u}}^T) \begin{pmatrix} U_u^\dagger [\ ] U_u & U_u^\dagger [\ ] U_{\bar{u}} \\ U_{\bar{u}}^\dagger [\ ] U_u & U_{\bar{u}}^\dagger [\ ] U_{\bar{u}} \end{pmatrix} \begin{pmatrix} \tilde{u} \\ \tilde{\bar{u}}^* \end{pmatrix}.$$

Note, if the boundary conditions at M_{GUT} are those of the CMSSM, then

$$A_u \equiv A_0 \mathbb{I}, \quad \tilde{m}_q^2 = \tilde{m}_{\bar{u}}^2 = m_0^2 \mathbb{I}. \tag{50.74}$$

As a consequence the squark and slepton mass matrices (see Eqn. 50.61) are diagonal in the SUSY flavor basis, defined at M_{GUT}. At low energies, however, there will still be flavor-violating terms in the squark mass matrices given by radiative corrections proportional to V_{CKM}.

Scalar Mass Eigenstates

Assume, for example, that the scalar masses are block diagonal within a generation and assume universal scalar masses at the GUT scale, then a large top Yukawa coupling drives the stop mass to be smaller than the other squark masses. The stop mass term in the Lagrangian is given by

$$-\mathcal{L} \supset (\tilde{t}^\dagger \ \tilde{\bar{t}}^T) \begin{pmatrix} \tilde{m}_{Q_3}^2 + m_t^2 + D_t & -m_t A_t + \mu m_t / \tan \beta \\ -m_t A_t + \mu m_t / \tan \beta & \tilde{m}_{\bar{t}}^2 + m_t^2 + D_{\bar{t}} \end{pmatrix} \begin{pmatrix} \tilde{t} \\ \tilde{\bar{t}}^* \end{pmatrix}. \tag{50.75}$$

The mass eigenstates are then $\tilde{t}_1$, $\tilde{t}_2$. If $m_t A_t > \tilde{m}_{Q_3}^2$, then we have one heavy and one light stop with $\tilde{m}_{t_2} > \tilde{m}_{t_1}$. The mass matrices for the down and charged lepton scalars are similar.

Quark and Lepton Masses

These renormalize similarly to non-SUSY theories. However, in SUSY $SU(5)$ the unification of the bottom and tau Yukawa couplings works well for specific values of $\tan \beta$, i.e. for $\tan \beta \sim 1$ or $\tan \beta \sim 50$.

Chargino and Neutralino Masses

The mass terms for the charginos are given by

$$\mathcal{L} \supset -\sqrt{2}ig\left(h_u^\dagger \frac{T_a}{2}\tilde{\omega}_a \tilde{h}_u + h_d^\dagger \frac{T_a}{2}\tilde{\omega}_a \tilde{h}_d\right) \tag{50.76}$$
$$- \sqrt{2}ig'\left(h_u^\dagger\left(+\frac{1}{2}\right)\tilde{b}\tilde{h}_u + h_d^\dagger\left(-\frac{1}{2}\right)\tilde{b}\tilde{h}_d\right)$$
$$+ \frac{1}{2}M_2(\tilde{\omega}_a\tilde{\omega}_a) + \frac{1}{2}M_1(\tilde{b}\tilde{b}) + h.c.$$
$$- \mu(\tilde{h}^+\tilde{\tilde{h}}^- - \tilde{h}^0\tilde{\tilde{h}}^0) + h.c.$$

Define the charged winos by

$$\tilde{\omega}^\pm \equiv \frac{\tilde{\omega}_1 \mp i\tilde{\omega}_2}{\sqrt{2}} \tag{50.77}$$

such that

$$\frac{1}{2}(\tau_1\tilde{\omega}_1 + \tau_2\tilde{\omega}_2) = \frac{1}{\sqrt{2}}(\tau^+\tilde{\omega}^+ + \tau^-\tilde{\omega}^-) \tag{50.78}$$

and also

$$\tilde{\omega}^+\tilde{\omega}^- \equiv \frac{1}{2}(\tilde{\omega}_1\tilde{\omega}_1 + \tilde{\omega}_2\tilde{\omega}_2). \tag{50.79}$$

In terms of these fields, the mass terms for charginos are given by

$$\mathcal{L} \supset -ig\Big[h_u^\dagger(\tau^+\tilde{\omega}^+ + \tau^-\tilde{\omega}^-)\tilde{h}_u \tag{50.80}$$
$$+ h_d^\dagger(\tau^+\tilde{\omega}^+ + \tau^-\tilde{\omega}^-)\tilde{h}_d\Big]$$
$$+ M_2(\tilde{\omega}^+\tilde{\omega}^-) - \mu(\tilde{h}^+\tilde{\tilde{h}}^-) + h.c.$$
$$= - gh^{+\dagger}(i\tilde{\omega}^+)\tilde{h}^0 - gh^{0\dagger}(i\tilde{\omega}^-)\tilde{h}^+$$
$$- g\bar{h}^{0\dagger}(i\tilde{\omega}^+)\tilde{\tilde{h}}^- - g\bar{h}^{-\dagger}(i\tilde{\omega}^-)\tilde{\tilde{h}}^0$$
$$- M_2(i\tilde{\omega}^+)(i\tilde{\omega}^-) - \mu\tilde{h}^+\tilde{\tilde{h}}^-.$$

Now when the neutral scalar Higgs bosons get VEVs, we obtain the mass matrix

$$\mathcal{L} \supset -(i\tilde{\omega}^- \quad \tilde{\tilde{h}}^-)\begin{pmatrix} M_2 & \frac{gv}{\sqrt{2}}\sin\beta \\ \frac{gv}{\sqrt{2}}\cos\beta & \mu \end{pmatrix}\begin{pmatrix} i\tilde{\omega}^+ \\ \tilde{h}^+ \end{pmatrix}. \tag{50.81}$$

We can diagonalize this mass matrix with two unitary transformations defined as follows.

$$\begin{pmatrix} i\tilde{\omega}^+ \\ \tilde{h}^+ \end{pmatrix}_a = V_{ab}^\dagger P_L \tilde{\chi}_b^+ \tag{50.82}$$

and

$$\begin{pmatrix} i\tilde{\omega}^- \\ \tilde{\tilde{h}}^- \end{pmatrix}_a = U_{ab}^\dagger P_L \tilde{\tilde{\chi}}_b^\dagger \tag{50.83}$$

such that the Dirac chargino spinor is given by

$$\psi_{\chi^+} = \begin{pmatrix} \tilde{\chi}^+ \\ i\sigma_2 \tilde{\chi}^{-*} \end{pmatrix}. \tag{50.84}$$

The mass term for the neutralinos is given by the terms in the Lagrangian

$$\mathcal{L} \supset -i\frac{g}{\sqrt{2}}(h_u^\dagger \tau_3 \tilde{\omega}_3 \tilde{h}_u + h_d^\dagger \tau_3 \tilde{\omega}_3 \tilde{h}_d) \tag{50.85}$$

$$- i\frac{g'}{\sqrt{2}}(h_u^\dagger \tilde{b}\tilde{h}_u - h_d^\dagger \tilde{b}\tilde{h}_d)$$

$$+ \frac{1}{2}M_2(\tilde{\omega}_3\tilde{\omega}_3) + \frac{1}{2}M_1(\tilde{b}\tilde{b}) + h.c.$$

$$+ \mu(\tilde{h}^0 \tilde{\tilde{h}}^0) + h.c.$$

In matrix form we have

$$\mathcal{L} \supset -\frac{1}{2}(i\tilde{b}\, i\tilde{\omega}_3\, \tilde{\tilde{h}}^0\, \tilde{h}^0) \begin{pmatrix} M_1 & 0 & -\frac{g'v}{2}\cos\beta & \frac{g'}{2}\sin\beta \\ 0 & M_2 & \frac{gv}{2}\cos\beta & -\frac{gv}{2}\sin\beta \\ -\frac{g'v}{2}\cos\beta & \frac{gv}{2}\cos\beta & 0 & -\mu \\ \frac{g'v}{2}\sin\beta & -\frac{gv}{2}\sin\beta & -\mu & 0 \end{pmatrix} \begin{pmatrix} i\tilde{b} \\ i\tilde{\omega}_3 \\ \tilde{\tilde{h}}^0 \\ \tilde{h}^0 \end{pmatrix}. \tag{50.86}$$

The mass eigenstates are then given by

$$\begin{pmatrix} i\tilde{b} \\ i\tilde{\omega}_3 \\ \tilde{\tilde{h}}^0 \\ \tilde{h}^0 \end{pmatrix}_i = N_{ij}^\dagger P_L \tilde{\chi}_j^0. \tag{50.87}$$

51

Spontaneous SUSY Breaking

In the Minimal Supersymmetric Standard Model (MSSM) we discussed a small number of soft supersymmetry (SUSY)-breaking parameters. In this chapter we focus on three simple mechanisms/models which exhibit spontaneous SUSY breaking. These models are applicable to four-dimensional supersymmetric theories. In higher dimensions, where the extra spatial dimensions are compactified into small balls, there are additional SUSY-breaking mechanisms which may be relevant for experiment. However, these are beyond the scope of this book.

51.1 O'Raifeartaigh Mechanism

In this section we consider the O'Raifeartaigh mechanism of supersymmetry breaking (O'Raifeartaigh, 1975). Consider the simple model with three superfields, A, B, C, and superpotential given by

$$W = A\left(B^2 + M^2\right) + \mu CB. \tag{51.1}$$

The Kahler potential is given by

$$K = A^*A + B^*B + C^*C. \tag{51.2}$$

The F-equations for a supersymmetric vacuum are given by

$$
\begin{aligned}
-F_A^* &= b^2 + M^2 = 0, \\
-F_B^* &= 2ab + \mu c = 0, \\
-F_C^* &= \mu b = 0.
\end{aligned}
\tag{51.3}
$$

Clearly, there is no simultaneous solution to all of these equations. Thus, SUSY must be spontaneously broken. Since the scalar potential is given by

$$V(A, B, C) = |F_A|^2 + |F_B|^2 + |F_C|^2, \tag{51.4}$$

we still want to minimize each term. We can easily obtain $F_B = 0$ along the flat direction,

$$c_0 = -\frac{2b_0}{\mu}a_0, \tag{51.5}$$

while b_0 is determined by minimizing the potential.

We have

$$V = |b^2 + M^2|^2 + \mu^2 |b|^2. \tag{51.6}$$

Then

$$\frac{\partial V}{\partial b} = 2b_0(b_0^{*2} + M^2) + \mu^2 b_0^* = 0 \tag{51.7}$$

$$\frac{\partial V}{\partial b^*} = 2b_0^*(b_0^2 + M^2) + \mu^2 b_0 = 0.$$

We then have

$$b^* \frac{\partial V}{\partial b^*} - b \frac{\partial V}{\partial b} = -2M^2(b_0^2 - b_0^{*2}) = 0. \tag{51.8}$$

Let $b_0 = re^{i\theta}$ and we find $\sin 2\theta = 0$ or $2\theta = n\pi$ with $n \in \mathbb{Z}$. Let $n = \pm 1$ and we find

$$r^3 = (M^2 - \frac{\mu^2}{2})r \tag{51.9}$$

or

$$b_0 = 0, \ \pm i\sqrt{M^2 - \frac{\mu^2}{2}}. \tag{51.10}$$

The first possibility is a maximum of the scalar potential, while the latter is self-consistent when $M \geq \mu/\sqrt{2}$. At the minimum of the potential, we have

$$V(b_0) = \frac{\mu^4}{4} + \mu^2(M^2 - \mu^2/2) = \mu^2(M^2 - \mu^2/4) > 0. \tag{51.11}$$

Let's now consider two cases, (1) $a_0 = c_0 = 0$ or (2) $a_0, \ c_0 \gg \mu, \ M.$[1]

(1) $a_0 = c_0 = 0$

Define the new superfields,

$$B = B' + b_0, \tag{51.12}$$

$$\mathcal{G} = \left(\langle \frac{\partial W}{\partial A} \rangle A + \langle \frac{\partial W}{\partial C} \rangle C \right) N^{-1/2},$$

$$L = \left(-\langle \frac{\partial W}{\partial C}^* \rangle A + \langle \frac{\partial W}{\partial A} \rangle^* C \right) N^{-1/2},$$

where we shall see that $\mathcal{G}$ is a Nambu–Goldstone superfield and the normalization constant, N, is given by

$$N = \left| \langle \frac{\partial W}{\partial A} \rangle \right|^2 + \left| \langle \frac{\partial W}{\partial C} \rangle \right|^2 \equiv V(a_0, \ b_0, \ c_0). \tag{51.13}$$

---

[1] Note, once SUSY is broken, the flat direction in the scalar potential will obtain radiative corrections. Hence it will no longer be flat. We shall not evaluate these corrections here, but just discuss two different possible scenarios.

The vacuum expectation value (VEV) of $\mathcal{G}$ is given by

$$\langle \mathcal{G} \rangle_0 = \left(\langle \frac{\partial W}{\partial A} \rangle F_A + \langle \frac{\partial W}{\partial C} \rangle F_C \right) N^{-1/2}\theta^2 = -N^{1/2}\theta^2; \tag{51.14}$$

whereas $\langle L \rangle_0 = \langle B' \rangle_0 = 0$. Thus, $\mathcal{G}$ contains the goldstino.

The Lagrangian is now given by

$$\mathcal{L} = \int d^4\theta \left(\mathcal{G}^*\mathcal{G} + L^*L + B'^*B' \right) + \int d^2\theta W + h.c., \tag{51.15}$$

with

$$W = A(B'^2 + 2b_0 B' + b_0^2 + M^2) + \mu C(B' + b_0). \tag{51.16}$$

The fields A, C are given by the orthogonal transformation

$$\begin{pmatrix} A \\ C \end{pmatrix} = N^{-1/2} \begin{pmatrix} \langle \frac{\partial W}{\partial A} \rangle^* & -\langle \frac{\partial W}{\partial C} \rangle \\ \langle \frac{\partial W}{\partial C} \rangle^* & \langle \frac{\partial W}{\partial A} \rangle \end{pmatrix} \begin{pmatrix} \mathcal{G} \\ L \end{pmatrix}. \tag{51.17}$$

Plugging this back into W we obtain

$$W = N^{-1/2} \left(\langle \frac{\partial W}{\partial A} \rangle^* \mathcal{G} - \langle \frac{\partial W}{\partial C} \rangle L \right) \underbrace{(B'^2 + 2b_0 B' + b_0^2 + M^2)}_{\langle \frac{\partial W}{\partial A} \rangle} \tag{51.18}$$

$$+ N^{-1/2} \left(\langle \frac{\partial W}{\partial C} \rangle^* \mathcal{G} + \langle \frac{\partial W}{\partial A} \rangle L \right) \underbrace{(\mu B' + \mu b_0)}_{\langle \frac{\partial W}{\partial C} \rangle}$$

$$= N^{-1/2}\mathcal{G} \left[N + \langle \frac{\partial W}{\partial A} \rangle^* B'^2 + B' \left(\langle \frac{\partial W}{\partial A} \rangle^* 2b_0 + \langle \frac{\partial W}{\partial C} \rangle^* \mu \right) \right]$$

$$+ N^{-1/2}L \left[-\langle \frac{\partial W}{\partial C} \rangle B'^2 + B' \left(-\langle \frac{\partial W}{\partial C} \rangle 2b_0 + \langle \frac{\partial W}{\partial A} \rangle \mu \right) \right].$$

We notice the following:

- The $\mathcal{G}B'$ term vanishes, since (Eqn. 51.7)

$$\frac{\partial V}{\partial b_0} = \left(\langle \frac{\partial W}{\partial A} \rangle^* 2b_0 + \langle \frac{\partial W}{\partial C} \rangle^* \mu \right) = 0. \tag{51.19}$$

- The term linear in L times a constant vanishes.
- The B', L states have a supersymmetric mass given by

$$m \equiv N^{-1/2} \left(-\langle \frac{\partial W}{\partial C} \rangle 2b_0 + \langle \frac{\partial W}{\partial A} \rangle \mu \right). \tag{51.20}$$

- B' states have SUSY-breaking masses given by the term

$$N^{-1/2}\langle \frac{\partial W}{\partial A} \rangle^* \langle \mathcal{G} \rangle_0 B'^2 = -\langle \frac{\partial W}{\partial A} \rangle^* B'^2 \theta^2 \equiv \frac{1}{2}\Delta m^2 B'^2 \theta^2. \tag{51.21}$$

The fermion mass matrix is given by

$$\frac{1}{2}\mathcal{M}_{1/2} \equiv \frac{1}{2} \begin{pmatrix} 0 & m \\ m & 0 \end{pmatrix} \begin{pmatrix} B' \\ L \end{pmatrix}, \tag{51.22}$$

while the scalar mass matrix is given by

$$\frac{1}{2}\mathcal{M}_0^2 \equiv \frac{1}{2} \begin{pmatrix} m^2 & \Delta m^2 & 0 & 0 \\ \Delta m^2 & m^2 & 0 & 0 \\ 0 & 0 & m^2 & 0 \\ 0 & 0 & 0 & m^2 \end{pmatrix} \begin{pmatrix} B' \\ B'^* \\ L \\ L^* \end{pmatrix}. \tag{51.23}$$

Note, the supertrace of the mass squared, is defined by

$$\text{STr}\mathcal{M}^2 \equiv \sum_J (-1)^{2J}(2J+1)\text{Tr}\mathcal{M}_J^2 = \text{Tr}\mathcal{M}_0^2 - 2\text{Tr}\mathcal{M}_{1/2}^2 = 0. \tag{51.24}$$

This result can be proven in general for direct SUSY breaking. Note also that if the superpotential was of the form $W \supset N^{1/2}\mathcal{G} + L^2 + L^3$, then SUSY is still broken. But no one would know, since the goldstino does not couple to any fields.

$$\textbf{(2)} \; c_0 = -\frac{2b_0}{\mu}a_0 \sim M_{large} \gg \mu, \; M$$

The SUSY-breaking scale is still given by

$$\Delta m^2 = N^{1/2} = \mu(M^2 - \mu^2/4)^{1/2}. \tag{51.25}$$

In this limit, the superpotential is given by

$$W = \Lambda_{SUSY}^2 \mathcal{G} + M_{large}B'^2 + mLB' + \lambda \mathcal{G}B'^2 + \lambda' LB'^2. \tag{51.26}$$

We then have the following:

- the B' mass is of order $M_{large} \gg \Lambda_{SUSY} \equiv \sqrt{\Delta m^2}$ at tree level,
- a supersymmetric mass term of the form $M_{large}B'^2 + mLB'$,
- dimensionless couplings, λ, λ'.

At tree level, only B' feels SUSY breaking of order $\Lambda_{SUSY} \ll M_{large}$. When we diagonalize the mass matrix, we find one light state with supersymmetric mass of order $m_{light} \sim m^2/M_{large}$ and light scalar masses of order $\tilde{m}_{light}^2 \sim (m^2 \pm \frac{1}{2}\Delta m^2)^2/M_{large}^2$.

51.2 Gravity-Mediated SUSY Breaking

Consider now an example of SUSY breaking in supergravity. For a review of gravity-mediated SUSY breaking, see Nilles (1984). The simplest example is known as the Polonyi model. We have the chiral superfields,

$$Z, \; Y_a, \; a = 1, \ldots N-1, \tag{51.27}$$

where Z is the Polonyi field and Y_a are all other fields. The Kahler potential is given by

$$K = \left(|Z|^2 + \sum_a |Y_a|^2 \right) / M_{Pl}^2, \tag{51.28}$$

where $M_{Pl} = 2.4 \times 10^{18}$ GeV is the reduced Planck mass scale. The superpotential is given by

$$W = m^2(Z + \beta M_{Pl}) + h(Y_a), \tag{51.29}$$

where $h(Y_a)$ is, for example, the superpotential of the MSSM. The factor β is chosen such that the potential at the minimum vanishes. We have $\beta = (2 - \sqrt{3})$. The Kahler covariant derivative

$$D_Z W = \frac{\partial W}{\partial Z} + \frac{Z^*}{M_{Pl}^2} W \neq 0 \tag{51.30}$$

at the minimum, and thus SUSY is spontaneously broken. We have the scalar potential

$$V(z, y_a) = e^K \left(|D_Z W|^2 + |D_a W|^2 - \frac{3}{M_{Pl}^2} |W|^2 \right) + \frac{1}{2} D_A^2, \tag{51.31}$$

where D_A is the D term associated with the low-energy gauge interactions. At the minimum of the potential we find

$$z_0 = (\sqrt{3} - 1)M_{Pl}, \quad y_{a0} = 0, \quad h(0) = 0, \tag{51.32}$$

with

$$W_0 = m^2 M_{Pl} \tag{51.33}$$

and

$$(D_Z W)_0 = \sqrt{3}\, m^2. \tag{51.34}$$

Finally, the gravitino mass is given by

$$m_{3/2} = e^{K/2} W_0 / M_{Pl}^2 = \exp[(\sqrt{3} - 1)^2 / 2] \frac{m^2}{M_{Pl}}. \tag{51.35}$$

For a gravitino mass $m_{3/2} \sim 1$ TeV we need a fundamental SUSY-breaking scale $\Lambda_{SUSY}^2 = \sqrt{3} m^2 \sim (10^{11} \text{ GeV})^2$.

The effective low-energy theory can be found in the limit $M_{Pl} \to \infty$, keeping $m_{3/2}$ fixed. For a flat Kahler potential one finds

$$V_{LE} = |\tilde{h}_a|^2 + m_{3/2}^2 |y_a|^2 + m_{3/2} \left[A\tilde{h}^{(3)} + (A - 1)\tilde{h}^{(2)} + (A - 2)\tilde{h}^{(1)} \right] + \frac{1}{2} D_A^2, \tag{51.36}$$

where $\tilde{h} = e^{K/2} h(y_a)$, $A = 3 - \sqrt{3}$ and $\tilde{h}^{(n)}$ are terms in the superpotential containing superfields to the power n. Generalizations of this simple model have been discussed in Chamseddine et al. (1982), Nilles et al. (1983), Hall et al. (1983) and Soni and Weldon (1983).

In general, the soft SUSY-breaking terms are arbitrary and they typically lead to large flavor-violating processes. One would need a fundamental theory of SUSY breaking to obtain a solution to this problem. Such theories of SUSY breaking have been presented which can solve this problem. They typically require extra dimensions, perhaps in the context of string theory, with SUSY breaking and matter localized in the extra spatial dimensions. Some more general solutions to the flavor problem include:

- family symmetries and the alignment of scalar and fermion mass matrices;
- heavy first- and second-family scalars and lighter third-family scalars since flavor-violating operators scale as $1/\tilde{m}^2$.

Another mechanism for SUSY breaking, which addresses the flavor problem, exists in four space–time dimensions. We shall discuss this next.

In order to generate gaugino masses one needs a term in the Lagrangian of the form[2]

$$\mathcal{L} \supset \frac{1}{8g^2} \int d^2\theta \varepsilon (f^{AB}(\Phi_i) W_A^\alpha W_{B\alpha}) \tag{51.37}$$

such that

$$(f^{AB}(\Phi_i) W_A^\alpha W_{B\alpha}) \supset \frac{\partial f^{AB}}{\partial \Phi_i} \langle F_{\Phi_i} \rangle \theta^2 (\lambda_A^\alpha \lambda_{B\alpha}) \tag{51.38}$$

and

$$\langle F_{\Phi_i} \rangle = e^{K/2} g^{ij^*} D_{j^*} W^* \neq 0. \tag{51.39}$$

For example, if $f^{AB} = \delta^{AB} \Phi/M_{Pl}$ and $D_\Phi W \sim m^2$, then $m_\lambda \sim m_{3/2}$.

To summarize, in gravity-mediated SUSY breaking, the fundamental SUSY-breaking scale can be of order $\Lambda_{SUSY} \sim 10^{11}$ GeV, but, since it is mediated by gravity, the effective SUSY-breaking scale of the low-energy theory is of order Λ_{SUSY}^2/M_{Pl}. Note, in local SUSY with a simple Kahler potential, the supertrace formula receives an extra term, i.e.

$$\text{STr}\mathcal{M}^2 \equiv \sum_J (-1)^{2J}(2J+1)\text{Tr}\mathcal{M}_J^2 = 4m_{3/2}^2. \tag{51.40}$$

51.3 Gauge-Mediated SUSY Breaking

In the case of gravity-mediated SUSY breaking, the low-energy SUSY-breaking scale was suppressed by a factor of $1/M_{Pl}$. We call gravity the messenger of SUSY breaking, since this is the only interaction which connects the Polonyi field to the Standard Model sector. In gauge-mediated SUSY breaking, it is the gauge interactions of the Standard Model which act as the messengers of SUSY breaking (Dimopoulos and Raby, 1981; Dine et al., 1981; Witten, 1981; Alvarez-Gaume et al.,

---

[2] ε is a gravitational superfield (Wess and Bagger, 1992).

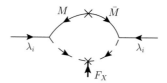

Fig. 51.1   One-loop contribution to gaugino masses.

1982; Dine and Fischler, 1982; Nappi and Ovrut, 1982; Dimopoulos and Raby, 1983; Dine and Nelson, 1993; Dine et al., 1995, 1996). For a review of gauge-mediated SUSY breaking, see Martin (1997) and Giudice and Rattazzi (1999).

Consider a chiral superfield

$$X(x_\mu, \theta) = X(x_\mu) + (\theta \psi_X(x_\mu)) + \theta^2 F_X(x_\mu) \tag{51.41}$$

and assume that X obtains a VEV given by[3]

$$\langle X(x_\mu, \theta) \rangle = X + \theta^2 F_X \neq 0. \tag{51.42}$$

Now consider the case when $X(x_\mu, \theta)$ couples in the superpotential to fields, $M, \bar{M}$, carrying Standard Model gauge charges. We shall call these fields messenger fields. We have

$$W \supset X \bar{M} M. \tag{51.43}$$

As a result of this coupling, the messengers obtain a supersymmetric mass of order X and a SUSY-breaking mass term of order F_Z. As an explicit example, let's assume that the fields $(M, \bar{M})$ transform as a $(5, \bar{5})$ of $SU(5)$ Grand Unified Theory (GUT). The Standard Model states do not feel SUSY breaking at the tree level, but at one loop the gauginos feel SUSY breaking, as in Fig. 51.1. They receive mass at one loop given by

$$m_{\lambda_i} = \frac{\alpha_i}{4\pi} \Lambda, \tag{51.44}$$

with the effective SUSY-breaking scale, Λ, given by

$$\Lambda = \frac{F_X}{X}, \tag{51.45}$$

when $X \gg \sqrt{F_X}$. For a gluino mass of order 1 TeV we might have $\Lambda \sim 10^5$ GeV. Note, only the SUSY-breaking scale, Λ, is relevant for the soft SUSY-breaking MSSM mass terms; the scale of X is arbitrary.

MSSM scalars obtain soft SUSY-breaking mass terms at two loops, as in Fig. 51.2. We obtain the soft scalar mass squared given by

$$\tilde{m}_a^2 = 2 \sum_i C_2^i(a) \left(\frac{\alpha_i}{4\pi} \right)^2 \Lambda^2. \tag{51.46}$$

The quadratic Casimirs are for the gauge group $i = 1, 2, 3$ and matter state, $\tilde{\phi}_a$. The

---

[3] This would require a SUSY-breaking potential for the X superfield, which we shall not consider here.

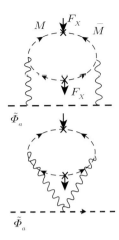

Fig. 51.2   Two-loop contribution to scalar masses.

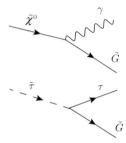

Fig. 51.3   The NLSP (neutralino or stau) can decay into a gravitino plus (photon or tau).

lightest scalars are the $\tilde{\bar{e}}_i$ since they only have $U(1)_Y$ gauge interactions. Note, also, in minimal gauge-mediated SUSY breaking, the cubic scalar soft SUSY-breaking A terms are not present. This makes it difficult to generate large enough radiative corrections to the Higgs mass. However, there is no problem with flavor violation, since the gauge bosons couple to all families equally.

When one combines gauge-mediated SUSY breaking in local SUSY, we obtain a massive gravitino with mass of order

$$m_{3/2} \sim \frac{F_X}{\sqrt{3}M_{Pl}} = \Lambda \left(\frac{X}{\sqrt{3}M_{Pl}} \right). \tag{51.47}$$

Hence, for $X \ll M_{Pl}$, the gravitino is typically the lightest supersymmetric particle. For example, in so-called low-energy gauge-mediated SUSY breaking, where $X \sim \sqrt{F_X} \sim 10^5$ GeV, we have $m_{3/2} \sim 10^{-8}$ GeV = 10 eV (see for example Dimopoulos et al., 1996; Martin, 1997). In this case the gravitino might be a dark-matter candidate. Moreover, the next-to-lightest SUSY particle (NLSP) can decay to the gravitino. For example, in Fig. 51.3, the lightest neutralino or stau can decay into a photon or tau and a gravitino. The decay rate in these processes is of order

$$\Gamma \sim \alpha \frac{\tilde{m}^5}{F_X^2}. \tag{51.48}$$

It is suppressed by F_X and not M_{Pl}^2, since the decay is to the goldstino component of the gravitino.

This could lead to processes such as $pp \to h + Z$ followed by $h \to \gamma\gamma + E_T^{miss}$, as in Fig. 52.3, or $pp \to \tau^+\tau^- + E_T^{miss}$ where, when the stau decays to the tau plus gravitino, there might be a visible kink in the charged-particle track.

In high-energy gauge-mediated SUSY breaking, with $X \sim M_{GUT}$, soft SUSY-breaking terms, due to gauge mediation, can be an order of magnitude larger than those due to gravity mediation, and again the SUSY flavor problem is ameliorated. Nevertheless, the spectrum is very much like that of the constrained MSSM with $A_0 = 0$. $A_0 \neq 0$ can be obtained in non-minimal models of gauge mediation, where the messengers couple to matter via Yukawa interactions (Dine et al., 1997; Giudice and Rattazzi, 1999).

MSSM Phenomenology

52.1 Gluino–Squark Detection at the LHC

The Large Hadron Collider (LHC) is a proton–proton collider. As such, physics at the LHC is most sensitive to the production of colored states, such as gluinos and squarks. In Section 48.1 we discussed supersymmetric quantum chromodynamics. The Lagrangian, in component form, is given in Eqn. 48.47. Let's focus on gluino–squark–quark interactions. The relevant terms in the Lagrangian are

$$\mathcal{L} \supset -\frac{1}{2}\mathrm{Tr}(G_{\mu\nu}G^{\mu\nu}) + 2i\mathrm{Tr}(\tilde{g}^\dagger\bar{\sigma}^\mu D_\mu\tilde{g}) + M_3\mathrm{Tr}(\tilde{g}\tilde{g}) + iq^\dagger\bar{\sigma}^\mu D_\mu q + i\bar{q}^\dagger\bar{\sigma}^\mu D_\mu\bar{q}$$
$$+ |D_\mu\tilde{q}|^2 + |D_\mu\tilde{\bar{q}}|^2 - \sqrt{2}ig_s\tilde{q}^*(\tilde{g}q) + \sqrt{2}ig_s\tilde{\bar{q}}^*(\tilde{g}^T\bar{q}) + h.c. \tag{52.1}$$

The covariant derivatives are given by

$$D_\mu\tilde{g} = \partial_\mu\,\tilde{g} + ig_s[g_\mu,\ \tilde{g}], \tag{52.2}$$
$$D_\mu q = (\partial_\mu + ig_s g_\mu)q,$$
$$D_\mu\tilde{q} = (\partial_\mu + ig_s g_\mu)\tilde{q},$$
$$D_\mu\bar{q} = (\partial_\mu - ig_s g_\mu^T)\bar{q},$$
$$D_\mu\tilde{\bar{q}} = (\partial_\mu - ig_s g_\mu^T)\tilde{\bar{q}},$$

and the gluon field strength is given by

$$G_{\mu\nu} = \partial_\mu g_\nu - \partial_\nu g_\mu + ig_s[g_\mu,\ g_\nu]. \tag{52.3}$$

The gluon and gluino fields are matrix valued with $g_\mu = T_A g_{\mu A}$, $\quad\tilde{g} = T_A\tilde{g}_A$ and $T_A = \frac{\lambda_A}{2}$. Hence,

$$2i\mathrm{Tr}(\tilde{g}^\dagger\bar{\sigma}^\mu D_\mu\tilde{g}) = 2i\tilde{g}_A^\dagger\bar{\sigma}^\mu\mathrm{Tr}(T_A D_\mu\tilde{g}) \tag{52.4}$$
$$= 2i\tilde{g}_A^\dagger\bar{\sigma}^\mu\mathrm{Tr}\left(T_A(\partial_\mu\tilde{g} + ig_s[g_\mu,\ \tilde{g}])\right)$$
$$= i\tilde{g}_A^\dagger\bar{\sigma}^\mu\partial_\mu\tilde{g}_A - ig_s f_{ABC}\tilde{g}_A^\dagger\bar{\sigma}^\mu g_{\mu B}\tilde{g}_C.$$

Define the four-component Majorana gluino by

$$\psi_{\tilde{g}} = \begin{pmatrix} \tilde{g} \\ i\sigma_2\tilde{g}^* \end{pmatrix}. \tag{52.5}$$

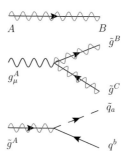

Fig. 52.1 The gluino propagator and vertex diagrams.

Then in Dirac notation, the first two terms in Eqn. 52.1 give the free gluino kinetic terms,[1]

$$\frac{1}{2}\bar{\psi}_{\tilde{g}}^A i\gamma^\mu \partial_\mu \psi_{\tilde{g}}^A - \frac{1}{2}m_{\tilde{g}}\bar{\psi}_{\tilde{g}}^A \psi_{\tilde{g}}^A = 2i\text{Tr}\left(\tilde{g}^\dagger \bar{\sigma}^\mu \partial_\mu \tilde{g}\right) + \frac{1}{2}M_3(\tilde{g}^A \tilde{g}^A) + h.c. \qquad (52.6)$$

and the gluon–gluino interaction term,

$$-i\frac{g_s}{2}f_{ABC}\bar{\psi}_{\tilde{g}}^A \gamma^\mu g_\mu^B \psi_{\tilde{g}}^C = +i\frac{g_s}{2}f_{ABC}g_\mu^A \bar{\psi}_{\tilde{g}}^B \gamma^\mu \psi_{\tilde{g}}^C. \qquad (52.7)$$

The Feynman propagators for the gluino is given by (see Fig. 52.1)[2]

$$\langle T(\psi_{\tilde{g}}^B \bar{\psi}_{\tilde{g}}^A)\rangle_0 = \frac{i\delta^{AB}}{\not{p} - m_{\tilde{g}}} \qquad (52.8)$$

and

$$\langle T(\psi_{\tilde{g}}^B \psi_{\tilde{g}}^{A\,T})\rangle_0 = \frac{i\delta^{AB}}{\not{p} - m_{\tilde{g}}}(i\gamma_2\gamma_0)^{-1} \qquad (52.9)$$

(where $(i\gamma_2)$ is the charge conjugation matrix (Eqn. 16.7) and we used the identity $\psi^T \equiv \bar{\psi}(i\gamma_2\gamma_0)^{-1}$) and the gluon–gluino and gluino–quark–squark vertices are as follows,

$$+ig_s f_{ABC}\gamma^\mu \qquad (52.10)$$

and

$$-\sqrt{2}ig_s(T^A)^a{}_b. \qquad (52.11)$$

The tree-level Feynman diagrams leading to gluino production at the LHC are as follows. In Fig. 52.2, quark–anti-quark annihilation produces a pair of gluinos. In Fig. 52.3, gluon–gluon fusion produces a pair of gluinos. In the latter, the production

[1] $M_3 = m_{\tilde{g}}$.
[2] For a nice discussion of propagators and wave-functions for Majorana fermions, see Haber and Kane (1985).

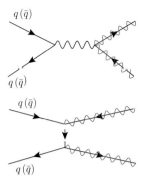

Fig. 52.2   Quark–anti-quark collisions produce gluinos in pairs.

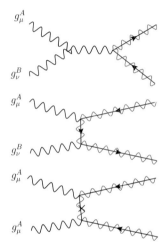

Fig. 52.3   The gluino production via gluon fusion.

cross-section due to gluon–gluon fusion dominates at small Feynman x. The cross-section has been calculated in Dawson et al. (1985). We have

$$\sigma_{tot} = \frac{3\pi\alpha_s^2}{2\hat{s}} \left[3 \left(1 + \frac{4m_{\tilde{g}}^2}{\hat{s}} - \frac{4m_{\tilde{g}}^4}{\hat{s}^2} \right) \ln \left(\frac{\hat{s} + \sqrt{\lambda(\hat{s}, m_{\tilde{g}}^2, m_{\tilde{g}}^2)}}{\hat{s} - \sqrt{\lambda(\hat{s}, m_{\tilde{g}}^2, m_{\tilde{g}}^2)}} \right) \right.$$
$$\left. - \left(4 + \frac{17m_{\tilde{g}}^2}{\hat{s}} \right) \frac{\sqrt{\lambda(\hat{s}, m_{\tilde{g}}^2, m_{\tilde{g}}^2)}}{\hat{s}} \right]. \tag{52.12}$$

For the LHC with the parton $\hat{s} = x_a x_b \, (13 \text{ TeV})^2$ and a gluino with mass $m_{\tilde{g}} = 2$ TeV, the total gluino production cross-section, integrated with the parton luminosity functions, gives $\sigma_{tot} \sim 0.3$ fb (see Fig. 52.4).[3] This corresponds to

---

[3] Note, this is about the same as the parton cross-section at $\sqrt{\hat{s}} \sim 13$ TeV, since $x_a \sim x_b \sim 0.3$ for this process, corresponding to the dominance of valence quarks. Moreover, we see in Fig. 52.4 that squark–squark and squark–gluino production dominates for squarks and gluinos with mass of order 2 TeV.

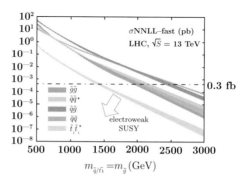

Fig. 52.4 Supersymmetric (SUSY)-particle production cross-section at the LHC (Masciovecchio, 2019). For a color version of this figure, please see the color plate section.

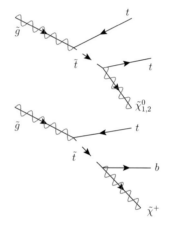

Fig. 52.5 Gluinos will typically decay, eventually into the lightest superparticle.

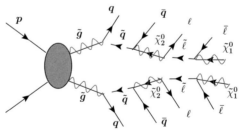

Fig. 52.6 Gluinos decay processes may have several sequential steps.

about 42 gluino events with $139\,\mathrm{fb}^{-1}$ of data.

Then the next question is, how does one see a gluino? Gluinos are unlikely to be the lightest superpartners. This is most likely a neutralino. Therefore, a gluino can decay as in Fig. 52.5 into a quark–anti-quark pair and either a neutralino or chargino. The latter can then decay further, unless it is the lightest superparticle

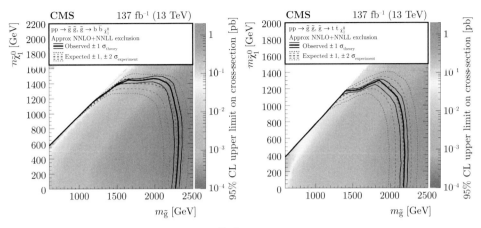

(left) Simplified model with $BR(\tilde{g} \to b\bar{b}\tilde{\chi}_1^0) = 1$, and (right) simplified model with $BR(\tilde{g} \to t\bar{t}\tilde{\chi}_1^0) = 1$ (Sirunyan et al., 2019). Published under the CC BY 4.0 license. For a color version of this figure, please see the color plate section.

(LSP). If the LSP is a neutralino, it will escape the detector leaving only missing $\not{E}_T$. A typical process for production and decay of gluinos may look as in Fig. 52.6

52.2 Simplified Models

The CMS and ATLAS detectors at CERN use simplified models to obtain bounds on the gluino mass. In the following searches, they are looking for jets plus missing E_T, and also vetoing events with isolated leptons. For example, in Fig. 52.7 with 137 fb^{-1} of data, CMS finds the gluino mass is greater than ~ 2.2 TeV, assuming the gluino decays 100% of the time to $b\bar{b}\tilde{\chi}_1^0$ (left) and, similarly, the bound is 2.25 TeV assuming 100% branching ratio for $\tilde{g} \to t\bar{t}\tilde{\chi}_1^0$ (right). ATLAS finds similar bounds with the same simplified models with only 79.8 fb^{-1} of data.

Finally, in a CMS search for gluinos in events with jets missing E_T and leptons, the results are a bit different. For example, in Fig. 52.8, the gluino mass bound is significantly lower. The LM signal regions, Fig. 52.9, are characterized by low p_T^{miss}: exactly two same-sign leptons, both with $p_T > 25$ GeV, and $p_T^{miss} < 50$ GeV. In addition, the different signal regions within this category are characterized by the following, Table 52.1. The LM signal region has ≥ 3 b jets and $H_T > 300$ GeV (where H_T is the scalar p_T sum of all jets in the event). There is a 2.3σ excess of events over background in this signal region. Considering these several excess bins, the gluino mass bound given in Fig. 52.8 is ~ 1700 GeV. This is significantly lower that the expected bound of ~ 1820 GeV. Nevertheless, overall the results are consistent with the Standard Model and only bounds can be obtained.

Table 52.1 The relevant cuts used to define the LM signal region (SR) (Sirunyan et al., 2020).

N_b	N_{jets}	$H_T \in [300, 1125]$	$H_T \in [1125, 1300]$	$H_T > 1300$
0	2–4	SR1		
	≥5	SR2	SR8 ($N_{jets} < 5$)	SR10 ($N_{jets} < 5$)
1	2–4	SR3		
	≥5	SR4		
2	2–4	SR5	SR9 ($N_{jets} \geq 5$)	SR11 ($N_{jets} \geq 5$)
	≥5	SR6		
≥3	≥2	SR7		

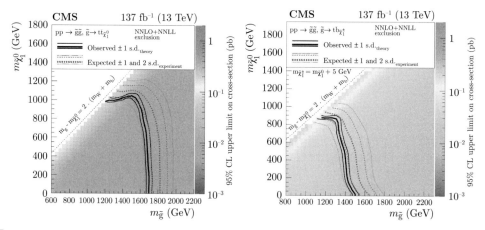

Fig. 52.8   Gluino bound with simplified models in the search for jets missing E_T and leptons. (left) $\tilde{g} \to t\bar{t}\tilde{\chi}_1^0$ and (right) $\tilde{g} \to tb\tilde{\chi}_1^{\pm}$, 100% of the time (Sirunyan et al., 2020). For a color version of this figure, please see the color plate section.

The bounds on bottom and top squarks from CMS data have been analyzed using the simplified models in Fig. 52.10. In Fig. 52.11 we see that the CMS bound on sbottom and stop masses using 137 fb^{-1} of data is of order 1220 GeV.

52.3 Electroweakinos

One can also search for the lighter electroweak gauginos. However the production cross-section for these electroweakinos is much smaller than for either gluinos or squarks, since they don't participate in strong interactions. Most studies conclude that it may take at least 3 ab^{-1} of data to obtain a discovery for an electroweakino with mass of order a few hundred GeV and the mass difference $m_{\tilde{\chi}^{\pm}} - m_{\tilde{\chi}_1^0} <$

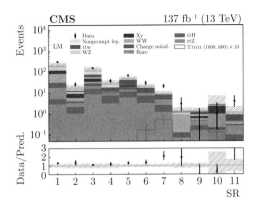

Fig. 52.9 The LM signal regions (Sirunyan et al., 2020). For a color version of this figure, please see the color plate section.

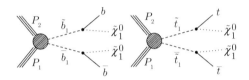

Fig. 52.10 Simplified models used by CMS for stop and sbottom bounds (Sirunyan et al., 2019). Published under the CC BY 4.0 license.

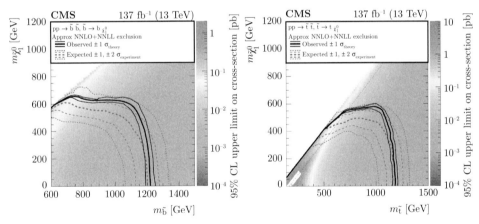

Fig. 52.11 (left) Simplified model with $BR(\tilde{b} \to b\tilde{\chi}_1^0) = 1$, and (right) simplified model with $BR(\tilde{t} \to t\tilde{\chi}_1^0) = 1$ (Sirunyan et al., 2019). Published under the CC BY 4.0 license. For a color version of this figure, please see the color plate section.

$100-200$ GeV. For example, see the latest results from ATLAS, Fig. 52.12, which gives a lower bound on the electroweakino mass of order $300-350$ GeV.

There is a completely separate search strategy for models with so-called *gauge-mediated SUSY breaking* (GMSB). In this case, the gravitino is generally the LSP

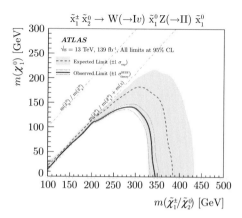

Fig. 52.12 $\tilde{\chi}_1^{\pm}, \tilde{\chi}_2^0$ mass bound with simplified model of Fig. 52.13 (Aad et al., 2020).

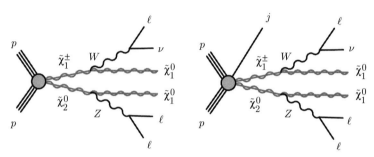

Fig. 52.13 Chargino–neutralino sequential decays (Aad et al., 2020).

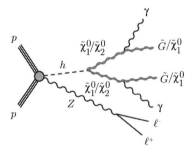

Fig. 52.14 Bounds on Higgs boson to two photons in a simplified model with gauge-mediated SUSY breaking (ATLAS Collaboration, 2018).

and the next to lightest SUSY particle (NLSP) might be the lightest neutralino. ATLAS has searched for the Higgs boson to decay to two photons and missing E_T, as described by the simplified model used in Fig. 52.14. In Fig. 52.15 one finds some excess events in the signal region (to the right of the arrow) obtained by ATLAS with 79.8 fb^{-1} of data. Finally, in Fig. 52.16 we find the ATLAS bounds on the branching ratio for Higgs to decay to two photons plus missing E_T. The observed

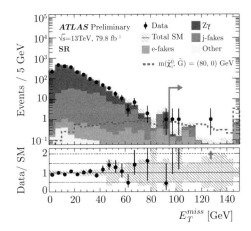

Fig. 52.15   ATLAS data for Higgs to two photons via GMSB. Note, the excess events in the signal region (to the right of the arrow) (ATLAS Collaboration, 2018). For a color version of this figure, please see the color plate section.

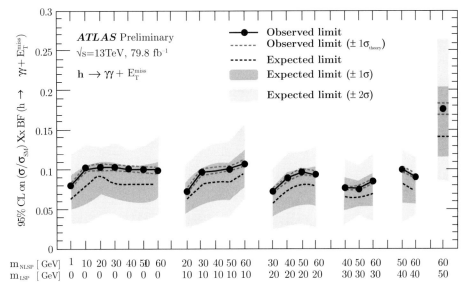

Fig. 52.16   Bounds on the branching fraction of Higgs boson decays to two photons plus E_T^{miss} depending on the NLSP and LSP masses (ATLAS Collaboration, 2018). For a color version of this figure, please see the color plate section.

upper limits on the branching fraction of Higgs boson decays to one or two photons plus E_T^{miss} range from 5% to 18% depending on the NLSP and LSP masses.

We should note that we have not presented all the data on SUSY searches and, moreover, we have chosen to show some results which might suggest that a discovery may lie just around the corner. Unfortunately, it is clear that any SUSY signature has a significant Standard Model background. As a consequence, the signal-to-noise ratios will only increase as the square root of the amount of data. Thus, even if SUSY is there, it may take quite a long time for a discovery, UNLESS, for example, a high-energy LHC (HE-LHC) is built. Then a discovery could, in principle, come sooner.

Summary

We have yet to observe the new SUSY particles, but the theory is so beautiful that it is hard to imagine that nature does not make use of it. It can solve the huge gauge hierarchy problem, eliminating the fine-tuning to one part in 10^{28} in order to obtain the Higgs mass so much smaller than a Grand Unified Theory (GUT) scale. It makes grand unification at the scale of order 10^{16} GeV very predictive, which then explains the low-energy charge assignments of quarks and leptons under the Standard Model gauge symmetry. Clearly, SUSY GUTs provide a natural extension of the Standard Model. It also provides a new high scale in nature, perhaps relevant for the see-saw mechanism to explain the light neutrino masses. And it leads to a natural candidate for dark matter. Unfortunately, we cannot know the origin or scale of SUSY breaking until we actually find SUSY particles. We look forward to the full analysis of Run 2 of the LHC and then to the high luminosity or possibly high-energy LHC phase. Of course, there may be new discoveries of physics beyond the Standard Model which have yet to be thought of in our theories.

PART XIII

SECOND-SEMESTER PROJECTS

53

Suggested Term Projects 1

Term projects compliment the lectures and also allow the students to do some independent research. There are many topics which cannot be covered in class due to a lack of time. However, they are very much a part of the saga of understanding the development of the Standard Model. In the following, I will present and motivate each possible topic. Note, some topics have been of more interest to high-energy theory, or nuclear theory or high-energy experimental students. The proposed references are only a starting point for the student research. Students should naturally expand their list of references as they get deeper into the subject matter.

Students choose one topic. Each person will have a different topic. Students can also suggest a topic and if it is relevant to the material to be discussed during the semester, then they may be able to take that topic. Reports should be submitted as a PDF file so that they can be put online for everyone to read. Students will also be expected to give a talk to the class, based on their reports.

1. Semi-leptonic weak decays of the baryon octet and the Cabibbo theory. Using

$$H_{weak} = \frac{G_F}{\sqrt{2}} J_\mu J^{\mu\dagger} \text{ with } J_\mu = \bar{\psi}\gamma_\mu(1 - \gamma_5)T^+\psi, \ \psi = \begin{pmatrix} u \\ d \\ s \end{pmatrix} \text{ and } T^+ = $$

$\begin{pmatrix} 0 & c_1 & s_1c_3 \\ 0 & 0 & 0 \\ 0 & 0 & 0 \end{pmatrix}$, find s_1c_3 by fitting weak decay rates. This analysis demonstrated that the weak flavor-violating decays can be described in terms of one mixing angle, θ_C. References: Cabibbo, *Phys. Rev. Lett.* **10**, 531 (1963); Shrock and Wang, *Phys. Rev. Lett.* **41**, 1692 (1978).

2. Determine the masses of the light quarks, u, d, s. The $\overline{MS}$ masses of light quarks have the largest uncertainties. Chiral Lagrangian analysis, perturbation theory for baryonic states and non-perturbative physics associated with instantons all play a role. References: Gasser and Leutwyler, *Nucl. Phys.* **B94**, 269 (1975); Leutwyler, *Phys. Lett.* **48B**, 431 (1974); Georgi, *Weak Interactions and Modern Particle Theory* (Addison-Wesley, 1984); Kaplan and Manohar, *Phys. Rev. Lett.* **56**, 2004 (1986).

3. Using the $SU(6)$ non-relativistic quark model, derive the magnetic moments for the baryon octet, the $\Sigma - \Lambda$ mass splitting and g_A. The non-relativistic quark model was a mysterious success, requiring the distinction between current

quark and constituent quark masses. References: Greiner and Muller, *Quantum Mechanics: Symmetries* (Springer, 2012); De Rujula, Georgi and Glashow. *Phys. Rev. D12*, 147 (1975).

4. Calculate the hadronic spectrum for vector mesons, pseudo-scalars and hadrons using the MIT bag model. Also calculate the magnetic moments, charge radius and g_A/g_V for nucleons, n, p. Prior to lattice gauge theory, the MIT and Stanford Linear Accelerator Center (SLAC) bag models were used to obtain a rough understanding of confinement. Reference: Lee, *Particle Physics and Introduction to Field Theory* (Harwood, 1981).

5. Using Heavy Quark Effective Theory, relate mass splittings and decay rates in B- and D-meson systems. When quark masses are much larger than the quantum chromodynamics (QCD) scale, their interaction with light quarks becomes independent of their mass. This realization leads to a new way to test strong-interaction dynamics. References: Isgur and Wise, *Phys. Lett.* **B232**, 113 (1989), *Phys. Lett.* **B237**, 527 (1990); Georgi, *Ann. Rev. Nucl. Part. Sci*, **43**, 209 (1993).

6. Using the $SU(3) \otimes SU(3)$ non-linear chiral Lagrangian, calculate the following semi-leptonic decay rates: $K^- \to \mu^- \bar{\nu}_\mu$, $K^- \to \pi^0 \mu^- \bar{\nu}_\mu$ and $K^- \to \pi^0 \pi^0 \mu^- \bar{\nu}_\mu$. The effective theory below the confinement scale must incorporate the symmetries of the fundamental theory. These spontaneously broken chiral symmetries and their consequences are tests of our understanding of the fundamental theory. Reference: Georgi, *Weak Interactions and Modern Particle Theory* (Addison-Wesley, 1984).

7. Using a non-relativistic potential, calculate the charmonium spectrum. Prior to the discovery of charm quarks, quarks were believed to be mathematical constructs, fictitious particles, which were useful for understanding the hadronic spectrum and also deep-inelastic scattering, but not real. However, the discovery of charm quarks and the description of their bound states using non-relativistic quantum mechanics led to a paradigm shift in our understanding. Quarks were now REAL ! Reference: Eichten and Gottfried, et al., *Phys. Rev.* **34**, 369 (1975).

8. Calculate the decay $\pi^0 \to 2\gamma$. Derive the Sutherland–Veltman theorem. Use the Adler–Bell–Jackiw anomaly to obtain the result. Note, the original calculation of the decay $\pi^0 \to 2\gamma$ by Steinberger (1949), via a triangle diagram with a proton in the loop, led to a quantitatively good result, in spite of the Sutherland–Veltman theorem. References: Cheng and Li, *Gauge Theory of Elementary Particle Physics* (Clarendon Press, 1984); Veltman, *R. Soc. London* **A301**, 107 (1967, https://doi.org/10.1098/rspa.1967.0193; Sutherland, *Nucl. Phys.* **B2**, 433 (1967).

9. Evaluate the η' mass using the $U(3) \otimes U(3)$ non-linear chiral Lagrangian. Compare with experiment and explain the result. Discuss the solution to this so-called $U(1)$ problem. This is another example where chiral anomalies and strong-interaction dynamics solved problems inherent in particle physics prior to QCD. Reference: Cheng and Li, *Gauge Theory of Elementary Particle Physics* (Clarendon Press, 1984); Georgi, *Weak Interactions and Modern*

Particle Theory (Addison-Wesley, 1984); Witten, *Ann. of Phys.* **128**, 363 (1980).

10. Read papers by Nambu and Jona-Lasinio and discuss the pion as a Nambu–Goldstone boson of spontaneously broken chiral symmetry. These papers are the seminal work on dynamical symmetry breaking, i.e. spontaneous symmetry breaking in a theory without fundamental scalars. Nambu and Jona-Lasinio, *Phys. Rev.* **122**, 1 (1961), *Phys. Rev.* **124**, 1 (1961).

11. Discuss **CP** violation in the K^0, $\bar{K}^0$ system. Define **CP** eigenstates, K_L, K_S and their dominant decay modes. Describe the experimental observation of **CP** violation by Cronin and Fitch. Also discuss experiments on K_s regeneration and on the measurement of the $K_L - K_S$ mass difference. The K_L, K_S mass difference is one of the most accurate measurements of strong-interaction dynamics. References: Christenson, Cronin, Fitch and Turlay, *Phys. Rev. Lett.* **13**, 138 (1964); Fetcher, Kokkas and Pavlopoulos, *Zeit. Phys.* **72**, 543 (1996); Angelopoulos, *Phys. Lett.* **B503**, 49 (2001).

12. Read the paper by Nambu on superconductivity and discuss spontaneous symmetry breaking and the Meissner effect. This is the first paper on the so-called Higgs mechanism in the context of the BCS theory of superconductivity. Also, when the electric coupling constant, $e \to 0$, it describes the phenomenon of spontaneous symmetry breaking of a global $U(1)$ symmetry and the occurrence of massless Nambu–Goldstone bosons. Reference: Nambu, *Phys. Rev.* **117**, 648 (1960).

Students choose one topic. Each person will have a different topic. Students can also suggest a topic and if it is relevant to the material we will discuss during the semester, then they may be able to take that topic. Reports should be submitted as a PDF file so that they can be put online for everyone to read. Students will also be expected to give a talk to the class, based on their report.

1. Calculate the differential cross-section for Higgs production in the process $pp \to h + X$ via gluon fusion at the LHC. Discuss the experiments which discovered the Higgs boson, i.e. the production processes and the relevant decay modes. References: Gunion, Haber, Kane and Dawson, *The Higgs Hunter's Guide* (Addison-Wesley, 1990); Vainshtein, Voloshin, Zakharov and Shifman, *Sov. J. Nucl. Phys.* **30**, 711 (1979); Wilczek, *Phys. Rev. Lett.* **39**, 1304 (1977).

2. Use the operator product expansion to calculate the differential cross-section for deep-inelastic eN scattering. Discuss scaling violations with this formalism and compare to the analysis of scaling violations using the Altarelli–Parisi equations. Reference: Peskin and Schroeder, *An Introduction to Quantum Field Theory* (Westview Press, 1995).

3. Discuss the quantum chromodynamics (QCD) Θ vacua and the strong CP problem. References: Belavin, Polyakov, Schwartz and Tyupkin, *Phys. Lett.* **59B**, 85 (1975); Jackiw and Rebbi, *Phys. Rev. Lett.* **37**, 172 (1976); Callan, Dashen and Gross, *Phys. Lett.* **63B**, 334 (1976); 't Hooft, *Phys. Rev. Lett.* **37**, 8 (1976), *Phys. Rev.* **14D**, 3432 (1976); M. Srednicki, *Quantum Field Theory* (Cambridge University Press, 2007).

4. Describe the Peccei–Quinn solution to the strong CP problem and axions. References: original papers, Peccei and Quinn, *Phys. Rev. Lett.* **37**, 1440 (1977), *Phys. Rev.* **D16**, 1791 (1977); S. Weinberg, *Phys. Rev. Lett.* **40**, 223 (1978); F. Wilczek, *Phys. Rev. Lett.* **40**, 279 (1978); J. E. Kim, *Phys. Rev. Lett.* **43**, 103 (1979); M. Dine, W. Fischler and M. Srednicki, *Phys. Lett.* **B104**, 199 (1981) there are also many review articles.

5. Calculate the gluino production cross-section at the LHC. Discuss the sensitivity of the LHC Run II to gluino discovery.

6. Derive the phase structure of the electroweak theory at high temperature. References: Le Bellac, *Thermal Field Theory* (Cambridge Press, 1996); Quiros, hep-ph/9901312.

7. Discuss partial wave unitarity violation in early weak-interaction theories and the recovery of partial wave unitarity in the Standard Model.

(a) Consider the Fermi effective theory of charged-current processes and discuss the process $\nu_\mu e^- \to \nu_e \mu^-$ and violation of partial wave unitarity.

(b) Add the $W^\pm$ to resolve this difficulty and then discuss the process $\nu\bar\nu \to W^+ W^-$ scattering and a violation of unitarity. Discuss possible solutions.

(c) Finally, discuss $W^+ W^-$ elastic scattering and a violation of partial wave unitarity. Discuss the solution (and the bound on the Higgs mass).

References: Langacker, *The Standard Model and Beyond* (CRC Press, 2017).

8. Discuss the process of baryogenesis in the early universe. Assume the solution via the process of leptogenesis followed by sphalerons in the Standard Model which violate baryon and lepton number via order one effects at high temperature. References: Fukugita and Yanagida, *Phys. Lett.* **B174**, 45 (1986); Kolb and Turner, *The Early Universe* (Addison-Wesley, 1990).

9. Discuss the WIMP solution to dark matter. Consider the bounds on Wino dark matter. Reference: Jungman, Kamionkowski and Greist, *Phys. Report* **267**, 195 (1996).

PART XIV

APPENDICES

Appendix A Gell-Mann–Low Theorem

A.1 Interaction Picture

Consider the generic field $\phi(x)$ which is fully interacting. The S matrix is then given in terms of the n-point Green's function

$$G(x_1, \ldots, x_n) \equiv (\psi_0|T(\phi(x_1) \cdots \phi(x_n))|\psi_0), \tag{A.1}$$

where ψ_0 is the Heisenberg ground state, satisfying

$$H\psi_0 = E_0\psi_0. \tag{A.2}$$

The Heisenberg-picture fields satisfy the equation

$$\phi(\vec{x}, t) \equiv e^{iHt}\phi(\vec{x}, 0)e^{-iHt}, \tag{A.3}$$

since H is the time translation operator. For free fields we have $H = H_0$, while for interacting fields we have $H = H_0 + H_I$ with $H_I \equiv H_{interaction}$.

We want to define the interaction picture in quantum mechanics. But first consider the Schrödinger picture. The operators are time independent, satisfying

$$[q_S, \ p_S] = i \tag{A.4}$$

and the wave functions satisfy the Schrödinger equation,

$$H(p_S, q_S)\psi_S(t) = i\frac{d}{dt}\psi_S(t). \tag{A.5}$$

Thus, we have

- operators are time independent,
- states are time dependent.

The expectation value of an operator A is given by

$$(\psi_S|A(p_S, q_S)|\psi_S) \equiv \hat{A}. \tag{A.6}$$

In the Heisenberg picture, we have

$$\psi_H \equiv e^{iHt}\psi_S(t) \tag{A.7}$$

such that

$$\frac{d\psi_H}{dt} \equiv e^{iHt}\left(iH\psi_S(t) + \frac{d\psi_S(t)}{dt}\right) \equiv 0. \tag{A.8}$$

Thus, we have the following.

- States are time independent.

The Heisenberg operators are given by

$$p_H(t) \equiv e^{iHt} p_S e^{-iHt}, \tag{A.9}$$
$$q_H(t) \equiv e^{iHt} q_S e^{-iHt}.$$

Hence, we have the Heisenberg equations of motion

$$\frac{dp_H(t)}{dt} \equiv i\,[H,\ p_H], \tag{A.10}$$
$$\frac{dq_H(t)}{dt} \equiv i\,[H,\ q_H].$$

- The operators are time dependent, but still satisfy

$$[q_H(t), p_H(t)] = i. \tag{A.11}$$

Note:

$$\hat{A} \equiv (\psi_S | A(p_S, q_S) | \psi_S) \tag{A.12}$$
$$= (\psi_S | e^{-iHt} e^{iHt} A(p_S, q_S) e^{-iHt} e^{iHt} | \psi_S)$$
$$\equiv (\psi_H | A(p_H, q_H) | \psi_H).$$

Thus, the value of observables, i.e. matrix elements of Hermitian operators, are independent of the picture.

Up until now in field theory we have been working in the Heisenberg picture, where ψ_0 is the Heisenberg ground state and $\phi(\vec{x}, t)$ is an Heisenberg operator. Let us now define the interaction picture. Given the Hamiltonian, $H = H_0 + H_I$, we define the interaction picture operators by

$$p_I(t) \equiv e^{iH_0(0)t} p_S e^{-iH_0(0)t}, \tag{A.13}$$
$$q_I(t) \equiv e^{iH_0(0)t} q_S e^{-iH_0(0)t}.$$

Then we have

$$p_H(t) \equiv e^{iHt} e^{-iH_0 t} p_I(t) \underbrace{e^{iH_0 t} e^{-iHt}}_{U(t,0)} \tag{A.14}$$

or

$$p_H(t) \equiv U^\dagger(t,0) p_I(t) U(t,0). \tag{A.15}$$

They satisfy the equations of motion,

$$\frac{dp_I(t)}{dt} \equiv i\,[H_0,\ p_I], \tag{A.16}$$
$$\frac{dq_I(t)}{dt} \equiv i\,[H,\ q_I],$$

i.e. the free-field equations of motion. The interaction-picture states are given by

$$\psi_I(t) \equiv e^{iH_0 t} \psi_S(t) \tag{A.17}$$

and satisfy the equations of motion

$$\frac{d\psi_I(t)}{dt} = e^{iH_0t}\left(iH_0\psi_S(t) + \frac{d\psi_S(t)}{dt}\right) \tag{A.18}$$

$$= -ie^{iH_0t}(H - H_0)\psi_S(t)$$

$$= -ie^{iH_0t}(H - H_0)e^{-iH_0t}\psi_I(t)$$

$$\equiv -iH_I(t)\psi_I(t),$$

where

$$H_I(t) \equiv e^{iH_0t}H_Ie^{-iH_0t}. \tag{A.19}$$

Thus, the time dependence of the interaction-picture state is due entirely to the perturbation. For field theory, we have

$$\phi(x) = U^\dagger(t,0)\phi_I(x)U(t,0) \tag{A.20}$$

and

$$\frac{d\phi_I}{dt} = i[H_0, \phi_I], \tag{A.21}$$

i.e. the free-field equations.

Now, all the physics of the interactions is in the unitary time development operator

$$U(t,0) \equiv e^{iH_0(0)t}e^{-iHt}. \tag{A.22}$$

We will now show that

$$U(t,0) \equiv T\left(e^{-i\int_0^t dt' H_I(t')}\right). \tag{A.23}$$

Proof

$$\frac{dU(t,0)}{dt} = ie^{iH_0t}H_0e^{-iHt} - ie^{iH_0t}He^{-iHt} \tag{A.24}$$

$$\equiv -ie^{iH_0t}(H - H_0)e^{-iH_0t}e^{iH_0t}e^{-iHt}$$

$$\equiv -iH_I(t)U(t,0),$$

with the boundary condition $U(0,0) \equiv 1$. Note, in field theory, the Hamiltonian

$$H(\phi_S) \equiv H_0(\phi_S) + H_I(\phi_S) \tag{A.25}$$

is time independent. On the other hand, the interaction-picture Hamiltonian is time dependent, i.e.

$$H_I(t) \equiv e^{iH_0t}(H(\phi_S) - H_0(\phi_S))e^{-iH_0t} \equiv H_I(\phi_I(t)). \tag{A.26}$$

The solution of the differential equation

For the first step in the solution, we have the integral equation with the correct boundary condition,

$$U(t,0) = 1 - i\int_0^t dt' H_I(t')U(t',0). \tag{A.27}$$

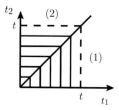

Fig. A.1   Integration regions $(\mathbf{1})$ and $(\mathbf{2})$.

We then solve the integral equation by iteration.

$$U(t,0) = 1 - i \int_0^t dt' H_I(t') \left(1 - i \int_0^{t'} dt'' H_I(t'') U(t'') \right) \tag{A.28}$$

$$= 1 - i \int_0^t dt' H_I(t') + (-i)^2 \int_0^t dt' \int_0^{t'} dt'' H_I(t') H_I(t'') U(t'')$$

$$= 1 - i \int_0^t dt' H_I(t') + (-i)^2 \int_0^t dt' \int_0^{t'} dt'' H_I(t') H_I(t'') \tag{A.29}$$

$$+ (-i)^3 \int_0^t dt' \int_0^{t'} dt'' \int_0^{t''} dt''' H_I(t') H_I(t'') H_I(t''') + \cdots$$

$$\equiv \sum_{n=0}^{\infty} (-i)^n \int_0^t dt_1 \int_0^{t_1} dt_2 \cdots \int_0^{t_{n-1}} dt_n T(H_I(t_1) \cdots H_I(t_n)).$$

Note:

$$T(H_I(t_1) \cdots H_I(t_n)) \tag{A.30}$$

is a symmetric function under the interchange $t_i \leftrightarrow t_j$.

Consider the third term above in line A.29.

In the integration region $(\mathbf{1})$, see Fig. A.1, we have

$$(\mathbf{1}) = \int_0^t dt_1 \int_0^{t_1} dt_2 \underbrace{H_I(t_1) H_I(t_2)}_{T(H_I(t_1) H_I(t_2))} \tag{A.31}$$

$$(\mathbf{2}) = \int_0^t dt_2 \int_0^{t_2} dt_1 \underbrace{H_I(t_2) H_I(t_1)}_{T(H_I(t_1) H_I(t_2))} \quad \text{just renamed variables 1 and 2.}$$

Hence $(\mathbf{1}) \equiv (\mathbf{2})$. As a result, we have

$$(\mathbf{1}) \equiv \frac{1}{2} \int_0^t dt_1 \int_0^t dt_2 T\left(H_I(t_1) H_I(t_2) \right). \tag{A.32}$$

Thus, in general, we have

$$U(t,0) \equiv \sum_{n=0}^{\infty} \frac{(-i)^n}{n!} \int_0^t dt_1 \cdots \int_0^t dt_n T(H_I(t_1) \cdots H_I(t_n)) \tag{A.33}$$

$$\equiv T\left(e^{-i \int_0^t dt' H_I(t')} \right).$$

$\square$

The overlap integral cancels.

Note $U^\dagger(t,0)U(t,0) \equiv 1 \equiv U(t,0)U^\dagger(t,0)$. Hence

$$U^\dagger(t,0) \equiv T\left(e^{i\int_0^t dt' H_I(t')}\right) = T\left(e^{-i\int_t^0 dt' H_I(t')}\right). \tag{A.34}$$

Show: for $t_1 > t_2$,

$$U(t_1,0)U^\dagger(t_2,0) = T\left(e^{-i\int_{t_2}^{t_1} dt' H_I(t')}\right) \equiv U(t_1,t_2). \tag{A.35}$$

Proof Consider first the case $t_1 > 0, \quad t_2 < 0$. Then we have

$$U(t_1,0)\, U^\dagger(t_2,0) \equiv T\left(e^{-i\int_0^{t_1} dt' H_I(t')}\right) T\left(e^{-i\int_{t_2}^0 dt' H_I(t')}\right). \tag{A.36}$$

Now, consider the case $t_1 > t_2 > 0$ (Fig. A.2). The overlap integral is given by

$$T\left(e^{-i\int_0^{t_2} dt' H_I(t')}\right) T\left(e^{-i\int_{t_2}^0 dt' H_I(t')}\right) \equiv U(t_2,0) \times U^\dagger(t_2,0) \equiv 1. \tag{A.37}$$

$\square$

A.2 Gell-Mann–Low Theorem

Consider the n-point Green's function

$$(\psi_0 \phi(x_1) \cdots \phi(x_n)\psi_0) \tag{A.38}$$
$$\equiv \left(\psi_0 U^\dagger(t_1,0)\phi_I(x_1)U(t_1,0)U^\dagger(t_2,0)\phi_I(x_2)\cdots \phi_I(x_n)U(t_n,0)\psi_0\right)$$
$$= \left(\psi_0 U^\dagger(t_1,0)\phi_I(x_1)U(t_1,t_2)\phi_I(x_2)\cdots \phi_I(x_n)U(t_n,0)\psi_0\right).$$

We now want to express

$$\psi_0 \equiv e^{iHt}\psi_S^0(t) \tag{A.39}$$

in terms of Φ_0, where Φ_0 is the free-field ground state, i.e. vacuum. Define

$$\psi_I^0(t) \equiv e^{iH_0 t}\psi_S^0(t) = e^{iH_0 t}e^{-iHt}\psi_0 \equiv U(t,0)\psi_0. \tag{A.40}$$

We expect that in the limit $t \to \pm\infty$, we have

$$\psi_I^0(t) \to \Phi_0. \tag{A.41}$$

Proof Consider Φ_1, Φ_2, two free-field states. Then we have

$$\left(\Phi_1 | e^{iHt}\Phi_2\right) \equiv \sum_n (\Phi_1 | \psi_n)e^{iE_n t}(\psi_n | \Phi_2) \tag{A.42}$$

such that

$$H\psi_n \equiv E_n\psi_n. \tag{A.43}$$

We then have

$$\left(\Phi_1|e^{iHt}\Phi_2\right) = e^{iE_0t}\left[\left(\Phi_1|\psi_0\right)\left(\psi_0|\Phi_2\right) + \sum_{n\neq 0} e^{i(E_n-E_0)t}\left(\Phi_1|\psi_n\right)\left(\psi_n|\Phi_2\right)\right], \tag{A.44}$$

where $E_n - E_0 > 0$. Then take the limit, $t \to \pm\infty(1 \pm i\epsilon)$. As a consequence, the second term vanishes and we have

$$\lim_{t\to\pm\infty(1\pm i\epsilon)}\left(\Phi_1|e^{iHt}\Phi_2\right) = \lim_{t\to\pm\infty(1\pm i\epsilon)} e^{iE_0t}\left(\Phi_1|\psi_0\right)\left(\psi_0|\Phi_2\right), \tag{A.45}$$

which is valid for all Φ_1, Φ_2. $\square$

This then implies the general result

$$\lim_{t\to\pm\infty(1\pm i\epsilon)} e^{iHt}\Phi_2 = \lim_t e^{iE_0t}\psi_0\left(\psi_0|\Phi_2\right) \tag{A.46}$$

$$= \lim_t \psi_0|\left(\psi_0 e^{iHt}\Phi_2\right).$$

Now let

$$\Phi_2 \equiv e^{-iH_0t}\Phi_0. \tag{A.47}$$

We then have

$$\lim_{t\to\pm\infty(1\pm i\epsilon)} \frac{U^\dagger(t,0)\Phi_0}{\left(\psi_0 U^\dagger(t,0)\Phi_0\right)} \equiv \psi_0. \tag{A.48}$$

Now plug this expression into the Green's function, Eqn. A.38, e.g.

$$\lim_{t\to+\infty(1+i\epsilon)} \frac{\langle\Phi_0|U(t,0)}{\langle\Phi_0|U(t,0)\psi_0\rangle} = (\psi_0|. \tag{A.49}$$

We have

$$\left(\psi_0\phi(x_1)\cdots\phi(x_n)\psi_0\right) \tag{A.50}$$

$$= \left(\psi_0 U^\dagger(t_1,0)\phi_I(x_1)U(t_1,t_2)\phi_I(x_2)\cdots\phi_I(x_n)U(t_n,0)\psi_0\right)$$

$$= \lim_{\substack{T\to+\infty(1+i\epsilon)\\T'\to-\infty(1-i\epsilon')}} \frac{\langle\Phi_0|U(T,t_1)\phi_I(x_1)U(t_1,t_2)\phi_I(x_2)\cdots\phi_I(x_n)U(t_n,T')|\Phi_0\rangle}{\langle\Phi_0|U(T,0)\psi_0)(\psi_0|U^\dagger(T',0)\Phi_0\rangle}.$$

Consider, the denominator. Show that it is equivalent to

$$\langle\Phi_0|U(T,T')|\Phi_0\rangle. \tag{A.51}$$

Proof   Note:

$$\lim_{\substack{T\to+\infty(1+i\epsilon)\\T'\to-\infty(1-i\epsilon')}} \langle\Phi_0|U(T,T')|\Phi_0\rangle \equiv \sum_n \langle\Phi_0|U(T,0)\psi_n\rangle\left(\psi_n U^\dagger(T',0)|\Phi_0\rangle\right). \tag{A.52}$$

However,

$$\lim_{T'\to-\infty(1-i\epsilon')} U^\dagger(T',0)|\Phi_0\rangle = \lim_{T'\to-\infty(1-i\epsilon')} |\psi_0)\left(\psi_0 U^\dagger(T',0)|\Phi_0\rangle\right) \tag{A.53}$$

and using the identity

$$(\psi_n|\psi_0) \equiv \delta_{n,0} \tag{A.54}$$

we have

$$\langle\Phi_0|U(T,T')|\Phi_0\rangle = \lim_{T,T'} \langle\Phi_0|U(T,0)\psi_0\rangle \left(\psi_0 U^\dagger(T',0)|\Phi_0\rangle \equiv \text{denominator.} \tag{A.55}\right.$$

$\square$

Thus,

$$(\psi_0\phi(x_1)\cdots\phi(x_n)\psi_0) \tag{A.56}$$

$$= \lim_{\substack{T\to+\infty(1+i\epsilon)\\T'\to-\infty(1-i\epsilon')}} \frac{\langle\Phi_0|U(T,t_1)\phi_I(x_1)U(t_1,t_2)\phi_I(x_2)\cdots\phi_I(x_n)U(t_n,T')|\Phi_0\rangle}{\langle\Phi_0|U(T,T')|\Phi_0\rangle} \ .$$

Finally, we have

$$G(x_1,\ldots,x_n) \equiv (\psi_0 T\left(\phi(x_1)\cdots\phi(x_n)\right)\psi_0) \tag{A.57}$$

$$\equiv \lim_{\substack{T\to+\infty(1+i\epsilon)\\T'\to-\infty(1-i\epsilon')}} \frac{\langle\Phi_0|T\left(U(T,T')\phi_I(x_1)\cdots\phi_I(x_n)\right)|\Phi_0\rangle}{\langle\Phi_0|U(T,T')|\Phi_0\rangle}$$

$$\equiv \lim_{\substack{T\to+\infty(1+i\epsilon)\\T'\to-\infty(1-i\epsilon')}} \frac{\langle\Phi_0|T\left(e^{-i\int_{T'}^T dt' H_I(t')}\phi_I(x_1)\cdots\phi_I(x_n)\right)|\Phi_0\rangle}{\langle\Phi_0|T\left(e^{-i\int_{T'}^T dt' H_I(t')}\right)|\Phi_0\rangle} \ .$$

B

Appendix B Wick's Theorem

Consider free-field theory and the operator

$$T\left(\phi(x_1)\cdots\phi(x_n)\right),\tag{B.1}$$

where

$$\phi(x)\equiv\phi_+ + \phi_-,\tag{B.2}$$

with

$$\phi_+(x)\equiv\int\frac{d^3\vec{p}}{(2\pi)^3 2E_p}e^{ip\cdot x}a^\dagger(\vec{p})\tag{B.3}$$

$$\phi_-(x)\equiv\int\frac{d^3\vec{p}}{(2\pi)^3 2E_p}e^{-ip\cdot x}a(\vec{p}).$$

Or in the case that $\phi(x)$ is replaced by $\psi_\alpha(x)$, we have

$$\psi_\alpha(x)\equiv\psi_{\alpha+} + \psi_{\alpha-}\equiv\bar{v}(x) + u(x),\tag{B.4}$$

with

$$\bar{v}(x)\equiv\psi_{\alpha+}(x)\equiv\int\frac{d^3\vec{p}}{(2\pi)^3 2E_p}\sum_s e^{ip\cdot x}v(\vec{p},s)d^\dagger(\vec{p},s)\tag{B.5}$$

$$u(x)\equiv\psi_{\alpha-}(x)\equiv\int\frac{d^3\vec{p}}{(2\pi)^3 2E_p}\sum_s e^{-ip\cdot x}u(\vec{p},s)b(\vec{p},s).$$

We also define

$$\bar{\psi}_\alpha(x)\equiv\bar{u}(x) + v(x).\tag{B.6}$$

Similarly, we divide vector bosons into their creation and annihilation parts.

Now define the normal product with all $+$ fields on the left and all $-$ fields on the right. For example

$$N\left(\phi(x_1)\phi(x_2)\right)\equiv\phi_+(x_1)\phi_+(x_2) + \phi_+(x_1)\phi_-(x_2) + \phi_+(x_2)\phi_-(x_1) + \phi_-(x_1)\phi_-(x_2).\tag{B.7}$$

We can now define the Wick contraction for two fields by

$$T\left(\phi(x_1)\phi(x_2)\right)\equiv N\left(\phi(x_1)\phi(x_2)\right) + \underbracket{\phi(x_1)\phi}(x_2).\tag{B.8}$$

Consider now the left-hand and right-hand sides of Eqn. B.8.

$$\text{LHS} = \phi(x_1)\phi(x_2)\theta(x_1^0 - x_2^0) + \phi(x_2)\phi(x_1)\theta(x_2^0 - x_1^0) \tag{B.9}$$
$$\equiv (\phi_+(x_1)\phi_+(x_2) + \phi_+(x_1)\phi_-(x_2)$$
$$+ \phi_-(x_1)\phi_+(x_2) + \phi_-(x_1)\phi_-(x_2)) + (1 \leftrightarrow 2)\,.$$

$$\text{RHS} = \{N\left(\phi(x_1)\phi(x_2)\right) + [\phi_-(x_1),\ \phi_+(x_2)]\}\theta(x_1^0 - x_2^0) + (1 \leftrightarrow 2)\,. \tag{B.10}$$

Thus

$$\text{LHS} \equiv \text{RHS} \tag{B.11}$$
$$T(\phi(x_1)\phi(x_2)) = N(\phi(x_1)\phi(x_2)) + [\phi_-(x_1),\ \phi_+(x_2)]\,\theta(x_1^0 - x_2^0)$$
$$+ [\phi_-(x_2),\ \phi_+(x_1)]\,\theta(x_2^0 - x_1^0)$$
$$\equiv N\left(\phi(x_1)\phi(x_2)\right) + i\Delta_F(x_1 - x_2),$$

i.e. the last two terms in the first line give us the Feynman propagator. Therefore,

$$\underbrace{\phi(x_1)\phi}(x_2) \equiv i\Delta_F(x_1 - x_2). \tag{B.12}$$

For fermions,

$$T\left(\psi(x_1)\bar{\psi}(x_2)\right) \equiv N\left(\psi(x_1)\bar{\psi}(x_2)\right) + iS_F(x_1 - x_2), \tag{B.13}$$

where

$$N\left(\psi(x_1)\bar{\psi}(x_2)\right) \equiv N\big((\bar{v}(x_1) + u(x_1))(\bar{u}(x_2) + v(x_2))\big) \tag{B.14}$$
$$\equiv \bar{v}(x_1)\bar{u}(x_2) - \bar{u}(x_2)u(x_1) + u(x_1)v(x_2) + \bar{v}(x_1)v(x_2).$$

Proof

$$T\left(\psi(x_1)\bar{\psi}(x_2)\right) \equiv \psi(x_1\bar{\psi}(x_2)\theta(x_1^0 - x_2^0) - \bar{\psi}(x_2)\psi(x_1)\theta(x_2^0 - x_1^0) \tag{B.15}$$
$$\equiv (\bar{v}(x_1) + u(x_1))(\bar{u}(x_2) + v(x_2))\theta(x_1^0 - x_2^0)$$
$$-(\bar{u}(x_2) + v(x_2))(\bar{v}(x_1) + u(x_1))\theta(x_2^0 - x_1^0)$$
$$= \big[N\left(\psi(x_1)\bar{\psi}(x_2)\right) + \{u(x_1),\ \bar{u}(x_2)\}\big]\theta(x_1^0 - x_2^0)$$
$$+ \big[N\left(\psi(x_1)\bar{\psi}(x_2)\right) - \{v(x_2),\ \bar{v}(x_1)\}\big]\theta(x_2^0 - x_1^0)$$
$$\equiv N\left(\psi(x_1)\bar{\psi}(x_2)\right) + iS_F(x_1 - x_2).$$

$\square$

In addition, we have

$$T(\psi(x_1)\psi(x_2)) \equiv N(\psi(x_1)\psi(x_2))\,. \tag{B.16}$$

Proof

$$T\left(\psi(x_1)\psi(x_2)\right) \equiv T\big((\bar{v}(x_1) + u(x_1))(\bar{v}(x_2) + u(x_2))\big) \tag{B.17}$$
$$= (\bar{v}(x_1) + u(x_1))(\bar{v}(x_2) + u(x_2))\theta(x_1^0 - x_2^0)$$
$$-(\bar{v}(x_2) + u(x_2))(\bar{v}(x_1) + u(x_1))\theta(x_2^0 - x_1^0)$$
$$= (\bar{v}(x_1)\bar{v}(x_2) + u(x_1)u(x_2) + \bar{v}(x_1)u(x_2) - \bar{v}(x_2)u(x_1))$$
$$\equiv N(\psi(x_1)\psi(x_2))\,.$$

$\square$

Hence we find the Wick contractions for fermions given by

-

$$\psi(x_1)\bar{\psi}(x_2) \equiv iS_F(x_1 - x_2), \tag{B.18}$$

-

$$\psi(x_1)\psi(x_2) \equiv 0. \tag{B.19}$$

In general, the normal product of n fields is given by ($\pm$ labels are suppressed)

$$N\left(\phi(x_1)\cdots\phi(x_n)\right) \equiv \delta_p \phi(x_{p_1})\cdots\phi(x_{p_n}), \tag{B.20}$$

where all annihilation operators are to the right; p is a permutation of the original order $1,\cdots,n$; and

$$\delta_p = \begin{cases} +1 & \text{if the permutation of fermions is even} \\ -1 & \text{if the permutation of fermions is odd} \end{cases}. \tag{B.21}$$

Now, generalize this definition further.

$$N(c\phi(x_1)\cdots\phi(x_n)) \equiv cN(\phi(x_1)\cdots\phi(x_n)), \tag{B.22}$$

where c is a c number. This implies that

$$N\left(\phi(x_1)\phi(x_2)\phi(x_3)\cdots\phi(x_n)\right) \equiv \phi(x_1)\phi(x_2)N(\phi(x_3)\cdots\phi(x_n)). \tag{B.23}$$

Finally, define

$$N\left(\phi(x_1)\phi(x_2)\phi(x_3)\cdots\phi(x_n)\right) \equiv \delta_p\phi(x_1)\phi(x_3)N(\phi(x_2)\cdots\phi(x_n)) \tag{B.24}$$

and

$$T\left(\phi(x_1)\cdots\phi(x_n)\right) \equiv \sum_{\text{all time orderings}} \delta_p\phi(x_{p_1})\cdots\phi(x_{p_n})\theta(x_{p_1}^0 - x_{p_2}^0)\theta(x_{p_3}^0 - x_{p_4}^0)\cdots. \tag{B.25}$$

We are now in a position to state Wick's theorem.

$$\begin{aligned} T\left(\phi(x_1)\cdots\phi(x_n)\right) \equiv\ & N(\phi(x_1)\cdots\phi(x_n)) \\ & + N\left(\phi(x_1)\phi(x_2)\cdots\phi(x_n)\right) \\ & + N\left(\phi(x_1)\phi(x_2)\phi(x_3)\cdots\phi(x_n)\right) \\ & + \text{all permutations of one contraction} \\ & + N\left(\phi(x_1)\phi(x_2)\phi(x_3)\phi(x_4)\cdots\phi(x_n)\right) \\ & + N\left(\phi(x_1)\phi(x_2)\phi(x_3)\phi(x_4)\cdots\phi(x_n)\right) \\ & + \text{all permutations of two contractions} \\ & + \ldots \text{until all pairs of fields are contracted} \\ & \quad\text{in all possible ways.} \end{aligned} \tag{B.26}$$

Thus, the vacuum-to-vacuum amplitude of the time-ordered product of free fields in the Fock vacuum is given by

$$\langle \Phi_0 | T(\phi(x_1) \cdots \phi(x_n)) | \Phi_0 \rangle = N \left(\underbrace{\phi(x_1) \phi(x_2) \phi(x_3) \cdots \phi(x_n)} \right) \quad \text{totally contracted.}$$

$$(B.27)$$

Proof   *Proof by induction*
We have proven it to be true for $n = 2$ fields. Now assume it is valid for $n - 1$ fields and prove that it is true for n fields. Let $x_1^0 > x_2^0 > x_3^0 > \cdots > x_n^0$. (If not then relabel the points and both sides are unchanged.)

$$T(\phi(x_1) \cdots \phi(x_n)) = \phi(x_1) \cdots \phi(x_n) \qquad\qquad (B.28)$$
$$\equiv \phi(x_1) T (\phi(x_2) \cdots \phi(x_n))$$
$$= \phi(x_1) \left[N (\phi(x_2) \cdots \phi(x_n)) + N \left(\underbrace{\phi(x_2) \phi(x_3)} \cdots \phi(x_n) \right) \right.$$
$$\left. + \text{ sum over all contractions} \right]$$
$$= (\phi_+(x_1) + \phi_-(x_1)) [\ldots]$$
$$= \left[N(\phi_+(x_1) \phi(x_2) \cdots \phi(x_n)) \right.$$
$$+ N \left(\phi_+(x_1) \underbrace{\phi(x_2) \phi(x_3)} \cdots \phi(x_n) \right) + \cdots \right]$$
$$+ \left[\phi_-(x_1) N (\phi(x_2) \cdots \phi(x_n)) \right.$$
$$\left. + \phi_-(x_1) N \left(\underbrace{\phi(x_2) \phi(x_3)} \cdots \phi(x_n) \right) + \cdots \right].$$

Consider the first term in the last line above. We have

$$\phi_-(x_1) N(\phi(x_2) \cdots \phi(x_n)) = \phi_-(x_1) \left[\sum_{\text{all combinations}} \delta_p \phi(x_{p_2}) \cdots \phi(x_{p_n}) \right] \quad (B.29)$$
$$= \phi_-(x_1) \left[\cdots \delta_p \phi_+(x_{p_2}) \phi_+(x_{p_3}) \right.$$
$$\left. \times \cdots \phi_-(x_{p_{i+1}}) \cdots \phi_-(x_{p_n}) \cdots \right]$$
$$= \left[\cdots \delta_p \left(\pm \phi_+(x_{p_2}) \phi_-(x_1) + \{\phi_-(x_1),\ \phi_+(x_{p_2})\}_\mp \right) \right.$$
$$\left. \times \phi_+(x_{p_3}) \cdots \phi_+(x_{p_i}) \phi_-(x_{p_{i+1}}) \cdots \phi_-(x_{p_n}) \cdots \right],$$

where in the last line we have a $+$ sign if $\phi(x_{p_2}), \phi_-(x_1)$ are both bosons or one boson and one fermion, and a $-$ sign if they are both fermions. For the bracket we have

$$\{\phi_-(x_1),\ \phi_+(x_{p_2})\}_\mp \, \theta(x_1^0 - x_{p_2}^0) = \{\text{either } 0 \text{ or } i\Delta_F(x_1 - x_{p_2})$$
$$\text{the propagator for appropriate fields}$$

$$(B.30)$$

Example 1.

We then continue this process on each term until the annihilation operator, $\phi_-(x_1)$, is to the right of all creation operators. As a result, we have

$$\phi_-(x_1)N(\phi(x_2)\cdots\phi(x_n)) \tag{B.31}$$

$$= \Big[\cdots\delta_{p'}\phi_+(x_{p_2})\phi_+(x_{p_3})\cdots\phi_+(x_{p_i})\phi_-(x_1)\phi_-(x_{p_{i+1}})\cdots\phi_-(x_{p_n})\cdots$$
$$+\cdots\delta_{p'}\underline{\phi_-(x_1)\phi_+(x_{p_2})}\phi_+(x_{p_3})\underline{\cdots\cdots}\ \cdots$$
$$+\cdots\delta_{p'}\underline{\phi_-(x_1)\phi_+(x_{p_2})}\phi_+\underline{\phi_-(x_1)\phi_+(x_{p_2})}\phi_+(x_{p_3})\underline{\cdots\cdots}\ \cdots$$
$$+\text{sum over all single contractions with this time ordering}\Big]$$

The first line, i.e. the sum over all non-contractions, just gives

$$N(\phi_-(x_1)\phi(x_2)\cdots\phi(x_n)). \tag{B.32}$$

The sum over all single contractions gives

$$N\Big(\underline{\phi(x_1)\phi(x_2)}\phi(x_3)\cdots\phi(x_n)\Big) \tag{B.33}$$
$$+N\Big(\underline{\phi(x_1)\phi(x_2)\phi(x_3)}\cdots\phi(x_n)\Big)$$
$$+\text{all single contractions of } \phi(x_1) \text{ with } \phi(x_i).$$

Upon adding all the terms in Eqns. B.28–B.31 we obtain, Eqn. B.26.   □

We finish this appendix with a few examples.

Example 1: see Fig. B.1.

$$\langle\Phi_0|T\big(\phi(x_1)\psi(x_2)\bar\psi(x_3)\phi^\dagger(x_4)\big)|\Phi_0\rangle \tag{B.34}$$

$$= \langle\Phi_0|N\Big(\underline{\phi(x_1)\psi(x_2)\bar\psi(x_3)\phi^\dagger(x_4)}\Big)|\Phi_0\rangle$$

$$= \langle\Phi_0|N\Big(\underline{\phi(x_1)\phi^\dagger(x_4)}\ \underline{\psi(x_2)\bar\psi(x_3)}\Big)|\Phi_0\rangle$$

$$= i\Delta_F(x_1-x_4)iS_F(x_2-x_3).$$

Example 2: see Fig. B.2.

$$\langle\Phi_0|T\big(\phi(x_1)\phi^\dagger(x_2)\phi(x_3)\phi^\dagger(x_4)\big)|\Phi_0\rangle \tag{B.35}$$

$$= \langle\Phi_0\Big[N\Big(\underline{\phi(x_1)\phi^\dagger(x_2)\phi(x_3)\phi^\dagger(x_4)}\Big)|\Phi_0\rangle$$

$$+\langle\Phi_0|N\Big(\underline{\phi(x_1)\phi^\dagger(x_2)}\ \underline{\phi(x_3)\phi^\dagger(x_4)}\Big)|\Phi_0\rangle\Big]$$

$$= i\Delta_F(x_1-x_4)i\Delta_F(x_3-x_2) + i\Delta_F(x_1-x_2)i\Delta_F(x_3-x_4).$$

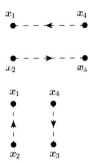

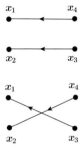

Fig. B.2   Example 2.

Fig. B.3   Example 3.

Example 3: see Fig. B.3.

$$\langle \Phi_0 | T\big(\psi(x_1)\psi(x_2)\bar{\psi}(x_3)\bar{\psi}(x_4)\big) | \Phi_0 \rangle \tag{B.36}$$

$$= \langle \Phi_0 \Big[N\Big(\psi(x_1)\psi(x_2)\bar{\psi}(x_3)\bar{\psi}(x_4) \Big) | \Phi_0 \rangle$$

$$+ \langle \Phi_0 | N\Big(\psi(x_1)\psi(x_2)\bar{\psi}(x_3)\bar{\psi}(x_4) \Big) | \Phi_0 \rangle \Big]$$

$$= iS_F(x_1 - x_4)iS_F(x_2 - x_3) - iS_F(x_1 - x_3)iS_F(x_2 - x_4).$$

Appendix C One-Loop Calculations in QED

C.1 Vacuum Polarization

Calculate the one-loop Feynman diagram, Fig. 30.1, using dimensional regularization. At one loop, the photon propagator is given by

$$iD_{\mu\nu}(q) = i\Delta_{\mu\nu}(q) + i\Delta_\mu{}^\lambda (i\Pi_{\lambda\rho}) i\Delta^\rho{}_\nu, \tag{C.1}$$

where the polarization tensor, $\Pi_{\lambda\rho}(q)$, is given by

$$i\Pi_{\lambda\rho}(q) \equiv -\int \frac{d^4p}{(2\pi)^4} \left[(iS_F(p))_{\beta\beta'} (-ie\gamma_\lambda)_{\beta'\alpha'} (iS_F(p+q))_{\alpha'\alpha} (-ie\gamma_\rho)_{\alpha\beta} \right] \tag{C.2}$$

$$= -e^2 \int \frac{d^4p}{(2\pi)^4} \text{Tr} \left[\frac{\not{p}+m}{p^2-m^2+i\epsilon} \gamma_\lambda \frac{\not{p}+\not{q}+m}{(p+q)^2-m^2+i\epsilon} \gamma_\rho \right]. \tag{C.3}$$

Note the minus sign in front is a result of the closed fermion loop.

Evaluating

$$q^\lambda \Pi_{\lambda\rho} = ie^2 \int \frac{d^4p}{(2\pi)^4} \text{Tr} \left[\frac{1}{\not{p}-m} [(\not{p}+\not{q}-m) - (\not{p}-m)] \frac{1}{\not{p}+\not{q}-m} \gamma_\rho \right] \tag{C.4}$$

$$= ie^2 \int \frac{d^4p}{(2\pi)^4} \text{Tr} \left[\left(\frac{1}{\not{p}-m} - \frac{1}{\not{p}+\not{q}-m} \right) \gamma_\rho \right] \equiv 0.$$

Note, also,

$$\Pi_{\lambda\rho}(q) = ie^2 \int \frac{d^4p}{(2\pi)^4} \frac{\{\text{Tr}(\gamma_\alpha \gamma_\lambda \gamma_\beta \gamma_\rho) p^\alpha (p+q)^\beta + \text{Tr}(\gamma_\lambda \gamma_\rho) m\}}{p^2-m^2+i\epsilon(p+q)^2-m^2+i\epsilon} \tag{C.5}$$

looks to be quadratically divergent. However, since $q^\lambda \Pi_{\lambda\rho} = 0$, we have

$$\Pi_{\lambda\rho}(q) \equiv (-g_{\lambda\rho} q^2 + q_\lambda q_\rho) \Pi(q^2). \tag{C.6}$$

Thus, $\Pi_{\lambda\rho}$ is only logarithmically divergent.

ASIDE: **Feynman parameters**

$$\frac{1}{a_1 a_2 \cdots a_n} = (n-1)! \int_0^1 \frac{dz_1 \cdots dz_n \delta(1 - \sum_{i=1}^n z_i)}{(a_1 z_1 + a_2 z_2 + \cdots + a_n z_n)^n} \tag{C.7}$$

$$\frac{1}{a_1^2 a_2 \cdots a_n} = n! \int_0^1 \frac{z_1 dz_1 \cdots dz_n \delta(1 - \sum_{i=1}^n z_i)}{(a_1 z_1 + a_2 z_2 + \cdots + a_n z_n)^{n+1}}.$$

Thus,

$$\frac{1}{p^2 - m^2 + i\epsilon} \frac{1}{(p+q)^2 - m^2 + i\epsilon} = \int_0^1 \frac{dz_1 dz_2 \delta(1 - z_1 - z_2)}{[(p^2 - m^2 + i\epsilon)z_1 + ((p+q)^2 - m^2 + i\epsilon)z_2]^2} \tag{C.8}$$

$$= \int_0^1 \frac{d\alpha}{[\ell^2 + a^2 + i\epsilon]^2},$$

where

$$\ell \equiv p + q\alpha, \quad z_2 = \alpha, \quad z_1 = 1 - \alpha \tag{C.9}$$

and

$$a^2 \equiv q^2 \alpha(1 - \alpha) - m^2. \tag{C.10}$$

Upon changing momentum integration variable from p_μ to ℓ_μ, we have

$$\Pi_{\lambda\rho}(q) = ie^2 \int_0^1 d\alpha \left\{ \text{Tr}(\gamma_\alpha \gamma_\lambda \gamma_\beta \gamma_\rho) \left(I^{\alpha\beta} - q^\alpha I^\beta \alpha + q^\beta I^\alpha (1 - \alpha) - q^\alpha q^\beta \alpha (1 - \alpha) I \right) \right. \tag{C.11}$$
$$\left. + \text{Tr}(\gamma_\lambda \gamma_\rho) m^2 I \right\},$$

where we have defined

$$I_{\lambda\rho} \equiv \int \frac{d^4\ell}{(2\pi)^4} \frac{\ell_\lambda \ell_\rho}{(\ell^2 + a^2 + i\epsilon)^2}, \tag{C.12}$$

$$I_\lambda \equiv \int \frac{d^4\ell}{(2\pi)^4} \frac{\ell_\lambda}{(\ell^2 + a^2 + i\epsilon)^2},$$

$$I \equiv \int \frac{d^4\ell}{(2\pi)^4} \frac{1}{(\ell^2 + a^2 + i\epsilon)^2}.$$

Let's now evaluate I. Consider the integral over ℓ_0 from $-\infty$ to $+\infty$ over the real axis. Note,

$$\ell_0^2 = \vec{\ell}^{\,2} - a^2 - i\epsilon = 0 \tag{C.13}$$

gives the position of the poles in the complex ℓ_0 plane. Let's analytically continue q_μ to Euclidean momentum, such that $q_0 \Rightarrow iq_4$ and

$$q^2 = q_0^2 - \vec{q}^{\,2} \Rightarrow -(q_4^2 + \vec{q}^{\,2}) \equiv -q_E^2. \tag{C.14}$$

Also,

$$a^2 \Rightarrow -a_E^2 \tag{C.15}$$

with

$$a_E^2 = q_E^2 \alpha(1 - \alpha) + m^2 > 0. \tag{C.16}$$

Then the poles are located at

$$\ell_0 = \pm \left(\sqrt{\vec{\ell}^{\,2} + a_E^2} - i\epsilon \right) \equiv \pm \tilde{\ell}. \tag{C.17}$$

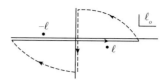

Fig. C.1   Integration along the real ℓ_0 axis is extended into the complex ℓ_0 plane.

We thus have

$$I \equiv \int_{-\infty}^{+\infty} \frac{d\ell_0 d^3\vec{\ell}}{(2\pi)^4} \frac{1}{(\ell_0 - \tilde{\ell})^2 (\ell_0 + \tilde{\ell})^2}, \tag{C.18}$$

where the integral is as in Fig. C.1.

According to Cauchy's theorem we have

$$\int_{-\infty}^{+\infty} f(\ell_0)d\ell_0 = i \int_{-\infty}^{+\infty} f(i\ell_4)d\ell_4 + (\text{integral over the contour at infinity} = 0). \tag{C.19}$$

Therefore, we have

$$I = i \int_{-\infty}^{+\infty} \frac{d\ell_4 d^3\vec{\ell}}{(2\pi)^4} \frac{1}{(i\ell_4 - \tilde{\ell})^2(i\ell_4 + \tilde{\ell})^2} \tag{C.20}$$

$$= i \int_{-\infty}^{+\infty} \frac{d^4\ell_E}{(2\pi)^4} \frac{1}{(\ell_E^2 + a_E^2)^2},$$

where $\ell_E^2 = \ell_4^2 + \vec{\ell}\,^2 = -\ell^2$. Now the integral

$$\int d^4\ell_E = \int_0^\infty \ell^3 d\ell \underbrace{\int_0^{2\pi} d\theta_1 \int_0^\pi \sin\theta_2 d\theta_2}_{4\pi} \underbrace{\int_0^\pi \sin^2\theta_3 d\theta_3}_{\pi/2} = 2\pi^2 \int_0^\infty \ell^3 d\ell. \tag{C.21}$$

Hence, the integral

$$\int \frac{d^4\ell_E}{\ell^4} \sim \int_0^\infty \frac{d\ell}{\ell} \tag{C.22}$$

is logarithmically divergent in the ultraviolet, i.e. as $\ell \to \infty$. We must thus regulate the integral, i.e. make it finite, and then renormalize the Green's function, i.e. subtract away the divergent term. We will use dimensional regularization. We extend space–time to d dimensions such that

$$\int d^d\ell_E = \int_0^\infty \ell^{d-1} d\ell \int_0^{2\pi} d\theta_1 \int_0^\pi \sin\theta_2 d\theta_2 \int_0^\pi \sin^2\theta_3 d\theta_3 \tag{C.23}$$

$$\times \cdots \times \int_0^\pi \sin^{d-2}\theta_{d-1} d\theta_{d-1} \equiv \frac{2\pi^{d/2}}{\Gamma(\frac{d}{2})} \int_0^\infty \ell^{d-1} d\ell.$$

Note, the integral is defined for integer d and then evaluated for d arbitrary. We have used the following identities,

$$\int_0^\pi \sin^m\theta d\theta \equiv \pi^{1/2} \frac{\Gamma(\frac{m+1}{2})}{\Gamma(\frac{m+2}{2})}, \quad \Gamma(m+1) \equiv m!, \quad 0! \equiv 1. \tag{C.24}$$

Finally, we find

$$I = i \int_0^\infty \frac{d\ell\, \ell^{d-1}}{(2\pi)^d} \frac{1}{(\ell^2 + a_E^2)^2} \frac{2\pi^{d/2}}{\Gamma(\frac{d}{2})}. \tag{C.25}$$

It can be shown that this is an analytic function in d for $Re(d) < 4$.

To evaluate the integral, we use the identity

$$\int_0^\infty \frac{dt\, t^{m-1}}{(t+a^2)^n} \equiv \frac{1}{(a^2)^{n-m}} \frac{\Gamma(m)\Gamma(n-m)}{\Gamma(n)}. \tag{C.26}$$

Then let $t = \ell^2$, $dt = 2\ell\, d\ell$ and we have

$$I = \frac{i\pi^{d/2}}{(2\pi)^d \Gamma(\frac{d}{2})} \int_0^\infty \frac{dt\, t^{\frac{d}{2}-1}}{(t+a_E^2)^2} \tag{C.27}$$

$$= \frac{i}{(4\pi)^{\frac{d}{2}}} \left(\frac{1}{a_E^2}\right)^{\frac{4-d}{2}} \Gamma(\frac{4-d}{2}).$$

Now let $d = 4 - \epsilon$ with $\epsilon > 0$. Using the identity $z\Gamma(z) \equiv \Gamma(z+1)$, we have

$$\Gamma(\frac{4-d}{2}) \equiv \Gamma(\frac{\epsilon}{2}) = \frac{2\Gamma(1+\frac{\epsilon}{2})}{\epsilon} = \frac{2}{\epsilon} - \gamma + O(\epsilon), \tag{C.28}$$

with a pole as $\epsilon \to 0$.

ASIDE: Some useful identities.

$$\psi(x) \equiv \frac{d\ln\Gamma(x)}{dx} = \frac{1}{\Gamma(x)} \frac{d\Gamma(x)}{dx}. \tag{C.29}$$

$$\Gamma(1 + \frac{\epsilon}{2}) = \Gamma(1) + \frac{\epsilon}{2} \frac{d\Gamma(x)}{dx}\big|_{x=1} + O(\epsilon^2) \tag{C.30}$$

$$= 1 + \frac{\epsilon}{2}\psi(1) + O(\epsilon^2).$$

$$\psi(1) \equiv -\gamma \quad \text{Euler constant} \tag{C.31}$$

$$\psi(n+1) = 1 + \frac{1}{2} + \cdots + \frac{1}{n} - \gamma \quad \text{for } n \geq 1.$$

$$\gamma = 0.57721\ldots \tag{C.32}$$

$$\Gamma(\epsilon - n) = \frac{(-1)^n}{n} \frac{1}{[\frac{1}{\epsilon} + \psi(n+1) + O(\epsilon)]} \quad \text{for } n \geq 1. \tag{C.33}$$

Using the above identities and expanding to order ϵ^0 (since in the end we will take the limit $\epsilon \to 0$), we find

$$I = \frac{i}{(4\pi)^2} \frac{(4\pi)^{\frac{\epsilon}{2}}}{(a_E^2)^{\epsilon/2}} \left(\frac{2}{\epsilon} - \gamma + O(\epsilon)\right)$$

$$= \frac{i}{(4\pi)^2} e^{-\frac{\epsilon}{2}(\ln\frac{a_E^2}{4\pi})} \left(\frac{2}{\epsilon} - \gamma + O(\epsilon)\right)$$

$$I = \frac{i}{16\pi^2} \left[\frac{2}{\epsilon} - (\ln a_E^2 - \ln 4\pi + \gamma) + O(\epsilon)\right]. \tag{C.34}$$

Similarly, for

$$I_\lambda \equiv \int \frac{d^4\ell}{(2\pi)^d} \frac{\ell_\lambda}{(\ell^2 + a^2 + i\epsilon)^2}, \tag{C.35}$$

we analytically continue to Euclidean momenta, i.e. Wick rotation, and find via symmetric integration

$$I_\lambda = 0. \tag{C.36}$$

Finally, we evaluate the tensor

$$
\begin{aligned}
I_{\lambda\rho} &= \frac{g_{\lambda\rho}}{d} \int \frac{d^d\ell}{(2\pi)^d} \frac{\ell^2}{(\ell^2 + a^2 + i\epsilon)^2} \\
&= \frac{-ig_{\lambda\rho}}{d} \int \frac{d^d\ell_E}{(2\pi)^d} \frac{\ell_E^2}{(\ell_E^2 + a_E^2)^2} \\
&= \frac{-ig_{\lambda\rho}}{(2\pi)^d d} \frac{\pi^{d/2}}{\Gamma(\frac{d}{2})} \underbrace{\int_0^\infty \frac{dt\, t^{d/2}}{(t + a_E^2)^2}}_{\left(\frac{1}{a_E^2}\right)^{1-\frac{2}{2}} \frac{\Gamma(\frac{d}{2}+1)\Gamma(1-\frac{d}{2})}{\Gamma(2)}} \\
&= \frac{-ig_{\lambda\rho}}{2(4\pi)^2} \frac{(4\pi)^{\epsilon/2}\Gamma(\frac{\epsilon}{2} - 1)}{(a_E^2)^{\frac{\epsilon}{2}-1}} \\
&= \frac{-ig_{\lambda\rho}}{2(16\pi^2)} a_E^2 e^{-\frac{\epsilon}{2}(\ln \frac{a_E^2}{4\pi})} \left(-\frac{2}{\epsilon} - \psi(2)\right), \\
I_{\lambda\rho} &= \frac{ig_{\lambda\rho}}{2(16\pi^2)} a_E^2 \left[\frac{2}{\epsilon} + 1 - (\ln a_E^2 - \ln 4\pi + \gamma) + O(\epsilon)\right].
\end{aligned}
$$

We now have $\Pi_{\lambda\rho}(q)$ given by the following equation (note, I and $I_{\lambda\rho}$ are still functions of q_E^2, which we must still continue back to Minkowski space, and in $d = 4 - \epsilon$ dimensions the coupling constant has dimensions of mass to the power $\frac{\epsilon}{2}$; thus the mass parameter μ is added so that e remains dimensionless):

$$
\begin{aligned}
\Pi_{\lambda\rho}(q) &= 4ie^2\mu^\epsilon \int_0^1 d\alpha \left\{(g_{\alpha\lambda}g_{\beta\rho} + g_{\alpha\rho}g_{\beta\lambda} - g_{\alpha\beta}g_{\lambda\rho})\right\} \tag{C.38} \\
&\quad \left(I^{\alpha\beta} - q^\alpha q^\beta \alpha(1-\alpha)I\right) + g_{\lambda\rho}m^2 I\} \\
&= 4ie^2\mu^\epsilon \int_0^1 d\alpha \left\{[2I_{\lambda\rho} - (g_{\alpha\beta}I^{\alpha\beta})g_{\lambda\rho}]\right\} \\
&\quad -2q_\lambda q_\rho \alpha(1-\alpha)I + (q^2\alpha(1-\alpha) + m^2)g_{\lambda\rho}I\} \\
&= \frac{4ie^2}{16\pi^2} \int_0^1 d\alpha \left\{-ig_{\lambda\rho}a_E^2 + ig_{\lambda\rho}(q^2\alpha(1-\alpha) + m^2)\right. \\
&\quad -2iq_\lambda q_\rho \alpha(1-\alpha)\} \left(\frac{2}{\epsilon} - (\ln(a_E^2/\mu^2) - \ln 4\pi + \gamma)\right),
\end{aligned}
$$

where in the last line the factor

$$[[2I_{\lambda\rho} - (g_{\alpha\beta}I^{\alpha\beta})g_{\lambda\rho}] = -i \frac{g_{\lambda\rho}}{16\pi^2} a_E^2 \left[\frac{2}{\epsilon} - (\ln a_E^2 - \ln 4\pi + \gamma)\right]. \tag{C.39}$$

Now continuing back to Minkowski space we have

$$a_E^2 \Rightarrow -q^2\alpha(1-\alpha) + m^2 \qquad (C.40)$$

and

$$\ln a_E^2/\mu^2 \Rightarrow \ln\left|\frac{q^2\alpha(1-\alpha)-m^2}{\mu^2}\right| + i\pi\theta(q^2\alpha(1-\alpha)-m^2) \quad \text{where} \quad \theta(x) = \begin{cases} 1 & x>0 \\ 0 & x<0 \end{cases}$$
$$(C.41)$$

At last we obtain

$$\Pi_{\lambda\rho}(q) = \frac{e^2}{2\pi^2}\left(q_\lambda q_\rho - g_{\lambda\rho}q^2\right)\left\{\frac{1}{3\epsilon} - \int_0^1 d\alpha\ \alpha(1-\alpha)\left[\ln\left|\frac{q^2\alpha(1-\alpha)-m^2}{\mu^2}\right|\right.\right.$$
$$\left.\left. + i\pi\theta(q^2\alpha(1-\alpha)-m^2) + \gamma - \ln 4\pi\right]\right\}. \qquad (C.42)$$

C.2 The Renormalized Polarization Tensor

At this order in perturbation theory, in order to cancel the divergence in $\frac{1}{\epsilon}$ as $\epsilon \to 0$, we must add a counter term to the Lagrangian density of the form

$$-\frac{1}{4}(Z_3 - 1)F_{\mu\nu}^2. \qquad (C.43)$$

The one-loop contribution to the polarization tensor plus a contribution from the counter term has the form[1]

$$i\Pi_{\lambda\rho}^r(q) = i\Pi_{\lambda\rho}^u(q) + i(Z_3 - 1)(-q^2 g_{\lambda\rho} + q_\lambda q_\rho) \qquad (C.44)$$
$$\equiv i(q_\lambda q_\rho - g_{\lambda\rho}q^2)\Pi^r(q^2)$$

and is, by definition, finite. The renormalization constant

$$Z_3 - 1 = -\frac{e^2}{6\pi^2\epsilon} + C, \qquad (C.45)$$

where C is a constant, finite correction. In the $\overline{MS}$ subtraction scheme we choose

$$C = \frac{e^2}{12\pi^2}(-\ln 4\pi + \gamma). \qquad (C.46)$$

This is equivalent to choosing the renormalization point $\mu^2 = \bar{\mu}^2\left(\frac{e^\gamma}{4\pi}\right)$. In the $\overline{MS}$ scheme, the renormalized polarization tensor is given by

$$\Pi_{\lambda\rho}^r(q) = -\frac{e^2}{2\pi^2}\left(q_\lambda q_\rho - g_{\lambda\rho}q^2\right)\int_0^1 d\alpha\ \alpha(1-\alpha) \qquad (C.47)$$
$$\left[\ln\left|\frac{q^2\alpha(1-\alpha)-m^2}{\mu^2}\right| + i\pi\theta(q^2\alpha(1-\alpha)-m^2)\right].$$

---

[1] The superscripts $u(r)$ denote unrenormalized (renormalized) Green's functions.

We can now obtain the full photon propagator by summing the infinite series of renormalized one-loop diagrams. We have

$$
\begin{aligned}
iD_{\mu\nu}(q) &= i\Delta_{\mu\nu}(q) + i\Delta_{\mu\lambda}(i\Pi^{\lambda\rho})i\Delta_{\rho\nu} + \cdots \\
&= i\Delta_{\mu\nu}(q)\left[g^{\lambda}{}_{\nu} - \Pi^{\lambda\rho}\Delta_{\rho\nu} + \cdots\right] \\
&= i\Delta_{\mu\lambda}(q)\left[g^{\lambda}{}_{\nu} + P^{\lambda}{}_{\nu}\sum_{n=1}^{\infty}(-\Pi^{r}(q^{2}))^{n}\right] \\
&\equiv i\left[\frac{-g_{\mu\nu} + \frac{q_{\mu}q_{\nu}}{q^{2}}}{q^{2}(1 + \Pi^{r}(q^{2}))} - \xi\frac{q_{\mu}q_{\nu}}{(q^{2})^{2}}\right],
\end{aligned}
$$
(C.48)

where the projection operator,

$$
P^{\lambda}{}_{\nu} \equiv g^{\lambda}{}_{\nu} - \frac{q^{\lambda}q_{\nu}}{q^{2}}
$$
(C.49)

satisfies

$$
P^{\lambda}{}_{\nu}P^{\nu}{}_{\mu} = P^{\lambda}{}_{\mu}
$$
(C.50)

and

$$
\Pi^{\lambda\rho}\Delta_{\rho\nu} \equiv (q^{\lambda}q^{\rho} - g^{\lambda\rho}q^{2})\Pi^{r}(q^{2}) \times \frac{\left(-g_{\rho\nu} + (1-\xi)\frac{q_{\rho}q_{\nu}}{q^{2}}\right)}{q^{2}} = P^{\lambda}{}_{\nu}\Pi^{r}(q^{2}).
$$
(C.51)

C.3   Vertex Function and $(g-2)$

Calculate the one-loop Feynman diagram, Fig. 14.2. This one-loop contribution to the vertex is given by

$$
\begin{aligned}
d\Gamma^{\mu}(p',p) &= \int\frac{d^{4}k}{(2\pi)^{4}}\frac{-ig_{\nu\rho}}{k^{2}-\mu_{0}^{2}+i\epsilon}(-ie\gamma^{\nu})\frac{i(\slashed{k}+\slashed{p}'+m)}{(k+p')^{2}-m^{2}+i\epsilon} \\
&\quad \gamma^{\mu}\frac{i(\slashed{k}+\slashed{p}+m)}{(k+p)^{2}-m^{2}+i\epsilon}(-ie\gamma^{\rho}) \\
&= -ie^{2}\int\frac{d^{4}k}{(2\pi)^{4}}\frac{\gamma^{\nu}(\slashed{k}+\slashed{p}'+m)\gamma^{\mu}(\slashed{k}+\slashed{p}+m)\gamma_{\nu}}{(k^{2}-\mu_{0}^{2}+i\epsilon)((k+p')^{2}-m^{2}+i\epsilon)((k+p)^{2}-m^{2}+i\epsilon)}.
\end{aligned}
$$
(C.52)

Note, $p' = p + q$ and μ_{0} is an infrared cut-off. Using the identity

$$
\begin{aligned}
\frac{1}{a_{1}a_{2}a_{3}} &= 2\int_{0}^{1}\frac{dz_{1}dz_{2}dz_{3}\delta(1-\sum_{i=1}^{3}z_{i})}{(a_{1}z_{1}+a_{2}z_{2}+a_{3}z_{3})^{3}} \\
&\equiv 2\int_{0}^{1}dx\int_{0}^{x}dy\frac{1}{(a_{1}y+a_{2}(x-y)+a_{3}(1-x))^{3}}
\end{aligned}
$$
(C.53)

with $x = z_1 + z_2$, $y = z_1$ and

$$a_1 = (k + p')^2 - m^2 + i\epsilon, \tag{C.54}$$
$$a_2 = (k + p)^2 - m^2 + i\epsilon$$
$$a_3 = k^2 - \mu_0^2 + i\epsilon, \tag{C.55}$$

we find (after some algebra) the denominator in the integrand for the "on-shell" vertex correction, i.e. $p^2 = p'^2 = m^2$, given by D^3 with

$$D = \ell^2 - m^2 x^2 - \mu_0^2(1 - x) + q^2 y(x - y) + i\epsilon, \tag{C.56}$$

with $\ell \equiv k + p'y + p(x - y)$.

We then have

$$d\Gamma^\mu(p', p) = -2ie^2 \int_0^1 dx \int_0^x dy \int \frac{d^4\ell}{(2\pi)^4} \frac{\gamma^\nu(\cdots)\gamma_\nu}{D^3}, \tag{C.57}$$

with

$$(\cdots) = [\slashed{\ell} + \slashed{p}'(1 - y) - \slashed{p}(x - y) + m]\gamma^\mu[\slashed{\ell} - \slashed{p}'y + \slashed{p}(1 - x + y) + m]. \tag{C.58}$$

Now consider

$$\bar{u}(p')\delta\Gamma^\mu u(p) \tag{C.59}$$
$$= -2ie^2 \int_0^1 dx \int_0^x dy \int \frac{d^4\ell}{(2\pi)^4} \frac{\bar{u}(p')[\cdots]u(p)}{(\ell^2 - m^2 x^2 - \mu_0^2(1 - x) + q^2 y(x - y) + i\epsilon)^3}$$

with

$$[\cdots] = \gamma^\nu \slashed{\ell}\gamma^\mu \slashed{\ell}\gamma_\nu + \gamma^\mu \left[2m^2(2 - 4x + x^2) - 2q^2(y(x - y) + 1 - x)\right] \tag{C.60}$$
$$+ 4m \left[p'^\mu(x - y - xy) + p^\mu(y + xy - x^2)\right].$$

Now, we can perform the integral using dimensional regularization in d dimensions. Using the identity and symmetric integration, with $\ell_\alpha \ell_\beta = g_{\alpha\beta}/d$ we have

$$\gamma^\nu \slashed{\ell}\gamma^\mu \slashed{\ell}\gamma_\nu = (2 - d)(\gamma^\alpha g^{\mu\beta} + \gamma^\beta g^{\mu\alpha} - \gamma^\mu g^{\alpha\beta})\ell_\alpha \ell_\beta = (2 - d)^2/d\gamma^\mu. \tag{C.61}$$

Thus, the on-shell vertex function becomes, in Euclidean space with $q^2 \to -q_E^2$, $\ell^2 \to -\ell_E^2$, $\ell_0 \to i\ell_4$ and the spinors u and $\bar{u}$ implicit,

$$\delta\Gamma^\mu \doteq +2e^2 \int_0^1 dx \int_0^x dy \int \frac{d^d\ell_E}{(2\pi)^d} \times \left\{ \gamma^\mu \left[\frac{(2 - d)^2}{d}\ell_E^2 - 2m^2(2 - 4x + x^2) \right. \right.$$
$$\left. - 2q_E^2(y(x - y) + 1 - x) \right]$$
$$- 4m \left[p'^\mu(x - y - xy) \right.$$
$$\left. \left. + p^\mu(y + xy - x^2) \right] \right\} / \left(\ell_E^2 + m^2 x^2 + \mu_0^2(1 - x) + q_E^2 y(x - y) \right)^3. \tag{C.62}$$

Notice the following symmetry of the integrand. We have a symmetry under the transformation $y \to x - y$, $x \to x$. Under this transformation, $y(x - y)$ is invariant and $y + xy - x^2 \leftrightarrow x - y - xy$. The latter terms mean that for the terms proportional to p, p' we can take

$$\frac{1}{2}(y + xy - x^2 + x - y - xy) = \frac{1}{2}x(1 - x). \tag{C.63}$$

We thus have

$$\delta\Gamma^\mu \doteq -2e^2 \int_0^1 dx \int_0^x dy \int \frac{d^d\ell_E}{(2\pi)^d} \tag{C.64}$$

$$\times \left\{ \gamma^\mu \left[-\frac{(2-d)^2}{d}\ell_E^2 + 2m^2(2 - 4x + x^2) + 2q_E^2(y(x - y) + 1 - x) \right] \right.$$
$$\left. + 2m(p'^\mu + p^\mu)x(1 - x) \right\} / \left(\ell_E^2 + m^2 x^2 + \mu_0^2(1 - x) + q_E^2 y(x - y) \right)^3.$$

Now using the following identities with $d \equiv 4 - \epsilon$,

$$\int \frac{d^d\ell_E \; \ell_E^2}{(\ell_E^2 + a^2)^3} \simeq \pi^2 \left(\frac{2}{\epsilon} - [\ln(\frac{a^2}{4\pi}) + \frac{1}{2} + \gamma] \right) \tag{C.65}$$

$$\int \frac{d^4\ell_E}{(\ell_E^2 + a^2)^3} = \frac{\pi^2}{2a^2}. \tag{C.66}$$

Note the second integral is finite. We then obtain

$$\delta\Gamma^\mu(p', p) \doteq -\frac{2e^2}{16\pi^2} \int_0^1 dx \int_0^x dy \left\{ \gamma^\mu \left[-\left(\frac{2}{\epsilon} - [\ln\frac{a^2}{4\pi\mu^2} + \gamma + 2] \right) \right. \right.$$
$$+ \frac{m^2}{a^2}(x^2 - 4x + 2)$$
$$\left. + \frac{q_E^2}{a^2}(y(x - y) + 1 - x) \right] + \frac{m(p'^\mu + p^\mu)}{a^2}x(1 - x) \right\}, \tag{C.67}$$

with

$$a^2 \equiv m^2 x^2 + \mu_0^2(1 - x) + q_E^2 y(x - y). \tag{C.68}$$

In Eqns. 15.11 and 15.23 we define the form factors

$$J^\mu(p', p) \equiv \bar{u}(\vec{p}\,', s_p') \, \Gamma_\mu(p', p) \, u(\vec{p}, s_p) \tag{C.69}$$
$$= \bar{u}(\vec{p}\,', s_p') \, [\gamma^\mu F_1(q^2) + i\Sigma^{\mu\nu}\frac{q_\nu}{m}F_2(q^2)]u(\vec{p}, s_p).$$

Using the Gordon identity, Eqn. 14.3, we can re-write this as

$$J^\mu(p', p) = \bar{u}(\vec{p}\,', s_p') \, [\gamma^\mu(F_1(q^2) + F_2(q^2)) - \frac{1}{2m}(p'^\mu + p^\mu)F_2(q^2)]u(\vec{p}, s_p) \tag{C.70}$$

where the anomalous magnetic moment is given by the equation

$$\frac{g - 2}{2} = F_2(0) = \frac{2e^2}{16\pi^2} \int_0^1 dx \int_0^x dy \frac{2m^2}{a^2}x(1 - x) \tag{C.71}$$

$$= \frac{\alpha}{2\pi} \int_0^1 dx \int_0^x dy \frac{2}{x^2}x(1 - x) = \frac{\alpha}{2\pi}$$

and we have identified the term proportional to $(p'^\mu + p^\mu)$ in Eqn. C.67 and set $\mu_0 = 0$. Note, since $\delta\Gamma^\mu$ is divergent, one must add a counter term to cancel the divergence with the on-shell condition that $F_1(0) = 1$.

C.4 One-Loop Fermion Self Energy

In Fig. C.2 we show the one-loop correction to the fermion two-point function which defines $-i\Sigma(k)$. The renormalized inverse propagator is given by

$$-i\Sigma(k) \equiv \left[\int \frac{d^4p}{(2\pi)^4} - ie(\gamma^\mu)_{\alpha\lambda}(iS_F)_{\lambda\delta}(k-p) - ie(\gamma^\nu)_{\delta\beta}i\Delta^\xi_{\mu\nu}(p) \right] + i(Z_2-1)\slashed{k} - i\delta m. \tag{C.72}$$

$$(\Sigma(k))_{\alpha\beta} \equiv ie^2 \int \frac{d^4p}{(2\pi)^4} \frac{[\gamma^\mu(\slashed{k} - \slashed{p} + m)\gamma^\nu]_{\alpha\beta}}{(k-p)^2 - m^2 + i\epsilon} \frac{\left(-g_{\mu\nu} + (1-\xi)\frac{p_\mu p_\nu}{p^2}\right)}{p^2 + i\epsilon} - (Z_2-1)\slashed{k} + \delta m. \tag{C.73}$$

ASIDE: The full one-loop fermion propagator is then given by summing an infinite geometric series

$$i\tilde{S}_F = iS_F + iS_F \left(\frac{\Sigma}{i}\right) iS_F + \cdots \tag{C.74}$$

$$= iS_F \left(\sum_{n=0}^{\infty} (\Sigma S_F)^n\right) = iS_F(1 - \Sigma S_F)^{-1}$$

or

$$\tilde{S}_F^{-1} = (1 - \Sigma S_F)S_F^{-1} = S_F^{-1} - \Sigma \equiv \slashed{k} - m - \Sigma(k). \tag{C.75}$$

Note:

$$\Sigma(k) \equiv A(k^2)\slashed{k} + B(k^2). \tag{C.76}$$

Now

$$\mathrm{Tr}\Sigma \equiv B(k^2)f(d) \quad \text{in } d \text{ dimensions} \tag{C.77}$$
$$\mathrm{Tr}(\slashed{k}\Sigma) \equiv k^2 A(k^2)f(d).$$

Integration along the real ℓ_0 axis is extended into the complex ℓ_0 plane as described in the figure.

Also

$$\operatorname{Tr}\left[\gamma^\mu(\not{k} - \not{p} + m)\gamma^\nu\right] \equiv mg^{\mu\nu}f(d) \tag{C.78}$$
$$\operatorname{Tr}\left[\not{k}\gamma^\mu(\not{k} - \not{p} + m)\gamma^\nu\right] \equiv [k^\mu(k-p)^\nu + k^\nu(k-p)^\mu - g^{\mu\nu}k\cdot(k-p)]\,f(d).$$

The factor $f(d) \to 4$ for $d = 4$. We shall just take $f(d) = 4$ and ignore order ϵ corrections for $d = 4 - \epsilon$.

Let's now calculate A, B.

$$B(k^2) = ie^2\mu^\epsilon \int \frac{d^dp}{(2\pi)^d} \frac{mg^{\mu\nu}}{(k-p)^2 - m^2 + i\epsilon} \frac{\left(-g_{\mu\nu} + (1-\xi)\frac{p_\mu p_\nu}{p^2}\right)}{p^2 + i\epsilon} + \delta m \tag{C.79}$$

$$B(k^2) = -ie^2\mu^\epsilon[m(d-1+\xi)] \int \frac{d^dp}{(2\pi)^d} \frac{1}{(p^2 + i\epsilon)((k-p)^2 - m^2 + i\epsilon)} + \delta m$$

$$k^2 A(k^2)$$
$$= ie^2\mu^\epsilon \int \frac{d^dp}{(2\pi)^d} \frac{[k^\mu(k-p)^\nu + k^\nu(k-p)^\mu - g^{\mu\nu}k\cdot(k-p)]}{(k-p)^2 - m^2 + i\epsilon} \frac{\left(-g_{\mu\nu} + (1-\xi)\frac{p_\mu p_\nu}{p^2}\right)}{p^2 + i\epsilon}$$
$$-(Z_2 - 1)k^2$$

$$k^2 A(k^2) = ie^2\mu^\epsilon \int \frac{d^dp}{(2\pi)^d} \frac{\left[(d-2)k\cdot(k-p) + (1-\xi)\left(\frac{2k\cdot p(k-p)\cdot p}{p^2} - k\cdot(k-p)\right)\right]}{(p^2 + i\epsilon)((k-p)^2 - m^2 + i\epsilon)}$$
$$-(Z_2 - 1)k^2.$$

Now continue to Euclidean space with $k^2 \to -k_E^2$, $p^2 \to -p_E^2$, $dp_0 \to idp_4$. We then have

$$B(k_E^2) = e^2\mu^\epsilon m[d-1+\xi] \int \frac{d^dp_E}{(2\pi)^d} \frac{1}{p_E^2((k-p)_E^2 + m^2)} + \delta m \tag{C.80}$$

$$-k_E^2 A(k_E^2)$$
$$= e^2\mu^\epsilon \int \frac{d^dp_E}{(2\pi)^d} \frac{\left[(d-2)k_E\cdot(k-p)_E + (1-\xi)\left(\frac{2k_E\cdot p_E(k-p)_E\cdot p_E}{p_E^2} - k_E\cdot(k-p)_E\right)\right]}{p_E^2((k-p)_E^2 + m^2)}$$
$$+ (Z_2 - 1)k_E^2.$$

Continuing with the evaluation of B, we have

$$\tag{C.81}$$
$$B = e^2\mu^\epsilon m[d-1+\xi] \int_0^1 d\alpha \int \frac{d^dp_E}{(2\pi)^d} \frac{1}{(\alpha((k-p)_E^2 + m^2) + (1-\alpha)p_E^2)^2} + \delta m$$
$$= e^2\mu^\epsilon m[d-1+\xi] \int_0^1 d\alpha \int \frac{d^dp_E}{(2\pi)^d} \frac{1}{((p_E - \alpha k_E)^2 + \alpha(1-\alpha)k_E^2 + \alpha m^2)^2} + \delta m.$$

Let

$$\ell \equiv p_E - \alpha k_E, \qquad \Delta \equiv \alpha(1-\alpha)k_E^2 + \alpha m^2. \tag{C.82}$$

Define

$$I \equiv \mu^\epsilon \int \frac{d^d\ell}{(2\pi)^d} \frac{1}{(\ell^2 + \Delta)^2} \tag{C.83}$$

and using the result of Eqns. C.25 and C.34, we have

$$I = \frac{1}{16\pi^2} \left(\frac{2}{\epsilon} - \left(\ln \frac{\Delta}{\mu^2} - \ln 4\pi + \gamma \right) \right). \tag{C.84}$$

Hence,

$$B(k_E^2) = e^2 m \, [3 + \xi - \epsilon] \int_0^1 d\alpha \left[\frac{1}{8\pi^2\epsilon} - \frac{1}{16\pi^2} \left(\ln \frac{\Delta}{\mu^2} - \ln 4\pi + \gamma \right) \right] + \delta m. \tag{C.85}$$

In the $\overline{MS}$ scheme, we have

$$\delta m \equiv -\frac{me^2}{16\pi^2}(3 + \xi) \left[\frac{2}{\epsilon} + \ln 4\pi - \gamma \right]. \tag{C.86}$$

Thus,

$$B(k_E^2) = -\frac{me^2}{16\pi^2} \left[(3 + \xi) \left(\int_0^1 d\alpha \ln \frac{\Delta}{\mu^2} \right) + 2 \right]. \tag{C.87}$$

Note, as a consequence of chiral symmetry, the radiative correction to the fermion mass, m, vanishes when $m = 0$.

Now, continuing with the evaluation of A, we have

$$k_E^2 A(k_E^2) = -e^2 \mu^\epsilon \int \frac{d^d p_E}{(2\pi)^d} \frac{[(d - 2 - (1 - \xi))k_E \cdot (k - p)_E - (1 - \xi)2k_E \cdot p_E]}{p_E^2((k - p)_E^2 + m^2)}$$

$$- e^2 \mu^\epsilon (1 - \xi) \int \frac{d^d p_E}{(2\pi)^d} \frac{2(k_E \cdot p_E)^2}{(p_E^2)^2((k - p)_E^2 + m^2)} - (Z_2 - 1)k_E^2 \tag{C.88}$$

$$= -e^2 \int_0^1 d\alpha \int \frac{d^d\ell}{(2\pi)^d}$$

$$\times \frac{[(d - 2 - (1 - \xi))(k_E^2(1 - \alpha) - \ell \cdot k_E) - (1 - \xi)(2\alpha k_E^2 + 2k_E \cdot \ell)]}{(\ell^2 + \Delta)^2}$$

$$- 4e^2(1 - \xi) \int_0^1 d\alpha(1 - \alpha) \int \frac{d^d\ell}{(2\pi)^d} \frac{(k_E \cdot (\ell + \alpha k_E))^2}{(\ell^2 + \Delta)^3} - (Z_2 - 1)k_E^2 \tag{C.89}$$

$$\equiv -e^2 \int_0^1 d\alpha \left[(d - 2 - (1 - \xi))(k_E^2(1 - \alpha) - (1 - \xi)(2\alpha k_E^2) \right] I$$

$$- 4e^2(1 - \xi) \int_0^1 d\alpha(1 - \alpha)I' - (Z_2 - 1)k_E^2. \tag{C.90}$$

The integral I' is given by

$$I' = \mu^\epsilon \int \frac{d^d\ell}{(2\pi)^d} \frac{k_E \cdot \ell)^2 + \alpha^2(k_E)^2)}{(\ell^2 + \Delta)^3} \tag{C.91}$$

$$= \mu^\epsilon k_E^2 \int \frac{d^d\ell}{(2\pi)^d} \frac{(\ell^2/d + \alpha^2 k_E^2)}{(\ell^2 + \Delta)^3}. \tag{C.92}$$

We find

$$I' = \frac{k_E^2}{32\pi^2\epsilon} - \frac{k_E^2}{64\pi^2}\left(\ln\frac{\Delta}{\mu^2} - ln4\pi + \gamma - \frac{2\alpha^2 k_E^2}{\Delta}\right). \tag{C.93}$$

We then find

$$A(k_E^2) = -\frac{e^2}{16\pi^2}\int_0^1 d\alpha\,[(2-\epsilon)(1-\alpha) - (1-\xi)(1+\alpha)]\left(\frac{2}{\epsilon} - \left(\ln\frac{\Delta}{\mu^2} - \ln 4\pi + \gamma\right)\right)$$
$$- \frac{e^2}{16\pi^2}(1-\xi)\int_0^1 d\alpha(1-\alpha)\left(\frac{2}{\epsilon} - \left(\ln\frac{\Delta}{\mu^2} - \ln 4\pi + \gamma - \frac{2\alpha^2 k_E^2}{\Delta}\right)\right) - (Z_2 - 1) \tag{C.94}$$

where

$$\Delta \equiv \alpha(1-\alpha)k_E^2 + \alpha m^2. \tag{C.95}$$

In the $\overline{MS}$ scheme we then have

$$(Z_2 - 1) = -\frac{e^2}{8\pi^2\epsilon}\left([1 - (1-\xi)] + \ln 4\pi - \gamma\right). \tag{C.96}$$

Fermion Pole Mass

The fermion pole mass is defined by the equation

$$\not{k} - m - \Sigma(k) = 0 \tag{C.97}$$

for

$$\not{k} \equiv m_{pole} \equiv M. \tag{C.98}$$

At one loop,

$$M = m(1 + O(e^2)). \tag{C.99}$$

We have

$$\not{k} = m + \not{k}A(k^2) + B(k^2) \tag{C.100}$$

and since A and B are of order e^2, we have

$$M = m + B(m^2) + m\,A(m^2) + O(e^4). \tag{C.101}$$

In Minkowski space

$$\Delta = \alpha(m^2 - (1-\alpha)k^2) = \alpha^2 m^2 \quad \text{for} \quad k^2 = m^2. \tag{C.102}$$

Define

$$\Sigma_1 \equiv B(m^2) + mA(m^2). \tag{C.103}$$

Then we have

$$
\frac{\Sigma_1}{m} = -\frac{e^2}{16\pi^2}\left[4\int_0^1 d\alpha\left(\ln\frac{\Delta}{\mu^2}+\frac{1}{2}\right)\right.
$$

$$
\left.-2\int_0^1 d\alpha(1-\alpha)\left(\ln\frac{\Delta}{\mu^2}+1\right)\right]
$$

$$
+\frac{e^2}{16\pi^2}(1-\xi)\left[\int_0^1 d\alpha\left(\ln\frac{\Delta}{\mu^2}\right)-\int_0^1 d\alpha(1+\alpha)\left(\ln\frac{\Delta}{\mu^2}\right)\right.
$$

$$
\left.+\int_0^1 d\alpha(1-\alpha)\left(\ln\frac{\Delta}{\mu^2}+\frac{2\alpha^2 m^2}{\Delta}\right)\right].
$$

(C.104)

Using the identities

$$
\int_0^1 d\alpha\ln\alpha^2 = -2,
$$

(C.105)

$$
\int_0^1 d\alpha(1-\alpha)\ln\alpha^2 = -3/2
$$

we find

$$
\frac{\Sigma_1}{m} = \frac{e^2}{4\pi^2}\left[1-\frac{3}{4}\ln\frac{m^2}{\mu^2}\right].
$$

(C.106)

Hence,

$$
M = m(\mu)\left(1+\frac{e^2}{4\pi^2}\left[1-\frac{3}{4}\ln\frac{m^2}{\mu^2}\right]\right).
$$

(C.107)

Note, both $m(\mu)$ and M are gauge-invariant quantities, i.e. the gauge-dependent term has dropped out of the result. Finally, at the point $\mu = m$ we have

$$
M = m(m)\left(1+\frac{e^2}{4\pi^2}\right).
$$

(C.108)

D Appendix D Renormalization in QED

D.1 The Un-renormalized and Renormalized Lagrangian

The un-renormalized Lagrangian, which we used to calculate the un-renormalized polarization tensor, is written as

$$\mathcal{L}_u = -\frac{1}{4}F_{\mu\nu}^{u}{}^{2} + \bar{\psi}_u(i\slashed{D}_u - m_0)\psi_u \tag{D.1}$$

with

$$D_\mu^u\psi_u \equiv (\partial_\mu + ie_0 A_\mu^u)\psi_u. \tag{D.2}$$

ψ_u, A_μ^u are un-renormalized fields and e_0, m_0 are (bare) un-renormalized parameters. In order to subtract the divergences, we must add counter terms (order-by-order in the perturbation series) to $\mathcal{L}$.

The counter terms are constrained by gauge invariance to be of the same form as terms in $\mathcal{L}$. In a renormalizable theory the counter terms have dimension ≤ 4. The renormalized Lagrangian (including these counter terms) gives (by definition) finite results. In general, we have

$$\mathcal{L}_r = -\frac{1}{4}F_{\mu\nu}^{r}{}^{2} + \bar{\psi}_r(i\slashed{D}_r - m_r)\psi_r$$
$$-\frac{1}{4}(Z_3 - 1)F_{\mu\nu}^{r}{}^{2} + (Z_2 - 1)\bar{\psi}_r i\slashed{\partial}\psi_r$$
$$-\delta m\bar{\psi}_r\psi_r - (Z_1 - 1)e_r\mu^{\frac{\epsilon}{2}}\bar{\psi}_r\gamma^\mu\psi_r A_\mu^r. \tag{D.3}$$

However, gauge invariance requires

$$D_\mu^r\psi_r \equiv (\partial_\mu + ie_r\mu^{\frac{\epsilon}{2}} A_\mu^r)\psi_r. \tag{D.4}$$

Hence,

$$Z_1 \equiv Z_2. \tag{D.5}$$

After rearranging the renormalized Lagrangian, it is then given by

$$\mathcal{L}_r = -\frac{1}{4}Z_3 F_{\mu\nu}^{r}{}^{2} + Z_2\bar{\psi}_r(i\slashed{D}_r)\psi_r - (m_r + \delta m)\bar{\psi}_r\psi_r. \tag{D.6}$$

Note, Green's functions calculated with $\mathcal{L}_u$ (un-renormalized) are divergent. However, Green's functions calculated using $\mathcal{L}_r$ (renormalized) are *finite*. The renormalization constants, Z_3, Z_2, Z_1, δm are determined order by order in perturbation theory. We say, the theory is *renormalizable* if there are a finite number of counter terms. Note, $\mathcal{L}_r$ has the same form as $\mathcal{L}_u$. In fact, we can obtain $\mathcal{L}_u$ from $\mathcal{L}_r$ (or vice versa) by a rescaling of the fields (wave-function renormalization) and redefining the charge and mass. Let

$$A_\mu^u \equiv Z_3^{1/2} A_\mu^r, \quad \psi_u \equiv Z_2^{1/2} \psi_r, \quad \bar\psi_u \equiv Z_2^{1/2} \bar\psi_r, \tag{D.7}$$

such that $\mathcal{L}_r \Rightarrow \mathcal{L}_u$ with

$$\mathcal{L}_u = -\frac{1}{4} F_{\mu\nu}^{u\,2} + \bar\psi_u i (\slashed{\partial} + i e_r \mu^{\frac{\epsilon}{2}} Z_3^{-1/2} \slashed{A}^u) \psi_u - (m_r + \delta_m) Z_2^{-1} \bar\psi_u \psi_u, \tag{D.8}$$

with

$$e_0 = e_r \mu^{\frac{\epsilon}{2}} Z_3^{-1/2}, \quad Z_2 m_0 - \delta m \equiv m_r. \tag{D.9}$$

D.2 Generic β functions

Let us now evaluate the β functions in the MS renormalization scheme.[1,2] Given

$$e_0 = \mu^{\frac{\epsilon}{2}} Z_3^{-1/2} e \tag{D.10}$$

with $e \equiv e_r$. We have to all orders in perturbation theory

$$e_0 \equiv \mu^{\frac{\epsilon}{2}} \left[e + \sum_{n=1}^\infty \frac{a_n(e)}{\epsilon^n} \right]. \tag{D.11}$$

At one loop, we have

$$Z_2 = Z_1 = 1 - \frac{e^2}{8\pi^2 \epsilon} \quad \text{for} \quad \xi = 1, \tag{D.12}$$

$$Z_3 = 1 - \frac{e^2}{6\pi^2 \epsilon}. $$

Thus, to order e^2 we have

$$a_1(e) = \frac{e^3}{12\pi^2}. \tag{D.13}$$

Note, e_0 is independent of the arbitrary scale μ. μ is only needed to regularize and define the renormalized Green's functions.

Now consider

$$\mu \frac{de_0}{d\mu} \equiv 0 = \frac{\epsilon}{2} \mu^{\frac{\epsilon}{2}} \left[e + \sum_{n=1}^\infty \frac{a_n(e)}{\epsilon^n} \right] \tag{D.14}$$

$$+ \mu^{\frac{\epsilon}{2}} \left[\mu \frac{de}{d\mu} + \sum_{n=1}^\infty \frac{1}{\epsilon^n} \frac{da_n}{de} \mu \frac{de}{d\mu} \right], \tag{D.15}$$

---

[1] In the MS scheme, only the pole in ϵ is subtracted.
[2] For the renormalization procedure using dimensional regularization, see ('t Hooft, 1973).

which gives the equation

$$\frac{\epsilon}{2}e + \left(\frac{1}{2}a_1 + \mu\frac{de}{d\mu}\right) + \sum_{n=1}^{\infty}\frac{1}{\epsilon^n}\left[\frac{da_n}{de}\mu\frac{de}{d\mu} + \frac{a_{n+1}}{2}\right] = 0. \tag{D.16}$$

The renormalized coupliing, $e(\epsilon, \mu)$, is finite as $\epsilon \to 0$. Therefore, assume

$$\mu\frac{de}{d\mu} = d_0 + d_1\epsilon + d_2\epsilon^2 + \cdots. \tag{D.17}$$

We then have

$$0 = \epsilon\left(\frac{1}{2}e + d_1\right) + \left(\frac{1}{2}a_1 + d_0\right) + \sum_{n=2}^{\infty}d_n\epsilon^n + \sum_{n=1}^{\infty}\frac{1}{\epsilon^n}\left[\frac{da_n}{de}\left(d_0 + \sum_{m=1}^{\infty}d_m\epsilon^m\right) + \frac{a_{n+1}}{2}\right]. \tag{D.18}$$

Re-expanding in ϵ^n we have

$$0 = \sum_{n=2}^{\infty}\epsilon^n\left(d_n + \sum_{s=1}^{\infty}\frac{da_s}{de}d_{s+n}\right) + \epsilon\left(\frac{e}{2} + d_1 + \sum_{s=1}^{\infty}\frac{da_s}{de}d_{s+1}\right) \tag{D.19}$$

$$+ \left(\frac{1}{2}a_1 + d_0 + \sum_{s=1}^{\infty}\frac{da_s}{de}d_s\right) + \sum_{n=1}^{\infty}\frac{1}{\epsilon^n}\left(\frac{da_n}{de}d_0 + \frac{a_{n+1}}{2} + \sum_{s=1}^{\infty}\frac{da_{s+n}}{de}d_s\right).$$

Consider the first line with terms of order ϵ^n with $n > 1$. We have

$$\epsilon^n\left(d_n + \sum_{s=1}^{\infty}\frac{da_s}{de}d_{s+n}\right) = 0 \tag{D.20}$$

or, equivalently,

$$\begin{pmatrix} 1 & \frac{da_1}{de} & \frac{da_2}{de} & \cdots \\ 0 & 1 & \frac{da_1}{de} & \cdots \\ 0 & 0 & 1 & \cdots \\ \vdots & \vdots & \vdots & \end{pmatrix}\begin{pmatrix} d_2 \\ d_3 \\ d_4 \\ \vdots \end{pmatrix} = 0 \equiv AD. \tag{D.21}$$

The terms $\frac{da_n(e)}{de}$ are a power series expansion in e and

$$D \equiv \begin{pmatrix} d_2 \\ d_3 \\ d_4 \\ \vdots \end{pmatrix}, \tag{D.22}$$

but

$$\det A \equiv 1; \tag{D.23}$$

hence,

$$D = 0. \tag{D.24}$$

Therefore,

$$d_n = 0 \quad \text{for} \quad n > 1. \tag{D.25}$$

Now consider the other terms. We have

$$\epsilon^1: \qquad\qquad d_1 + \tfrac{e}{2} = 0 \qquad\qquad\qquad\qquad \Rightarrow \quad d_1 = -\frac{e}{2} \qquad\qquad \text{(D.26)}$$

$$\epsilon^0: \qquad\qquad \tfrac{1}{2}a_1 + d_0 + \tfrac{da_1}{de}d_1 = 0 \qquad\qquad \Rightarrow \quad d_0 = -\frac{1}{2}a_1 + \frac{e}{2}\frac{da_1}{de}$$

$$\epsilon^{-n}: \qquad \tfrac{da_n}{de}d_0 + \tfrac{a_{n+1}}{2} + \tfrac{da_{n+1}}{de}d_1 = 0, \quad n = 1,\ldots,\infty \quad \Rightarrow \quad a_2,\ a_3,\cdots \text{ are given}$$

$$\text{in terms of } a_1.$$

The last equation is due to the fact that higher-order powers in $\frac{1}{\epsilon}$ are due to multiple contributions of the order loop divergences.

We now define the β function,

$$\beta(e) \equiv \mu\frac{de}{d\mu}\Big|_{\epsilon=0} \equiv d_0 = -\frac{1}{2}a_1 + \frac{e}{2}\frac{da_1}{de}. \qquad\qquad \text{(D.27)}$$

With $a_1(e)$, Eqn. D.13, we find

$$\beta(e) = \frac{e^3}{12\pi^2}. \qquad\qquad \text{(D.28)}$$

Anomalous Dimension for the Mass Operator

We define

$$\gamma_m \equiv \lim_{\epsilon\to 0} \frac{d\ln m}{d\ln\mu} \equiv \lim_{\epsilon\to 0} \frac{\mu}{m}\frac{dm}{d\mu}. \qquad\qquad \text{(D.29)}$$

In an arbitrary R_ξ gauge we have (at one loop)

$$\delta m \equiv (Z_m - 1)m \qquad\qquad \text{(D.30)}$$

with

$$Z_m - 1 = -\frac{e^2}{8\pi^2\epsilon}[4 - (1-\xi)] \qquad\qquad \text{(D.31)}$$

and

$$Z_2 - 1 = -\frac{e^2}{8\pi^2\epsilon}[1 - (1-\xi)]. \qquad\qquad \text{(D.32)}$$

Following Eqn. D.9 we have

$$m_0 \equiv Z_2^{-1}Z_m m, \qquad\qquad \text{(D.33)}$$

with $m \equiv m_r$. But

$$Z_2^{-1}Z_m \simeq 1 - \frac{3e^2}{8\pi^2\epsilon} + O(e^4). \qquad\qquad \text{(D.34)}$$

In general, we have

$$Z_2^{-1}Z_m = 1 + \sum_{n=1}^{\infty} \frac{b_n(e)}{\epsilon^n}, \qquad\qquad \text{(D.35)}$$

where at one loop

$$b_1 = -\frac{3e^2}{8\pi^2}. \qquad\qquad \text{(D.36)}$$

Using the same analysis as before, we find

$$\gamma_m = \frac{e}{2}\frac{db_1}{de} \tag{D.37}$$

and at one loop

$$\gamma_m = -\frac{3e^2}{8\pi^2}. \tag{D.38}$$

Note, $\gamma_m(e)$ is gauge invariant, since the mass operator is gauge invariant.

Given the anomalous dimension for the mass operator, we could define the running mass, $m(\mu)$. At one loop we have

$$\gamma_m = \gamma_0\alpha, \tag{D.39}$$

with

$$\gamma_0 = -\frac{3}{2\pi}, \quad \alpha = \frac{e^2}{4\pi}. \tag{D.40}$$

Define $t \equiv \ln\frac{\mu}{\mu_0}$ and recall $\frac{d\alpha}{dt} = b_0\alpha^2$ with $b_0 = \frac{2}{3\pi}$. Then,

$$\int \frac{dm}{m} = \gamma_0\alpha dt, \tag{D.41}$$

$$\int \frac{d\alpha}{\alpha} = b_0\alpha dt. \tag{D.42}$$

Thus,

$$\int \frac{dm}{m} = \frac{\gamma_0}{b_0}\int \frac{d\alpha}{\alpha}. \tag{D.43}$$

The solution is

$$m(\mu) = m(\mu_0)\left[\frac{\alpha(\mu)}{\alpha(\mu_0)}\right]^{\frac{\gamma_0}{b_0}}. \tag{D.44}$$

Anomalous Dimensions

We define

$$\gamma_i \equiv \frac{1}{2}\mu\frac{d\ln Z_i}{d\mu} = \frac{1}{2}\frac{\mu}{Z_i}\frac{dZ_i}{d\mu}, \tag{D.45}$$

where Z_i are the wave function renormalization constants. Let

$$Z_i^{-1} \equiv 1 + \sum_{n=1}^{\infty}\frac{b_n^i(e)}{\epsilon^n}, \tag{D.46}$$

then

$$\gamma_i \equiv \frac{c}{4}\frac{db_1^i}{de}. \tag{D.47}$$

We then have for electrons (at one loop),

$$b_1^2 = \frac{e^2}{8\pi^2}[1 - (1-\xi)] \tag{D.48}$$

and thus

$$\gamma_2 = \frac{e^2}{16\pi^2}[1 - (1 - \xi)]. \tag{D.49}$$

Note, this is gauge dependent.

D.3 Callan–Symanzik Equations

Consider an n-point Green's function (in momentum space) with $n = n_{fermions} + n_{photons}$. We have

$$G_0^{(n)}(p_i; e_0, m_0), \tag{D.50}$$

calculated with the un-renormalized Lagrangian $\mathcal{L}_u$. In terms of renormalized fields and couplings, we obtain the renormalized Green's function given by

$$G_0^{(n)}(p_i; e_0, m_0) \equiv Z_3^{\frac{n_\gamma}{2}} Z_2^{\frac{n_f}{2}} G^{(n)}(p_i; e, m, \mu). \tag{D.51}$$

Note,

$$\mu \frac{d}{d\mu} G_0^{(n)} \equiv 0. \tag{D.52}$$

We then obtain the Callan–Symanzik (CS) equations for the renormalized Green's function by applying $\frac{d}{d\mu}$ to the right-hand side of the equation (Callan, 1970; Symanzik, 1970a,b, 1971; Coleman, 1985). We find

$$\left\{ \frac{n_\gamma}{2} \left(\frac{\mu}{Z_3} \frac{dZ_3}{d\mu} \right) + \frac{n_f}{2} \left(\frac{\mu}{Z_2} \frac{dZ_2}{d\mu} \right) + \mu \frac{de}{d\mu} \frac{\partial}{\partial e} + \mu \frac{dm}{d\mu} \frac{\partial}{\partial m} + \mu \frac{\partial}{\partial \mu} \right\} G^{(n)}(p_i; e, m, \mu) \equiv 0. \tag{D.53}$$

Given the definitions from the last section, we have

$$\beta(e) \equiv \mu \frac{de}{d\mu} \qquad \gamma_m(e) \equiv \frac{\mu}{m} \frac{dm}{d\mu} \tag{D.54}$$

$$\gamma_f(e) \equiv \frac{1}{2} \mu \frac{d\ln Z_2}{d\mu} \qquad \gamma_\gamma(e) \equiv \frac{1}{2} \mu \frac{d\ln Z_3}{d\mu}$$

and thus the CS equation becomes

$$\left\{ \mu \frac{\partial}{\partial \mu} + \beta(e) \frac{\partial}{\partial e} + \gamma_m(e) m \frac{\partial}{\partial m} + (n_\gamma \, \gamma_\gamma(e) + n_f \, \gamma_f(e)) \right\} G^{(n)}(p_i; e, m, \mu) \equiv 0. \tag{D.55}$$

The meaning of the CS equation is that physics remains unchanged if one changes the renormalization scale μ, if at the same time one rescales e, m and the fields. Physical observables should be both gauge invariant and renormalization-group invariant, i.e. independent of the arbitrary scale μ.

Using the CS equation (or, equivalently, the renormalization group equation [RGE]) we can evaluate the Green's function $G^{(n)}(\sigma p_i; e, \, m, \, \mu)$, i.e. at a rescaled value of the momentum, given the value of $G^{(n)}(p_i; e, m, \mu)$. We define the running charge by

$$\frac{d\bar{e}}{dt} \equiv \beta_e(\bar{e}), \quad t \equiv \ln \sigma, \tag{D.56}$$

with boundary condition

$$\bar{e}(0) \equiv e(\mu). \tag{D.57}$$

The solution to the CS equation is given by

$$G^{(n)}(\sigma p_i; e, m, \mu) = \exp\left[\left(n_\gamma + \frac{3}{2}n_f - 4(n-1)\right)t\right] \tag{D.58}$$

$$\times \exp\left[n_\gamma \int_0^t \gamma_e(\bar{e}(t'))dt' + n_f \int_0^t \gamma_f(\bar{e}(t'))dt'\right]$$

$$\times G^{(n)}(p_i; \bar{e}(t), \bar{m}(t), \mu).$$

Proof   Using dimensional analysis, we have

$$\dim[G^{(n)}(p_i, \dots)] = \dim[\psi](n_\psi + n_{\bar{\psi}}) + \dim[A_\mu]n_\gamma - 4n + 4 = \frac{3}{2}n_f + n_\gamma - 4n + 4, \tag{D.59}$$

where $(n_\psi + n_{\bar{\psi}}) \equiv n_f$ and the factor of $4n - 4$ takes into account the Fourier transform to momentum space. Therefore, we have

$$G^{(n)}(p_i; e, m, \mu) \equiv \mu^{\frac{3}{2}n_f + n_\gamma - 4n + 4} \times \bar{G}^{(n)}\left(\frac{p_i}{\mu}; e, \frac{m}{\mu}\right), \tag{D.60}$$

which explicitly takes into account the scale dependence.

Now define the Green's function of the rescaled momenta,

$$G^{(n)}(\sigma p_i; e, m, \mu) \equiv \mu^{\frac{3}{2}n_f + n_\gamma - 4n + 4} \times \bar{G}^{(n)}\left(\frac{\sigma p_i}{\mu}; e, \frac{m}{\mu}\right). \tag{D.61}$$

Let $A \equiv \frac{\sigma p_i}{\mu}$ and $B \equiv \frac{m}{\mu}$. Consider the following

$$\left[\sigma\frac{\partial}{\partial\sigma} + m\frac{\partial}{\partial m}\right]\bar{G}^{(n)}(A; e, B) \equiv \frac{\partial\bar{G}}{\partial A}\left(\sigma\frac{\partial A}{\partial\sigma}\right) + \frac{\partial\bar{G}}{\partial B}\left(m\frac{\partial B}{\partial m}\right) \tag{D.62}$$

$$= -\left(\frac{\partial\bar{G}}{\partial A}\left(\mu\frac{\partial A}{\partial\mu}\right) + \frac{\partial\bar{G}}{\partial B}\left(\mu\frac{\partial B}{\partial\mu}\right)\right) \equiv -\mu\frac{\partial}{\partial\mu}\bar{G},$$

where in the last line we used the identities

$$\sigma\frac{\partial A}{\partial\sigma} \equiv A \equiv -\mu\frac{\partial A}{\partial\mu}, \tag{D.63}$$

$$m\frac{\partial B}{\partial m} \equiv B \equiv -\mu\frac{\partial B}{\partial\mu}.$$

Define the differential operator

$$\theta \equiv \sigma\frac{\partial}{\partial\sigma} + m\frac{\partial}{\partial m} + \mu\frac{\partial}{\partial\mu}, \tag{D.64}$$

then

$$\theta \bar{G} \equiv 0. \qquad (D.65)$$

Also,

$$\theta G^{(n)} = \theta \left(\mu^{\frac{3}{2} \ n_f + n_\gamma - 4n + 4} \ \bar{G}^{(n)} \right) \qquad (D.66)$$

$$= \theta \left(\mu^{\frac{3}{2} \ n_f + n_\gamma - 4n + 4} \right) \bar{G}^{(n)}$$

$$= \left(\frac{3}{2} \ n_f + n_\gamma - 4n + 4 \right) \bar{G}^{(n)}.$$

Therefore, we have the revised CS equation given by

$$\left\{ \sigma \frac{\partial}{\partial \sigma} - \beta(e) \frac{\partial}{\partial e} - \left[\gamma_m(e) - 1 \right] m \frac{\partial}{\partial m} \right. \qquad (D.67)$$

$$\left. - \left(\left(\frac{3}{2} + \gamma_f(e) \right) n_f + n_\gamma (1 + \gamma_\gamma(e)) - 4n + 4 \right) \right\} G^{(n)}(\sigma p_i; e, m, \mu) \equiv 0.$$

Note, it should be clear that the quantities γ_f and γ_γ are known as anomalous field dimensions.

Define $t \equiv \ln \sigma$ and $\sigma \frac{\partial}{\partial \sigma} \equiv \frac{\partial}{\partial t}$. We then have

$$\left\{ \frac{\partial}{\partial t} - \beta(e) \frac{\partial}{\partial e} - \left[\gamma_m(e) - 1 \right] m \frac{\partial}{\partial m} \right. \qquad (D.68)$$

$$\left. - \left(\left(\frac{3}{2} + \gamma_f(e) \right) n_f + n_\gamma \left(1 + \gamma_\gamma(e) \right) - 4n + 4 \right) \right\} G^{(n)}(e^t p_i; e, m, \mu) \equiv 0.$$

ASIDE: Consider a particle with position $\vec{x}(t)$ satisfying the differential equation

$$\frac{d\vec{x}}{dt} = \vec{v}(\vec{x}(t)). \qquad (D.69)$$

The trajectory for the particle's motion is given by

$$\vec{x} = \vec{x}(t, \vec{x}_0), \qquad (D.70)$$

with boundary condition

$$\vec{x}(0, \vec{x}_0) = \vec{x}_0. \qquad (D.71)$$

Assume, the trajectory passes through a point $\vec{x}_a$ at time $t = t_a$. We then have the equation

$$\vec{x}(t, \vec{x}_0) \equiv \vec{x}(t - t_a, \vec{x}_a). \qquad (D.72)$$

This gives the equation

$$\vec{x}(dt, \vec{x}_0) = \vec{x}_0 + \underbrace{\frac{d\vec{x}}{dt}\bigg|_{t=0}}_{\vec{v}(\vec{x}_0) \equiv \vec{v}_0} + \cdots = \vec{x}_0 + \vec{v}_0 \, dt + \cdots \equiv \vec{x}(0, \vec{x}_0 + \vec{v}_0 dt). \qquad (D.73)$$

This then generalizes to the equation

$$\vec{x}(t + dt, \vec{x}_0) = \vec{x}(t, \vec{x}_0 + \vec{v}_0 \, dt), \tag{D.74}$$

$$\vec{x}(t) + \frac{\partial \vec{x}(t, \vec{x}_0)}{\partial t} \, dt + \cdots = \vec{x}(t) + \vec{v}_0 \cdot \frac{\partial}{\partial \vec{x}_0} \, \vec{x}(t, \vec{x}_0) + \cdots$$

and gives the final equation

$$\left(\frac{\partial}{\partial t} - \vec{v}_0(t) \cdot \frac{\partial}{\partial \vec{x}_0} \right) \vec{x}(t, \vec{x}_0) \equiv 0. \tag{D.75}$$

Now in the case that we had several dimensionless coupling constants, λ_i, we have $\vec{x}_i \equiv \bar{\lambda}_i$ and $\vec{v}_i \equiv \beta_{\lambda_i}(\bar{\lambda}_i)$ and the differential operator

$$D_t \equiv \frac{\partial}{\partial t} - \sum_i \beta_i(\lambda_j) \frac{\partial}{\partial \lambda_i}. \tag{D.76}$$

In our case we have just one coupling constant, e, and we find

$$D_t \bar{e}(t, e) = \left(\frac{\partial}{\partial t} - \beta_e(e) \frac{\partial}{\partial e} \right) \bar{e}(t, e) \equiv 0. \tag{D.77}$$

Now define

$$z_i(t) \equiv \exp \int_0^t dt' \gamma_i(\bar{e}(t')). \tag{D.78}$$

Then

$$D_t z_i(t) = \gamma_i(\bar{e}(t)) z_i(t) - [\beta_e(e) \frac{\partial}{\partial e} \int_0^t dt' \gamma_i(\bar{e}(t'))] z_i(t) \tag{D.79}$$

$$= [\gamma_i(\bar{e}(t)) - \int_0^t dt' \underbrace{(\beta_e \frac{\partial}{\partial e}) \bar{e}(t')}_{\frac{\partial \bar{e}(t')}{\partial t'}} \frac{\partial \gamma_i}{\partial \bar{e}(t')}] z_i(t)$$

$$= \left\{ \gamma_i(\bar{e}(t)) - \int_0^t dt' \frac{\partial}{\partial t'} \gamma_i(\bar{e}(t')) \right\} z_i(t)$$

or, finally,

$$D_t z_i(t) \equiv \gamma_i(e) z_i(t). \tag{D.80}$$

As the final step in the proof, define

$$G^{(n)}(\sigma p_i; e, \, m, \, \mu) = e^{[\frac{3}{2} n_f + n_\gamma - 4n + 4]t} z_\gamma^{n_\gamma} z_f^{n_f} F(e^t p_i; e, \, m, \mu) \tag{D.81}$$

such that

$$\left(D_t - (\gamma_m - 1) \, m \frac{\partial}{\partial m} \right) F(e^t p_i; e, \, m, \mu) \equiv 0. \tag{D.82}$$

If we define the running mass, $\bar{m}(\mu,\ e,\ m)$, satisfying the equation

$$0 = \mu \frac{d}{d\mu}\ \bar{m} \equiv \left(\mu \frac{\partial}{\partial \mu} + \mu \frac{\partial e}{\partial \mu} \frac{\partial}{\partial e} + \mu \frac{\partial m}{\partial \mu} \frac{\partial}{\partial m} \right) \bar{m} \tag{D.83}$$

$$\equiv \left(\mu \frac{\partial}{\partial \mu} + \beta_e(e) \frac{\partial}{\partial e} + \gamma_m(e)\ m \frac{\partial}{\partial m} \right) \bar{m},$$

then this is equivalent to the trajectory in terms of $\sigma = e^t$,

$$\left(D_t - (\gamma_m - 1) \frac{\partial}{\partial m} \right) \bar{m}(t,\ e,\ m) \equiv 0. \tag{D.84}$$

Using this we then find the solution for F given by

$$F \equiv F(p_i; \bar{e}(t),\ \bar{m}(t),\ \mu). \tag{D.85}$$

We have

$$G^{(n)}(\sigma p_i; e,\ m,\ \mu) = \exp\left(\left[\frac{3}{2}\ n_f + n_\gamma - 4n + 4 \right] t \right) e^{\int_0^t dt'\, n_\gamma \gamma_\gamma(t')} e^{\int_0^t dt'\, n_f \gamma_f(t')}$$

$$\times F(p_i; \bar{e}(t),\ \bar{m}(t),\ \mu), \tag{D.86}$$

which must satisfy the boundary condition

$$G^{(n)}(p_i; e,\ m,\ \mu) = F(p_i; e,\ m,\ \mu). \tag{D.87}$$

Finally, we have

$$G^{(n)}(\sigma p_i; e,\ m,\ \mu) = \exp\left(\left[\frac{3}{2}\ n_f + n_\gamma - 4n + 4 \right] t \right) e^{\int_0^t dt'\, n_\gamma \gamma_\gamma(t')} e^{\int_0^t dt'\, n_f \gamma_f(t')}$$

$$\times G^{(n)}(p_i; \bar{e}(t),\ \bar{m}(t),\ \mu). \tag{D.88}$$

$\square$

Three Simple Examples

1 As an example of the solution to the CS equation, consider a toy model where $\beta(e^*) = 0$, see Fig. D.1. The quantities $\gamma_\gamma(e^*)$, $\gamma_f(e^*)$ are called, anomalous dimensions, since they add to the naive scaling dimensions of fields. At the fixed point, the n-point Green's function scales as follows.

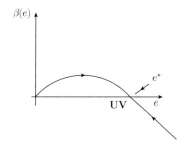

Fig. D.1   In this toy model the $\beta(e)$ function has a fixed point at $e = e^*$.

$$G^{(n)}(\sigma p_i; e, \, m, \, \mu) = \exp\left[\left(n_\gamma(1 + \gamma_\gamma(e^*)) + n_f\left(\frac{3}{2} + \gamma_f(e^*)\right) - 4(n-1)\right)t\right]$$
$$\times \, G^{(n)}(p_i; \bar{e}(t), \, \bar{m}(t), \, \mu). \tag{D.89}$$

For example, consider the case $n = n_f = 2$, $n_\gamma = 0$, i.e. the fermion propagator. We have

$$G^{(2)}(\sigma p_i; e, \, m, \, \mu) \sim \frac{\left(\frac{\sigma p_i}{\mu}\right)^{2\gamma_f(e^*)}}{\sigma p_i}, \tag{D.90}$$

with power-law scaling plus logarithmic corrections when not exactly at the fixed point.

2 In QCD the fixed point is at $g^* = g_s^* = 0$. Then, the integral over the anomalous dimensions gives terms as follows.

$$\int_0^t \gamma(\bar{g}(t'))dt' \equiv \int_g^{\bar{g}(t)} \frac{\gamma(x)}{\beta_g(x)}dx, \tag{D.91}$$

where we used

$$\beta_g(g) = -\beta_0\frac{g^3}{16\pi^2}, \quad \beta(\bar{g})dt = d\bar{g}, \quad \gamma(g) = \gamma_0\frac{g^2}{4\pi^2}. \tag{D.92}$$

Thus, we obtain

$$\int_0^t \gamma(\bar{g}(t'))dt' = -\int_g^{\bar{g}(t)} \frac{4\gamma_0}{\beta_0}\frac{dx}{x} = -\frac{4\gamma_0}{\beta_0}\ln\left(\frac{\bar{g}(t)}{g}\right) \tag{D.93}$$

and

$$\exp\int_0^t \gamma(\bar{g}(t'))dt' = \left(\frac{\bar{g}(t)}{g}\right)^{-\frac{4\gamma_0}{\beta_0}}, \tag{D.94}$$

which gives logarithmic scaling violations.

3 Also, in QCD, consider the quantity

$$\Lambda_{QCD} \equiv \mu e^{-\frac{2\pi}{\alpha\beta_0}}. \tag{D.95}$$

We can now show that Λ_{QCD} is an RG-invariant quantity. We have

$$\left(\mu\frac{\partial}{\partial\mu} + \beta(g)\frac{\partial}{\partial g}\right)\Lambda_{QCD} = \lambda_{QCD} + \mu\beta(g)\frac{\partial}{\partial g}e^{-\frac{8\pi^2}{\beta_0 g^2}} \tag{D.96}$$

$$= \Lambda_{QCD}\left(1 + \beta(g)\frac{16\pi^2}{\beta_0 g^3}\right) \equiv 0$$

to order g^2.

Appendix E Triangle Anomaly

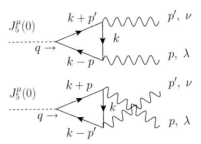

$J_5^\mu(0)$

$k + p'$ p', ν

$q \rightarrow$ k

$k - p$ p, λ

$J_5^\mu(0)$

$k + p$ p', ν

$q \rightarrow$ k

$k - p'$ p, λ

Fig. E.1   The triangle diagram with the $U(1)_{axial}$ current, with momentum, q_μ, at one vertex and photons at the other two vertices. Electrons are in the loop.

In Fig. E.1 we have the triangle graph which results in the Adler-Bell-Jackiw (ABJ) anomaly (Adler, 1969; Bell and Jackiw, 1969) with $q = p + p'$. Define the Green's functions[1]

$$\tilde{R}^{\mu\nu\lambda}(p',p) = \int d^4y d^4z e^{ip\cdot z} e^{ip'\cdot y} \langle 0|T(J_5^\mu(0)J^\nu(y)J^\lambda(z)|0\rangle \qquad (\text{E.1})$$

and

$$\tilde{R}^{\nu\lambda}(p',p) = \int d^4y d^4z e^{ip\cdot z} e^{ip'\cdot y} \langle 0|T(\bar{\psi}\gamma_5\psi(0)J^\nu(y)J^\lambda(z)|0\rangle. \qquad (\text{E.2})$$

The classical Ward identities are given by

$$p_\lambda \tilde{R}^{\mu\nu\lambda}(p',p) = 0 \quad = p'_\nu \tilde{R}^{\mu\nu\lambda}(p',p), \qquad (\text{E.3})$$
$$q_\mu \tilde{R}^{\mu\nu\lambda}(p',p) = 2m\tilde{R}^{\nu\lambda}(p',p).$$

We can check these identities explicitly. We have

$$\tilde{R}^{\mu\nu\lambda}(p',p) = (-1)\int \frac{d^4k}{(2\pi)^4}\text{Tr}\left[\gamma^\mu\gamma_5 i\frac{\slashed{k} - \slashed{p} + m}{(k-p)^2 - m^2 + i\epsilon}\right.$$
$$\left. \times\ \gamma^\lambda i\frac{\slashed{k} + m}{k^2 - m^2 + i\epsilon}\gamma^\nu i\frac{\slashed{k} + \slashed{p'} + m}{(k+p')^2 - m^2 + i\epsilon}\right] + \left\{\begin{smallmatrix}p\leftrightarrow p'\\ \lambda\leftrightarrow\nu\end{smallmatrix}\right\} \qquad (\text{E.4})$$

---

[1] We follow the very nice discussion in Cheng and Li (1984).

and

$$\tilde{R}^{\nu\lambda}(p',p) = (-1)\int\frac{d^4k}{(2\pi)^4}\,\mathrm{Tr}\left[\gamma_5 i\frac{\slashed{k}-\slashed{p}+m}{(k-p)^2-m^2+i\epsilon}\right.$$

$$\left.\times\gamma^\lambda i\frac{\slashed{k}+m}{k^2-m^2+i\epsilon}\gamma^\nu i\frac{\slashed{k}+\slashed{p}'+m}{(k+p')^2-m^2+i\epsilon}\right]+\left\{{p\leftrightarrow p'\atop\lambda\leftrightarrow\nu}\right\}. \quad (E.5)$$

Let's first check the second identity. Using the identities

$$\slashed{q}\gamma_5 \equiv (\slashed{k}+\slashed{p}'-m)\gamma_5 + \gamma_5(\slashed{k}-\slashed{p}-m) + 2m\gamma_5 \quad\quad (E.6)$$

$$\equiv (\slashed{k}+\slashed{p}-m)\gamma_5 + \gamma_5(\slashed{k}-\slashed{p}'-m) + 2m\gamma_5, \quad\quad (E.7)$$

we have

$$q_\mu\tilde{R}^{\mu\nu\lambda}(p',p) = i\Delta^{\nu\lambda}_{(1)} + i\Delta^{\nu\lambda}_{(2)} + 2m\tilde{R}^{\nu\lambda}(p',p), \quad\quad (E.8)$$

where

$$\Delta^{\nu\lambda}_{(1)} = \int\frac{d^4k}{(2\pi)^4}\left\{\mathrm{Tr}\left[\gamma_5\gamma^\nu i\frac{\slashed{k}-\slashed{p}+m}{(k-p)^2-m^2+i\epsilon}\gamma^\lambda i\frac{\slashed{k}+m}{k^2-m^2+i\epsilon}\right]\right. \quad (E.9)$$

$$\left.-\,\mathrm{Tr}\left[\gamma_5\gamma^\nu i\frac{\slashed{k}+m}{k^2-m^2+i\epsilon}\gamma^\lambda i\frac{\slashed{k}+\slashed{p}+m}{(k+p)^2-m^2+i\epsilon}\right]\right\}$$

and

$$\Delta^{\nu\lambda}_{(2)} = \int\frac{d^4k}{(2\pi)^4}\left\{\mathrm{Tr}\left[\gamma_5\gamma^\lambda i\frac{\slashed{k}-\slashed{p}'+m}{(k-p')^2-m^2+i\epsilon}\gamma^\nu i\frac{\slashed{k}+m}{k^2-m^2+i\epsilon}\right]\right. \quad (E.10)$$

$$\left.-\,\mathrm{Tr}\left[\gamma_5\gamma^\lambda i\frac{\slashed{k}+m}{k^2-m^2+i\epsilon}\gamma^\nu i\frac{\slashed{k}+\slashed{p}'+m}{(k+p')^2-m^2+i\epsilon}\right]\right\}.$$

ASIDE: Linearly divergent integrals

Note, if we can shift momenta such that $k \to k-p$ or $k \to k-p'$, then $\Delta^{\nu\lambda}_{(1)}$ and $\Delta^{\nu\lambda}_{(2)}$ vanish. But the integrals are linearly divergent. Consider

$$\Delta(a) \equiv \int_{-\infty}^{+\infty}[f(x+a)-f(x)] \quad\quad (E.11)$$

and expand the integrand in a Taylor series. We have $[f(x+a)-f(x)] = af'(x) + \frac{a^2}{2}f''(x) + \cdots$. Hence

$$\Delta(a) = a[f(\infty)-f(-\infty)] + \frac{a^2}{2}[f'(\infty)-f'(-\infty)] + \cdots. \quad\quad (E.12)$$

If the integral converges or diverges at most logarithmically, then

$$f(x) \sim \frac{1}{(x^2)^{1/2+\epsilon/2}} \quad\text{as}\quad x \to \pm\infty. \quad\quad (E.13)$$

Thus,

$$f(\pm\infty) = 0 = f'(\pm\infty) = \ldots \quad\quad (E.14)$$

and

$$\Delta(a) = 0. \tag{E.15}$$

However, for a linearly divergent integral we have

$$\Delta(a) = a[f(\infty) - f(-\infty)], \tag{E.16}$$

i.e. a surface term. We can generalize this to four dimensions. We have

$$\Delta(a) = \int d^4x[f(x+a) - f(x)] = \int d^4x \, a^\mu \frac{\partial}{\partial x^\mu} f(x) + \cdots \tag{E.17}$$

$$= a^\mu \int_S f(R) R^3 d\Sigma_\mu.$$

Using $d\Sigma_\mu = \frac{R_\mu}{R} d\Omega$ in Euclidean space, we have

$$\Delta(a) = i2\pi^2 a^\mu \lim_{R\to\infty} \left(\frac{R_\mu}{R}\right) R^3 f(R). \tag{E.18}$$

In addition, we have

$$\int \frac{R_\mu R_\nu}{R^2} d\Omega = (2\pi^2)\frac{g_{\mu\nu}}{4}. \tag{E.19}$$

Using Eqns. E.17 and E.18, we can evaluate the anomaly. We find

$$\Delta_{(1)}^{\nu\lambda} = \frac{i^3 2p^\tau}{(2\pi)^4} \int d^4k \frac{\partial}{\partial k^\tau} \mathrm{Tr}\left[\gamma_5\gamma^\nu \frac{\slashed{k} - \slashed{p} + m}{(k-p)^2 + m^2}\gamma^\lambda \frac{\slashed{k} + m}{k^2 + m^2}\right] \tag{E.20}$$

$$= \frac{2p^\tau}{(2\pi)^4} \lim_{k\to\infty} \int d\Omega \frac{k_\tau k^2}{k^4} \mathrm{Tr}\left[\gamma_5\gamma^\nu(\slashed{k} - \slashed{p})\gamma^\lambda \slashed{k}\right]$$

$$= \frac{2p^\tau}{(2\pi)^4} \lim_{k\to\infty} \int d\Omega \frac{k_\tau}{k^2} \mathrm{Tr}\left[\gamma_5\gamma^\nu\gamma^\alpha\gamma^\lambda\gamma^\beta\right] (k_\alpha - p_\alpha)k_\beta.$$

Using

$$\mathrm{Tr}\left[\gamma_5\gamma^\nu\gamma^\alpha\gamma^\lambda\gamma^\beta\right] = 4i\epsilon^{\nu\alpha\lambda\beta} \tag{E.21}$$

and Eqn. E.19, we have

$$\Delta_{(1)}^{\nu\lambda} = \frac{-2p^\tau}{(2\pi)^4} \lim_{k\to\infty} \int d\Omega \frac{k_\tau}{k^2} 4i\epsilon^{\nu\alpha\lambda\beta}p_\alpha k_\beta = \frac{-i}{4\pi^2}\epsilon^{\nu\alpha\lambda\beta}p_\alpha p_\beta \equiv 0. \tag{E.22}$$

Equivalently, we have

$$\Delta_{(2)}^{\nu\lambda} \equiv 0. \tag{E.23}$$

But since the integral is linearly divergent, and thus it is not invariant under a shift of integrals, consider taking $k \to k + a$. We define

$$\Delta^{\mu\nu\lambda}(a) = \tilde{R}^{\mu\nu\lambda}(a) - \tilde{R}^{\mu\nu\lambda}(0) \tag{E.24}$$

$$= (-1) \int \frac{d^4k}{(2\pi)^4} \left\{ \mathrm{Tr}\left[\gamma^\mu \gamma_5 i \frac{\slashed{k} + \slashed{a} - \slashed{p} + m}{(k+a-p)^2 - m^2 + i\epsilon} \gamma^\lambda \right.\right.$$

$$\left. \times i \frac{\slashed{k} + \slashed{a} + m}{(k+a)^2 - m^2 + i\epsilon} \gamma^\nu i \frac{\slashed{k} + \slashed{a} + \slashed{p}' + m}{(k+a+p')^2 - m^2 + i\epsilon} \right]$$

$$- \mathrm{Tr}\left[\gamma^\mu \gamma_5 i \frac{\slashed{k} - \slashed{p} + m}{(k-p)^2 - m^2 + i\epsilon} \gamma^\lambda i \frac{\slashed{k} + m}{k^2 - m^2 + i\epsilon} \gamma^\nu \right.$$

$$\left.\left. \times i \frac{\slashed{k} + \slashed{p}' + m}{(k+p')^2 - m^2 + i\epsilon} \right]\right\} + \left\{ \begin{smallmatrix} p \leftrightarrow p' \\ \lambda \leftrightarrow \nu \end{smallmatrix} \right\}$$

$$\equiv \Delta^{\mu\nu\lambda}_{(1)} + \Delta^{\mu\nu\lambda}_{(2)},$$

where $\Delta^{\mu\nu\lambda}_{(1)}$ is the first term and $\Delta^{\mu\nu\lambda}_{(2)}$ is gotten by $\left\{ \begin{smallmatrix} p \leftrightarrow p' \\ \lambda \leftrightarrow \nu \end{smallmatrix} \right\}$.

Using Eqns. E.18 and E.19, we have

$$\Delta^{\mu\nu\lambda}_{(1)} = \int \frac{d^4k}{(2\pi)^4} a^\tau \frac{\partial}{\partial k^\tau} \mathrm{Tr}\left[\gamma^\mu \gamma_5 \frac{\slashed{k} - \slashed{p} + m}{(k-p)^2 + m^2} \gamma^\lambda \frac{\slashed{k} + m}{k^2 + m^2} \gamma^\nu \frac{\slashed{k} + \slashed{p}' + m}{(k+p')^2 + m^2} \right]$$

$$= \frac{a^\tau}{(2\pi)^4} \lim_{k \to \infty} \int d\Omega \, k_\tau k^2 \frac{\mathrm{Tr}\left[\gamma^\mu \gamma_5 \slashed{k} \gamma^\lambda \slashed{k} \gamma^\nu \slashed{k} \right]}{k^6}$$

$$= \frac{a^\tau}{(2\pi)^4} \lim_{k \to \infty} \int d\Omega \frac{k_\tau}{k^4} 4i\epsilon^{\alpha\lambda\beta\mu}(2k^\nu k_\beta - g^\nu{}_\beta k^2) k_\alpha$$

$$= -\frac{2\pi^2 a^\tau}{(2\pi)^4} \lim_{k \to \infty} 4i\epsilon^{\alpha\lambda\nu\mu} \frac{g_{\tau\alpha}}{4} \tag{E.25}$$

and finally,

$$\Delta^{\mu\nu\lambda}_{(1)} = \frac{i}{8\pi^2} a_\alpha \epsilon^{\alpha\mu\nu\lambda}. \tag{E.26}$$

Let $a \equiv up + vp'$. Then

$$\Delta^{\mu\nu\lambda} = \frac{i}{8\pi^2} \left[(up_\alpha + vp'_\alpha)\epsilon^{\alpha\mu\nu\lambda} + (up'_\alpha + vp_\alpha)\epsilon^{\alpha\mu\lambda\nu} \right] \tag{E.27}$$

$$= \frac{i}{8\pi^2} \hat{u} \, (p - p')_\alpha \, \epsilon^{\alpha\mu\nu\lambda},$$

where $\hat{u} \equiv u - v$. Hence, we find

$$\tilde{R}^{\mu\nu\lambda}(\hat{u}) = R^{\mu\nu\lambda}(0) + \frac{i}{8\pi^2} \hat{u} \epsilon^{\alpha\mu\nu\lambda}(p - p')_\alpha, \tag{E.28}$$

i.e. there is an ambiguity which must be resolved. The Ward identity for the axial current becomes

$$q_\mu \tilde{R}^{\mu\nu\lambda}(\hat{u}) = 2m\tilde{R}^{\nu\lambda} + \frac{i}{8\pi^2} \hat{u} \epsilon^{\alpha\mu\nu\lambda} q_\mu (p - p')_\alpha \tag{E.29}$$

$$= 2m\tilde{R}^{\nu\lambda} + \frac{i}{4\pi^2} \hat{u} \epsilon^{\nu\lambda\alpha\beta} p_\alpha p'_\beta.$$

In order to resolve the ambiguity, let's check the vector Ward identity. The electromagnetic current is necessarily conserved and we should have

$$p_\lambda R^{\mu\nu\lambda}(0) = 0. \qquad (E.30)$$

$$
\begin{aligned}
p_\lambda \tilde{R}^{\mu\nu\lambda}(0) = (-1) \int \frac{d^4k}{(2\pi)^4} \Bigg\{ & \mathrm{Tr}\left[\gamma^\mu \gamma_5 i \frac{\slashed{k} - \slashed{p} + m}{(k-p)^2 - m^2 + i\epsilon} \right. \\
& \times \slashed{p} i \frac{\slashed{k} + m}{k^2 - m^2 + i\epsilon} \gamma^\nu i \frac{\slashed{k} + \slashed{p}' + m}{(k+p')^2 - m^2 + i\epsilon} \Bigg] \\
& + \mathrm{Tr}\left[\gamma^\mu \gamma_5 i \frac{\slashed{k} - \slashed{p}' + m}{(k-p')^2 - m^2 + i\epsilon} \gamma^\nu i \frac{\slashed{k} + m}{k^2 - m^2 + i\epsilon} \slashed{p} \right. \\
& \times i \frac{\slashed{k} + \slashed{p} + m}{(k+p)^2 - m^2 + i\epsilon} \Bigg] \Bigg\}.
\end{aligned}
\qquad (E.31)
$$

Using $\slashed{p} \equiv (\slashed{k} + \slashed{p} - m) - (\slashed{k} - m) = (\slashed{k} - m) - (\slashed{k} - \slashed{p} - m)$ we have

$$(E.32)$$

$$
\begin{aligned}
p_\lambda \tilde{R}^{\mu\nu\lambda}(0) = i \int \frac{d^4k}{(2\pi)^4} \Bigg\{ & \mathrm{Tr}\left[\gamma^\mu \gamma_5 \frac{\slashed{k} - \slashed{p} + m}{(k-p)^2 - m^2 + i\epsilon} \gamma^\nu \frac{\slashed{k} + \slashed{p}' + m}{(k+p')^2 - m^2 + i\epsilon} \right] \\
& - (k \to k + p - p') \\
& + \mathrm{Tr}\left[\gamma^\mu \gamma_5 \frac{\slashed{k} - \slashed{p}' + m}{(k-p')^2 - m^2 + i\epsilon} \gamma^\nu \frac{\slashed{k} + m}{k^2 - m^2 + i\epsilon} \right] \\
& - (k \to k + p') \Bigg\} \\
= & -\frac{2\pi^2}{(2\pi)^4} (p - p')^\tau \lim_{k\to\infty} \frac{k_\tau k^2}{k^4} \mathrm{Tr}\left[\gamma^\mu \gamma_5 \gamma^\alpha \gamma^\nu \gamma^\beta \right] (k_\alpha p'_\beta - p_\alpha k_\beta) \\
& -\frac{2\pi^2}{(2\pi)^4} p'^\tau \lim_{k\to\infty} \frac{k_\tau k^2}{k^4} \mathrm{Tr}\left[\gamma^\mu \gamma_5 \gamma^\alpha \gamma^\nu \gamma^\beta \right] (-p')_\alpha k_\beta.
\end{aligned}
$$

Note, the last term vanishes identically. Hence, we have

$$
\begin{aligned}
p_\lambda \tilde{R}^{\mu\nu\lambda}(0) &= -\frac{i}{8\pi^2} (p - p')_\alpha (p + p')_\beta \epsilon^{\alpha\nu\beta\mu} \qquad (E.33) \\
&= -\frac{i}{4\pi^2} \epsilon^{\mu\nu\alpha\beta} p_\alpha p'_\beta.
\end{aligned}
$$

Also,

$$
\begin{aligned}
p_\lambda \tilde{R}^{\mu\nu\lambda}(\hat{u}) &= p_\lambda \tilde{R}^{\mu\nu\lambda}(0) - \frac{i}{8\pi^2} \hat{u} \epsilon^{\alpha\mu\nu\lambda} p_\lambda p'_\alpha \qquad (E.34) \\
&= -\frac{i}{4\pi^2} \epsilon^{\mu\nu\alpha\beta} p_\alpha p'_\beta \left(1 - \frac{\hat{u}}{2} \right).
\end{aligned}
$$

Gauge invariance thus requires $\hat{u} = 2$ and we obtain the axial anomaly given by

$$q_\mu \tilde{R}^{\mu\nu\lambda}(\hat{u} = 2) = 2m\tilde{R}^{\nu\lambda} + \frac{i}{2\pi^2} \epsilon^{\nu\lambda\alpha\beta} p_\alpha p'_\beta. \qquad (E.35)$$

Hence, the axial vector is not conserved.

If we now couple two photons to the electromagnetic currents in the Green's function, Eqn. E.1, we find

$$-iq_\mu \tilde{R}^{\mu\nu\lambda}|_{\text{anomaly}} (ie)^2 \epsilon_\nu^*(p')\epsilon_\lambda^*(p) = -\frac{e^2}{2\pi^2} \epsilon^{\alpha\lambda\beta\nu} ip_\alpha \epsilon_\lambda^*(p) ip'_\beta \epsilon_\nu^*(p'). \qquad (E.36)$$

Using

$$\langle \gamma(p'), \gamma(p)|F_{\mu\nu}\,{}^*F^{\mu\nu}(0)|0\rangle = 4\epsilon^{\mu\nu\alpha\beta}(ip_\mu)\epsilon_\nu^*(p)(ip'_\alpha)\epsilon_\beta^*(p'), \qquad (E.37)$$

with

$$^*F^{\mu\nu} \equiv \frac{1}{2}\epsilon^{\mu\nu\alpha\beta}F_{\alpha\beta}, \qquad (E.38)$$

we find

$$\partial_\mu J_5^\mu = -2im\bar{\psi}\gamma_5\psi - \frac{e^2}{8\pi^2}F_{\mu\nu}\,{}^*F^{\mu\nu}. \qquad (E.39)$$

References

Aad, G. et al. 2012. Observation of a new particle in the search for the Standard Model Higgs boson with the ATLAS detector at the LHC. *Phys. Lett. B*, **716**, 1–29.

Aad, G. et al. 2020. Search for chargino–neutralino production with mass splittings near the electroweak scale in three-lepton final states in $\sqrt{s}$=13 TeV pp collisions with the ATLAS detector. *Phys. Rev. D*, **101**(7), 072001.

Aaij, R. et al. 2015. Measurement of the forward–backward asymmetry in $Z/\gamma^* \to \mu^+\mu^-$ decays and determination of the effective weak mixing angle. *JHEP*, **11**, 190.

Abdurashitov, J. N. et al. 2009. Measurement of the solar neutrino capture rate with gallium metal. III: Results for the 2002–2007 data-taking period. *Phys. Rev.*, **C80**, 015807.

Abe, F. et al. 1995. Observation of top quark production in $\bar{p}p$ collisions. *Phys. Rev. Lett.*, **74**, 2626–2631.

Abe, K. et al. 2006. A measurement of atmospheric neutrino flux consistent with tau neutrino appearance. *Phys. Rev. Lett.*, **97**, 171801.

Abe, K. et al. 2011. Solar neutrino results in Super-Kamiokande-III. *Phys. Rev.*, **D83**, 052010.

Abe, K. et al. 2014. Search for proton decay via $p \to \nu K^+$ using 260 kilotonyear data of Super-Kamiokande. *Phys. Rev.*, **D90**(7), 072005.

Abe, K. et al. 2017. Search for nucleon decay into charged antilepton plus meson in 0.316 megaton·years exposure of the Super-Kamiokande water Cherenkov detector. *Phys. Rev.*, **D96**(1), 012003.

Abe, Y. et al. 2012. Indication of reactor $\bar{\nu}_e$ disappearance in the Double Chooz experiment. *Phys. Rev. Lett.*, **108**, 131801.

Abers, E. S. and Lee, B. W. 1973. Gauge theories. *Phys. Rept.*, **9**, 1–141.

Abramowicz, H. et al. 1986. A precision measurement of $\sin^2\theta_W$ from semileptonic neutrino scattering. *Phys. Rev. Lett.*, **57**, 298.

Abrams, G. S. et al. 1974. The discovery of a second narrow resonance in e^+e^- annihilation. *Phys. Rev. Lett.*, **33**, 1453–1455.

Abreu, P. et al. 2000. Cross-sections and leptonic forward backward asymmetries from the Z^0 running of LEP. *Eur. Phys. J.*, **C16**, 371–405.

Acciarri, R. et al. 2015. Long-Baseline Neutrino Facility (LBNF) and Deep Underground Neutrino Experiment (DUNE). arXiv:1512.06148.

Achasov, M. N. et al. 2001. Direct measurement of the phi(1020) leptonic branching ratio. *Phys. Rev. Lett.*, **86**, 1698–1701.

Acker, A. and Pakvasa, S. 1994. Solar neutrino decay. *Phys. Lett.*, **B320**, 320–322.

Adamson, P. et al. 2016. Limits on active to sterile neutrino oscillations from disappearance searches in the MINOS, Daya Bay, and Bugey-3 experiments. *Phys. Rev. Lett.*, **117**(15), 151801. [Addendum: Phys. Rev. Lett. 117,no.20,209901(2016)].

Adamson, P. et al. 2019. Search for sterile neutrinos in MINOS and MINOS+ using a two-detector fit. *Phys. Rev. Lett.*, **122**(9), 091803.

Adler, S. L. 1965. Sum rules for the axial vector coupling constant renormalization in beta decay. *Phys. Rev.*, **140**, B736.

Adler, S. L. 1969. Axial vector vertex in spinor electrodynamics. *Phys. Rev.*, **177**, 2426–2438.

Adler, S. L. and Dashen, R. F. 1968. *Current Algebra and Application to Particle Physics*. New York: Benjamin.

Agafonova, N. et al. 2019. Final results on neutrino oscillation parameters from the OPERA experiment in the CNGS beam. *Phys. Rev.*, **D100**(5), 051301.

Aghanim, N. et al. 2020. Planck 2018 results. VI. Cosmological parameters. *A&A*, **641**, A6.

Agostini, M. et al. 2019. Probing Majorana neutrinos with double-β decay. *Science*, **365**, 1445.

Aguilar-Arevalo, A. A. et al. 2018. Significant excess of electronlike events in the MiniBooNE short-baseline neutrino experiment. *Phys. Rev. Lett.*, **121**(22), 221801.

Aharmim, B. et al. 2013. Combined analysis of all three phases of solar neutrino data from the Sudbury Neutrino Observatory. *Phys. Rev.*, **C88**, 025501.

Ahmad, Q. R. et al. 2002. Direct evidence for neutrino flavor transformation from neutral current interactions in the Sudbury Neutrino Observatory. *Phys. Rev. Lett.*, **89**, 011301.

Ahn, J. K. et al. 2012. Observation of reactor electron antineutrino disappearance in the RENO experiment. *Phys. Rev. Lett.*, **108**, 191802.

Ahn, M. H. et al. 2003. Indications of neutrino oscillation in a 250 km long baseline experiment. *Phys. Rev. Lett.*, **90**, 041801.

Albert, J. B. et al. 2018. Search for neutrinoless double-beta decay with the upgraded EXO-200 detector. *Phys. Rev. Lett.*, **120**(7), 072701.

Alduino, C. et al. 2018. First results from CUORE: A search for lepton number violation via $0\nu\beta\beta$ decay of ^{130}Te. *Phys. Rev. Lett.*, **120**(13), 132501.

Alexandrou, C. et al. 2014. Baryon spectrum with $N_f = 2 + 1 + 1$ twisted mass fermions. *Phys. Rev.*, **D90**(7), 074501.

Allaby, J. V. et al. 1987. A precise determination of the electroweak mixing angle from semileptonic neutrino scattering. *Z. Phys.*, **C36**, 611.

Altarelli, G. 1974. The physics of deep inelastic phenomena. *Riv. Nuovo Cim.*, **4**, 335.

Altarelli, G. and Parisi, G. 1977. Asymptotic freedom in parton language. *Nucl. Phys.*, **B126**, 298–318.

Altarelli, G. et al. 2000. From minimal to realistic supersymmetric $SU(5)$ grand unification. *JHEP*, **11**, 040.

Altmann, M. 1998. GALLEX solar neutrino observations: Results from the total data set. *Proceedings, 33rd Rencontres de Moriond 98: Electroweak Interactions and Unified Theories Les Arcs, France, Mar 14-21, 1998.* Paris: Edition Frontieres, pp. 345–352.

Altmann, M. et al. 2005. Complete results for five years of GNO solar neutrino observations. *Phys. Lett.*, **B616**, 174–190.

Alvarez-Gaume, L. et al. 1982. Low-energy supersymmetry. *Nucl. Phys.*, **B207**, 96.

Alvis, S. I. et al. 2019. A search for neutrinoless double-beta decay in ^{76}Ge with 26 kg-yr of exposure from the Majorana Demonstrator. *Phys. Rev.*, **C100**(2), 025501.

Amaldi, U. et al. 1991. Consistency checks of GUTs with LEP data. *Conf. Proc. C*, **910725V1**, 690–693.

An, F. P. et al. 2012. Observation of electron-antineutrino disappearance at Daya Bay. *Phys. Rev. Lett.*, **108**, 171803.

Angrik, J. et al. 2005. KATRIN design report 2004. INIS volume 36 (26).

Anselmino, M. et al. 1994. Polarized deep inelastic scattering at high-energies and parity violating structure functions. *Z. Phys.*, **C64**, 267–274.

Aoki, S. et al. 2009. 2+1 Flavor lattice QCD toward the physical point. *Phys. Rev.*, **D79**, 034503.

Aoki, S. et al. 2014. Review of lattice results concerning low-energy particle physics. *Eur. Phys. J.*, **C74**, 2890.

Aoyama, T. et al. 2018. Revised and improved value of the QED tenth-order electron anomalous magnetic moment. *Phys. Rev.*, **D97**(3), 036001.

Appelquist, T. and Carazzone, J. 1975. Infrared singularities and massive fields. *Phys. Rev.*, **D11**, 2856.

Armbruster, B. et al. 2002. Upper limits for neutrino oscillations muon–anti-neutrino to electron–anti-neutrino from muon decay at rest. *Phys. Rev.*, **D65**, 112001.

Arnison, G. et al. 1983a. Experimental observation of isolated large transverse energy electrons with associated missing energy at s**(1/2) = 540-GeV. *Phys. Lett.*, **122B**, 103–116.

Arnison, G. et al. 1983b. Experimental observation of lepton pairs of invariant mass around 95-GeV/c**2 at the CERN SPS collider. *Phys. Lett.*, **126B**, 398–410.

Arnison, G. et al. 1983c. Further evidence for charged intermediate vector bosons at the SPS collider. *Phys. Lett.*, **129B**, 273–282.

Arnison, G. et al. 1984a. Observation of muonic Z^0 decay at the anti-p p collider. *Phys. Lett.*, **147B**, 241–248.

Arnison, G. et al. 1984b. Observation of the muonic decay of the charged intermediate vector boson. *Phys. Lett.*, **134B**, 469–476.

Ashie, Y. et al. 2005. A measurement of atmospheric neutrino oscillation parameters by Super-Kamiokande I. *Phys. Rev.*, **D71**, 112005.

Astier, P. et al. 2003. Search for nu(mu) –> nu(e) oscillations in the NOMAD experiment. *Phys. Lett.*, **B570**, 19–31.

ATLAS collaboration. 2018. Search for exotic decays of the Higgs boson to at least one photon and missing transverse momentum using 79.8 fb^{-1} -1 of proton-proton collisions collected at $\sqrt{s} = 13$ s=13 TeV with the ATLAS detector. Report number: ATLAS-CONF-2018-019

Aubert, B. et al. 2002. Measurement of the $B^0 - \overline{B}^0$ oscillation frequency with inclusive dilepton events. *Phys. Rev. Lett.*, **88**, 221803.

Aubin, C. B. et al. 2004. Light hadrons with improved staggered quarks: Approaching the continuum limit. *Phys. Rev.*, **D70**, 094505.

Augustin, J. E. et al. 1974. Discovery of a narrow resonance in e^+e^- annihilation. *Phys. Rev. Lett.*, **33**, 1406–1408.

Azzolini, O. et al. 2018. First result on the neutrinoless double-β decay of ^{82}Se with CUPID-0. *Phys. Rev. Lett.*, **120**(23), 232502.

Babu, K. S. et al. 2000. Fermion masses, neutrino oscillations, and proton decay in the light of Super-Kamiokande. *Nucl. Phys.*, **B566**, 33–91.

Bagnaia, P. et al. 1983. Evidence for $Z^0 -> e^+e^-$ at the CERN anti-p p Collider. *Phys. Lett.*, **129B**, 130–140.

Bagnaia, P. et al. 1984. A study of high transverse momentum electrons produced in anti-p p collisions at 540-GeV. *Z. Phys.*, **C24**, 1.

Bahcall, J. N. 2000. Solar neutrinos: An overview. *Phys. Rept.*, **333**, 47–62.

Bahcall, J. N. 2004. Solar neutrinos. www.sns.ias.edu/~jnb/Papers/Popular/Wiley/paper.pdf.

Bahcall, J. N. 2005. Solar models and solar neutrinos: Current status. *Phys. Scripta*, **T121**, 46–50.

Bahcall, J. et al. 1987. The MSW effect in electron–neutrino scattering experiments. *Phys. Rev.*, **D35**, 2976.

Bahcall, J. N. et al. 1998. How uncertain are solar neutrino predictions? *Phys. Lett. B*, **433**, 1–8.

Banks, T. et al. 1977. Strong coupling calculations of the hadron spectrum of quantum chromodynamics. *Phys. Rev.*, **D15**, 1111.

Banner, M. et al. 1983. Observation of single isolated electrons of high transverse momentum in events with missing transverse energy at the CERN anti-p p Collider. *Phys. Lett.*, **122B**, 476–485.

Barber, D. P. et al. 1979a. Discovery of three jet events and a test of quantum chromodynamics at PETRA energies. *Phys. Rev. Lett.*, **43**, 830.

Barber, D. P. et al. 1979b. Tests of quantum chromodynamics and a direct measurement of the strong coupling constant α_S at $\sqrt{s} = 30$-GeV. *Phys. Lett.*, **89B**, 139–144.

Bardeen, W. et al. 1972. Light cone current algebra, π^0 decay, and e^+e^- annihilation. In Scale and Conformal Symmetry in Hadron Physics: Topical Meetings on Outlook for Broken Conformal Symmetry in Elementary Particle Physics. Frascati, Italy, 4 - 5 May.

Barger, V. et al. 1993. Supersymmetric grand unified theories: Two loop evolution of gauge and Yukawa couplings. *Phys. Rev.*, **D47**, 1093–1113.

Barnes, V. E. et al. 1964. Observation of a hyperon with strangeness -3. *Phys. Rev. Lett.*, **12**, 204–206.

Bartel, W. et al. 1980. Observation of planar three jet events in e^+e^- annihilation and evidence for gluon Bremsstrahlung. *Phys. Lett.*, **91B**, 142 147.

Bazavov, A. et al. 2010. Nonperturbative QCD simulations with 2+1 flavors of improved staggered quarks. *Rev. Mod. Phys.*, **82**, 1349–1417.

Becchi, C. et al. 1976. Renormalization of gauge theories. *Annals Phys.*, **98**, 287–321.

Bell, J. S. and Jackiw, R. 1969. A PCAC puzzle: $\pi^0 \to \gamma\gamma$ in the σ model. *Nuovo Cim.*, **A60**, 47–61.

Bellini, G. et al. 2010. Measurement of the solar ^{8}B neutrino rate with a liquid scintillator target and 3 MeV energy threshold in the Borexino detector. *Phys. Rev.*, **D82**, 033006.

Bellini, G. et al. 2011. Precision measurement of the ^{7}Be solar neutrino interaction rate in Borexino. *Phys. Rev. Lett.*, **107**, 141302.

Benaksas, D. et al. 1972. pi$^+$ pi$^-$ production by e^+e^- annihilation in the rho energy range with the Orsay storage ring. *Phys. Lett.*, **39B**, 289–293.

Bennett, G. W. et al. 2006. Final report of the muon E821 anomalous magnetic moment measurement at BNL. *Phys. Rev.*, **D73**, 072003.

Berger, C. et al. 1979. Evidence for gluon Bremsstrahlung in e^+e^- annihilations at high energies. *Phys. Lett.*, **86B**, 418–425.

Beringer, J. 2002. CP violation, B mixing and B lifetime results from the BaBar experiment. *Proceedings, 36th Rencontres de Moriond on QCD and High Energy Hadronic Interactions: Les Arcs, France, Paris: Edition Frontieres*, pp. 197–200.

Bernard, C. et al. 2011. Tuning Fermilab heavy quarks in 2+1 flavor lattice QCD with application to hyperfine splittings. *Phys. Rev.*, **D83**, 034503.

Bethe, H. A. and Salpeter, E. E. 1957. *Quantum Mechanics of One- and Two-Electron Atoms*. Berlin: Springer.

Bietenholz, W. et al. 2011. Flavour blindness and patterns of flavour symmetry breaking in lattice simulations of up, down and strange quarks. *Phys. Rev.*, **D84**, 054509.

Bjorken, J. D. 1969. Asymptotic sum rules at infinite momentum. *Phys. Rev.*, **179**, 1547–1553.

Bjorken, J. D. and Paschos, E. A. 1969. Inelastic electron proton and gamma proton scattering, and the structure of the nucleon. *Phys. Rev.*, **185**, 1975–1982.

Bjorken, J. D. and Dunietz, I. 1987. Rephasing invariant parametrizations of generalized Kobayashi–Maskawa matrices. *Phys. Rev.*, **D36**, 2109.

Blazek, T. et al. 1997. A global chi**2 analysis of electroweak data in $SO(10)$ SUSY GUTs. *Phys. Rev.*, **D56**, 6919–6938.

Bloom, E. D. et al. 1970. Recent results in inelastic electron scattering. Proceedings of the *15th International Conference on High Energy Physics*, Kiev, August 26–September 4.

Bodek, A. et al. 1973. Comparisons of deep inelastic e p and e n cross-sections. *Phys. Rev. Lett.*, **30**, 1087.

Bodek, A. et al. 1974. Observed deviations from scaling of the proton electromagnetic structure functions. *Phys. Lett.*, **52B**, 249–252.

Bodek, A. et al. 1979. Experimental studies of the neutron and proton electromagnetic structure functions. *Phys. Rev.*, **D20**, 1471–1552.

Bogert, D. et al. 1985. A determination of $\sin^2 \theta_W$ and ρ in deep inelastic neutrino–nucleon scattering. *Phys. Rev. Lett.*, **55**, 1969.

Brandelik, R. et al. 1979. Evidence for planar events in e^+e^- annihilation at high energies. *Phys. Lett.*, **86B**, 243–249.

Brandelik, R. et al. 1980. Evidence for a spin one gluon in three jet events. *Phys. Lett.*, **97B**, 453–458.

Breidenbach, M. et al. 1969. Observed behavior of highly inelastic electron–proton scattering. *Phys. Rev. Lett.*, **23**, 935–939.

Brown, H. N. et al. 2001. Precise measurement of the positive muon anomalous magnetic moment. *Phys. Rev. Lett.*, **86**, 2227–2231.

Buchalla, G. et al. 1990. The anatomy of epsilon-prime/epsilon in the Standard Model. *Nucl. Phys.*, **B337**, 313–362.

Bumiller, F. et al. 1961. Electromagnetic form factors of the proton. *Phys. Rev.*, **124**, 1623–1631.

Buras, A. et al. 1994. Waiting for the top quark mass, $K^+ -> \text{pi}^+$ neutrino anti-neutrino, $B(s)^0$–anti-$B(s)^0$ mixing and CP asymmetries in B decays. *Phys. Rev.*, **D50**, 3433–3446.

Cabibbo, N. 1963. Unitary symmetry and leptonic decays. *Phys. Rev. Lett.*, **10**, 531–533.

Cabibbo, N. et al. 2003. Semileptonic hyperon decays. *Ann. Rev. Nucl. Part. Sci.*, **53**, 39–75.

Cabibbo, N. et al. 2004. Semileptonic hyperon decays and CKM unitarity. *Phys. Rev. Lett.*, **92**, 251803.

Callan, C. G., Jnr. 1970. Broken scale invariance in scalar field theory. *Phys. Rev. D*, **2**, 1541–1547.

Callan, C. G., Jr. and Gross, D. J. 1969. High-energy electroproduction and the constitution of the electric current. *Phys. Rev. Lett.*, **22**, 156–159.

Carruthers, P. A. 1966. *Introduction to Unitary Symmetry.* New York: Interscience Publishers.

Casher, A. et al. 1973. Vacuum polarization and the quark parton puzzle. *Phys. Rev. Lett.*, **31**, 792–795.

Caswell, W. E. 1974. Asymptotic behavior of nonabelian gauge theories to two loop order. *Phys. Rev. Lett.*, **33**, 244.

Chambers, E. E. and Hofstadter, R. 1956. Structure of the proton. *Phys. Rev.*, **103**, 1454–1463.

Chamseddine, A. H. et al. 1982. Locally supersymmetric grand unification. *Phys. Rev. Lett.*, **49**, 970.

Chatrchyan, S. et al. 2012. Observation of a new boson at a mass of 125 GeV with the CMS Experiment at the LHC. *Phys. Lett. B*, **716**, 30–61.

Cheng, T. P. and Li, L. F. 1984. *Gauge Theory of Elementary Particle Physics.* New York, NY: Oxford University Press.

Christ, N. H. et al. 2010. The η and η' mesons from lattice QCD. *Phys. Rev. Lett.,* **105**, 241601.

Christenson, J. H. et al. 1964. Evidence for the 2π decay of the K_2^0 meson. *Phys. Rev. Lett.,* **13**, 138–140.

Chuang, S. H. 2004. Standard-Model Higgs searches at CDF Run II. BEACH 04, Chicago, IL, June 27–July 3,

CKMfitter Collaboration. Unitarity Triangle. http://ckmfitter.in2p3.fr/www/results/plots_summer19/ckm_res_summer19.html

Cleveland, B. T. et al. 1998. Measurement of the solar electron neutrino flux with the Homestake chlorine detector. *Astrophys. J.,* **496**, 505–526.

Cohen, A. G. 1993. Selected topics in effective field theories for particle physics. Proceedings of Theoretical Advanced Study Institute (TASI 93) Elementary Particle Physics: The Building Blocks of Creation - From Microfermius to Megaparsecs, Boulder, Colorado, June 6–July 2, pp. 53–100.

Coleman, Sidney. 1985. Aspects of Symmetry: Selected Erice Lectures. Cambridge, U.K.: Cambridge University Press.

Coleman, S. R. and Glashow, S. L. 1961. Electrodynamic properties of baryons in the unitary symmetry scheme. *Phys. Rev. Lett.,* **6**, 423.

Collins, J. et al. 1985. Factorization for short distance hadron–hadron scattering. *Nucl. Phys.,* **B261**, 104–142.

Cravens, J. P. et al. 2008. Solar neutrino measurements in Super-Kamiokande-II. *Phys. Rev.,* **D78**, 032002.

Crawford, S. et al. 1959. Experimental determination of the Λ spin. *Phys. Rev. Lett.,* **2**(Feb), 114–116.

Dalitz, R. H. 1954. Decay of tau mesons of known charge. *Phys. Rev.,* **94**, 1046–1051.

Dalitz, R. H. 1994. The τ–θ puzzle. *AIP Conf. Proc.,* **300**, 141–158.

Danby, G. et al. 1962. Observation of high-energy neutrino reactions and the existence of two kinds of neutrinos. *Phys. Rev. Lett.,* **9**, 36–44.

Davier, M. et al. 2017. Reevaluation of the hadronic vacuum polarisation contributions to the Standard Model predictions of the muon $g-2$ and $\alpha(m_Z^2)$ using newest hadronic cross-section data. *Eur. Phys. J.,* **C77**(12), 827.

Davoudiasl, H. and Marciano, W. J. 2018. A tale of two anomalies. *Phys. Rev. D,* **98**, 075011.

Dawson, S. et al. 1981. Radiative corrections to $\sin^2\theta_W$ to leading logarithm in the W boson mass. *Phys. Rev.,* **D23**, 2666.

Dawson, S. et al. 1985. Search for supersymmetric particles in hadron–hadron collisions. *Phys. Rev.,* **D31**, 1581.

de Baere, W. et al. 1967. Three body final States in $K+p$ Interactions at 3.5 GeV/c. *Nuovo Cim. A,* **51**, 401.

DeGrand, T. et al. 1977. Jet structure in e^+e^- annihilation as a test of QCD and the quark-confining string. *Phys. Rev.,* **D16**, 3251.

Dermisek, R. et al. 2001. SUSY GUTs under siege: Proton decay. *Phys. Rev.*, **D63**, 035001.

Dimopoulos, S. et al. 1981. Supersymmetry and the scale of unification. *Phys. Rev.*, **D24**, 1681–1683.

Dimopoulos, S. and Georgi, H. 1981. Softly broken supersymmetry and SU(5). *Nucl. Phys.*, **B193**, 150.

Dimopoulos, S. and Raby, S. 1981. Supercolor. *Nucl. Phys.*, **B192**, 353–368.

Dimopoulos, S. and Raby, S. 1983. Geometric hierarchy. *Nucl. Phys.*, **B219**, 479.

Dimopoulos, S. et al. 1982. Proton decay in supersymmetric models. *Phys. Lett.*, **112B**, 133.

Dimopoulos, S. et al. 1996. Experimental signatures of low-energy gauge mediated supersymmetry breaking. *Phys. Rev. Lett.*, **76**, 3494–3497.

Dine, M. Fischler, W. 1982. A phenomenological model of particle physics based on supersymmetry. *Phys. Lett.*, **110B**, 227–231.

Dine, M. and Nelson, A. E. 1993. Dynamical supersymmetry breaking at low-energies. *Phys. Rev.*, **D48**, 1277–1287.

Dine, M. et al. 1981. Supersymmetric technicolor. *Nucl. Phys.*, **B189**, 575–593.

Dine, M. et al. 1995. Low-energy dynamical supersymmetry breaking simplified. *Phys. Rev.*, **D51**, 1362–1370.

Dine, M. et al. 1996. New tools for low-energy dynamical supersymmetry breaking. *Phys. Rev.*, **D53**, 2658–2669.

Dine, M. et al. 1997. Variations on minimal gauge mediated supersymmetry breaking. *Phys. Rev.*, **D55**, 1501–1508.

Donoghue, J. F. et al. 1992. Dynamics of the standard model. *Camb. Monogr. Part. Phys. Nucl. Phys. Cosmol.*, **2**, 1–540.

Dowdall, R. J. et al. 2012. Precise heavy–light meson masses and hyperfine splittings from lattice QCD including charm quarks in the sea. *Phys. Rev.*, **D86**, 094510.

Dreiner, H. K. et al. 2006. What is the discrete gauge symmetry of the MSSM? *Phys. Rev.*, **D73**, 075007.

Drell, S. D. and Yan, T.-M. 1971. Partons and their applications at high energies. *Annals Phys.*, **66**, 578.

Dudek, J. J. et al. 2011. Isoscalar meson spectroscopy from lattice QCD. *Phys. Rev.*, **D83**, 111502.

Durbin, R. et al. 1951. The spin of the pion via the reaction $\pi^+ + D \leftrightarrow P + P$. *Phys. Rev.*, **83**, 646–648.

Durr, S. et al. 2008. Ab-initio determination of light hadron masses. *Science*, **322**, 1224–1227.

Eichten, E. et al. 1975. The spectrum of charmonium. *Phys. Rev. Lett.*, **34**, 369–372. [Erratum: Phys. Rev. Lett.36,1276(1976)].

Eichten, E. et al. 1978. Charmonium: The model. *Phys. Rev.*, **D17**, 3090. [Erratum: Phys. Rev.D21,313(1980)].

Einhorn, M. B. et al. 1981. Heavy particles and the rho parameter in the Standard Model. *Nucl. Phys.*, **B191**, 146–172.

Einhorn, M. B. and Jones, D. R. T. 1982. The weak mixing angle and unification mass in supersymmetric $SU(5)$. *Nucl. Phys. B*, **196**, 475–488.

Ellis, J. 2014. The discovery of the gluon. *Int. J. Mod. Phys.*, **A29**(31), 1430072.

Ellis, J. R. and Karliner, I. 1979. Measuring the spin of the gluon in e^+e^- annihilation. *Nucl. Phys.*, **B148**, 141–147.

Ellis, J. R. et al. 1976. Search for gluons in e^+e^- annihilation. *Nucl. Phys.*, **B111**, 253. [Erratum: Nucl. Phys.B130,516(1977)].

Ellis, J. R. et al. 1982. GUTs 3: SUSY GUTs 2. *Nucl. Phys.*, **B202**, 43–62.

Ellis, J. R. et al. 1991. Probing the desert using gauge coupling unification. *Phys. Lett. B*, **260**, 131–137.

Englert, F. and Brout, R. 1964. Broken symmetry and the mass of gauge vector mesons. *Phys. Rev. Lett.*, **13**, 321–323.

Esteban, I. et al. 2019. Global analysis of three-flavour neutrino oscillations: Synergies and tensions in the determination of $\theta_{23}, \delta_C P$, and the mass ordering. *JHEP*, **01**, 106.

Faddeev, L. D. and Popov, V. N. 1967. Feynman diagrams for the Yang–Mills field. *Phys. Lett.*, **B25**, 29–30.

Farrar, G. R. and Fayet, P. 1978. Phenomenology of the production, decay, and detection of new hadronic states associated with supersymmetry. *Phys. Lett.*, **76B**, 575–579.

Feinberg, G. and Weinberg, S. 1959. On the phase factors in inversions. *Nuovo Lim.* **14**, 571.

Ferrara, S. (ed). 1987. *Supersymmetry.* Amsterdam: North-Holland.

Ferrara, S. et al. 1974. Supergauge multiplets and superfields. *Phys. Lett.*, **51B**, 239.

Feynman, R. P. 1963. Quantum theory of gravitation. *Acta Phys. Polon.*, **24**, 697–722.

Feynman, R. P. 1969a. The behavior of hadron collisions at extreme energies. *Conf. Proc.*, **C690905**, 237–258.

Feynman, R. P. 1969b. Very high-energy collisions of hadrons. *Phys. Rev. Lett.*, **23**, 1415–1417.

Feynman, R. P. and Gell-Mann, M. 1958. Theory of fermi interaction. *Phys. Rev.*, **109**, 193–198.

Field, R. D. and Feynman, R. P. 1977. Quark elastic scattering as a source of high transverse momentum mesons. *Phys. Rev.*, **D15**, 2590–2616.

Field, R. D. and Feynman, R. P. 1978. A parametrization of the properties of quark jets. *Nucl. Phys.*, **B136**, 1.

Forte, S. et al. 2001. Polarized parton distributions from charged current deep inelastic scattering and future neutrino factories. *Nucl. Phys.*, **B602**, 585–621.

Friedman et al. 1972. Deep inelastic electron scattering. *Ann. Rev. Nucl. Part. Sci.*, **22**, 203–254.

Fritzsch, H. and Gell-Mann, M. 1971. Light cone current algebra. Proceedings, International Conference on Duality and Symmetry in Hadron Physics: Tel Aviv, Israel, April 5–7, pp. 317–374.

Fritzsch, H. and Minkowski, P. 1975. Unified interactions of leptons and hadrons. *Annals Phys.*, **93**, 193–266.

Fritzsch, H. et al. 1973. Advantages of the color octet gluon picture. *Phys. Lett.*, **47B**, 365–368.

Fujikawa, K. 1979. Path integral measure for gauge invariant fermion theories. *Phys. Rev. Lett.*, **42**, 1195–1198.

Fujikawa, K. 1980. Path integral for gauge theories with fermions. *Phys. Rev.*, **D21**, 2848. [Erratum: Phys. Rev.D22,1499(1980)].

Fukuda, S. et al. 2000. Tau neutrinos favored over sterile neutrinos in atmospheric muon–neutrino oscillations. *Phys. Rev. Lett.*, **85**, 3999–4003.

Fukuda, Y. et al. 1998. Evidence for oscillation of atmospheric neutrinos. *Phys. Rev. Lett.*, **81**, 1562–1567.

Gaillard, M. K. and Lee, B. W. 1974. Rare decay modes of the K-mesons in gauge theories. *Phys. Rev.*, **D10**, 897.

Gaillard, M. et al. 1999. The Standard Model of particle physics. *Rev. Mod. Phys.*, **71**, S96–S111.

Gaisser, T. K. and Honda, M. 2002. Flux of atmospheric neutrinos. *Ann. Rev. Nucl. Part. Sci.*, **52**, 153–199.

Gando, A. et al. 2011. Constraints on θ_{13} from a three-flavor oscillation analysis of reactor antineutrinos at KamLAND. *Phys. Rev.*, **D83**, 052002.

Gando, A. et al. 2016. Search for Majorana neutrinos near the inverted mass hierarchy region with KamLAND-Zen. *Phys. Rev. Lett.*, **117**(8), 082503. [Addendum: Phys. Rev. Lett.117,no.10,109903(2016)].

Gasiorowicz, St. 1966. *Elementary Particle Physics.* New York: Wiley.

Gasser, J. and Leutwyler, H. 1985. Chiral perturbation theory: Expansions in the mass of the strange quark. *Nucl. Phys.*, **B250**, 465–516.

Gates, S. J. et al. 1983. Superspace or one thousand and one lessons in supersymmetry. *Front. Phys.*, **58**, 1–548.

Gell-Mann, M. 1956. The interpretation of the new particles as displaced charge multiplets. *Nuovo Cim.*, **4**(S2), 848–866.

Gell-Mann, M. 1961. The Eightfold Way: A theory of strong interaction symmetry. Technical Report CTSL-20, TID-12608.

Gell-Mann, M. 1962. Symmetries of baryons and mesons. *Phys. Rev.*, **125**, 1067–1084.

Gell-Mann, M. 1964. A schematic model of baryons and mesons. *Phys. Lett.*, **8**, 214–215.

Gell-Mann, M. and Levy, M. 1960. The axial vector current in beta decay. *Nuovo Cim.*, **16**, 705.

Gell-Mann, M. and Rosenbaum, E. P. 1957. Elementary particles. *Sci. Am.*, **197**(1), 72–92.

Gell-Mann, M. et al. 1978. Color embeddings, charge assignments, and proton stability in unified gauge theories. *Rev. Mod. Phys.*, **50**, 721.

Georgi, H. 1975. The state of the art gauge theories. *AIP Conf. Proc.*, **23**, 575–582.

Georgi, H. 1982. *Lie Algebras in Particle Physics: From Isospin to Unifed Theories.* Benjamin Cummings, Fontiers in Physics, volume 54.

Georgi, H. 1984. *Weak Interactions and Modern Particle Theory.* Menlo Park, CA: Benjamin–Cummings

Georgi, H. and Glashow, S. L. 1974. Unity of all elementary particle forces. *Phys. Rev. Lett.*, **32**, 438–441.

Georgi, H. et al. 1974. Hierarchy of interactions in unified gauge theories. *Phys. Rev. Lett.*, **33**, 451–454.

Geweniger, C. et al. 1974. Measurement of the kaon mass difference m(L)-m(S) by the two regenerator method. *Phys. Lett.*, **52B**, 108–112.

Girardello, L. and Grisaru, M. T. 1982. Soft breaking of supersymmetry. *Nucl. Phys.*, **B194**, 65.

Giudice, G. F. and Rattazzi, R. 1999. Theories with gauge mediated supersymmetry breaking. *Phys. Rept.*, **322**, 419–499.

Glashow, S. L. 1961. Partial symmetries of weak interactions. *Nucl. Phys.*, **22**, 579–588.

Glashow, S. L. et al. 1970. Weak interactions with lepton-hadron symmetry. *Phys. Rev.*, **D2**, 1285–1292.

Glashow, S. L. and Krauss, L. M. 1987. "Just so" neutrino oscillations. *Phys. Lett.*, **B190**, 199–207.

Gohn, Wesley. 2019. The muon g-2 experiment at Fermilab. Proceedings, 18th Lomonosov Conference on Elementary Particle Physics: Moscow, August 24–30, pp. 232–236.

Goldberger, M. L. and Treiman, S. B. 1958. Decay of the pi meson. *Phys. Rev.*, **110**, 1178–1184.

Goldstone, J. 1961. Field theories with superconductor solutions. *Nuovo Cim.*, **19**, 154–164.

Goldstone, J. et al. 1962. Broken symmetries. *Phys. Rev.*, **127**, 965–970.

Gregory, E. B. et al. 2011. Precise B, B_s and B_c meson spectroscopy from full lattice QCD. *Phys. Rev.*, **D83**, 014506.

Gregory, E. B. et al. 2012. A study of the eta and eta$'$ mesons with improved staggered fermions. *Phys. Rev.*, **D86**, 014504.

Gross, D. J. and Jackiw, R. 1972. Effect of anomalies on quasirenormalizable theories. *Phys. Rev.*, **D6**, 477–493.

Gross, D. J. and Wilczek, F. 1973a. Asymptotically free gauge theories – I. *Phys. Rev.*, **D8**, 3633–3652.

Gross, D. J. and Wilczek, F. 1973b. Ultraviolet behavior of nonabelian gauge theories. *Phys. Rev. Lett.*, **30**, 1343–1346.

Gross, D. J. and Wilczek, F. 1974. Asymptotically free gauge theories. 2. *Phys. Rev.*, **D9**, 980–993.

Guralnik, G. S. et al. 1964. Global conservation laws and massless particles. *Phys. Rev. Lett.*, **13**, 585–587.

Haber, H. n.d. Physics 215 course notes. http://scipp.ucsc.edu/haber/ph215/djpi18.pdf.

Haber, H. E. and Kane, G. L. 1985. The search for supersymmetry: probing physics beyond the standard model. *Phys. Rept.*, **117**, 75–263.

Hall, L. J. et al. 1983. Supergravity as the messenger of supersymmetry breaking. *Phys. Rev.*, **D27**, 2359–2378.

Hanneke, D. et al. 2011. Cavity control of a single-electron quantum cyclotron: measuring the electron magnetic moment. *Phys. Rev.*, **A83**, 052122.

Hanson, G. et al. 1975. Evidence for jet structure in hadron production by e^+e^- annihilation. *Phys. Rev. Lett.*, **35**, 1609–1612.

Hasert, F. J. et al. 1973a. Observation of neutrino like interactions without muon or electron in the Gargamelle neutrino experiment. *Phys. Lett.*, **B46**, 138–140.

Hasert, F. J. et al. 1973b. Search for elastic ν_μ electron scattering. *Phys. Lett.*, **B46**, 121–124.

Herb, S. W. et al. 1977. Observation of a dimuon resonance at 9.5-GeV in 400-GeV proton-nucleus collisions. *Phys. Rev. Lett.*, **39**, 252–255.

Higgs, P. W. 1964. Broken symmetries and the masses of gauge bosons. *Phys. Rev. Lett.*, **13**, 508–509.

Higgs, P. W. 1966. Spontaneous symmetry breakdown without massless bosons. *Phys. Rev.*, **145**, 1156–1163.

Hirata, K. S. et al. 1989. Observation of B-8 solar neutrinos in the Kamiokande-II detector. *Phys. Rev. Lett.*, **63**, 16.

Hirata, K. S. et al. 1991. Real time, directional measurement of B-8 solar neutrinos in the Kamiokande-II detector. *Phys. Rev.*, **D44**, 2241. [Erratum: Phys. Rev.D45,2170(1992)].

Hosaka, J. et al. 2006. Solar neutrino measurements in Super-Kamiokande-I. *Phys. Rev.*, **D73**, 112001.

Ibanez, L. E. and Lopez, C. 1983. N=1 Supergravity, the breaking of $SU(2) \times U(1)$ and the top quark mass. *Phys. Lett.*, **126B**, 54–58.

Ibanez, L. E. and Ross, G. G. 1981. Low-energy predictions in supersymmetric grand unified theories. *Phys. Lett.*, **B105**, 439.

Ibanez, L. E. and Ross, G. G. 1982. $SU(2)$-L x $U(1)$ symmetry breaking as a radiative effect of supersymmetry breaking in guts. *Phys. Lett.*, **110B**, 215–220.

Ibanez, L. E. and Ross, G. G. 1992a. Discrete gauge symmetries and the origin of baryon and lepton number conservation in supersymmetric versions of the Standard Model. *Nucl. Phys.*, **B368**, 3–37.

Ibanez, L. E. and Ross, G. G. 1992b. Electroweak breaking in supersymmetric models. arXiv:hep-ph/9204201v1.

Inami, T. and Lim, C. S. 1981. Effects of superheavy quarks and leptons in low-energy weak processes $k(L) ->$ mu anti-mu, $K^+ ->$ pi$^+$ neutrino anti-neutrino and $K^0 ->$ anti-K^0. *Prog. Theor. Phys.*, **65**, 297.[Erratum: Prog. Theor. Phys.65,1772(1981)].

Itzykson, C. and Zuber, J. B. 1980. *Quantum Field Theory.* International Series In Pure and Applied Physics. New York: McGraw-Hill.

Jegerlehner, F. 2017. Variations on photon vacuum polarization. arxiv.org/abs/1711.06089.

Jones, D. R. T. 1974. Two loop diagrams in Yang–Mills theory. *Nucl. Phys.*, **B75**, 531.

Kajita, T. et al. 2016. Establishing atmospheric neutrino oscillations with Super-Kamiokande. *Nucl. Phys. B*, **908**, 14–29.

Kamioka Observatory, ICRR (Institute for Cosmic Ray Research), The University of Tokyo. *Super-Kamiokande images.*

Kane, G. L. et al. 1994. Study of constrained minimal supersymmetry. *Phys. Rev.*, **D49**, 6173–6210.

Kaplan, D. B. and Manohar, A. V. 1986. Current mass ratios of the light quarks. *Phys. Rev. Lett.*, **56**, 2004.

Kazakov, D. I. 2000. Beyond the standard model: In search of supersymmetry. *Proceedings of the 2000 European School of High-Energy Physics, Caramulo, Portugal, Aug 20–Sept 2, pp. 125–199.*

Kim, C. S. et al. 1990. Impact of new V_{ub}/V_{cb} and epsilon-prime/epsilon measurements on weak mixing angles. *Phys. Rev.*, **D42**, 96–111. [Erratum: Phys. Rev.D45,389(1992)].

Kobayashi, M. and Maskawa, T. 1973. CP violation in the renormalizable theory of weak interaction. *Prog. Theor. Phys.*, **49**, 652–657.

Kogut, J. B. and Susskind, L. 1974. Parton models and asymptotic freedom. *Phys. Rev.*, **D9**, 3391–3399.

Kroll, N. M. and Wada, W. 1955. Internal pair production associated with the emission of high-energy gamma rays. *Phys. Rev.*, **98**, 1355–1359.

Lamb, W. E. and Retherford, R. C. 1947. Fine structure of the hydrogen atom by a microwave method. *Phys. Rev.*, **72**, 241–243.

Lamb, W. E. and Retherford, R. C. 1950. Fine structure of the hydrogen atom. Part I. *Phys. Rev.*, **79**, 549–572.

Landau, L. D. 1948. On the angular momentum of a system of two photons. *Dokl. Akad. Nauk Ser. Fiz.*, **60**(2), 207–209.

Langacker, P. et al. 1987. On the Mikheev–Smirnov–Wolfenstein (MSW) mechanism of amplification of neutrino oscillations in matter. *Nucl. Phys.*, **B282**, 589–609.

Langacker, P. and Luo, M-X. 1991. Implications of precision electroweak experiments for M_t, ρ_0, $\sin^2\theta_W$ and grand unification. *Phys. Rev. D*, **44**, 817–822.

Lee, B. W. 1972. Renormalizable massive vector meson theory. Perturbation theory of the Higgs phenomenon. *Phys. Rev.*, **D5**, 823–835.

Lee, B. W. 1974. A Theory of electromagnetic and weak interactions. *Princeton 1971, Local Currents and Their Applications*, Amsterdam: North Holland Publishers, pp. 131–141.

Lee, T. D. 1981. Particle physics and introduction to field theory. *Contemp. Concepts Phys.*, **1**, 1–865.

Lee, T. D. and Yang, C.-N. 1956a. Charge conjugation, a new quantum number G, and selection rules concerning a nucleon anti-nucleon system. *Nuovo Cim.*, **10**, 749–753.

Lee, T. D. and Yang, C-N. 1956b. Question of parity conservation in weak interactions. *Phys. Rev.*, **104**, 254–258.

Lee, T. D. and Yang, C.-N. 1958. Possible determination of the spin of Λ^0 from its large decay angular asymmetry. *Phys. Rev.*, **109**, 1755–1758.

Lehmann, H. et al. 1957. On the formulation of quantized field theories. II. *Nuovo Cim.*, **6**, 319–333.

Leith, D. W. G. S. 1978. High-energy photoproduction: diffractive processes. In *Electromagnetic Interactions of Hadrons*, Volume 1. Donnachie A. and Shaw G. (eds). New York: Plenum Press, 1978, 345–441.

Leutwyler, H. 1990. How about m (u) = 0? *Nucl. Phys.*, **B337**, 108–118.

Leutwyler, H. 1996. The ratios of the light quark masses. *Phys. Lett.*, **B378**, 313–318.

Llewellyn Smith, C. H. 1983. On the determination of $\sin^2(\theta_W)$ in semileptonic neutrino interactions. *Nucl. Phys.*, **B228**, 205–215.

Loomis, W. A. 1981. B Meson physics at CESR. *AIP Conf. Proc.*, **72**, 432–444.

Lucas, V. and Raby, S. 1996. GUT scale threshold corrections in a complete supersymmetric $SO(10)$ model: Alpha-s $(M(Z))$ versus proton lifetime. *Phys. Rev.*, **D54**, 2261–2272.

Lucas, V. and Raby, S. 1997. Nucleon decay in a realistic $SO(10)$ SUSY GUT. *Phys. Rev. D*, **55**, 6986–7009.

Machacek, M. E. and Vaughn, M. T. 1983. Two loop renormalization group equations in a general quantum field theory. 1. Wave function renormalization. *Nucl. Phys.*, **B222**, 83–103.

Maki, Z. et al. 1962. Remarks on the unified model of elementary particles. *Prog. Theor. Phys.*, **28**, 870–880.

Marciano, W. J. and Senjanovic, G. 1982. Predictions of supersymmetric grand unified theories. *Phys. Rev. D*, **25**, 3092.

Martin, S. P. 1997. A Supersymmetry primer. arXiv:hep-ph/9709356.

Masciovecchio, M. 2019. Strong SUSY results in ATLAS and CMS LHC Run II. Proceedings, 54rd Rencontres de Moriond on Electroweak Interactions and Unified Theories (Moriond EW 2019): La Thuile, Italy, March 16–23.

McDonald, A. B. 2016. Nobel Lecture: The Sudbury Neutrino Observatory: Observation of flavor change for solar neutrinos. *Rev. Mod. Phys.*, **88**(3), 030502.

McFarland, K. S. et al. 1998. A Precision measurement of electroweak parameters in neutrino–nucleon scattering. *Eur. Phys. J.*, **C1**, 509–513.

McFarland, K. S. et al. 2002. A departure from prediction: Electroweak physics at NuTeV. *Frascati Phys. Ser.*, **27**, 283–289.

McKeown, R. D. and Vogel, P. 2004. Neutrino masses and oscillations: Triumphs and challenges. *Phys. Rept.*, **394**, 315–356.

Michael, C. et al. 2013. η and η' mixing from lattice QCD. *Phys. Rev. Lett.*, **111**(18), 181602.

Michel, L. 1950. Interaction between four half spin particles and the decay of the μ meson. *Proc. Phys. Soc.*, **A63**, 514–531.

Mikheyev, S. P. and Smirnov, A. Y. 1985. Resonance amplification of oscillations in matter and spectroscopy of solar neutrinos. *Sov. J. Nucl. Phys.*, **42**, 913–917.

Mikheyev, S. P. and Smirnov, A. Yu. 1986. Resonant amplification of ν oscillations in matter and solar-neutrino spectroscopy. *Il Nuovo Cimento C*, **9**(1), 17–26.

Miller, G. et al. 1972. Inelastic electron–proton scattering at large momentum transfers. *Phys. Rev.*, **D5**, 528.

Mohler, D. and Woloshyn, R. M. 2011. D and D_s meson spectroscopy. *Phys. Rev.*, **D84**, 054505.

Mohr, P. J. et al. 2016. CODATA recommended values of the fundamental physical constants: 2014. *Rev. Mod. Phys.*, **88**(3), 035009.

Morel, L. et al. 2020. Determination of the fine-structure constant with an accuracy of 81 parts per trillion. *Nature*, **588**(7836), 61–65.

Mori, T. 2017. Final results of the MEG experiment. *Nuovo Cim.*, **C39**(4), 325.

Nachtmann, O. 1972a. Inequalities for structure functions of deep inelastic lepton–nucleon scattering giving tests of basic algebraic structures. *Nucl. Phys.*, **B38**, 397–417.

Nachtmann, O. 1972b. Tests for the internal quantum numbers of partons. *Phys. Rev.*, **D5**, 686–689.

Nambu, Y. 1960. Quasiparticles and gauge invariance in the theory of superconductivity. *Phys. Rev.*, **117**, 648–663.

Nambu, Y. and Jona-Lasinio, G. 1961a. Dynamical model of elementary particles based on an analogy with superconductivity. I. *Phys. Rev.*, **122**, 345–358.

Nambu, Y. and Jona-Lasinio, G. 1961b. Dynamical model of elementary particles based on an analogy with superconductivity. II. *Phys. Rev.*, **124**, 246–254.

Nappi, C. R. and Ovrut, B. A. 1982. Supersymmetric extension of the $SU(3) \times SU(2) \times U(1)$ Model. *Phys. Lett.*, **113B**, 175–179.

Ne'eman, Y. 1961. Derivation of strong interactions from a gauge invariance. *Nucl. Phys.*, **26**, 222–229.

Nilles, H. P. 1984. Supersymmetry, supergravity and particle physics. *Phys. Rept.*, **110**, 1–162.

Nilles, H. P. et al. 1983. Weak interaction breakdown induced by supergravity. *Phys. Lett.*, **120B**, 346.

Okubo, S. 1962. Note on unitary symmetry in strong interactions. *Prog. Theor. Phys.*, **27**, 949–966.

O'Raifeartaigh, L. 1975. Spontaneous symmetry breaking for chiral scalar superfields. *Nucl. Phys.*, **B96**, 331–352.

Ovrut, B. A. and Schnitzer, H. J. 1981a. Gauge theories with minimal subtraction and the decoupling theorem. *Nucl. Phys.*, **B179**, 381–416.

Ovrut, B. A. and Schnitzer, H. J. 1981b. The decoupling theorem and minimal subtraction. *Phys. Lett.*, **100B**, 403–406.

Parker, R. H. et al. 2018. Measurement of the fine-structure constant as a test of the Standard Model. *Science*, **360**, 191.

Paschos, E. A. and Wolfenstein, L. 1973. Tests for neutral currents in neutrino reactions. *Phys. Rev.*, **D7**, 91–95.

Pati, J. C. and Salam, A. 1974. Lepton number as the fourth color. *Phys. Rev.*, **D10**, 275–289. [Erratum: Phys. Rev.D11,703(1975)].

Perkins, D. H. 1981. Experimental tests of quantum chromodynamics. Proceedings on the Conference on Nuclear Structure and Particle Physics, Hanko, Finland, 6–19 June, pp. 29–74.

Perl, M. L. 1976. Interpretation of anomalous e mu events produced in e^+e^- annihilation. Proceedings, International Neutrino Conference 1976: Aachen, Germany, June 8–12, pp. 0147.

Perl, M. L. et al. 1975. Evidence for anomalous lepton production in $e^+ - e^-$ annihilation. *Phys. Rev. Lett.*, **35**, 1489–1492.

Peskin, M. E. and Schroeder, D. V. 1995. *An Introduction to Quantum Field Theory.* Reading: Addison-Wesley.

Petratos, G. et al. 2000. Measurement of the $F^n(2)/F^p(2)$ and d/u ratios in deep inelastic electron scattering off H-3 and He-3. *Workshop on Nucleon Structure in High x-Bjorken Region (HiX2000) Philadelphia, Pennsylvania, March 30-April 1, 2000.*

Pilkuhn, H. (1967). *The Interaction of Hadrons,* Amsterdam: North Holland Publishing Company; see in particular ch. 1.

Poh, Z. et al. 2017. Pati–Salam SUSY GUT with Yukawa unification. *Phys. Rev.*, **D95**(11), 115025.

Politzer, H. D. 1973. Reliable perturbative results for strong interactions? *Phys. Rev. Lett.*, **30**, 1346–1349.

Pontecorvo, B. 1968. Neutrino experiments and the problem of conservation of leptonic charge. *Sov. Phys. JETP*, **26**, 984–988.

Poucher, J. S. et al. 1974. High-energy single-arm inelastic e - p and e - d scattering at 6-degrees and 10-degrees. *Phys. Rev. Lett.*, **32**, 118.

Press, W. H. and Spergel, D. N. 1985. Capture by the sun of a galactic population of weakly interacting massive particles. *Astrophys. J.*, **296**, 679–684.

Qian, X. and Vogel, P. 2015. Neutrino mass hierarchy. *Prog. Part. Nucl. Phys.*, **83**, 1–30.

Raby, S. 2017. Supersymmetric grand unified theories. *Lect. Notes Phys.*, **939**, 1–308.

Reutens, P. G. et al. 1985. Measurement of $\sin^2(\theta_W)$ and ρ in deep inelastic neutrino–nucleon scattering. *Phys. Lett.*, **152B**, 404–410.

Riordan, E. M. et al. 1974. Extraction of $r = \mathrm{sigma}(l)/\mathrm{sigma}(t)$ from deep inelastic $e\,p$ and $e\,d$ cross-sections. *Phys. Rev. Lett.*, **33**, 561.

Rojo, J. et al. 2015. The PDF4LHC report on PDFs and LHC data: Results from Run I and preparation for Run II. *J. Phys.*, **G42**, 103103.

Rosen, S. P. and Gelb, J. M. 1986. Mikheev–Smirnov–Wolfenstein enhancement of oscillations as a possible solution to the solar neutrino problem. *Phys. Rev.*, **D34**, 969.

Sachs, R. G. 1962. High-energy behavior of nucleon electromagnetic form factors. *Phys. Rev.*, **126**, 2256–2260.

Sakai, N. 1981. Naturalness in supersymmetric GUTs. *Z.Phys.*, **C11**, 153.

Sakai, N. and Yanagida, T. 1982. Proton decay in a class of supersymmetric grand unified models. *Nucl. Phys.*, **B197**, 533.

Sakai, Y. 2002. Results of time evolution analyses of B decays at BELLE. *Nucl. Phys. Proc. Suppl.*, **111**, 14–19.

Salam, A. 1968. Weak and electromagnetic interactions. *Conf. Proc.*, **C680519**, 367–377.

Salam, A. and Strathdee, J. A. 1974. Supergauge transformations. *Nucl. Phys.*, **B76**, 477–482.

Salam, A. and Strathdee, J. A. 1975. On superfields and Fermi–Bose symmetry. *Phys. Rev.*, **D11**, 1521–1535.

Samios, N. P. et al. 1962. Parity of the neutral pion and the decay $\pi^0 \rightarrow 2e^+ + 2e^-$. *Phys. Rev.*, **126**, 1844–1849.

Schael, S. et al. 2006. Precision electroweak measurements on the Z resonance. *Phys. Rept.*, **427**, 257–454.

Schildknecht, D. 2006. Vector meson dominance. *Acta Phys. Polon.*, **B37**, 595–608.

Schopper, H. 1981. Two years of PETRA operation. *Comments Nucl. Part. Phys.*, **10**(1), 33–54.

Schwinger, J. S. 1948. On quantum electrodynamics and the magnetic moment of the electron. *Phys. Rev.*, **73**, 416–417.

Sikivie, P. et al. 1980. Isospin breaking in technicolor models. *Nucl. Phys.*, **B173**, 189–207.

Sirunyan, A. M. et al. 2019. Search for supersymmetry in proton–proton collisions at 13 TeV in final states with jets and missing transverse momentum. *JHEP*, **10**, 244.

Sirunyan, A. M. et al. 2020. Search for physics beyond the standard model in events with jets and two same-sign or at least three charged leptons in proton–proton collisions at $\sqrt{s} = 13$ TeV. arXiv:2001.10086CMS-SUS-19-008CERN-EP-2020-001.

Soding, P. and Wolf, G. 1981. Experimental evidence on QCD. *Ann. Rev. Nucl. Part. Sci.*, **31**, 231–293.

Soni, S. K. and Weldon, H. A. 1983. Analysis of the supersymmetry breaking induced by N=1 supergravity theories. *Phys. Lett.*, **126B**, 215–219.

Steinberger, J. 1949. On the Use of subtraction fields and the lifetimes of some types of meson decay. *Phys. Rev.*, **76**, 1180–1186.

Stirling, J. *13/8 TeV LHC luminosity ratios.* www.hep.ph.ic.ac.uk/∼wstirlin/ plots/plots.html.

Sturm, S. et al. 2013. Electron g-factor determinations in Penning traps. *Annalen Phys.*, **525**(8-9), 620–635.

Susskind, L. 1979. Dynamics of spontaneous symmetry breaking in the Weinberg–Salam theory. *Phys. Rev.*, **D20**, 2619–2625.

Sutherland, D. G. 1967. Current algebra and some nonstrong mesonic decays. *Nucl. Phys.*, **B2**, 433–440.

Symanzik, K. 1970a. Renormalizable models with simple symmetry breaking. 1. Symmetry breaking by a source term. *Commun. Math. Phys.*, **16**, 48–80.

Symanzik, K. 1970b. Small distance behavior in field theory and power counting. *Commun. Math. Phys.*, **18**, 227–246.

Symanzik, K. 1971. Small distance behavior analysis and Wilson expansion. *Commun. Math. Phys.*, **23**, 49–86.

't Hooft, G. 1971a. Renormalizable Lagrangians for massive Yang–Mills fields. *Nucl. Phys.*, **B35**, 167–188.

't Hooft, G. 1971b. Renormalization of massless Yang–Mills fields. *Nucl. Phys.*, **B33**, 173–199.

't Hooft, Gerard. 1973. Dimensional regularization and the renormalization group. *Nucl. Phys. B*, **61**, 455–468.

't Hooft, G. and Veltman, M. J. G. 1972. Regularization and renormalization of gauge fields. *Nucl. Phys.*, **B44**, 189–213.

Tanabashi, M. et al. 2018. Review of particle physics. *Phys. Rev.*, **D98**(3), 030001.

Teubner, T. 2017. International workshop on e^+e^- collisions from Phi to Psi 2017, Mainz, Germany. https://indico.mitp.uni-mainz.de/event/86/contribution/17/material/slides/0.pdf.

Treiman, S. et al. 1972. *Lectures on Current Algebra and Its Applications*. Princeton, NJ: Princeton University Press.

Tsai, Y-S. 1971. Decay correlations of heavy leptons in $e^+e^- ->$ lepton$^+$ lepton$^-$. *Phys. Rev.*, **D4**, 2821. [Erratum: Phys. Rev.D13,771(1976)].

Tyutin, I. V. 1975. Gauge invariance in field theory and statistical physics in operator formalism. P.N. Lebedev Physical Institute, No. 39, arxiv.org/abs/0812.0580.

UTfit Collaboration. *Unitarity Triangle fit.* www.utfit.org/UTfit/.

Vilain, P. et al. 1994. Precision measurement of electroweak parameters from the scattering of muon-neutrinos on electrons. *Phys. Lett.*, **B335**, 246–252.

Weinberg, S. 1967. A model of leptons. *Phys. Rev. Lett.*, **19**, 1264–1266.

Weinberg, S. 1975. The $U(1)$ problem. *Phys. Rev.*, **D11**, 3583–3593.

Weinberg, S. 1980. Effective gauge theories. *Phys. Lett.*, **91B**, 51–55.

Weinberg, S. 1982. Supersymmetry at ordinary energies. 1. Masses and conservation laws. *Phys. Rev.*, **D26**, 287.

Weinberg, S. 1996. *The Quantum Theory of Fields*, Cambridge: Cambridge University Press.

Weisberger, W. I. 1965. Renormalization of the weak axial vector coupling constant. *Phys. Rev. Lett.*, **14**, 1047–1051.

Weisskopf, V. and Wigner, E. 1930a. Over the natural line width in the radiation of the harmonius oscillator. *Z. Phys.*, **65**, 18–29.

Weisskopf, V. and Wigner, E. P. 1930b. Calculation of the natural brightness of spectral lines on the basis of Dirac's theory. *Z. Phys.*, **63**, 54–73.

Wess, J. and Bagger, J. 1992. *Supersymmetry and Supergravity*. Princeton, NJ: Princeton University Press.

Wess, J. and Zumino, B. 1974a. A Lagrangian model invariant under supergauge transformations. *Phys. Lett.*, **49B**, 52.

Wess, J. and Zumino, B. 1974b. Supergauge invariant extension of quantum electrodynamics. *Nucl. Phys.*, **B78**, 1.

Wilson, K. G. 1976. Quarks on a lattice, or, the colored string model. *Phys. Rept.*, **23**, 331–347.

Wilkes, 2011. Results from Super-Kamiokande. https://agenda.infn.it/event/3101/contributions/44523/attachments/31309/36805/superk-wilkes.pdf.pdf.

Witten, E. 1981. Dynamical breaking of supersymmetry. *Nucl. Phys.*, **B188**, 513.

Wolf, G. et al. 1979. Tasso results on e^+e^- annihilation between 13-GeV and 31.6-GeV and evidence for three jet events. *eConf*, **C790823**, 34.

Wolfenstein, L. 1978. Neutrino oscillations in matter. *Phys. Rev.*, **D17**, 2369–2374.

Wolfenstein, L. 1983. Parametrization of the Kobayashi–Maskawa matrix. *Phys. Rev. Lett.*, **51**, 1945.

Wu, C. S. et al. 1957. Experimental test of parity conservation in beta decay. *Phys. Rev.*, **105**, 1413–1414.

Yang, C-N. 1950. Selection rules for the dematerialization of a particle into two photons. *Phys. Rev.*, **77**, 242–245.

Yang, C.-N. and Mills, R. L. 1954. Conservation of isotopic spin and isotopic gauge invariance. *Phys. Rev.*, **96**, 191–195.

Yukawa, H. 1935. On the interaction of elementary particles I. *Proc. Phys. Math. Soc. Jap.*, **17**, 48–57.

Zweig, G. 1964a. An $SU(3)$ model for strong interaction symmetry and its breaking. Version 1.

Zweig, G. 1964b. An $SU(3)$ model for strong interaction symmetry and its breaking. Version 2. Published in Lichtenberg, D. B. and Rosen, S. P. (eds) 1980, *Developments in the Quark Theory of Hadrons*. Nonantum, MA: Hadronic Press, pp. 22–101.

Index